The Handbook of Environmental Chemistry

Volume 73

More information about this series at http://www.springer.com/series/698

Groundwater in the Nile Delta

Volume Editor: Abdelazim M. Negm

With contributions by

I. Abd-Elaty · H. F. Abd-Elhamid · S. Abdel-Fattah ·
A. T. Abdelhameed · A. M. Abdel Sattar · H. Ali · A. M. Armanuos ·
S. A. Atta · M. Attwa · H. Bonakdari · A. S. El Hassanein ·
M. Elhakeem · A. ElNahrawy · G. El-Qady · G. Elsaiedy ·
M. G. A. Eltarabily · O. M. Gaame · B. Gharabaghi · T. M. Hassan ·
I. M. Ibraheem · K. M. A. Khallaf · W. E. Mahmod · M. A. Mahmoud ·
N. N. Mohamed · A. Nahrawy · A. M. Negm · E. E. Omran ·
O. M. Osman · S. Sakr · Z. E. Salem · A. M. Sharaky ·
A. M. M. Soliman · M. M. Solimsn

Springer

Editor
Abdelazim M. Negm
Faculty of Engineering
Zagazig University
Zagazig, Egypt

ISSN 1867-979X ISSN 1616-864X (electronic)
The Handbook of Environmental Chemistry
ISBN 978-3-319-94282-7 ISBN 978-3-319-94283-4 (eBook)
https://doi.org/10.1007/978-3-319-94283-4

Library of Congress Control Number: 2018964916

This Springer imprint is published by the registered company Springer Nature Switzerland AG
The registered company address is: Gewerbestrasse 11, 6330 Cham, Switzerland

Editors-in-Chief

Advisory Editors

The Handbook of Environmental Chemistry Also Available Electronically

The Handbook of Environmental Chemistry is included in Springer's eBook package *Earth and Environmental Science*. If a library does not opt for the whole package, the book series may be bought on a subscription basis.

For all customers who have a standing order to the print version of *The Handbook of Environmental Chemistry*, we offer free access to the electronic volumes of the Series published in the current year via SpringerLink. If you do not have access, you can still view the table of contents of each volume and the abstract of each article on SpringerLink (www.springerlink.com/content/110354/).

You will find information about the

- Editorial Board
- Aims and Scope
- Instructions for Authors
- Sample Contribution

at springer.com (www.springer.com/series/698).

All figures submitted in color are published in full color in the electronic version on SpringerLink.

Aims and Scope

Since 1980, *The Handbook of Environmental Chemistry* has provided sound and solid knowledge about environmental topics from a chemical perspective. Presenting a wide spectrum of viewpoints and approaches, the series now covers topics such as local and global changes of natural environment and climate; anthropogenic impact on the environment; water, air and soil pollution; remediation and waste characterization; environmental contaminants; biogeochemistry; geo-ecology; chemical reactions and processes; chemical and biological transformations as well as physical transport of chemicals in the environment; or environmental modeling. A particular focus of the series lies on methodological advances in environmental analytical chemistry.

Series Preface

With remarkable vision, Prof. Otto Hutzinger initiated *The Handbook of Environmental Chemistry* in 1980 and became the founding Editor-in-Chief. At that time, environmental chemistry was an emerging field, aiming at a complete description of the Earth's environment, encompassing the physical, chemical, biological, and geological transformations of chemical substances occurring on a local as well as a global scale. Environmental chemistry was intended to provide an account of the impact of man's activities on the natural environment by describing observed changes.

While a considerable amount of knowledge has been accumulated over the last three decades, as reflected in the more than 70 volumes of *The Handbook of Environmental Chemistry*, there are still many scientific and policy challenges ahead due to the complexity and interdisciplinary nature of the field. The series will therefore continue to provide compilations of current knowledge. Contributions are written by leading experts with practical experience in their fields. *The Handbook of Environmental Chemistry* grows with the increases in our scientific understanding, and provides a valuable source not only for scientists but also for environmental managers and decision-makers. Today, the series covers a broad range of environmental topics from a chemical perspective, including methodological advances in environmental analytical chemistry.

In recent years, there has been a growing tendency to include subject matter of societal relevance in the broad view of environmental chemistry. Topics include life cycle analysis, environmental management, sustainable development, and socio-economic, legal and even political problems, among others. While these topics are of great importance for the development and acceptance of *The Handbook of Environmental Chemistry*, the publisher and Editors-in-Chief have decided to keep the handbook essentially a source of information on "hard sciences" with a particular emphasis on chemistry, but also covering biology, geology, hydrology and engineering as applied to environmental sciences.

The volumes of the series are written at an advanced level, addressing the needs of both researchers and graduate students, as well as of people outside the field of "pure" chemistry, including those in industry, business, government, research establishments, and public interest groups. It would be very satisfying to see these volumes used as a basis for graduate courses in environmental chemistry. With its high standards of scientific quality and clarity, *The Handbook of Environmental Chemistry* provides a solid basis from which scientists can share their knowledge on the different aspects of environmental problems, presenting a wide spectrum of viewpoints and approaches.

The Handbook of Environmental Chemistry is available both in print and online via www.springerlink.com/content/110354/. Articles are published online as soon as they have been approved for publication. Authors, Volume Editors and Editors-in-Chief are rewarded by the broad acceptance of *The Handbook of Environmental Chemistry* by the scientific community, from whom suggestions for new topics to the Editors-in-Chief are always very welcome.

Damià Barceló
Andrey G. Kostianoy
Editors-in-Chief

Preface

This volume adds deep insights to the picture of the Nile Delta in Egypt due to the fact that groundwater in Egypt is the second source of freshwater. The demand for groundwater is increasing in Egypt due to its limited water resources. The volume consists of 23 chapters divided into 11 parts and contributed by more than 33 scientists, groundwater specialists, and researchers from Egypt.

Part I is an introduction to the volume and it consists of two chapters. The first chapter is titled "An Overview of Groundwater Resources in Nile Delta Aquifer" and presents a comprehensive overview of groundwater in the Nile Delta as a freshwater resource in Egypt. The authors covered almost all topics related to groundwater in the Nile Delta to provide a background for the volume subject. In the chapter "Land and Groundwater Resources in the Egypt's Nile Valley, Delta and Its Fringes," the author assesses the land and groundwater resources and its sustainable development in the Nile valley, delta, and its fringes.

Part II consists of two chapters which deal with agriculture and water uses. The chapter "Use of Groundwater in Nile Alluvial Soils and Their Fringes" provides a comprehensive overview of the available water resources including groundwater and their usage particularly in irrigation and agricultural activities. In the last chapter of Part II, which is titled "Groundwater and Agriculture in the Nile Delta," the author provides an overview of the exchangeable relationships between groundwater and agriculture in the Nile Delta region.

Part III consists of two chapters to cover the sedimentology of the northern Nile Delta and hydrogeochemical characteristics of the Nile Delta. In the chapter "Sedimentological Characteristics of the Quaternary Groundwater Aquifer, Northwestern Nile Delta, Egypt," the authors focus on the grain size analysis and mineralogical study of the Plio-Pleistocene sediments of the area under consideration. In the chapter "Hydrogeophysical Characteristics of the Central Nile Delta Aquifer," the authors estimate the aquifer parameters as clay volume, porosity, clay thickness, permeability, total dissolved solids, and hydraulic conductivity of the formation for typical study areas in the central Nile Delta, namely, Bliqas and Mit Ghamer cities using electrical resistivity and gamma ray logs for 34 wells.

Part IV consists of two chapters which deal with groundwater investigations, namely, resistivity characterization and hydrogeophysical investigations. In the chapter "Resistivity Characterization of Aquifer in Coastal Semi-Arid Areas: An Approach for Hydrogeological Evaluation," the authors present the fundamentals of direct current resistivity (DCR) and induced polarization (IP) as groundwater investigation methods and demonstrate the efficiency of DCR method for hydrogeological assessment in the Nile Delta, Egypt, with an emphasis on technical constraints to achieve sustainable development in coastal and semi-arid areas. On the other hand, the chapter "Hydrogeophysical Investigations at El-Nubariya-Wadi El-Natrun Area, West Nile Delta, Egypt" is planned to address and evaluate the hydrogeological regime in the area of West Nile Delta using geophysical electrical and electromagnetic methods.

Also, Part V consists of two chapters which dealing with groundwater contamination and degradation. The chapter "Salinization and Origin of the Coastal Shallow Groundwater Aquifer, Northwestern Nile Delta, Egypt" discusses the originality of brackish water in the coastal areas, the chemical analysis of pollutants and determination of the source of pollution, the hydrochemical characterizations of groundwater in the northwestern Nile Delta, and the evaluation of the suitability of groundwater for different purposes. The chapter "Soil Aquifer Treatment System Design Equation for Organic Micropollutant Removal" presents a novel model for more accurate prediction of the removal efficiency of the soil aquifer treatment (SAT) system for the organic micropollutants (OMPs) and the fate of the OMPs trapped within the vadose zone of the aquifer.

Part VI comprises of two chapters to investigate the seawater intrusions in the Nile Delta aquifer and how to control this intrusion. The first chapter titled "Investigation of Saltwater Intrusion in Coastal Aquifers" reviews briefly the history of saltwater intrusion in coastal aquifers and also presents the traditional methods that are applied to investigate saltwater intrusion in coastal aquifers around the world and in the Nile Delta aquifer. The chapter titled "Control of Saltwater Intrusion in Coastal Aquifers" discusses the different methods and approaches for saltwater intrusion in the coastal aquifers.

Part VII consists of three chapters to explore the usage of geophysical methods to delineate the groundwater flow and saltwater intrusion. The chapter "Use of One-Dimensional Subsurface Temperature Profiles to Characterize the Groundwater Flow System in the Northwestern Part of the Nile Delta, Egypt" presents the usage of the borehole temperatures to trace the groundwater stream lines and to assess the groundwater vertical flux rates in the study region which is located at northwestern part of the Nile Delta. In the chapter "Use of Geoelectrical Resistivity to Delineate the Seawater Intrusion in the Northwestern Part of the Nile Delta, Egypt," the authors present a detailed vertical electrical sounding survey in the northern part of the Nile Delta to delineate and follow seawater intrusion with interpretation of results to help groundwater manager to take proper decision and design a good management startegy. In the chapter "Integrated Subsurface Thermal Regime and Hydrogeochemical Data to Delineate the Groundwater Flow System and Seawater Intrusion in the Middle Nile Delta, Egypt," the authors utilize the temperature and water chemistry for following the groundwater stream framework and seawater intrusion in the middle Nile Delta.

Part VIII consists of three chapters which deal with modeling and assessment of groundwater in the Nile Delta and its fringes. In the chapter titled "Integrated Groundwater Modeling for Simulation Saltwater Intrusion in the Nile Delta Aquifer, Egypt," the authors build an integrated 3-D groundwater model for the Nile Delta aquifer to simulate saltwater intrusion under different climate change scenarios using the MODFLOW and SEAWAT softwares including the rescharge from the existing irrigation canals network in the Nile Delta region. Also, they proposed different scenarios for management and control of saltwater intrusion in the Nile Delta aquifer. In the chapter titled "Groundwater Modeling and Assessment Under Uncertain Hydrological Conditions for Egyptian Sahara," the author performs groundwater flow analysis using the developed modified grey model (MGM) which combines the finite element method (FEM) and the new developed modified genetic algorithms (MGA) for the Kharga Oasis southwest of the Nile Delta in the Egyptian Sahara. The chapter "Groundwater Potential in the New Valley South West of the Nile Delta in Egypt" provides the benefits of applying the modeling techniques. Numerous valuable inputs for the national development plan in Egypt are presented.

Part IX consists of two chapters to cover the usage, assessment, and management. In the chapter "Hydrogeochemistry and Quality Assessment of Groundwater Under Some Central Nile Delta Villages, Egypt," the authors present the hydrogeochemistry of groundwater in some village in the central Nile Delta. The quality of groundwater in these villages is assessed too. The chapter "Assessment of the Groundwater Quality for Drinking and Irrigation Purposes in the Central Nile Delta Region, Egypt" provides an integrated approach of water quality indices to evaluate groundwater for irrigation and drinking purposes in the central Nile Delta.

Part X has two chapters dealing with groundwater management for sustainability. The chapter "Groundwater Management for Sustainable Development East of the Nile Delta Aquifer" presents budget analysis for the study area and provides some strategies for sustainable management. In the chapter "Groundwater Management for Sustainable Development Plans for the Western Nile Delta," the authors integrated the GIS and MODFLOW models to simulate the groundwater flow in the western Nile Delta of the Quaternary aquifer. The groundwater potentiality is evaluated, and different management scenarios are analyzed for groundwater prediction.

The volume ends with the conclusions chapter which briefly summarizes the most significant findings and recommendations of the volume.

The editor would like to express his great thanks and his special appreciation to all the authors who have contributed to this volume. Without their patience and effort in writing and revising the different versions to satisfy the high-quality standards of Springer, it would not have been possible to produce this volume and make it a reality. All appreciation and thanks must be extended to include all members of the Springer team who have worked long and hard to produce this volume and make it a reality for the researchers, graduate students, and scientists around the world. I must thank all the experts who contributed to the review processes of the volume chapters.

Also, the editor would like to thank all those who contributed to make the book entitled *Groundwater in the Nile* a great source of information and knowledge for all those who are interested in it. Much appreciation and great thanks are also owed to the editors of the HEC book series and the Springer team who worked hard during the production of this volume.

The volume editor would be so happy to receive any comments to improve future editions. Comments, feedback, suggestions for improvement, or new chapters for next editions are much welcome and can be sent directly to the volume editor.

Zagazig, Egypt
18 April 2018

Abdelazim M. Negm

Contents

Part I
Introduction

An Overview of Groundwater Resources in Nile Delta Aquifer

Abdelazim M. Negm, Sameh Sakr, Ismail Abd-Elaty, and Hany F. Abd-Elhamid

Abstract Egypt's renewable water resources are limited to its share from the Nile River as well as some minor rainfall along the coastal areas. The Delta and the narrow valley of the Nile comprise 5.5% of the total area of Egypt. The Nile Delta aquifer, Egypt, which is replenished directly by the Nile water, is considered another renewable resource. However, this aquifer system is in direct contact with the Mediterranean Sea from the north and the Suez Canal from the east. Groundwater resources in Nile Delta are an essential source of freshwater due to increasing water demand and shortage in supply as a result of increasing agriculture, domestic, and industrial consumption. It provides about 85% of total groundwater abstractions in Egypt by 6.1 BCM per year. It is obvious that since 1981 the groundwater potentiality is decreasing annually in a linear fashion by 0.1 billion cubic meters. On the other hand, water quality in the irrigation and drainage canals is deteriorating as we move downstream due to the increased pollution load from the heavy agricultural activities and high population density. Agriculture, domestic, and industrial effluents are the main sources of contamination of the Nile Delta aquifer. The water quality of the Nile Delta aquifer has been assessed through testing some groundwater samples, surface canal water, and drainage water. All the studies showed that the concentration of most water quality parameters and heavy metals exceeded the WHO standards for drinking water, and some studies indicated the areas where the groundwater is not suitable for irrigation. There are damage and shortage in aquifer water quantity and degradation in the quality due to the gradual

A.M. Negm (✉), I. Abd-Elaty, and H.F. Abd-Elhamid
Water and Water Structures Engineering Department, Faculty of Engineering, Zagazig University, Zagazig, Egypt
e-mail: Amnegm@zu.edu.eg; Amnegm85@yahoo.com; Eng_abdelaty2006@yahoo.com; Eng_abdelaty@zu.edu.eg; Hany_farhat2003@yahoo.com

S. Sakr
Groundwater Institute, National Water Research Center, Ministry of Water Resources and Irrigation, Cairo, Egypt
e-mail: Sameh_nw@hotmail.com; sameh_nw@yahoo.com

A. M. Negm (ed.), *Groundwater in the Nile Delta*, 3
Hdb Env Chem (2019) 73: 3–44, DOI 10.1007/698_2017_193,

increase of groundwater pumping especially in the delta fringes. Also, problems were reported as a result of the continuous decline of the piezometric surface and the increase of the water salinities. This chapter gives an overview of groundwater resources in the Nile Delta aquifer. Detailed investigation of groundwater in the Nile Delta aquifer was presented.

Keywords Irrigation, Pollution, Wastewater, Water quality, Water resources

Contents

1 Introduction

Groundwater is considered the main source of water supply in many coastal regions. It represents about 22% of all freshwater on the earth; polar ice represents 77%, while another freshwater in rivers and lakes represents about 0.30%. Population growth increases the water requirements leading to increasing pumping from the aquifers [1]. Groundwater is an important natural resource. Many agricultural, domestic, and industrial water users rely on groundwater as the sole source because of its low cost and high quality. However, in recent years it has become clear that human activities and climate change have a negative impact on both quantity and quality of groundwater resources. The depletion of groundwater may occur due to excessive pumping and contamination of the groundwater by waste disposal or other activities [2]. The Mediterranean region is characterized by a strong development of coastal areas with a high concentration of water-demanding human activities, resulting in weakly controlled withdrawals of groundwater which accentuate the saltwater intrusion phenomenon [3]. Groundwater or subsurface water is a term used to denote all the waters found beneath the surface of the ground and is considered as part of the hydrologic cycle (Fig. 1).

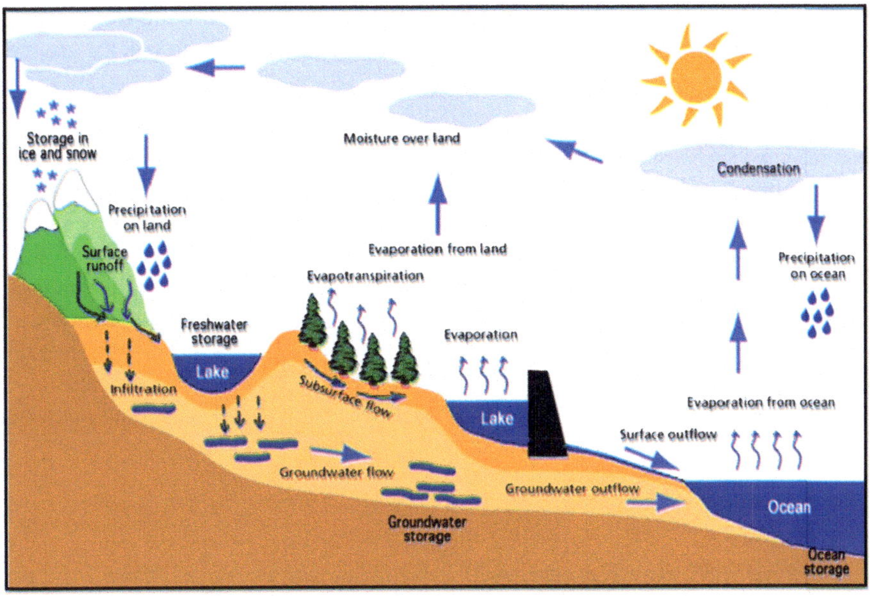

Fig. 1 Schematic diagram of hydrologic cycle [4]

2 Groundwater Problems

Groundwater quality is an important issue in the development and management of water resources. In fact, with the increasing demand for water in most parts of the world and with the intensification of water utilization, the quality problem becomes the limiting factor in the development and management of water resources in many parts of the world. The quality of both surface and groundwater resources deteriorates as a result of pollution. Special attention should be devoted to the pollution of aquifers. Groundwater pollution is usually traced back to four sources [5]:

(a) *Environmental*: this type of pollution is due to the characteristics of the environment through which the flow of groundwater takes place, for example, flow through carbonate rock, seawater intrusion, and invasion by brackish water from the adjacent aquifer.
(b) *Domestic*: domestic pollution may be caused by the accidental breaking of sewers, percolation from septic tanks, rain infiltrating through sanitary landfills, acid rains, and artificial recharge using sewage water after being treated to different levels and biological contaminants (e.g., bacteria and viruses).
(c) *Industrial*: industrial pollution may come from sewage disposal, which contains heavy metals, non-deteriorating compounds, and radioactive materials.

(d) *Agriculture*: this is due to irrigation water and rainwater dissolving and carrying fertilizers, salts, herbicides, pesticides, etc., as they infiltrate through the ground surface and replenish the aquifer.

Groundwater quality in the Nile system is fairly reasonable; however pollution had affected some shallow groundwater bodies. Almost 20% of groundwater in the Nile aquifer does not meet drinking water standards, especially at the fringes where there is little or no protective clay cap. Groundwater quality in the Nile Delta has better quality than that in the Nile valley. Fertilizers are a large source of pollution for soil and water resources. Egyptian farmers consume more than 1.8 million tons of fertilizers annually [6], mainly using nitrogen, phosphorus, and potassium in different forms [7]. Industrial wastewater constitutes 39% of the environmental problems of the industrial sector [8]. It contains dissolved industrial organic and inorganic wastes, solids, and metals, all having negative and hazardous impacts with direct reflection on human health [9].

3 Egypt's Water Resources

The Nile River is considered the main water resource for Egypt as it provides Egypt with 55.5 BCM/year. Groundwater in the Western Desert and Sinai is the second water source. The annual abstraction from aquifers in these areas is 0.5 BCM/year. Rainfall in the coastal zones of Egypt provides 1.5 BCM/year [10]. Table 1 shows the current water budget of Egypt 2010, as given in the 2050 water strategy of the ministry of water resources and irrigation [11]. The water sector in Egypt is facing a lot of challenges such as: limited water resources from the Nile River, non-renewable groundwater and rainfall, population increase and demographic distribution, water quality degradation, costs for new projects, lack of awareness, need for more coordination of different stakeholders, more than 90% of water supply of Egypt comes from outside of its borders, impacts of world trade agreements, and globalization on local economy [12].

Egypt is the most populous country in the Middle East and the third-most populous on the African continent, nearly 97% of the country's 82.5 million (2012 estimate). Projected population for Egypt is estimated at 104 million by 2025, 146 million by 2050, and 237 million by 2100. The annual per capita water share of Nile in Egypt has decreased from 2,500 m^3 per capita per year in the 1950s to about 680 m^3 per capita per year in 2012 and is projected to drop to about 350 m^3 per capita per year in 2050. The agricultural sector is the largest user and consumer of water in Egypt, with its current allocation (in 2010) exceeding 68% of the total freshwater supplies or 82% of the total used water (after recycling). About 80% of the drinking water supplies use surface water sources from the Nile or the main canals, and 20% rely on groundwater, while the desalination is limited to remote areas on the coasts. The total generated energy from hydropower is 14,630 GWH, or about 11.5% of the gross national generated electricity which is about 132,000

Table 1 Current water budget of Egypt (2010), all sources and allocation/usage [11]

Water supply	Volume (BCM/year)	Demand by sector	Consumption (BCM/year)	Usage/allocation (BCM/year)
Conventional water sources		Drinking (fresh-water only)	1.80	9.00
Nile (HAD)	55.00	Industry	1.40	2.00
Deep groundwater	2.50	Agriculture	40.40	67.00
Rainfall and flashfloods	1.30	Drainage to sea	12.20	
Desalination	0.70	Evap. losses	3.00	3.00
Total supply conventional	59.50	Env. balance	0.20	0.20
Unconventional sources		Total consumption	59.00	
Shallow groundwater (Delta)	6.50			
Reuse of agricultural drainage water	13.50			
Total supply non-conventional	20.00			
Total water supply	79.50	Total water use or allocation		81.20

GWH (2050 water strategy). This hydropower is generated mainly from the Aswan High Dam, the Aswan dams, and the new Esna and new Naga Hammadi Barrages, in addition to a small plant on Bahr Yussef canal. In addition to that, there are some hydropower generation projects at the new Assiut Barrage and some other canals and drains in the Nile valley and Delta. Again, reduced flow due to climate change and increased demands would certainly impact the hydropower generation and will have direct economic impacts on the country development plans [7].

4 Main Aquifers in Egypt

The major groundwater system in Egypt consists of several aquifers as shown in Fig. 2. These are:

(a) *Nile Delta* aquifer is considered a semi-confined aquifer due to the upper clay layer. Also, it is considered a shallow aquifer that is recharged mainly by infiltration of excess irrigation water or from the irrigation network. "It is composed of a thick sand and gravel layer covered by a clay cap of varying depth up to 50 m. It provides about 85% of total groundwater abstractions in Egypt. About 6.1 BCM/year are annually extracted from the aquifer" [7].

(b) *Nubian Sandstone* Aquifer System (NSAS). According to CEDARE/IFAD [13], the NSAS is the Africa's largest fossil aquifer system with its reserves estimated at 259.293 km^3 (372,960 km^3 with post-Nubian). The NSAS is shared

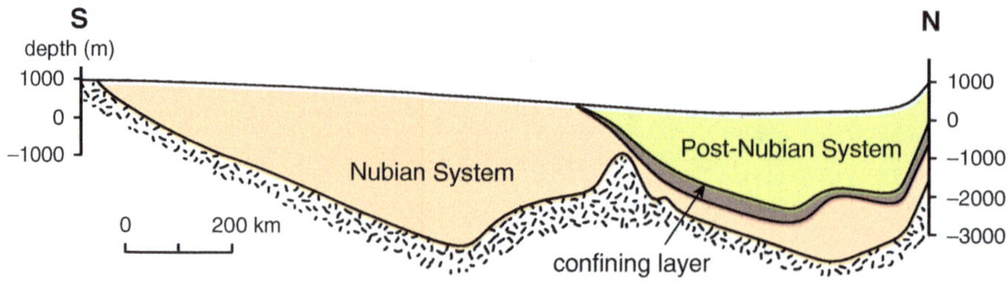

Fig. 2 The Nubian sandstone aquifer system and the post-Nubian aquifer system

by Libya, Egypt, Sudan, and Chad and covers approximately 2.2 million km^2, of which 235,000 km^2 are in Chad (11%), 828,000 km^2 in Egypt (38%), 760,000 km^2 in Libya (34%), and 376,000 km^2 in Sudan (17%) (CEDARE [14] cited in [15, 16]). Table 1 represents the main characteristic of the aquifer. The maximum depth of NSAS reaches 4,500 m with hydraulic head ranges from 570 m above sea level west of Darfur to 78 m in the Qatar depression [17]. "The thickness of the fresh layer ranges from 200 m in East Owinat to 3,500 m North West of El-Farafra Oasis" [7]. Some studies concluded the existence of recharge to the NSAS from the Nile River in a few areas at the eastern boundary by groundwater influx from the Blue Nile and by precipitation in some mountain regions. Low values of infiltration rate were estimated compared to the natural groundwater flow due to discharge into depressions, evaporation in areas with low depth to the groundwater table, and leakage into confining beds. It is thus considered to be a nonrenewable groundwater resource under current climate conditions [18] (Table 2, Fig. 3).

(c) *Fissured carbonate rock aquifer* extends from Sinai to Libya and occupies more than 50% of the Egypt's area. It has many natural springs and acts as a confining layer on top of the Nubian Sandstone Aquifer System. The aquifer recharge is unknown, and there is no reliable information regarding its potential [7].
(d) *Moghra* aquifer is located in the northwestern desert, and groundwater is directed toward Qattara Depression. It is recharged by rainfall and lateral inflow from the Nile aquifer. It contains freshwater only with salinity increases toward the north and west [7].
(e) *Coastal* aquifer "system is recharged by rainfall, and the abstractions are limited due to the presence of saline water underneath the freshwater lenses" [7].
(f) *Fissured hard* rock aquifer system in the Eastern Desert and Southern Sinai is also recharged by small quantities of infiltrating rainwater [7].

Table 2 Recoverable volumes and present extraction from the NSAS

	Nubian sandstone aquifer system		Post-Nubian aquifer system						
Region	Area (km^2)	Volume in storage (km^3)	Area (km^2)	Volume in storage (km^3)	Volume of freshwater in storage (km^3)[a]	Total recoverable groundwater volume (km^3)[b]	Present extraction from post-NSAS (km^3)	Present extraction from NSAS (km^3)	Total present extraction from NSAS (km^3)
Egypt	815,670	154,720	494,040	35,867	190,587	5,367	0.306	0.200	0.506
Libya	754,088	136,550	426,480	48,746	185,296	4,850	0.264	0.567	0.831
Chad	232,980	47,810	–	–	47,810	1,630	–	0.000	0.000
Sudan	373,100	33,880	–	–	33,880	2,610	–	0.833[c]	0.833
Total	2,175,838	372,960	920,520	84,613	457,573	14,457	0.570	1.600	2.170

Source: Bakhbakhi [15]

[a]Assuming a storativity of 10^{-4} for the confined part of the aquifer and a 7×10^{-2} for the unconfined part

[b]Assuming a maximum water decline of 100 m in the unconfined aquifer areas and 200 m in the confined aquifer areas

[c]Most of this water is extracted in the Nile Nubian Basin (833 ucm) which is not considered to be part of the Nubian Basin

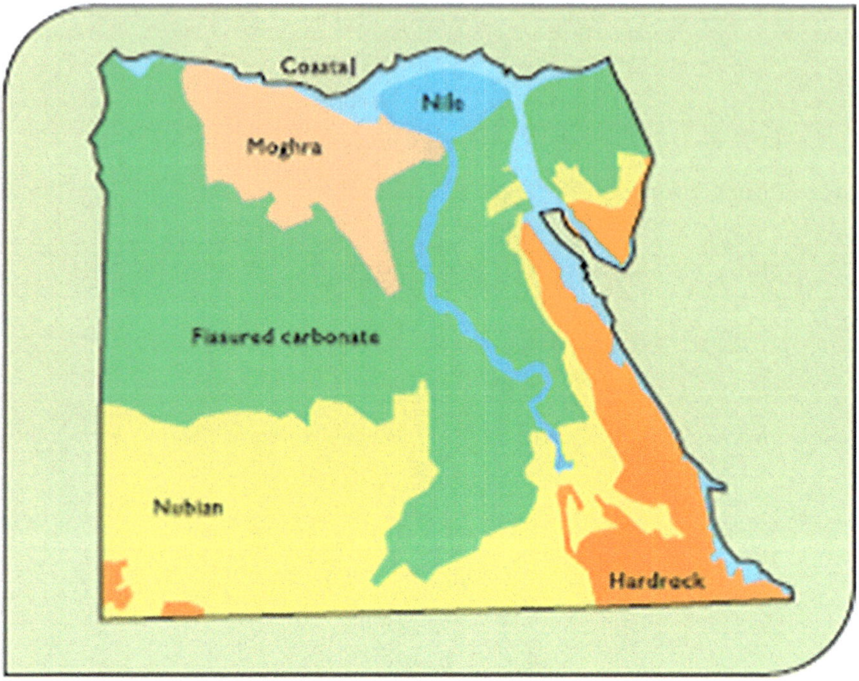

Fig. 3 Major sources of aquifer in Egypt: National Water Resources Plan 2017 [19]

5 Investigation of Groundwater in Nile Delta Aquifer

Nile Delta aquifer is one of the largest groundwater aquifers in the world due to the extension in area and thickness of its layer. The Nile Delta and the narrow valley of the Nile comprise 5.5% of the total area of Egypt. Much of Egypt's infrastructure and development are concentrating on the deep coastal areas and dependence on the Nile Delta for agricultural land. The sea level rise (SLR) will have a direct and critical impact on Egypt's entire economy by coastal inundation or saline intrusion caused by anthropogenic climate change [20]. Also, a number of studies were carried out on Nile Delta aquifer under different scenarios of extraction rate [21].

5.1 Physical Setting

5.1.1 Location

Nile Delta is one of the largest river deltas in the world bounded by the Mediterranean Sea in the north, Nile River in the south, Ismailia Canal in the east, and Nubaria Canal in the west. It is located between latitudes 30° 00′ and 31° 45′N and longitudes 29° 30′ and 32° 30′E as shown in Fig. 4. The Nile Delta consists of flat, low-lying areas, where most areas are used for agriculture. It is about 200 km from the south to

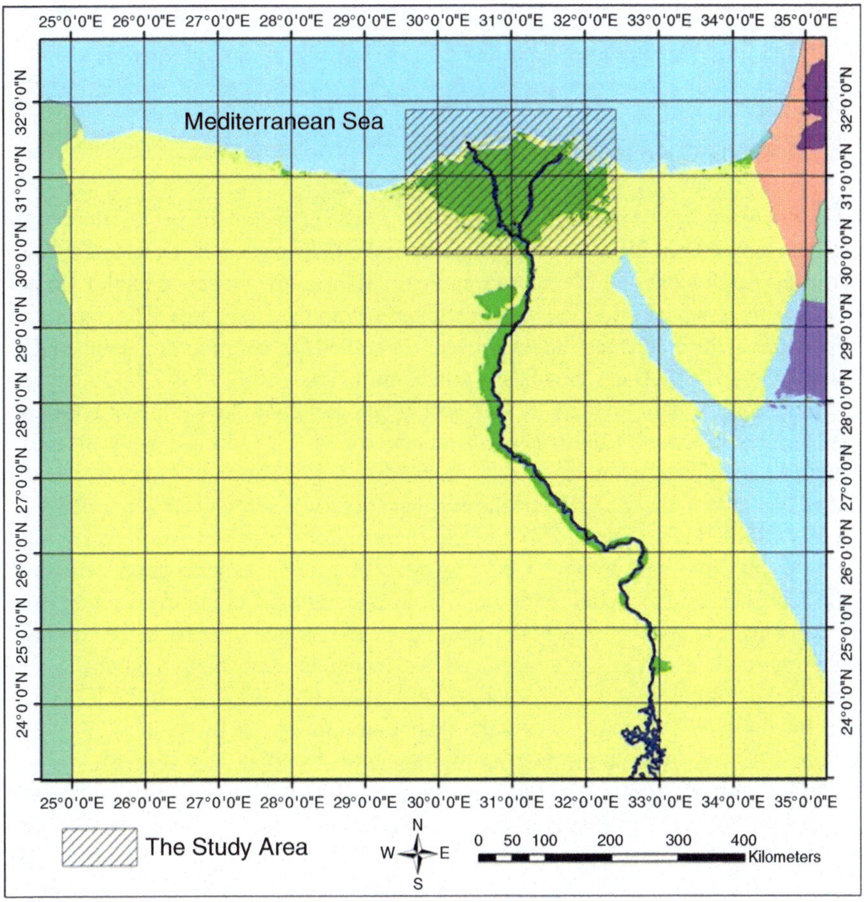

Fig. 4 Location map of the Nile Delta aquifer [22]

north, and the coastline is about 300 km long, with an area of about 25,000 km^2. It is considered among the most densely populated agricultural areas in the world. It contains 10 governorates of which there are about 25 large cities and 4 brackish lagoons or lakes. Nile Delta aquifer system is considered as one of the most important water resources in Egypt following the surface water source. It is needed essentially to face the increase in water requirements. This aquifer belongs mainly to the Quaternary age; it is formed from fluviomarine deposits and Deltaic deposits. The aquifer is generally composed of unconsolidated sand and gravel with occasional clay lenses that increases toward Northern direction [7]. The area is bounded by the Mediterranean Sea in the north, the apex of the Delta in the south, the Suez Canal in the east, and the Wadi El-Natrun fault in the west. The two branches of the

Nile River form a triangle with its base along the Mediterranean Sea in the north and the apex at the Delta Barrage in the south.

5.1.2 Climate Conditions

The study area is located in an arid climate region; the average rainfall in the delta is very small and ranges from 25 mm/year in the south and middle part of the Delta to 200 mm/year in the north along the Mediterranean Sea [23]. The low values of rainfall do not contribute by any means to the aquifer recharge within the study area. The maximum temperature in the Delta area ranges between 26 and 34 °C and minimum temperature between 6 and 13 °C. The average daily temperature ranges from 17 to 20 °C along the Mediterranean to more than 25 °C in Upper Egypt along the Nile [8]. From 1961 to 2000, the mean maximum air temperature increased 0.34 °C per decade, while the mean minimum air temperature increased 0.31 °C per decade [8]. The annual mean values for relative humidity in the morning and the evening are between 60 and 80%, respectively. The Met Office Hadley Centre [24] projected temperature increases over Egypt around 1–1.5 °C by 2050 and around 3–3.5 °C by 2100 with a general good consistent agreement between the models over the Middle East region. Current evaporation rates in Egypt range between 7 mm/day in Upper Egypt and about 4 mm/day in the Northern Mediterranean coast [25]. It is recognized that increased temperature and changes in wind and humidity will affect these values. As reported by Eid [26], and IPCC [27], a temperature rise of 1 °C may increase the evapotranspiration rate by about 4–5%, while a rise of 3 °C may increase the evapotranspiration rate by about 15% (Fig. 5).

The Intergovernmental Panel on Climate Change (IPCC) [27] indicated that the global warming, due to increasing the concentrations of greenhouse gases by 0.13° per decade in the atmosphere, is expected to have a full range of temperature projection from 1.1 to 6.4° over the period 1990 to 2100. Scientists use climate models to describe how various factors affect the temperature of the Earth's surface. Such models help climatologists to estimate the future changes in temperature and sea levels. SLR for some reasons includes thermal expansion of oceans and seas, melting of glaciers and ice caps, and melting of Greenland and Antarctic ice sheets [28]. However, some researchers are still discussing the accuracy of the magnitudes of sea level rise. The sea level rise was estimated between 18 and 58 cm by the end of the century. However, recently published estimates have greatly expanded these values to reach more than 180 cm [29]. Currently, the Nile Delta region is subject to climate changes, including increase in sea level, and shoreline changes, due to erosion and accretion [30]. The Nile Delta coastal zone is highly vulnerable to the impacts of sea level rise through direct inundation and saltwater intrusion as shown in Fig. 6.

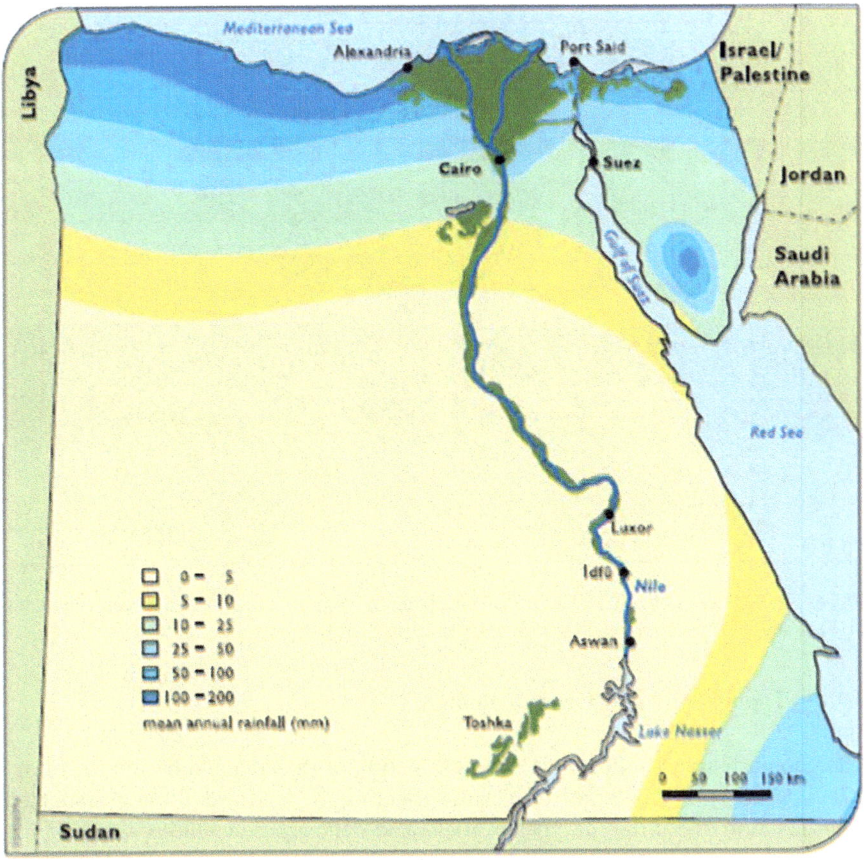

Fig. 5 Average annual precipitation in Egypt (mm/year) [19]

5.1.3 Population

Egypt is the most populous country in the Middle East and the third-most populous on the African continent. Nearly 97% of the country's 82.5 million (2012 estimate) people live in three major regions of the country: Cairo, Alexandria, and elsewhere along the banks of the Nile valley and the Nile Delta and along the Suez Canal. These regions, which occupy about 4% of the country's area, are among the world's most densely populated regions, where the population density is about 1,500 inhabitants per km^2. Projected population for Egypt is estimated as 104 million by 2025, 146 million by 2050, and 237 million by 2100 [7].

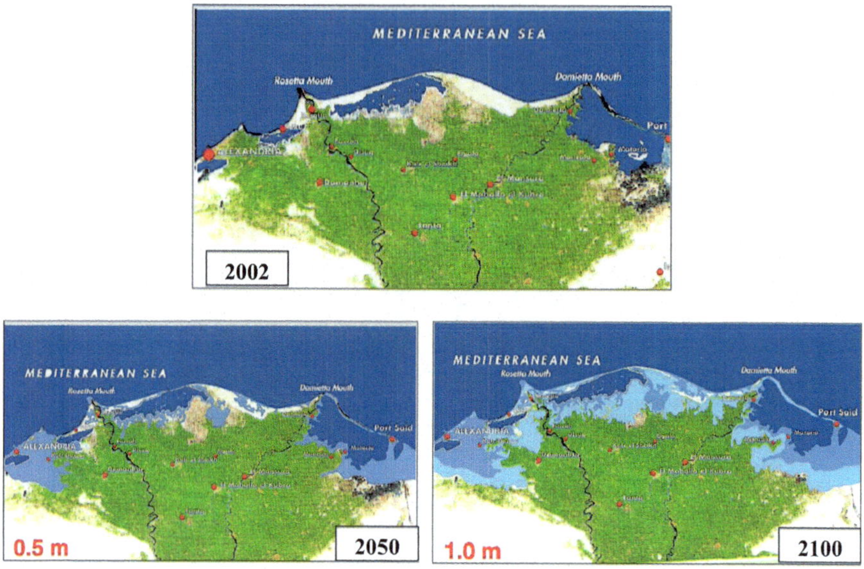

Fig. 6 Satellite maps of Nile Delta, showing the potential impact of SLR with the occurring status in 2002 and the coastal inundation with a 0.5 and 1 m SLR [31]

5.1.4 Topography and Geomorphology

The geomorphology reflects the geologic setting of the area and has an important effect on the type and degree of hazards affecting the land use of the investigated area including the different types of rocks and deposits, constituting the different landscapes and landform units varying in relief and extension [32]. The Nile Delta region is a morphotectonic depression open to the Mediterranean Sea in the north. The Nile enters the Delta 20 km north of Cairo, where it divides into two branches, Rosetta in the west and Damietta in the east. The Delta region is bounded on the east and southeast by two main watershed areas rising to more than 500 m above mean sea level. To the west, the watershed is less developed morphologically, being only 100 m above mean sea level. The ground elevation ranges between about 18 m above mean sea level (AMSL) in the south at El Qanater El Khayreya and about 5 m (AMSL) near Tanta sloping very gently in the northward direction by the average value of 1 m/10 km [33]. The geomorphologic features of the Nile Delta region are shown in Fig. 7 and are discussed below in details:

The Foreshore Plain

The foreshore plain occupies the area determined by the coastal lakes and their inland extension into brackish water lagoons. Landforms existing in that plain include the wetland areas of the main lakes and the sabkhas.

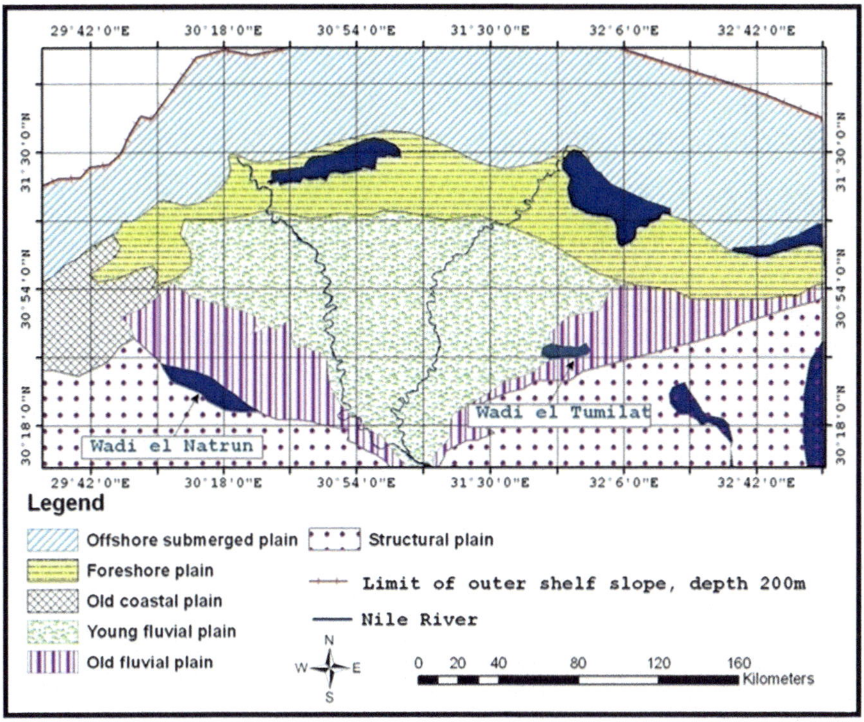

Fig. 7 Geomorphologic units for the Nile Delta region [22]

The Old Coastal Plain

The old coastal plain occupies a portion of the northwest corner of the Nile Delta depression and may be a remnant of the offshore plain, which was submerged in early Holocene times.

The Young Alluvial Plains

The young alluvial plains dominate the cultivated lands bordering the channel of the Nile River and its two branches. These plains cover most of the Nile Delta region [34]. The irrigation canals and drains break through these plains. The ground surface elevation ranges between 18 m above mean sea level (MSL) in the south and about 5 m near Tanta sloping gently in the northward direction by an average value of 10 cm/km [33]. Furthermore, the Nile Delta slopes also from the east to west making Damietta Branch higher in elevation than Rosetta Branch by about 3 m. The area is flat and covered by recent and Quaternary sediments that were formed from the disintegrated igneous and metamorphic rocks of the Ethiopian Plateau and South of Sudan. These sediments were transported by the Nile River and its tributaries to the Delta during the flood seasons for more than 10,000 years. The old and traditionally cultivated lands are dominant in these areas.

The Old Alluvial Plains

The old alluvial plains are present along both sides of the Nile Delta and are exposed at various heights above the young alluvial plains. They occupy the areas on the eastern and western fringes of the Nile Delta. The surface of this plain is gently undulating and displays classical examples of landforms by wind deflation. The surface is also incised by the downstream portions of a good number of dry channels (wadis), which acted as active rivers in Middle Pleistocene.

5.1.5 Land Use in the Nile Delta Region

The Nile Delta region is divided into three subregions with respect to the two Nile Branches, Damietta and Rosetta. These subregions are mainly the Western Delta, the Middle Delta, and the Eastern Delta. As shown in Fig. 8, the Nile Delta can be distinguished into three main regions with respect to land use: (1) the agricultural land, (2) the wetland, and (3) the desert. The agricultural land, 20,000 km^2, includes the traditionally cultivated areas and the newly reclaimed areas. The wetland portion, 1,000 km^2, includes the coastal lakes and the marshlands. The desert portion, 20,000 km^2, borders the Nile Delta from both east and west sides. Geographically, the traditional cultivated lands are predominated in the Middle Delta, while the reclaimed lands are present in the eastern and western Fringes of the Nile Delta. Surface water, groundwater, or conjunctive use of irrigation water is applied when appropriate. Many towns and villages and some industrial zones are scattered in the Delta with an extensive population.

5.2 Geology of the Nile Delta Region

5.2.1 Geologic Setting of the Nile Delta Region

The Nile Delta area is covered by the Quaternary deposits consisting of Nile silt, clay, sandy clay, sands, and gravels (Fig. 9). Schlumberger [35] published a generalized litho-stratigraphic column of the Nile Delta as shown in Fig. 10. Deep drilling in the Nile Delta region revealed a very thick sedimentary succession. The litho-stratigraphic cross sections in the Nile Delta are shown in Fig. 11. There are two main geological units in the Nile Delta region which are:

(a) The Quaternary deposits that include the Holocene and Pleistocene sediments. The Holocene comprises sand dunes, coastal deposits, sabkha deposits, and silty

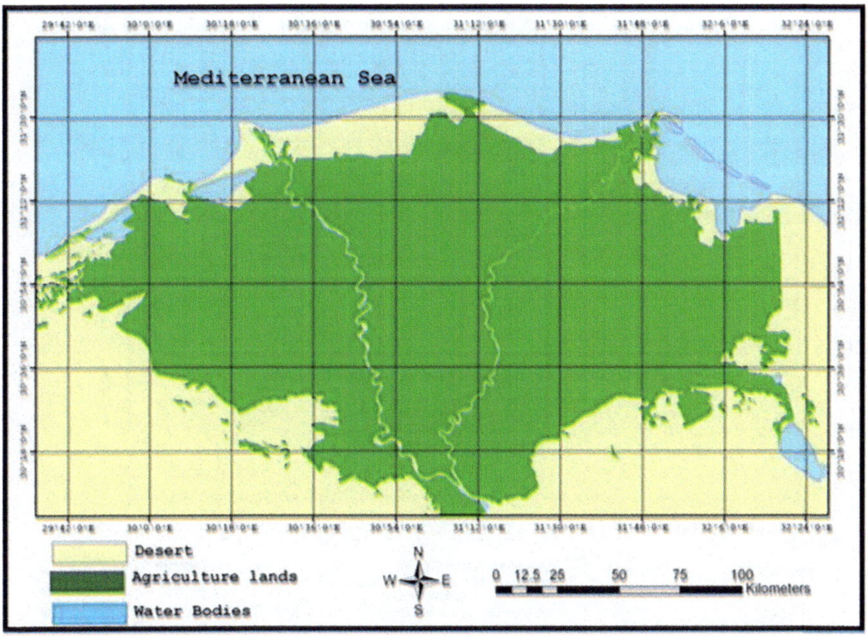

Fig. 8 Land use map for the Nile Delta region [22]

clay sediments capping the floodplain. The Pleistocene comprises desert crusts, kurkar ridges, and the graded sand and gravel that contain the main water-bearing formation. The thickness of the Quaternary aquifer is about 100 m at Cairo, reaches to 1,000 m at the coast, and decreases to zero to the east and west of the Nile Delta fringes. Figure 12 shows the contour lines of the base of the Quaternary aquifer relative to the mean sea level [36].

(b) The Tertiary deposits include the Pliocene, the Miocene, the Oligocene, the Eocene, and the Paleocene sediments. The Pliocene forms the lower boundary of the main water-bearing formation. The Miocene deposits (e.g., the Moghra) exist underneath the surface with a thickness that reaches up to 2,000 m. The Oligocene and Eocene are of a little hydrogeological interest due to their small contribution to groundwater.

5.2.2 Geological Evolution of the Nile Delta Region

History of the Nile Sediment Deposition

The Nile passed through five main periods which are the Eonile (T_{mu}), Paleonile (T_{plu}), Protonile (Q_1), Prenile (Q_2), and Neonile (Q_3).

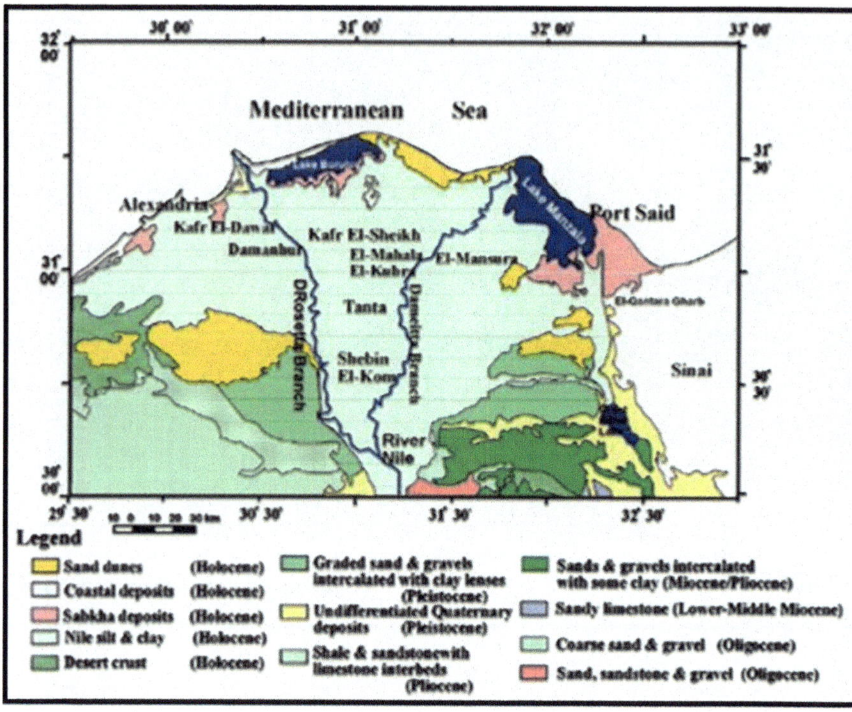

Fig. 9 Geologic map of the Nile River Delta (Based on CONOCO Coral 1987 and TM Satellite image, taken in 2003)

Eonile: The Eonile predominated in the Late Miocene. The depth of the Eonile canyon in Northern Egypt reached 2,500 m. The available drilling data indicates the existence of evaporite in both North Delta Embayment and the Mediterranean Sea. Miocene Eonile is alternating beds of coarse-grained quartz and poorly sorted sandstone with pebbles.

Paleonile: The Early Pliocene sediments in the North Delta Embayment consist of thick sand-shale deposits. The sediments include more sand members in their lower part and are made up of shale in their upper part. The Early Pliocene sediments in the South Delta Block vary in thickness and lithological composition from one place to another, depending on their position relative to the Eonile canyon. The sediment thickness in the South Delta Block is in the range of 250–500 m, whereas in the North Delta Embayment, the Paleonile sediment thickness approaches 1,000 m.

Protonile: The fluviatile deposits occupying the Protonile basin were in the form of complex gravel, coarse sand, and loamy deposits. The loose gravel layer ranges in thickness from 2 to 4.5 m and reaches 6 m in some places. These deposits could also be found in the Nile valley.

Prenile: The Prenile represented a various and competent river with a huge supply of water and a wide floodplain. Sediment is coarse, massive, and thick. The source of these sediments is from the Ethiopian Highlands. In the Delta, a consistent

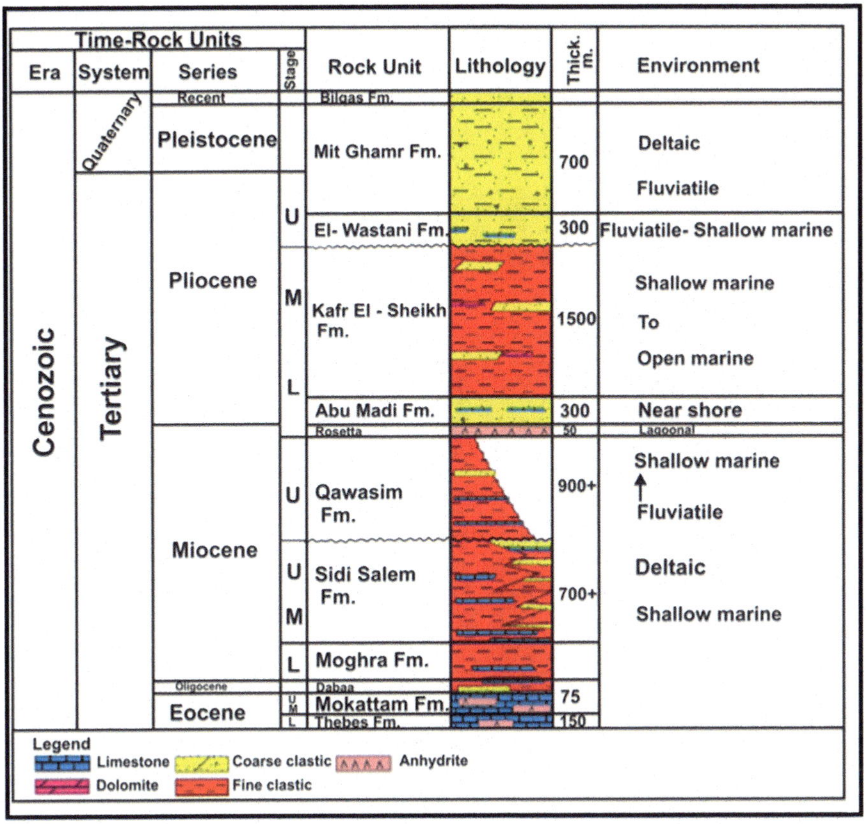

Fig. 10 Generalization litho-stratigraphic column of the Nile Delta [35]

unit of thick layers of quartzose sands and pebbles occurred above El-Wastani formation (Mit Ghamr formation). The sands are medium to coarse-grained. The pebbles are mainly quartz. In some locations, interbeddings of coquina of marine shells exist. Some peat layers are also observed. The upper levels of this formation are marked by the appearance of thin beds of clays, silt, and peat containing coastal or lagoonal faunas. Said [37] concluded that the Delta extended into the Mediterranean Sea and had an area at least three times as large as that of the present Delta.

Neonile: The deposits of the Neonile are made up of silts and clays similar to those which were deposited over the land of Egypt by the modern Nile up to the very recent past. Seven major branches of the Nile are mentioned in various documents and in ancient maps. Five of them silted up in the course of history, whereas two, the present-day Damietta and Rosetta Branches, remained active. Prior to 1900, the sediment supply by the Nile was estimated to be more than 100 million tons per year [38]. Old maps showed that between 1800 and 1900, the Rosetta and Damietta promontories advanced 3.6 and 3 km, respectively. From 1900 up to the present time, the river sediment discharge has decreased due to various reasons, such as control works constructed on

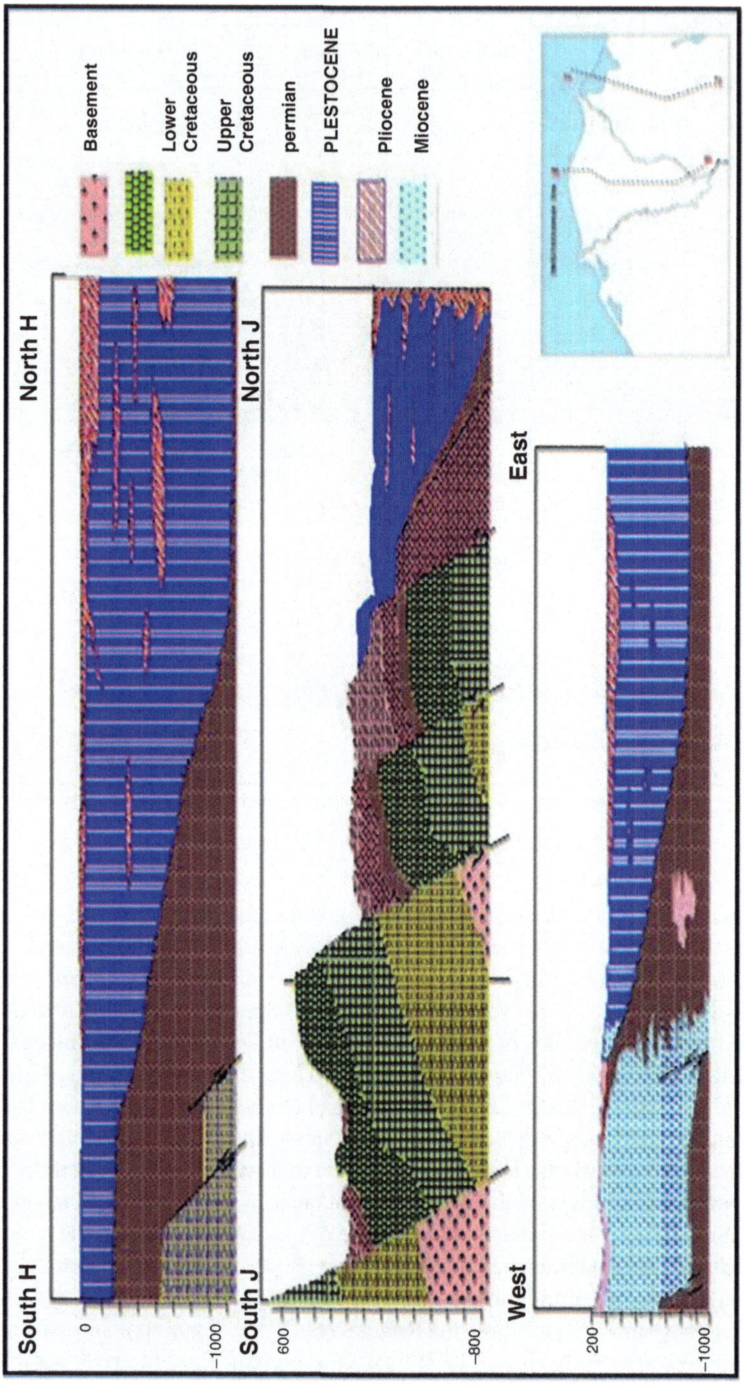

Fig. 11 Geological cross sections in the Nile Delta region [23]

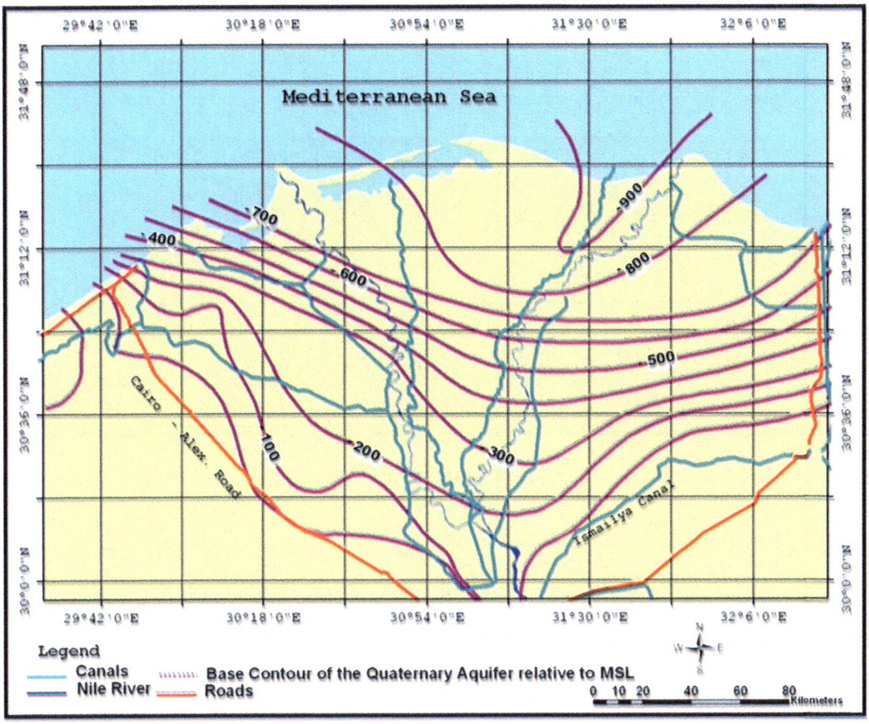

Fig. 12 Contour map of the base of the Quaternary aquifer relative to the mean sea level [36]

the river itself, the use of water for permanent irrigation, and interaction with the shoreline morphology [39]. The rate of sedimentation in the Delta would be 12.5 cm/century or 20 times that are prevailing today in the perennially irrigated Delta lands [38].

History of the Nile Delta Shoreline and the Mediterranean Sea Level

The shoreline of the Nile Delta extends for 240 km along the Mediterranean Sea from Alexandria City in the west to Port Said City in the east. Regional variations in the Holocene deposits predominate the location of the coastal flexure zone. This zone of Holocene sediments deviates from the present line with a distance varying from 15 km (in the north-central sector) to 22 km in the northwestern plain and reaching 45 km in the Manzala lagoon region [40]. The coastal region of the Delta has been affected by the sea level fluctuation regime. Lowering the sea level exposes the submarine formations to aeration and leaves sanded beaches and shore lands exposed forming a new shoreline. Figure 13 indicates that ancient beach and dune formations which are now submerged wedges should be found above the level of −70, −50, and −30 m. The ancient shore formation caused by lowering the sea level was later

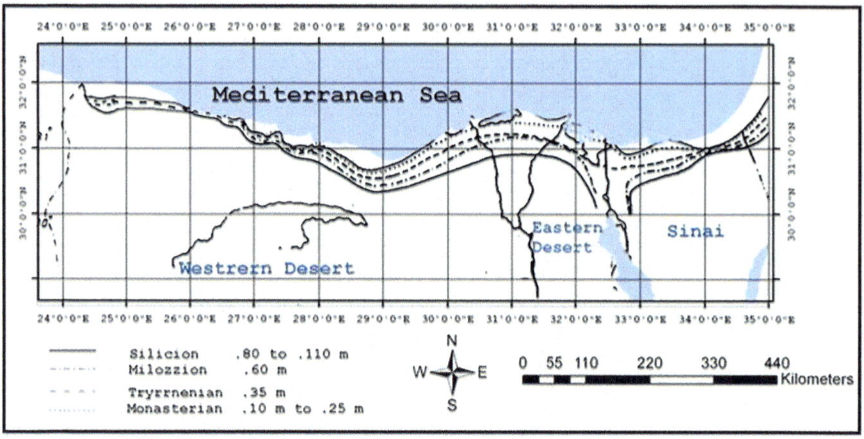

Fig. 13 Ancient Nile Delta's shoreline of the Mediterranean Sea [34]

subjected to a transgressive sea. At the beginning of the Holocene, the Delta must have had a coast that probably lay 50 km to the north of the modern coast as shown in Fig. 13. The regime of the river has been more or less constant for the past 9,000 years and that the Nile in Egypt must have assumed at the inception of the Holocene its modern gradient.

The Early Miocene depositional facies range from nonmarine in the south shelf and slope to the north. Due to sea level changes, the thickness of Early Miocene beds is highly influenced by rotational block faulting in the east and west parts of the Delta. Middle Miocene depositional environments are similar to the Early Miocene, in that nonmarine deposits occupied the southern part of the Delta. The thickness patterns are related mainly to structural movement in central part of the Delta, Late Miocene. During the Tortonian, this river may have built a small delta, which progressed eastward beginning nonmarine or marginal marine conditions to San El Hagar and eventually to Matariya. The thickness of Late Miocene beds is shown in the northeast part of the Delta region. The Late Miocene is thin in the north-central Delta area partly because of the low rates of subsidence. Rotational faulting affected thickness variations in the east and west-central parts of the Delta area.

During the period 1981–1990, the analysis of 230 profiles along the Nile Delta coast revealed that the shoreline retreated along almost the entire coast except in some short regions between Abu Madi and west of Rosetta [39]. The rate of the shoreline retreat varied between 70 m/year at Rosetta and about 5 m/year at Burullus Lake.

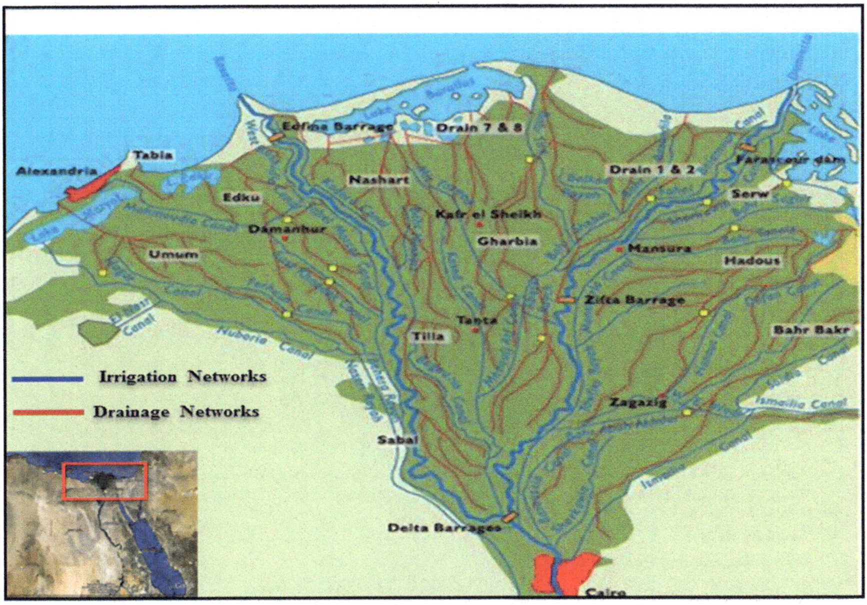

Fig. 14 Main canals and surface drains of the Nile Delta (http://entroportal.nilebasin.org)

5.3 *Hydrogeology of the Nile Delta Region*

5.3.1 Aquifer Geometry

The hydrogeologic investigation of the Nile Delta aquifer revealed the existence of well-defined hydraulic boundaries that delimit the aquifer geometry. As shown in Fig. 14, the aquifer is bounded by the Mediterranean Sea in the north and the Suez Canal in the east. Ismailia Canal in the southeast is considered as the limiting boundary of the aquifer, while in the west and southwest, Wadi El-Natrun fault represents another distinct boundary of the aquifer. In the vertical dimension, the Nile Delta aquifer has variable thicknesses as shown earlier in Fig. 14.

5.3.2 Irrigation: Drainage Network in the Nile Delta Region

The Nile River represents an outstanding feature that has shaped not only the physical characteristics of the area but also its history and the nature of its settlers. The water resources in the Nile Delta consist of complex irrigation and drainage networks which are hydraulically connected to the underlying aquifer system. The Damietta and Rosetta branches of the Nile River, the irrigation system, and drains are formed through the area. These hydraulic features are directed to the natural discharging areas, namely, the northern lakes and the Mediterranean Sea. The

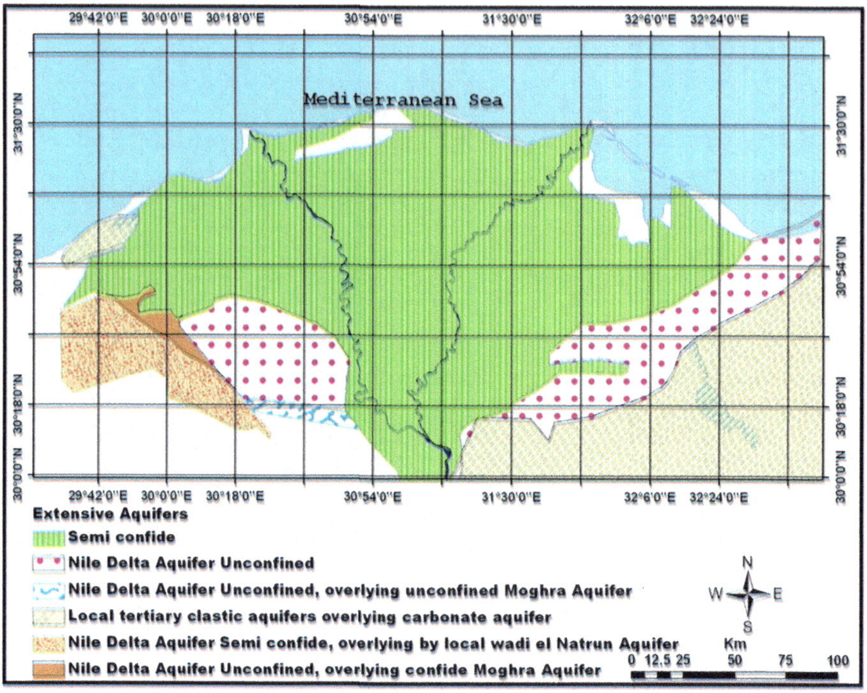

Fig. 15 Configuration of aquifer systems in the Nile Delta region [22]

surface water levels in the Nile River and its irrigation canals are controlled by the Delta Barrages and the other hydraulic control structures scattered everywhere. At the tail ends of the two branches, hydraulic dams were constructed to prevent the flow of the salt water from the Mediterranean Sea to the Nile River. Figure 15 presents an overview of the Nile valley and Delta where the irrigation network and main structures are pointed out. The irrigation and drainage system is complicated, and there is a large portion of the agricultural drainage water is reused to supplement the shortage of the freshwater especially in the low reaches of the canals [7].

Since the construction of the Aswan High Dam (AHD), the flow rates in the irrigation system vary according to the variable water demand. This water demand does not change significantly from year to year unless new expansion projects are being implemented. During the low flood periods, there is a chance to have less flow in the Nile River, and hence the water levels in all the irrigation-drainage system are affected. The great barrages built on the Nile are controlling the surface water levels in the Nile branches and the irrigation canals. The lands of the delta were converted to perennial irrigation at the beginning of the last century.

Open drains are extensively constructed in the Nile Delta region and used principally to drain excess irrigation water. Nowadays, most of the Nile Delta lands are served by tile drains to improve the agricultural drainage. These tile drains are constructed at a depth of 1.5 m below ground surface. Near the coastal line, the water levels in the open drains are less than the sea level. Therefore, water has to be discharged mechanically to the northern lakes or directly to the Mediterranean Sea.

5.3.3 Surface Water and Groundwater Relationship

There is a distinct hydrogeological relationship between the groundwater and surface water in the area of the study, where this area is dissected by many surface freshwater canals and drains at the west, east, and south and saltwater bodies at the north and east. The contact between surface water and groundwater in the area of the study varies according to the nature of sediments, infiltration rate, groundwater water table relative to the surface water level, hydraulic conductivity of the soil, and the difference in hydraulic properties of both surface and groundwater [41].

5.3.4 Groundwater Aquifer System in the Nile Delta Region

On the basis of geomorphology, the hydrogeological and hydrochemical features of the Nile Delta region are divided into three major regions which are the floodplain (area of 9,126 km^2), the eastern Nile Delta fringes (area of 10,220 km^2), and the western Nile Delta fringes (area of 11,042 km^2). There are different groundwater aquifers with different importance for exploitation in the Nile Delta region. These aquifers are the semi-confined Quaternary aquifer, phreatic sandy aquifer, Pliocene aquifer, Moghra aquifer, and sand dune aquifer as shown in Fig. 15.

The Moghra aquifer is located in the Western Delta region. It has an area of about 50,000 km^2. The aquifer consists of Lower Miocene sand and gravel. The aquifer is found in the west of Wadi El-Natrun and extends toward the Qattara Depression. The aquifer is phreatic south of latitude 30°N but confined by Pliocene deposits in the northern direction. Oligocene rocks (basalts or shales) underlie the Moghra aquifer. The base of the Moghra aquifer slopes from ground level near Cairo to 100 m above sea level near Borg El Arab area. The saturated thickness ranges from 70 to 700 m [42]. The groundwater flow in the Moghra aquifer is directed westward, toward the Qattara Depression. Inter-aquifer flow is a minor component of recharge, which occurs from the Nile Delta toward the Moghra and Pliocene aquifers. The groundwater salinity is good to brackish with a maximum value of 7,000 ppm. The salinity values increase from very low in Wadi El-Farigh to high in the north and western part. West of Wadi El-Natrun, the groundwater quality is brackish. The potentiality of Moghra aquifer varies between low and moderate according to the Research Institute for Groundwater (RIGW) [22]. The outflow of the Nile Delta aquifer also occurs toward the Moghra aquifer along the fringes of the western Nile Delta, with a transfer estimated by RIGW/IWACO (1990) between 50 and 100 Mm^3/year [42].

The Pliocene aquifer is present in Wadi El-Natrun depression. The aquifer is considered a local low productive aquifer. It is a multilayered aquifer consisting of an alternation of sand and clayey layers belonging to the Pliocene age. The aquifer

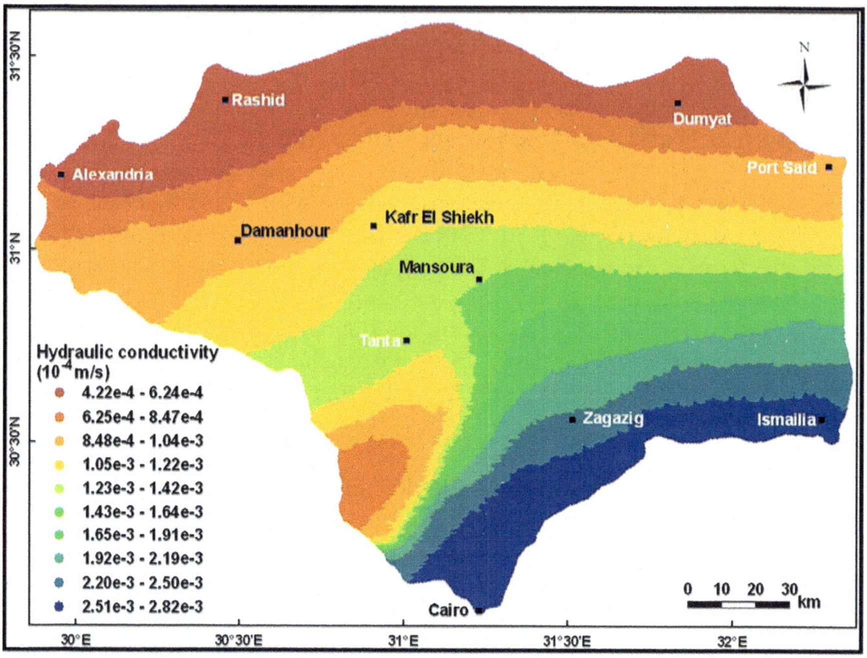

Fig. 16 Calibrated hydraulic conductivity for the clay layer [43]

is underlain by the Moghra aquifer but separated from it by layers of lower Pliocene age. In this area, the groundwater is discharged through a great number of seepage zones into small lakes and ponds. Groundwater is lost by direct evaporation with an annual rate of 70 million cubic meters according to RIGW (1998) [22].

The most important regional aquifer in the Nile Delta is the Quaternary aquifer. This aquifer consists of Pleistocene graded sand and gravel, changing to fine and clayey facies in the north. The aquifer is found along the entire Nile Delta floodplain. The clay cap of the Nile aquifer is a semi-confining layer and has a thickness up to 20 m. The Nile Delta aquifer is underlain by Pliocene marine clay in the Central Delta and wedges out toward the fringes.

Data were collected and interpreted to outline the hydrogeological environment of the Nile Delta groundwater aquifer. Simplified latitudinal and longitudinal hydrogeological cross sections, representing the Nile Delta aquifer, were developed as shown in Fig. 16. According to these cross sections, the hydrogeological units of interest in studying the Nile Delta aquifer are as follows: a top unit of Holocene clay aquitard, Quaternary and Late Tertiary gravel and sand unit (aquifer), and basal unit of Pliocene clay aquiclude (Fig. 17).

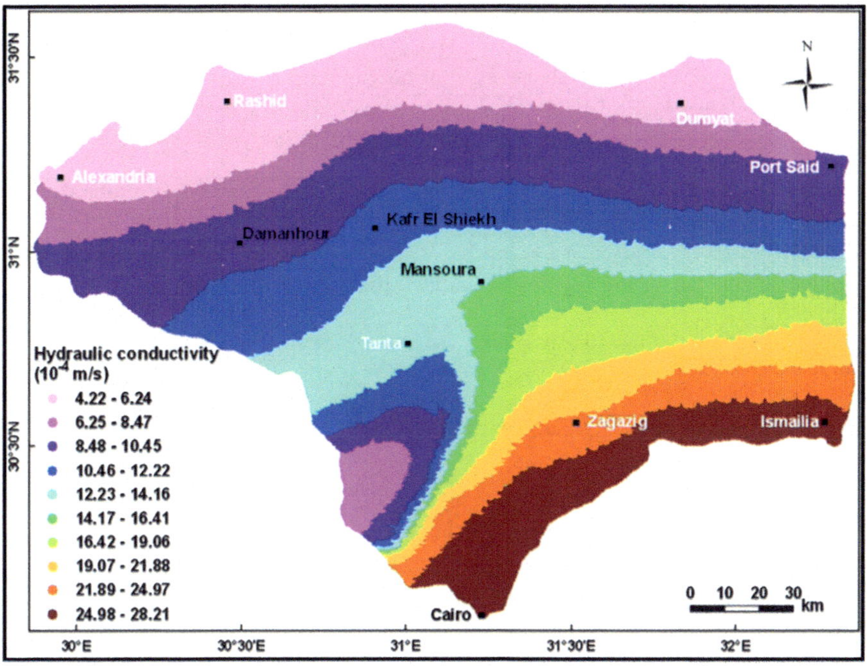

Fig. 17 Calibrated hydraulic conductivity for quaternary layers [43]

5.3.5 Hydrogeologic Characteristics of the Nile Delta Aquifer Units

Top Unit of Holocene Clay and Fine Sand (Aquitard)

The top boundary of the deltaic deposits is a formation that belongs to the Holocene (10,000 years). This formation is made up of a semi-pervious clay and silt. It acts as a cap for the main Quaternary aquifer. It is generally heterogeneous and anisotropic. "This unit mainly consists of Nile silt, sandy clay, clayey sand, occasionally with fine sand intercalations" [36]. The clay content in the topsoil layer ranges between 5 and 50%, while the silt content varies between 5 and 95% in most of the area. The thickness of this top layer varies from 20 m in the north and totally absent at some localities such as the Nile Delta fringes and turtlebacks. According to the laboratory experiments made by RIGW [36], the average vertical hydraulic conductivity of the clay cap is 2.5 mm/day, and the average horizontal hydraulic conductivity is between 50 and 500 mm/day.

The aquifer porosity ranges between 25 and 40%. According to the thickness and the hydraulic conductivity of the top clay layer, the second aquifer unit exhibits unconfined and semi-confined conditions. For the unconfined aquifer condition, the specific yield is

Table 3 Hydraulic parameters of the Quaternary aquifer in the Nile Delta

Main hydraulic units	Hydraulic conductivity K (m/day)	Transmissivity T (m^2/day)	Storage coefficient S	Specific yield Ss (1/m)	Porosity n (%)	Effective porosity n_{eff} (%)
RIGW [23]	75	15,000–75,000	10^{-4}–10^{-3}	–	25–40	–
Farid [45]	112	–	2.35 10^{-3}	–	40	37.35
Mabrook et al. [46]	72–108	–	–	–	21–30	–
Zaghloul [47]	119	–	10^{-4}–10^{-3}	0.15	30	–
Leaven [48]	150	10,350–59,800	–	–	25–30	–
Bahr [49]	75	–	1.1 10^{-3}	–	25	18
Sherif et al. [43]	36–240	2,000–15,000	–	–	25–40	–

0.20, and for the semi-confined aquifer condition, the storage coefficient ranges between 10^{-3} and 10^{-4} [44].

The water in this aquitard layer is in contact with the main underlying aquifer through downward or upward leakage. A large area of Nile Delta covers with a clay cap. This cap keeps the Nile Delta aquifer to semi-confined aquifer as shown in Table 3. The thickness of the clay layer varies from 5 to 20 m in the south and the middle part of the Delta and reaches 50 m in the north [50].

Quaternary and Late Neogene Gravel and Sand Unit (Aquifer)

This is a thick unit that consists of coarse sand and gravel with occasional clay lense intercalations. It underlies the Holocene top clay layer and overlies the lower marine clay deposits of Neogene impervious clay. It belongs to Pleistocene and its lower part to Late Pliocene time [36]. The thickness of these strata increases northward. It ranges between 150 m at El Qanater El Khayreya in the south and more than 500 m near Tanta. It increases further in northward to reach 1,000 m near the coast. Also, the aquifer decreases in thickness toward the southeastern and western fringes of the Nile Delta, Fig. 18. The hydraulic conductivity of this aquifer unit increases northward and eastward. It ranges between 50 m/day south of El-Bagur and increases northward to more than 100 m/day [36].

This unit forms the main groundwater aquifer in the area, which is considered a semi-confined (leaky) aquifer in most areas where the clay cap covers the aquifer. Unconfined (phreatic) groundwater aquifer is present in the fringes of the Nile Delta where the Quaternary sediments are not covered with clay cap. Leakage and infiltration from the surface water system are the main processes of recharging the aquifer with a very great recharge rate. On the other hand, the groundwater in this aquifer is discharged naturally, along with some areas on the Rosetta Branch and to

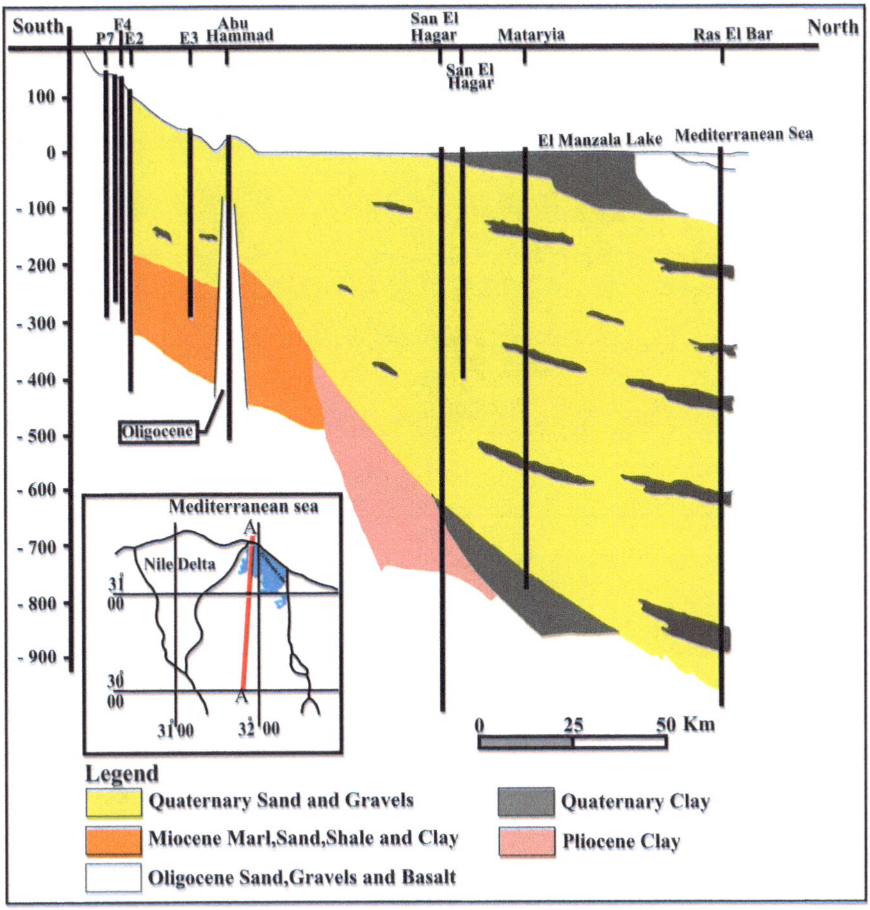

Fig. 18 Longitudinal cross section showing the thickness and litho-facies variation of the Quaternary aquifer in Nile Delta [51]

the Western Desert, the Suez Canal, and the Mediterranean Sea, or mechanically by pumping for irrigation, drinking, and/or industrial purposes. The thickness and lithological facies change of Quaternary aquifer are illustrated Fig. 18.

Basal Unit of Pliocene Clay (Aquiclude)

The base of the deltaic deposits rests unconformably on a thick and dense clay section, which belongs to the Pliocene age. This clay section acts as an aquiclude that does not have any potential.

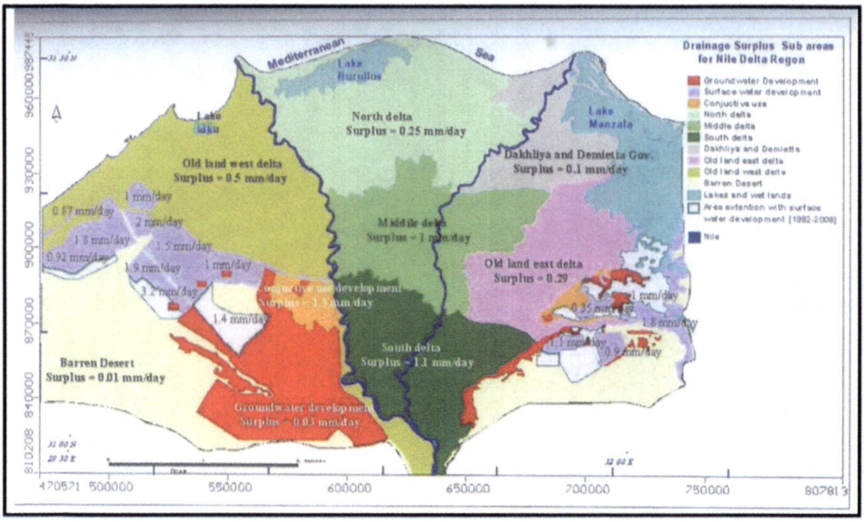

Fig. 19 Distribution of drainage surplus in the Nile Delta region in 2008 [54]

5.3.6 Recharge-Discharge Sources of the Nile Delta Aquifer

Aquifer Recharge

The Nile Delta aquifer in the floodplain is continuously recharged by irrigation water in the southern and central portions. Therefore, the aquifer acts as a storage reservoir that can be used conjunctively with the other surface water supplies. Recharge of groundwater is taking place in the Nile Delta area by three processes: (1) infiltration of rainfall; (2) downward leakage of the excess irrigation and seepage from the Nile River, irrigation canals, and drains; and (3) inter-aquifer flow of groundwater.

Recharge of the Nile Delta aquifer occurs mainly through direct seepage from the irrigation canals and drains. In the central and southern parts of the Nile Delta floodplain, the downward leakage toward the aquifer varies between 0.25 and 0.80 mm/day [52]. In the desert fringes, high leakage rates of 1–2.5 mm/day are recorded for basin irrigation, whereas low rates of 0.1–0.5 mm/day are occurring for drip irrigation according to RIGW [53]. Figure 19 shows the distribution of drainage surplus in the Nile Delta region in 2008.

Aquifer Discharge

Discharge of groundwater takes place by four components: outflow into the drainage system, direct evaporation, extraction, and inter-aquifer flow of groundwater.

Groundwater discharge to the drainage system occurs in the northern portions of the Delta through upward seepage with a daily rate of 0.2–0.9 mm/day

[52]. Discharge of groundwater through evaporation may occur in low-lying areas with a shallow groundwater table. In Wadi El-Natrun depression, lakes, and sabkhas, groundwaters are discharged naturally by evaporation.

Inter-aquifer flow is a minor component of discharge, which occurs between the Nile Delta aquifer on the one hand and the Moghra aquifer and the Wadi El-Natrun aquifer on the other hand. The total amount of groundwater flowing from the Nile Delta to the Moghra aquifer is estimated to vary between 50 and 106 million m^3/year [36]. Groundwater extraction from the Nile Delta aquifer will be presented in a subsequent section. Abstraction rates over the last 30 years increase during the period of 1980–2010. The total annual abstraction rate in 1980 was estimated at 1.6×109 m^3/year according to the Research Institute for Groundwater (RIGW) in Egypt, and the net recharge rate to the Quaternary aquifer was estimated to be 2.60×109 m^3/year. RIGW report confirmed significant increase patterns of abstraction, which reached around 2.6×109 m^3/year in 1991. In 2003, the total annual abstraction reached 3.5×109 m^3/year [55]. In 2010, it reached about 4.6×109 m^3/year. The increase of abstraction in Nile Delta over the period from the year 1981 to the year 2010 is shown in Fig. 20. It can be noticed that it increases linearly by about 0.10×109 m^3 per year, except from the period of 2003 till 2010 where the abstraction increases dramatically by the rate of 0.20×109 m^3 per year [56] (Fig. 21).

5.3.7 Groundwater Levels in the Nile Delta Aquifer

Depth to Groundwater Surface

The depth to groundwater surface depends mainly on the ground surface elevation at a particular point. The field survey indicates that the depth to the groundwater surface is less than 5 m in most of the Nile Delta region, Fig. 22. It is obvious from the map that the depth to groundwater decreases northward and northeast. As a result, waterlogging problems are encountered in such areas. In the Nile Delta fringes, the depth to groundwater increases to more than 50 m due to high topography in these areas.

Fluctuations of the Piezometric Heads

Groundwater levels in the Nile Delta aquifer fluctuate in response to the stage of the Nile River, aquifer recharge from excess irrigation water and groundwater pumping. Since the construction of the Aswan High Dam, it was observed that the piezometric head increased within the Nile Delta aquifer. The agriculture projects that rely on the groundwater resources caused a severe decline in the piezometric head as noticed in the delta fringes. On the contrary, the agriculture projects that depend on the Nile water caused an increase in the water table and the development of water mounds at some localities.

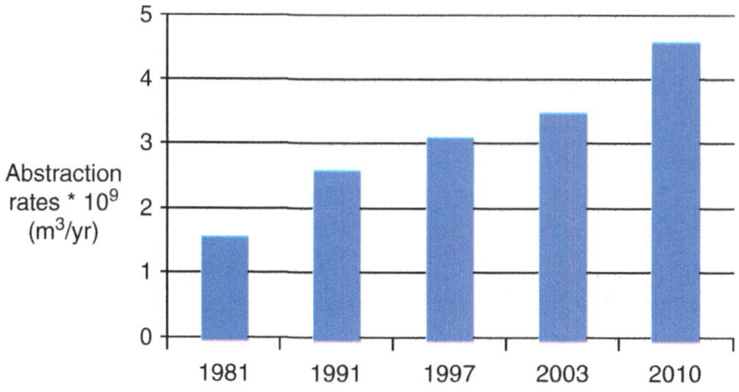

Fig. 20 Abstraction rates versus time in Nile Delta (RIGW, 1980, 1992, 1999, 2003, and 2010) [56]

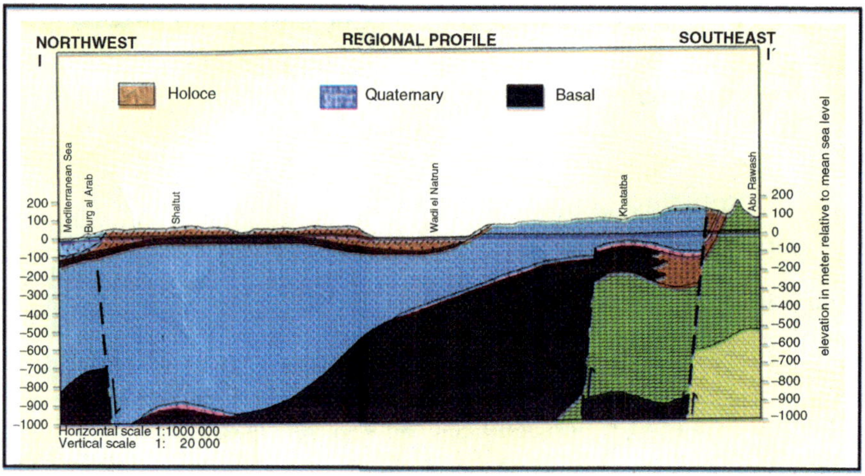

Fig. 21 Hydrogeological cross section in the Nile Delta area [22]

The piezometric head map for the year 2002 (Fig. 23) shows that the groundwater level decreases gradually toward the north and northeast direction with an average gradient of 11 cm/km. The groundwater head decreases from 15 m at Cairo to less than 1 m near the Mediterranean coast.

Records of the groundwater levels in the Nile Delta region are obtained from observation wells by RIGW. These wells have well screens at a depth between 20 and 75 m below groundwater surface. On the other hand, the production wells tap the aquifer generally at a depth between 20 and 150 m below groundwater surface, whereas the production wells tap the aquifer generally at a depth between 20 and 150 m. The groundwater levels in the study area range between 16.96 m above (MSL) at the southern border and 0 m above (MSL) at the northern. Figure 24

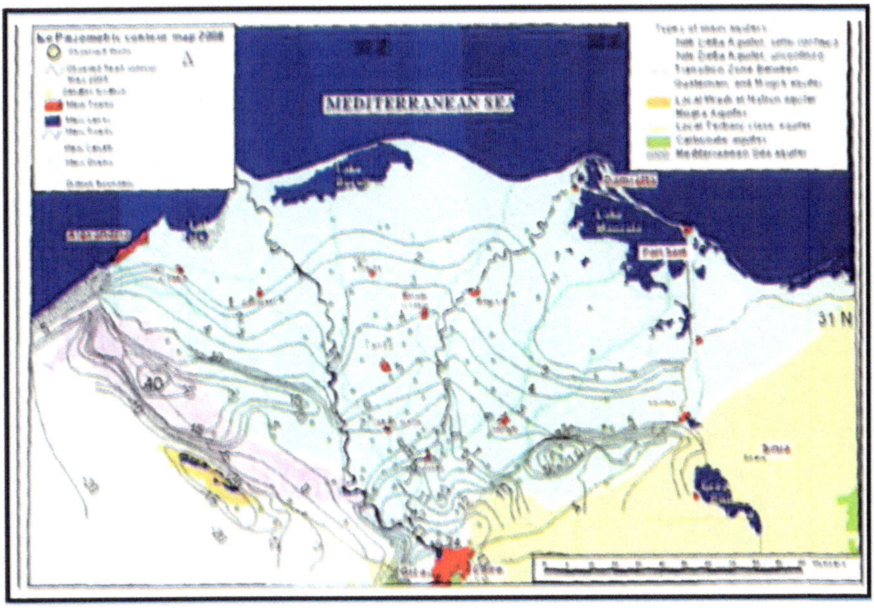

Fig. 22 Contour map of groundwater in the Quaternary aquifer recorded in 2008 [54]

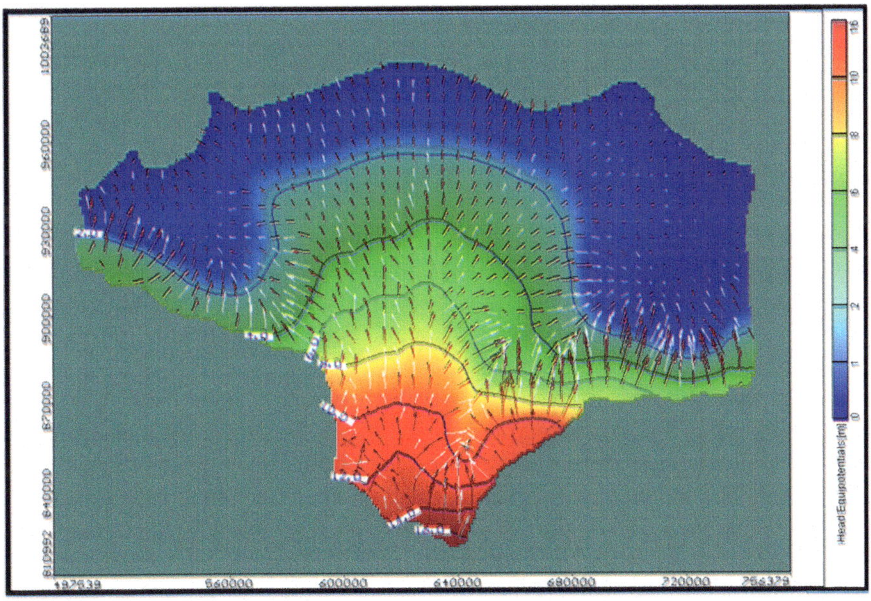

Fig. 23 Contour map of groundwater and velocity direction using numerical model for Quaternary aquifer [57]

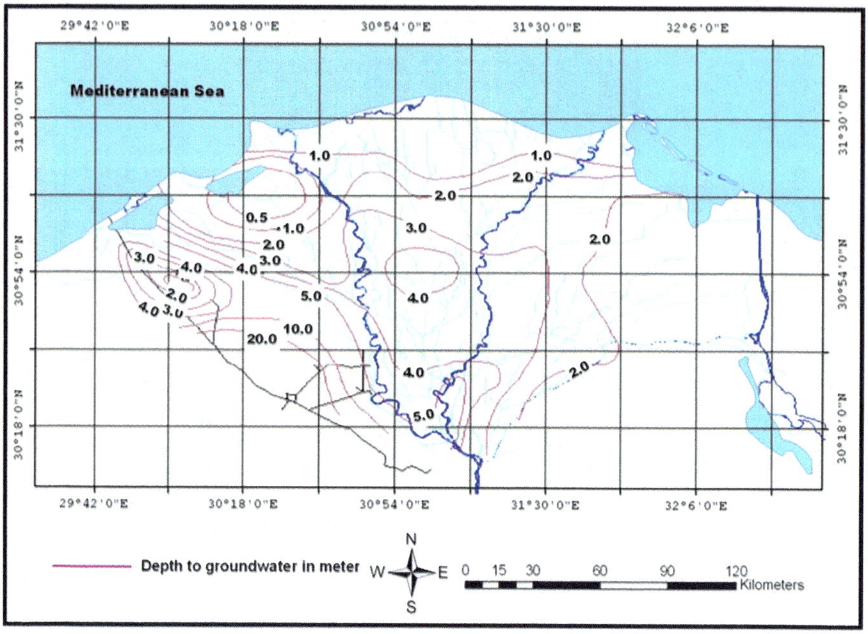

Fig. 24 Depth to groundwater in the Quaternary aquifer [58]

shows groundwater depth in the aquifer, and also Fig. 25 shows the observed piezometric head contour lines for 2008.

5.3.8 Groundwater Use in the Nile Delta

Before the construction of the Aswan High Dam, groundwater in the Nile Delta was used to supplement irrigation water at the tail ends of the irrigation canals. In 1957, the first well inventory was launched and revealed that a total of 0.2 billion cubic meters was pumped annually from the aquifer through 5,600 wells [59]. It was reported that most of the production wells are 70 m deep and the screen length ranges between 20 and 30 m. Shallow wells operated by hand pumps are still extensively used for domestic purposes especially in the rural communities of the Nile Delta. Freshwater outflow to the sea was calculated, and it was found that a total of 0.37 billion cubic meters was flowing directly to the sea in 1958 compared to 0.283 billion cubic meters in 1962. The reduction in the outflow of the freshwater was referred to as the illegal drilling of new wells by farmers. For coastal aquifers such as the Nile Delta, such freshwater outflow to the sea is required to keep the balance of the interface between the salt water and the freshwater.

In 1980 an extensive study was conducted to evaluate the safe yield of the Nile Delta aquifer [36]. It was reported that the total annual extraction rate from the aquifer was 1.6 billion cubic meters, while the net recharge rate to the aquifer from

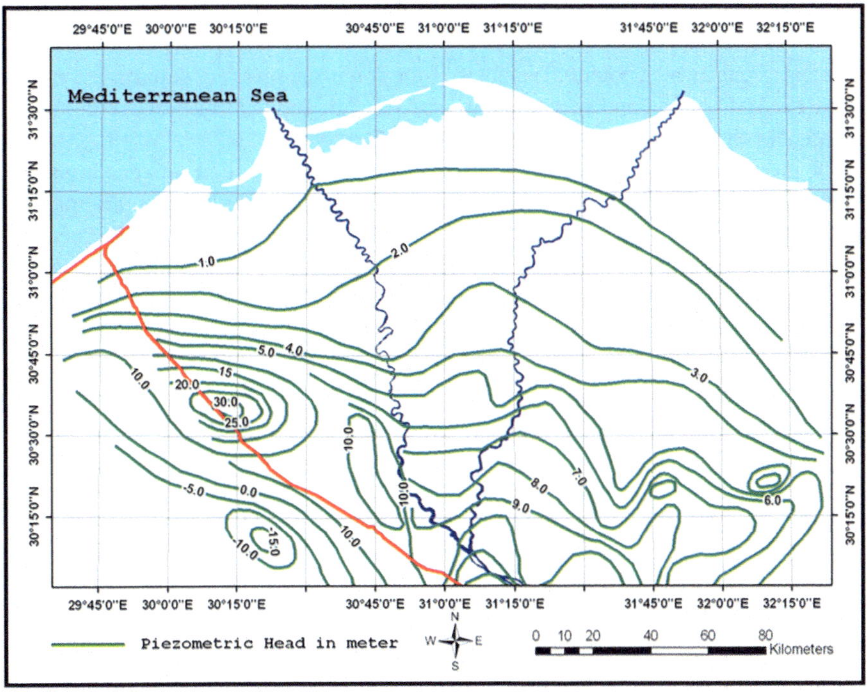

Fig. 25 Piezometric head map of the Quaternary aquifer [58]

the Nile River and the irrigation canals was 2.645 billion cubic meters. The annual outflow of freshwater to the sea was estimated as 97 million cubic meters. Despite the reduction in the outflow to the sea and the increase in the groundwater extraction, the chemical analysis of the groundwater did not show a significant increase in the salinity level. This concluded the fact that the salt water and freshwater were in dynamic equilibrium. The study recommended increasing the annual extraction by 500 million cubic meters to lower the piezometric surface to a level that prevents waterlogging and soil salinization. This recommendation was based on the results of a two-dimensional finite difference model that did not account for the seawater intrusion phenomenon.

In 1991, the total annual groundwater extraction from the Nile Delta aquifer was 2.77 billion cubic meters [23]. The total number of wells was 13,000 compared to 5,600 wells in 1958. In 1997, the annual groundwater extraction was reported as 0.86, 1.6, and 0.56 billion cubic meters for the western, middle, and eastern Nile Delta regions, respectively [60]. Recently in 2003, the annual extraction rate was 0.9, 2.0, and 0.6 for the 3 regions of the Delta [61]. From the above, it is obvious that since 1981 the groundwater extraction is increasing annually in a linear fashion by 0.1 billion cubic meters.

5.3.9 Groundwater Potential in the Nile Delta

The total groundwater potential in each of the three regions was determined using the TRIWACO model and was calculated as 1.2, 2.4, and 0.71 billion cubic meters per year for the Western, Middle, and Eastern Delta [58].

The groundwater potentiality of the floodplain can be classified between high and moderate. Areas with high groundwater potential cover the south and central part of the Nile Delta. The characteristics of the aquifer (semi-confined) and type and continuity of recharge (surface Nile water) enable the extraction of large quantities of groundwater of good quality from shallow depths (0–5 m). Areas with moderate potential are found in the north where the water salinity is increasing.

In the western fringes, areas with moderate potential are located near the Nile floodplain, which are characterized by its good water quality. These areas are irrigated with surface water and/or groundwater. However, the depth to groundwater and the aquifer productivity are classified as moderate to low. Low potential areas in the western Nile Delta fringes are characterized by the deep groundwater and the poor water quality where salinity is more than 5,000 ppm. In areas where the groundwater depth exceeds 80 m from the ground surface, groundwater exploitation is not economically feasible.

Around the Eastern Delta projects which are irrigated using surface water, private farmers have reclaimed their land using groundwater. The aquifer has a medium to low potentiality in these areas. The aquifer is characterized by its limited recharge and over-abstraction of groundwater.

5.4 Hydrogeochemistry of the Nile Delta Aquifer

5.4.1 Groundwater Origins

The main source of groundwater in Nile Delta aquifer is the Nile, which flows from the Ethiopian Plateau. Therefore, the groundwater is of meteoric origin. The meteoric water of Pleistocene sediments covers all Delta except the coastal area up to N 31° 00′ which is occupied by the saline water of marine origin coming from either the Mediterranean Sea or from the old marine deep aquifers. In the Miocene aquifer of the Western Delta, the groundwater is of meteoric origin, whereas the same aquifer in Eastern Delta is of marine origin as well. The groundwater of Pliocene aquifer in Wadi El-Natrun is of marine origin. Groundwater of old marine origin was observed in areas with old active tectonic movements ascending along fault plains in the Northern Delta, in the southeastern fringe, and in the northwestern fringe. This type of groundwater is characterized by the considerable presence of $CaCl_2$. Old meteoric groundwater was observed along the old passages of Nile tributaries in the eastern and western fringes. This type of groundwater is characterized by the presence of Na_2SO_4.

The main sources of contamination of the Nile Delta aquifer are agriculture, domestic and industrial effluent. High concentrations of nitrate (No_3^-), sulfate (So_4^{-2}), potassium (K^+), and phosphate (PO_3^{-4}) are caused by the widespread use of chemical fertilizers in agriculture. The high concentration of trace elements such as iron, manganese, and aluminum in the Nile Delta was due to the dumping of industrial wastes into the drainage system. Moreover, septic tanks and the lack of sewage systems in many rural areas in the Nile Delta and disposal of domestic sewage directly into irrigation canals and drains have increased ammonium and BOD concentration in water. A number of testing samples from the Nile Delta have been taken from groundwater, surface water, and drainage water. The results revealed that the concentration of most water quality parameters and heavy metals exceeded the WHO standards. Also, some studies showed the areas where groundwater is unsuitable for irrigation [62].

5.4.2 Groundwater Types in the Nile Delta Aquifer

The water type is classified according to the chemical composition and the hydrochemical process encountered in the aquifer. Figure 26 shows the different water-type zones in the Nile Delta aquifer. In the southern part of the Nile Delta (Pleistocene and Moghra aquifer), fresh Ca $(HCO_3)_2$ and Mg $(HCO_3)_2$ groundwater types are found. The chemical composition and the hydrochemical process indicate that groundwater replenishment is taking place as a result of the continuous recharge of the aquifer by excess irrigation water. To the northwest, north, and east part of the Nile Delta, zones of fresh, brackish Na HCO_3 and Na mix water type are present. Refreshing of the aquifer is taking place as a result of the downward seepage of irrigation water. More to the north and in the eastward and west direction, a zone with NaCl-type groundwater is found. In this zone, the groundwater is brackish to saline, but to some extent, the aquifer is still flushed by freshwater. Near the coast, NaCl to seawater type is found, indicating that the groundwater is invaded by the saline water from the Mediterranean Sea or affected by the saline groundwater flowing upward from the deeper aquifers.

The present water-type distribution in the northern Nile Delta has been affected by the encountered shift of shoreline 20,000 years ago. At this time, the shoreline was at 125 m below its present level and located about 50 km north of the present coast. During this period, the coastal region was formed of saline lagoons and depressions. Following this period, around 10,000 years ago, a rapid sea level rise and a southward movement of the shoreline occurred. Later, the rate of sea level rise started to decline, and the Holocene clay cap started to accumulate as an overbank deposit of the Nile River system. As a result of the Holocene deposits and its depositional evolution, the salinity distribution in the Northern Delta is highly variable and complex [63].

The salinity of groundwater is changing with changing water levels of the canals based on analyzing the historical records. From 1957 till 1984, the water salinity records showed that it was enhanced and the freshwater was dominating and overcoming seawater intrusion. Also, the groundwater heads were increasing during this period, and he attributed that to the construction of Aswan High Dam. After 1984, the groundwater salinity started to increase due to extensive abstraction and

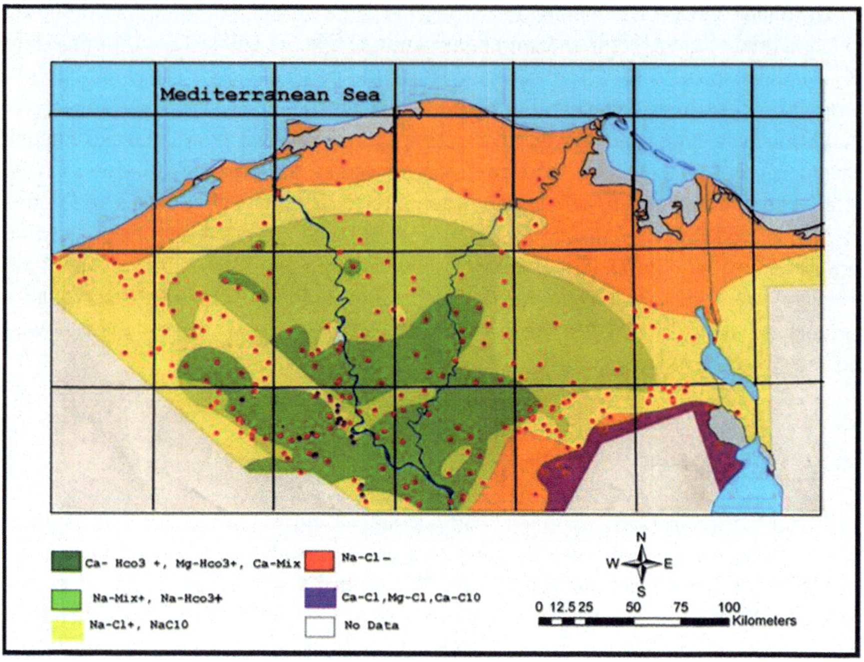

Fig. 26 Groundwater type in the Quaternary aquifer [58]

reduction in the flow of the Nile. When the Nile water flow increased in 1990, the salinity of groundwater reduced again to its former levels. However, in 2000, the salinity of groundwater increased again due to extensive abstraction and new reclamation projects [64]. From period 1992 to 2008 the contour line of 1,000 ppm became more north in the middle delta. There is an decrease in salinity in the eastern and northwestern Nile Delta fringes due to increasing recharge from Nubaria and Ismailia canal as shown in Figs. 26 and 27. There is also a increasing salinity in Southwestern fringes due to the presence of salty clay layers [54].

In middle Nile Delta: Hussein [65] assessed the impact of industrial activities in the eastern part of the middle Nile Delta. A number of 75 samples from drinking water wells, industrial production wells, sewage water and drains of the industrial area (Quesna) have been tested. The major cations and anions and ammonium (NH_4^+), nitrite (NO_2^-), phosphate (PO_4^{-3}), boron (B^+), and trace elements have been analyzed. The results showed deterioration in groundwater quality in all production wells located inside or close to the industrial area due to increasing levels of TDS, Fe, and Al concentrations. This deterioration of groundwater quality could be related to a number of causes: factories' illegal sewage disposal, dump disposal sites, leakage from transportation sewage pipes, leakage from sewage storage tanks, and leakage from El Khadrawya drain that receives the Mubarak industrial area wastewater. Ghoraba et al. [66] investigated groundwater quality in the central part of the middle Nile Delta (El-Gharbia Governorate) through the hydrochemical analysis of 34 groundwater

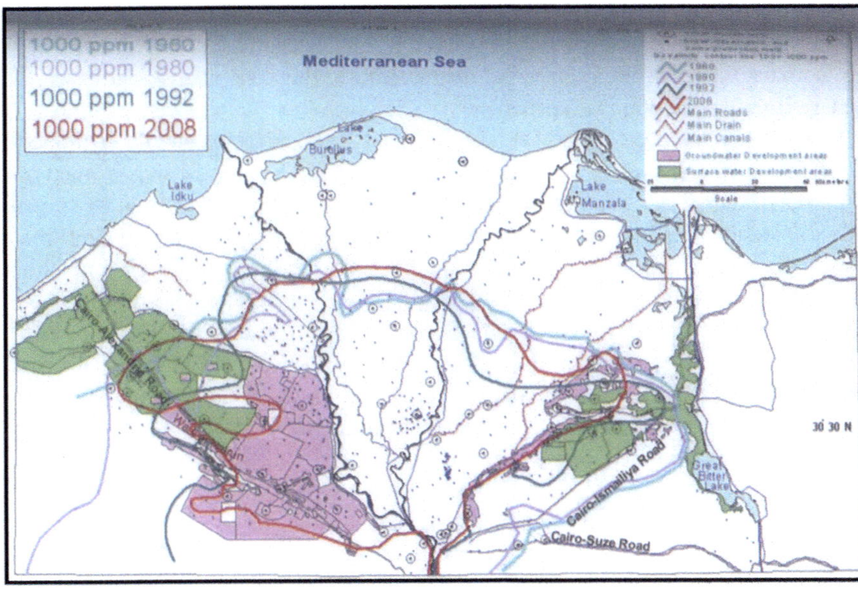

Fig. 27 Salinity (TDS) in groundwater during the periods 1960, 1980, 1992, and 2008 [54]

samples (with depths ranging between 13 and 60 m) plus 14 samples from drains and 10 samples from canals. The study concluded that the aquifer in the studied area is vulnerable to pollution. The deterioration of groundwater quality indicated that human activities caused serious pollution problems, including high nitrate concentrations within the area except in the northeastern part. The study indicated high concentration of ammonium in a village located in the central part of El-Gharbia governorate such as Birma village. The values of ammonium concentration reached 3.3 mg/L which exceeded by far the drinking water standards (0.5 mg/L). Also, nitrogen pollution was noticed due to the use of effluents from a sewage treatment plant for irrigation, sludge and animal manure, septic tanks, soil nitrogen and chemical fertilizers. They suggested a management scenario to reduce the contamination in the study area by drilling eight extraction wells to pump the contaminated groundwater out of the aquifer for treatment. Abd El-Fattah [67] conducted another study on groundwater in the southern part of the middle Nile Delta by testing the chemical properties of 80 samples. The study showed that salinity had values ranging from 296 mg/L to 810 mg/L. A salinity diagram was used by Abd El-Fattah [67] to test the suitability of water for drinking and irrigation. The result reflected the suitability of groundwater for irrigation purposes under normal conditions, while he recommended a set of mitigations for the safe use of groundwater for drinking, and also highlighted the need for treatment of wastewater before discharge into the Rosetta branch [68].

Ghoraba [69] and Zeidan et al. [70] studied the groundwater quality management in middle Nile Delta through a combination of laboratory and numerical modeling works. Environmental isotope techniques were used to investigate the recharge sources and the nitrogen compound pollution sources. MODFLOW and MT3DMS were employed numerically by applying the method of finite differences for solving the three-dimensional problem. The obtained results include the prediction of water levels and solute concentration value distribution in the area at different times. A scenario for remediation is proposed.

In west Nile Delta: Sharaky et al. [71] analyzed and assessed the hydrogeochemistry of groundwater in the western Nile Delta aquifers, using a total of 108 groundwater wells, varying in depth from 27.5 to 120 m. It has been examined and sampled to carry out the physicochemical parameters and chemical compositions of the groundwater and to obtain additional information on the possible contamination with major elements, nutrients (NO_3^-, PO_4^{-2}, NH_4^+) north and westward due to seawater intrusion and mixing with the Miocene aquifer, respectively. The freshwater is mainly concentrated in the central-eastern part, close to Rosetta Branch. The higher concentrations of the anions are chloride (Cl^-) and sulfate (SO_4^{-2}) ions, while the cations are sodium (Na^+), calcium (Ca^{+2}), and magnesium (Mg^{+2}). The concentrations of the major ions are higher than the maximum standard limits according to the World Health Organization [72]. Masoud [73] examined 451 shallow groundwater samples for the same area. Most hydrochemical parameters showed very wide ranges. Concentrations of TDS (201–24,400 mg/l), pH (6.72–8.65), Na^+ (28.30–7,774 mg/l), and Cl^- (7–12,186 mg/l) suggest complex hydrochemical processes of multiple sources. He explained that the high concentrations of Fe^{+2}, Mn^{+2}, Zn^{+2}, Cu^{+2}, and Ni^{+2} were mostly related to their natural presence in water-bearing sediments and/or to contamination from industrial leakage. Also, he reported very high nitrate concentrations exceeding the permissible limit (50 mg/L) due to wastewater leakage. Morsy [54] noted that salinity increased by 50–650 ppm from the year 1992 to 2008 in areas suffering from lowering groundwater levels, such as Khatatba road, Wadi el Faregh, and Dina farm in the west of the Nile Delta. Increasing levels of groundwater salinity were also found in the water logging areas in El Nubaria (west of Delta), and Mullak, Wadi Tumilat, Salhiya (east of the Delta) within a range of 500 to 1,200 ppm.

In east Nile delta: Morsy [54] indicated that the water samples showed contamination with reuse of agricultural drainage water in Sharqia Governorate (Bahr El-Baqar, Belbeis, and Qalyubeya drain).

6 Summary and Conclusions

Several studies have been carried out to investigate groundwater in the Nile Delta aquifer.

Groundwater represents about 22% of all freshwater on the Earth; polar ice represents 77%, while another freshwater in rivers and lakes represents about 0.30%. Groundwater is considered the second main source of water supply in Egypt after the Nile River. The Nile Delta aquifer is among the largest underground freshwater reservoirs in the world. The Nile Delta aquifer is replenished directly by the Nile water. The Nile Delta aquifer has been extensively utilized and conjunctively used with the Nile water to cope with the increased demands due to implementing economic development plan in Egypt. Increasing groundwater pumping from the Nile Delta aquifer has led to the decline of the piezometric surface and the increase in water salinity. This chapter presented an overview of groundwater resources in the Nile Delta aquifer and investigation of groundwater in the Nile Delta aquifer. It is recommended to identify all the locations where groundwater wells are operating both official and nonofficial. Also, evaluation of the efficiency of these wells should be known. Therefore, the concerning authority can take the needed measure to regulate the use of the groundwater wells and find the suitable measure to control the excessive use of the Nile Delta groundwater. This will help to protect the Nile Delta aquifer against severe saltwater intrusion.

The authors recommend a re-distribution of abstraction wells from Nile aquifer. Also, the crop pattern should be rearrangement to keep the shortage in the Delta Nile aquifer to its minimum and to help in keeping groundwater in the Nile Delta sustainable.

References

1. Bear J, Cheng AH, Sorek S, Quazar D, Herrera I (1999) Seawater intrusion in coastal aquifers, concepts, methods and practices. Kluwer Academic Publisher, Dordrecht. ISBN 0-7923-5573-3
2. Istok JD (1989) Groundwater modelling by the finite element. American Geophysical Union, Washington
3. De Filippis G, Giudici M, Negri S, Margiotta S, Cattaneo L, Vassena C (2014) Numerical modeling of groundwater flow in the coastal aquifer system of Taranto (Southern Italy). Geophysical Research Abstracts 16 EGU2014-393-1
4. Abd-Elhamid HF (2010) A simulation-optimization model to study the control of seawater intrusion in Coastal Aquifers. PhD thesis, College of Engineering, Mathematics and Physical Sciences, University of Exeter, Exeter
5. Bear J (1979) Hydraulics of groundwater. Mc Graw-Hill, New York
6. FAO (2006) World reference base for soil resources. A framework for international classification, correlation and communication. Food and Agriculture Organization, Rome
7. MWRI (2013) Adaptation to climate change in the Nile Delta through integrated coastal zone management. Ministry of Water Resources and Irrigation, Giza
8. SNC (2010) Egypt's second national communication. Egyptian Environmental Affairs Agency (EEAA-May 2010) under the United Nations framework convention on climate change
9. EEAA (2004) Environmental status report for 2004
10. El Arabi N (2012) Environmental management of groundwater in Egypt via artificial recharge extending the practice to soil aquifer treatment (SAT). Int J Environ Sustain 1(3):66–82
11. MWRI (2010) Water resources development and management strategy in Egypt – 2050. Ministry of Water Resources and Irrigation, Cairo
12. El-Atfy H (2007) Integrated national water resources plan in Egypt. Ministry of Water Resources and Irrigation Alexandria Governorate, Cairo
13. Centre for Environment & Development for the Arab Region and Europe (CEDARE) and International Fund for Agricultural Development (IFAD) (2002) Regional strategy for utilisation of the Nubian Sandstone Aquifer system. http://water.cedare.int/cedare.int/files15/File2919.pdf

14. CEDARE (2002) Centre for Environment and Development for the Arab Region and Europe (CEDARE) on activities to support African country parties under UNCCD, Cairo. http://water.cedare.int/cedare.int/files15/File2919.pdf
15. Bakhbakhi M (2006) Nubian sandstone aquifer system. In: Foster S, Loucks DP (eds) Non-renewable groundwater resources: a guidebook on socially sustainable management for water-policy makers. IHP-VI series on groundwater 10. United Nations Educational, Scientific and Cultural Organization, Paris, pp 75–81
16. Abu-Zeid KM (n.d.) Regional management of the Nubian Sandstone Aquifer. Potential Arab Region & Latin America Cooperation on Large Aquifers. http://www2.mre.gov.br/aspa/semiarido/data/Palestr%20Khaled% 20Abu-Zeid.doc [online]
17. Sefelnasr A, Gossel W, Wycisk W (2007) GIS-based groundwater flow modeling of the Nubian aquifer system, western desert Egypt. University Halle, Halle
18. Mirghani M (2012) Groundwater need assessment Nubian Sandstone Basin. Watertrac, book, Nile IWRM-Net
19. NWRP (2005) National water resources plan, facing the challenge. Ministry of Water Resources and Irrigation Integrated Water Resources Management Plan for 2017
20. MWRI (Ministry of Water Resources and Irrigation) (2009) Proposed climate change adaptation strategy for the Ministry of Water Resources and Irrigation in Egypt. Ministry of Water Resources and Irrigation. Coastal Research Institute, the Egyptian Shore Protection Authority, United Nations, Development Programme, Government of Egypt, project document
21. Sherif MM, Al-Rashed MF (2001) Vertical and horizontal simulation of seawater intrusion in the Nile Delta aquifer. In: Proceeding of the 1st international conference and workshop on saltwater intrusion and coastal aquifers, monitoring, modelling, and management, Essaouira, Morocco
22. Sakr SA (2005) Impact of the possible sea level rise on the Nile delta aquifer. A study for Lake Nasser flood and drought control project (LNFDC/ICC), Planning Sector, Ministry of Water Resources and Irrigation
23. RIGW (1992) Hydrogeological map of Egypt. Nile Delta; scale 1:500,000, 1st edn. El-Qanatir, Egypt
24. Met Office Hadley Centre (2011) Evidence: the state of the climate [Internet]. http://www.metoffice.gov.uk/media/pdf/m/6/evidence.pdf. Accessed 31 Jul 2012
25. WMRI-NWRC (2002) Unpublished report under the matching supply and demand project. Water Management Research Institute, National Water Research Center, Ministry of Water Resources and Irrigation, Cairo
26. Eid HM (2001) Climate change studies on Egyptian agriculture. Soils, Water and Environment Research Institute (SWERI) ARC, Ministry of Agriculture, Giza
27. IPCC (2007) An assessment of the Intergovernmental Panel on Climate Change. In: Adopted section by section at IPCC Plenary XXVII, represents the formally agreed statement of the IPCC concerning key findings and uncertainties contained in the Working Group contributions to the fourth assessment report, Valencia, Spain, 12–17 November 2007
28. IPCC (1996) Climate change 1995: the science of climate change. In: Houghton JT, Meira Filho LG, Callander BA, Harris N, Kattenberg A, Maskell K (eds) Contribution of working Group I to the second assessment report of the intergovernmental panel on climate change. Cambridge University Press, Cambridge, p 572
29. El-Raey M (2010) Impact and implications of climate change for the coastal zones of Egypt. In: Michel D, Pandya A (eds) Coastal zones and climate change
30. Agrawala S, Moehner A, El Raey M, Conway D, Aalst M, Hagenstad M, Smith J (2004) Development and climate change in Egypt: focus on coastal resources and the Nile. In: Environment policy committee, working party on global and structural policies and working party on development co-operation and environment, Organization for Economic Co-operation and Development (OECD)

31. Simmons CT, Narayan KA, Woods JA, Herczeg AL (2002) Groundwater flow and solute transport at the mourquong saline-water disposal basin, Murray Basin, southeastern Australia. Hydrogeol J 10:278–295
32. Zaid SM (2006) Geo-environmental study of eastern Nile Delta, Egypt. PhD thesis, Faculty of Science, Zagazig University, Zagazig, p 339
33. Saleh MF (1980) Some hydrological and hydro-chemical studies on the nile delta. MSc thesis, Faculty of Science, Ain Shams University, Cairo
34. Shata A, El-Fayoumey I (1970) Remarks on the regional geological structure of the Nile Delta. Bucharest symposium, UNESCO publication chapter, Catalog Number 14859, pp 189–197
35. Schlumberger (1995) Well evaluation conference of Egypt. Schlumberger Technical Editing Services, Chester
36. RIGW (1980) Safe use studies for groundwater reservoirs in the Nile Delta and Upper Egypt. Research Institute for Groundwater, Cairo
37. Said R (1981) Geologic evolution of the River Nile. Springer, New York
38. El-Fishawy N (1989) Coastal erosion in relation to sea level changes, subsidence and river discharge, Nile Delta coast. Acta Mineralogica-Petergraphica, Szeged
39. Frihy O, Fanos A (1995) The importance of marine investigation in implementing coastal projects. In: Proceeding of the annual conference of the National Water Research Center, Cairo, Egypt
40. Stanley DJ, Warne AG (1993) Nile Delta: recent geological evolution and human impact. Science 260:628–634
41. Nossair AM (2011) Climate changes and their impacts on groundwater occurrence in the northern part of east Nile Delta. MSc thesis, Faculty of Science, Zagazig University, Zagazig
42. RIGW/IWACO (1990) Hydrological inventory and groundwater development plan western Nile Delta region. TN77. 01300-9-02 Research Institute for Groundwater, Kanater El-Khairia
43. Sherif MM, Sefelnasr A, Javad A (2012) Incorporating the concept of equivalent freshwater 15 head in successive horizontal simulations of seawater intrusion in the Nile Delta aquifer, Egypt. J Hydrol 464:186–198
44. Shahin M (1985) Hydrology of the Nile basin development in water science 21. Elsevier Science Publishers, B.V., Amsterdam, p 575
45. Farid MS (1980) Nile Delta groundwater study. MSc thesis, Faculty of Engineering, Cairo University, Giza
46. Mabrook B, Swailem F, El Sheikh R, El Dairy F (1983) Shallow aquifer parameters and its influence on groundwater flow, Nile Delta Egypt. Australian water resources council conference, series 8, pp 187–197
47. Zaghloul MG (1985) Flow distribution through groundwater aquifer of the Nile Delta. MSc thesis, Faculty of Engineering, Alexandria University, Alexandria
48. Leaven MT (1991) Hydrogeological study of the Nile Delta and adjacent desert areas Egypt, with emphasis on hydrochemistry and isotope hydrology. Thesis, Free University, Amsterdam, also published by RIGW/IWACO as Technical note TN 77.01300-91-01
49. Bahr B (1995) Nile Delta aquifer with emphasis on saltwater intrusion in the northern area. MSc thesis, Technical University of Berlin, Institute for Applied Geoscience, Berlin
50. Diab MS, Dahab K, El Fakharany M (1997) Impacts of the paleohydrological conditions on the groundwater quality in the northern part of Nile Delta. Geol J 4112B:779–795
51. El-Fayoumy IF (1968) Geology of groundwater supplies in the eastern region of the Nile Delta and its extension in North Sinai. PhD thesis, Faculty of Science, Cairo University, Cairo, pp 1–207
52. DRI (1989) In: Amer MH, de Ridder NA (eds) Land drainage in Egypt. Drainage Research Institute, Cairo
53. Environics (2010) EIA study for greeter cairo metro line number 4 phase one. National Authority of Tunnel, Final report
54. Morsy WS (2009) Environmental management to groundwater resources for Nile Delta Region. PhD thesis, Faculty of Engineering, Cairo University, Cairo

55. RIGW (2003) Monitoring of groundwater microbiological activities in the Nile Delta Aquifer. In: A study completed for the National Water Quality and Availability Management project (NAWQAM), Kanater El-Khairia
56. Mabrouk MB, Jonoski A, Solomatine D, Uhlenbrook S (2013) A review of seawater intrusion in the Nile Delta groundwater system the basis for assessing impacts due to climate changes and water resources development. J Hydrol Earth Syst Sci 10:10873–10911
57. Abd-Elaty IM (2014) Numerical and experimental study for simulating climatic changes effects on Nile Delta aquifer. PhD thesis, Faculty of Engineering, Zagazig University, Zagazig
58. RIGW (2002) Nile Delta groundwater modeling. Research Institute for Groundwater, El Qanater El Khayreya
59. GWR (1966) Groundwater research along the Nile River. Ministry of Irrigation, Giza
60. Hefny K (1998) Water use in Egypt. In: Report submitted to the National Water Quality and Availability Management project (NAWQAM), El Qanater El Khayreya
61. Sakr SA (2005) Enhancing groundwater availability in the Nile Delta Aquifer of Egypt. Ain Shams University-Faculty of Engineering. Sci Bull 40(3)
62. Al-Ag DE, Closas A, Molle F (2015) Survey of groundwater use in the central part of the Nile Delta. Activity report (Draft), Water and salt management in the Nile Delta: Report No. 6
63. RIGW/IWACO (1999) Environmental management of groundwater resources (EMGR), Final Technical Report TN/70.0067/WQM/97/20, Research Institute for Groundwater (RIGW), El Kanater El-Khairia
64. Sakr SA, Attia FA, Millette JA (2004) Vulnerability of the Nile Delta aquifer of Egypt to seawater intrusion. International conference on water resources of arid and semi-arid regions of Africa. Issues and challenges, Gaborone, Botswana
65. Hussien MM (2007) Environmental Impacts of new settlements on the ground water in a region in Delta. Msc thesis, Zagazig University, Faculty of Engineering, Zagazig
66. Ghoraba SM, Zyedan BA, Rashwan IM (2013) Solute transport modeling of the groundwater for quaternary aquifer quality management in middle delta, Egypt. Alex Eng J 52:197–207
67. Abd El-Fattah EG (2014) Evaluation of the environmental and chemical impacts of banana plantation on the groundwater aquifer in the southern part of the Nile Delta and its vicinities. PhD thesis, Submitted to Environmental Studies and Research Institute (ESRI), University of Sadat City, Sadat City
68. El-Agha D, Closas A, Molle F (2015) Survey of groundwater use in the central part of the Nile Delta. Water and salt management in the Nile Delta project report
69. Ghoraba SA (2009) Ground water quality management in the Middle Delta utilizing environmental isotopes. PhD thesis, Faculty of Engineering, Tanta University, Tanta
70. Zeidan BA, Aly AI, Rashwan IM, Ahmed MA, Ghoraba SM (2015) Scenarios for groundwater remediation using N15 In Nile Delta. In: 18th international water technology conference, IWTC2015, Sharm El-Shiekh, Egypt, March 12–14
71. Sharaky M, Atta SA, El Hassanein AS, Khallaf KM (2007) Hydrogeochemistry of groundwater in the western Nile Delta aquifers, Egypt. In: Proceedings of the 2nd international conference on the geology of tethys, 19–21 March, Cairo University, Cairo
72. World health Organization (WHO) (1996) Fighting disease, fostering development, report, Geneva. http://www.who.int/whr/1996/en/
73. Masoud AA (2014) Groundwater quality assessment of the shallow aquifers west of the Nile Delta (Egypt) using multivariate statistical and geostatistical techniques. J African Earth Sci 95:123–137

Land and Groundwater Resources in the Egypt's Nile Valley, Delta, and Its Fringes

El-Sayed Ewis Omran

Abstract The current Egyptian situation is framed by land and water scarcity, which are under severe pressure. The Nile Delta is one of the most densely populated deltas in the world. Soil and water resources are at the center of sustainable development and are critical for socio-economic development.

Nile Delta branches gain water from the aquifer in some reaches and lose water to the aquifer in other reaches. The flow directions between groundwater and surface water can change seasonally with variations of the water table level with respect to the level in nearby waterways. Available data on the evolution of the salinity of groundwater in the Delta indicate that the construction of the High Aswan dam resulted in a shift of salinity isolines towards the seashore, and that current pumping rates have not yet critically affected this balance. Possible localized over-pumping, however, results in "up-coning" of salinity from deeper layers. Contamination of groundwater by agriculture and more prominently by seepage from domestic and industrial effluents had already attained worrying levels. This may jeopardize the quality of the Nile Delta aquifer in the end. The increasing use of groundwater for irrigation poses a serious threat to food security and could lead to unaffordable prices of staple foods. Therefore, groundwater overuse rising could hit food prices. Aquifer depletion can induce significant environmental degradation, such as land subsidence and seawater intrusion. The amount of non-renewable groundwater used for irrigation was doubled in Nile Valley and Delta. The annual groundwater abstraction in the Nile aquifer system and fringes is about 4.6 billion m^3. Another 0.5 billion m^3 is abstracted from the desert aquifers and the coastal areas. Groundwater abstraction is expected to increase to 11.4 billion m^3. Model output revealed that groundwater recharge has not changed significantly over time, while

E. E. Omran (✉)
Soil and Water Department, Faculty of Agriculture, Suez Canal University, Ismailia 41522, Egypt
e-mail: ee.omran@gmail.com

A. M. Negm (ed.), *Groundwater in the Nile Delta*,
Hdb Env Chem (2019) 73: 45–104, DOI 10.1007/698_2017_64,

pumping has. Because of these trends, groundwater was estimated to be in a deficit of approximately 24 billion m^3 (±15%) in year 2011, compared to year 1957.

Most of the Nile Delta soils are recent alluvial soils. The soils generally have a light to heavy clay texture. The clay content varies from 40% in the south to nearly 70% in the north. The soils located near the north coast and lakes are of marine and alluvial deposits. Close to the desert fringe on both sides of the Delta occurs the desert sandy plains, which are flat to undulating topography. The soils of the coastal plains and beaches are sandy with some low to medium longitudinal sand dunes. The salinity problem becomes more severe in the Delta as we approach the seacoast and lakes, due to the effect of the shallow, saline groundwater and the brackish water intrusion from the sea and lake. The old and young terraces of both western and eastern sides of the Nile Delta are of alluvial origin and non-uniform in nature. Most of the soils are originated from the ancient Nile sediments, which are mostly derived from igneous and metamorphic rocks of the Abyssinian Plateau. The soils are alluvial deposits of the Nile Delta (Qatabeya) and valley (Qena), swamps and fluviomarine-lacustrine deposits (El-Manzala), beach sands (Edku). The old alluvial soil is more developed than other soils.

Keywords Egypt, GIS, Groundwater, Land, Virtual Water, Nile Delta, Nile Valley, RS

Contents

1 Introduction

Climate change is likely to increase the stress on resources, especially soil and water, which are the most imperative and central factors in agriculture development. The current Egyptian situation is framed by land and water scarcity, which is under severe pressure from increasing populations, fast development, agriculture intensification, and the degrading environment in Egypt. The contested agricultural land makes up only 4% of the country. About 95% of the people (93 million in 2016 [1]) actually live on the scarce agricultural land in the Nile Valley and Delta in a country that is otherwise a desert. With a population density of 1,500 inhabitants per square kilometer when including the capital [2], the Nile Delta is one of the most densely populated delta in the world. On the one hand, soil and water resources are at the center of sustainable and socio-economic development. On the other hand, water and soil resources are under threat because of the way we treat it. High Dam has freed the Egyptian Nile Valley and Delta from floods, but also dramatically confines sediment transport and water availability. Since the 1960s, the Egyptian Nile is completely controlled by the High Dam and a series of barrages along its course to the Mediterranean Sea. A Grand Ethiopian Renaissance Dam (GERD) will put additional pressure on the territory [3]. As our soil and water resources come under increasing pressure, hard decisions should be made so that resources are not degraded or tipping points reached.

Water is an irreplaceable resource and it is only renewable if well managed. As the demand on freshwater increases, consideration regarding non-conventional water resources consequently increases in Egypt. The demand for adequate and safe supplies of water is becoming crucial especially in the overpopulated of Egypt. Water resources in Egypt are the Nile, rainfall, deep groundwater, and desalinated water complimented by shallow groundwater in the Nile Delta and both re-used agricultural drainage and treated wastewater as non-conventional water resources [4]. Eighty-six percent of that water comes from the Nile and around 11% from underground sources, 2% is recycled [5]. Groundwater plays a vital role in water supply everywhere throughout Egypt. In the desert regions, groundwater is considered as one of the key sources for both rural and agriculture water supplies. Although several aquifer systems exist in Egypt, the most important is the Nile aquifer system, which covers the Nile floodplain and the desert fringes where 90% of Egypt's population lives [6]. In the center of the floodplain, the aquifer is semi-confined (10 m of silt-clay). However, on the edges of the floodplain and the desert fringes, the aquifer becomes phreatic. The thickness of the aquifer ranges from 300 m in the south to 800 m in the Delta [6]. Groundwater in the Nile aquifer and desert fringes is not a resource in itself as it is replenished from the river Nile by seepage from canals and percolation from irrigation application and infiltration. The flow direction is, with a few local exceptions, south north. Recharge varies according to hydrogeological conditions. Extraction from the aquifer is by seepage to the river and extraction from wells. Increasing abstraction to meet rising demand has raised concerns about the sustainability of the groundwater resource. The annual groundwater abstraction in the Nile aquifer system and fringes is about 4.6 billion m^3 [7]. Another 0.5 billion m^3 is abstracted from the desert aquifers and the coastal areas. Groundwater abstraction is expected to increase to 11.4 billion m^3 [8], thus these are posing a challenging question to researchers: Are the land and groundwater resources in the Nile Valley and Delta sustainable? Accordingly, the main scope of the present chapter is to assess the land and groundwater resources and its sustainable development in the Nile Valley and Delta of Egypt and its fringes.

2 Description of the Nile Valley and Delta

2.1 Egypt Gift of the Nile and Egyptians

The Nile in Egypt was drilled as a result of unusual circumstances caused the Mediterranean basin drying since about 6 million years ago, due to higher Gibraltar (which linking the Mediterranean with the Atlantic Ocean), and thus the arrival of water to the Mediterranean sea was stopped. It transforms into a lake, which its water was evaporated until dried [9]. It was connected as a result of earth movements and volcanic eruptions associated with the formation of the great African Rift Valley, which led to the formation of lakes Tana and Victoria. In addition, the grace of God and a gift from God to Egypt caused these ground movements in turning the water from the Ethiopian highlands to the Nile instead of from the Red Sea as it was in the past. With this shift in the earth movements, which have formed the Rift Valley, the Nile River was connected to Africa and the water has been arrived from the sources. Otherwise, the Nile River has dried up and become like other desert valleys such as the Wadi Qena and Assiouti and the Egyptian man was not able to settle down in the valley and not to establish a stable agricultural civilization [10].

Herodotus has said, “Egypt is the Gift of the Nile.” While others believe (and I am one of them) that “Egypt is the Gift of the Nile and Egyptians together.” The river passes through several countries starting from the upstream and downstream. There was no giant civilization influenced in human like ancient civilization. Nile water cannot make a civilization, but it is the Egyptian man, who controls runoff of the Nile and digging canals and drains. It has monitored the Egyptian priests’ stars to calculate the flood, and thus created astronomy. Is there found the people other than the Egyptian have a calendar for 4,000 years BC?

2.2 Nile History

The Neonile was arrived to Egypt in the declining snow era of the last ice and thus increased rainfall on the plateau of tropical lakes and the Ethiopian plateau especially between 12,500 and 10,000 years ago. It has led to an increase in the water, which the Nile carries them to Egypt, and began to form its fertile soil in the valley and delta. It became in its modern form after a series of changes over the Nile before reaching its current form, and can be summarized as follows [9]:

1. The Eonile (5.4–6.0 million years ago) is being stuffy Nile because of drought Mediterranean Sea. The first Nile River was created because of the existence of a great fold (6 million years ago). The Mediterranean Sea dried up and turned into a desolate desert 6 million years ago because of the closure of the Straits of Gibraltar. It was then filled with water after a million years of drought (since 5 million years). These changes have had a significant impact on the pre-Nile. It was not connected to tropical Africa until only 800 thousand years ago.
2. Gulf Sea Development (3.3–5.4 million years ago).
3. The Paleonile or old Nile (1.8–3.3 million years ago) is a local river occupies the Gulf and followed by the desert developed (0.8–1.8 million years old) turned Egypt into the desert, the Nile stop the flow.
4. The Prenile (400,000–800,000 years) is the first Nile in Egypt liaise tropical Africa.
5. The Neonile (400,000 years till now) is a period dominated by a river less able and Communication Africa has become transient. Lake Victoria has been without any connection to the Nile for a long time after its inception and has been connected to the Nile River only 12,500 years after the height of its level by 26 m. Then the modern Nile was appeared in the Paleolithic and then another river in the Middle Stone Age and finally the back of the last Nile, which is currently called the Neonile. It was followed by the present Nile with a sustainable flow (12,000 years ago). The current Nile is a modern river born with rains that followed the decline of the last Ice Age snow (10,000 years ago). The ice age declined and the water level of the river increased and began to sink the silt that has been known for 7,500 years.

The River Nile has two sources of water: plateau tropical lakes and Ethiopian highlands with summer rains. Plateau lakes have five large lakes connected to the Nile. Some of these lakes are great rift (groove) lakes in the center of the west (Edward, George, and Albert), and others are large shallow (downward) lakes such as Victoria and Kyoga [11]. Lake Victoria (Fig. 1) is the largest lake with an area of about 69,000 km^2 (more than double the Delta area and the valley of the River Nile). Lakes Victoria, Kyoga, and Albert have its present form because of earth movements that made up the Great African Rift. This is one of the most prominent geographical phenomena on the Earth's surface, where a length of nearly 3,000 km, and its depth varies from place to place [9]. The rivers of the core of the Nile, which originates from the Ethiopian highlands, include the Blue, Atbara, and Sobat (Fig. 1), which stems from the volcanic area. Igneous rocks (basalt) are covered in the Ethiopian highlands, which cover the bulk of the plateau. The average rainfall is between 1,000 and 1,400 mm a year. The Ethiopian highlands are the source of most of the Nile water. However, the other water sources, even though it low, it makes the river flow sustained throughout the year. The lack of water arriving from tropical plateau was due to the loss of river water in the marshes and the various

Fig. 1 Nile River Basin

bodies in the region, where the biggest losses of the water were to the sea. Nile River enters the Egyptian territory near Adindan on the Egyptian–Sudanese border and holds the river for a distance of about 1,536 km until emptying into the Mediterranean Sea. The total length of the river from Aswan to Ras Delta at Cairo is about 981 km.

2.3 The Nile Delta and Valley Land

The land of Egypt was (before the arrive Nile) covered by the sea from the north since tens of millions of years (about 60 million years ago). It covered a large part of the north of Sudan. The sea began to retreat until its beach became the line running between El-Faiyum and Siwa, and thereafter seaside has become close to the current position. Egypt with its lands extending over 1 million km^2 is endowed with varied agro-ecological zones, which represents the greater majority of cultivated lands of the Nile Valley, as well as, most of the reclaimed desert lands, mainly, on the western and eastern fringes of the Delta in addition to relatively limited areas at on fringes of the Valley in Upper Egypt (total areas over 7.5 million feddans). The Nile Valley system extends from the Mediterranean shores of the Nile Delta to the north till Aswan in the south over an area extending from 22° to 32° latitude north under arid and hyperarid conditions.

Egypt has started many national plans to increase its cultivated areas to resettle the population outside the narrow valley and to build new communities to diminish hostility of its vast deserts. These potential areas included Nile Delta margins, the Western Desert, and North Sinai. Nile Delta reclamation projects included El-Tahrir (west of the Delta) to reclaim 0.2 million feddan, Maryout (near Alexandria) to reclaim 0.1 million feddan, and Nubariah (south of Alexandria) to reclaim 0.2 million feddan [12]. To the east of the Nile Delta, large reclamation projects were represented by the El-Salam (peace) Canal project, which aimed to reclaim about 0.6 million feddan in the Eastern Delta and Sinai regions by mixing drainage water from the Delta agricultural drains with the Nile water in a 1:1 ratio [13]. Further to the south, a mega reclamation project; the South Valley Development (known as Toshka project) was implemented to convey water from Lake Nasser behind the High Dam at southern Egypt to cultivate 0.5 million feddan by 2017 and resettle 7 million people in the Western Desert [14].

Soils varied including the fertile deep alluvial soils of the old Nile Valley, soils of the river terraces at different reliefs which are deep soils with gravelly and reddish subsoils, in addition to the soils of the fringes including desert calcareous soils of varied textures and non-calcareous soils characterized with low soil fertility and inferior soil physical, chemical, and biological properties.

2.4 The Nile Delta History

Herodotus since about 500 BC was the first one to name the Delta, because of the Latin Delta (▼) character [15]. The Nile basin by nearly 10,000 years has several branches (nine branches) and then reduced to seven branches, then five, then three, and finally the existing branches of Damietta and Rosetta. Toussoun [16] collects and achieves much information about the branches of the Nile. Maps and ancient manuscripts show that the seven branches were formed during the period leading up to the big increase happen in the sea level (5,000 years BC), in which the surface of the sea was low. These branches have been silting in times where the river was acting a little bit, and therefore the rate of deposition of silt over these branches. As for the failure silting of Damietta and Rosetta branches that it may be due to the branches that were ending directly into the sea are destined to inflation and survival, but the one which was aimed at the lakes are those which have as much atrophy and silting [17]. Figure 2 shows the old branches of the Nile, which were drawn by

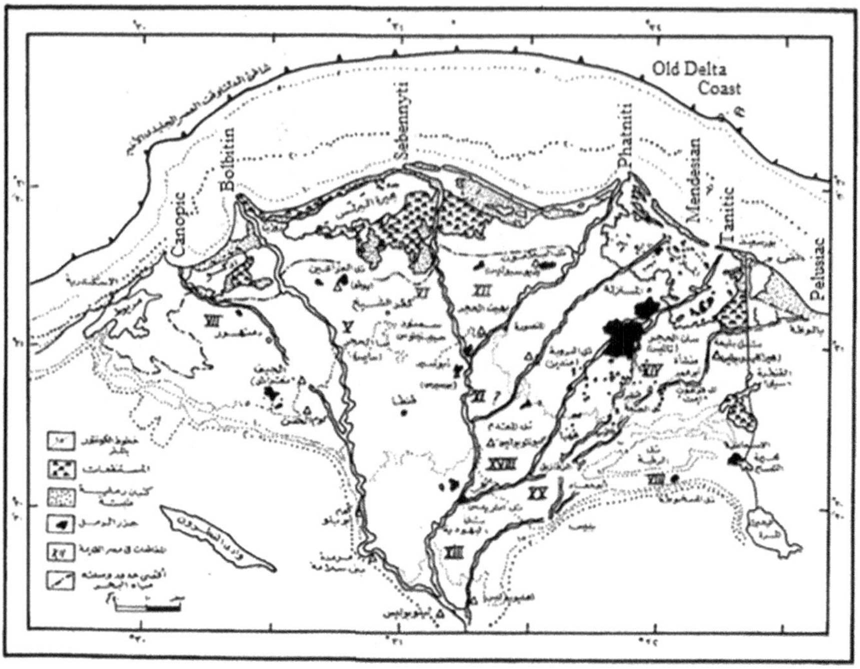

Fig. 2 Old branches and historic cities of the Delta [9]

geographers and historians ancients. The following is a brief on these branches: (Fig. 2)

1. Pelusiac branch was the main branch, relative to the town of Bellows (Farma).
2. Tanitic branch was the secondary branch and takes named relative to Sais (Sun Sea). The Tanitic branch deposited material underneath what is now Manzala Lagoon, and prograded across to present day Port Said.
3. Mendesian branch was the secondary branch, which was taken from the branch of Alsabenity its course and the original part of the small sea. The Mendesian branch formed a depositional center under the western portion of Manzala Lagoon and prograded (advance towards the sea) to the northeast to merge with the depositional center of the Tanitic branch. The supply of sediments by these two branches (Mendesian and Tanitic) gave rise to up to 40 m thick sections of Holocene deposits, currently underlying Manzala Lagoon. Around 1,500–2,500 years ago. These two branches silted up allowing the Damietta branch to become the largest contributor of Nile sediment flux [18].
4. Phatnitic branch was the secondary branch of Sebennytic, which was taking its course in line with the lower part of the Damietta branch.
5. Sebennytic branch was the main branch in the center of the Delta and starts at the top Delta and ends at Borollos tower and its name relative to Spenatios (Samanoud).

6. Bolbitine branch was the secondary branch, which was subdivided from canopic near Damanhur being in the lower part of the Rosetta branch.
7. Canopic branch was the third main western branch and was ended at Canopus (Abu Qir).

2.5 Current Nile Delta Description

Nile Delta starts in branching north of the City of Cairo (Fig. 3). The river is divided into Rashid (239 km) and Damietta (245 km) branches. The average width of the Rosetta branch 500 m, while the Damietta branch of about 270 m. Damietta branch and its drained water much lower than the discharge of the Rosetta branch, which discharges water up to one and half the Damietta branch. Rosetta is erosional branch, while the Damietta is depositional branch [19]. Environmental conditions have helped on the speed of the Nile Delta composition. These factors are leveling the surface of the deposition area, clay aggregates, which facilitates deposition by salts in seawater, and finally the lack influenced the Mediterranean Sea coast marine currents because it is one of the semi-enclosed seas. The length of the Delta from the south to the north (about 170 km), while the base up to about 220 km, an area of about 22,000 km^2 (Fig. 3), twice the area of the valley.

Average rainfall changes from 25 mm/year in the south and central part of the Delta to 200 mm/year in the northwestern part [20]. Excess rainfall is largely intercepted by the drainage system and cannot reach the aquifer [21]. The Nile Delta is among the most populated agricultural areas in the world, with 1,080–1,500 inhabitants per km^2 [2, 22], covering nine governments with about 43% of the Egyptian population (about 39 million inhabitants) [1]. The "old" cultivated lands

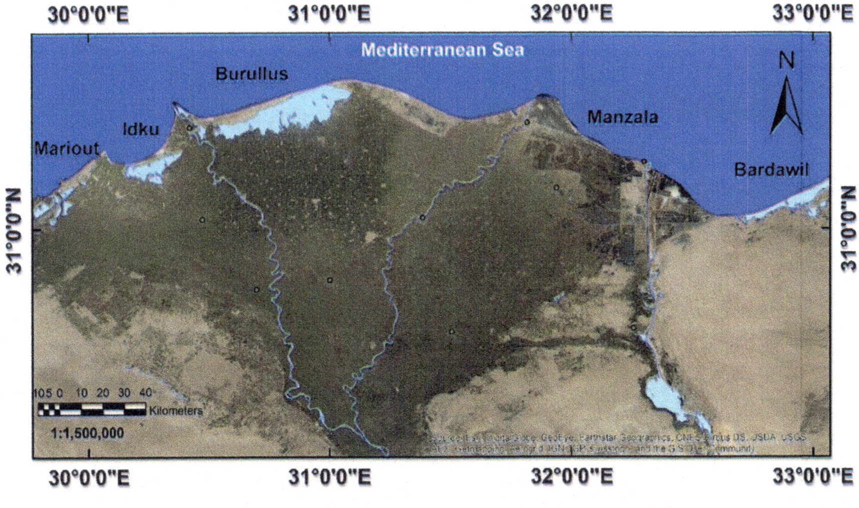

Fig. 3 The Nile Delta and its governments and coastal lakes

are in the Delta and "new" commonly reclaimed desert in the east and the west of the Delta. Coastal lagoons and lakes (Manzala, Borollus, Edku, and Mariout), salt marshes, and sabkhas are comprised the northern part of the Nile Delta. They are usually a low surface of depression composed by the removal of dry and loose material down to the groundwater level or the capillary water zone [23].

3 Groundwater Resources in the Nile Delta

Of six groundwater aquifers in Egypt (Fig. 4), the Nile aquifer is viewed as a fundamental source of groundwater abstraction as it represents 87% of the total groundwater abstraction in the country [25]. The Nile Delta aquifer is one of the biggest aquifers in Egypt with a total capacity of 500 Bm^3 [26], around 15 times the amounts of water that enters the Delta every year through the river flow itself. The recharge source of the Nile Delta aquifer is the Nile water leakage from canals, drains, and the irrigated plots themselves [27]. The conjunctive utilization of groundwater and surface water plays an essential role during the period of peak irrigation demand. The amount of water abstraction from the Nile aquifer (Delta and Valley) was estimated in 2010 at 6.2 Bm^3, which is under the safe yield (8.4 Bm^3) estimated by MWRI [28].

The Nile Delta aquifer serves nine governments, and according to FAO [29] the total irrigated area by groundwater in the Nile Delta would amount to 414, 240 feddan. Little or no land areas are being irrigated with groundwater in Damietta, Port Said, and Kafr El Sheikh Provinces due to the effect of saline water intrusion from the Mediterranean Sea into those provinces, and that the higher share of the groundwater use occurs in El-Beheira governorate. This is due to the inclusion of western new lands (outside the Delta) in the statistics.

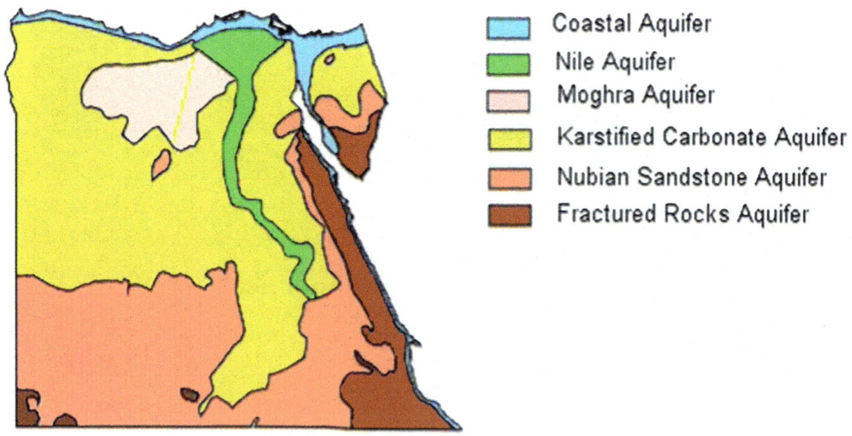

Fig. 4 Groundwater aquifers in Egypt [24]

3.1 The Nile Delta Aquifer Depth

The elements of the Nile Delta aquifer have been perceived by various studies (e.g., [30–32]). The Nile Delta aquifer is viewed as a semi-confined aquifer shaped by Quaternary deposits. Figure 5 demonstrates the contour map of the depth underneath sea level of the lower boundary of the main Delta aquifer. The highest point of the aquifer is secured by a thin clay layer, which contrasts in thickness and disappears in some places (Fig. 5). In such places, the aquifer is thought to be phreatic since its free water surface is exposed to atmospheric pressure. The thin clay layer changes from 5 m in the south to 20 m in the center and achieves 50 m in the north of the Nile Delta (Fig. 5). The thickness and lithological contrasts of the clay layer have a great effect on the degree of hydraulic connection among ground and surface water [34]. The saturated thickness of the aquifer changes from 200 m in the southern parts to around 1,000 m in the northern parts [20].

The depth of the groundwater table in the Nile Delta goes between 1 and 2 m in the north, 3–4 m in the center, and 5 m in the south [35]. Abd El-Fattah [36] studied the southern part of the central area of the Nile Delta and reasoned that the depth of water declines towards Rosetta branch (2.5–5 m). Thus, the flow of groundwater movement is from southeast to northwest towards Rosetta branch in the central part of the Nile Delta. Morsy [21] drew a regional contour map of groundwater depth and the piezometeric contour map in the Nile Delta region in 2008 in light of field observations. Areas with very shallow groundwater incorporate the region between Damanhur and Lake Edku and the saline depressions south of Lake Manzala, and near Ismailia. The ground elevation ranges between about 18 m above mean sea level (AMSL) in the south at Qanater El-Khairia to about 5 m (AMSL) near Tanta sloping down very gently in a northward direction by an average value of 1 m/10 km [34]. Besides, the Nile Delta slopes from east to west making the Damietta branch 2 m higher than the Rosetta branch [37]. According to Attia [38] and Farid

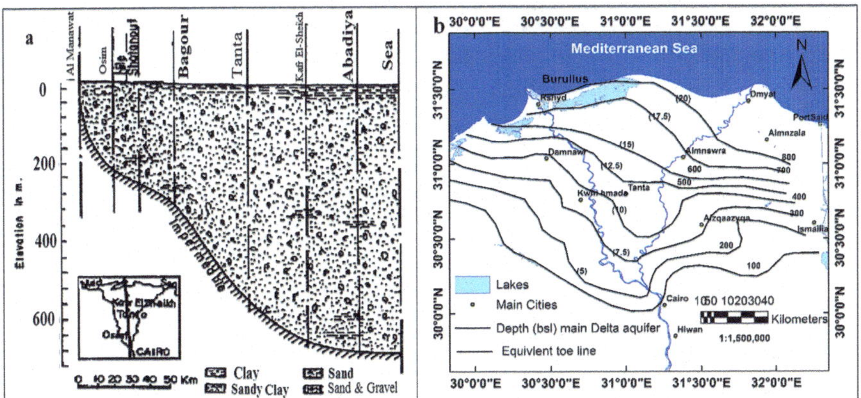

Fig. 5 Vertical cross section (**a**) at the center of Nile Delta [33]. Contour map (**b**) of the depth below sea level of the lower boundary of the main Delta aquifer

[39], the Rosetta branch is as a discharge stream for groundwater at Kafr El-Ziyat and Tamalay cities, while it recharges the aquifer in the Edfina and Desouq region. The Damietta branch gathers water from the aquifer as the groundwater level of the aquifer is higher than the surface water level in the Rosetta branch [29].

Natural discharge of groundwater by evaporation and/or evapotranspiration occurs in the low lying areas with shallow groundwater tables as well as small depressions in the eastern Nile Delta, where water can be taken up by plants or moves upwards through capillary rise. Mabrouk [35] indicated that the total evapotranspiration rate of the Nile Delta using both the irrigation inflow and the aquifer storage is about 2,000 mm/year. Outflow of the Nile Delta aquifer also occurs towards the Moghra aquifer, along the fringes of the western Nile Delta. The total amount of groundwater flowing from the Nile Delta aquifer into the Moghra aquifer was estimated between 50 and 100 Mm^3/year [40]. However, Morsy [21] observed that the drawdown of the groundwater levels in the Moghra aquifer was between 5 and 15 m due to the extensive groundwater exploitation.

An estimation of the amount of groundwater used in irrigation by Abu-Zeid [41] was about 906 Mm^3/year in the Nile Delta area (326 Mm^3/year in the East Delta, 73 Mm^3/year in the Middle Delta, and 506 Mm^3/year in the West Delta). The total annual groundwater abstraction for irrigation and drinking in the Nile Delta area was estimated by Morsy [21] to be 4.9 Bm^3 in 2008. She indicated that the beginning of the increase in abstraction rates was 3 Bm^3 in 1992 [40], which was due to the uncontrolled land cultivation in the reclaimed lands of the western Nile Delta. Mabrouk et al. [35] indicated that annual abstraction increased linearly by about 0.1 Bm^3 per year, except from the period of 2003 till 2010, when abstraction increased by 0.2 Bm^3 per year.

Fattah and Ragab [42] assess the depth to groundwater in the southern part of Nile Delta, Egypt (Fig. 6). Depth to groundwater decreases towards Rosetta branch and decreases close main irrigation canals and El-Rayahat. It ranges between 2.0 m at the northwest and 10 m at the southeast. Potentiometric heads decrease from the southeastern part to the north and northwestern parts. They range from 17 m above MSL at the eastern part to 6 m above MSL close to northwestern part around Rosetta branch. Groundwater shows low depths (<2.5 m) in areas close to the Rosetta branch, which indicates high vulnerability to pollution in this area. Areas of high groundwater depth (> 4 m) indicate moderately vulnerable to pollution. The general direction of groundwater movement is from southeast to northwest towards Rosetta branch which acts as a main drain for drainage water of most drains along its course and that leads to high vulnerability to pollution in the areas adjacent to the branch.

3.2 *Management of the Nile Delta Aquifer*

Many researchers have studied the Nile Delta aquifer, using several numerical models to simulate groundwater flows and to define and analyze the hydrological

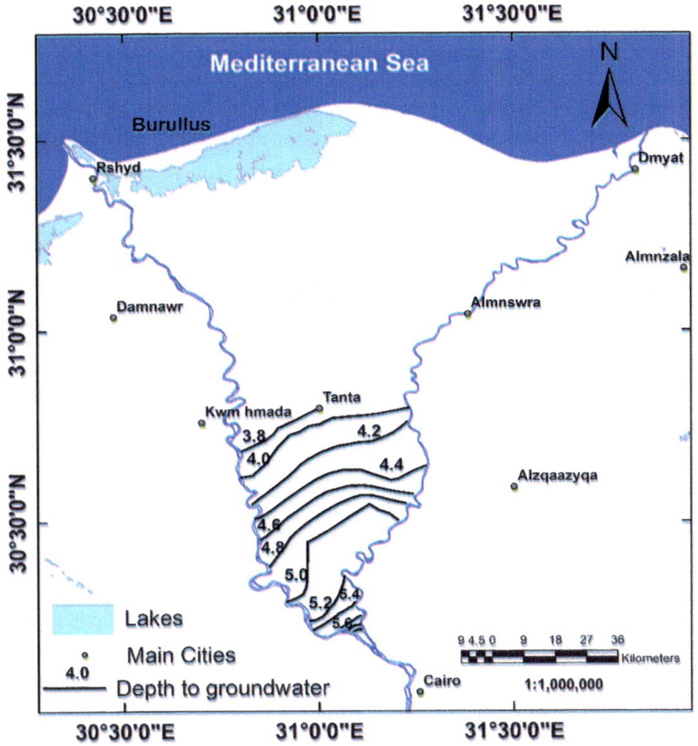

Fig. 6 Depth to groundwater in Pleistocene aquifer of the southern part of the Nile Delta, Egypt

and topographic characteristics of the aquifer. A change in the groundwater level and salinity has been indicated by several researchers due to the rapid development of the land reclamation in the Western Nile Delta region. A whole set of studies testing several development scenarios has been undertaken by different researchers (e.g., [24, 43–46]). Different combinations of simulations and optimization models have been used to recommend a plan for the conjunctive use to optimize the benefits of using both ground and surface water [47, 48]. RIGW/IWACO [40] evaluates the potential of groundwater in the Western Nile Delta's New Lands. They proposed increasing groundwater abstraction from 0.5 to 2.5 Bm^3 for the period from 1990 to 2000 [21]. Mohammed [43] used satellite images to assess the increase in the area under cultivation from 1987 to 1998, and the area developed with groundwater in the Western Nile Delta between 1990 and 2000. They found an increase in the cultivated area exceeding that of the development plan of 1990, predicting that critical areas will face severe groundwater drawdown.

Khater et al. [44] studied the groundwater levels in Sadat City (one of the urban areas in the Western Nile Delta region), which depends on groundwater as the main source for irrigation. They indicated that continuous groundwater extraction exceeded the planned level, causing some hazards on the groundwater potentiality.

Dawoud et al. [49] test two alternative management scenarios using GIS-based model for preserving the aquifer system in the Western Nile Delta. They proposed reducing the surface water inflow while increasing the annual abstraction of groundwater by about 450 Mm^3. Alternatively, they recommended constructing a canal to increase the annual aquifer storage potential by about 23%. Likewise, Moll et al. [50] studied different management scenarios for the western Nile Delta aquifer to satisfy the need for 460,000 feddan of the new land reclamation. These scenarios studied alternative conjunctive uses for available water resources in the Western Nile Delta (surface water, groundwater, and drainage water reuse) to prevent aquifer depletion. They also recommended constructing a canal with 2.1 Bm^3 of surface water and the increase of groundwater abstraction must not exceed 1.0 Bm^3/year to avoid aquifer exploitation. Mohamed and Hua [51] developed a model to simulate the water resources in the Western Nile Delta. Unsustainable use of groundwater was indicated as the total amount of groundwater abstraction for irrigation (1.9 Bm^3), which was larger than groundwater recharge (1.54 Bm^3). They commented that this situation in the long term will undermine the capacity and sustainability of the aquifer and will lead to saltwater intrusion into this aquifer. They recommended reducing groundwater abstraction by 20% in order to reach a sustainable situation.

Switzman et al. [52] present an approach for generating transient groundwater model inputs (Fig. 7) to assess the groundwater resources in a highly data-sparse context of Wadi El Natrun. An ensemble of groundwater model inputs was generated and used in a 3D groundwater flow (MODFLOW) of Wadi El Natrun's multilayer aquifer system to analyze trends in water levels and water budgets over time. Model output revealed that groundwater recharge has not changed significantly over time, while pumping has. Because of these trends, groundwater was estimated to be in a deficit of approximately 24 billion m^3 ($\pm$15%) in 2011, compared to 1957. In the 1990s, a significant trend in water level declines has been observed in monitoring records, which are directly attributed to abstraction. Over abstraction of groundwater is directly linked to declines in water levels and depletion of storage from all aquifers in the area. The problem is most severe in the Pleistocene and Pliocene aquifers due to their more limited storage capacity and longer history of being exploited.

Morsy [21] studied the impact of groundwater development plan from the year 1992 to 2008 on the water table and highlighted waterlogging problems in both western and eastern fringes of the Nile Delta. She noted a drawdown of groundwater levels in the southern part of the western fringes of the Nile Delta (5–15 m) due to extensive exploitation of groundwater and lack of recharge. El-Fakharany [46] tested three strategies for reducing irrigation water supply from canals to El-Saidiya area (East Nile Delta) and substituting the reduction with groundwater. She examined the effect of different pumping rates in the long term, seasonal, and interannual strategies and recommended to satisfy 25% of the study area irrigation requirements with groundwater. In addition, Samak [53] simulated the groundwater flow at Abu-Kabir district in El-Sharqia governorate (East Nile Delta) using a numerical model (MODFLOW) to show the effect of the utilization of both

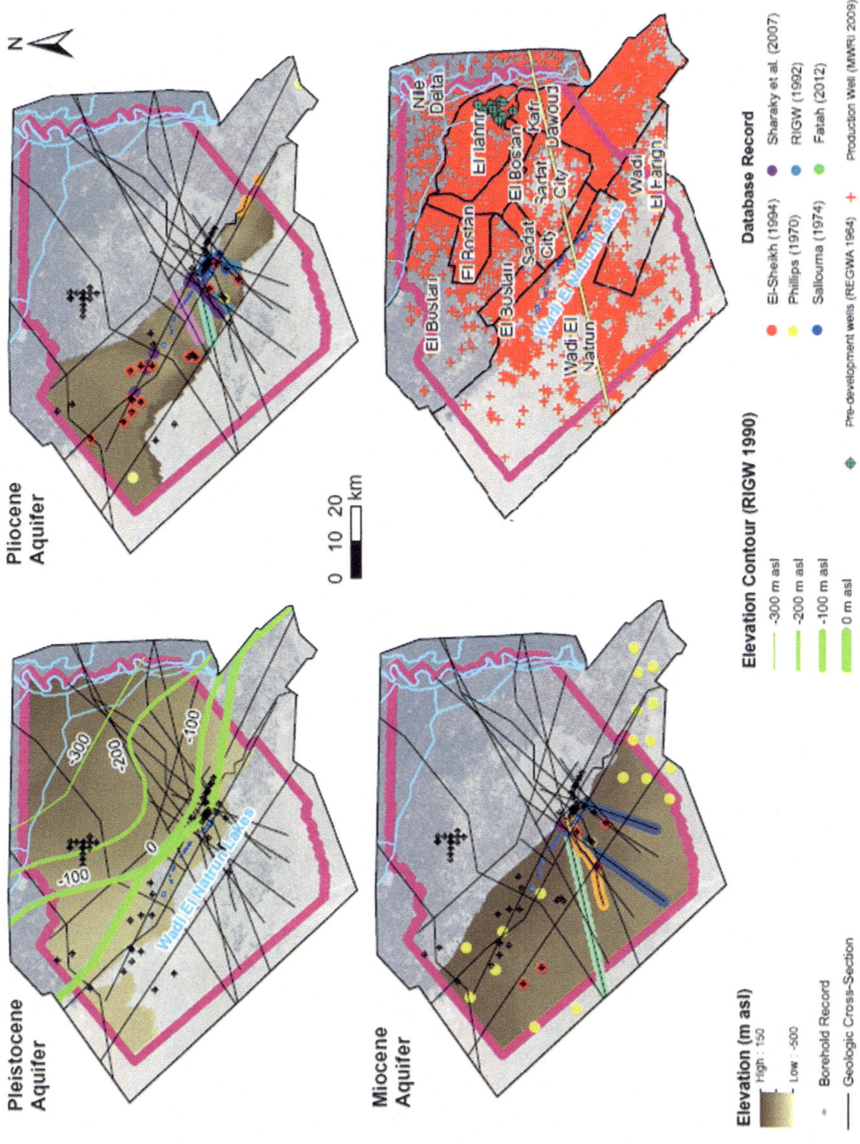

Fig. 7 Datasets records used to map the bottom elevation of each aquifer layer implemented in MODFLOW [52]

groundwater and surface water to overcome irrigation water shortages. Ten scenarios of groundwater management were evaluated to determine the optimal conjunctive ground and surface water and the impact of continuous extraction of groundwater. The best and safest scenario recommended for water management in the studied area was able to satisfy 30% of agricultural demand from groundwater for a period of 20 years.

3.3 Groundwater Quality of the Nile Delta

Groundwater in the Nile Delta aquifer is in hydraulic connection with surface water from Nile River branches, irrigation canals, and drainage network. Therefore maintaining groundwater quality requires an integrated approach to both ground and surface water. Many factors contribute to the deterioration of groundwater quality in the Nile Delta aquifer including: leaching of chemicals and pesticides from agriculture areas; disposal of industrial waste; over-pumping of groundwater to meet increasing demand; and seawater intrusion. Saline water intrusion occurs where saline water displaces or mixes with freshwater [54]. Mathematical and numerical models have been used to predict salinity distribution within the Nile Delta aquifer. Atta [55] analyzed the groundwater salinity and found that the range of groundwater salinity is between 227 and 15,264 mg/l. Farid [56] indicated that the northern zone is highly saline due to seawater intrusion. Iso-salinity lines distribution in the Delta range from 640 to 45,000 mg/l (Fig. 8).

FAO [29] produced groundwater salinity contour maps for year 2012 (Fig. 8). It shows a noticeable increase in groundwater salinity all over the Nile Delta and especially in the middle of the Nile Delta (around Tala City) which has TDS values up to 3,000 mg/l in 1990, decreasing after 2000 and reaching less than 1,000 mg/l in 2012. This phenomenon is called up coning and happened due to the over extraction of groundwater and mobilization of the deeper layers of the aquifer.

Farid [56] presented a cross section of the middle of the Nile Delta aquifer (Fig. 9) which shows the interface between the salty groundwater below and fresh groundwater above with iso-salinity lines ranging from 1,000 to 35,000 mg/l. This interface contains gradually mixing fresh and saltwater and is called a transition zone. Sherif et al. [57] described the Nile Delta aquifer as one of the few cases where seawater migrated to a distance of more than 100 km from the shoreline. Laeven [58] indicated that the saline water of the Mediterranean Sea intrudes the aquifer at a depth of 175–225 m. Ebraheem [59] described that the depth of the fresh–brackish interface (1,000–3,000 mg/l) exceeds 150 m in Tanta and decreases northward to 40 m or less between Qotur and Kafr El-Sheikh. South of Kafr El-Sheikh, the depth of the brackish saline interface exceeds 180 m and decreases northward to 70 m near Hadadi village. Sakr et al. [60] studied the salinity distribution at the middle of the Nile Delta and indicated that the transition zone of brackish water (concentration from 2,000 to 10,000 mg/l) and saltwater

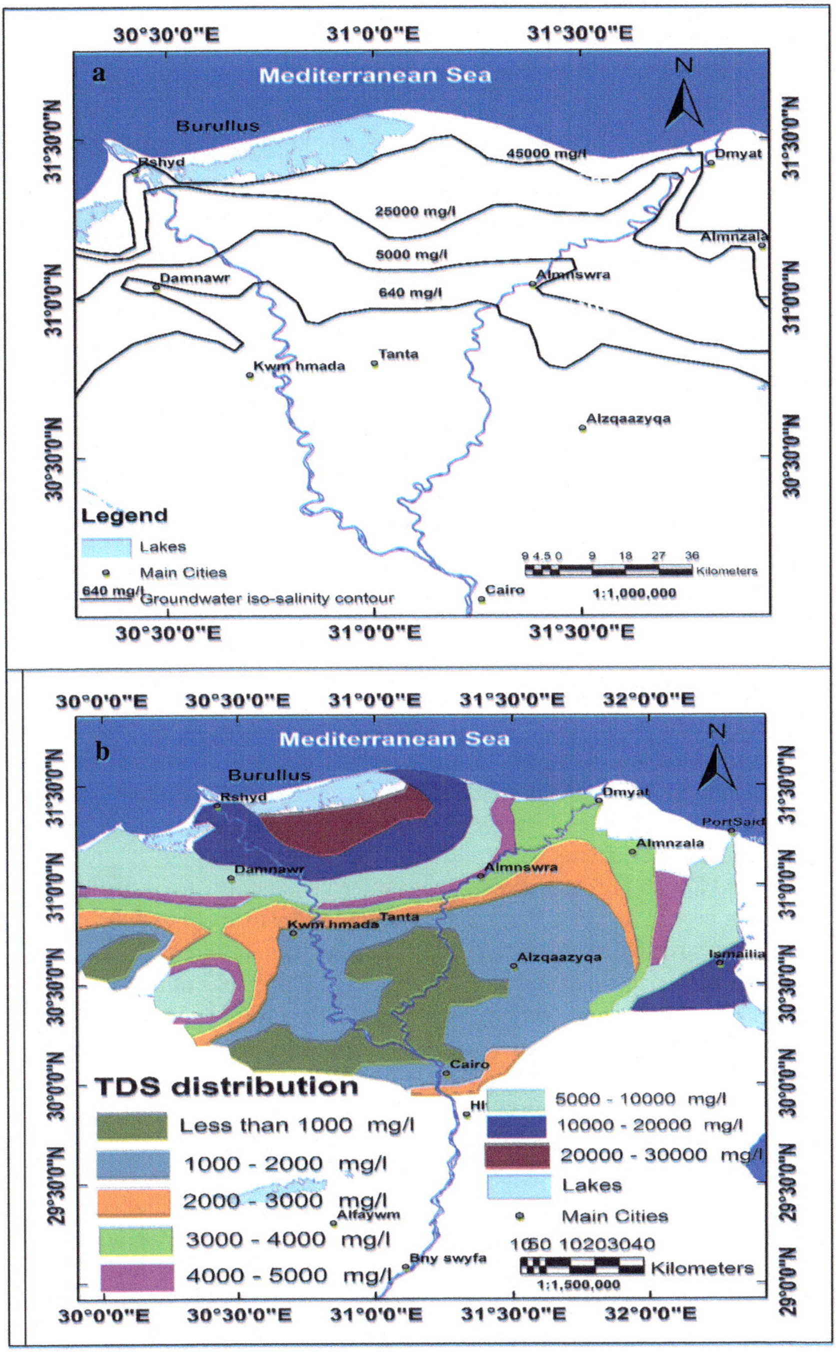

Fig. 8 Groundwater iso-salinity contour (**a**) map of the Nile Delta aquifer [39] and TDS distribution (**b**) map in the Nile Delta in year 2012 [29]

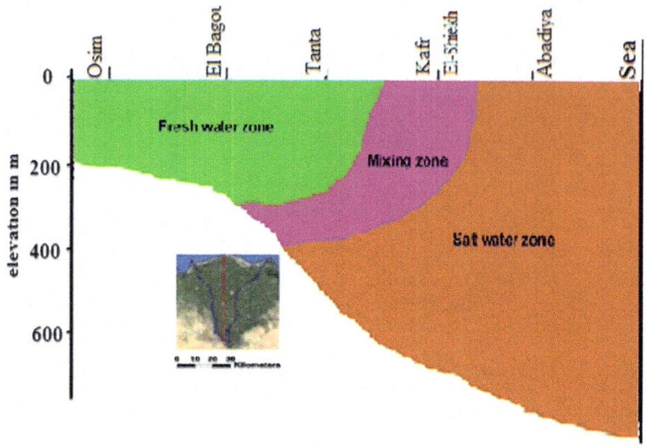

Fig. 9 Schematic presentation of a cross section in the middle Nile Delta aquifer [56]

(concentration >10,000 mg/l) forms a wedge extending into the aquifer to a distance of 90 km from the coast.

Higher concentrations of major ions such as nitrate, sulfate, potassium, and phosphate can be caused by the extensive use of chemical fertilizers in agriculture. Also high concentrations of trace elements like manganese, iron, and aluminum were found in the old lands of the Nile Delta aquifer due to the dumping of industrial effluents into the drainage system. Morsy [21] noticed that concentrations of iron and manganese were higher in the old lands due to the general natural characteristics of the Nile Delta aquifer and more pronounced in the areas having a clay cap. She indicated that local contamination from industrial areas also led to high concentrations of lead (Pb) in the City of Tenth Ramadan (Eastern Nile Delta) and Sadat City (Western Nile Delta). Also high concentrations of cadmium (Cd) were found in Mobarak City (middle of the Nile Delta).

Fattah and Ragab [42] assess the groundwater vulnerability to contamination in the southern part of the Nile Delta. Its vulnerability assessment to delineate areas that are more susceptible to contamination has become an important element for water resource management and land use planning. The results of this study can be utilized to figure out where communities ought to undertake aggressive protection of the groundwater. These values were reclassified into three classes (Fig. 10): low, moderate, and highly vulnerable zones.

The high vulnerability zones of the study area are located in the western part of the study area beside the Rosetta Nile branch. Highly vulnerable rate in such areas is due to the low thickness of the clay cap layer of Holocene aquifer, coarse texture of the soil zone, and high downward flows, which allow infiltration of more pollutants from the upper Holocene to the underlying Pleistocene aquifer. The risk of vulnerability of groundwater contamination in that area is high. Regional development planners will benefit from knowledge of local sensitive aquifers. Moderate vulnerable zones are situated in the center of the study area. Low vulnerable zones are very much dispersed and are for the most part situated in the center

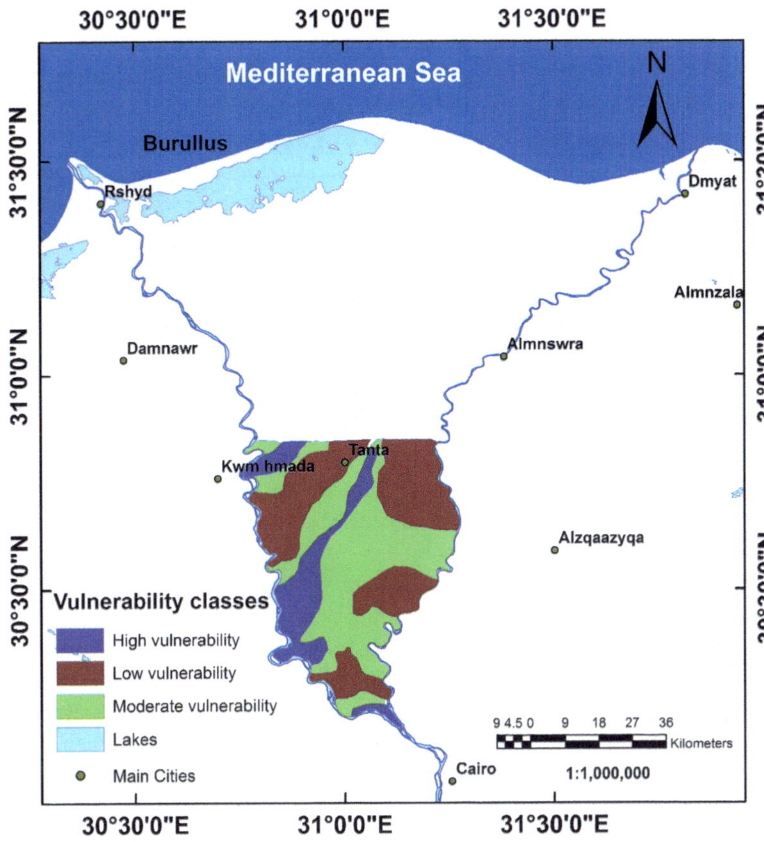

Fig. 10 Vulnerability classification map of the southern part of Nile Delta, Egypt

and eastern parts of the study area. Some low vulnerable zones can also be found in the northwestern part.

For the western fringe of the Nile Delta (Newlands), Sharaky et al. [61] examined 108 groundwater samples from depths between 27.5 and 120 m. The results indicated that the concentrations of major ions were higher than the maximum standard limits according to the World Health Organization. The nutrient content in phosphates was also higher than the maximum standard values, while nitrate levels were critical. They concluded that groundwater had good quality and was suitable for irrigation for all types of soils but not suitable for drinking purposes. Masoud [62] examined 451 shallow groundwater samples for the same area. Most hydrochemical parameters showed very wide ranges. Concentrations of TDS (201–24,400 mg/l), pH (6.72–8.65), Na^+ (28.30–7,774 mg/l), and Cl^- (7–12,186 mg/l) suggest complex hydrochemical processes of multiple sources. The high concentrations of Fe^{2+}, Mn^{2+}, Zn^{2+}, Cu^{2+}, and Ni^{2+} were related to their natural presence in water-bearing sediments and/or to contamination from industrial leakage. In addition, very high nitrate concentrations exceeding the permissible limit (50 mg/l) exist due to wastewater leakage. Groundwater was unsuitable to irrigate regular crops due to high salinity in most areas close to the sea (Abis, North-Tahrir, Mariut) and west of the Cairo-Alexandria road (with only the track of land between this road and the Nubaria canal having access to reasonably good groundwater).

Abd El-Fattah [36] conducted a study on groundwater in the southern part of the Nile Delta by testing the chemical properties of 80 samples. Salinity had values ranging from 296 to 810 mg/l. Ghoraba et al. [63] investigated groundwater quality in the central part of the Middle Nile Delta (El-Gharbia Governorate) through the hydrochemical analysis of 34 groundwater samples (depths between 13 and 60 m) plus ten samples from canals and 14 samples from drains. They concluded that the Quaternary aquifer is quite vulnerable to pollution. The deterioration of groundwater quality clearly indicated that human activities caused serious pollution problems, including high nitrate concentrations within the area (except the northeastern part). They indicated high concentration of ammonium in a village located in the central part of El-Gharbia governorate (Birma village). The values of ammonium concentration reached 3.3 mg/l, which exceeded by far the drinking water standards (0.5 mg/l).

Salem et al. [64] investigated the water quality parameters (iron and manganese) of the central part of the Middle Nile Delta in El-Gharbia Governorate from 34 well samples. The results indicated concentrations of iron (0.1–1.33 mg/l) and manganese (0.5–1.45 mg/l) in most samples higher than World Health Organization and Egyptian Standard. They concluded that the main polluting sources were from chemicals and fertilizers used in fruit farms, as well as from petrochemical and industrial activities. Hussein [65] assessed the impact of industrial activities in the eastern part of the middle of the Nile Delta. Seventy-five samples from drinking water wells, industrial production wells, sewage water, and drains in the industrial area (Quesna) were tested. The major cations and anions and the ammonium (NH_4^+), nitrite (NO_2^-), phosphate (PO_4^{3-}), boron (B^+), and trace elements were analyzed. Also, Morsy [21] indicated that the water samples showed contamination with reuse of agricultural drainage water in Sharqia Governorate (Bahr El-Baker, Belbeis, and Qalyubeya drain).

4 Resources and Quality of the Nile Valley Groundwater

The alluvium underlying the Nile Valley may be subdivided into two units, a clay-silt layer underlain by a graded sand layer. Graded sand layer ranges in thickness along the valley from a minimum of 20 m to a maximum of 300 m. It wedges out laterally at the escarpments, which mark the limits of the valley. Its width ranges from about 2 km at Aswan, up to about 20 km in some reaches. The layer consists almost exclusively of sand and includes lenses of both predominantly coarse and fine sand. The clay-silt layer overlies the graded sand layer. It consists predominantly of silt and clay, though some lenses of fine sand also occur.

4.1 Esna Aquifer

The aquifer underlying Esna has been simulated as two layers. The silty-clay cap represents the upper layer, while the lower aquifer is represented by the sand aquifer. Interchange between the two aquifers has been simulated through the specification of a leaky layer, having a vertical permeability of one-tenth the horizontal permeability of the upper layer. Ghanem et al. [66] study the groundwater lowering in aquifer systems of Esna City, which is underlain by two layers: a shallow silty-clay layer and a deep sand aquifer. Both layers are hydraulically interconnected.

About half of the city's area is subjected to a high groundwater level, 1 m or less below ground surface. The areas are strongly affected by elevated groundwater levels lay in the western and southwestern parts of the city. Results of the simulation model showed that the sanitary drainage system will have a positive impact on the groundwater situation of the city, but some locations will remain with elevated groundwater table. Thus, a groundwater lowering system consisting of a secondary tile drainage system and a main covered drain is recommended. The expected lowering of groundwater levels in the critical areas of the city with the groundwater lowering system in operation range between 0.5 and 1.4 m. The critical area (depth to groundwater <1.0 m) is expected to be reduced to about 10% of its present area. The simulation of the combined effect of the groundwater drainage system and the sanitary drainage system showed that expected lowering of groundwater levels ranges between 0.6 and 1.7 m, with a minimum depth to groundwater table of 1.50 m. El-Fakharany and Fekry [67] assess the impact of the barrage effect on groundwater aquifer in Esna. The minimum depth to shallow groundwater table reaches about 1.0 m and the maximum one reaches about 5.94 m (Fig. 11).

Faid [68] detects and identifies groundwater aquifer in Esna. The study area is covered with Quaternary deposits of sand and gravel with thin clay interbeds. The study reveals that the typical stratigraphic column and geophysical survey (geoelectric) in the region shows various aquifers. Distinction can be made for the Quaternary and Plio-Pleistocene (Fig. 12). The maximum thickness of the first aquifer is 180 m, while for Plio-Pleistocene is 90 m. The Plio-Pleistocene aquifers vertically recharged from infiltration of irrigation water in the reclaimed areas, and some from wadis and rainfall during the occasional cloud bursts. Lateral inflow from limestone aquifer and vertical one from the deeper aquifer recharge it also.

4.2 Luxor Aquifer

Ismail et al. [69] integrate geological/geophysical/hydrological study to identify and map different geoelectric (geologic/hydrologic) units in the shallow subsurface (<100 m), and determine the rise of groundwater and its accompanying increase in salinity in Luxor. Groundwater flow paths were from the newly irrigated areas

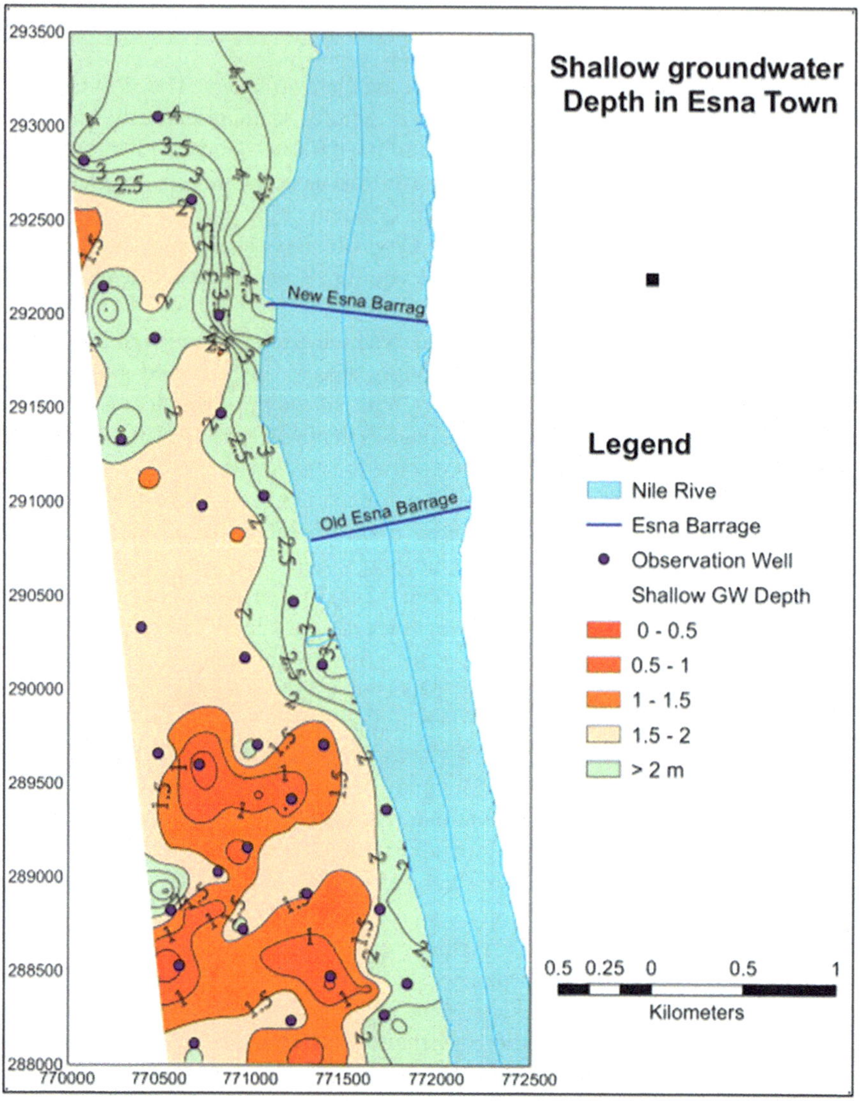

Fig. 11 Depth to shallow groundwater in Esna [67]

outside Luxor back towards the River Nile, which causes the groundwater to elevate in the area (Fig. 13). The regional increase in groundwater salinity was found to coincide with the flow paths, towards the River Nile in the same direction of the groundwater flow.

The thickness of the second geoelectric unit increases towards the River Nile, and thins and pinches out eastward (Fig. 13). However, this unit is anomalously thicker (12–28 m) underneath the area of Luxor Temples, where it may have been

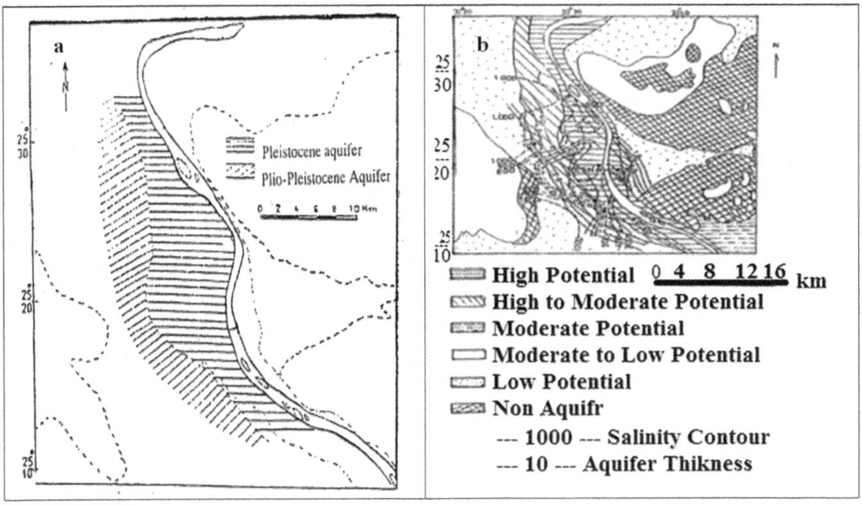

Fig. 12 Aquifer distribution map (**a**) and aquifer potentiality map (**b**) in Esna [68]

deposited in a paleo-meander of the River Nile. This anomaly is significant and must affect the local hydrologic environment. This could include reducing the potential capillary water rise, decreasing the groundwater lateral flow, reducing the vertical drainage of surface water, and increasing the salinity near the paleo-meander. The third geoelectric unit is characterized by relatively high resistivity (17–95 Ω-m; Fig. 13) and significant thickness (10–60 m). It represents the sand, silt, and gravels forming the Quaternary aquifer.

This geoelectric unit exhibits significantly higher resistivity than the overlying silty clay and the underlying sand and clay of the Plio-Pleistocene aquifer. These resistivity contrasts allowed for the determination of the thickness and resistivity of the unit. Resistivity values within the third unit generally decrease towards the River Nile. This decrease in resistivity could be caused by a general increase in groundwater salinity in this direction and/or lateral lithologic changes. It was verified using the groundwater salinity data.

4.3 *Qena Aquifer*

Selim [70] studied areas located at the east of the Nile. It is located within the transitional zone between the eastern desert and the Nile Valley. Two major wadis in the south valley region (Wadi Qena and Wadi El-Mathula) bound it. The Quaternary sediments in the area between Wadi Qena and Wadi Mathula represent an important aquifer both in the floodplain area and in the desert fringes. Nubian Sandstone rocks represent another aquifer, in which groundwater exists under high artesian condition. The potentiometric map reveals that the Quaternary aquifer,

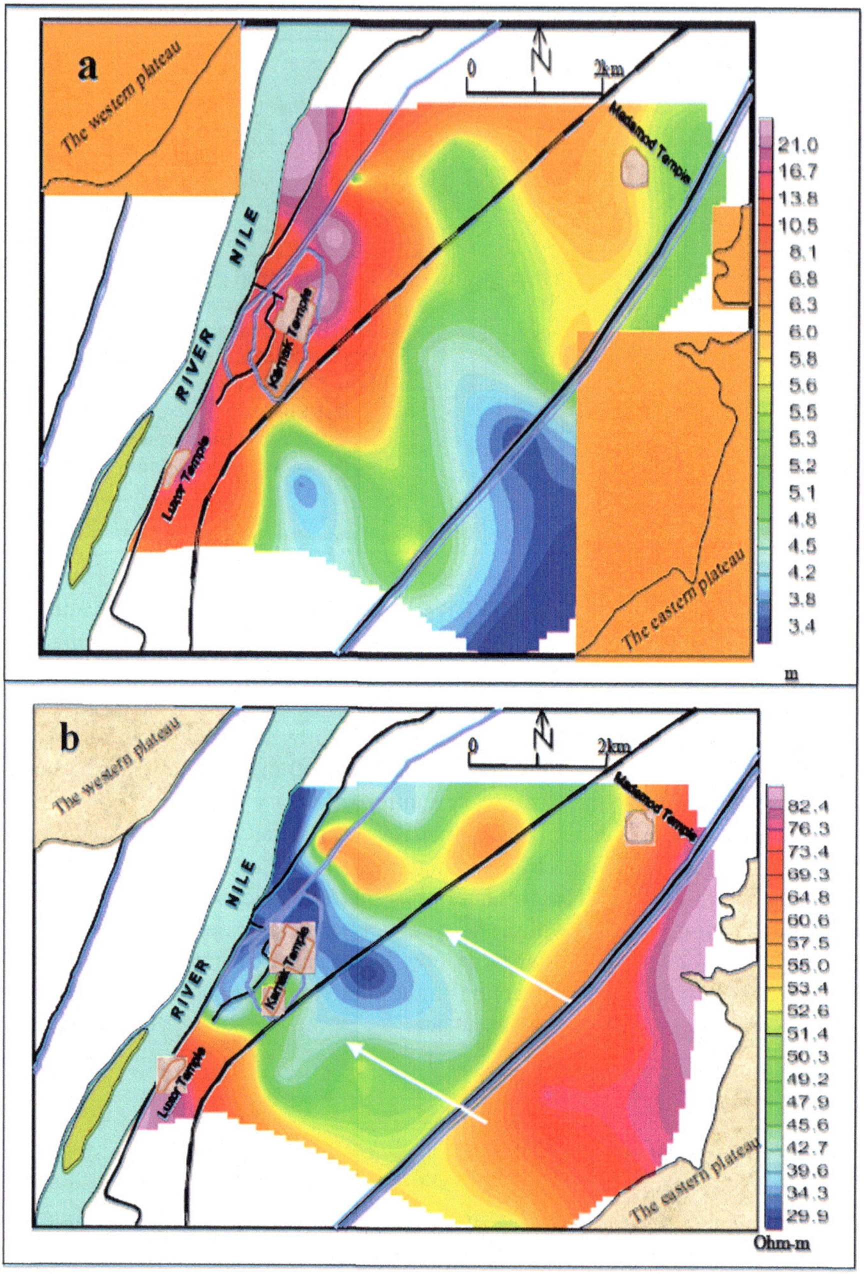

Fig. 13 (**a**) Isopach map of the second geoelectric unit (moist silty clay); (**b**) Resistivity map of the third geoelectric unit (graded sand and gravel of the Quaternary aquifer) in Luxor area [69]

under the floodplain area, has water level in the range from 65 to 70 m (above sea level), while it varies from 70 to 115 m under the desert fringes and along the wadis. The flow direction of groundwater is from the desert fringes to the floodplain areas. The source of groundwater recharge is mainly from the irrigation system in the Nile Valley and secondary from the underflow from the catchments areas of Wadi Qena and Wadi El-Mathula, which receive heavy rainfalls during the winter months every few years. The aquifers are discharged by pumping from the drilled wells, which used for irrigation purposes, as well as by seepage to the Nile River. Groundwater salinity of the Quaternary aquifer varies from fresh to brackish, which has an average TDS value from 728 to 3,308 mg/l. In the Nubian Sandstone aquifer, groundwater salinity ranges from 1,531 to 2,187 mg/l. The water chemistry of the study area indicates that sodium ions represent the main dominating cation, while chloride and sulfate are the main dominating anions. The high concentration of these ions in the groundwater of the desert fringes may be related to leaching processes of highly soluble minerals, which have high effect in the geochemistry of groundwater in its flow path.

Abdalla et al. [71] study the quality of the Quaternary aquifer at Qena. This aquifer is extensive and highly productive. It is distinguished into semi-confined conditions under the cultivated areas and unconfined conditions under the new reclaimed areas at the desert fringes on both sides of the Nile Valley. The aquifer thickness decreases from 300 m at the northern boundary to a few meters in the southwestern boundary of the study area. This aquifer can be categorized into the Holocene aquifer, which composed of clay, silty-clay, and clayey-silt deposits and graded sand and gravel intercalated with clayey lenses (Fig. 14). The Holocene aquifer including the phreatic groundwater that constitutes the base of the cultivated lands with thickness varies from 12.5 to 26 m in the western bank of the River Nile. The main component of recharge of the aquifer is the seepage from irrigation canals, subsurface drainage from irrigated lands, and upward leakage from the deep aquifers through fault planes. The depth to groundwater in the Quaternary aquifer as measured from some available wells varies from a few meters to about 30 m below ground level. The groundwater flow in the aquifer decreases gradually towards the Nile. The River Nile is acting as a discharging line for the Quaternary aquifer as the groundwater levels are higher than those of the River Nile except at the upstream of the barrages.

Many locations in the study area are highly polluted and need urgent measures to protect the Quaternary aquifer and mitigate further pollution. Lack of sewage water systems at rural areas, improper application of wastewater treatment plants as well as uncontrolled use of agrochemicals is the main factors responsible for groundwater pollution. Nitrates, phosphates, ammonia, and *E. coli* bacteria pollute groundwater sources. Groundwater quality varied from locality to another concerning its suitability for drinking purposes according to the World Health Organization and by Egyptian standards. Most of the collected groundwater samples are not suitable for drinking purposes. It is suitable for irrigation purposes with limited restriction. The study recommended management of pollution sources including application of

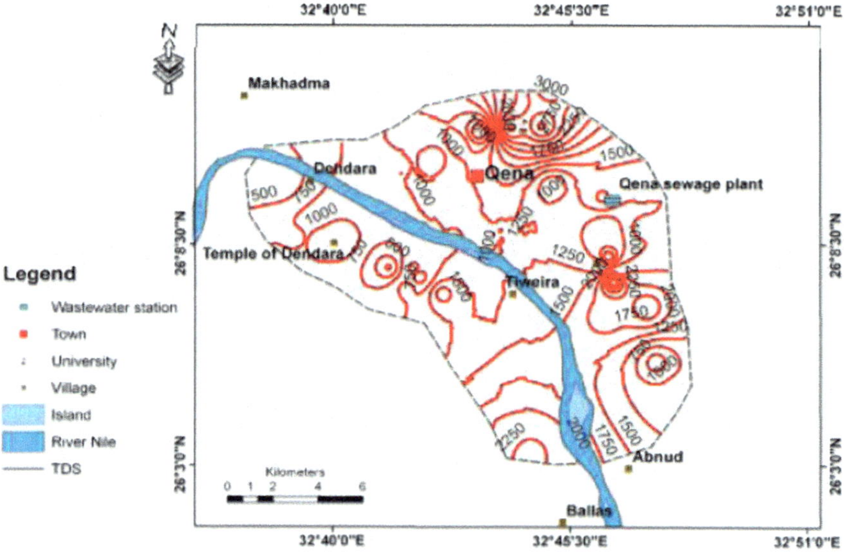

Fig. 14 TDS distribution in the groundwater quality of the Quaternary aquifer at Qena

agrochemicals and controlling the seepage of wastewater from rural areas as well as management of wastewater treatment plants.

Present conditions indicate that groundwater of Quaternary aquifer will become more worsened unless urgent measures are taken. To protect the Quaternary groundwater aquifer and mitigate the risk of pollution, the following recommendations should be followed:

1. Managing the application and use of the hazardous fertilizers and pesticides.
2. Controlling the use of groundwater for irrigation at the desert lands to avoid salinity hazards.
3. Using surface water for irrigation and leaching soils at the desert lands to mitigate salinity and hardness of groundwater.
4. Management of wastewater treatment plants by selecting the best safe locations and managing the processes of wastewater treatment.
5. Determining the vulnerability of Quaternary aquifer to pollution to determine the priority areas to be managed.
6. Continuous monitoring of pollution and degree of water quality degradation to help in decision-making.

4.4 Sohag Aquifer

Esam et al. [72] compare of the groundwater in the West Tahta, Upper Egypt in 1989 and 2011 where they found the Nile Valley characterized by one aquifer

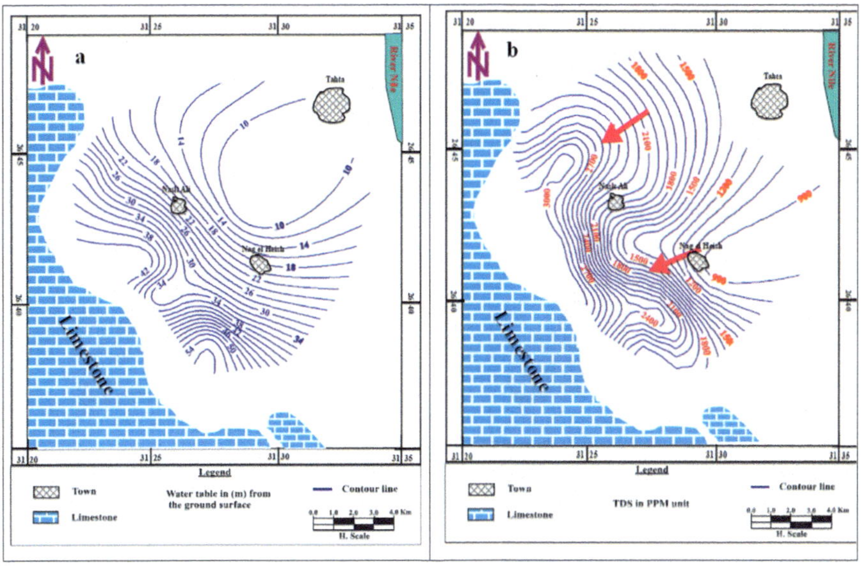

Fig. 15 (**a**) Water table contour map 2011; (**b**) Iso-salinity contour map 2011 of West Tahta [72]

system (Quaternary aquifer), which consists of fluvial sands with minor conglomerate and clay. It is capped with the Neonile silt and fine-grained sands that constitute the base of the cultivated lands. Along the eastern and western fringes, the recent sediments replace the Neonile silty layer. Therefore, the aquifer system in the floodplain is under semi-confined condition, but in the desert fringes, it is under unconfined condition. The penetrated thickness of the aquifer system varies from 150 m in the central part of the floodplain to about 50 m in the desert fringes. The water depth was increased in the study area from 1989 (1–48 m) to 2011 (5–60 m). The extensive withdrawal of groundwater with rare recharge as well as the dissolution of limestone leads to increase in the water salinity. In 1989, it varied between 205 and 2,880 mg/l, while in 2011 it reached to 4,448 mg/l (Fig. 15).

4.5 Assiut Aquifer

El Tahlawi et al. [73] used two DRASTIC models, which are generic and pesticide to get groundwater vulnerable levels to pollution in the Nile aquifer along Assiut governorate. It reaches the northern edge of Sohag Governorate at latitude 27° 37′ N and extends the southern edge of El-Minia Governorate at latitude 26° 47′ N. It is bordered between longitudes 30° 37′–31° 34′ E, as shown in Fig. 16. River Nile divides the study area into a western and an eastern part. In general, the land surface in the fringes of both parts of slopes towards the River Nile. The sharp declination is the main feature of the fringes because of the limestone plateau,

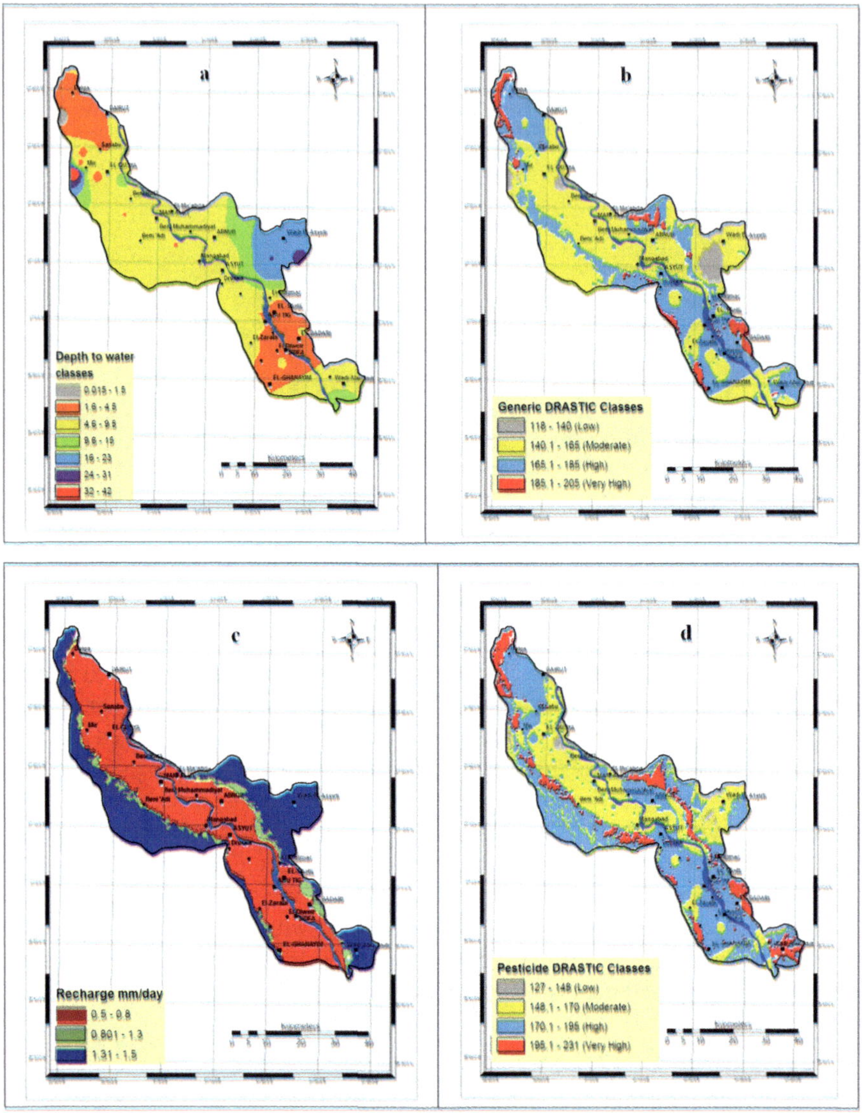

Fig. 16 Groundwater vulnerable levels to pollution in the Nile aquifer along Assiut governorate. Depth to water map (**a**), vulnerability map of Nile aquifer (generic DRASTIC model) (**b**), net recharge of the Nile aquifer map (**c**), and vulnerability map of Nile aquifer (pesticide DRASTIC model) (**d**) [73]

which limits the area from the east and the west, except of the northwestern part that has a moderate slope [19]. There are some wadis joined with the study area like: Wadi El-Assiuti and Wadi El Ibrahimi in the central east, Wadi Abu Shih in the southeast.

The purpose of the vulnerability study is to spot the light on those vulnerable areas by decision makers to protect groundwater. Groundwater vulnerability maps of generic DRASTIC model show that moderate vulnerable and highly vulnerable covers the majority of the study area; where 55.2% of the area are moderately vulnerable and 35.4% has high level of vulnerability. However, in the vulnerability map generated by pesticide DRASTIC model, the results concluded that about 64% of the study area has an extreme to high vulnerability to contamination, 34.6% has a moderate vulnerability, and small areas occupy about 1.4% and has a low vulnerability. The water depth of the majority of the study area ranges from 1.6 to 4.5 m and from 4.5 to 9.5 m (Fig. 17).

Waleed et al. [74] evaluate the groundwater aquifer in the area between El-Qusiya and Manfalut, which represent a large part of the Nile Valley in Assiut governorate using vertical electric soundings technique. Maps of the aquifer base and its thickness were constructed to display the variations of the aquifer thickness. Based on data of true depths deduced from the interpretation of the vertical electrical soundings a contour map of the aquifer base is constructed and shown in (Fig. 17). The interpretation of this map indicates that the greater depth to the aquifer base is located at the central parts of the study area and decreases outward (i.e., to the East and West). The maximum depth is recorded near Manfalut City, where it reaches 300 m from the ground surface. While the minimum depth is recorded at the east of El-Qusiya City, where the depth ranging between 150 and 100 m below the ground surface. The thickness of the Quaternary aquifer ranges between 75 and 300 m, in which the maximum thicknesses are detected around Manfalut and at the west of El-Qusiya. Generally, the central parts of the map have greater thickness than the outside parts and the western parts than the Eastern also.

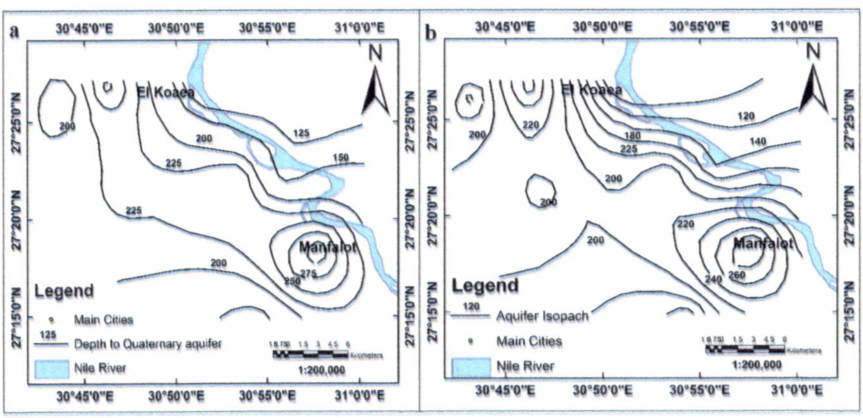

Fig. 17 The depth to the base of the Quaternary aquifer (**a**) and aquifer Isopach map (**b**) in the area [74]

4.6 El-Minia Aquifer

Salem [75] studied the hydrogeological settings of the topmost aquifers along the area lying west of Samalot (El-Minia Governorate). The area of study extends between longitudes 30° 20′ and 30° 45′ E, and latitudes 28° 10′ and 28° 30′ N. The middle Eocene (Samalot Formation) fractured limestone is the available water-bearing rocks in the investigated area. The maximum penetrated thickness reaches 225 m. It occurs under free water table conditions. The fracture density, caving and fissuring, controls the geometry of the aquifer. The depth to water ranges between 109.43 m to the west and 14.30 m to the east (Fig. 18). The groundwater flows to the east and northeast towards Abo Edahab irrigation canal, which acts as a discharging area rather than recharging one. An annual drop in water level occurs. It ranges between 0.5 and 2.45 m (Fig. 18). The aquifer transmissivity ranges between 4394.9 m^2/day and 3515.929 m^2/day (high potentiality) along the eastern parts. However, it ranges between 274.68 m^2/day and 15.13 m^2/day (moderate

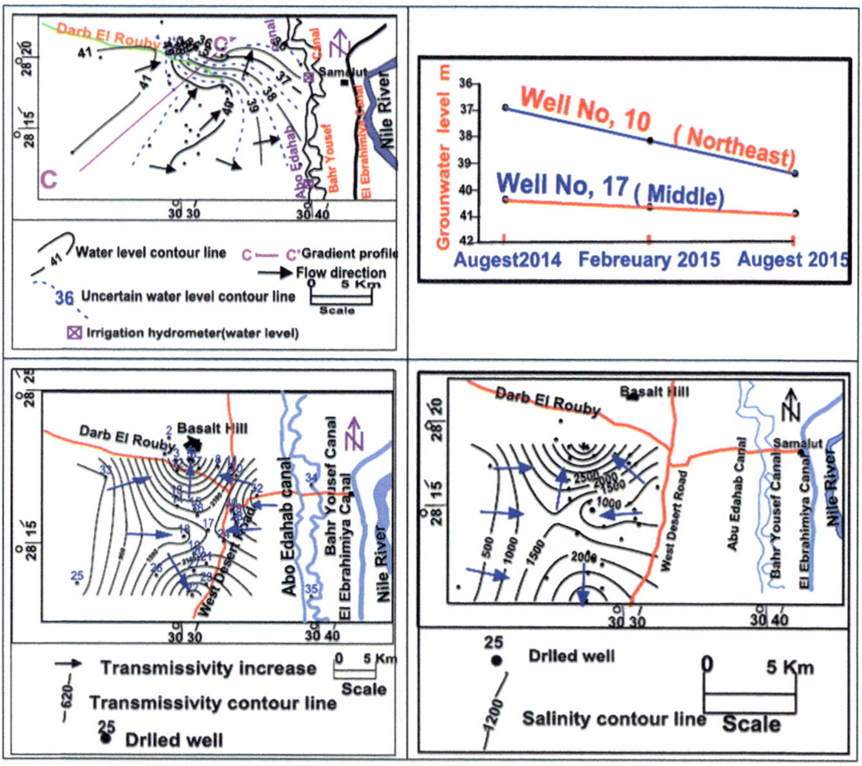

Fig. 18 Hydrogeological settings of the Samalot Formation aquifer (El-Minia Governorate). Water table contour map (records of august 2015), annual drop in groundwater level, transmissivity isolines map, and iso-salinity contour map [75]

potentiality) in the western parts (Fig. 18). Most of the analyzed groundwater samples are of sodium chloride and some of sodium sulfate water type.

4.7 *Giza Aquifer*

Gedamy et al. [76] detect pollutants in groundwater and their impact on human health at El-Saff area which occupies the southern position of Giza governorate. El-Saff area extends along the eastern side of the Nile River between latitudes 29° 30′–29° 45′ N and longitudes 31° 15′–31° 25′ E. Its location is just to the south of Helwan area, which includes many factories for iron, steel, and cements production. These factories constitute sources for water resources pollution besides El-Saff wastewater canal and polluted agricultural drainage water. The surface water system in the study area comprises Nile River, El Khashab, and El-Hagar canals. The groundwater is represented by the shallow Quaternary aquifer. The major threat to the water resources in El-Saff area is coming from the polluted sewage water and drainage water, which are considered very serious to people's health.

Groundwater has Hg^{2+} (0.003–0.033 mg/l), which is greater than the acceptable limit of pollution. Also, 22% of the groundwater samples are contaminated with Fe^{3+} (0.4321–1.77 mg/l), 14% of the water samples are polluted with Mn^{2+} (0.4028–0.5156 mg/l) and Al^{3+} and Zn^{2+} ions (0.337 and 9.84 mg/l, respectively), which exceeded the permissible limit of pollution. These results confirm that there is a seepage from El Khashab and El Hager canals as well as from El-Saff wastewater canal that contains relatively high soluble metals as well as downward infiltration of excess amounts of irrigation water rich in fertilizers and pesticides to the groundwater aquifer. Quaternary aquifers recharge sources are the seepage from Nile water and downward infiltration from the excess irrigation water of cultivated lands as well as the seepage from canals and drain. El-Saff wastewater canal was generally dug in the Pliocene clay sediments and sometimes cutting the foot slope of the eastern limestone plateau. These rocks traversed by many wadi courses. Wadi deposits that fill these courses facilitate the arrival of the canal water not only to the old cultivated lands but also to the Nile course itself. The consequent water level contour map shows that the movement of groundwater takes place essentially from east to west, i.e., from the new reclaimed area around El-Saff canal (+71 m) to the old cultivated lands (+26 to +24 m) near the Nile River (+23 m) since the latter acts as a discharging area for the former. In addition, the results of the geoelectrical survey in Ghamaza area reveal that the groundwater exists at shallow depths ranging from 7.5 to 10.8 m and the sedimentary section above the water table consists of gravel and sand, which permits seepage and filtration from the disposal site to the groundwater.

5 Land Resources in the Nile Valley and Delta

Egyptian Nile River basin includes Delta and Valley, El-Faiyum, and Lake Nasser. Fluvial sediments and soil that covers the Delta and Valley and some parts of El-Faiyum are the result of the annual precipitation for the silt from African headwaters of the Nile before the construction of the High Dam. Fluvial sedimentary soils of El-Faiyum were considered the Egyptian Nile River Basin, due to contact El-Faiyum with Nile River by Joseph Sea. As for the area of Lake Nasser the situation is different, it has been put the Lake Nasser area within the Egyptian Nile River basin as a geographic and not because the soils are sedimentary, because most soils around Lake Nasser is not sedimentary soils except soils of khors and wadis, such as Khor Karkar, Khor Kalabsha, Khor Toshka, and Khor Allaqi where these khors were filled with water and mud during a flood before the construction of the High Dam. Therefore, there are the fluvial sedimentary soils. As for the soils in areas around the lake are mostly composed of local rock deposits that are not affected by the presence of the Nile at high levels.

5.1 Soils of the Nile Valley and Delta

Soils of Delta and Valley consider the most fertile soils. The ancient Egyptian civilization was placed on the banks of the Nile River. Most of the agricultural areas are concentrated in the Delta and Valley. The majority of the planting areas were located west of the Nile. The cultivated areas east of the Nile were much lower than in the west Bank of the Nile (Fig. 19). The river between Aswan and Cairo always has a tendency to adhere to the right side of the valley. It does not rotate to the left side of the river until it reverts to the right side. This phenomenon is not clear in Qena, where the river runs away from east to west. However, it is apparent from a distance in particular between Manfalut (Assiut) and north of Cairo. In Minya and Beni Suef, for example, the sedimentary plain is situated on the left side. However, on the right side, the river is hardly separated from the desert. This implies that the river through its sediments on the left side dissolves in the carving of a small part of the right side so that the plain on the left. On the right side, there is a cliff and a desert plateau.

Delta and Valley soil properties (Fig. 20) vary depending on its geographical location. For example, the northern parts of the Delta and near the northern lakes are high salinity, unlike the central and southern areas in the delta is less salty. Also, the eastern and western edge parts of the valley and adjacent to the desert differ in their properties on the parts located in the center of the valley. Delta and the valley soils are very poor in organic matter content (0.5–2%), and contain very small

Fig. 19 Satellite image shows the majority of the planting areas were located west of the Nile. The river between Aswan and Cairo always tends to adhere to the right side of the valley

amounts of $CaCO_3$ (1–4% or a bit more), and clay vary from 40% or less to 70%. The following is the different soil mapping units for the Nile Valley and Delta [78]:

1. Recent fluvial sedimentary soils include light to heavy very dark brown clay soils and clay loam to sandy loam soils, sometimes coarse sand in the surface layer (the shoulders of the river).
2. Fulvio marine sedimentary soils include heavy, very dark brown clay soils, and clay to clay loam brown soils at the depth of 50–120 cm above the shells mixed with bluish layers of sand and clay.
3. Under Deltaic soils include yellow sandy soils and gray sandy or faint soils, sometimes sandy loam to clay loam in the middle layers.
4. Desert plain soils include sandy soils and loam sand to clay loam dark brown over yellow sand, flat.
5. River terraces soils include faint sandy limestone soils mixed with pebble and gravel of various sizes and undulating surface.
6. Marine sandy beach soils include sandy often submerged and hills of sand deep mid-rise.

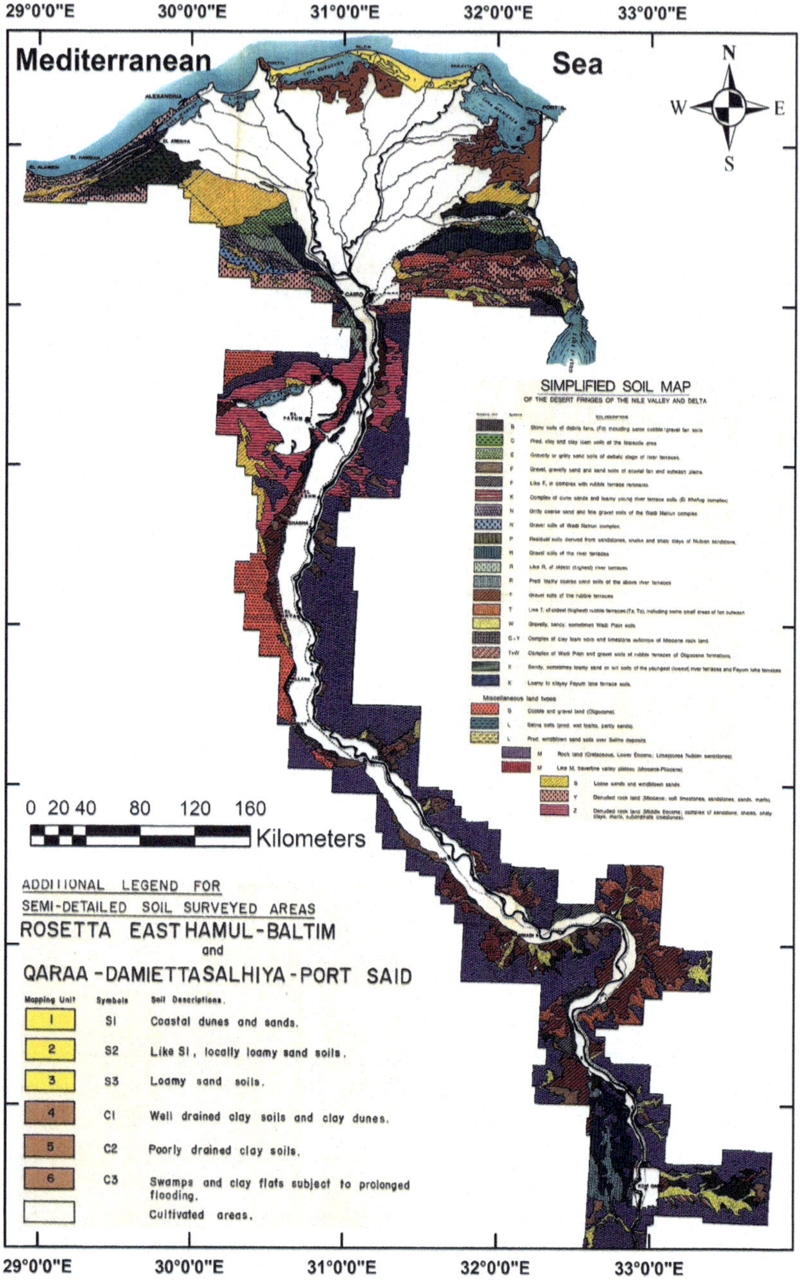

Fig. 20 Simplified soil mapping units adjacent areas of the delta and valley [77]

However, the delta soils were divided into eight mapping units based on soil data and surface terrain database [79]: (1) Sandy to clay soils, high salinity, slope 2% – *Aquolic Salorthids*. (2) Sand dunes. (3) Sedimentary marine, clay, high salinity – *Vertic Ustifluvents*. (4) Fine clay, high salinity, slope 1% – *Typic Ustifluvents*. (5) Sand, high salinity, slope 2% – *Typic Torripsamments*. (6) Clay, non-saline, slope 1–2% – *Typic Torrerts*. (7) Sandy loam, non-saline, slope 1–2% – *Typic Torrifluvents*. (8) Clayey, high salinity, slope 2% – *Typic Salitorrents*. Figure 20 shows the soil mapping units adjacent areas of the delta and valley [77].

The laboratory analyzes showed that the Delta soil ranging from light to heavy clay, and that the percentage of clay ranging from 40% in the south to about 70% in the north (Fig. 21a). Most Delta soils are non-saline (Fig. 21b), but the salinity problem becomes more serious as we get closer to the sea, lakes, beach due to high groundwater level as a result of salty seawater and lakes nominated in the surrounding soils. Figure 21c shows the groundwater depth of Delta soils, where the groundwater depth of up to more than 150 cm in most of the southern part. The shallower northern part is near the coast and lakes. In general, the layers beneath the soil in the northern areas are wet, exhibiting the Mottling phenomenon, and have the gley horizon because of poor ventilation and poor drainage.

5.2 The Soils of El-Faiyum Depression

El-Faiyum depression lies in the western desert of Egypt close to the Nile Valley with a distance of 40 km. It lies to the southwest of Cairo at a distance of about 90 km. The depression is situated between the altitudes 20° 34′ and 29° 2′, while the City of El-Faiyum – the capital – lies between 29° 18′ altitude to the north and 30° 50′ longitude to the east. The depression is bordered by Qattrani Mountain, to the south of the Libyan desert. On the eastern border, it is connected with the Nile by Bahr Hassan Wassif and the main desert highway (about 92 km) joining the northeast of the depression with Cairo (Fig. 22). The depression is surrounded by the Libyan desert except for a very narrow cultivated strip connecting it with the Nile Valley and with lake Qarun to the north. El-Faiyum depression has, in general, an extremely arid climate characterized by long dry and hot summers and short nearly rainless and cold winters [80]. The Nile water is considered the main source of irrigation in El-Faiyum by Joseph Sea. The water of agricultural drainage disposes in Lake Qarun while remaining part disposes in El-Rayan Valley (Fig. 22).

El-Faiyum is a deep circular depression in the limestone of the northern part of the western desert. The soils of the area comprised in the El-Faiyum fan, the recent lake terraces, the old lake terraces, the older lake terraces, the depression edges, El-Faiyum plain, the Ghrak basin, and the old wadis. These soils have been developed from three types of parent materials, namely the Nile deposits, lacustrine deposits, and the calcareous deposits. Because of the differences in the nature of soil parent materials, these soils reflect different physical, chemical, and morphological characteristics. These soils are young and show weak horizon differentiation

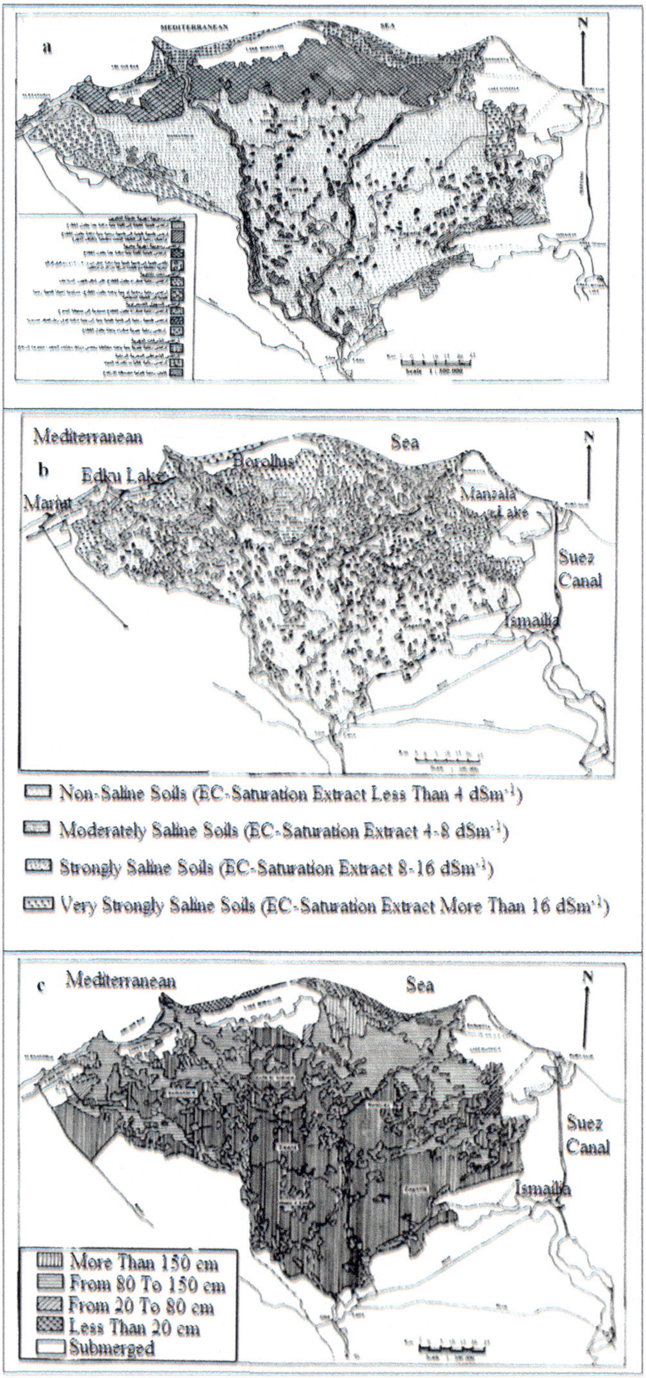

Fig. 21 The morphological properties (**a**), soil salinity (**b**), and groundwater depth (**c**) of the Nile Delta [78]

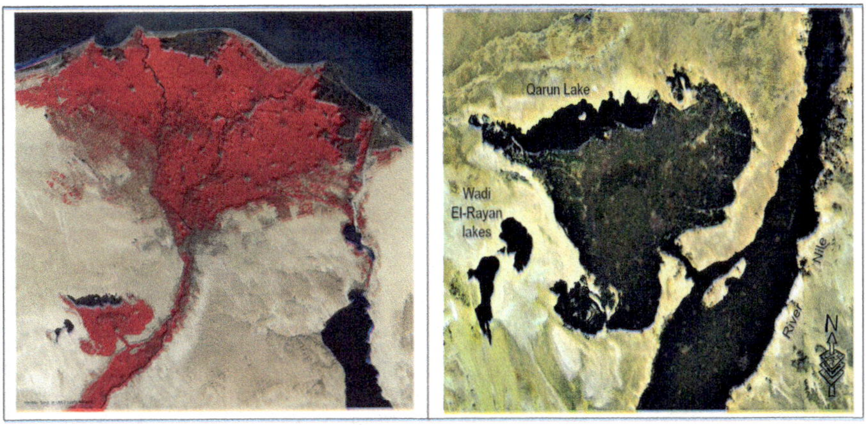

Fig. 22 Landsat color image of El-Faiyum depression

with the absence of diagnostic horizons other than salic, gypsic, and calcic horizons. Dominant pedogenetic processes are salinization, translocation of iron oxides, and probably transformation of minerals [81].

With regard to the main soil characteristics of El-Faiyum Governorate, many authors studied these properties [44, 80, 82–87]. El-Faiyum soils have divided into the following units [88]:

1. Soils with diagnostic horizons, such as the salic horizon, gypsic horizon, and the calcic horizon, which include: (a) Fluvial sedimentary soil, and have accumulation of calcium carbonate horizon. (b) Soil containing gypsum and $CaCO_3$ horizons, located near the eastern edge of the depression. (c) Soil with salic horizon, mostly on the coast of Lake Qarun, which is the lowest part in El-Faiyum, consists of fluvial deposits, high water table, salt crust on the surface, and most of which is not cultivated.
2. Soils do not have diagnostic horizons, which include: (a) Sandy soils are at low edges and mostly not cultivated due to the high water level. (b) Fluvial sedimentary soils, occupy a relatively large area in the center of the depression, have silty sand to silty clay texture, low salinity, and a large part is cultivated with citrus and other.
3. Heavy clay soils, most of low depression, consist of river sediment and salinity varies from region to region.

From the soil association map of Egypt [89], the Nile alluvium soils are the major in El-Faiyum depression as well as stony and loamy sand lithosols on rough to undulating denuded terrain that developed mainly from limestone. These soils could be grouped, in general, into Nile alluvial fluvio-lacustrine, residual regosols and lithosols (Fig. 23). The soil parent materials are of two kinds: transported or residual. In both types, there is not much difference and it is only confined to the formation of salic, gypsic, calcic, and gley horizons. El-Faiyum soils belong to the three orders vertisols, entisols, and aridisols. Abu El Einane [91] stated that the

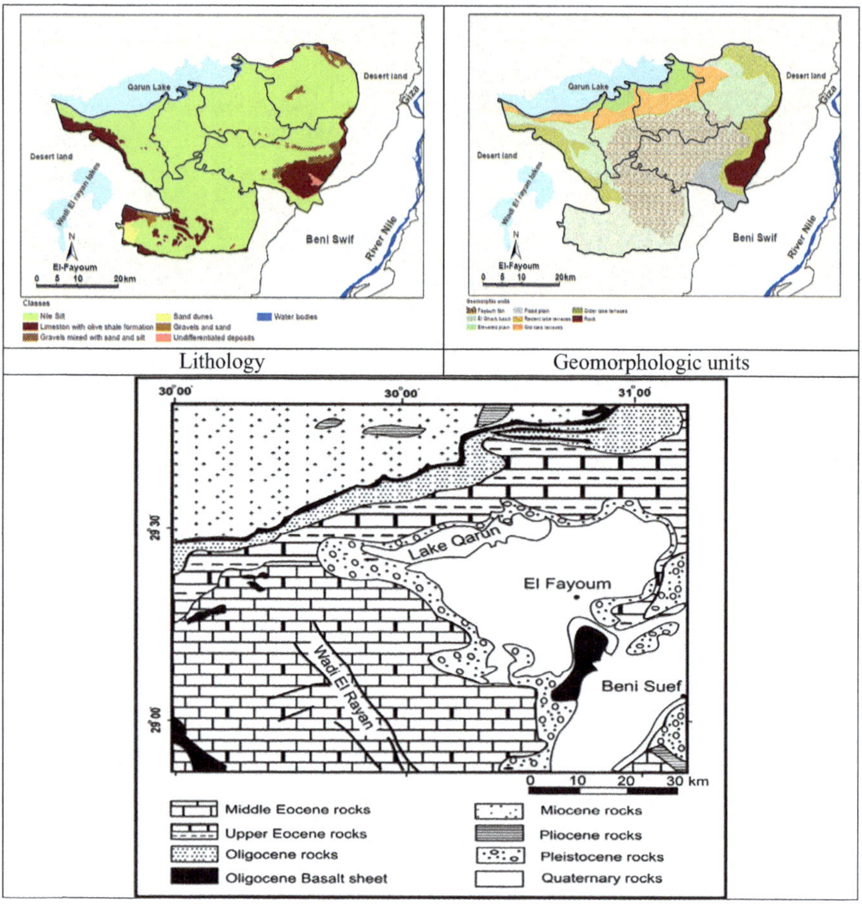

Fig. 23 Geological map of El-Faiyum and surrounding areas [90]

calcareous parent material is the prevailing arid climatic conditions and relief is mainly considered as the limiting factors affecting the soil forming development processes in El-Faiyum.

Abdel Aal [92] studied the soils of different geomorphic features of El-Faiyum area. Morphological and analytical studies reveal that, on geomorphical basis, four geomorphic units El-Faiyum depression were recognized, namely (a) the El-Faiyum fan, (b) El-Faiyum plain including the flood and elevated features, (c) the El-Faiyum lake terraces including the recent, old, and older features, and (d) El Gharak basin. There exists a mutual relation between geomorphic aspects and either geogenic processes or pedological characteristics. Most of the soil properties can be interpreted by the climatic fluctuation between humid and arid conditions in both Pleistocene and Holocene. Therefore, some areas are characterized by recently formed and weakly developed soil without any diagnostic horizons due to the dominance of cumulization processes, which causes additions of new Aeolian or aqueous sediments under arid and humid conditions. Most of the studied soils are apparently formed from aqueous media, together with Aeolian origin, which was seemingly obvious in some localities lying on middle part (Faiyum fan) or at the edges of the depression (older lake terraces). The general conclusion, however, is that soil development as expected is weakly or not so pronounced, except for soil, which achieve some pedological diagnostic horizons. Hanna and Labib [93] showed that the soils located near the desert fringes of El-Faiyum depression are formed from fluvial-colluvial-aeolian sediments, while those near Qarun Lake are mainly formed from saline lacustrine deposits. They added that these soils have a sandy clay loam to loamy sand and clayey textures for the desertic formations and those near the Lake, respectively.

Soils adjacent to Qarun Lake include three geomorphic units, which are Nile alluvial deposits, desertic formation, and interference zone between lacustrine and desertic deposits. The soils south of Qarun Lake have been developed under arid conditions with different parent materials. Drainage played a major role in the soil formation through the effect on the soil salinization. In addition, Shendi [85] noted that the soil forming processes are salinization, calcification, alkalinization, illuviation, and gleization in these soil. The main soil forming factors in this area are parent material, climate, and topography. Studies of Kassem and Elwan [94] on these soils showed that these soils were developed under either stratified condition or multi-depositional regime. They also indicated that the amount and type of minerals assemblage reveals the origin of these soils, which could be related to lacustrine as well as post-lower Paleolithic Nile deposits.

The geographic location of the different soil types can be attributed to the distance to Qarun Lake in the northwest and thereby the frequency and severity of flooding [9]. Figure 24 displays the soil map of El-Faiyum. *Vertic Torrifluvents* are the dominant soil subgroup in El-Faiyum, covering an area of 760 km^2 (43% of the study area). Other important subgroups are *Typic Haplocalcids* (421 km^2, 24%) and *Typic Torrifluvents* (141 km^2, 8%). The rest of the soil subgroups (*Typic Haplogypsids*, *Typic Haplosalids*, and *Typic Torripsamments*) are found in small areas and all together cover approximately 11% of the study area. Most of the cultivated soils in El-Faiyum province are deep Nile alluvial loam to clayey soils. In addition, calcareous clayey and a part from the sandy soils are found as patches in the depression edges. A general discussion of such types was provided by Shendi [85] who classified the soil types in this area in five major units: Nile alluvial soils, fluvio-lacustrine soils, fluvio-desertic soils, lacustrine soils, and desert soils.

The main identified soil deposits in El-Faiyum depression are: (a) Nile alluvial, (b) fluvio-lacustrine, (c) fluvio-desertic, (d) recent and old lacustrine, and (e) aqueous and Aeolian desertic formation. Nile alluvial deposits are mainly occupied the eastern-south and middle parts of El-Faiyum depression. Fluvio-lacustrine deposits occupied a large area, which exists in the eastern and southern rims of Qarun Lake. The general mineral assemblage and ratios of resistant minerals may be interpreted the heterogeneity of the parent materials, which prior to inception to geogenetic weathering, as well as the contribution of a basic source

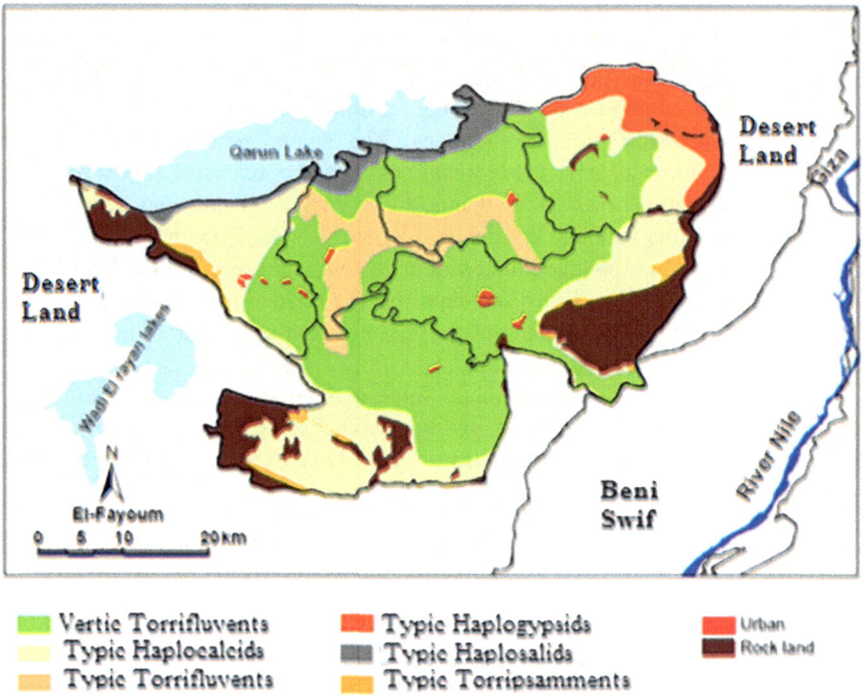

Fig. 24 Soil map of El-Faiyum [95]

rock. It is rich in the ferro-magnesium minerals, together with the Nile suspended matter in forming a multi-origin profiles. Fluvio-desertic deposits occupied a small area at the outer rims of El-Faiyum depression, especially the western and eastern ones. Representative soils are characterized by heterogeneity nature of the parent materials; therefore, these deposits possess quite different textural classes that reflect the fluctuation in the lithological and environmental conditions during the profile formation. Recent lacustrine deposits are located at the present shoreline area of Qarun Lake. The frequency distribution of minerals and ratios of resistant minerals within the profile layers indicates multi-depositional stages, as well as a multi-origin profile and advanced stages of geochemical weathering. The suite of minerals of the old lacustrine deposits reveals that these deposits are subjected to an intensive chemical weathering for a long time and formed under multi-depositional regimes. Aqueous desertic deposits are located in the northern-east area. Unique depositional and mineralogical features that are differed from those of Nile and El-Faiyum deposits characterize these deposits. Aeolian desertic deposits occupied the northern-east and west area. These deposits are mainly derived from acidic plutonic rock with a little contamination from a metamorphic source rock. Also, these deposits have a relatively low index figure, an elementary weathering stage, almost uniform origin, and similar depositional environments [96].

From the study of soils that covered the main fluctuation in the environmental conditions of El-Faiyum, Awadalla [97] explained the mode of soil formation, soil origin, and profile uniformity. The soil deposits are of a poorly sorted nature, indicating that these sediments are mainly transported and deposited in aqueous media by water agent. The exception condition of the uppermost layers at fluvio-desertic formation soil exhibits an improvement in sorting coefficient, as it is qualified as a moderately well sorted or moderately sorted nature, indicating the combination effect of the Aeolian deposits, which transported by wind from the adjacent desertic formation. The assemblage and frequency of heavy minerals identified in the soil sediments developed on the Nile alluvial denote the low value of the index figure and the moderate content of parametamorphic minerals. These deposits are originated from igneous and metamorphic rocks, mostly related to Abyssinian plateau, basaltic-Sudan, and granitic-Aswan regions. However, soil sediments of both recent lacustrine and fluvio-lacustrine, which are dominated by ferromagnetism silicate minerals together and less pronounced amounts of parametamorphic ones, may be due to a basic source rock. Also, the high values of both index figure and opaques together with moderate content of ferro-magnesium minerals in the deposits of old lacustrine soil may have stress on their calcareous in nature, as well as, the possibilities of enrichments from a basic source rock. Concerning the deposit of the fluvial-old lacustrine, obtained data indicated a moderate content of ultra-stable minerals (zircon, rutile, and tourmaline) that occurred in the surface layer, which confirms the characters of the Aeolian desertic formation. Regarding uniformity ratios and weathering values, results obtained indicate that soil sediments of Nile alluvial, recent lacustrine and desertic formation soils exhibit an apparent homogeneity.

5.3 Lake Nasser Area

Lake Nasser (22° 31′–23° 45′ N and 31°30′–33°15′ E) reached its operating level of 175 m (asl) in 1975, with a total amount of 121×10^9 m^3 of stored water. Lake Nasser is a vast reservoir in southern Egypt and northern Sudan. Lake Nasser refers only to the much larger portion of the lake that is in Egyptian territory (83% of the total). Lake Nasser is currently the largest man-made lake in the World. The region around the reservoir is a potential area for land reclamation and new settlements, urgently needed by Egypt's expanding population [98]. Lake Nasser (Fig. 25) has a number of side extensions known as khors. All khors have a "U" shape in cross section, with a flat sandy central belt. There are 100 important khors in Lakes Nasser and Nubia combined. Their total length when the lake is full is nearly 3,000 km and their total surface area is 4,900 km^2 (79% of total lake surface). In volume, they contain 86.4 km^3 water (55% of total lake volume). Khors Allaqi, Kalabsha, and Toshka are the largest [99].

The soils, which lie on a level between 180 and 200 m above sea level, were described only, because the soils that fall under this may be attributed to the

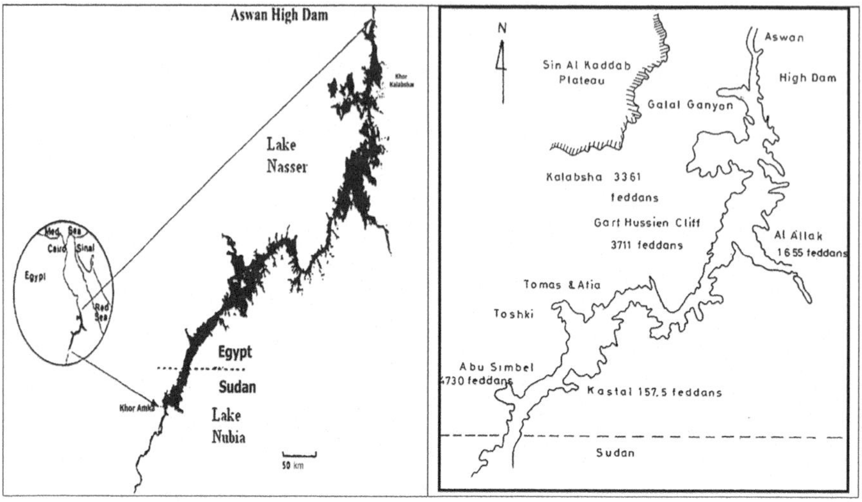

Fig. 25 Map of the High Dam (Lake Nasser) soils

submerged during the flood season, and thus these soils were excluded. The parent materials of these soils are wadis sediments, shales, and Nubian sandstones [100]. On the western side of the lake, soils each of Karkar, Kalabsha, Aldka, Sarah, Toshka, Abu Simbel, and Adindan have consisted mainly of sandstone and shales. The soils of the Wadi Allaqi side east of the lake were established from igneous and metamorphic rocks.

Karkar valley is the Nuba part west of Lake Nasser and runs until the Sinn El-Kedab Plateau (Fig. 26) bordering the valley from the west. The plateau extends to south Kharga with the area of 800 km^2 and a height of 180–550 m. The soils consist of different parent materials such as sandstone, limestone, clay, and sand transported by wind. The sandy soils are dominant in the region with a low content of salt.

Kalabsha is located south of the Aswan City about 75 km. It extends from the Sinn El-Kedab Plateau (Fig. 26) in the direction of the northwest of Lake Nasser. Kalabsha Khor holds the lower part of the Valley Kalabsha, starts from Lake Nasser, and extends towards the west for a distance of 10.5 km with an average width of about 3 km. It is divided into two branches: the northern branch continues its extension towards the northwest for a distance of 5.5 km with an average width of 1 km, and the southern branch continues its extension towards the southwest for a distance of 11.5 km with an average width of about 1.1 km.

There are between Kalabsha and Aldka Kaolin mine (Kaolinite-mineral), with thickness ranges from a few centimeters to over 5 m, covering these deposits a thin veneer of wadis deposits. Some single hills exist such as Marawah Mount (270+ m). Soft sediments found in a low circular shape like Aburihwa Playa, located north of this Marawah Mount, with an area of about 30 km^2. The soil texture is often clay loam with a thickness of about 10 m.

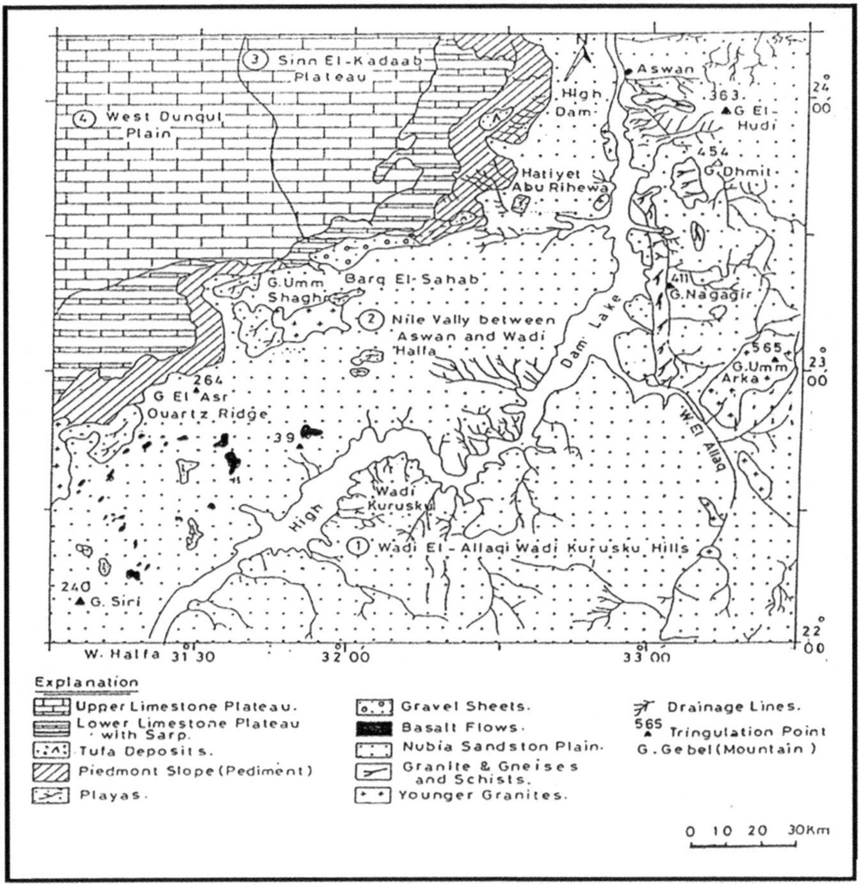

Fig. 26 The geomorphological units of the Lake Nasser area and adjacent areas in southern Egypt [101]

Sarah Khor soils are located on the western bank of Lake Nasser, and lie about 350 km south of Aswan, which is part of the plain of Lower Nubia. There are some sand dunes and rocky bumps on the surface in some locations, and there are hills and plateaus separate (Mesas) of the Nubian. Many wadis cut the area in which the most important and the longest are Sarah Khor.

Toshka (Fig. 27) is Nubian word with two syllables. The first is "Touche," which is a plant that grows in depression, "Key" meaning home or place. Toshka means plant or flowers places [102]. The Toshka area is the first sites to gather the population in the desert, by prehistoric man, and the region of archaeological value, including the effects of and information about the origin of human civilization. Many dry valleys cut surface of Lower Nubia, which discharged, into the lake, and the valleys are Toshka and Wadi Aldka, where the headwaters are located in the plains of Nubia. The region is surrounded by hills in some parts of the Nubian

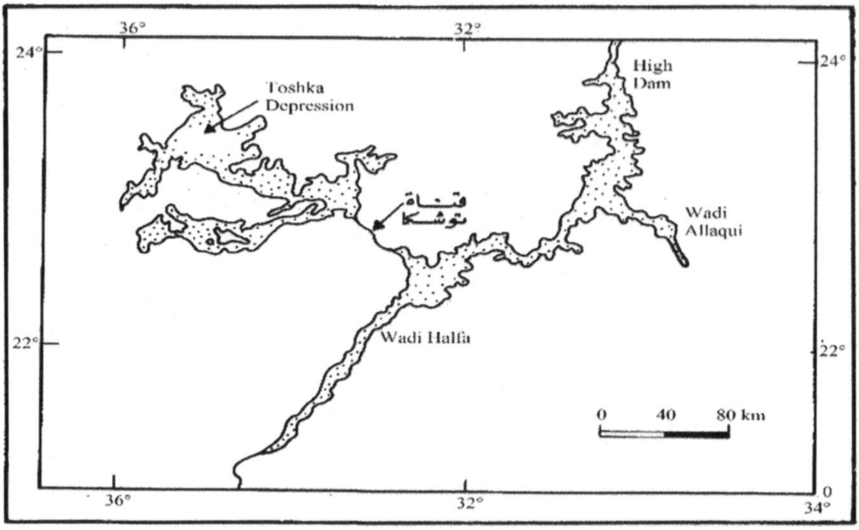

Fig. 27 A map showing the flood of Toshka in the desert

Sandstone, where hilly landscape forms a hierarchical area shape or continuous sandstone ridges.

Shallow sandy soils occupy part of the region, while the deep sandy soils occupy relatively small areas. Most of the soils (sandy or shallow sandy) are less than 50 cm depth. The shallow sandy soil area of about 89,000 acres. The surface is covered with small size pebbles, low salinity, which rarely exceeds 1.3 dS/m. The amount of $CaCO_3$ ranges from 1.1 to 4%, and up to 15% or more in the some sites [103]. While the deep sandy soils occupy a small area (13,000 acres), small gravel on the surface, very low salinity (<1 dS/m), and few $CaCO_3$.

The dominant clay minerals [100] in the Lake Nasser consist of kaolinite, montmorillonite, illite, and chlorite, but the kaolinite is dominant in sandstone soil, a good kind of crystallization (well crystalline), while the transported sedimentary soils have weak crystallization kaolinite.

The Adindan soils are located in the far south of Egypt, where the Egyptian–Sudanese border. Many valleys cut the region including Khor Adindan. The following soils are identified:

1. Deep sandy soil (11,448 acres) of which the soil surface is covered with a thin layer of sand, gravel, and mixed soils. Soil salinity ranges from 6 to 11 dS/m and $CaCO_3$ ranges from 2 to 14%.
2. Medium deep sandy soil (19,788 acres) ranges from 50 to 100 cm of which soil surface is covered by desert pavement, mixed with gravel sand. Salinity is ranging from 7 to 10 dS/m and $CaCO_3$ ranges from 7 to 14%.
3. Shallow sandy soils (2009 acres) at the depth of 50–25 cm, with some of fine layered, rock Nubian sandstone at depths close to the surface. The salinity ranges

from 3 to 11 dS/m, and $CaCO_3$ ranges from 2 to 14%, being depending on location.
4. Very shallow sandy loam soils (15,605 acres) at the depth of 25 cm. It is bounded on the bottom of the Nubian, mixed with gravel with sand through the layers of the soil at different rates depending on the location. The salinity ranges from 7 to 61 dS/m and $CaCO_3$ ranges from 3 to 7%.

5.4 Evaluation of the Soils from Lake Nasser to the Delta

Although the Nile Valley and delta soils occupy the smallest region in the country, Egyptian peoples depend on these soils for agriculture and food since thousands of years. According to the soil association map of Egypt (Fig. 7) [89], there are five soil types: (a) Nile alluvium that formed the soils of the valley and delta except of its northern part, (b) salt affected soils of the lower Nile Delta areas, (c) salt marshes of the lower delta plain, (d) sandy clay loam with calcareous crusts and sand dunes of the delta lacustrine complexes, and (e) gravels and gravelly sand soils of deltaic phase with sand dunes in the eastern and western sides of delta. Alluvial soils are constituted by river sediment deposits that formed from the disintegration of the eruptive and metamorphic rocks of the Ethiopian plateau through physical, chemical, and biological weathering processes.

5.4.1 Soils Surrounding of Lake Nasser

The soils surrounding of Lake Nasser, i.e., Kalabsha, Tushka, El-Dakka, Abu-Simble at the western side of the lake and Allaqui area on the eastern side (Fig. 28) were formed as a result of water action while wind action is neglected [100]. In addition, the parameters show that Wadi soils are heterogeneous and stratified, while sandstone has a degree of uniformity. Wadi soils are composed of subrounded grains; this may be related to the distance of transportation. The origin of the clay minerals varies according to the lithological origin of soils. Kaolinite could be diagenetic formed in sandstone rocks and inherited in soils. Smectite may be derived from mica in the soils at the western side of the lake, while from mica or volcanic ash at the eastern side. Chlorite of Wadi Allaqui soils was most probably derived from the primary chlorite of the drainage basin of Wadi Allaqui. Illite may be formed because of changes on mica. Abu Al-Izz [37] reported that the Nile river terraces on both sides of the valley formed from sediments belonging to the Pliocene and Pleistocene. The Pliocene sediments in the southern part of the valley between Kum-Umbu and Beni Suef comprise of conglomerates, gravel, and sand. Those are distributed in some parts of the valley between Pleistocene and Holocene sediments of the floodplain and the two scarps bordering the valley. The Pleistocene deposits comprise of sand and gravel originating in the Red Sea Mountains. Moreover, the formation of the river terraces is related to three main factors, namely

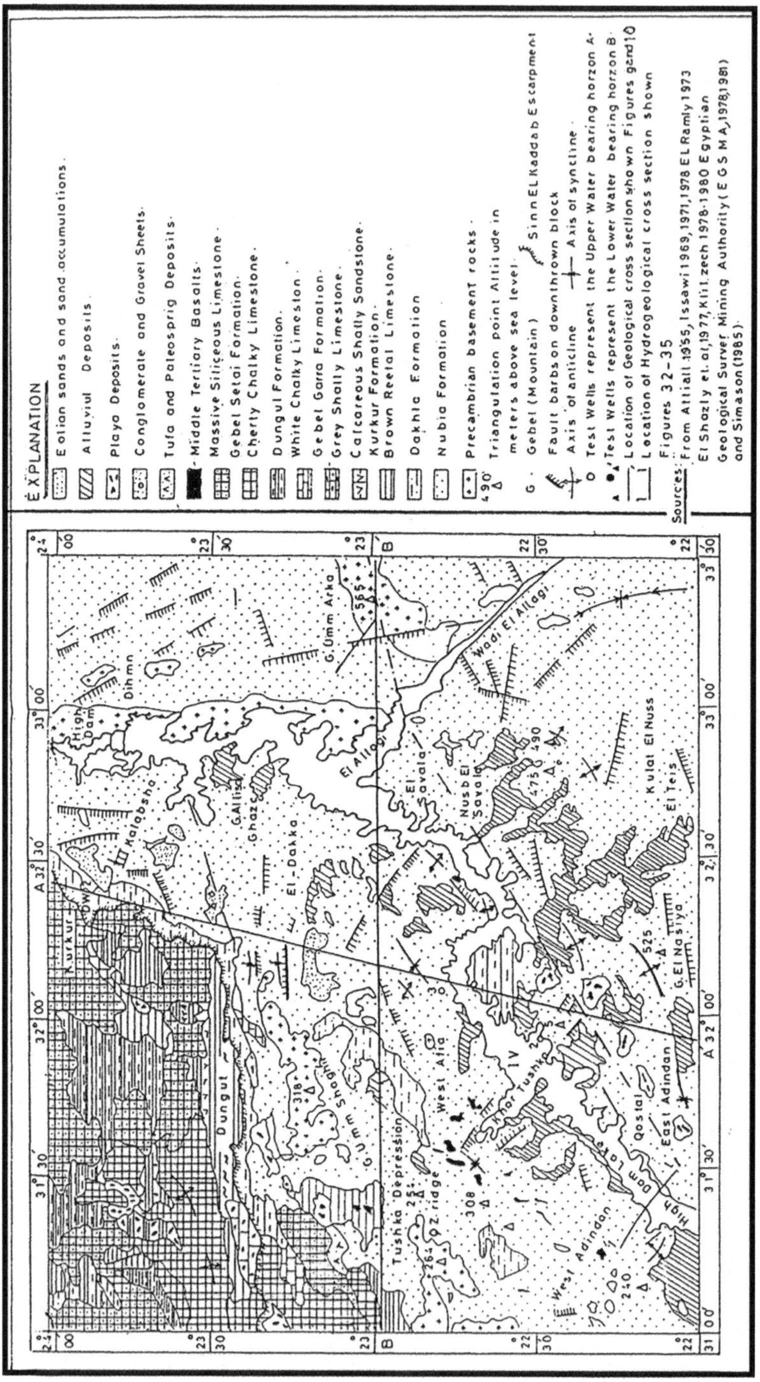

Fig. 28 Geological map of Lake Nasser area and adjacent areas south of the valley [101]

changes in base level, changes in water volume and load, and changes in the hydrographic system of the Nile.

5.4.2 Aswan: Qina Soils

Soil formation of the Nile alluvial in Upper Egypt (ldfu at Aswan and Nag Hammadi at Qina) was investigated by Wahab et al. [104]. Soils features and subsequently their formation are deeply affected by the origin of their parent material (fluvial and colluvial origin). The major part of the valley soils is of fluvial origin. The soils near the western and eastern desert fringes are of fluvial-colluvial origin due to the continuous addition of colluvial and Aeolian sandy materials from the near desert. The effect of man on the soil formation in this area could be identified in the formation of an Ap horizon in the cultivated soils due to old cultivation and addition of organic manures. The presence of high underground water table in some soil profiles is generally associated with the formation of gley and natric horizons.

Studies of Sawy et al. [105] on the soils of the eastern area of the Nile river at Qena government showed that the different deposits occurred in this area are alluvial deposits (Island and river bank deposits, and floodplain deposits), alluvial colluvial deposits (relatively high and low parts), desertic deposits (relatively high, moderately high, low parts), and dry valleys. The soil texture is mainly affected by distance from the main Nile river course, the slope and relief, and the landform. However, the desertic soils differ in their texture due to the differences of their parent materials. The $CaCO_3$ fractionation showed that the highest amounts occurred in the finer fractions in the alluvial soils of the Nile river valley and interference zone (alluvial-colluvial soils), while the opposite is true for desertic soils. The former is mainly due to the secondary form of $CaCO_3$ in alluvial soils and the finer transported particles in the interference zone, while the latter is due to the products of the physical disintegration of desertic deposits, which are rich in $CaCO_3$ content.

5.4.3 Nile Terraces at Assiut Soils

Faragallah and Essa [106] studied the soils of the Nile terraces at Assiut and found that the origin of these soils derived from different provenances, i.e., the acidic and basic igneous and metamorphic rocks, which are related to the igneous and metamorphic rocks of Ethiopian Plateau. In addition, the presence of calcite mineral in these soils suggests their derivation from lime-rich plateau surrounding the Nile Valley during their transportation and soil formation. The abundance of relatively unstable minerals, i.e., pyroxenes epidotes, amphiboles, micas, and feldspar suggests that these soils are young and weakly developed and/or recent deposition. These soils were stratified and/or mostly formed under multi-sedimentation regimes. The soil stratifications of the studied Nile terraces could be attributed to the variable water conditions occurred during their transportation and sedimentation together with the effect of the paleo-topography. The abundance of the relatively unstable minerals such as epidotes, amphiboles, pyroxenes, and feldspars (plagioclase, microcline, and orthoclase) in the alluvial soils of the Nile Valley and the interference zone between the desert and the valley at Assiut suggests that these soils are young and weakly developed. The presence of opaques, pyroxenes, and rutile reflects basic igneous rocks; the sphene, zircon, apatite represent the acidic igneous rocks; the epidotes represent the metamorphic rocks, and the amphiboles, biotite, tourmaline as well as garnet represent igneous and/or metamorphic rocks. The characterization of the sand fractions suggested that the deposition mode was by one source, namely Nile River water.

5.4.4 Kalubiah Governorate Soils

Soils of Kalubiah government are mostly formed from the Nile alluvium deposits, while some soils are developed over sub-deltaic deposits or over a mixture of wind-blown sand and calcareous deposits. Soils of recent Nile alluvium are usually clayey except the soils adjacent to the Nile River and the main canals which having a medium texture (loam or clay loam). Soils are developed on sub-deltaic deposits located in Shebin El-Kanater District and they consist mainly of coarse sand. Soils of calcareous deposits mixed with wind-blown sand are located in El-Khanka District. Soils that irrigated with sewage water for a long time are well developed, where horizons are easily recognized. (a) Horizon is friable and free of $CaCO_3$, whereas (b) horizon is compact because of the cementation of $CaCO_3$ and gypsiferous material. However, soils irrigated with Nile water are young. Horizons are weakly developed as showed in the Nile alluvial soils [107]. From soil development studies of some zones between the Nile Delta and the desert in the same government, there are clear differences between the older soils (*Typic Torrerts*), the recent soils (*Vertic Torrifluvents, Typic Torrifluvents, Typic Torriorthents, and Typic Torripsamments*), and Aridisols (*Typic Calciorthids*). Soils of *Typic Torrerts* had developed more than the other soils of Entisols and Aridisols [108].

5.4.5 Eastern and Western Side of the Nile Delta Soils

Some soils representing the main geomorphic units in the eastern (young deltaic plain, old deltaic plain, fluviomarine plain, wadi El-Tumulat, and the structural plain) and western (young deltaic plain, old deltaic plain, fluviomarine plain, alluvial deposits of desert wadi, and inland dunes) sides of the Nile Delta were evaluated by Morsi [109]. From the grain size parameters, the data of sorting Skewness and Kurtosis reveal that water and wind or both share together in the sedimentation of the soils. The sediments forming the investigated soils are mostly deposited under aqueous environment. Moreover, soils constituting each profile are heterogeneous either due to their multi-origin or due to the subsequent variation along the course of sedimentation. Therefore, they are still considered young from the pedological viewpoint.

At the end of the Pliocene period and during the formation of river terraces in the Nile Valley, the river was depositing great amounts of sand and gravels in the Mediterranean forming the delta base. These old sediments have almost disappeared under the new alluvium and are accordingly called sub-deltaic soils. Some of the sediments that have relatively high relief appear on the surface a sandy island underlying the recent fine silt and clay deposits. They occur in small scattered areas mainly in the eastern side and in the south central part of the delta. Sub-deltaic soils are distributed between 30°–31° N and longitudes 31°–32° E [37]. The application of the morphological rating scale on the cultivated and uncultivated sub-deltaic soils indicates that the cultivated soils are relatively more developed than the virgin ones especially in the surface layers. The deposits in the virgin areas as well as in the deepest layers of cultivated soils are mainly due to both water and wind actions. The water action is the common effective factor in the transportation and deposition of the surface layers in cultivated soils. Virgin soils seem to have one homogenous parent material and one origin, while the cultivated soils have apparently stratified layers of multi-origin [110]. Sandy hillocks within the recent alluvial soils of the Nile Delta constitute the distinguished phase of the old sub-deltaic deposits, which occupy considerable areas, especially in the south of the central and eastern parts of the Nile Delta. These deposits comprise the parent material of the sub-deltaic soils, which have a very different nature from the recent alluvial soils. The surrounding interference zones soils are characterized by a gradual change, from the sand texture (the sub-deltaic soils) to the clay texture (the recent alluvial soils) through the development of pedon characteristics.

Data of the study of the soil terraces on both sides of Ismailia canal, east delta shows that these terraces are of very poorly sorted nature transported in a rolling and suspension and deposited in an aqueous medium and characterized by the fine materials derived from fluvial origin, but non-uniform in nature. The soils located along the north and south Ismailia canal between Tell El-Kebier and Ismailia are the river terraces of fluvial origin, occupying the former Wadi El-Tumillat. The stratified conditions observed in most cases are mostly attributed to local changes during sedimentation together with the side effect of the natural formation closed to such terraces. The heavy minerals assemblage in the soil terraces indicate that these soils are weakly developed. Soils of the terraces are formed of either multi-origin or multi-depositional regime, due to the heterogeneity of the parent materials prior to inception to geogenetic weathering or resulted in from the geochemical weathering prevailed in the source sediments [111].

Soil formations in the area located at the eastern part of the Nile Delta, between 30° 18′ and 31° 42′ N and longitudes 31° 40′ and 32° 5′ E, are of different sources. The western part is from the Nile alluvium deposits. To the east, the soil becomes finer due to the effect of the Aeolian sand deposits. The northern part is mostly of fluviomarine deposits. The topography of the area plays an important role in soil formation through its elevation differences and the nature of its slopes. The mineral assemblages are characterized by an abundance of iron ores, pyroxenes, amphipoles, and epidotes. The slight difference in frequencies than that of the main Nile sediments is related to the relative increase in the frequencies of the iron oxides, metamorphic, and ultra-stable minerals, which are thought to be derived from the Oligocene and younger sediments exposed to the east and south of the area. Additions to these fluviatile sediments from another source rocks took place possibly by wind as evidenced from the much lower index figure than that of the Nile sediments, and the presence of rounded to subrounded frosted quartz grains. Due to the prevailing aridity in the area, the mechanical weathering is dominant. Uniformity and weathering ratios indicate that the top layers have been subjected to a relatively higher weathering than in the subsoils. In conclusion, the mineral suites of the examined soil have been resulted from the mixing of the Nile sediments and the older sediments in the area, the provenance is considered an igneous and metamorphic complex [112].

The salt efflorescence occurs covering the surface of most the soils of fluviomarine marshes located at the eastern site of the Nile Delta as well as on the western site of the Suez Canal coupled with gypsum formation and shell remnants. This is due to high water table and salinity of these soils that are adjacent the southern border of El-Manzala Lake. The soils of fluviomarine marshes and depressions contain high contents of $CaCO_3$ and gypsum especially in the lower layers and this could be attributed to the marine origin of these stratified layers, i.e., lacustrine deposits. These soils have shallow salic horizon; however, the majority of the soils of the plains and deltaic stage of the river terraces does not have any diagnostic horizon in the upper 1 m [113].

5.4.6 Northern Zone of the Nile Delta Soils

The soils of the northern zone of the Nile Delta are characterized by the presence of four main geomorphic units: (1) Nile alluvial deposits, (2) fluvio-lacustrine deposits, (3) coastal barrier plains, and (4) coastal sand dunes [114]. The northern part of the Nile Delta that extends 170 km along the coast of the Mediterranean Sea from Damietta branch in the east to Edku Lake in the west, with 60 km inland from the coastal shoreline in the north to the old cultivated lands in the south localities was studied by Hagag [115]. The clay fraction of soils of Nile alluvial deposits is dominated by montmorillonite, kaolinite that are mostly inherited from parent materials (alluvium). Soils of fluvio-lacustrine are sandy clay loam which contain some calcareous and dominated by the montmorillonite, hydrous mica, kaolinite, and palygroskite. These soils come from the original matter (fluvio-lacustrine deposits). Concerning the soils of coastal barrier plains are loamy sand, which contains a high percentage of $CaCO_3$ and salts and dominated by the montmorillonite, mica, kaolinite, vermiculite, and palygroskite. These deposits come from the pedogenic or inherited from parent origin sediments derived wholly or in part, from lacustrine or calcareous. However, the soils of coastal sand dunes are loamy sand and contain a moderate percentage of $CaCO_3$ and salts. These soils are formed by

Aeolian and marine deposits. The effect of the low topography appeared by the effect of the salinity and the marine nature of the depositions on the clay minerals which exist especially in the coastal barrier plains. These conditions helped to turn the original clay minerals to mixed layers as a transformational stage. In addition, the effect of the high topography appeared in the sand dunes movement and the creping of sands from north to south played an important role in the distribution of clay minerals. The depositional environment which formed the area contains the soils of Nile alluvial plains formed by Nile deposits materials, but the soils of fluvio-lacustrine, though derived from two different parent sediments, namely lacustrine and alluvium. The soils of the coastal barrier plains formed by the sea, and so the soils of coastal sand dunes formed by the Aeolian and marine deposits.

Khater and Mansour [116] reported that the grain size image of the soil sediments in the northern part of the Nile Delta between Damiatia and Rossitta cities showed that the transporting agent and environment of deposition were mainly fluviatile in nature, except for the beach sandy sediments, which have dual sources between the water and wind action, i.e., related to Aeolian and fluviatile environment. The beach sandy sediments contain a relatively high content of heavy minerals due to the influence of Aeolian sea sediments. Degree of minerals weathering is mainly affected by the drainage conditions of the soils. The soil materials were originated from basaltic rocks of Sudan, igneous and metamorphic rocks of Abyssinian plateau, and granitic rocks of Aswan region. Most of the soil sediments have quite different features due to lithological discontinuity, which is indicated by the distribution of minerals within the successive layers of studied soils. This difference in distribution is attributed either to heterogeneity of parent material or to a more advanced stage of weathering. Most of the sediments in the area are of multi-origin or formed due to multi-depositional regimes.

Elmaaz [117] studied some soils adjacent to lakes at the north of Egypt and found the soils of recent Nile alluvium, marine alluvium, and desert plain have mainly heavy texture (clay), whereas the others of sub-deltaic and sandy beaches are lighter (loam). These soils are non-saline to moderately saline and total carbonate content varies widely from one profile to another. Many lime concretions and broken shells especially in the downstairs layers of profiles south of Lake Mariout. The different statistical size parameters indicated that the most soil materials of recent Nile alluvium, marine alluvium, and desert plains deposits are poorly sorted with platy to very platy kurtic pattern, this indicates that the water is the main factor responsible for transportation and formation of soil materials of these deposits. The cumulative curves of these soils are nearly similar and symmetrical reflecting almost uniform and homogenous soil materials. However, results of sub-deltaic and sandy beaches deposits indicate that their soil materials are formed under a combined effect of water and wind action. Their soil materials are nearly heterogeneous and formed under different depositional regime. Most soils are considered young from the pedological point of view. The majority of recent Nile alluvium, marine alluvium, and desert plain are formed under similar depositional regime. The soils of other landforms are deposited under multi-depositional regime. Abu-Agwa and Amira [118] stated that the soils adjacent to

Borullus, Manzala, and Qarun salty lakes differ from sandy loam to clay texture affecting mainly by natural sedimentation pattern and circumstances of each area. The soils situated close to the lakes have moderate horizontation may be due to the intermixing between recent alluvium and lacustrine deposits in these areas.

The alluvial salt affected soils, the northern part of the Nile Delta were investigated in governorates of El-Beheara (Ferhash and El-Lakana), Alexandria (Abees), Kafr El-Sheikh (Burg Megasal and Shalma), Damietta (Kafer Soliman El-Bahri), El-Dakahlia (Mear Meraga salcil), and Sharqia (El-Monaga El Kobra and El-Tal El-Kebir) [119]. The obtained results showed the significance of the Mediterranean Sea water, Lakes, and groundwater as sources of salinity to adjacent soils. However, the action of seasonal wetting and drying under the environmental arid climate and the low elevation from surrounding areas adds some extra salts to these soils. Soils were categorized as saline soils, saline alkali soils, and alkali soils. Generally, most of the salt affected soils of the Nile Delta are of the saline alkali nature [120]. The study of soil morphology and sedimentation pattern showed that these soils vary in their components in different locations based on soil relief, the Nile, sea, lakes, and western and eastern desert deposits. Accordingly, the different sediments of these parent materials interfere with each other, but in different deposits rates in the different locations [119].

Abo El-Ennan et al. [121] studied the genesis of clay minerals in some saline and alkaline soils in the northern part of the Nile Delta. They stated that the clay minerals of the investigated soils could have been carried either by the Nile River from its upper sources or formed in situ by the weathering of primary silicates. Smectite, kaolinite, and illite minerals may have been transported in the suspended matter carried by the Nile. The presence of interstratified clay minerals in some soil samples may be explained by pedogenic formation and transformation processes. An inadequate drainage system and a high water table possible promoted such a transformation. The soils are influenced by saline water and relatively high temperature, which enable the alteration of the lattice framework of the clay minerals. These soils have high Mg/Ca or Na/Ca ratios because of the inadequate drainage system and seepage of saline water from the Mediterranean Sea and the northern lakes. Seawater rich in Mg-ions leads to the transformation of hydrous mica into montmorillonite whereas the K-ions between the layers have been replaced and due to this the attractive forces were weakened and water molecules entered the lattice resulting in its expansion.

6 Conclusions and Outlook

This chapter however, brings to light the land and water resources development in the Nile Valley and Delta. Analysis of the hydrogeological features of the Nile Delta starts from the recognition that surface and groundwater are tightly intertwined. Nile Delta branches (Rosetta and Damietta) gain water from the aquifer in some reaches and lose water to the aquifer in other reaches. The flow directions between groundwater and surface water can change seasonally with variations of the water table level with respect to the level in nearby waterways. Available data on the evolution of the salinity of groundwater in the Delta indicate that the construction of the High Aswan dam resulted in a shift of salinity isolines towards the seashore (because of a constant supply of water year round), and that current pumping rates have not yet critically affected this balance (but more recent data are needed to analyze the evolution in the past 10 years). Possible localized over-pumping, however, results in "up-coning" of salinity from deeper layers. This preliminary review on groundwater in the Nile Delta has also shown that contamination of groundwater by agriculture and more prominently by seepage from domestic and industrial effluents had already attained worrying levels. This may jeopardize the quality of the Nile Delta aquifer in the long run and requires more research and attention.

The future study should be focused on cases where underground reservoirs or aquifers are overused. The increasing use of groundwater for irrigation poses a serious threat to food security and could lead to unaffordable prices of staple foods. Recent hydrological modelling and Earth Observations have located and quantified alarming rates of groundwater depletion, which is due to water withdrawals for irrigation. Aquifer depletion can also induce significant environmental degradation, such as land subsidence and seawater intrusion. Studies of water resources embedded in food trade – virtual water trade – have sometimes distinguished between blue water (surface water and groundwater) and green water (soil moisture) sources, which can provide additional information on the potential environmental impact of their use. Globally, countries irrigating crops from over exploited aquifers export them in various proportions. Eleven percent of groundwater extraction for irrigation is linked to agricultural trade. India keeps most of its large GWD-based crop production for domestic use (only 4% of GWD exported), while the USA, Pakistan, and Mexico export significant portions of their GWD-based crop production (42%, 26%, and 23%, respectively). The major importers of GWD via crops include China (9% of global GWD trade), the USA, Iran, Mexico, Japan, Saudi Arabia, Canada, Bangladesh, UK, and Iraq. Individual crops contributing most to global GWD transfers are rice (29%), followed by wheat (12%), cotton (11%), maize (4%), and soybeans (3%). Citrus crops and sugar crops account for 5% of GWD transfers each.

In Egypt, which is irrigation-intense country, agriculture is the leading user of groundwater, making up more than 80–90% of withdrawals. There is not sufficient precipitation or surface water available to grow crops like maize or rice and so farmers use water from the underground to irrigate. Based on the current chapter, the amount of non-renewable groundwater used for irrigation was doubled in Nile Valley and Delta. The annual groundwater abstraction in the Nile aquifer system and fringes is about 4.6 billion m^3. Another 0.5 billion m^3 is abstracted from the desert aquifers and the coastal areas. Groundwater abstraction is expected to increase to 11.4 billion m^3. Model output revealed that groundwater recharge has not changed significantly over time, while pumping has. Because of these trends, groundwater was estimated to be in a deficit of approximately 24 billion m^3 (±15%) in 2011, compared to 1957.

Crops such as rice, wheat, cotton, maize, and sugar crops are most reliant on this unsustainable water use. Egypt was locally most affected due to groundwater depletion and exporting agricultural products grown with non-sustainable groundwater. When a country imports maize grown with this non-renewable water, it virtually imports non-renewable groundwater. Therefore, groundwater overuse rising could hit food prices. Putting water labels, along the lines of food labels is important for creating awareness. These labels show how much water is used domestically and internationally in the production and whether these water amounts are from sustainable or non-sustainable sources. Water labels and caps on extraction, linked to agricultural trade, are among measures proposed tackling the trend.

References

1. CAPMAS (2016) Central Agency for Public Mobilization and Statistics. http://www.capmasgoveg/
2. Tamburelli P, Thill O (2013) The Nile metropolitan area. Berlage-Institute, TU Delft
3. Kantoush SA (2013) The downstream impacts of Ethiopia's Cascade dams in the upper Blue Nile on Egypt. Proceedings of regional sustainable building conference SB13 Cairo Fairmont Towers Hotel Cairo, Egypt
4. El-Din MMN (2013) Climate change risk Management in Egypt proposed climate change adaptation strategy for the Ministry of Water Resources & Irrigation in Egypt. UNESCO Office, Cairo
5. CAPMAS (2012) Central Agency for Public Mobilization and Statistics. http://www.capmasgoveg/
6. Wahaab R, Badawy M (2004) Water quality assessment of the River Nile system: an overview. Biomed Environ Sci 17:87–100
7. Elnashar W (2014) Groundwater management in Egypt. IOSR J Mech Civil Eng (IOSR-JMCE) 11(4):69–78
8. El Arabi N (2012) Environmental management of groundwater in Egypt via artificial recharge extending the practice to soil aquifer treatment (SAT). Int J Environ Sustain 1 (3):66–82
9. Said R (1993) The River Nile; geology, hydrology and utilization. Pergamon Press, Oxford
10. Hozaien S (1991) Egyptian civilization. Dar Al Shorouk ISBN: 9770900252
11. Awad M (1998) The Nile River. Egyptian General Book Authority
12. Beaumont P, Blake G, Wagstaff J (1988) The Middle East: a geographical study. David Fulton, London
13. Elarabawy M, Attia B, Tosswell P (2000) Integrated water resources management for Egypt. J Water Supply Res Technol AQUA 49:111–125
14. Lonergan S, Wolf A (2001) Moving water to move people. Water Int 26:589–596
15. Harms JC, Wray JL (1990) Nile Delta. In: Said R (ed) The geology of Egypt. A.A. Balkema, Rotterdam & Brookfield, Vermont, pp 329–343
16. Toussoum O (1922) Memire sur les annciennes branches du Nil. Imprimeric d'Instit Francais Epoque ancienne TIVD'archeologie Orientale cairo
17. Shahin AAW (1978) Some of the geological phenomena in the Nile Delta. Arab Geogr Mag (11):9–26

18. Stanley DJ, Warne AG (1993) Nile delta: recent geological evolution and human impact. Sci Technol 260:628–634
19. Hamdan G (1984) The personality of Egypt. Dar Al Hilal
20. RIGW (1992) Groundwater resources and projection of groundwater development. Water Security Project, (WSP), Cairo
21. Morsy WS (2009) Environmental management to groundwater resources for Nile Delta region. PhD thesis, Faculty of Engineering, Cairo University, Egypt
22. Aquastat (2013) FAO http://www.FAO.org/nr/water/aquastat/irrigationmap/egy/index
23. Al-Agha DE, Closas A, Molle F (2015) Survey of groundwater use in the central part of the Nile Delta. Water and salt management in the Nile Delta: Report No 6
24. RIGW (2003) Impacts of future surface water development on groundwater aquifer system in the West Nile Delta Region, Cairo
25. MWRI (2005) National Water Resources Plan for Egypt 2017. Ministry of Water Resources and Irrigation
26. Sherif M (1999) The Nile Delta aquifer in Egypt. Chapter 17 in seawater intrusion in coastal aquifers, concepts methods and practices. In: Bear J, Cheng A, Sorek S, Ouazar D, Herrera A (eds) Theory and application of transport in porous media, vol 14. Kluwer Academic Publishers, Dordrecht, The Netherlands, pp 559–590
27. Sefelnasr A, Sherif M (2013) Impacts of seawater intrusion in the Nile Delta Aquifer, Egypt. Ground Water 52(2):264–276. National Groundwater Association
28. MWRI (2012) Strategy of water resources of Egypt till 2050. Ministry of Water Resources and Irrigation
29. FAO (2013) Monitoring of climate change risk impacts of sea level rise on groundwater and agriculture in the Nile Delta
30. Zaghloul MG (1985) Flow distribution through groundwater aquifer of the Nile Delta. MSc thesis, Faculty of Engineering, Alexandria University, Egypt
31. Amer A, Sherif M (1996) An integrated study for seawater intrusion in the Nile Delta aquifer. WRSR publication series no 14, NWRC, MWRI, Egypt
32. Diab MS, Dahab K, El Fakharany M (1997) Impacts of the paleohydrological conditions on the groundwater quality in the northern part of Nile Delta. The geological society of Egypt. J Geol 4112B:779–795. Cairo
33. Fadlelmawla A, Dawoud MA (2006) Protection areas at the Nile Delta. Egypt J Environ Manage 79(2):140–149
34. Saleh MF (1980) Some hydrogeological and Hydrochemical studies on the Nile Delta. MSc thesis, Faculty of Science, Ain Shams University
35. Mabrouk MB, Jonoski A, Solomatine D, Uhlenbrook S (2013) A review of seawater intrusion in the Nile Delta groundwater system – the basis for assessing impacts due to climate changes and water resources development
36. Abd El-Fattah EM (2014) Evaluation of the environmental and chemical impacts of banana plantation on the groundwater aquifer in the southern part of nile delta and its vicinities. PhD thesis in Environmental Science University of Sadat City, Egypt, 190 p
37. Abu Al-Izz MS (1971) Land forms of Egypt. The American University in Cairo press, Dar Al Maaref, Cairo, Egypt
38. Attia MI (1954) Deposits in the Nile Valley and the Delta Mines and Quarrie Department. Geological Survey, Egypt, Cairo, p 356
39. Farid MSM (1980) Nile Delta groundwater study. MSc thesis, Cairo University
40. RIGW/IWACO (1990) Hydrogeological inventory and groundwater development plan, Western Nile Delta Region. Main Report, vol 4, Technical Note 7701300–90-02RIGW, 64 p
41. Abu-Zied M, Abdel-Dayem G (1991) Sail load in irrigation and drainage water in the Nile Delta. In: African Regional Symposium on techniques for environmentally sound water resources development
42. Fattah M, Ragab E (2014) Assessment of groundwater vulnerability to pollution in the southern part of Nile Delta, Egypt. Stand Sci Res Essays 2(13):725–738

43. Mohamed MAE (2002) Study of groundwater development in the Western Delta by using the remote sensing and GIS. Msc thesis, Faculty of Engineering, Ain Shams University, Egypt
44. Khater A, Fadlelmawla A, Fekry A (2002) Assessment of groundwater quality in El Sadat City and proposed action for groundwater protection. Proceedings of the III international symposium on environmental hydrology American Society of Civil Engineering & Egyptian Society of Irrigation Engineering, Cairo, Egypt
45. Abd el Salam M (2004) Remedial solutions for the impacts of the existing groundwater development plan in Western Nile Delta. Second regional conference on Arab Water, action plans for integrated development
46. El Fakharany ZMA (2002) Management of conjunctive use of surface water and groundwater. Msc thesis, Helwan Univerisity, Faculty of Engineering, Mataria, Egypt
47. Darweesh WMK (2001) Optimal management of underground water in west delta zone. PhD thesis, Faculty of Engineering, Ain Shams University
48. El-Beshri MZ, Labadie JW (1994) Optimal conjunctive use of surface and ground water resources in Egypt. International water resources association VIII IWRA world congress on water resources, Cairo, Egypt
49. Dawoud M, Darwish M, El-Kady M (2005) GIS-based groundwater management model for western Nile Delta. Water Resour Manag 19:585–604
50. Molle F, Rap E, Ezzat Al-Agha D, Ismail A, Abou El Hassan W, Freeg M (2015) Irrigation improvement projects in the Nile Delta: promises, challenges, surprises. Water and salt management in the Nile Delta project report no 4 IWMI, WMRI, Cairo
51. Mohamed RF, Hua CZ (2010) Regional groundwater flow modeling in Western Nile Delta, Egypt. World Rural Observ 2(2):37–42
52. Switzman H, Coulibaly P, Zafar A (2015) Modeling the impacts of dryland agricultural reclamation on groundwater resources in Northern Egypt using sparse data. J Hydrol 520: 420–438
53. Samak SAMA (2007) Groundwater management in delta. Msc thesis, Faculty of Engineering, Zagazig University, Egypt
54. Todd DK, Mays LW (2005) Groundwater hydrology. 3rd edn. Wiley, New York, NY, pp 589–606
55. Atta SA (1979) Studies on the groundwater properties of the Nile Delta, Egypt. MSc, Faculty of Science, Cairo University, Cairo
56. Farid MSM (1985) Management of groundwater systems in the Nile Delta. PhD thesis, Cairo Univ, Giza, Egypt
57. Sherif M, Sefelnasr A, Javadi A (2012) Incorporating the concept of equivalent freshwater head in successive horizontal simulations of seawater intrusion in the Nile Delta aquifer, Egypt
58. Leaven MT (1991) Hydrogeological study of the Nile Delta and adjacent desert areas, Egypt, with emphasis on hydrochemistry and isotope hydrology. MSc thesis, Free University, Amsterdam also published by RIGW/IWACO as technical note TN 7701300-91-01
59. Ebraheem AM, Senosy MM, Dahab KA (1997) Geoelectrical and hydrogeochemical studies for delineating ground-water contamination due to salt-water intrusion in the northern part of the Nile Delta, Egypt. Ground Water 35(2):216–222
60. Sakr SA, Attia FA, Millette JA (2004) Vulnerability of the Nile Delta aquifer of Egypt to seawater intrusion. International conference on water resources of arid and semi-arid regions of Africa, Issues and challenges, Gaborone, Botswana
61. Sharaky M, Atta SA, El Hassanein AS, Khallaf KMA (2007) Hydrogeochemistry of groundwater in the Western Nile Delta aquifers, Egypt. 2nd international conference on the geology of Tethys, Cairo University
62. Masoud A (2014) Groundwater quality assessment of the shallow aquifers west of the Nile Delta (Egypt) using multivariate statistical and geostatistical techniques. J Afr Earth Sci 95:123–137

63. Ghoraba SM, Zyedan BA, Rashwan MH (2013) Solute transport modeling of the groundwater for quaternary aquifer quality Management in Middle Delta, Egypt. Alex Eng J 52: 197–207
64. Salem MG, El-Awady MH, Amin E (2012) Enhanced removal of dissolved iron and manganese from nonconventional water resources in Delta District. Egypt Energ Procedia 18: 983–993
65. Hussien MM (2007) Environmental impacts of new settlements on the ground water in a region in Delta. Msc thesis, Zagazig University, Faculty of Engineering, Egypt
66. Ghanem AHM, Zaghloul AS, El-Ayouti S (2011) Groundwater lowering in multi-layer aquifer systems: case study of Esna City – Egypt. A paper presented at the 64th CWRA National Conference, "Our Water – Our Life – The Most Valuable Resource," in St John's NL Canada, June, 2011
67. El-Fakharany Z, Fekry F (2014) Assessment of New Esna barrage impacts on groundwater and proposed measures. Water Sci 28(1):65–73
68. Faid AM (2003) Interpretation of goelectrical data to identify aquefers and pale-channel of river Nile in Esna area, upper Egypt. Sci J Fac Sci Minufiya Univ XMI:89–118
69. Ismail A, Anderson N, Rogers D (2005) Hydrogeophysical investigation at Luxor, Southern Egypt. J Environ Eng Geophys 10(1):35–49
70. Selim S (2009) Hydrogeochemical assessment of groundwater at the area between Wadi Qena and Wadi El-Mathula, Upper Egypt. Sci J Fac Sci Min ujia Univ XXIII(2):81–101
71. Abdalla F, Ahmed A, Omer A (2009) Degradation of groundwater quality of quaternary aquifer at Qena, Egypt. J Environ Stud 1:19–32
72. Esam I, Abdalla F, Erich N, Hermann M (2012) Comparison of the groundwater quality in the West Tahta Area, Upper Egypt in 1989 and 2011. J Environ Prot 03:1442–1457
73. El Tahlawi M, Abo-El Kassem M, Baghdadi G, Saleem H (2016) Assessment of groundwater vulnerability – a case study. Int J Adv Rem Sens GIS 5(2):1561–1579
74. Waleed SS, Abd El-Monaim AE, Mansour MM, El-Karamany MF (2009) Evaluation of groundwater aquifer in the area between El-Qusiya and Manfalut using vertical electric soundings (ves) technique. J Eng Sci Assiut Univ 37(5):1193–1207
75. Salem A (2015) Hydrogeological studies on the shallow aquifers in the area West Samalot, El-Minia Governorate, Egypt. Egypt J Pure Appl Sci 53(4):49–60
76. Gedamy Y, El-Aassar A, Abdel-Gawad A (2012) Pollutants detection in water resources at El Saff Area and their impact on human health, Giza Governorate, Egypt. Int J Environ 1 (1):1–14
77. FAO (1965) Simplified map of the desert fringes of the Nile Valley and Delta. High Dam Soil Survey Project Ministry of Agricultuie
78. El-Nahal MA, Abd El-Aal RM, Aid El-Wahed AA, Raafat I (1977) Soil studios on the Nile Delta. Egypt J Soil Sci 17(1):55–65
79. Rahim SI (2006) Compilation of a soil and terrain data base of the Nile Delta at Scale 1:100000. J Appl Sci Res 2(44):226–231
80. Ghabbour TK (1988) Soil salinity mapping and monitoring using remote sensing and a geographical information system (some applications in Egypt. PhD thesis, Fac of Sciences, State Univ of Ghent
81. Abu El-Einane SM (1985) Pedogenesis of the Fayoum area PhD thesis, Fac of Agric, Al-Azhar Univ, Cairo, Egypt
82. Khater EA (1973) Pedological studies on Fayoum Governorate, A.R.E. MSc thesis, Fac of Agric, Cairo Univ, Egypt
83. Abdel-Hady MA, El-Kassas IA, Ayob AS (1982) Automatic classification of lake Qarun water by digital processing of land sat Mss data. In: Proceeding of the international symposium of remote sensing of environment first thematic conference: remote sensing of arid and semi-arid lands, Cairo, Egypt, pp 1143–1165
84. Abd El-Aal TS (1984) Pedological and physical studies on some soil of the expansion areas in Fayoum Governorate, Egypt. MSc thesis, Fac Of Agric, Cairo Univ, Egypt

85. Shendi MM (1984) Pedological studies on soils adjacent to Qarun lake – Fayoum Governorate – EGYPT. MSc thesis, Cairo Univ, Egypt
86. El-Naggar MA (2004) Land Evaluation of some soils in Arab Republic of Egypt. PhD thesis, Fac of Agric, Moshtohor, Zagazig University, Banha Branch
87. Harun MRO (2004) Soil evaluation systems as a guide to identify an economical feasibility study for agricultural purposes in El-Fayoum Governorate. PhD thesis, Fac of Agric Cairo University, Egypt
88. Hamdy H (1982) Soil map of Egypt. Final report academy of scientific research and Technology – ASRT, Cairo, Egypt
89. Hammad MA (2011) Soil association map of Egypt. Updated legend The Soil Survey Institute Wageningen the Netherland & Digitized by AAAbdel-Ghany SWR ARC Giza Egypt
90. Said R (1962) The geology of Egypt. El-Sevier Publishing Company, Amesterdam, NY
91. Abu El Einane SM (1977) Pedology and geomorphology relationship in El Fayoum governorate. MSc thesis, Fac of Agric, Al Azhar Univ
92. Abd El-Aal TS (1990) Cyclic formation of soil on different geomorphic features of Fayoum area. Egypt. PhD thesis, Faculty of Agriculture at Fayoum, Cairo University, Egypt
93. Hanna FS, Labib FB (1977) The soils of the Fayoum depression. Egypt J Soil Sci 17(1):33–43
94. Kassem YS, Elwan AA (1980) Origin and uniformity of the soils adjacent to Qarun lake, Fayoum Governorate. Egyptian J Soil Sci 20(1):57–63
95. Shendi MM (1990) Some mineralogical aspects of soil sediments with special reference to both lithology and environmental conditions of formation in Fayoum area, Egypt. PhD thesis, Facof Agricat El-Fayoum, Cairo University, Egypt
96. USDA (2010) Keys to soil taxonomy. United State Department of agriculture: natural resources conservation service (NRCS), 3rd edn
97. Awadalla AA (1998) Studies on the main soil pedogenic aspects in El-Fayoum depression as related to the dominant soil formation processes. PhD thesis, Fac of Agric at Fayoum, Cairo University, Egypt
98. Abu Zeid M (1987) Environmental impact assessment for Aswan High Dam. In: Biswas AK, Geping Q (eds) Environmental impact assessment for developing countries, London, pp 168–190
99. El-Shabrawy MG (2009) Lake Nasser–Nubia. In: Dumont HJ (ed) The Nile: origin, environments, limnology and human use, © Springer Science + Business Media BV
100. Reda M (1978) Some pedological features in some arid soils of Egypt. PhD Fac Agric Al-Azhar Univ
101. DRS and LERI (1999) Land resources South Valley and Toshka South Valley and Toshka. Desert Research Center and Land and Environment Research Institute. Encyclopedia Volume II
102. Saleh I (1998) Project Toshka, Human – Investment – Development. Family Library
103. Abdel-Salam AM, El-Kadi HA, Reda M (1974) Soils of wadi kalabsha (Lake Nasser region). The Regional Planning of Aswan and the Desert Institute
104. Wahab MA, Hanan FS, Hady F, Abdel Samie AG, Hamdi H (1972) Studies on the formation of some soils in the Nile Valley. Egypt J Soil Sci 12(1):31–40
105. Sawy S, Magd MH, Hanna F, Abd El-Hady AA (1990) Physiography and soil map of the eastern area of the River Nile at Qena governorate, Egypt. Egypt J Soil Sci 30(3):457–479
106. Faragallah MA, Essa MA (2011) Sand and clay mineralogical composition in relation to origin, sedimentation regime, uniformity and weathering rate of Nile terrace soils at Assiut, Egypt. Aust J Basic Appl Sci 5(10):239–256
107. Boctor S, Abo Gelayel AMA, Yaeoub EW, Ibrahim AM (1979) The soil classification of Kalubiah governorate, Egypt. Agric Reach Rev 57(5):147–163
108. Mousa AM, Naguib BH, Khalil MA (2000) Soil development studies of some zones of encroachment between the Nile Delta and the desert in Qalubeya governorate. J Agric Sci Mansoura Univ 25(10):6567–6583
109. Morsi MA (2003) Pedological studies on the soil located in the eastern and western sides of the Nile Delta. PhD thesis, Fac of Agric, Zagazic Univ (Benha Branch), Egypt

110. Amira MS (1996) Pedological, mineralogical studies and evaluation of sub-deltaic soils in the Nile Delta, Egypt. Minufiya I Agric Res 21(1):237–253
111. Noaman KI (1989) Origin, mode of formation and uniformity of soil terraces on both sides of Ismailiya canal, east Delta, Egypt. Egypt J Soil Sci 29(2):121–131
112. Khail JB, Noman KI (1980) Mineralogical studies on the sand fraction of the river terraces, and its relation to soil genesis, in the eastern part of the Nile Delta. ARE Desert Instil Bull 30(1):81–97
113. Gobran OA, Abou Agwa FE, Shehata RB (1992) Morphology, physical and chemical changes associated with physiographic units of selected soils in Egypt. Egypt J Soil Sci 32(2):265–286
114. FAO (1963) The semidetaile soil survey. High Dam Soil Survey, UAR 111 (Report, 16)
115. Hagag AA (2005) Mineralogical composition of the clay fraction, in relation to lithology, topography and depositional environments in the northern coastal zone of the Nile Delta. Zagazig JAgric Res 32(1):125–142
116. Khater E, Mansour A (1985) Texture characterisics and mineral composition the main soil sediments in the northern part of the Nile Delta, Egypt. Bull Fac Agric Univ Cairo 36(1): 467–474
117. Elmaaz EIM (2005) Pedological and mineralogical on soils adjacent to some lacks at the north of Egypt. PhD thesis, Fac of Agric, Minufiya Univ, Egypt
118. Abu-Agwa FE, Amira MS (1998) Characteristics and evaluation of soils adjacent to salty lakes in Egypt. Minufiya Agric Res 23(4):111–1128
119. Kandil MF, Hanna F, Abd El-Aal SI (1980) Diagnostic features of Egyptian salt affected soils in the Nile Delta. Agric Reach Rev 58(4):115–133
120. Kandil MF, Hanna F, Abd El-Aal SI (1980) Sources and natures of the salinity and alkalinity in the salt affected soils of the northern part of Nile Delta, Egypt. Agric Reach Rev 58(4): 99–114
121. Abo El-Ennan SM, Salem MZ, MM EL-B (1990) Genesis of the clay minerals of some soils in the Nile Delta, A.R.E. Egypt. Soil Sci 30(3):445–456

Part II
Groundwater Use

Use of Groundwater in Nile Alluvial Soils and Their Fringes

Nader Noureldeen Mohamed

Abstract Groundwater reportedly provides drinking water to at least 50% of the global population and accounts for 43% of all water used for irrigation. Food production requires the largest quantities of water, with groundwater resources providing more than 40% of all water used globally for irrigated agriculture. According to the Egyptian Ministry of Irrigation and Public Works the annual water resources in Egypt depend mainly on the Nile water (55.5 BCM), 5.5 BCM groundwater, and 1.3 BCM of rain water that falls on the agricultural land in the Delta. Most of the groundwater in Egypt is non-renewable except for the shallow groundwater in the Nile valley and Delta lands and its fringes in addition to some depression sources and oasis like Wadi El-Natrun in the west Delta (the Valley of Sodium salts) and Siwa oasis south of the northwest coast of Mediterranean. The main aquifers are generally formed of granular rocks (sand and gravel) or fissured limestone and rocks. The deep-lying aquifers systems is comprised of the regional Nubian Sandstone aquifer System, occupying much of the area of Egypt. The thickness of the sediments varies from a few hundred meters in the south, to 4,000 m west of Abu Mongar. Carbonate Aquifers occupy at least 50% of Egypt. The Moghra aquifer system has a broad geographical distribution in the region west of the Nile Delta and south of the Qattara depression. The Nile valley and Delta aquifer are the most productive, containing around 200×103 million m^3 of water that is renewable by seepage from the Nile river irrigation systems. The thickness of this aquifer decreases from 300 m at Sohag Governorate in Upper Egypt to few meters near Great Cairo (Cairo, Giza, and Qalyubia governorates) and also in the south near Aswan. The coastal aquifer lies 35 km from the seashore, 45 km north of Cairo and is recharged mainly from rainwater and from high-pressure water in the Nubian Sandstone aquifer. Rose basement rock has the same characteristics as the

N.N. Mohamed (✉)
Department of Soil and Water Sciences, Faculty of Agriculture, Cairo University, Giza, Egypt
e-mail: nadernour@hotmail.com

A. M. Negm (ed.), *Groundwater in the Nile Delta*,
Hdb Env Chem (2019) 73: 107–140, DOI 10.1007/698_2017_61,

Carbonate aquifer but is difficult to explore since it is very deep (1,200–2,000 m depth). The main problem of the Siwa oasis depression is the poor drainage and lack of a drainage outlet, thus causing water logging. The second problem is the shallow and under pressure groundwater that pops up to the ground creating wetlands. In Wadi-El-Natrun depression, the water table depth is almost of 3–5 m but has a high concentration in sodium carbonates and bi-carbonates. This type of composition is completely different in other delta fringes such as in Nubaria (west delta) or in Salhia (east delta) in which it ranges between 30 and 60 m with a medium quality of maximum salinity of 2,000 ppm. Most of these areas in Nubaria or Salhia are irrigated with Nile water through El-Nasr canal in Nubaria and Salhia canal in the east Delta, but the wells of groundwater are stationed as stand-by or alternative resources when Nile irrigation water is not sufficient or in case of a delay in its delivery.

Keywords Deep and shallow groundwater, Groundwater, Irrigated agriculture, Nile valley and delta aquifer, Saline and sodic water, Siwa Oasis, Wadi El-Natrun depression, Water logging

Contents

1 Introduction

Around 30% of the world's freshwater resources are stored underground in the form of groundwater (shallow and deep groundwater basins up to 2,000 m, soil moisture, swamp water, and permafrost). This constitutes about 97% of all the freshwater that is potentially available for human use [1].

Groundwater is the lifeline for many rural populations around the globe and a cornerstone of global food production. Groundwater constitutes nearly half the world's drinking water and much of the world's irrigation water supply. Population growth, overexploitation, salinization, nonpoint source pollution from agricultural activities, degradation of surface water, and groundwater quality and quantity conflicts at the urban–rural interface have reached global dimensions and threaten water availability and food security [3].

Groundwater is inherently a local and regional resource. Access is mostly through individual landowners and also through local and regional water purveyors. Groundwater overdraft and/or groundwater degradation from agricultural activities have broadly affected regions, states, and nations. Agricultural practices and management, groundwater management, groundwater use and degradation policies, regulations, and groundwater resource assessment and forecasting occurs at the farm, community, region, or state level are delineated by groundwater basin and jurisdictional boundaries. Yet, around the globe, many groundwater users and benefactors share similar experiences in their struggle to control or curtail overdraft and groundwater quality degradation [3] (Table 1).

Groundwater provides drinking water to at least 50% of the global population and accounts for 43% of all water used for irrigation [4]. Groundwater also sustains the base flows of rivers and important aquatic ecosystems. Uncertainty over the availability of groundwater resources and their replenishment rates pose a serious challenge to their management and in particular to their ability to serve as a

Table 1 Groundwater irrigation worldwide [4]

		Volume used	
Region	Area (Mha)	km^3/a	%
Global Total	112.9	545	43
South Asia	48.3	262	57
East Asia	19.3	57	34
South East Asia	1.0	3	5
Middle East and Central Asia	11.9	76	38
Europe	7.3	18	38
North Africa	2.5	16	24
Sub-Saharan Africa	0.4	2	7
North America	19.1	100	54
Latin America	2.2	88	19
Australia	0.9	3	21

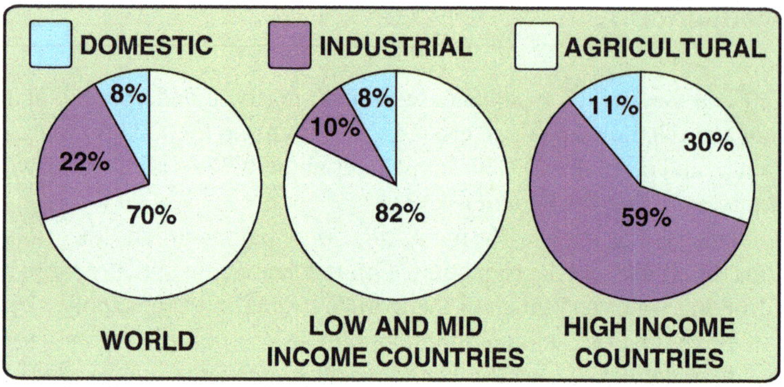

Fig. 1 Competing water uses in the different sectors [2]. Source: World Water Development Report, www.unesco.org/water/wwap/facts_figures/water_industry.shtml

buffer to offset periods of surface water scarcity [5]. Groundwater supplies are diminishing, with an estimated 20% of the world's aquifers being over-exploited [6, 7] leading to serious consequences such as land subsidence and saltwater intrusion in coastal areas [8]. Groundwater levels are declining in several of the world's intensely used agricultural areas and around numerous mega-cities.

Agriculture sector consumes an average of 70% of the total global water supply, which represents 85% of global water resources consumptions. Groundwater provides 43% of the blue water used in agriculture [1] (Fig. 1).

According to the Egyptian Ministry of Irrigation and Public Works [9, 10], the annual water resources in Egypt depend mainly on Nile water (55.5 Billion Cubic Meters (BCM)) in addition to 5.5 BCM groundwater and 1.3 BCM of rain water that falls on the agricultural land in the Delta. The remaining water resources in Egypt are water reuse sources such as recycled agricultural drainage water of 10 BCM, and primary treated sanitary wastewater of 5 BCM and 1 BCM from industrial wastewater. Accordingly, the total water resources in Egypt of fresh and reused water are almost 78.5 BCM. There is currently a shortage of 12.7 BCM since Egypt's population is 92 million per capita which would require 92 BCM for a share per capita equal to 1,000 m^3 to be above the water scarcity level. Figures 2 and 3 show water scarcity around the world, including Egypt.

Most of the groundwater in Egypt is non-renewable except in the shallow groundwater in the Nile valley and Delta lands and its fringes in addition to some depressions and oasis like Wadi El-Natrun in the west delta (The Valley of Sodium salts) and Siwa oasis south in the northwest coast of the Mediterranean, as seen in Fig. 4.

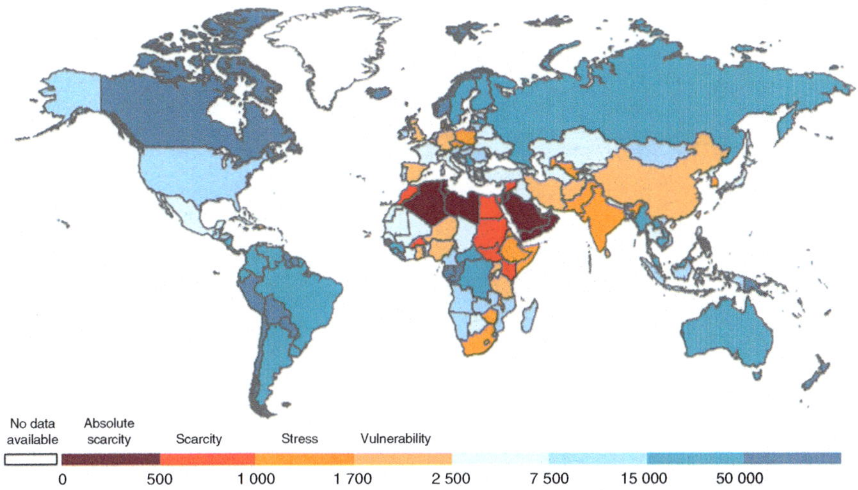

Fig. 2 Total renewable water per capita (2013) and water scarcity [11]. Source: WWAP, with data from the FAO AQUASTAT database (http://www.fao.org/nr/water/aquastat/main/index.stm) (aggregate data for all countries except Andorra and Serbia, external data), and using UN-Water category thresholds. Note: The figures indicate total renewable water resources per capita in m^3

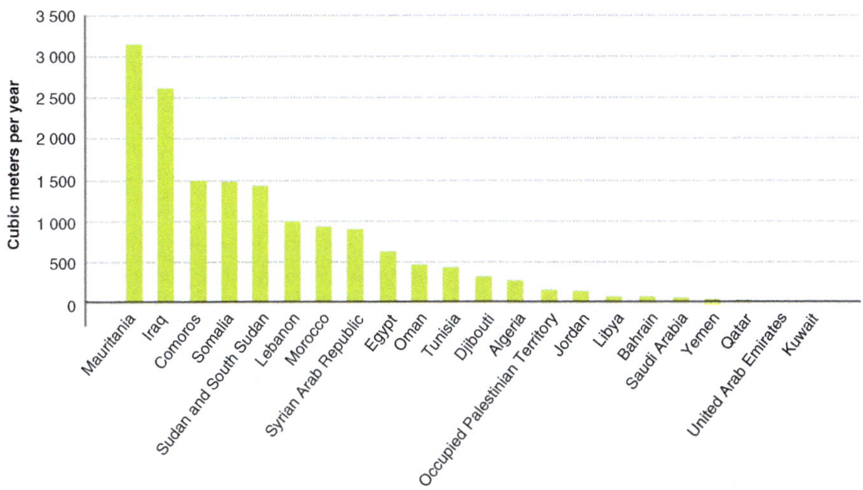

Fig. 3 Renewable water in the Arab Countries (2011) shows the water scarcity in Egypt [12]. Source: UNESCWA, prepared with data from FAO AQUASTAT. Note: The statistics for the year 2011 cover Sudan and South Sudan, as South Sudan did not become a country since July 2011

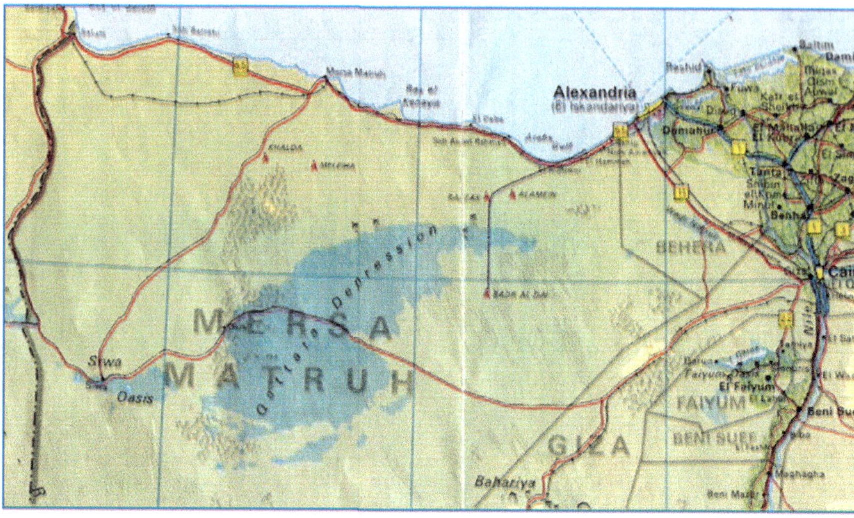

Fig. 4 Siwa oasis and Wadi El-Natrun depression (Google Earth)

2 Key Messages of Using Groundwater in Agriculture

- Food production requires the largest quantities of water, with groundwater resources providing more than 40% of all water used globally for irrigated agriculture [1, 11].
- Groundwater has proved to be a critical input for securing improved crop yields and enabling the 250% increase in food production achieved during the "green revolution" of 1970–2000 [1, 11].
- During the past 30–40 years there has been a remarkable level of investment in private water well construction for agricultural irrigation, because groundwater availability allows higher crop yields and high farmer returns per unit of water.
- Groundwater storage is very large but current withdrawal rates for irrigated agriculture in more arid areas are not physically sustainable, resulting in long-term (semi-permanent) depletion of aquifer reserves at rates in excess of 120 km^3/a [1].
- Land-use practices affect ground in the form of water recharge rates and quality; with intensification of cropping diffuse pollution of groundwater by plant nutrients, salinity and some residual pesticides is common.
- There is a pressing need to mobilize groundwater professionals, together with water-resource managers and irrigation engineers, to identify trans-sectoral governance and management responses for improving resource sustainability.

3 Priority Actions

- Elaboration of sustainable groundwater management plans for aquifers under pressure from irrigated agriculture, including the identification of improved irrigation-water management measures [1].
- Integrated evaluation and conjunctive management of groundwater and surface-water in major alluvial areas to enhance agricultural productivity and to avoid land drainage problems.
- Careful evaluation and monitoring to ensure that crop irrigation practices and new irrigation developments reduce groundwater salinization problems.
- Promotion of land management measures by farmers to enhance groundwater recharge rates and to reduce nutrient, salinity, and pesticide leaching to groundwater.
- Re-aligning government finances (such as crop guarantee-prices, pumping-energy subsidies, water well, and irrigation hardware grants) so as to reflect limited groundwater availability and the value of lost ecological services – thereby supporting initiatives for sustainable resource management.

4 Water Supply in Egypt

4.1 Nile River

The main (almost exclusive) source of water in Egypt is the Nile River. Egypt is unique among other countries in its dependence on water from one deterministic source. The Nile water agreement with Sudan in (1959) allocates 55.5 BCM/year to Egypt and 18.5 BCM to Sudan according to Sudan's low population and because they have a heavy rain in east and south Sudan as an additional water resource. This amount is guaranteed by the multi-year regulatory capacity provided by the High Aswan Dam (HAD) to catch and store the flooding water of the Nile to use it in the dry years. The Regular Cycle (rotation) of the Nile flooding water (before climate change) is 7 years high and heavy flooding; followed by another 7 years of low and small flooding and 6 years on average as not heavy not low flooding. The stored water in the front of High Aswan Dam helps Egypt to generate 2.1 Mega Wat/year which increases the percent of access electricity especially in rural area and village, in addition to complete covering of all cities. This amount of hydropower represents 50% of the Egyptian demands in year 1971 (the time of official opening of the dam). Figure 5 shows the Nile water distribution systems and its net canals.

4.2 Precipitation

Egypt has a winter rainfall in the form of scattered showers, usually beginning in November and ending in early April. The rain in Egypt rarely is heavy enough to

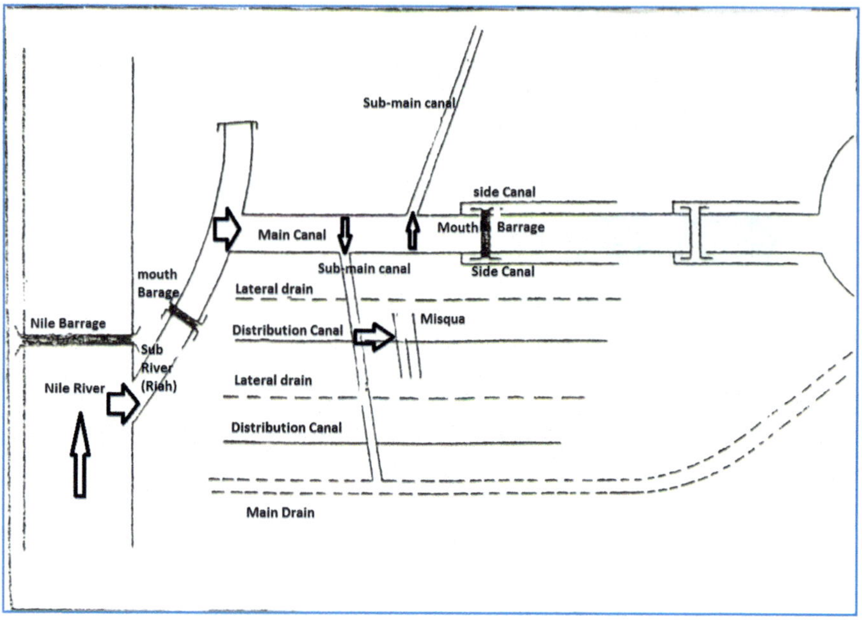

Fig. 5 Nile water distribution system (irrigation canal system; author artwork)

cause dangerous flooding except in the desert of Sinai and red sea cities and also in some governorates in Upper Egypt located close to high mountains (series of red sea and eastern desert mountains "Lime stone composition"). The average annual amount of effectively utilized rainfall water used in agriculture in the old soils of the Delta is estimated to be 1.3 BCM/year because most rains fall on the uninhabited desert especially in the Northwest cost of Mediterranean from west Alexandria city to the border with Libya (see Fig. 6).

The same trend is occuring to the North east coast of the Mediterranean in the north Sinai Peninsula up to the border with Gaza and Israel. In addition, there is a possibility of some dangerous flooding along the coast of Red Sea cities. The FAO [12] determined the all precipitation in Egypt to be 51 BCM/year but mostly in un-accessible areas and there is still no planning of projects for water harvesting (Table 2).

Rain-fed agriculture requires a rain (enough to saturate the root zone area and soil profile each time) in addition to continuous and steady rains along the growth season along the plant life cycle (from seed to seed) which usually extends for 150–180 days. But, excess Heavy rain and flash floods may cause runoff and soil erosion [13]. Most rain in Egypt is scattered and non-continuous through the plant life cycle and usually stops within 90 days only; thus, the amount of rain in the Mediterranean coastal lands (old and desert soils) is not considered a viable area for rain-fed agriculture due to its high spatial and temporal variability.

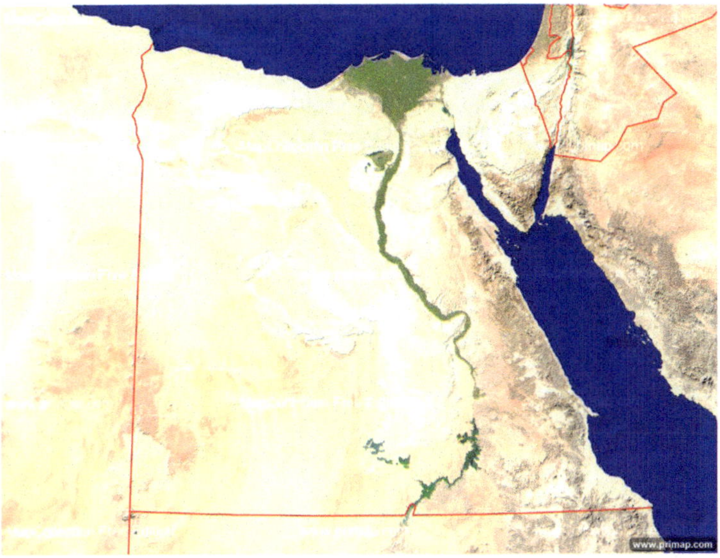

Fig. 6 The map of Egypt (Google map)

Table 2 Renewable water resources in Egypt (downstream country) compared with Ethiopia (upper stream country) [12]

Aspect of water use	Egypt	Ethiopia
Average precipitation in volume (billion m^3/year)	51.07	936.4
Total actual renewable water resources (km^3/year)	57.5/85.8	122.00
Total renewable water resources per capita (m^3/inhab/year)	702.80	1,512.00
% of withdrawals of total actual renewable water resources	117.20	5.10
% of total actual renewable water resources by agriculture in 2002	103.00	4.3
Dependency ratio (%)	96.86	0.00
Irrigation potential (1,000 ha)	4,420.00	2,700.00
Area equipped for irrigation (1,000)	3,422.00	290.00

4.3 Groundwater

Groundwater is stored in aquifers, which are water-bearing rock formations that hold water in the inter-particle pore space and cracks within rock material. There are two basic types of aquifers including the unconfined aquifer and the confined aquifer. An unconfined aquifer (also called a water table aquifer) has an extensive water table open to recharge by precipitation. A confined aquifer does not have an extensive water table. Water is pressurized and some of it can flow from the well without pumping [14]. This water, generally described as fossil water, is mostly nonrenewable except for coastal and Delta aquifers. The chemical quality of the water is generally suitable for irrigation and domestic uses with an average total dissolved solids (TDS) less than 1,000 ppm. Salinity increases approaching the east/

west extremities, and ranges from 500 to 3,000 ppm. High salinity water could be made suitable for crop irrigation by combining two parts of surface water to one part of groundwater [15]. This kind of groundwater is suffering from the residues of both pesticides and fertilizers in addition to salts that leached from the soil above.

Groundwater exists in the Western Desert and Sinai in aquifers that are mostly deep and non-renewable. The total groundwater volume has been estimated at about 40,000 BCM. However, current abstraction is estimated to be 2.0 BCM/year. The main obstacles in utilizing this huge resource are the great depths (up to 1,500 m in some areas) of these aquifers and deteriorating water quality at the increasing depths [9].

Shallow Groundwater in the Nile aquifer cannot be considered a separate source of water. The aquifer is recharged only by seepage losses from the Nile; the irrigation canals and drains and percolation losses from irrigated lands. Hence, its yield must not be added to Egypt's total water resources. Therefore, it is considered as a reservoir in the Nile River system with a huge capacity, but only 7.5 BCM/year is rechargeable live storage.

The total available storage of the Nile aquifer was estimated at about 500 BCM but the maximum renewable amount (the aquifer safe yield) was estimated to be only 7.5 BCM. The existing rate of groundwater abstraction in the Valley and Delta regions is about 4.5–6.5 BCM/year (2013), which is still below the potential safe yield of the aquifer [9].

4.4 Seawater Desalination

Desalination of seawater in Egypt has been given low priority as a water resource because the cost of treatment is high compared with other sources. Desalination is actually practiced in the Red Sea coastal area to supply tourist villages and resorts with adequate domestic water supply where the economic value of the water is high enough to cover the treatment costs (almost 0.1 BCM). Other groundwater desalination units are constructed at several locations in the Sinai such as a water supply for Bedouins. It may be crucial to use such resources in the future if the growth of the demand for water exceeds all other available water resources. However, its use will depend on technological development in this field. Desalination technologies are used extensively in wealthy countries like Saudi Arabia, United States, and the United Arab Emirates (see Fig. 7). One major factor causing the limited use of desalinating technologies in unwealthy countries like Egypt is the high demand for energy which constitutes 44% of the total cost of the desalination system (see Fig. 8).

In 2009, Food and Water Watch in the United State of America [17] and Pacific Institute, California 2006 [16] stated the drawbacks of desalination:

1. Seawater desalination is expensive.
2. Seawater desalination will provide little benefit and has a poor track record.
3. Seawater desalination invites corporate control and abuse of our water supply.
4. Seawater desalination endangers the environment and public health.

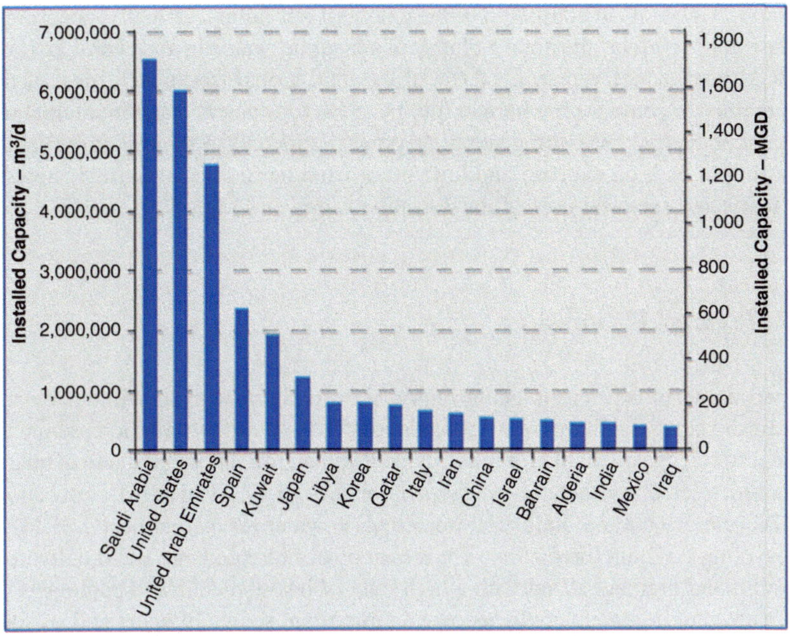

Fig. 7 The top list of countries producing desaline water [16]

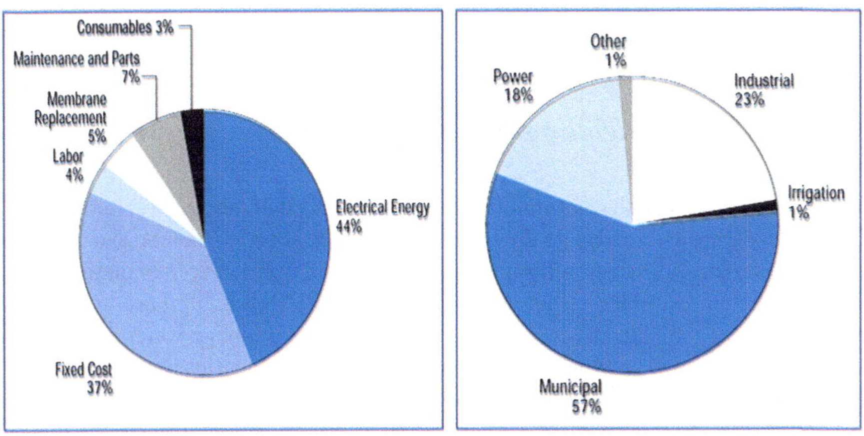

Fig. 8 The cost items of de-saline water production and its sectorial uses [16]

5. Seawater desalination could contribute to global warming.
6. Seawater desalination threatens fisheries and marine environments.
7. Seawater desalination poses a risk to human health.
8. Seawater desalination promotes social and environmental injustice.

At the end of both reports the following statement was summarized: "Desalination of the sea is not the answer to our water problems. It is survival technology, a life support system, an admission of the extent of our failure." Finally, according to the last conservative, the total volume of de-saline water in the world is low as 24 BCM/year, which represents 0.6% of the total global freshwater, most of them used mainly to conserve the human life, i.e. 57% for domestic and municipal uses, 23% for industry, 18% for generating power, and only 1% used in agriculture according to its high expense and low return from agriculture. Figure 8 shows the cost items and sectorial uses of the de-saline water.

4.5 Treated Wastewater

Treated domestic sewage is being reused for irrigation with or without blending with freshwater. The increasing demands for domestic water will increase the total amount of sewage available for reuse. It is estimated officially that the total quantity of reused treated wastewater in Egypt is about 0.3 BCM in 2013, with another 5 BCM primary treated. Industrial treated wastewater reached almost a BCM/year by recycling it within the facility. There is another untreated one BCM delivered in irrigation and drainage canal with a high load of heavy metals and pathogens [9].

Wastewater treatment could become an important source of water and should be considered in any new water resource development policy. However, proper attention must be paid to the associated issues with such reuse. The major issues include public health and environmental hazards as well as technical, institutional, socio-cultural, and sustainability aspects.

4.6 Reuse of Agricultural Drainage Water

The amount of water that returns to drains from irrigated lands is relatively high (about 25–30%). This drainage flow comes from three sources: tail end and seepage losses from canals, surface runoff from irrigated fields, and deep percolation from irrigated fields (partially required for leaching salt). None of these sources is independent of the Nile River. The first two sources of drainage water are considered to be freshwater with relatively good quality. The agricultural drainage of the southern part of Egypt returns directly to the Nile River where it is mixed automatically with Nile freshwater which can be used for different purposes downstream [9]. The total amount of such direct reuse was estimated to be about 4.07 BCM/year in 1995/1996. In addition, it is estimated that 0.65 BCM/year of drainage water is pumped to the El-Ibrahimia and Bahr Youssef canals for further reuse. Another 0.235 BCM/year of drainage water is reused in Fayoum while about 0.65 BCM/year of Fayoum is drained to Lake Qarun. Moreover, drainage pumping stations transport about 0.60 BCM/year of Giza drainage water to the Rossita Branch just downstream of the Delta Barrages for further downstream reuse.

According to the Ministry of Irrigation and Public Works 2016 [10], the total amounts of agricultural drains in the Delta cultivated lands (4.5 Million Feddans

(MF)) reached 15 MCM. Officially the amount of reused agricultural drainage water is 8 BCM but in reality it exceeds 10 BCM. On the other hand, Upper Egypt which has nearly 1.5–2 MF (4,200 m^2) of cultivated land is fed through both Nile River irrigation and drainage canals concurrently. The amount of agricultural drainage water that returns to the Nile course is expected to be between 4–5 BCM/year. Thus, the actual reuse of agricultural drainage water reached 10 BCM/year in the Delta with an additional 5 BCM/year in Upper Egypt for a total of 15 BCM/year [10].

5 Water Scarcity in Egypt

Water security is the availability of an acceptable quantity and quality of water for health, livelihoods, ecosystems, and production, coupled with an acceptable level of water-related risks to people, environment, and economies [14]. The share water in Egypt currently is 680 m per capita; this is below the scarcity level of 1,000 m per capita. Egyptian citizens should have renewable freshwater resources not less than 92 BCM for its 92 million population; however, only 62 BCM is available, leaving a water shortage of 28 BCM. Thus, Egypt must use the marginal and poor quality water and reuse and recycle when available. The reuse of agricultural drainage water, treated sanitary water, and treated industrial water, produces 17 BCM, helps to decrease the actual water shortage in Egypt from 31 BCM/year to 14 BCM/year. Improving the existing distribution system through the open non-cementing, non-piping irrigation canal, running along 30,000 km from the High Aswan Dam lake south Egypt to the last point of the irrigated soil in North Delta may add another 19 BCM/year usually lost by evaporation and seepage within this long wide mud irrigation canals net. The reuse of marginal poor quality water in Egypt has already been established and will continue to be developed in water policy in the future in the face of rising water scarcity.

6 Sectoral Water Demand

Water requirements in Egypt are continuously increasing due to population increases and the improving standards of living as well as the governmental policy to encourage industrialization. Demands for water can be categorized in classes representing the demand and water balance. Water requirements of the agricultural sector represent the largest component of the total water demand in Egypt.

6.1 Agriculture

Agriculture consumes more than 85% of Egypt's water resources or about 103% of the shared water from the Nile river by almost of 57 BCM (see Table 1). Although the country has lost a part of its fertile land due to urbanization, this has been balanced by expansion of agricultural areas into new regions. Expansion in agriculture is carried out horizontally and vertically through intensive cultivation in the different crops to produce food, forage, leaf, and medicinal plants. In 1990 cultivated lands occupied 6.92 MF with a cropped area of about 12.43 MF (two seasons a year), while in 2010 cultivated areas and cropped lands were 8.6 MF and 17.50 MF, respectively [18].

6.2 Municipal and Domestic

Municipal water requirements include water supply for major urban and rural villages. Part of this water comes from the Nile system and the other part comes from groundwater resources. Municipal water demand was estimated to be 10 BCM in 2010, where approximately 97% of the urban population and 70% of the rural population of Egypt access piped clean drinking water supply. Sanitation facilities are less developed where approximately 90% of the urban population and 40% of the rural population was planned to be connected to a sewerage system in 2016. Municipal water is supplied by about 83% from Nile water and 17% from groundwater.

6.3 Industry

The industrial sector, according to the Ministry of Irrigation and Public Works, required 3 BCM during the year 2016. A small portion of that water is consumed through evaporation during industrial processes especially in the cement and steel stoves (only 0.7 BCM) while the rest of water returns to the system in a contaminated form. These numbers must be reconciled when conducting an accurate assessment of financial aspects of industrial water use and its effects on the economics (Fig. 9).

7 Allocation of Groundwater in Egypt

The chemical quality of the water is generally suitable for irrigation and domestic uses with an average total dissolved solids (TDS) less than 2,000 ppm. Salinity increases approaching the east/west extremities, and ranges from 2,500 to 3,000 ppm. Water of high salinity could be made suitable for crop irrigation by combining two parts of surface water to one part of groundwater [20]. In Egypt, the main aquifers are generally formed of granular rocks (sand and gravel) or fissured and limestone rocks comprise the following items:

- The regional Nubian Sandstone Aquifer System, occupying much of Egypt and continuing across the border in a westward direction into Libya, in the south and

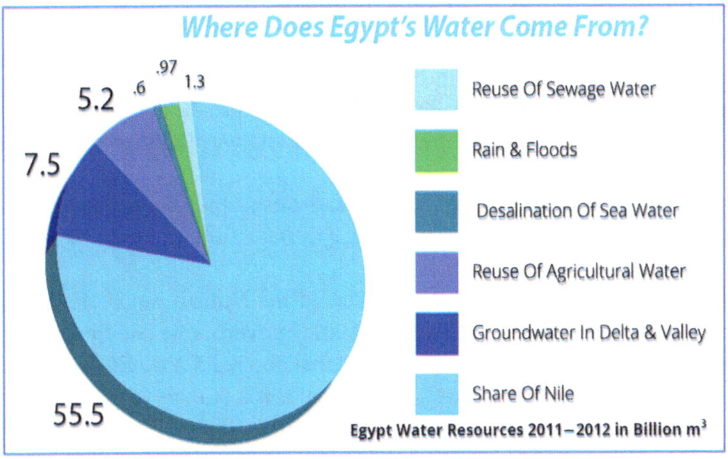

Fig. 9 The distribution of freshwater in the different sectors in Egypt (2010) [19]

southwestward direction into the Sudan and Chad, and in the eastward direction into Israel, Jordan, and the Arabian Peninsula [21]. The thickness of the sediments varies from a few hundred meters in the south, to 4,000 m west of Abu Mongar.

- Carbonate Aquifers occupy at least 50% of Egypt. They are made of fissured limestone, giving rise to natural springs with a total flow of 200,000 m^3. Salinity varies between 1,500 and 7,000 ppm. Lower salinity was reported in Siwa Oases at 200 ppm [22].
- The Moghra Aquifer system has a broad geographical distribution in the region west of the Nile Delta and south of the Qattara Depression [21], Fig. 4. It is characterized by its moderate to high salinity, which ranges between 1,000 and 5,000 ppm with a present dominant average of 4,000. The aquifer water is a mixture of Paleo water and renewable water.
- The Nile valley and Delta aquifer is the most productive, containing around 200×103 million m^3, renewable by seepage from Nile river irrigation systems. Thickness of this aquifer decreases from 300 m at Sohag Governorate in the Upper Egypt to few meters near Great Cairo (Cairo, Giza, and Qalyubia governorates) and also in the south near Aswan. This aquifer loses its water through the Rosetta Branch into the Mediterranean and Suez Canal [22]. The Nile receives a total amount of 1.6×103 million m^3/year as drainage from aquifers. The total amount of groundwater already used for agriculture and domestic purposes is about 2.6×109 m^3/year and is extracted from about 9,000 wells [18]. Approximately 1.0×103 million m^3/year of groundwater is extracted through productive wells. In 1990, total extraction of groundwater amounted to 3.3 bm^3 to meet irrigation and domestic water requirements. It is estimated that 4.8 bm^3 of groundwater is currently used in the Valley and Delta, as well as 0.57 bm^3 in the desert and Sinai for a total of 5.37 bm^3 of groundwater.

Groundwater extraction can be increased to 11 bm^3 without depletion of underground reservoirs. There is a possible future increase of 8.3 bm^3 [18].

- The Coastal Aquifer lies 35 km from the seashore, 45 km north of Cairo and is recharged mainly from rainwater and from high-pressure water in the Nubian Sandstone aquifer. It is renewable, but deeper and with salinity varying between 3,000 and 5,000 ppm [22].
- Rose basement rock has the same characteristics as the Carbonate Aquifer but is difficult to explore since it is very deep, (1,200–2,000 m depth) and has a high cost of digging (least L.E 5 million) [22].
- Groundwater occurs in the western desert in the Nubian sandstone aquifer that extends below the vast area of the New Valley governorate and the region east of Owaynat Mountain. It has been estimated that about 200,000 BCM of freshwater is stored in this aquifer. However, groundwater is available at greater depths and the aquifer is generally non-renewable.
- Utilization of such water depends on pumping costs and its depletion rates versus the potential economic return in the long run.
- Groundwater in the Sinai is mainly encountered in three different water-bearing formations: the shallow aquifers in northern Sinai, the valley aquifers, and the deep aquifers. The shallow aquifers in the northern part of Sinai are composed of sand dunes that hold the seasonal rainfall. The aquifers in the coastal area are subject to salt-water intrusion. The total dissolved solids in this water range from 2,000 to 9,000 ppm which can be treated to reach a suitable salinity level for use to irrigate certain crops.
- The groundwater aquifers in the valleys of the Sinai are recharged from rainfall and particularly heavy storms. The annual rainfall in the Sinai varies from 40 to 200 mm/year. Most of the rainfall water recharges the shallow groundwater aquifers in northern Sinai such as the Delta of Wadi El-Arish and the El-Beqaa flood-plain, while such aquifers are absent in southern Sinai. Although most of the shallow aquifers are renewable, only 10–20% of the deep aquifers are renewable by rainfall and flash floods.
- The total groundwater abstraction in the western desert in 1995/1996 was estimated to be about 0.48 BCM while it is only 0.09 BCM/year in Sinai.

8 Groundwater Potential

Groundwater development started in the early 1960s and hasbeen continuing. It is observed that the present use is concentrated in the Nile aquifer system, followed by the desert fringes and the Nubian sandstone.

Other aquifer systems are still underdeveloped. Moreover, within the developed systems, only part of the potential is utilized mainly where groundwater is fresh. Table 3 *shows the salinity and some other properties of some groundwater in Egypt* (Table 4).

Table 3 Major aquifers, renewability, salinity, and productivity in Egypt [18]

Aquifer	Confined/non-confined/ semi-confined	Renewable or non-renewable	Salinity	Productivity
Nile valley and Delta	Confined	Renewable	Low	Very high
Moghra	Semi-confined	Non-renewable	High	Low
Nubian sand stone	Unconfined	Non-renewable	Low	Medium
Coastal	Unconfined	Renewable	High	Medium
Carbonate	Non-confined	Non-renewable	High	Low
Rose	Non-confined	Non-renewable	High	Low

Table 4 Available water resources in Egypt, 2000 and 2017 (% of total) [18]

Water supply source	Supply in year 2000 (BCM)	Estimated supply 2017 (BCM)
River Nile	55.5	55.5
Rainfall	1.0	1.5
Desert and Sinai groundwater	0.6	3.5
Valley and Delta groundwater	4.8	7.5
Municipal reused drainage water	0.7	2.5
Agricultural reused drainage water	5.1	8.4
Improved irrigation system	0.0	7.0
Total available water 2000/2001	67.6	85.9

8.1 Nubian Sandstone Aquifer

Because the Nubian sandstone aquifer is the largest store of groundwater in Egypt, we will discuss it in detail. The Nubian Sandstone Aquifer System underlies virtually all of Egypt, much of eastern Libya, and significant areas of both northern Chad and northern Sudan [18]. It can be broadly described as two distinct aquifer systems vertically separated by layers of lower permeability that allow some upward leakage [23]. The deeper Nubian Sandstone Aquifer System is older and extends to the entire area while the Post Nubian Aquifer System lies above that in more recent geological formations and covers roughly the northern half of the larger system [18].

People have been extracting groundwater to a limited degree for thousands of years in the North African desert [24]. Historical rates of use, however, are relatively insignificant compared to the current rate of abstraction in the Nubian Sandstone Aquifer System, which has increased by roughly 500% since the early 1960s when large-scale development began [25]. Models of the aquifer's water budget show that the current scale of groundwater use and the increased rate of groundwater decline will significantly exceed historical rates [26]. While it is clear that the aquifer system holds an enormous reserve of water, estimates vary considerably as to the amount – from as little as 15,000 km^3 [22] to 135,000 km^3 [27], to as much as 457,570 km^3 [25]. It is generally believed that the aquifer's water dates from wetter climates in the past (5,000–10,000 and 20,000–25,000 years BP) and that no significant recharge is taking place under current climate conditions [18, 23–27]. Experts generally accept that the system has not been in equilibrium for thousands of years and that groundwater levels were already declining well before artificial extraction began [27, 28]. Thus, any water withdrawal from the aquifer under the current climate would be considered "water mining," or a rate that exceeds recharge [29]. Figures 10 and 11 show the Nubian sandstone aquifer.

8.2 *Precipitation in the Nubian Sandstone Aquifer*

Arid and hyper-arid climate conditions in most of the region magnify the aquifer's importance. Almost all of the water used by people living above the aquifer comes from either groundwater abstraction or diversion from the Nile. The average annual

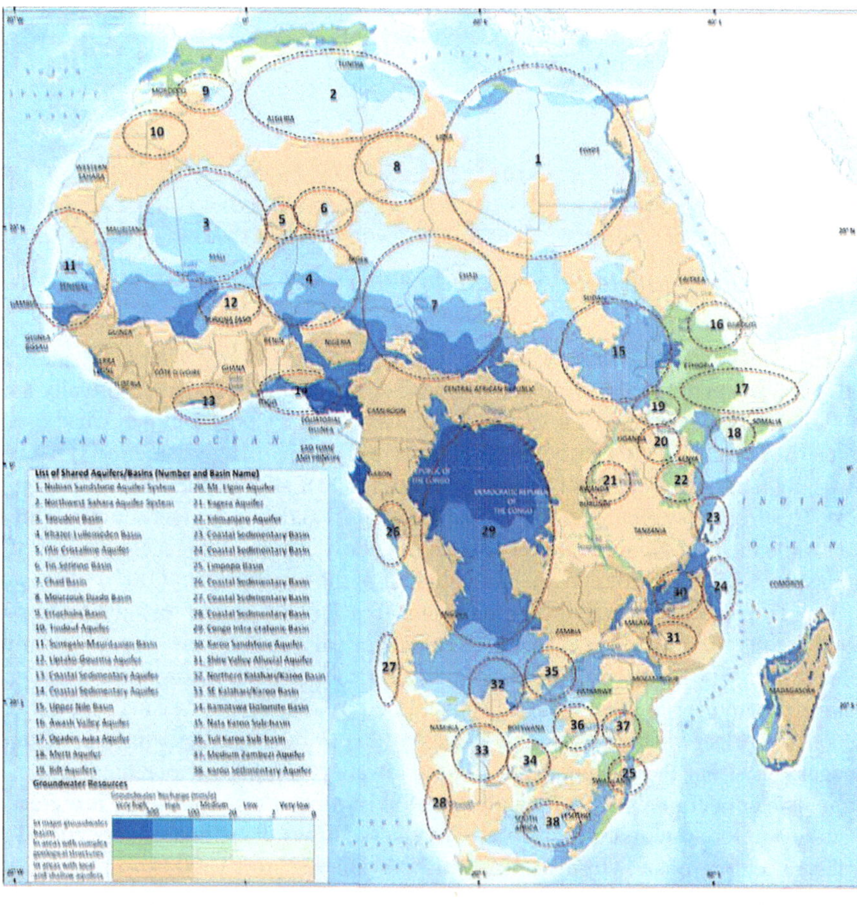

Fig. 10 Transboundary aquifers [29]

Fig. 11 Nubian sandstone aquifer [29]

mean rainfall is less than 50 mm and on average vast areas over the aquifer system receive no measurable precipitation. The wettest area at the northern most tip of Libya receives around 425 mm of rain. Northern Darfur in Sudan is the wettest area to the south but sees only an average of around 200 mm each year [30].

8.3 Desert Oasis Development: Dakhla Oasis

Dakhla Oasis lies 300 km west of the Nile and is surrounded by the driest of desert landscapes. Its location over the southern edge of the Post Nubian Aquifer, however, provides access to both shallow and deep wells within the Nubian Sandstone Aquifer System. While it is currently home to fewer than 100,000 people, archeologists believe that the Dakhla Oasis has been continuously settled for around 8,000 years. Water extraction from the deeper Nubian Sandstone Aquifer at the Oasis has grown tenfold since 1960 [18]. The related growth in agriculture can be seen in the pair of satellite images that span just a few of those years – 1986–2010, as seen in Fig. 12. Faced with increasingly crowded populations along the Nile, the Egyptian government has been further developing settlement and agriculture in the Western Desert's oasis. Some studies suggest that the planned rate of abstractions for these areas is unsustainable, however, because it will lead to local depressions in the water table level which increases the cost and difficulties of lifting this deep water [26].

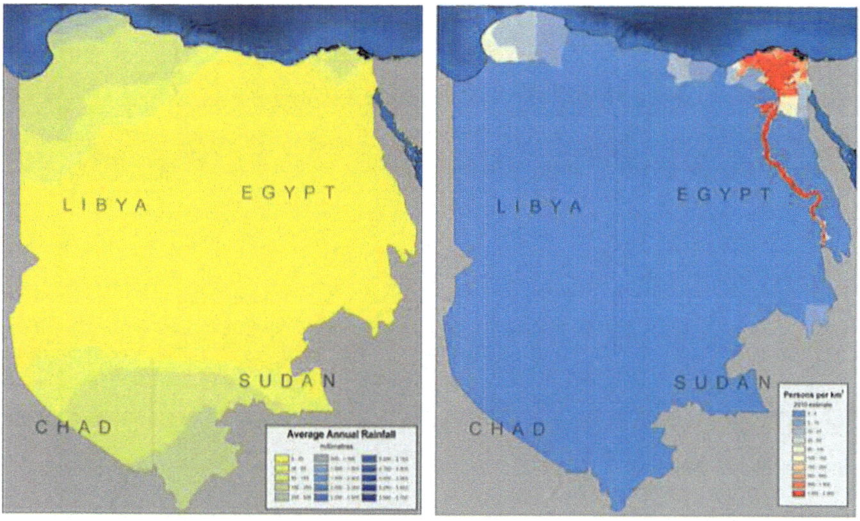

Fig. 12 Nubian sandstone system [29]; average annual rainfall (*left*), and population density (*right*)

9 Benefits of Groundwater Use for Agricultural Irrigation

Groundwater is a "very popular commodity" to farmers [30] since it:

- Is usually found close to the point-of-use (often only a well's depth away)
- Can be developed quickly at low capital cost by individual private investment
- Is available directly on-demand for crop needs (given a reliable energy source for pumping) and thus affords small-holders a high level of control year-round
- Is well-suited to pressurized irrigation and high productivity precision agriculture
- Has "democratized" irrigation by permitting irrigated agriculture outside canal command areas.

This situation has led to widespread depletion of groundwater resources, with the following collateral effects, which vary considerably in occurrence and intensity with hydrogeological setting:

1. Counterproductive competition between irrigation users.
2. Conflicts with rural and/or urban drinking-water provision, making it more difficult to achieve MDGs.
3. Impacts on natural aquifer discharge (spring flow, riverbed flows), which result cumulatively in an unacceptable impact on "downstream" surface water-flows.

Degradation of groundwater depends on aquatic ecosystems. Conceptual misunderstandings about groundwater resources tend to occur rather widely and there is need to substitute myth with reality [30] *by:*

- Making a clear distinction between "groundwater-only irrigation areas" and "conjunctive-use irrigation areas," since these present very different prospects, approaches, and challenges for resource optimization.
- Focusing resource-management efforts on constraining consumptive use (rather than just groundwater withdrawals), especially in groundwater-only irrigation areas
- Assessing groundwater-surface water connectivity in alluvial environments, as a basis for taking advantage of the opportunities of "conjunctive management" whilst avoiding the risk of "double-resource accounting."

9.1 Hazards of Excessive Groundwater Exploitation

Continuous groundwater depletion resulting from long-term excessive resource exploitation can in some cases result in a number of other serious consequences:

- The salinization of aquifers – which is a very insidious and often complex process arising from a variety of physical mechanisms.

- Troublesome land subsidence due to the settlement of interbedded aquitards in alluvial and/or lacustrine formations increasing (and in some cases spiraling) electrical energy costs for pumping, especially where use is "buffered" by subsidies or flat-rate tariffs – with serious implications for many electricity utilities and for the unit energy consumption and carbon footprint of irrigated agricultural production [31, 32].

9.2 Climate Change

Climate change will affect the natural water balance and water availability in several ways. First, changes in spatiotemporal patterns and variability of precipitation affect the replenishment of water resources. Second, increase in water and soil salinity due to high temperature of global warming. Third, the contamination of irrigation water will decrease water efficiency, water return benefits, and causing land degradation which leads to increase in the water requirements needed to produce food. This is not easy under the water deficiency in Egypt. Thus, much pressure will hit the ground and surface water to meet the increasing demands of different sectors of agriculture, industry, domestic, and municipal.

Groundwater reportedly provides drinking water to at least 50% of the global population and accounts for 43% of all water used for irrigation [33]. Groundwater also sustains the base flows of rivers and important aquatic ecosystems. Uncertainty over the availability of groundwater resources and their replenishment rates pose a serious challenge to their management and in particular to their ability to serve as a buffer to offset periods of surface water scarcity [34]. Groundwater supplies are diminishing, with an estimated 20% of the world's aquifers being over-exploited [35], leading to serious consequences such as land subsidence and saltwater intrusion in coastal areas [6]. Groundwater levels are declining in several of the world's intensely used agricultural areas and around numerous mega-cities [33]. In the Arabian Peninsula, freshwater withdrawal, as a percentage of internal renewable water resources, was estimated at 505% in 2011 [6], with significant volumes of groundwater reserves being transboundary in nature [12].

9.3 The Diagnosis of Groundwater Salinization

In major areas of agricultural irrigation, the salinization threat to groundwater varies widely with overall hydrogeological setting and climatic regime, and even down-the-length of major river basins. It arises through a number of distinct and independent mechanisms:

- Rising water-table due to excessive canal seepage and/or field application in head-water areas leading to soil water-logging and phreatic salinization, or sometimes naturally saline shallow groundwater becoming mobilized.
- Leaching of soil salinity across irrigation areas on first habilitation of arid soils and/or salt fractionation by "efficient" irrigation, with accumulation in tail-end sections of canal commands if no groundwater outflow occurs.
- More frequent coastal lateral intrusion or inland up-coning of saline groundwater due to excessive abstraction of fresh groundwater.
- Additionally, there are hyper-arid areas in which virtually all groundwater is naturally saline, except where there exists some infiltration from surface watercourses and irrigation canals forming "freshwater lenses." The implication is that groundwater salinization threats need sound diagnosis, close monitoring, and careful management.

9.4 Groundwater Quality Impacts of Irrigated Agriculture

Agricultural land-use practices in general also exert a major influence on groundwater recharge quality [8, 36] through:

- Leaching of soil nutrients is a problem that has been exceptionally widespread in the industrialized nations with (largely successful) attempts to increase grain, oil-seed, vegetable, fruit and milk production per unit area through the replacement of traditional crop rotations with near monocultures, but as yet has been less severe in the developing world where inorganic fertilizer applications have generally been much lower – in theory at least the problem of soil nutrient leaching should also be more manageable in irrigated than rain-fed agriculture.
- Contamination with pesticides is a potentially serious problem but one more confined geographically to recharge areas of aquifers exhibiting high vulnerability to pollution from the land-surface, where the more "mobile" pesticides (mainly certain herbicides and soil insecticides) have been regularly used at high application rates.
- Mobilization of salinity is an issue of very serious concern in arid and hyper-arid areas where the "irrigation frontier" has (or is) being extended through clearing of native desert scrubland with high salinity levels retained in the sub-soil profile.

10 Use of Groundwater in Agriculture in Egypt

FAO [37] stated the following five principles of sustainable agriculture:

Principle 1: Improving efficiency in the use of resources is crucial to sustainable agriculture.
Principle 2: Sustainability requires direct action to conserve, protect, and enhance natural resources.
Principle 3: Agriculture that fails to protect and improve rural livelihoods, equity, and social well-being is unsustainable.
Principle 4: Enhanced resilience of people, communities, and ecosystems is a key to sustainable agriculture.
Principle 5: Sustainable food and agriculture requires responsible and effective governance mechanisms.

The numbers in Fig. 13 indicate the following:

1. *Environmental services include*: Climate, Nutrient cycling, Biodiversity, Water cycles, Coastal protection…
2. *Natural resources include*: Land, Oceans, Water, Genetic resources, Forest resources, Aquatic systems, Nutrients, and Energy.
3. *Agriculture includes*: Crops, Livestock, Forestry, Fisheries, and Aquaculture

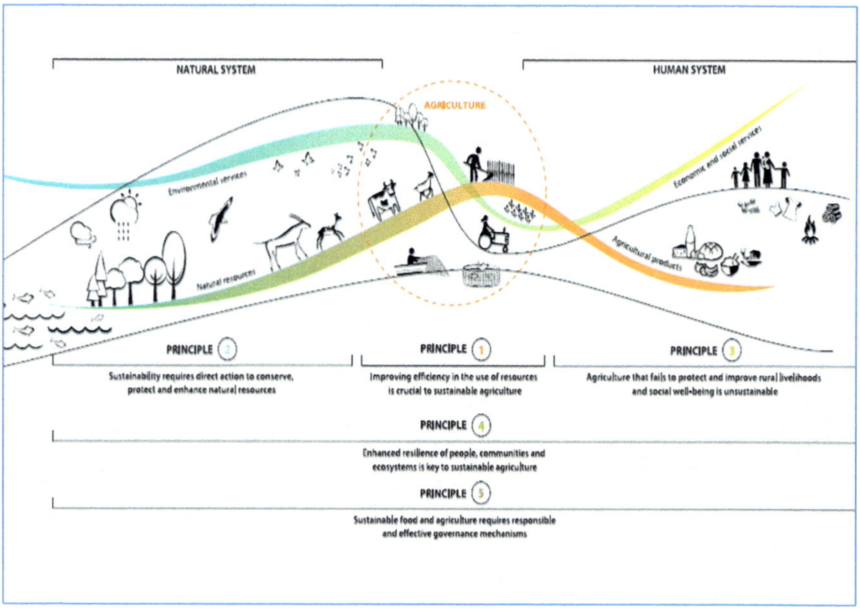

Fig. 13 The five principles of sustainable agriculture [37]. Source: FAO ([38], Fig. 3, pp. 18–19)

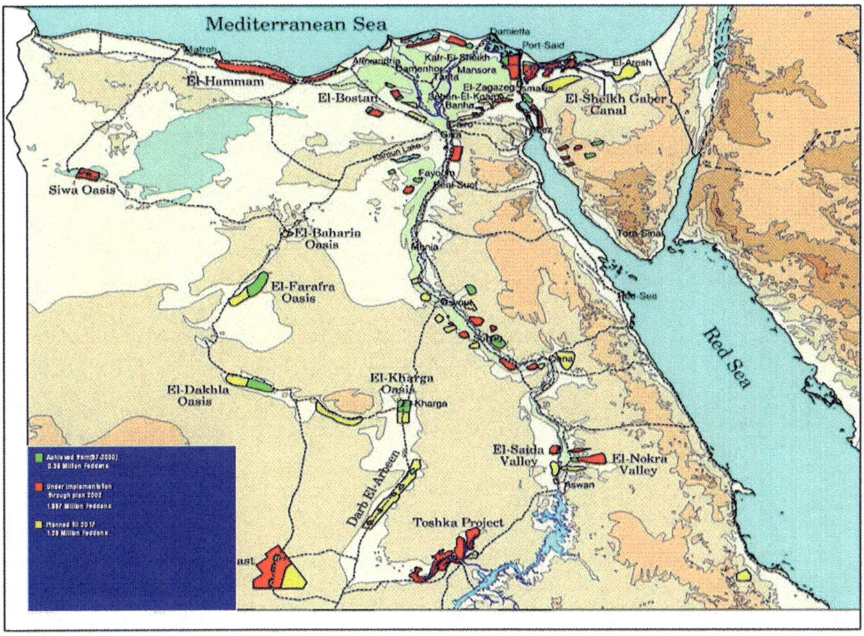

Fig. 14 The master plane of Egypt for agriculture extension up to year 2017 (nothing achieved) [39]

4. *Agricultural products include*: Food, Feed, Fiber, Fuel, and Medicinal crops
5. *Economic and social services include*: Growth, Poverty reduction, Employment, Stability, Health, and nutrition.

The use of groundwater in agriculture is effective in the new reclaimed lands in the deserts where the Nile water is not available or accessible (Cairo Alexandria desert road, Cairo-Al-Ismailia desert road, the new valley governorates and its Oasis, The east of Owenate mountain in the far south west close to Sudan and Libian boarder, Wadi El-Natrun depression, North and South Sinai in addition to the desert extension of all valley and delta governorates fringes). Figure 14 shows the master plan of agriculture extension projects in Egypt from the years 2000 up to year 2017, and Fig. 15 shows the location of the new project (year 2016) for reclaiming 1.5 MF within the next 5 years (0.63 million ha) that completely counts on the groundwater.

10.1 Use of Groundwater in Irrigation in the Nile Valley and Delta and Their Fringes

The total dissolved salts in Nile water is 500 g/m^3 measured at Giza Nile [37], and this concentration will add 3.5 tons of hazardous soluble salts every year to each

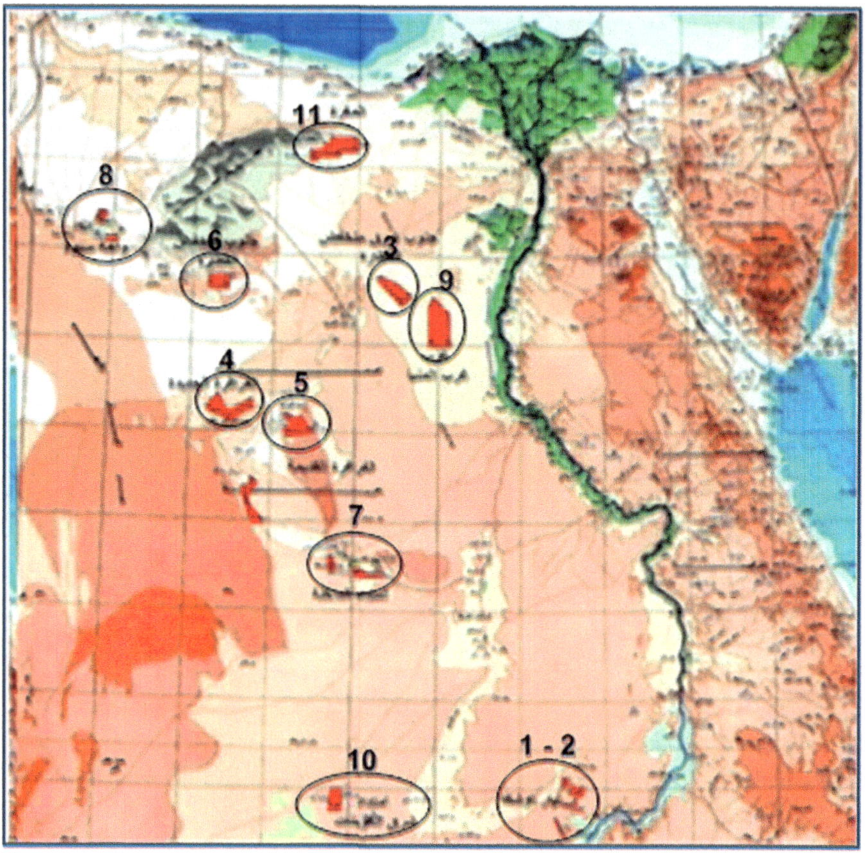

Fig. 15 The location of the reclamation project (2016) for reclaiming 1.5 MF [10]

feddan (4,200 m^3) at the rate of total annual irrigation water and consumptive use of 7,000 m^3. This number will increase threefold when using agricultural drainage water of 10 BCM/year (this amount represents about one third of total irrigation water in the delta soils with total area of 4.5 MF). The same trend of salt accumulation will exist when using shallow, poor quality water table in irrigation. Actually the Egyptian farmers obliged to frequently and continuous use of this poor quality water because the Ministry of Water Resources and irrigation only delivers 5,000 m^3/acre/year, instead of 7,000 m^3/acre/year, the actual water requirements per acre due to high temperature weather in Egypt. Thus, the farmers are obliged to manage and find a way to get an additional 2,000 m^3 from any near water resources to complete their field requirement of irrigation water. Most of the farmers in the delta areas or in the Giza Governorates make groundwater wells at a depth usually ranging between three meters and not to exceed 20 m to help them in completing their water needs.

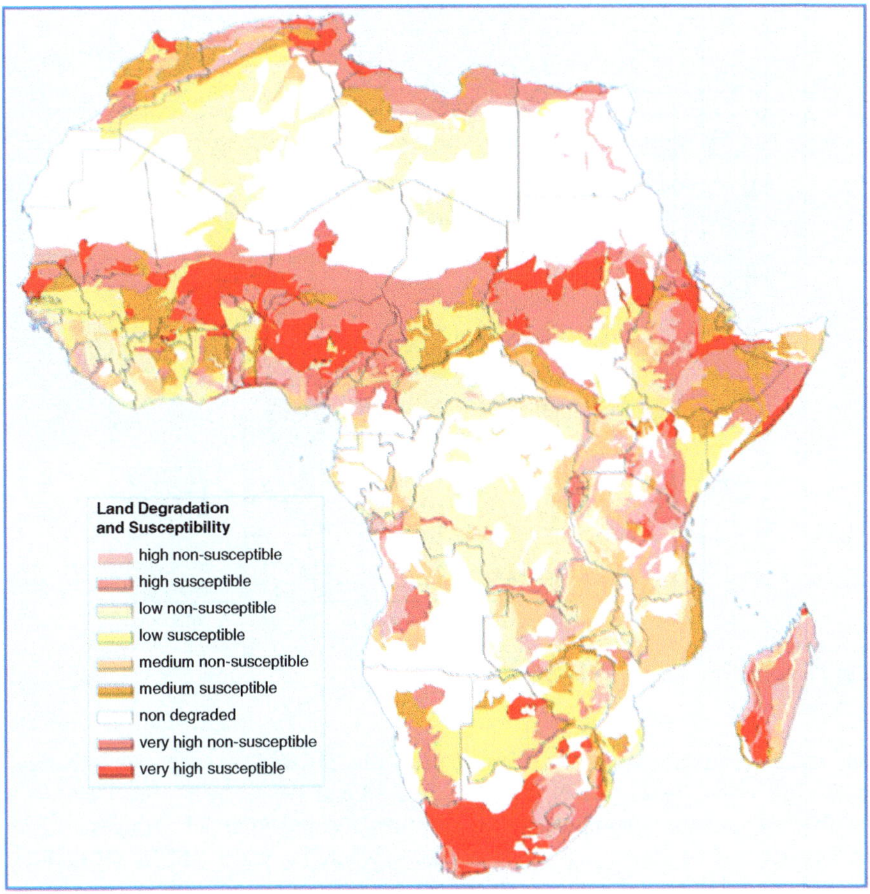

Fig. 16 Land degradation susceptibility in Africa [40]

The poor water quality causes a high salinity and severe land degradation in old alluvial soils, as shown in Figs. 16 and 17.

10.2 Wadi El-Natrun

Wadi El-Natrun (the Valley of Sodium) depression is located mid-way between Cairo-Alexandria on the western desert, the water table depth is 3–5 m but has a high concentration in sodium carbonates and bi-carbonates. The soils in this area are sandy soils and thus not affected by sodicity and high pH according to its low water holding capacity (6–12%) and low cation exchange capacity with low organic matter contents (sandy soils usually warm soils and that accelerates its processing to organic matter). The poor quality irrigation water causes yield reduction for salinity

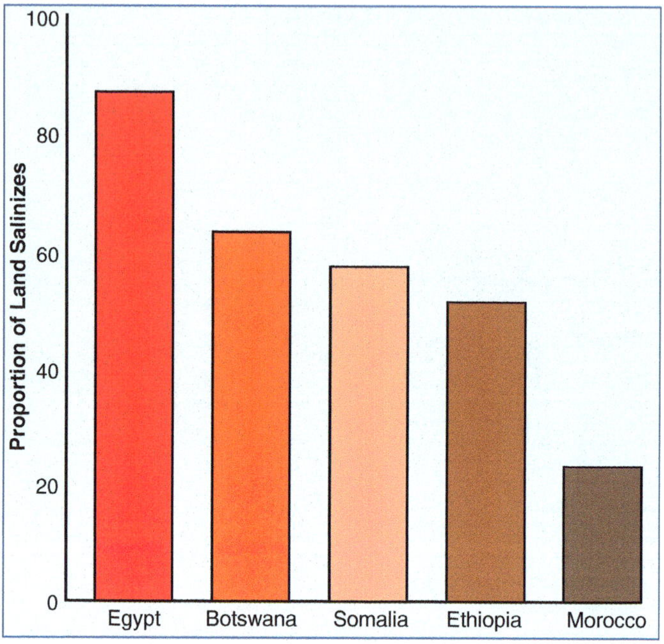

Fig. 17 Salinization in some African countries [41]

and sodicity sensitive crops and needs to use with only tolerant crop such as barley, wheat, olive tree, okra, peas and others (Figs. 18 and 19).

The last situation is completely different than in some other delta fringes such as in Nubaria (west delta) or in Salhia (east delta) where the water table is not shallow but not deep, and ranges between 30 and 60 m with a medium quality of maximum salinity of 2,000 ppm. Most of these areas in Nubaria or Salhia are irrigated with Nile water through El-Nasr canal in Nubaria and the Salhia canal in east delta, but the wells of groundwater work as stand-by or alternative resources when the Nile irrigation water is not enough or in case of delaying on its delivery period (Figs. 20 and 21).

10.3 Siwa Oasis

The Siwa Oasis is a natural depression about 18 m below sea level. It is located in the northern part of the Western Desert of Egypt (about 90 km east of the Libyan border) and 300 km south of the Mediterranean Sea. Groundwater is the only source of drinking and irrigation water in region. Old artesian wells originated from the top

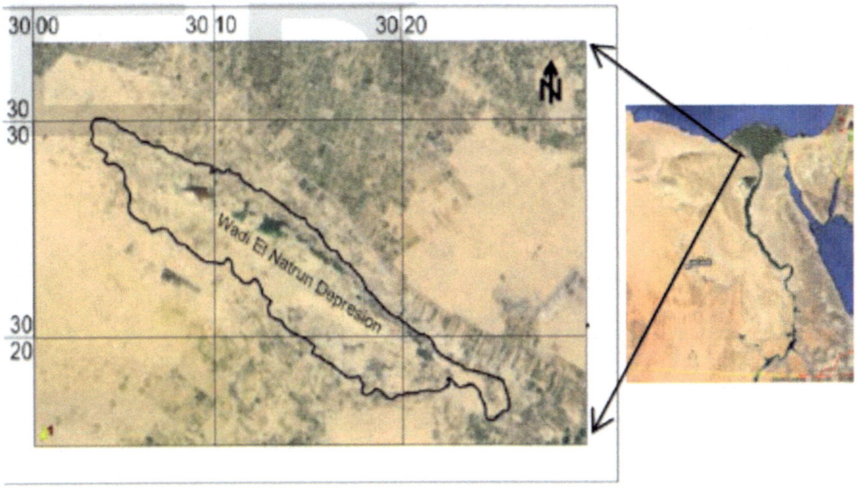

Fig. 18 Wadi El Natrun depression

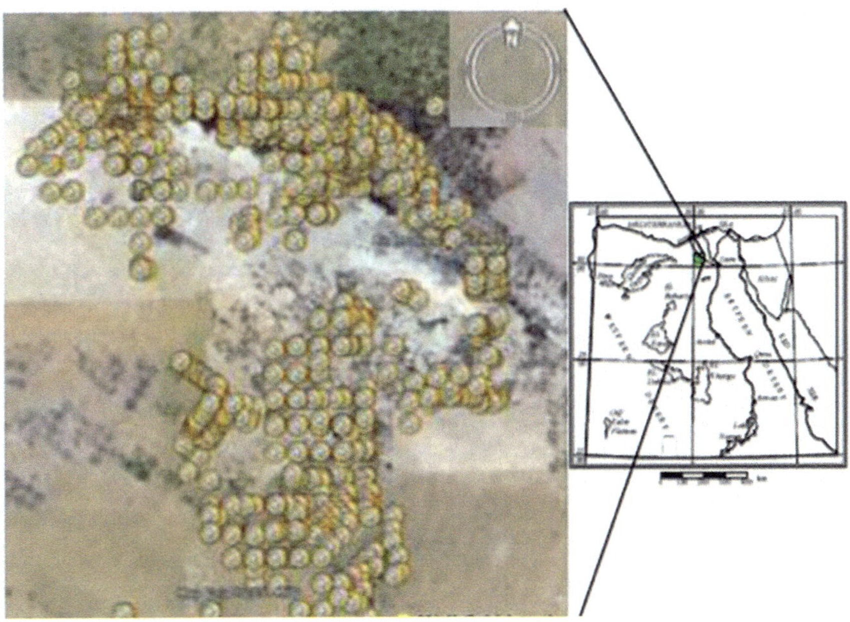

Fig. 19 Haphazard wells in Wadi-El-Faregh includes Wadi El-Natrun that led to depletion of groundwater

shallow aquifer (220 wells) are the traditional source of irrigation water in Siwa Oasis. Under the need for more water source, the upper limestone aquifer has been tapped by hand-drilled boreholes to add new source of irrigation water. For

Fig. 20 Salt and white crust in some delta soils resulting from using shallow groundwater in irrigation (Author 2015)

Fig. 21 Shallow groundwater in Wadi El-Natrun (*right*) and in the delta lands (*left*) [Author 2015]

historical, demographic, and economic reasons as well as its high potential for agricultural development, the Siwa Oasis is considered the most important oasis in the western desert of Egypt. In addition to the existing cultivated area, there are more than 17,000 feddans that were determined to be suitable for agricultural development. The area under cultivation has been gradually increased in recent years as the population of the oasis is on the rise. New land reclamation development projects aimed at the exploitation of water resources have been in progress.

Therefore, there is an urgent need for monitoring water quality in the Siwa Oasis [38, 42]. It is noted that the water quality in the Siwa Oasis has been deteriorating over time, and there is an urgent need for long-term monitoring of water quality of the available water resources in the Siwa Oasis.

A scientific study concluded that the Siwa Oasis is suffering from high salinity, soil as a result of the exchange, which will negatively impact agricultural areas where irrigation water is misused, leading to lower productivity of crops [43, 44].

The Siwa Oasis (1,000 km^2) is today suffering from an excessive rise in the subsoil water levels. The rate of groundwater has risen from 1.33 to 4.6 cm/year. Consequently, the fertile soils are subjected to deterioration and salinization. These events will require new chain water management scenarios that make use of suitable water (multiple reuses) before it reaches the last disposal point to increase the cultivated area (17,000 feddans). The salinity of groundwater representing the area changed from fresh to saline groundwater, 168.8–7,472.8 mg/L and total hardness from soft to hard, 30.7–1,829.7 mg/L as $CaCO_3$ [44].

The groundwater in the Siwa Oasis lies in two different aquifers: Nubian sandstone and fractured dolomite limestone. The fresh groundwater and brackish and saline contained Nubian sandstone and fractured dolomite limestone. Chemical analysis of the groundwater shows the total dissolved solutes changed from fresh to saline 168.8–7,472 mg/L, total hardness ranges from soft to hard, 30.7–1,829.7 mg/L as $CaCO_3$. The effective ions that cause an increase of water salinity are in descending order: As cations; $Na^+ > Mg^{2+} > Ca^{2+}$ while the anions $Cl^- > SO_4^{2-} > HCO^{3-}$ in both brackish and saline groundwater samples and $HCO^{3-} > Cl^- > SO_4^{2-}$ in fresh. Also, the ion ratios indicated that the groundwater has a mixed mineralization that is possibly pure meteoric water affected by leaching and dissolution and cation exchange of both terrestrial and marine salts [45].

Because Siwa Oasis is a depression inside a vast depression of western desert, their main problems are poor drainage with no drainage outlet, and water logging. The second problem is the shallow under pressure groundwater that pops up to the ground. Many scientists call Siwa oasis "the dead oasis," but others see Siwa for its potential for agricultural development.

11 Conclusion and Recommendation

Egypt relies most on Nile water (55.5 BCM/year), then the reuse of agricultural drainage water (10 BCM/year), and finally the groundwater (5.5 BCM/year) as an additional water source for drinking (first) and irrigation (second). There are also the uses of reclaimed and primary treated wastewater (5.5 BCM/year) with little amounts of rain that precipitate on the delta lands which has not exceeded 1.3 BCM/year. Use of groundwater in agriculture activity is effective in the new reclaimed lands in the deserts where the Nile water is not available or access. Most of the groundwater in Egypt is non-renewable water except the shallow groundwater in the Nile valley and delta lands and its fringes in addition to some famous depression and oasis like Wadi El-Natrun in west delta (the Valley of Sodium salts) and Siwa oasis south the northwest coast of Mediterranean.

The share water per capita in Egypt nowadays is 666 m; below the scarcity level of 1,000 m per capita/year. That means the Egypt should have additional renewable freshwater resources to raise the total water resource to 93 BCM/year for the 93 million capita to reach the scarcity level (1,000 m^3/capita.year). The actual situation refers to, the total available water resources is only 62 BCM with a deep gap of water shortage by 31 BCM. Thus, Egypt policy has no choice to use and reuses the marginal and poor quality water. The reuse of both agricultural drainage water and treated sanitary wastewater in addition to recycle of treated industrial wastewater offering about 17 BCM helps to minimize the water shortage to 11 BCM instead of 28 BCM. However, the improvement of water delivery system through the open mud irrigation canal extended for 30,000 km long may add another 19 BCM/year lost by evaporation and seepage from these outdated canals. Finally we can say, the reuse of poor quality water in Egypt is vital, essential and will be continuing until found another renewable water sources to meet the increasing water demands.

References

1. IAH (2015) Food security and groundwater. International Association of Hydrogeologists strategic overview series. www.iah.org
2. World Water Development (2010) Report. www.unesco.org/water/wwap/facts_industry.shtml
3. Davis UC, Water Education Foundation (2010) Abstracts of toward sustainable groundwater in agriculture of the International Conference on Liking Science and Policy, June 15-17. UC Davis and Water Education Foundation, Davis and Sacramento
4. FAO (2010) Statistic on groundwater use for agricultural irrigation, FAO: AQUASTAT – FAO's global information system on water and agriculture. FAO, Rome. http://www.fao.org/nr/aquastat
5. Van der Gun J (2012) Groundwater and global change: trends, opportunities and challenges. WWDR4 side ublication series No. 01. UNESCO, Paris
6. Gleeson T, Wada Y, Bierkens MFP, van Beek LPH (2012) Water balance of global aquifers revealed by groundwater footprint. Nature 488(7410):197–200. doi:10.1038/nature11295
7. Siebert S, Burke J, Faures JM, Frenken K, Hoogeveen J, Doell P, Portman FT (2010) Groundwater use for irrigation – a global inventory. Hydrol Earth Syst Sci 14(10):1863–1880
8. USGS (United States Geological Survey) (2013) Land subsistence. http://ga.water.usgs.gov/edu/earthgwlandsubside.html
9. Ministry of Water Resources and Irrigation, Egypt (2014) Water scarcity in Egypt: the urgent need for regional cooperation among the Nile Basin countries. http://www.mfa.gov.eg/SiteCollectionDocuments/Egypt%20Water%20Resources%20Paper_2014.pdf. Accessed 7 Nov 2016
10. Ministry of Agriculture and Land Reclamation (2016) The reclamation project to reclaim 1.5 million feddan. https://olx.com.eg/ad/130-ID6TZBk.html; https://www.facebook.com/ezrmisr/photos/a.706234316197524.1073741828.466575000163458/706234302864192/?type=1&theater
11. UNESCO (2015) Water for a sustainable world. World Water Development Report 2015. ISBN: 978-92-3-100071-3; ePub ISBN: 978-92-3-100099-7

12. FAO AQUASTAT (2012) Online database. Food and Agriculture Organization of the United Nations (FAO), Rome. http://www.fao.org/nr/water/aquastat/main/index.stm
13. Noureldden N (2016) Notes in water resource and the use of water resources in Egypt for the senior student. Soil Sciences Department, Faculty of Agriculture, Cairo University, Giza
14. Thompson SA (1999) Water use, management, and planning in the United States. Academic, San Diego
15. Mervat D, Grant M (2011) In: Beijer workshop on property rights structures and environmental resource management, Egypt, Mar
16. Cooley H, Gleick P, Wolff G (2006) Desalination, with a grain of salt. Pacific Institute, Oakland
17. Food and Water Watch (2009) Desalination an ocean of problems. www.foodandwaterwatch.org; https://www.foodandwaterwatch.org/sites/default/files/desalination_fs_oct_2009.pdf. Accessed 9 Dec 2016
18. Ministry of Water Resources (2000) Irrigation, planning sector. http://www.mwri.gov.eg
19. Environmental Justice Programme Officer (2014) A Publication by the Egyptian Center for Economic & Social Rights (ecesr.org). Environmental Justice Programme by: Isabel Bottoms – Environmental Justice Programme Officer March water pollution in Egypt. In: Causes and concerns
20. Farid S, Tuinhof A (1999) Groundwater development planning in the desert fringes of the Nile Delta
21. Shata A (1987) Management problems of the major regional aquifer in N.E. Africa. In: UN tech workshop, Khartoum
22. Hefny K (1999) Groundwater assessment in Egypt. MWRI, NWRC, Cairo
23. CEDARE (2001) Regional strategy for the utilisation of the Nubian Sandstone Aquifer System – executive summary. Centre for Environment and Development for the Arab Region and Europe – International Fund for Agricultural Development, Cairo
24. Alker M (2008) The Nubian Sandstone Aquifer System. In: Scheumann W, Herrfahrdt-Pähle E (eds) Conceptualizing cooperation on Africa's transboundary groundwater resources. German Development Institute. http://www.isn.ethz.ch/isn/Current-Affairs/Security-atch/Detail/?q51=Mali&ots591=0c54e3b3-1e9c-be1e-2c24-a6a8c7060233&lng=en&id=103285
25. Shahin M (1987) Groundwater resources in Egypt: potentials and limitations. Water for the future: hydrology in perspective. IAHS publ no. 164
26. Bakhbakhi M (2006) Nubian Sandstone Aquifer System. In: Foster S, Loucks D (eds) Non-renewable groundwater resources – a guidebook on socially-sustainable management for water-policy makers. UNESCO, Paris
27. Ebraheem A, Riad S, Wycisk P, Seif El-Nasr A (2003) Simulation of impact of present and future groundwater extraction from the non-replenished Nubian Sandstone Aquifer in southwest Egypt. Environ Geol 43:188–196
28. Gossel W, Ebraheem A, Wycisk P (2004) A very large scale GIS-based groundwater flow model for the Nubian Sandstone Aquifer in Eastern Sahara (Egypt, northern Sudan and eastern Libya). Hydrogeol J 12(6):698–713
29. Heinl M, Brinkman P (1989) A groundwater model of the Nubian Aquifer System. Hydrol Sci J 34(4):425–447
30. United Nations Environment Programme (UNEP) (2010) Africa Water Atlas. ISBN: 978-92-807-3110-1
31. Shah T, Villholth K, Burke J (2007) Groundwater: a global assessment of scale and significance. Water for food, water for life – a comprehensive assessment of water management in agriculture. IWMI, Colombo, pp 395–423
32. Shah T, Verma S (2008) Co-management of electricity and groundwater: an assessment of Gujurat's Jyoti-Gram Scheme. Indian Econ Pol Weekly 43(7):59–66
33. Garduno H, Foster S (2010) Sustainable groundwater irrigation–approaches to reconciling demand with resources. World Bank/GWP GW-MATE strategic overview series SO-4. World Bank, Washington

34. Groundwater Governance (n.d.) http://www.groundwatergovernance.org/
35. Van der Gun J (2012) Groundwater and global change: trends, opportunities and challenges. WWDR4 side publication series No. 01. UNESCO, Paris
36. Foster SSD, Chilton PJ, Lawrence AR (2000) Processes of diffuse groundwater pollution by agricultural land-use. In: Fu R, Yi Q, Shoemaker CA (eds) Groundwater contamination and its control in China. UNEP-SCOPE, Tsinghua University Press, Beijing, pp 1–11
37. Foster S, Candela L (2008) Diffuse groundwater quality impacts from agricultural land-use: management and policy implications of scientific realities. Groundwater science and policy: an international overview. RSC Publishing, London, pp 454–470
38. FAO (2014) Building a common vision for sustainable food and agriculture: principles and approaches. FAO, Rome. http://www.fao.org/3/a-i3940e.pdf
39. Ministry of Agriculture and Land Reclamation (2000) Reclamation projects up to 2017. http://www.agricultureegypt.com/NewsDetails.aspx?CatID=3d653fad-5feb-4d77-8eb8-15ef0cfa9348&ID=21f3cf52-e187-4b6a-9bc2-d6e8fdda61a6#.WVEUOWiGPIU
40. UNEP (2006) Hydro-political vulnerability and resilience along international waters: Africa. In: United Nations environment program
41. Barr J, Mafuta C (2007) Regional perspectives. Global environment outlook 4. United Nations Environment Programme, Nairobi. Chapter 6
42. Aly A, Benaabidate L (2010) Salinity of water resources in the Siwa Oasis: monitoring and diagnosis. In: Brikle P, Torres Alvaro IS (eds) Water-rock interaction. Taylor & Francis, London. ISBN: 978-0-415-60426-0
43. Anwar A, Abdulrasoul A, Mohamed A, Abdullah A, Mohammed A (2013) Hydrochemical and quality of water resources in Saudi Arabia groundwater: a comparative study of Riyadh and Al-Ahsa regions. Proc Int Acad Ecol Environ Sci 3(1):42–51
44. Hesham ME (2010) Development of low-cost technology for the removal of iron and manganese from ground water in Siwa Oasis. J Egypt Public Health Assoc 55(3):169–188
45. Fathy A, Traugott S (2012) Hydrochemistry of surface water and groundwater from a fractured carbonate aquifer in the Helwan area. Egypt J Earth Syst Sci 121(1):109–124

Groundwater and Agriculture in the Nile Delta

M.A. Mahmoud

Abstract Egypt is located in the arid and semiarid region, where the limited availability of renewable freshwater is the main challenge in future agriculture and urban development. The main water resource in Egypt is the River Nile; Nile water alone is no longer sufficient for the increasing water requirements for the different developmental activities in Egypt due to a rapid increase in population and expected impacts of climate change especially on the agriculture sector. The agriculture sector in Egypt is the main consumption of freshwater; it consumes more than 80% of the total water resources in Egypt. The role of groundwater is steadily increasing especially in the newly reclaimed areas along the desert fringes of the Nile Delta and Valley. Abstraction from groundwater in Egypt is dynamic in nature as it grows rapidly with the expansion of irrigation activities, industrialization and urbanization.

The quality of the groundwater in this area may be strongly affected by the impact of the sea level rise combined with changes of Nile River flows, leading to an increase in the salinity levels of groundwater. In addition, the current and future human activities, especially extensive and unplanned groundwater abstraction, are resulting in deterioration of the available groundwater resources. Serious negative socioeconomic impacts can follow as a consequence. In the Nile Delta, extensive groundwater abstraction is also a very significant factor that increases seawater intrusion. Groundwater wells which were beyond salinization zones in the past are consequently showing upconing of saline or brackish water.

There are many efforts from researchers to control groundwater level on farm via controlled drainage which contributes to water requirements for some crops like rice. On the other hand, shallow groundwater may cause soil salinization, waterlogging and damage to crop roots. Agriculture activity may cause pollution

M.A. Mahmoud (✉)
Water Requirements and Field Irrigation Research Department, Soils, Water and Environment Research Institute, Agricultural Research Center, Giza, Egypt
e-mail: mahmoud_abdalla96@yahoo.com

A. M. Negm (ed.), *Groundwater in the Nile Delta*,
Hdb Env Chem (2019) 73: 141–158, DOI 10.1007/698_2017_94,

of groundwater with fertilizers and pesticides through seepage so integrated management for sustainable use of groundwater is a very important issue in the Nile Delta, so in this chapter the author will provide an overview of the exchangeable relationships between groundwater and agriculture in the Nile Delta region.

Keywords Agriculture, Contamination, Groundwater, Nile Delta, Salinization

Contents

1 Introduction

Globally, the continuous increase of freshwater demand led to more attention for nonconventional water resources as groundwater which play an essential role in water supply. In Egypt, the main constraint of future agriculture is the limited availability of renewable freshwater. Nile River is the main water resource; it is no longer sufficient for the increasing water requirements for the different developmental activities, so the role of groundwater is steadily increasing and is expected to cover about 20% of the total water supply in the upcoming decades especially in the newly reclaimed areas along the desert fringes of the Nile Delta and Valley [1]. There are six groundwater aquifers in Egypt, and the Nile aquifer represents 87% of the total groundwater pumping in Egypt. However, the groundwater aquifer of the Nile Delta is not considered as an additional or separate water resource from the Nile because it is directly connected to its river channels [2]. Aquifers generally are refilled by effective rainfall, lakes and rivers. This water may reach aquifers rapidly via macro-pores or fissures or slowly through soil infiltration and permeable rocks [3]. The direct seepage of Nile water from drainage and irrigation systems and, moreover, irrigated and cultivated lands are the main sources of Nile Delta aquifer recharge [4–6].

There is a hydraulic connection between the groundwater aquifer and the Nile Delta branches especially in the Middle and Southern parts of the Nile Delta due to the thinner and irregular clay layer [7]. The results of SF6 tracer study of groundwater in reclamation areas in South-west of the Nile Delta indicated that the wells that are located nearest to the artificial canals of irrigation (El-Rayah El-Behery and

El-Rayah El-Nasiry) have a young age of between 0.5 and 12 years, which means that the wells close to the surface water are directly recharged by water derived from the river. With increasing the space from the surface water, the age of groundwater increases. This may be clarified by mixing between old water that is present in the aquifers and the infiltrated young water from the surface resources [8]. Drainage practices, soil type and irrigation method have a direct influence on the amount of water recharging the aquifer [9]. The Nile Delta aquifer thickness decreases from more than 900 m at the Mediterranean Sea to about 200 m in the South, near Cairo. The base is a clay aquiclude with an average slope of about 4 m/km [10].

2 Contribution of Groundwater in Agriculture

Groundwater is an essential source of freshwater. It could significantly contribute to agriculture through direct abstraction to use in irrigation mainly or supplemental irrigation. Also, groundwater could be used in irrigation via indirect way through control drainage system, so water table depth will decrease and, thus, contribute to crop water requirement.

2.1 Abstraction of Groundwater to Irrigate Newly Reclaimed Land

In arid and semiarid regions, coastal aquifers represent an essential source of freshwater. In these regions, groundwater resources are overexploited to meet the development and urbanization of coastal regions [6]. In Egypt, to face the challenge of high population density and constant freshwater resource from Nile River, Egypt adopted many policies to establish new agricultural communities outside the overpopulated Nile Delta and Nile Valley. These newly reclaimed areas depend almost exclusively on groundwater as water resource [8]. The growth of irrigation activities, urbanization and industrialization lead to rapid growth of groundwater abstraction. In most of the newly reclaimed areas in the Western Nile Delta region, groundwater is the main source for agriculture, industrial and domestic uses, which affect the quality of groundwater [1, 11].

In Egypt, the continuous increase of groundwater abstraction especially in newly reclaimed lands leads to knowing and studying the groundwater aquifer especially in the Nile Delta which is characterized as shown in Table 1 according to [12], in addition, estimating the safe abstraction yield to protect groundwater deterioration. Using groundwater besides surface water has an essential role during the peak irrigation demand period. The amount of water from the Nile aquifer (Delta and Valley) was estimated in 2010 at 6.2 Bm^3, which is under the safe yield (8.4 Bm^3) estimated by MWRI [13].

Table 1 Characteristics of the groundwater aquifers in the Nile Delta and its fringes [12]

Aquifer	Age	Rock type	Extent	Location	Transmissivity (m^2/day)
Mediterranean coastal aquifer	Pleistocene	Granular	Local	West of Alexandria	500–1,500
Nile Delta aquifer	Pleistocene	Granular	Extensive	Nile Delta	500–5,000
Northwest Sinai aquifer	Pleistocene	Granular	Extensive	Western Sinai	<500
Wadi El-Natrun aquifer	Pliocene	Granular/ clay	Local	Western Delta fringe	<500
Moghra aquifer	Miocene	Granular	Extensive	Northwest Desert	500–5,000
Cairo-Suez aquifer	Plio-Pleistocene/ Miocene/ Oligocene	Granular	Extensive	Cairo-Suez district	<500
Abu Zaabal volcanics	Oligocene	Fissured/ volcanic	Local	South portion	<500
Upper carbonate aquifer	Middle Miocene	Fissured and karstified	Extensive	Almost the entire area	Largely unexplored
Lower carbonate aquifer	Upper Cretaceous	Fissured and karstified	Extensive	Almost the entire area	
Nubian Sandstone aquifer	Cretaceous to Paleozoic	Mixed granular and fissured	Extensive	Almost the entire area	Largely unexplored
Basement	Precambrian	Fissured/ igneous	Extensive	Almost the entire area	Unexplored

2.2 Controlling Groundwater to Contribute Some Crop Water Requirements

A controlled subsurface drainage system allows farmers to adjust groundwater levels in their fields, simply by adjusting the overflow level in a control structure at the end of the lateral line. Conventional drainage is designed solely to remove excess water from the fields. The goal of controlled drainage is to reduce total drainage flow, reduce contaminant load, improve irrigation efficiency or some combination of these outcomes. In controlled drainage system, the groundwater table is maintained at shallower depth by a control structure which reduces deep percolation below the root zone by reducing hydraulic gradients and increases potential capillary upflow through careful water management at depth acceptable for the purpose for plant water [14].

In arid areas, in irrigated agriculture, controlled drainage could be used to improve water management and reduce the environmental impacts of subsurface drainage flow. Controlling the water table position makes shallow groundwater available for crop water use by conserving soil water content through a capillary rise in the root zone as shown in Fig. 1. Shallow groundwater is a resource that is routinely overlooked when water management alternatives are being considered in irrigated agriculture, even though it has the potential to provide significant quantities of water for crop use under the proper conditions and management. In shallow water table areas less than 3 m in depth, some crops can extract water from the groundwater, so irrigation water applied can be reduced significantly, and crop yields can often be enhanced. Under very shallow water table situations 0.5 m in depth, wheat extracted almost all its required water from the groundwater, whereas sunflower extracted more than 80% of its requirement. When the water table is less than 2 m in depth, most crops can obtain a big amount of their water requirements from the groundwater, and therefore, irrigation rates can be decreased below the evapotranspiration rates [15]. Shallow saline groundwater is contributing a part of a crop's water requirement by subirrigation when supplemented by surface irrigation or rainfall with good-quality water [16].

Numerous studies indicated that controlled drainage significantly contributes to reducing drainage volumes from agricultural land and enhancing water use efficiency [17]. The drainage volume reduction under controlled drainage was 11% [18]. Also, the volume reduction ranges from 17 to 85% [19]. Water use efficiency

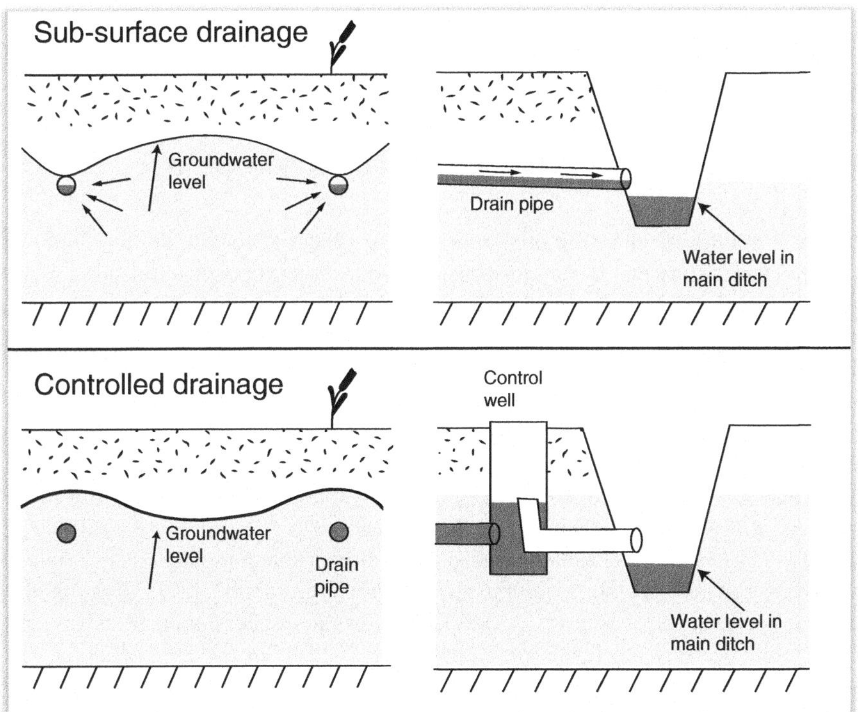

Fig. 1 Conceptual flow paths in controlled and uncontrolled subsurface drainage systems

on controlled drained plots with subirrigation was greater compared to controlled drained plots without subirrigation over a period of 6 years [20]. Application of controlled drainage has the potential to maintain and even increase yields per unit land whilst increasing the irrigation water use efficiency by 15–20% [21]. Water management techniques may be used to reduce drainage outflow during the growing season of rice. The use of controlled drainage and other water management practices plays an important role in reducing the amounts of irrigation water. Irrigation water requirements can be reduced to 80% of the total crop ET without reducing crop yield and increasing soil salinization in areas where water tables are shallow [22]. Introduced Modified drainage for rice areas in the Nile Delta by closing the main collector line during the rice season lead to water savings of up to 50% depending on local conditions. Thus, Farmers saved considerable irrigation time, and direct pumping costs were reduced by as much as 43% of total seasonal pumping costs [23]. Water requirements of rice under controlled drainage was 25% less than under conventional drainage system in Northern Western Nile Delta of Egypt [17].

Controlled drainage increases yields per unit land and increases water use efficiency by 15–20% [21]. Using controlled drainage systems (and combined subirrigation and controlled drainage systems) has positive results on crop yields of soybean and corn cropping systems in Canada and the United States [24].

3 Problems of Shallow Groundwater on Agriculture

3.1 Secondary Salinization of the Soils in Nile Delta

"Salinity management in the crop root zone is essential for the sustainability of irrigated agriculture and is a major consideration when proposing controlled drainage practices" [25]. Research has demonstrated that water and salt will move upwards from shallow groundwater and may result in the salinized soil in a short time. However, there have also been many studies over the years demonstrating that salinity in the root zone can be effectively managed from year to year by irrigation. Most investigators believed that shallow groundwater plays a major role in the soil salinization [26]. In shallow water table areas, due to applying controlled drainage, water and salt will move upwards from shallow groundwater and may result in the salinized soil in a short time, and it is highly dependent on water table position and salinity [25]. "High water tables are also a common problem in intensively irrigated areas and can lead to the development of salinity problems" [27]. "Results also show a positive correlation ($r^2 = 0.84$) between the concentration of salt on the surface and water table levels of 70 cm or less" [27]. In the area of study, salt accumulation, groundwater and inadequate drainage conditions are the major causes of salinization [27].

Excessive irrigation and lack of adequate drainage cause rises in the ground table and consequently irrigation salinity [28–30]. The depth of groundwater, a distance of irrigation canals, subsoil and groundwater salinity and land use are the mean reasons of salinity risk [31].

Salinity levels in the soil and groundwater are increasing due to continuous irrigated crop production. Over time, crop production will be negatively affected due to shallow saline water table which causes salt accumulation in the soil surface through the capillary rise and/or directly as a result of waterlogging [32]. The salinity in wetlands of arid/semiarid areas will vary naturally due to sporadic rainfall, high evaporative conditions, groundwater inflows and freshening after floods or rains. However, wetlands are often at particular risk of secondary salinization which may be due to lower elevation in the landscape that causes an increase in saline groundwater inflows by rising water tables [33].

In areas that has shallow water table and suitable soil properties of water movement through the surface layers, water moves from groundwater to the soil surface bringing dissolved salts in it. Evapotranspiration is the driving force for upward movement of water and salts [34]. There are mechanisms which explain the soil processes leading to transient salinity in root zone layers as shown in Fig. 2.

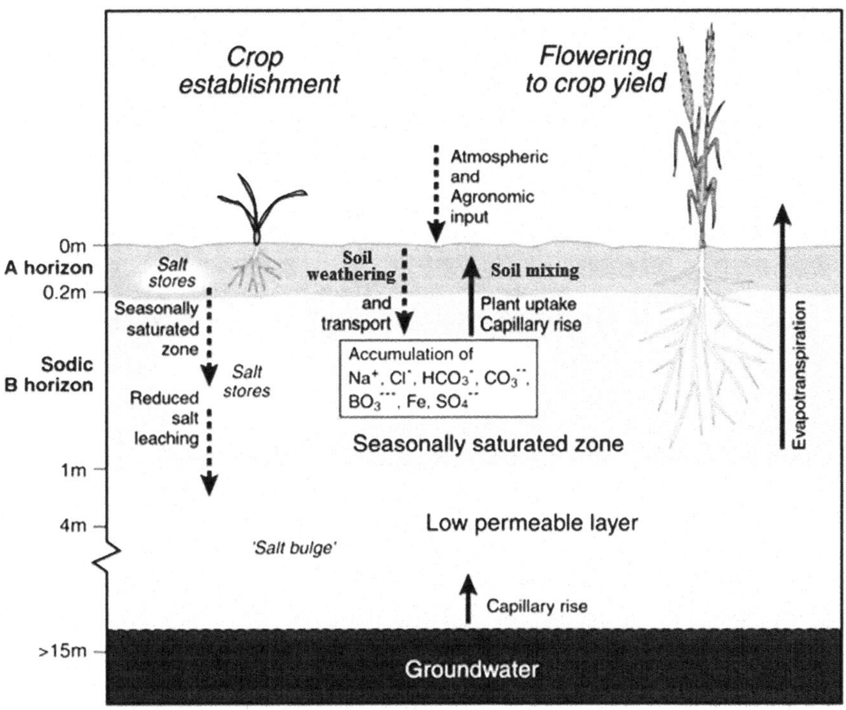

Fig. 2 Soil processes lead to transient salinity accumulation in the root zone layers [35] (Reproduced by kind permission of CSIRO Publishing)

3.2 Waterlogging and Its Solution

Irrigation in arid and semiarid zones inevitably leads to water table variations and, thus, to salinization and waterlogging problems. About one-third of the world's irrigated lands have decreased productivity as a consequence of poorly managed irrigation that has caused salinization and waterlogging [36]. In the irrigated lands such as Delta and Nile Valley, Egypt, groundwater levels have risen to produce waterlogging. This process has caused excessive salinity build-up in crop root zones and created yield reductions or caused land abandonment in severe cases [37]. Shallow groundwater table may have negative effects on crops; if water table is too shallow, crop yield could decrease due to waterlogging and root anoxia [38, 39]. "When the water table is very shallow, soil waterlogging limited the root growth of winter wheat due to the reduced oxygen concentration of the soil" [40].

Over the past 25 years, the role of drainage has improved from only purpose measure for controlling salinity and/or waterlogging to an important element of integrated water management under multiple land use [41]. In arid and humid zones, subsurface drainage is used to prevent waterlogging, increase the trafficability of soil and enhance aeration for better crop growth, thus allowing soil preparation for planting and harvest on time. In addition, in arid areas drainage contributes significantly to leaching capability to control salinity build-up in crop root zone [25]. Subsurface drainage systems have proved to be a technically feasible and cost-effective practice to combat the problems of root zone salinity and waterlogging in irrigated land and thus increase crop yields and rural income [42–47]. Drainage is one of the most effective solutions to reduce waterlogging and salinity problems [48].

4 Salinization of Groundwater and Seawater Intrusion

The coastal aquifers are essential sources of freshwater in coastal zones, but salinity intrusion can be the main problem in these areas. Saltwater intrusion is the process that saltwater moves from the ocean into coastal aquifers as a result of the over-pumping of groundwater, and it is a dynamic equilibrium of groundwater movement and subsequently may cause polluting of the groundwater [3, 49]. Saltwater intrusion is a natural phenomenon that occurs almost in all coastal aquifers which are connected to the sea as a result of the higher density of the seawater compared to the groundwater in the aquifer [50]. Due to the extraction of groundwater from aquifers which are in hydraulic connection with the sea, the gradients that are set up may cause a flow of seawater from the sea towards the well [51]. Increasing water abstraction from coastal aquifers lead to decreasing the movement of freshwater to the sea and accordingly, increasing seawater intrusion inland [6, 52]. The intrusion of seawater in coastal aquifers is the main problem and is encountered, with different degrees, in almost all coastal aquifers. It is considered one of the main factors that decrease water quality through raising salinity. It may happen due to human activities as overpumping and/or by natural events like sea level rise or decreasing recharge from surface water which quickens saltwater intrusion [3, 10, 52]. Saltwater intrusion has a direct influence on the salinity of the soil, resources of groundwater, coastal zone quality and agricultural productivity [53].

In Egypt, Nile Delta aquifer is one of these aquifers which are subjected to face intrusion of seawater from the Mediterranean Sea [52]. The sea level rise and fluctuation of Nile River flow may have a significant impact on groundwater quality in the Nile Delta, which lead to an increase in the salinity levels of groundwater [54]. Moreover, the current and future human activities, mainly unplanned and extensive groundwater abstraction, are resulting in degradation of the available groundwater resources [55]. Previous studies indicated that seawater intrusion in the Nile Delta aquifer had extended inland more than 100 km from the Mediterranean coast [56, 57]. Sea level rising might have significant long-term impacts on the Nile Delta, which include the distribution of groundwater salinity and erosion of the narrow and low-lying barriers of the Manzala and Burullus lagoons [58]. Rising the Mediterranean Sea level by 0.5 m, it will cause additional intrusion of 9.0 km in the Nile Delta aquifer [3].

Groundwater salinity in the Nile Delta is classified into two zones; the first one in the Southern Nile Delta that is a low salinity level (total dissolved solids $<$1,000 ppm) and the second zone in the central Nile Delta, in which the salinity increases sharply to a total dissolved solids $>$10,000 ppm [4]. The projected seawater rise would cause a reduction in a hydraulic gradient of the water table and/or piezometric head in the Nile Delta aquifer. So, under climate change condition and seawater rise, the deep coastal aquifers with mild hydraulic gradients would be more vulnerable. In Nile Delta aquifer, increasing Mediterranean Sea level by 0.5 m would cause movement of the equi-concentration lines 35, 5, and 1 g/L inland by distances of 1.5, 4.5 and 9 km, respectively [59]. Rising seawater as a result of climate change will have many impacts on groundwater. First, additional pressure heads will be forced at the seaside causing more seawater intrusion in land. Second, seawater will submerge the low lands along the shoreline, and the groundwater below these lands will become saline. Third, climate change may cause differences in rainfall which would affect the natural replenishment of groundwater. Fourth, as a result of the expected reduction in surface water resources and rainfall particularly in arid and semiarid regions, the dependence on, and exploration of, groundwater resources would increase to contribute for water needs of the different sectors [10, 60].

The overextraction of groundwater leads to an increase in salinity concentration and covers about large area [61]. The most extremely serious result indicated that overpumping of the groundwater especially in the Nile Delta aquifers, which is followed by salinization by seawater intrusion, thus decreases the obtainability of freshwater sources [62]. The combination of overpumping and sea level rise results in the movement of the transition zone further towards land. Rising the Mediterranean Sea water level by 100 cm will cause extra intrusion in the Nile Delta aquifer by 10 km. Overpumping will cause an extra intrusion by 8 km and reduction in piezometric head water at the land side. The combination of sea level rise and overpumping will cause an extra intrusion of 15 km [6].

5 Contamination of Groundwater

Pollution can be defined as the modification of chemical, physical and biological properties of water preventing or restricting its use in the various uses. Water is polluted naturally due to saltwater intrusion and others without human intervention or artificially by human activities [63]. There are many groundwater pollution sources which include oil and mining fields (petroleum exploration and development, abandoned oil wells and test wells, buried pipelines and storage tanks, disposal of oil field brine and mining activities), water wells (disposal, drainage and abandoned wells, overpumping, well construction, river infiltration and seawater encroachment), cultivated lands (animal wastes, dryland farming and evapotranspiration from vegetation, feedlots, fertilizers and pesticides, irrigation return flow and sewage treatment plant discharge), urban areas (urban and industrial landfills, surface impoundments, solid wastes, natural pollution and septic tanks and cesspools) and others (lakes and spills, natural leaching, water from fault zones and volcanic origin) [64]. Contamination of the Nile Delta aquifers has resulted from saltwater intrusion, the absence of sewage systems in most of villages and towns in Delta and oxidation and dissolution of applied pesticides, in addition, fertilizers into infiltrating irrigation water. These contaminations are characterized by high concentrations of total dissolved solids [65]. The degradation of groundwater quality in the Eastern part of the Nile Delta could be related to the leakage from sewage storage tanks, transportation sewage pipes and El Khadrawya drain which receives the Mobarak industrial area wastewater. Moreover, dump disposal sites and factories' illegal sewage disposal [66]. Land use and related human activities have significant correlation on groundwater quality, type and extent of pollution. Use and reuse of water for agricultural, domestic and industrial purposes result in the discharge of liquid or solid wastes into the geologic environment [63].

Shallow aquifers are more sensitive to degradation due to temporal and spatial changes in recharge and discharge. Extensive urban, industrial and agricultural expansions on the western fringe of the Nile Delta have applied much load on the water needs and lead to groundwater quality deterioration. These deteriorations may be due to salinization and contamination from nutrients and trace metals [67]. When the water table is shallow and the aquifers have low buffering capacity and are highly permeable, the risk of pollution increased [68]. There is hydraulic connection between groundwater in the Nile Delta aquifer and surface water from Nile River branches, drainage system and irrigation canals. Therefore there are many factors contributing to the decline of groundwater quality in the Nile Delta aquifer including leaching of fertilizers, pesticides and chemicals from agriculture areas, saline groundwater upconing and seawater intrusion, disposal of industrial waste and overpumping of groundwater [9].

The main pollutant of groundwater is saline water. Saline water intrusion takes place where saline water relocates or mixes with freshwater. The groundwater in most areas close to the sea in the western fringe of the Nile Delta, as North-Tahrir, Abis and Marriott, was unsuitable to irrigate regular crops. Also, West of the Cairo-Alexandria road except for the track of land between Nubaria Canal and this road has access to reasonably good groundwater [67]. The increase of salinity in the Middle part of the Nile Delta is caused by vertical seepage of sanitary, agricultural wastewater and the dissolving of salts from the sediments of the aquifer itself [69]. The excessive extraction of groundwater may affect groundwater quality by upward seepage of saline water [65]. Mobilization of stored salts in the unsaturated zone, local pollution and evaporites, which resulted from ancient marine intrusion, wind-driven sea spry and marine aerosols deposited at the topsoil, may further contribute to groundwater salinization [70, 71]. The salinization resulted from seawater mixing, dissolution of evaporite minerals in the water-bearing formation and the sabkhas and saline soils [67].

Agriculture, industrial and domestic discharges are the main sources of pollution of the Nile Delta aquifer. The overuse of chemical fertilizers in agriculture causes higher concentrations of nitrate, potassium, sulphate and phosphate. Also, higher concentrations of iron, manganese and aluminium were found in the old lands of the Nile Delta aquifer due to the dumping of industrial effluents into the drainage system. In addition, due to the lack of sewage systems in most rural areas and popularity of septic tanks in the Nile Delta [9]. Contamination with nutrients and trace metals is usually from agricultural, urban and industrial wastes [67]. Manganese and iron concentrations were higher in the old lands due to the general natural characteristics of the Nile Delta aquifer and more pronounced in the areas having a clay cap. Local contamination from industrial areas led to high concentrations of lead in Eastern Nile Delta and Western Nile Delta. Also, high concentrations of cadmium were found in the Middle of the Nile Delta [72]. The agricultural activities as the excessive use of pesticides and fertilizers particularly in the old land and using wastewater in irrigation which cause degradation of the groundwater quality are mainly attributing to enrich groundwater with nitrate, pesticides, phosphate and bacteria. In addition, domestic and industrial activities increase heavy metals, industrial pollutant and bacteria in groundwater [63, 73]. The maximum concentration of nutrients was mainly recorded in the old cultivated lands, indicating the contamination from irrigation water [11]. The shallow brackish coastal aquifer in northern Nile Delta is partially polluted with phosphates (0.24 mg/L) and nitrates (7.35 mg/L). Furthermore, groundwater contains excess amount of some trace elements (Fe, Mn, Ni) that are higher than the standard levels for drinking water [62]. Herbicides which were used to control submerged weeds in irrigation canals and water hyacinths in drains have caused serious environmental hazards [74].

6 Integrated Management for Sustainable Use of Groundwater

Maintaining groundwater quality and quantity for sustainable use requires an integrated management approach for both groundwater and surface water. For sustainable use of groundwater in areas where groundwater is renewable, the abstraction rate should not exceed the recharge rate. If this is not the case, the unplanned exploitation will lead to short term of use [75]. Many wells were dug on the Nubian aquifer in the Oases area in the Western Desert were stopped to produce naturally due to overextraction [64, 76]. Increasing recharge of groundwater by 10% and decreasing pumping by 10% are predicted to cause a 3 and 6 m increase in groundwater head in the heavily exploited coastal aquifer [77]. Under the condition of 0.5 m seawater level rise, it would sustain the freshwater resources when reducing the groundwater pumping by 50% [10].

The Nile Delta aquifer is strongly exposed to seawater intrusion from the Mediterranean Sea producing serious environmental impacts. Groundwater resources should thus be developed and managed carefully to avoid any further water quality deterioration [6]. In Western Nile Delta, the integrated and sustainable management plans for groundwater are essential to avoid the deterioration of the aquifer system in this area. The inclusive database that includes the characteristics of the aquifer system and modelling tools to achieve the impacts of decision alternatives are very important factors for efficient integrated and sustainable management of water resources [1].

Sustainability of groundwater is a very important issue, and it is not possible without water managers, monitoring and characterizing groundwater resources; local communities, management and hydrogeologists should work together to devise measure and policies by backcasting and, moreover, adapt future measures in achieving the long-term sustainable targets [78]. Decreasing, controlling or forbidding well drilling, treatment of sewage and drainage water before using in irrigation, integrated management of applying fertilizers in agriculture areas to prevent groundwater pollution and monitoring of surface and groundwater are very important actions to improve groundwater quality, avoid saline groundwater supply and prevent groundwater pollution of the Eastern Nile Delta aquifer [73].

7 Conclusions and Recommendations

Groundwater is a very important source of freshwater in Egypt besides River Nile. It is expected to cover about 20% of the total water supply in the upcoming decades especially in the newly reclaimed areas along the desert fringes of the Nile Delta and Valley. In Egypt, there are six groundwater aquifers, and the Nile aquifer represents 87% of the total groundwater pumping in Egypt. The direct seepage of Nile water from irrigation and drainage systems and the irrigated and cultivated lands are the main source of Nile Delta aquifer recharge. This is due to a hydraulic connection between the groundwater aquifer and the Nile Delta branches especially in the Middle and Southern parts of the Nile Delta due to the thinner and irregular clay layer.

Groundwater could significantly contribute to agriculture through direct abstraction to use in irrigation mainly or supplemental irrigation in rainfed areas. Also, it could be used in irrigation in an indirect way through controlling drainage system, so water table will decrease and, thus, contribute to crop water requirement. But, there are some problems with shallow water table which negatively affect agriculture production as secondary soil salinization and waterlogging. These two problems could be overcome through using surface and subsurface drainage systems.

Nile Delta aquifer is exposed to many challenges which decrease the quantity and/or deteriorate its quality as seawater intrusion. It may happen due to expected rising of seawater level due to projected climate change and/or the current and future human activities, mainly unplanned and extensive groundwater abstraction which lead to an increase in the salinity levels of groundwater, and also the transport of pollutant from surface sources to groundwater. So, to maintain groundwater resources in Nile Delta for sustainable use, an integrated management plan should be prepared. The plan should include preparing digital maps characteristic for groundwater distribution and depth and define the safe extraction. Also, controlling or forbidding drilling wells especially in North Nile Delta close to the Mediterranean Sea, increasing ground recharge and/or reducing ground abstraction, treatment of sewage and drainage water before using in irrigation, integrated management of applying fertilizers in agriculture areas to prevent groundwater pollution and applying controlled drainage in agricultural areas in Nile Delta because it reduces the transport of agricultural pollutant as pesticides, herbicides, nutrients and some of heavy metals in addition enhancing water use efficiency.

Finally, monitoring of surface and groundwater is a very important action to improve groundwater quality, avoid saline groundwater supply and prevent groundwater pollution in Nile Delta region.

References

1. Dawoud MA, Darwish MM, El-kady MM (2005) GIS-based groundwater management model for western Nile Delta. Water Resour Manag 19:585–604. https://doi.org/10.1007/s11269-005-5603-z
2. MWRI (2005) National water resources plan for Egypt 2017. Ministry of Water Resources and Irrigation, Giza
3. Kumar CP (2012) Climate change and its impact on groundwater resources. Int J Eng Sci 1:43–60
4. Geirnaert W, Laeven MP (1992) Composition and history of groundwater in the western Nile Delta. J Hydrol 138:169–189
5. Sefelnasr A, Sherif M (2013) Impacts of seawater intrusion in the Nile Delta aquifer, Egypt. Natl Groundwater Assoc 52:264–276

6. Abd-Elhamid H, Javadi A, Abdelaty I, Sherif M (2016) Simulation of seawater intrusion in the Nile Delta aquifer under the conditions of climate change. Hydrol Res 47:1198–1210. https://doi.org/10.2166/nh.2016.157
7. Kashef AI (1983) Harmonizing Ghyben-Herzberg interface with rigorous solutions. Groundwater 21:153–159
8. El-gamal H, Dahab K, Aeschbach-hertig W (2004) A multi-tracer study of groundwater in reclamation areas south-west of the Nile Delta, Egypt. In: International workshop on the application of isotope techniques in hydrological and environmental studies UNESCO, Paris, France, 6–8 Sept 2004
9. Al-Agha DE, Closas A, Molle F (2015) Survey of groundwater use in the central part of the Nile Delta. In: Water and salt management in the Nile Delta: Report No. 6
10. Sefelnasr A, Sherif M (2014) Impacts of seawater rise on seawater intrusion in the Nile Delta aquifer, Egypt. Groundwater 52:264–276. https://doi.org/10.1111/gwat.12058
11. Sharaky A, Atta SA, El Hassanein AS, Khallaf KMA (2007) Hydrogeochemistry of groundwater in the Western Nile Delta Aquifers, Egypt. In: 2nd international conference on the geology of the Tethys, Cairo University, 19–21 Mar 2007, pp 19–21
12. RIGW/IWACO (1988/1993) Hydrogeological map of Egypt, scale 1:2,000,000. Research Institute for Groundwater, Cairo
13. MWRI (2012) Strategy of water resources of Egypt till 2050. Ministry of Water Resources and Irrigation, Giza
14. Christen EW, Ayars JE (2001) Subsurface drainage system design and management in irrigated agriculture: best management practices for reducing drainage volume and salt load. CSIRO Land and Water, Clayton South
15. Kahlown MA, Ashraf M, Zia-ul-Haq (2005) Effect of shallow groundwater table on crop water requirements and crop yields. Agric Water Manag 76:24–35. https://doi.org/10.1016/j.agwat.2005.01.005
16. Rose DA, Ghamarnia HM, Gowing JW (2010) Development and performance of wheat roots above shallow saline groundwater. Aust J Soil Res 48:659–667
17. Khalil BM, Abdel-Gawad T, Millette JA (2004) Impact of controlled drainage on rice production, irrigation water requirement and soil salinity in Egypt. In: Drainage VIII proceedings of the eighth international symposium, Sacramento, CA, USA, 21–24 Mar 2004, p 1
18. Ng HYF, Tan CS, Drury CF, Gaynor JD (2002) Controlled drainage and subirrigation influences tile nitrate loss and corn yields in a sandy loam soil in Southwestern Ontario. Agric Ecosyst Environ 90:81–88
19. Skaggs RW, Youssef MA (2008) Effect of drainage water management on water conservation and nitrogen losses to surface waters. In: 16th national nonpoint source monitoring workshop, Columbus, OH, pp 14–18
20. Bonaiti G, Borin M (2010) Efficiency of controlled drainage and subirrigation in reducing nitrogen losses from agricultural fields. Agric Water Manag 98:343–352
21. Wahba MS, El-Ganainy MA, Amer MH (2008) Water table management strategies for irrigation water saving. In: Twelfth international water technology conference, IWTC12 2008, Alexandria, Egypt
22. Prathapar SA, Qureshi AS (1999) Modelling the effects of deficit irrigation on soil salinity, depth to water table and transpiration in semi-arid zones with monsoonal rains. Int J Water Resour Dev 15:141–159
23. DRI (2001) Drainage criteria study at Mashtul pilot areas part 3: fluctuation in the groundwater table. Technical Report No. 59, pilot area and drainage technology project. Advisory Panel on Land drainage in Egypt, Giza
24. Ballantine DJ, Tanner CC (2013) Controlled drainage systems to reduce contaminant losses and optimize productivity from New Zealand pastoral systems. N Z J Agric Res 56:171–185. https://doi.org/10.1080/00288233.2013.781509

25. Ayars JE, Christen EW, Hornbuckle JW (2006) Controlled drainage for improved water management in arid regions irrigated agriculture. Agric Water Manag 86:128–139. https://doi.org/10.1016/j.agwat.2006.07.004
26. Cary L, Trolard F (2006) Effects of irrigation on geochemical processes in a paddy soil and in groundwaters in Camargue (France). J Geochem Explor 88:177–180
27. Mohamed ES, Morgun EG, Bothina SMG (2011) Assessment of soil salinity in the Eastern Nile Delta (Egypt) using geoinformation techniques. Moscow Univ Soil Sci Bull 66:11–14. https://doi.org/10.3103/S0147687411010030
28. Houk E, Frasier M, Schuck E (2006) The agricultural impacts of irrigation induced waterlogging and soil salinity in the Arkansas Basin. Agric Water Manag 85:175–183
29. Corwin DL, Rhoades JD, ŠimYuunek J (2007) Leaching requirement for soil salinity control: steady-state versus transient models. Agric Water Manag 90:165–180
30. Herrero J, Pérez-Coveta O (2005) Soil salinity changes over 24 years in a Mediterranean irrigated district. Geoderma 125:287–308
31. Zhou D, Lin Z, Liu L, Zimmermann D (2013) Assessing secondary soil salinization risk based on the PSR sustainability framework. J Environ Manag 128:642–654. https://doi.org/10.1016/j.jenvman.2013.06.025
32. Houk E, Frasier M, Schuck E (2006) The agricultural impacts of irrigation induced waterlogging and soil salinity in the Arkansas Basin. Agric Cult Water Manag 85:175–183. https://doi.org/10.1016/j.agwat.2006.04.007
33. Jolly ID, Mcewan KL, Holland KL (2008) A review of groundwater – surface water interactions in arid/semi-arid wetlands and the consequences of salinity for wetland ecology. Ecohydrology 1:43–58. https://doi.org/10.1002/eco
34. Rengasamy P (2006) World salinization with emphasis on Australia. J Exp Bot 57:1017–1023. https://doi.org/10.1093/jxb/erj108
35. Rengasamy P (2002) Transient salinity and subsoil constraints to dryland farming in Australian sodic soils: an overview. Aust J Exp Agric 42:351–361
36. Fernández-Cirelli A, Arumí JL, Rivera D, Boochs PW (2009) Environmental effects of irrigation in arid and semi-arid regions. Chilean J Agric Res 69:27–40
37. De Wrachien D, Feddes R (2003) Drainage development in a changing environment: overview and challenges. In: 9th international drainage workshop – drainage for a secure environment and food supply, pp 1–17
38. Ibrahim SM (1999) Wheat cultivation under limited irrigation and high water table conditions. Egypt J Soil Sci 39:361–372
39. Nosetto MD, Jobbágy EG, Jackson RB, Sznaider GA (2009) Reciprocal influence of crops and shallow groundwater in sandy landscapes of the Inland Pampas. Field Crop Res 113:138–148
40. Brisson N, Rebiere B, Zimmer D, Renault P (2002) Response of the root system of a winter wheat crop to waterlogging. Plant Soil 243:43–55
41. Schultz B, Zimmer D, Vlotman WF (2007) Drainage under increasing and changing requirements. Irrig Drain 56:S1
42. Bhutta MN, van der Sluis TA, Wolters W (1995) Review of pipe drainage projects in Pakistan. In: Proceedings of the national workshop on drainage systems performance in plain and future strategies, pp 10–18
43. Ali AM, Van Leeuwen HM, Koopmans RK (2001) Benefits of draining agricultural land in Egypt: results of five years' monitoring of drainage effects and impacts. Int J Water Resour Dev 17:633–646
44. Nijland H, Croon FW, Ritzema HP (2005) Subsurface drainage practices: guidelines for the implementation, operation and maintenance of subsurface pipe drainage systems. ILRI, Wageningen
45. Ritzema HP, Satyanarayana TV, Raman S, Boonstra J (2008) Subsurface drainage to combat waterlogging and salinity in irrigated lands in India: lessons learned in farmers' fields. Agric Water Manag 95:179–189

46. Ritzema H (2009) Drain for gain – making water management worth its salt. Subsurface drainage practices in irrigated agriculture in semi-arid and arid regions. PhD thesis, Wageningen University and UNESCO-IHE, Delft
47. Ritzema H, Schultz B (2011) Optimizing subsurface drainage practices in irrigated agriculture in the semi-arid and arid regions: experiences from Egypt, India and Pakistan. Irrig Drain 60:360–369
48. Valipour M (2014) Drainage, waterlogging, and salinity. Arch Agron Soil Sci 60:1625–1640. https://doi.org/10.1080/03650340.2014.905676
49. Fang H-Y, Daniels J (1997) Introduction to environmental geotechnology. CRC press, Boca Raton
50. Ciaran H (2002) The effect of basement on heterogeneity on saltwater wedge a physical and numerical modelling approach. The University of Western Australasia, Crawley. Text Book
51. Freeze RA, Cherry JA (1979) Groundwater. Prentice Hall, Englewood Cliffs, pp 375–379
52. Abdelaty IM, Abd-Elhamid HF, Fahmy MR, Abdelaal GM (2014) Investigation of some potential parameters and its impacts on saltwater intrusion in Nile Delta aquifer. J Eng Sci Assiut Univ Fac Eng 42:931–955
53. El Raey M (2009) Vulnerability assessment of the coastal zone of the Nile Delta, Egypt, to the impacts of sea level rise. Ocean Coast Manag 37:29–40
54. Dawoud MA (2004) Design of national groundwater quality monitoring network in Egypt. Environ Monit Assess 96:99–118
55. Mabrouk MB, Jonoski A, Solomatine D, Uhlenbrook S (2013) A review of seawater intrusion in solid the Nile Delta groundwater system – the basis for assessing impacts due to climate changes and water resources development. Hydrol Earth Syst Sci Discuss 10:10873–10911. https://doi.org/10.5194/hessd-10-10873-2013
56. Faried MS (1985) Management of groundwater system in the Nile Delta. PhD thesis, Faculty of Engineering, Cairo University, Egypt
57. Sherif M, Sefelnasr A, Javadi A (2012) Incorporating the concept of equivalent freshwater head in successive horizontal simulations of seawater intrusion in the Nile Delta aquifer, Egypt. J Hydrol 464:186–198
58. Frihy OE (2003) The Nile Delta-Alexandria: vulnerability to sea-level rise, consequences and adaptation. Mitig Adapt Strateg Glob Chang 8:115–138
59. Sherif MM, Singh VP (1999) E ect of climate change on sea water intrusion in coastal aquifers. Hydrol Process 13:1277–1287
60. Sherif M (1999) Nile delta aquifer in Egypt. Seawater intrusion coast. Aquifers? Concepts, methods and practices. Kluwer, Dordrecht, pp 559–590
61. Wassef R, Schüttrumpf H (2016) Groundwater for sustainable development impact of sea-level rise on groundwater salinity at the development area western delta, Egypt. Groundwater Sustain Dev 2–3:85–103. https://doi.org/10.1016/j.gsd.2016.06.001
62. Atta SA, Sharaky AM, EL Hassanein AS, Khallaf KMA (2007) Salinization of the groundwater in the coastal shallow aquifer, Northwestern Nile Delta, Egypt. ISESCO Sci Technol Vis 3:112–123
63. Taha AA (2004) Pollution sources and related environmental impacts in the new communities Southeast Nile Delta, Egypt. Hydrol Earth Syst Sci 9:35–49
64. El Tahlawi MR, Farrag AA, Ahmed SS (2008) Groundwater of Egypt: an environmental overview. Environ Geol 55:639–652. https://doi.org/10.1007/s00254-007-1014-1
65. Ebraheem A-AM, Senosy MM, Dahab KA (1997) Geoelectrical and hydrogeochemical studies for delineating ground-water contamination due to salt-water intrusion in Northern part of the Nile Delta, Egypt. Groundwater 35:216–222
66. Hussien MM (2007) Environmental impacts of new settlements on the groundwater in a region in Delta. MSc thesis, Zagazig University, Faculty of Engineering, Egypt
67. Masoud AA (2014) Groundwater quality assessment of the shallow aquifers west of the Nile Delta (Egypt) using multivariate statistical and geostatistical techniques. J African Earth Sci 95:123–137. https://doi.org/10.1016/j.jafrearsci.2014.03.006

68. El Alfy M, Merkel B (2004) Assessment of human impact on quaternary aquifers of Rafah area, NE Sinai, Egypt. Int J Econ Environ Geol 1:1–9
69. Dahab KA (2003) Influence of hydrogeologic flow pattern and aquifer material on water quality and mineral contents: case study, Nile Delta Egypt. Sci J Fac Sci Menoufia Univ 17:65–86
70. Cruz JV, Silva MO (2000) Groundwater salinization in Pico Island (Azores, Portugal): origin and mechanisms. Environ Geol 39:1181–1189
71. Cartwright I, Weaver TR, Fulton S, Nichol C, Reid M, Cheng X (2004) Hydrogeochemical and isotopic constraints on the origins of dryland salinity, Murray Basin, Victoria, Australia. Appl Geochem 19:1233–1254
72. Morsy WS (2009) Environmental management to groundwater resources for Nile Delta region. PhD thesis, Faculty of Engineering, Cairo University, Egypt
73. Abo-El-Fadl M (2013) Possibilities of groundwater pollution in some areas, East of Nile Delta, Egypt. Int J Environ 1:1–21
74. Grwp (2004) Parys Underground Group. http://www.parysmountain.co.uk/
75. Nasr P, Sewilam H (2015) The potential of groundwater desalination using forward osmosis for irrigation in Egypt. Clean Technol Environ Policy 17:1883–1895. https://doi.org/10.1007/s10098-015-0902-4
76. Nashed A, Sproul AB, Leslie G (2014) Water resources and the potential of brackish groundwater extraction in Egypt: a review. J Water Supply Res Technol 63:399–428
77. Pandian RS, Nair IS, Lakshmanan E (2016) Finite element modelling of a heavily exploited coastal aquifer for assessing the response of groundwater level to the changes in pumping and rainfall variation due to climate change. Hydrol Res 47:42–60
78. Gleeson T, Alley WM, Allen DM, Sophocleous MA, Zhou Y, Taniguchi M, VanderSteen J (2012) Towards sustainable groundwater use: setting long-term goals, backcasting, and managing adaptively. Groundwater 50:19–26

Part III
Sedimentology and Hydrogeophysical Characteristics

Sedimentological Characteristics of the Quaternary Groundwater Aquifer, Northwestern Nile Delta, Egypt

Abdelmonem T. Abdelhameed, Zenhom E. Salem, and Osman M. Osman

Abstract A total of 110 sediment samples were collected at various levels during the well drilling. Mechanical analysis of 70 sediment samples was done in order to study the sedimentological and depositional environments of the studied aquifer. Another 40 samples were subjected to X-ray diffraction (XRD) analysis to detect the mineralogical composition of the aquifer sediments. The sedimentary work revealed that the majority of the samples are medium to coarse clayey sand, moderately sorted and coarse skewed. This may support the multi-directional depositional currents. The present-day, wind action influences skin of the sediments. The samples lie within the field of river processes. Mineralogical analysis by X-ray diffractometry revealed that smectite is the most abundant clay mineral, followed by kaolinite, whereas illite is the less abundant clay mineral. The essential carbonate minerals are calcite and dolomite whereas non-carbonate minerals include quartz, feldspar, and hematite.

Keywords Depositional environment, Groundwater aquifer, Mineralogy, Sedimentology, Western Nile Delta

Contents

A.T. Abdelhameed and Z.E. Salem (✉)
Geology Department, Faculty of Science, Tanta University, Tanta, Egypt
e-mail: zenhomsalem@yahoo.com

O.M. Osman
Geology Department, Faculty of Science, Damanhour University, Damanhour, Egypt

A. M. Negm (ed.), *Groundwater in the Nile Delta*,
Hdb Env Chem (2019) 73: 161–186, DOI 10.1007/698_2017_153,

1 Introduction

Sedimentological studies are of expanding significance in environmental research, especially contamination studies, where past patterns in ecological procedures should be joined with information on current conditions to foresee likely future changes [1]. Sedimentology is a vital branch of geology and its utilization in water resources administration, contamination investigations of subsurface water bodies and depositional condition cannot be over underscored. The presumptions that groundwater is perfect and drilling ought to be done blindly are demonstrations of ignorance and the outcomes far exceed the benefits. Basic sedimentological information should shape some portion of any evaluation of possibly polluted destinations and part of examinations concerning the scattering and catching of contaminants in fluvial frameworks [2]. These information are additionally required for sound ecological administration to guarantee that planning policies are good with common environmental limitations [1].

Sedimentological examination of subsurface borehole sediments is a noteworthy part in groundwater study and aid to predict the depositional condition utilizing fundamental standards of geology [1, 3–5]. The stratigraphy of a landscape assumes an imperative part in groundwater pollution particularly when the aquifer geometry is unconfined [6–8]. When the aquifer is unconfined and shallow, there is a high possibility of contaminants whatever source and kind, particularly when they are near a highly populated area with high anthropogenic effect [9]. Omorogieva and Imasuen [10], Joshua and Oyebanjo [11] estimated that aquifer deposits come from the close surface, igneous outcrops, volcanic and sedimentary rocks. Some of these are effectively disintegrated, though others, particularly the crystalline rocks, are influenced by streams just when adjusted in the surface [12].

It has been perceived that grain-size distribution of granular permeable media influences its hydraulic conductivity [13]. This interrelationship is helpful for the estimation of conductivity properties where coordinate permeability information is inadequate in the beginning times of aquifer investigation. In groundwater hydrology, the learning of soil conductivity is vital for demonstrating the water flow both in the saturated and unsaturated zone, and transportation of water-dissolvable contaminations in the aquifer. The textural properties of an aquifer are all the more effectively acquired, a potential option for evaluating hydraulic conductivity of soils is from grain-distribution [14].

The sedimentary clastic rocks are characterized by high porosity and permeability and forming most of the groundwater aquifers. In the Nile Delta area in general

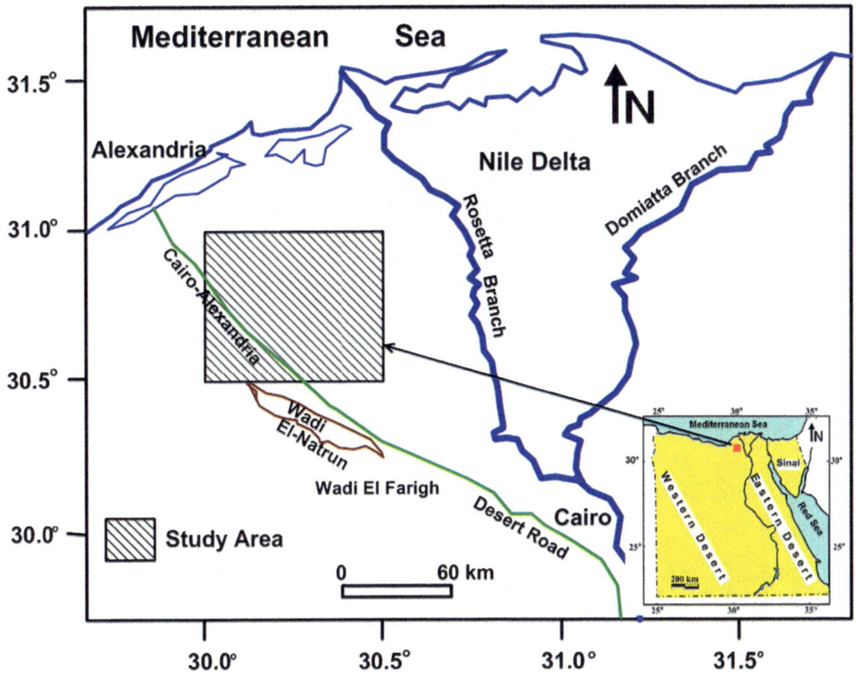

Fig. 1 Location map of the study area

and particularly the study area (Fig. 1), the main aquifer is the Pleistocene which consisted of sand and gravels with clay lenses intercalations. So, the sedimentary formation (i.e., Pleistocene aquifer) analysis for the study area is very important to reveal the aquifer properties as the grain size analyses and mineralogical study. This study deals with the grain size analysis and mineralogical study of the Plio-Pleistocene sediments of the area under consideration.

2 Study Area

West of the Nile Delta is one of the biggest groundwater territories in northern Egypt. Groundwater potentiality and dispersion are controlled by the geographic location, the sedimentary succession, the geologic structure, and physiographic setting. The study area is situated in the west of the Nile Delta in northern Egypt (Fig. 1). Its geographical boundaries are 30° 30′ and 31° 00′ N and longitudes 30° 00′ and 30° 30′ E. The common territories in this area are north of Wadi El Natrun, El Bustan extension reclamation project, west and east of El Nubaria canal, Abu El Matamir area, and Housh Eissa area.

Many authors studied the stratigraphy of the study area; among them El Fayoumy [15]; Shata et al. [16]; Omara and Sanad [17]; El Ghazawi [18]; Abdel

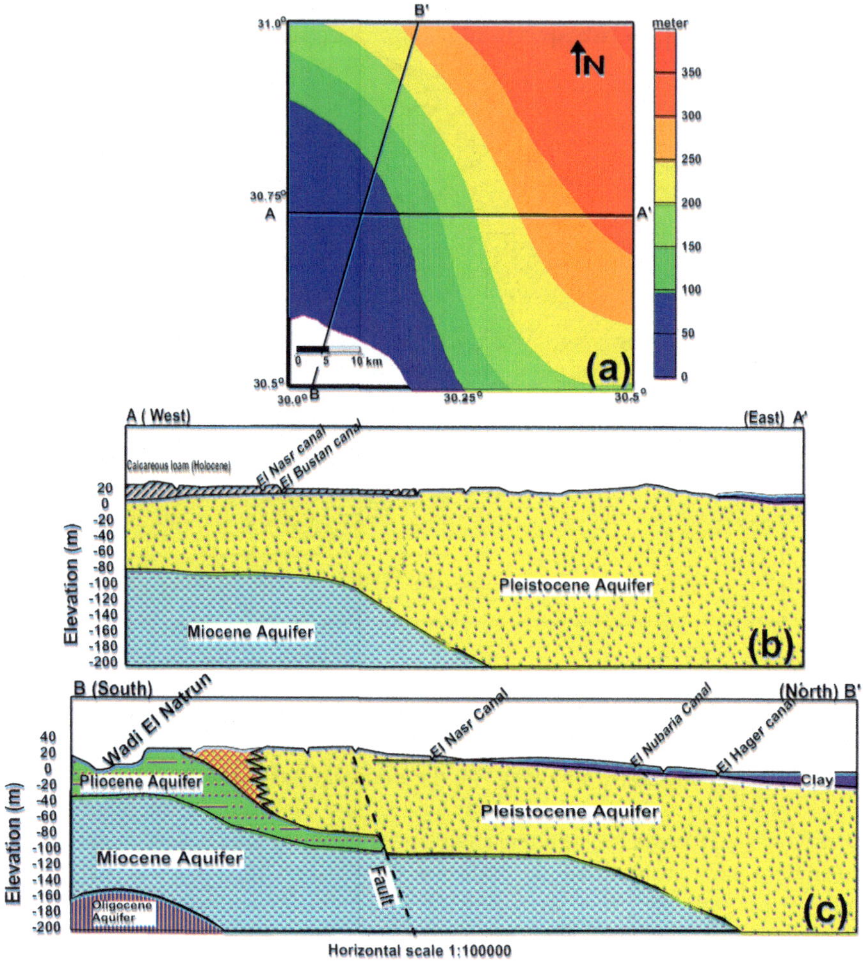

Fig. 2 Shows the hydrogeology of the study area. Pleistocene aquifer thicknesses increase to the eastern and northern directions is shown in (**a**). The hydro-stratigraphic succession of the study area along the cross-sections AA′ and BB′ is shown in (**b**) and (**c**), respectively (after [21])

Baki [19] and Abdel Wahab [20]. As disused in Salem and Osman [21, 22], sediments of the study aquifers are of Quaternary (Recent and Pleistocene), Neogene (Pliocene and Miocene), and Oligocene ages (Fig. 2). The Pleistocene aquifer is the most vital aquifer that has a greatest thickness of 300 m to the east to the northeast (Fig. 2a). The Holocene deposits are located in the western part of the investigated area and are composed of eolian sand (4–10 m thick) with calcareous intercalations (Fig. 2b). The Pleistocene aquifer is composed of gravels and sands with thin clay intercalations (Fig. 2b, c). This Quaternary aquifer overlays the clay-rich sediments of Pliocene and Miocene ages (Fig. 2b, c). The Pleistocene aquifer is mostly unconfined but of semiconfined characters at the northern direction where clays cover is predominant. Surface leakages from the watering framework trenches are the principal recharge source for the Pleistocene aquifer [19].

Structures including faults, folds, and unconformities influence the groundwater movement. The West Nile Delta region comprises the northern portion of the great Marmarican Homocline. The strata dip gently to the north and eastward reflecting the effect of the structural elements of the Taphro-geosynclinal basin of the Nile Delta basin [16]. The structural map of the area compiled by different authors is shown in Fig. 3. Major faults with downthrown sides due east have been delineated along the western margin of the Nile Delta. These faults have led, among other geologic features, to the facing of the highly permeable Pleistocene gravel of the Nile Delta opposite to older impermeable sediments to the west. This situation

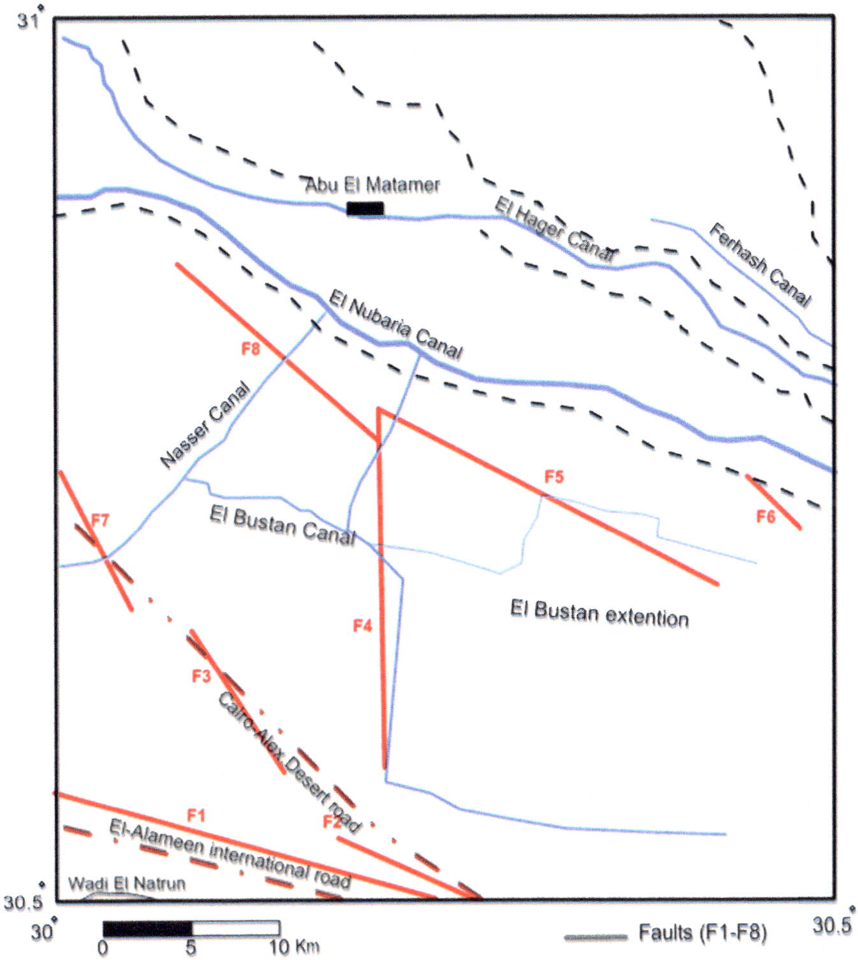

Fig. 3 Compiled structure map of the study area (after [18, 23, 24])

accelerates the subsurface westward flow of Nile water to the older sediments [23]. Two subsurface faults in the area were detected by El Ghazawi [18] comprising El Galala fault (F7) with downthrown to east and displacement reaches about 65 m, and El Nubaria fault (F8) with downthrown also to east (Fig. 3). They are related to the clysmic system during Neogene times. However, the rapid vertical and lateral facies changes of the Quaternary sediments hide this displacement. On the other hand, four probably subsurface faults (F3, F4, F5, and F6) have been detected with downthrown to the Nile Delta [24]. They greatly affect the thickness of the Pleistocene aquifer.

Wadi El Natrun anticlinal structure trends E 35° W and extends for about 60 km from El Ralat depression in the north to Beni Salama depression in the south [20]. It is a symmetrical structure affected by parallel and diagonal faults, among them F1 and F2 that led to the formation of the main central depression (Fig. 3). Unconformities are recognized in the area between the Quaternary deposits and Pliocene deposits, where Wadi El Natrun Formation (Early Pliocene) is covered directly by Quaternary clastics, and El Hagif Formation is absent (Late Pliocene) [18].

3 Methodology

The performed working methodology of this work is shown in Fig. 4. Thirty four boreholes were drilled in the study area (Fig. 5). A total of 110 sediment samples were collected at various levels during well drilling. Lithologic log of each well was constructed and then 2D and 3D lithological correlations were performed (Figs. 6 and 7). Mechanical analysis of 70 sediment samples was done to study the

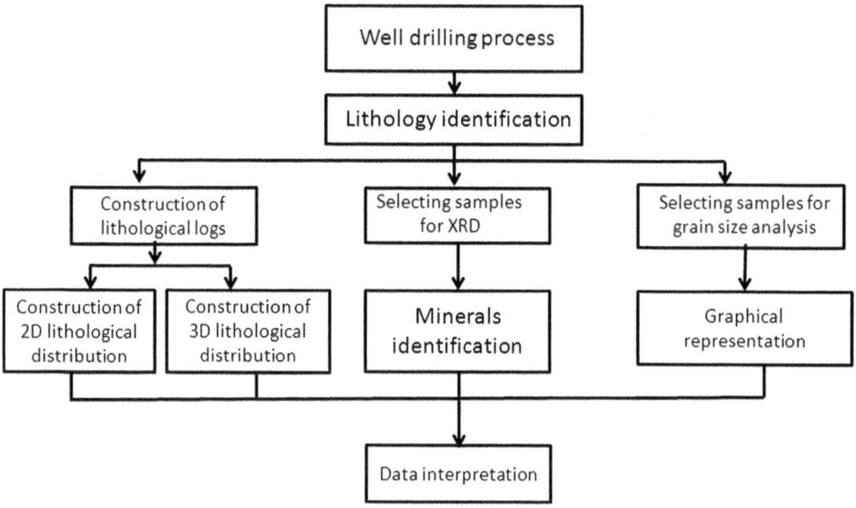

Fig. 4 Flowchart shows the working methodology

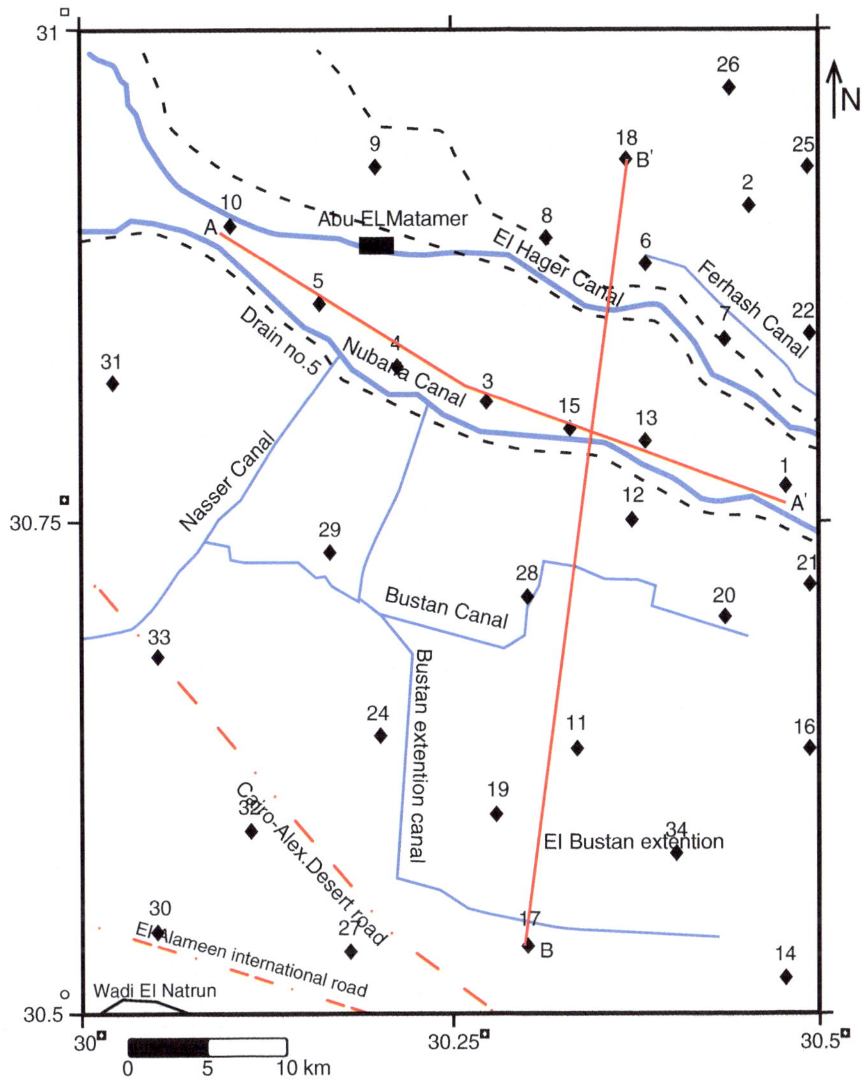

Fig. 5 Borehole and the 2D cross-sections location map

sedimentological and depositional environments of the studied aquifer. Results of the grain size analysis were represented in Table 1 and Figs. 8 and 9. The grain size analysis was carried out on the sand fraction using the standard sieving method. Another 40 samples were subjected to X-ray diffraction (XRD) analysis to detect the mineralogical composition of the aquifer sediments. Identified mineralogical compositions for the selected samples are shown in Table 3 and some representative X-Ray Diffractograms are plotted in Fig. 10. The mineralogical analysis was done on bulk samples by X-ray diffraction (XRD) analysis to delineate the clay and non-clay minerals of the sediment.

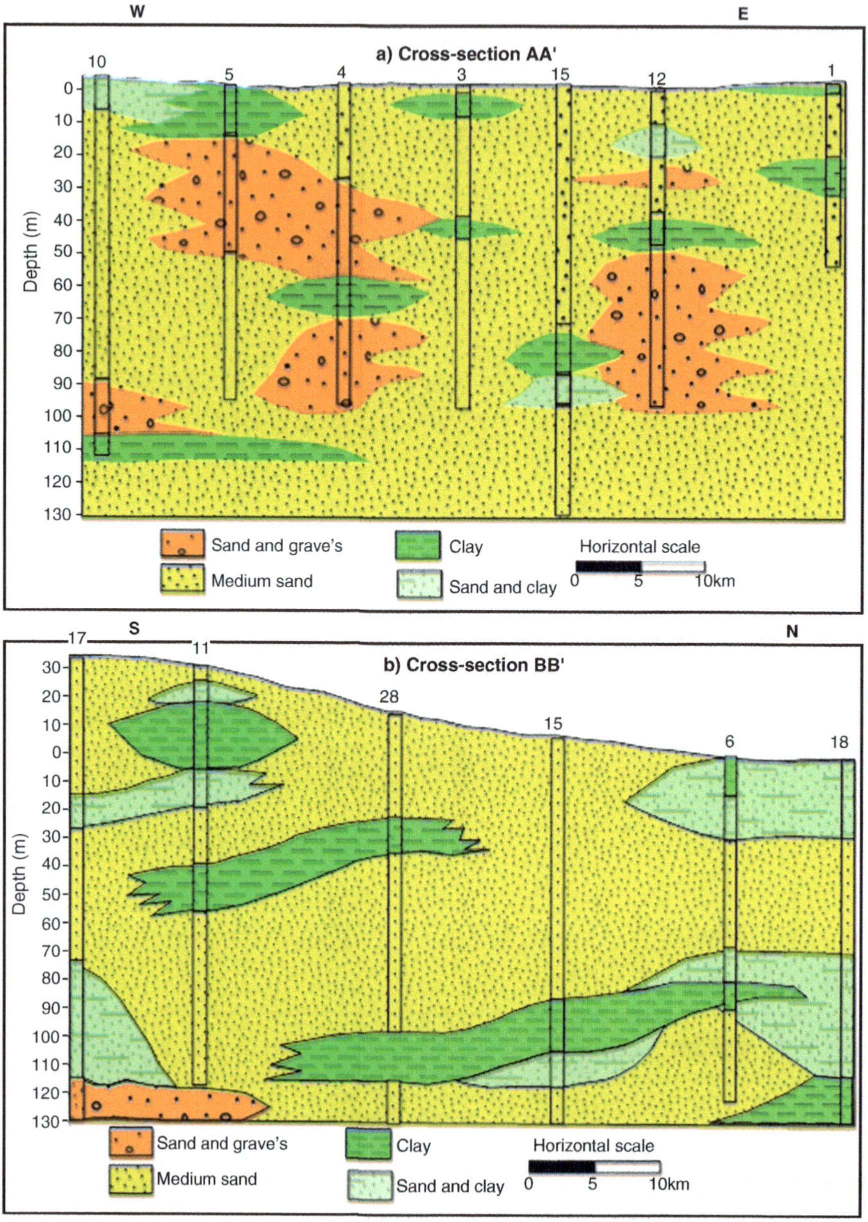

Fig. 6 Vertical lithological distribution along the West-East cross-section (AA′, **a**) and the South-North cross-section (BB′, **b**)

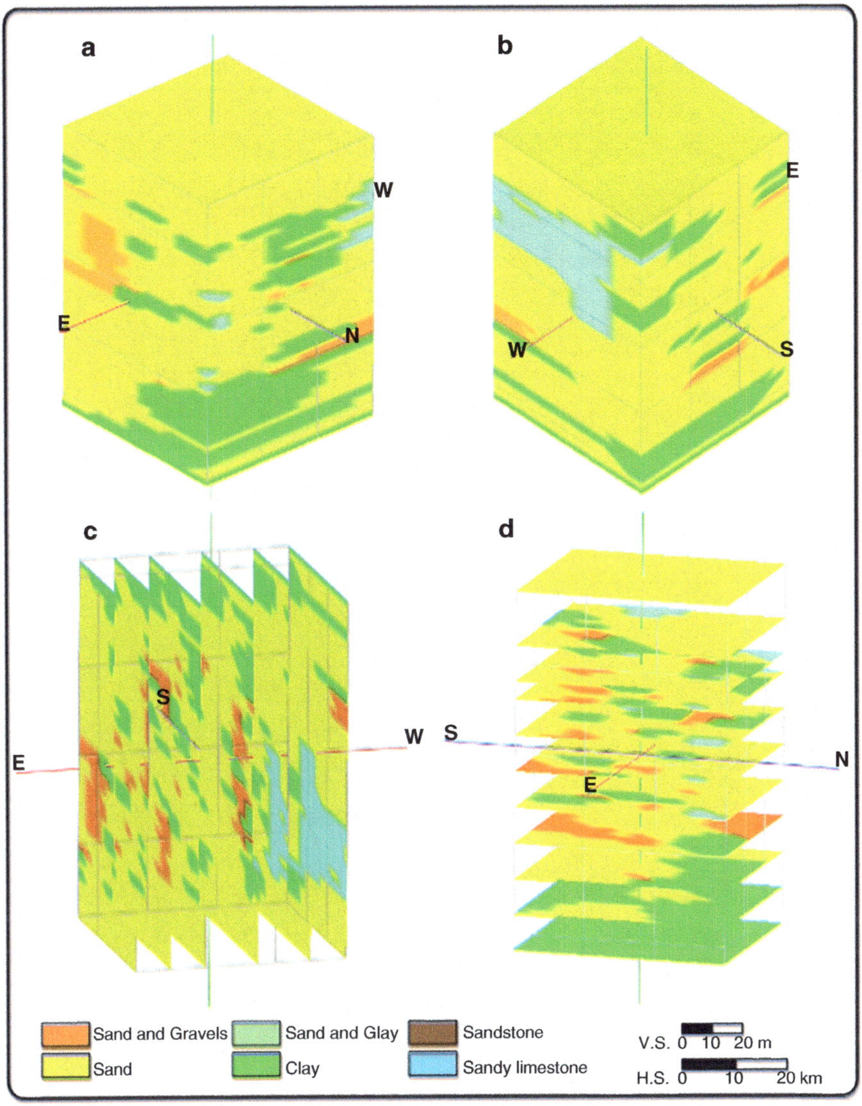

Fig. 7 3D lithology models of the study area

4 Result and Discussion

4.1 Lithological Correlation

The samples from different wells drilled in the area (Fig. 5) are distributed overall the area at various depths (Table 1). The cross-sections AA′ and BB′ (EW and SN, respectively) were established (Fig. 6) to show the vertical and lateral lithology

Table 1 Statistical grain size parameters of the collected samples

Well no.	Sample depth (m)	Md	Mz	S_I	SK_I	KG_I
1	1	1.10	1.43	1.07	0.40	2.32
	5	0.80	−0.37	3.37	−0.59	1.61
	10	0.60	0.20	2.21	−0.35	1.26
	15	0.50	0.33	2.21	−0.31	2.00
	25	0.90	0.67	1.91	−0.27	1.27
	35	−2.80	−3.27	4.27	−0.19	1.16
2	40	0.70	0.57	0.75	−0.33	1.76
	45	1.05	1.08	0.93	−0.02	1.19
	55	0.60	0.43	1.34	−0.18	1.06
	60	1.70	1.67	0.83	−0.14	1.69
	70	0.95	0.98	0.98	−0.02	1.19
	75	0.80	0.53	1.12	−0.25	1.55
	90	1.30	1.13	1.07	−0.35	1.45
	100	0.70	0.63	1.12	−0.14	1.53
	125	1.30	1.27	0.96	−0.15	1.19
	140	1.40	1.33	0.90	−0.16	1.12
3	20	1.10	0.93	1.31	−0.37	1.64
	28	1.00	1.00	1.03	−0.10	1.20
	48	0.90	0.77	1.27	−0.24	1.39
4	20	1.30	0.90	1.24	−0.42	1.07
	25	1.00	0.77	1.29	−0.32	1.13
	30	1.30	1.03	1.14	−0.40	1.23
	35	1.30	1.10	1.08	−0.36	1.20
	40	1.30	1.03	1.13	−0.39	1.30
	50	1.00	0.73	1.40	−0.30	1.18
	55	0.30	0.03	1.37	−0.26	1.06
	60	1.00	0.90	1.10	−0.23	1.20
	65	1.30	1.03	1.13	−0.39	1.20
	70	1.20	0.80	1.62	−0.48	1.56
	80	0.90	0.77	1.31	−0.24	1.28
	90	1.50	1.43	1.11	−0.17	1.08
	100	0.30	0.80	1.04	0.46	1.26
	110	1.30	1.20	0.96	−0.24	1.01
	130	1.50	1.33	0.99	−0.25	1.16
5	1	−0.70	−0.87	2.07	−0.07	1.23
	100	0.90	0.80	1.18	−0.17	1.02
	110	0.80	0.67	1.36	−0.17	1.13
6	10	0.90	0.57	0.94	−0.51	1.23
	30	0.70	0.50	1.57	−0.25	1.23
	50	−1.00	−1.80	3.78	−0.38	0.86
	80	0.30	0.17	0.93	−0.17	1.09
	120	1.00	0.90	1.16	−0.17	1.32
	150	1.00	0.97	1.01	−0.04	1.10

(continued)

Table 1 (continued)

Well no.	Sample depth (m)	Md	Mz	S_I	SK_I	KG_I
7	27	1.00	−0.30	3.33	−0.72	1.43
	35	1.20	1.10	1.29	−0.20	1.20
	43	1.00	0.93	1.08	−0.13	1.11
	48	0.10	−0.23	1.04	−0.31	1.01
	97	1.70	1.63	0.56	−0.25	1.97
	106	0.80	0.67	1.26	−0.14	0.84
7	120	1.20	0.93	1.25	−0.29	1.26
	125	0.70	0.57	1.30	−0.17	1.45
8	22	1.40	1.20	1.48	−0.24	0.77
	133	1.40	1.20	1.24	−0.28	0.67
9	0	1.00	1.27	1.24	0.27	0.89
	40	1.00	1.40	1.28	0.25	0.98
	70	1.40	1.27	0.90	−0.20	1.23
	75	1.60	1.47	0.80	−0.24	1.37
	80	0.90	1.03	1.07	0.00	1.27
	85	1.60	1.60	1.02	−0.12	1.16
	102	1.90	1.97	0.65	0.02	1.89
11	12	−2.60	−3.10	4.16	−0.17	1.28
	25	1.00	0.87	1.05	−0.24	1.13
	27	1.60	1.47	1.06	−0.25	1.17
	30	1.00	1.10	1.01	0.09	1.10
	35	1.40	1.27	0.88	−0.19	1.09
	45	1.20	0.63	1.58	−0.56	1.86
	55	−1.00	−0.97	2.28	−0.12	1.04
	60	0.40	0.13	1.46	−0.32	1.28
12	4	1.40	1.23	1.10	−0.29	1.40
	13	1.60	1.50	0.86	−0.14	1.19
	30	0.80	0.77	1.20	−0.06	1.05
	44	1.30	1.17	1.45	−0.28	1.55
13	10	1.20	1.20	1.21	−0.05	1.26
	30	1.30	1.17	1.30	−0.23	1.26
	40	1.30	1.07	1.10	−0.34	1.29
	50	1.10	1.13	1.53	−0.21	1.52
	60	1.40	1.23	1.02	−0.26	1.45
	120	1.30	1.23	0.88	−0.09	1.19
	135	1.00	0.83	1.29	−0.19	1.13
14	4	1.30	1.20	1.83	−0.16	1.17
	10	1.80	1.67	1.24	−0.32	1.08
	30	1.6	1.47	1.11	−0.28	1.37
14	85	0.80	0.37	1.81	−0.42	1.64
	15	1.20	1.20	1.55	−0.23	1.84
15	65	0.20	−0.30	2.53	−0.33	1.36

(continued)

Table 1 (continued)

Well no.	Sample depth (m)	Md	Mz	S_I	SK_I	KG_I
	75	1.40	1.43	1.14	−0.05	1.17
	80	1.30	1.23	1.27	−0.12	1.05
	85	0.80	0.90	1.23	0.11	0.96
	90	1.40	1.30	1.24	−0.22	1.29
	120	0.90	0.90	0.92	−0.15	1.15
	130	1.20	0.90	1.47	−0.43	1.64
	140	0.90	1.00	1.30	0.04	1.05
	150	0.40	0.37	1.33	−0.03	1.26
	155	0.20	−0.03	1.51	−0.27	1.07

Mz mean size, *Md* median size, *σl* standard deviation, *Sk* skewness, *KG* kurtosis

changes. These sections revealed that the sediments of the area are formed of fine to medium sand, gravels, and clay lenses at different levels. Cross-section AA′ (Fig. 6a) extends from E-W direction and contains seven lithology wells [1, 6, 19, 27, 29–31]. The general cross-section composition is sand from fine to coarse size. The sand and gravel sediments which are considered as an ideal groundwater aquifer are recorded in wells 12, 4, 5, and 10. Intercalations of some clay lenses are recorded all over this cross-section. These lenses might affect the level and quality of the groundwater in the investigated aquifer.

Cross-section BB′ contains six wells [23, 27, 32–35] (Fig. 6b) and shows that, the ground level is decreased toward the northern direction. It also indicates that the sand thickness increased to the central part and decreased to the southern and northern directions. In contrast, the number and thickness of the clay lenses increased to the southern and northern directions. Also, the constructed 3D models (Fig. 7) show the lithology variations as represented from AA′ and BB′ cross-sections in all directions. The 3D model shows the increasing of sand thickness toward the central part. Increasing in number and thickness of the clay lenses toward south, north and deeper zones led to decrease in sand thickness.

4.2 Granulometric Analysis

The granulometric study of sediment is an important textural parameter which measures the energy of the depositional environment. Both the friable and semi-friable samples from different wells were selected for the grain size analysis by sieving method. The statistical parameters of Folk and Ward's [31] were calculated from the grain size data. The data were then plotted on the cumulative curves.

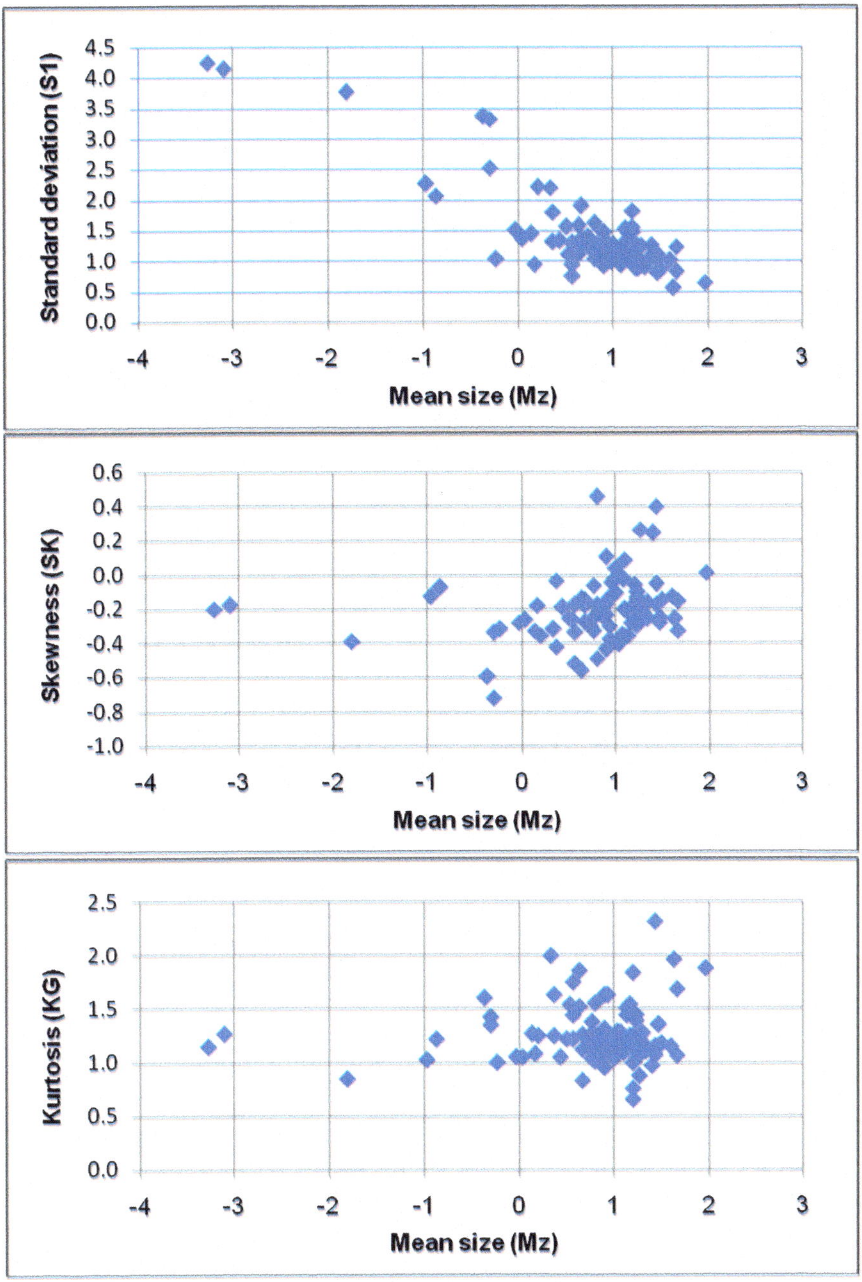

Fig. 8 The relationship between mean size (Mz) on one hand and standard deviation (σ1), skewness (SK), and kurtosis (KG) on the other hand of the Pleistocene sediments

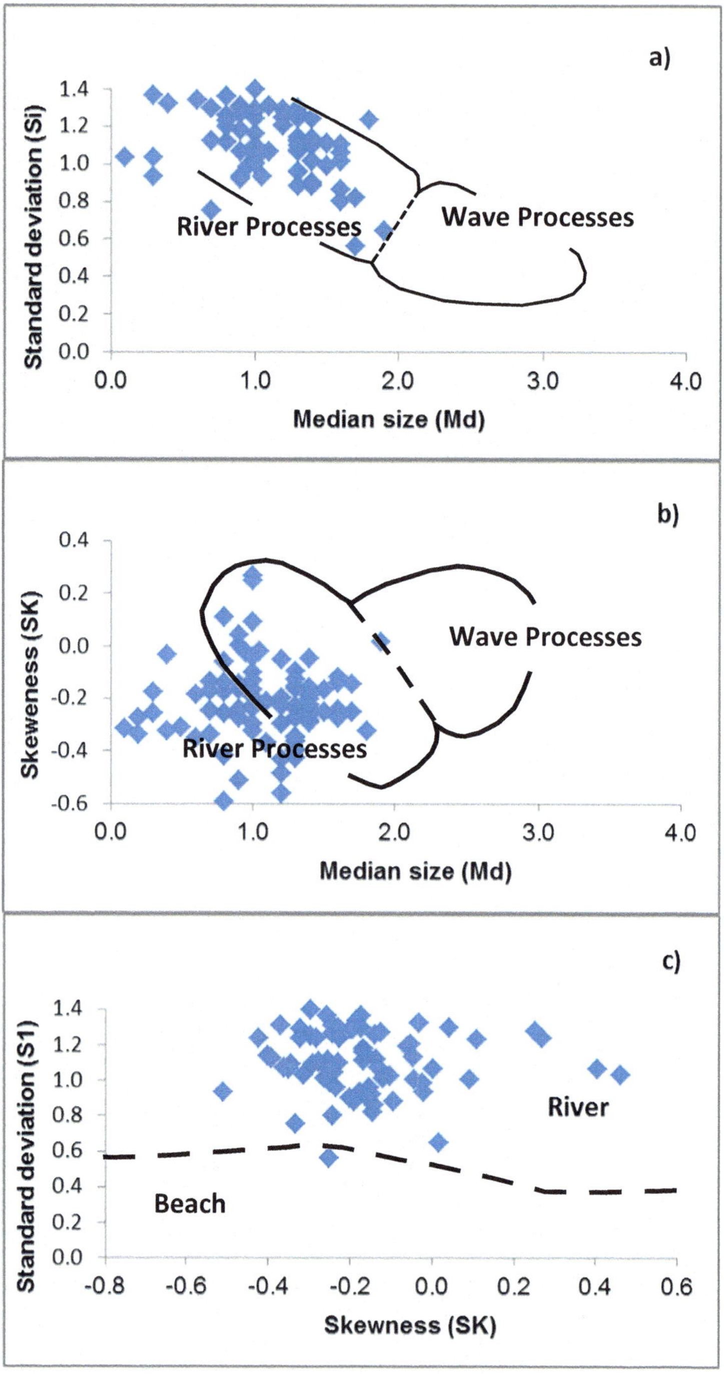

Fig. 9 Bivariate plots of the statistical grain-size parameters for the studied sediments (**a** and **b** boundaries after [25]; **c** after [26]). (**d**) Boundary after Friedman [27], (**e**) and (**f**) after Moiola and Weiser [28]

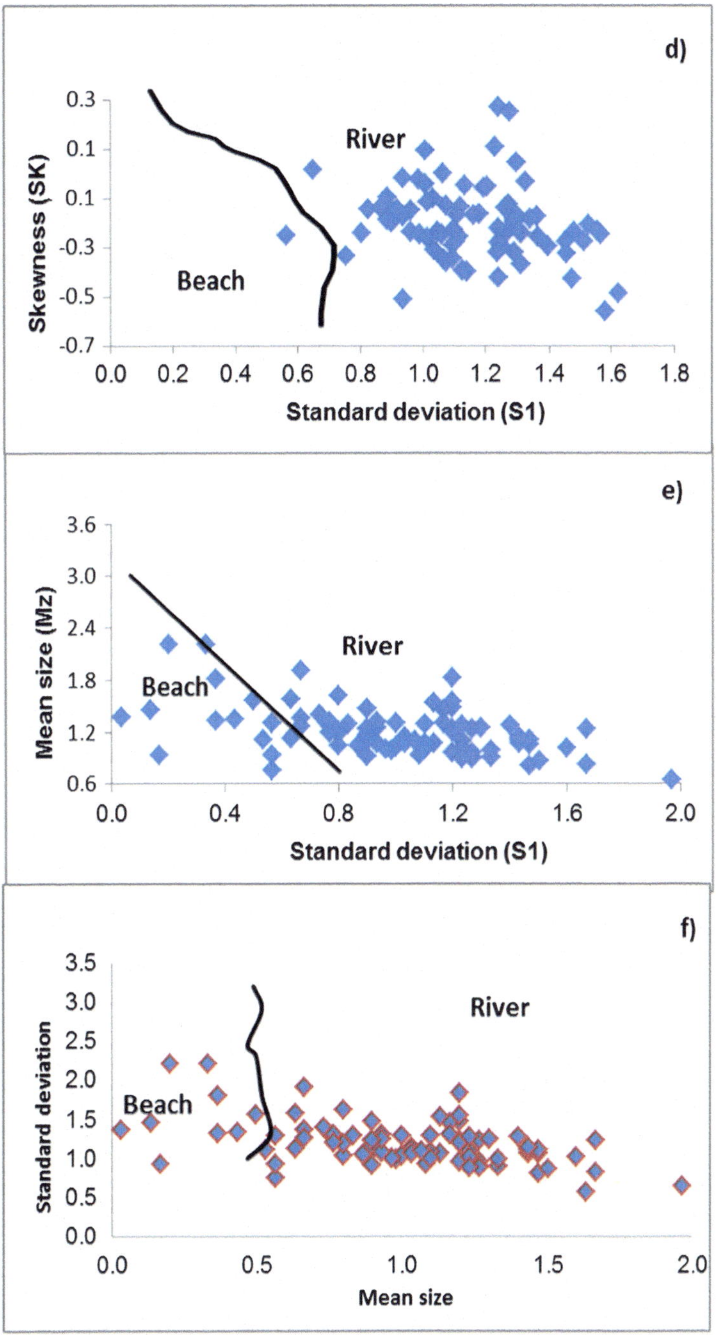

Fig. 9 (continued)

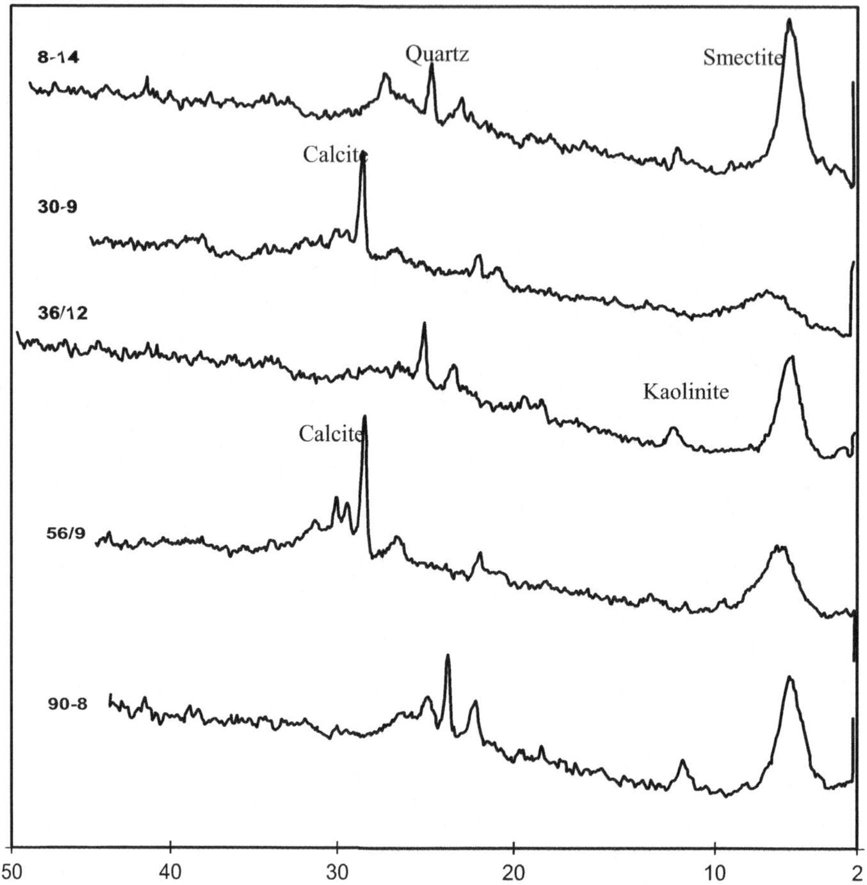

Fig. 10 X-Ray Diffractograms of some selected samples in the area. From up to down, these diffractograms are for samples 8 (well 14), 30 (well 9), 36 (well 12), 56 (well 9) and 90 (well 8). Horizontal axis represents two-theta (degree) and the vertical axis represents the relative peak intensity (counts) for each diffractogram

The grain-size parameters include graphic mean (Mz), median size (Md), inclusive graphic standard deviation (Si), skewness (SK) and kurtosis (KG). The diagrams proposed by Stewart [25], Friedman [26, 27], Moiola and Weiser [28], and Mason and Folk [35] have been applied to recognize the depositional environments of the sediments. The data of grain size distribution of the analyzed samples were plotted on cumulative curves, and the calculated parameters are shown in Table 1, and the summary of the grain size parameters is shown in Table 2.

Table 2 Summary of the grain size parameters of the collected samples

Parameter	Phi value (φ)		Description
Mz	Range	−3.27 to 1.97	Medium pebbles to medium sand
	Average	−0.6	Very coarse sand
Md	Range	−2.8 to 1.9	Fine pebbles to medium sand
	Average	−0.4	Very coarse sand
σ1	Range	0.56–4.27	Medium well sorted to extreme poorly sorted
	Average	2.45	Very poorly sorted
SK	Range	−0.72 to 0.46	Strongly coarse to strongly fine skewed
	Average	−0.13	Coarse skewed
KG	Range	0.67–2.32	Platykurtic to Leptokurtic
	Average	1.5	Leptokurtic

4.2.1 Univarient Grain Size Parameters

The mean size of sediments is controlled by the average energy distribution in the deposition media, the initial size and the source material [31, 36, 37]. It can be determined by using Eq. (1):

$$Mz = (\phi_{16} + \phi_{50} + \phi_{84})/3 \tag{1}$$

in which Mz is the mean size and ϕ_{16}, $\phi_{50,}$ and ϕ_{84} represent the size at 16, 50, and 84% of the sample by weight, respectively.

The mean size value of the Pleistocene samples (Table 2) ranges from −3.27 to 1.97 Φ (medium pebbles to medium sand) with an average value of −0.6 Φ (very coarse sand). The values of mean size fluctuate frequently reflecting the change of the depositing medium.

Grain sorting depends on the sediment source, grain size, and depositional regime [31]. The inclusive graphic standard deviation (σ) for the studied sediment varies from 0.56 to 4.27 Φ (medium well sorted to extreme poorly sorted) with an average value of 2.45 Φ (very poorly sorted). It can be calculated by using Eq. (2):

$$\sigma = (\phi_{84} - \phi_{16})/4 + (\phi_{95} - \phi_5)/6.6 \tag{2}$$

where ϕ_{16}, $\phi_{84,}$ $\phi_{5,}$ and ϕ_{95} represent the size at 16, 84, 5, and 95% of the sample by weight, respectively.

Skewness (SK) is a measure of symmetry of grain size distribution, and it is very useful descriptive term for the depositional processes of the sediment. The values of (SK) in these samples range from −0.72 to 0.46 Φ (strongly coarse sorted to strongly fine skewed), with an average value of −0.26 Φ (coarse skewed). It is calculated by [31], using Eq. (3):

$$\begin{aligned} SK = &\{(\phi_{16} + \phi_{84} - 2\phi_{50})/2(\phi_{84} - \phi_{16})\} \\ &+ \{(\phi_5 + \phi_{95} - 2\phi_{50})/2(\phi_{95} - \phi_5)\} \end{aligned} \quad (3)$$

According to [31], Kurtosis (KG) is a measure for the peakedness of the grain-size distribution. The obtained values of (KG) range from 0.67 to 2.32 Φ (platykurtic to leptokurtic) with a mean value of 1.5 Φ (leptokurtic). The graphic Kurtosis is given by Eq. (4):

$$KG = (\phi_{95} - \phi_5)/2.44(\phi_{75} - \phi_{25}) \quad (4)$$

where ϕ_{75} and ϕ_{25} represent the size at 75 and 25% of the sample by weight, respectively.

4.2.2 Bivariate Grain-Size Parameters

The bivariate plots between grain-size parameters are very useful in discriminating the depositional environments of rock [31]. The relationships between the calculated statistical parameters are discussed below, and their bivariate plots are given in Fig. 8.

The scatter plot diagram of the mean size versus the inclusive graphic standard deviation (Fig. 8a) shows that the majority of the samples are scattered around the field of moderately sorted, except some poorly sorted and well sorted samples. It further indicates that the decrease in the mean size of sediment is associated with better sorting.

Skewness has successfully been used by Mason and Folk [35], Friedman [14] and Moiola and Weiser [28] to differentiate between various environments of deposition. Strong skewed samples reflect zones of mixing environments [38]. The plot of skewness versus mean size shows that the majority of the samples are coarsely skewed (Fig. 8b). Such behavior may support the multi-directional depositional currents.

The relationship between mean size and kurtosis (Fig. 8c) further indicates that the present sand is leptokurtic. The leptokurtic character corresponds to what Mason and Folk [35] termed “aeolin flat.” This term does not mean that the sediments were deposited by aeolian agent but means that the present day skin of the sediments is influenced by wind action [30]. This explains the more pronounced leptokurtic characters of the studied samples.

Stewart [25] suggested that the plot of median diameters versus each of standard deviation and skewness is important to differentiate between river and wave processes. The plot of sorting vs. median diameter (Fig. 9a, b) indicates that the total samples lie within the field of river processes. Also, the plot of skewness versus median diameter shows that the total samples lie within the same field.

The plot of skewness versus standard deviation [26] shows the studied samples are distributed in the field of fluvial environments. The total samples belong to the river sands (Fig. 9c, d). The discriminating fields of Friedman [27] show the total samples are distributed within the river processes.

Moiola and Weiser [28] used various combinations of Folk and Ward's [31] textural parameters for differentiation between different environments. The plot of standard deviation versus mean diameters (Fig. 9e, f) shows that the most samples are distributed within the river processes, while 18% of the total samples lie within the beach environment.

4.3 Mineralogical Analysis

The mineralogical analysis was carried out on some selected bulk samples using X-ray diffractometry. The data of semi-quantitative determination of clay minerals are given in Table 3. X-Ray Diffractogram of the fine fraction of the studied sediments suggests that the clay mineral associations are dominated by smectite, kaolinite, and illite (see Fig. 10).

4.3.1 Clay Minerals

Smectite is the most abundant clay mineral in the studied sediments (Fig. 10). Its average value reaches up to 24%. The highest value (41.1%) is recorded in the north-central part of the area at 8 m depth, while the lowest value (5.4%) is obtained in the central part at 50 m depth. Smectite may be formed under authigenic and/or detrital conditions [39, 40]. Authigenic smectite may be formed in both continental and marine environments. Authigenic smectites formed in the continental aqueous sedimentary environment are essentially restricted to saline alkaline lakes and characterize the unusual water chemistries that may develop in these settings [32].

In the marine environment, authigenic smectites may be originated from the alteration of volcanics, mixing of ocean waters and plumes of hydrothermal fluid or halmrolysis (submarine weathering) and direct precipitation on the sea floor [41]. Detrital smectites may be formed under exogenous conditions chiefly upon weathering of basic igneous rocks, in alkaline medium. They also may be formed by secondary detrital degraded lattices (mica or chlorite), where their inter-sheet K^+ or Mg^{++} ions are replaced by water molecules [42].

Kaolinite is the second abundant clay mineral encountered in the studied sediments. The average value of kaolinite reaches 12.6%. The highest value (22.4%) is obtained in the northern part at 55 m depth, while the lowest value (5.1%) at 73 m depth in the central part. Kaolinite usually has a detrital origin, although authigenic kaolinite may also be precipitated from circulating groundwaters that are supersaturated with alumina and silica [33]. Typically, kaolinite develops in continental areas subjected to strong chemical weathering. It usually favors moderately warm

Table 3 X-ray diffraction data of the collected bulk samples

Well no.	Sample depth	Clay minerals			Non-clay minerals				
		Smectite	Illite	Kaolinite	Quartz	Feldspars	Calcite	Hematite	Dolomite
1	45	24.0	2.1	12.5	38.5	12.5	8.3	2.1	0
	50	5.4	1.8	8.1	12.6	4.5	63.1	4.5	0
	80	23.7	7.9	11.8	28.9	9.2	13.2	5.3	0
	85	17.3	5.1	7.1	45.9	9.2	10.2	5.1	0
2	6	30.6	3.3	16.5	20.7	9.9	12.4	0.0	6.6
	17	30.5	3.8	16.8	21.4	13.0	11.5	3.1	0
3	10	34.3	3.7	13.9	23.1	11.1	9.3	4.6	0
	37	40.7	5.5	11.0	18.7	11.0	9.9	3.3	0
4	5	5.9	5.9	16.1	16.9	5.9	47.5	1.7	0
	10	16.5	12.1	5.5	38.5	11.0	11.0	5.5	0
	15	27.9	6.3	9.0	31.5	10.8	14.4	0.0	0
	40	25.8	6.5	12.9	30.3	10.3	10.3	3.9	0
5	75	33.3	0.0	16.2	18.9	14.4	14.4	2.7	0
	90	22.4	5.1	12.2	45.9	7.1	5.1	2.0	0
7	45	15.6	6.5	9.1	37.7	15.6	10.4	5.2	0
	87	29.3	5.4	10.9	32.6	10.9	5.4	5.4	0
	96	11.8	2.7	11.8	25.5	10.9	28.2	9.1	0
	98	36.3	4.9	27.5	15.7	9.8	2.9	2.9	0
	104	20.7	3.7	17.1	24.4	14.6	12.2	7.3	0
	107	15.1	5.4	9.7	39.8	17.2	7.5	5.4	0
	118	16.9	7.0	11.3	23.9	14.1	18.3	8.5	0
	130	26.7	4.0	21.3	20.0	18.7	0.0	9.3	0
8	55	32.7	5.1	22.4	25.5	0.0	10.2	4.1	0
	90	29.8	3.5	14.2	25.5	14.2	9.2	3.5	0
	135	10.8	4.3	11.8	48.4	21.5	0.0	3.2	0

9	5	29.7	5.4	12.2	27.0	10.8	9.5	5.4	0
	10	18.2	3.6	7.3	38.2	25.5	0.0	7.3	0
	30	16.0	5.6	7.2	45.6	12.0	9.6	4.0	0
	56	19.9	4.0	8.5	41.5	15.3	8.5	2.3	0
	73	13.3	3.1	5.1	58.2	10.2	7.1	3.1	0
12	15	38.5	3.3	13.2	24.2	9.9	11.0	0.0	0
	36	38.3	0.0	13.8	28.7	10.6	8.5	0.0	0
13	80	23.5	5.9	17.6	36.8	16.2	0.0	0.0	0
	100	36.3	4.4	14.3	24.2	8.8	8.8	3.3	0
	102	23.4	4.7	12.5	31.3	17.2	10.9	0.0	0
14	8	41.1	4.0	11.3	17.9	7.9	14.6	3.3	0
	30	30.0	7.3	9.1	24.5	13.6	12.7	2.7	0
Average		24.7	4.7	12.7	30	12	12.1	3.8	0.2

to humid and well-drained acidic continental areas, where the rate of precipitation is much higher than the evaporation rate. Under these conditions, chemically soluble elements are preferentially removed leaving residual silica and alumina which are involved in the formation of kaolinite [43]. As kaolinite formation requires acidic leaching conditions, it cannot be formed under marine settings [44].

In the alkaline marine environment, there is no leaching, and the water contains a good deal of dissolved calcium. These environmental conditions favor the formation of smectite rather than kaolinite. A study of weathering processes indicates that kaolinite does not form calcareous parent material until all the carbonate has been removed. This kaolinite is likely to be transported to the marine environment by rivers and/or by the wind. It may also have a diagenetic origin. Velde [45] has shown that kaolinite is generally abundant in the early diagenesis, but as depth and maturity of rocks increase, kaolinite becomes less abundant.

Robert and Kennett [46] linked kaolinite abundances to increased precipitation and deposition during high stand of sea level. The great relief of drainage basins, established during low stand of sea level, favors the formation of kaolinite, especially during warm and humid conditions. However, Robert and Chamley [47] found no significant correlation between the abundances of kaolinite and sea level fluctuations.

Kaolinite is a typical low latitude clay mineral [48] and its concentration increases in tropical regions. Millot [43] assumed that kaolinite is dominant in fluviomarine and lacustrine deposits. Kaolinite formation requires strong leaching environment, an acid reaction medium and the presence of aluminum silicates [40]. The formation of kaolinite requires a favorable climate and an environment of high depleted cations other than Al and Si. Perrin [49] considered kaolinite as the most stable clay mineral that does not undergo appreciable change during transportation. However, diagenetic alteration of kaolinite in the marine environment has been noted by Biscaye [29].

Illite is the less abundant clay mineral encountered in the studied sediments. Its average value reaches up to 24%. Illite has a continental origin according to age determination [34]. It is formed by the weathering or partial hydrolysis of muscovite. Thus, it occurs often in soils derived from the weathering of mica-schist, gneisses, acid and medium igneous rocks in association with smectite and other minerals. Weaver [50] indicated that illite could often be formed from smectite in marine environments.

Diagenetic alteration of montmorillonite and kaolinite into illite generally occurs where sedimentation rates are low and time is also important for the chemical interaction between seawater and clay minerals [51]. Illitization of kaolinite is mentioned by Rumeau and Kulbicki [52]. Potassium and magnesium in seawater would be expected to promote the formation of the more stable clay minerals at the expense of the less stable ones.

4.3.2 Non-clay Minerals

The mineralogical examination has been carried out on some selected carbonate rocks using XRD technique. The data of bulk samples (Table 3) show that the essential carbonate minerals are calcite and rarely dolomite whereas noncarbonate minerals include quartz, feldspar, and hematite. The essential carbonate minerals in the studied samples are calcite and rare dolomite. These minerals are detected by their basal reflection at 3.04A and 2.88A, respectively, and their abundances are calculated by measuring peak height above the background. Calcite is the most common carbonate mineral in the studied samples. It is easily identified by the X-ray diffraction technique where the major peak appears at (3.04A). The average percent of calcite reaches about 11.8%. Dolomite is recorded only in two samples. Its identification is aided by X-ray diffraction technique, where the major peak appears at (2.88A). The average percent of dolomite reaches about 5.9%.

Quartz, feldspar, and hematite are detected in the studied samples. These minerals were detected by their basal reflections at 3.4A, 3.25A, and 2.69A, respectively, and their abundances are calculated by measuring the peak heights above the background. The most common detrital minerals in the samples are quartz and feldspar. Quartz is one of the most common detrital minerals in the samples. It is identified by X-ray diffraction technique where the major peak appears at (3.35A). Table 3 shows that the percentage of quartz ranges from 0 to 62% with a mean value of 10.8%. Feldspar is the second abundant detrital mineral in the samples as its chemical stability is lower than that of quartz. Its mean value is 11.8%. Hematite is the lowest mineral in the studied samples, and it is identified by X-ray diffraction where the major peak appears at (2.69A), its average value is 3.7%.

5 Conclusions

This study gave basic information for the future groundwater dynamics and water–rock interaction research in the study area. The grain size analysis data could help for hydraulic conductivity calculations, and the mineralogical description could answer many questions about the change in groundwater salinity and chemistry along the flow system.

The sedimentological study revealed that

1. The main composition of the aquifer in the area is medium sand to medium pebbles, poorly sorted with clay intercalations.
2. The majority of sediments belong to river sands and river processes with multi-directional depositional currents.
3. The main clay minerals in the intercalated clay lenses of the aquifer are smectite and kaolinite. The essential carbonate minerals include calcite and dolomite whereas noncarbonate minerals are represented by quartz, feldspar, hematite, and gypsum.

6 Recommendations

Lithological, grain size, and mineralogical analysis of the clastic sediments help to explain the water hydrodynamic and salinity behavior in an aquifer. Therefore the result of this work might be used for future groundwater investigation in the study area.

Acknowledgments The authors are grateful to Tanta University for the financial support offered during the course of this research work. The authors thank the editor Prof. Dr. Abdelazim Negm for his constructive remarks.

References

1. Brandly E, David M (1993) Implication of sedimetological studies for environment and management: examples from fluvial systems in North Queensland and West Australia. Sediment Geol 85(1–4):235–252
2. Imasuen OI, Omorogieva OM, Nwokoloh NJ (2016) Grain size and heavy mineral analyses of two boreholes in recent to miocene aquifer in benin formation. Niger J Technol 35:979–986
3. Nwajide CSA (2006) Guide for geological field trip to Anambra and related sedimentary basins in southeastern. University of Nigeria, Nsukka, p 28
4. Suzuki T (1975) Heavy minerals composition of the recent sediments in 3 different environments. In: Geological survey of Japan part, vol 1, p 501
5. Tawari GS (2008) Variation of heavy minerals in different lithofacies and Geomorphic Units of Ganga – Yamuna River near their confluence at Allahabad, Uttar Pradesh. Geol Soc India 1:133–140
6. Akpoborie I (2011) Aspect of the hydrology of the western Niger Delta wetlands: groundwater conditions in the Neogene deposits of the Ndokwa area. Afr Geosci Rev 18:25–36
7. Onyeobi TUS, Akujieze CN (2014) Characterization of soil and sediment parameters of Jisike-Izombe Upper aquifer system for assessment of potential of groundwater pollution. J Appl Sci Environ Manag 18:674–683
8. Oteze GE (2011) Water supply, groundwater and flood control in Benin City. In: Presented at the NMGS, Benin chapter sensitization workshop on water resources management and its implications in Benin City, Mar 22
9. Isikhuemen MI, Omorogieva OM (2015) Hydrogeochemical and biophysical characterization of groundwater in Eastern Nigeria; a case study of Onisha and Environ. Niger J Technol 34
10. Omorogieva OM, Imasuen OI (2016) Factors contributing to the concentration of heavy metals in stream sediment along Ikpoba river tributary in Oluku (Upstream) to Ikpoba River Dam (Downstream) and their implication. Niger J Appl Sci 34:187–193
11. Joshua EO, Oyebanjo OA (2010) Grain-size and heavy metals mineral analysis of river Osun sediments. Aust J Basic Appl Sci 4:498–501
12. Mahavir S (2001) Heavy mineral assemblage of the pinjor formation of the Northwestern Himalaya and its significance in deciphering the provenance of the sediments. J Geol 54:65–87
13. Freeze RA, Cherry JA (1979) Groundwater. Prentice Hall, Englewood Cliffs
14. Rosas J, Lopez O, Missimer TM, Dehwah AHA, Sesler K, Coulibaly K (2014) Hydraulic conductivity and grain size distribution: comparison of methods for different sedimentary depositional environments. Groundwater 52:399–413
15. El Fayoumy IF (1964) Geology of groundwater supplies in Wadi El-Natrun area. MSc thesis, Faculty of Science, Cairo University

16. Shata AA, El Fayoumy IF, Tamer M (1970) Geology, hydrogeology and soil of Wadi El-Natrun-Maryut agriculture project. Int. Rep., Desert Institute, Cairo, p 19
17. Omara SM, Sanad S (1975) Rock stratigraphy and structural features of the area between Wadi El-Natrun and Moghra depression, western desert, Egypt. Geol Jb B16:45–73
18. El Ghazawi MM (1982) Geological studies of the Quaternary-Neogene aquifers in the area northwest Nile Delta. MSc thesis, Al-Azhar University, Cairo
19. Abdel Baki AA (1983) Hydrogeological and hydrochemical studies on the area west of Rosette branch and south of El-Nasr Canal. PhD thesis, Faculty of Science, Ain Shams University, Cairo
20. Abdel Wahab S (1999) Hydrogeological and isotope assessment of groundwater in Wadi El-Natrum and Sadat city, Egypt. MSc thesis, Faculty of Science, Ain Shams University
21. Salem ZE, Osman MO (2016) Shallow subsurface temperature in the environs of El-Nubaria canal, northwestern Nile Delta of Egypt: mplications for monitoring groundwater flow system. Environ Earth Sci 75:1241. https://doi.org/10.1007/s12665-016-6046-y
22. Salem ZE, Osman OM (2017) Use of major ions to evaluate the hydrogeochemistry of groundwater influenced by reclamation and seawater intrusion, West Nile Delta, Egypt. Environ Sci Pollut Res 24:3675–3704. https://doi.org/10.1007/s11356-016-8056-4
23. El Shazly EM, Abdel-Hady MA, El-Ghawaby MA, El Kassas IA, Khawasil SM, El Shazly MM, Sanad S (1975) Geological interpretation of landsat satellite images for west Nile delta area, Egypt. Remote sensing research project, Academy of Scientific Research Technology, Egypt
24. Mabrouk MA (1978) Electrical prospecting on the groundwater in the area west of Cairo-Alexandria desert road (between Wadi El-Natron and El-Nasr Canal). MSc thesis, Faculty of Science, Ain Shams University, Cairo
25. Stewart HB (1958) Sedimentary reflections of depositional environment in San Miguel lagoon, Baja California, Mexico. Bull Am Assoc Petrol Geol 42:2567–2618
26. Friedman GM (1961) Distinction between dune, beach and river sands from their textural characteristics. J Sediment Petrol 31:514–529
27. Friedman GM (1967) Dynamic processes and statically parameters compared for size frequency distribution of beach and river sands. J Sediment Petrol 37:327–345
28. Moiola RJ, Weiser D (1968) Textural parameters and evaluation. J Sediment Petrol 38:45–53
29. Biscaye PE (1965) Mineralogy and sedimentation of recent deep sea clay in the Atlantic ocean and adjacent seas and oceans. Geol Soc Am Bull 76:803–831
30. El Hinnawi EE, Kabesh ML, Zahran I (1973) Mineralogy and chemistry of Nubian sandstone from the central Eastern Desert of Egypt. N Jb Miner Abh 118:211–234
31. Folk RL, Ward WC (1957) Brazos River bar, a study in the significance of grain size parameters. J Sediment Petrol 27:3–26
32. Chamley H (1989) Clay sedimentology. Springer, Berlin, p 623
33. Hendriks F (1985) Upper Cretaceous to Lower Tertiary sedimentary environments and clay mineral associations in the Kharga Oasis area (Egypt). N Jb Geol Paläont Mh 10:579–591
34. Hurley PM, Heezen BC, Pinson WH, Fairbairn HW (1963) K-Ar age values in pelagic sediments of the north Atlantic. Geochim Cosmochim Acta 27:393–399
35. Mason CC, Folk RL (1958) Differentiation of beach, dune and aeolian flat environments by size analysis. J Sediment Petrol 28:211–226
36. Pettijohn FJ, Potter PE, Siever R (1987) Sand and sandstone, 2nd edn. Springer, New York, p 553
37. Sahu WD (1964) Depositional mechanisms of clastic deposition. Am Assoc Petrol Geol Bull 34:73–83
38. Inman DL (1952) Measure for describing the size distribution of sediments. J Sediment Petrol 22:124–145
39. Keller WD (1970) Environmental aspect of clay minerals. J Sediment Petrol 40:788–813
40. Weaver CE (1958) Geologic interpretation of argillaceous sediments. Bull AAPG 42:254–309
41. Lewis WD, McConchie D (1994) Practical sedimentology. Chapman and Hall, New York, p 190

42. Musa GA (1985) Sedimentological and micropaleontological studies on Tariff well northeast Nile Delta, Egypt. Unpublished MSc thesis, Faculty of Science, Tanta University
43. Millot C (1970) Geology of clays. Springer, Berlin, p 429
44. Keller WD (1982) Kaoline a most diverse rock in genesis, texture, physical properties and uses. Geol Soc Am Bull 93:27–36
45. Velde B (1995) Compaction and diagenesis. In: Velde B (ed) Origin and mineralogy of clays: clays and the environment. Springer, New York, pp 220–246
46. Robert C, Kennett JP (1992) Paleocene and Eocene kaolinite distribution in south Atlantic and southern oceanic Antaractic climate and paleoanographic implication. Mar Geol 103:99–110
47. Robert C, Chamley H (1991) Development of early Eocene climates as inferred from clay mineral variation in oceanic sediments. Glob Planet Chang 89:315–331
48. Rateev MA, Gorbunova ZN, Lisitzin AP, Nosov GL (1969) The distribution of clay mineral in the oceans. Sedimentology 13:21–43
49. Perrin RMS (1971) The clay mineralogy of British sediments. Mineralogical Society, London, p 406
50. Weaver CE (1956) The distribution and identification of mixed layer clays in sedimentary rocks. Am Mineral 14:202–221
51. Grim RE, Diet RS, Bradly WF (1949) Clay mineral composition of some sediment from the Pacific Ocean of the California coast and the Gulf of California. Geol Soc Am Bull 60:1785–1808
52. Rumeau JL, Kulbicki G (1969) Evolution des mineraux agileux dans les dolomites etles calcaries du cretace superieur de lar platforme d' Aquitaine Intern. In: Clay Conf. 2nd, pp 103–117

Hydrogeophysical Characteristics of the Central Nile Delta Aquifer

Zenhom E. Salem, Abdelazim M. Negm, and Abdelaziz Nahrawy

Abstract The present study is carried out in Nile Delta aquifer, where the data of electrical resistivity and gamma ray logs of the 34 well ranging in depth between 80 and 140 m were used to calculate the aquifer parameters. It is aimed to estimate the spatial variability of the formation lithology, porosity, permeability, groundwater salinity, and the hydraulic conductivity. Bilqas Formation showed an increase in the thickness, porosity, and water salinity to the north and northeast directions. Permeability and hydraulic conductivity values decrease in the same direction. Bilqas Formation ranges in thickness from 3 m in the southwest direction to 31 m in the northeast direction. The shale content and porosity ranges are 54% to 97% and 21% to 55%, respectively. This layer has low values of permeability (16×10^{-9} to 78×10^{-9} mD) and hydraulic conductivity ($<2 \times 10^{-9}$ cm/s). The water salinity of this layer ranges from 200 mg to 1,600 mg/l.

In Mit Ghamr Formation, average shale content ranges from 4.5 to 22%. Numbers of scattered clay lenses are detected in different places with high intensity in the northeastern direction. Porosity ranges from 19 to 39%. High permeability values are recorded in this formation and ranged from 0.1×10^{-2} to 8.7×10^{-2} mD. The water salinity average values in this aquifer range from 220 mg to 2,100 mg/l. The calculated hydraulic conductivity values for this formation are of range 5.082×10^{-10} to 2.134×10^{-8} cm/s. In this layer, the increase in the shale content, the increase in porosity, decrease in the permeability and hydraulic conductivity, as well as the increase in salinity, are to the northern and northeastern directions.

Z.E. Salem and A. Nahrawy
Geology Department, Faculty of Science, Tanta University, Tanta, Egypt
e-mail: zenhomsalem@yahoo.com

A.M. Negm (✉)
Department of Water and Water Structure Engineering, Faculty of Engineering, Zagazig University, Zagazig 44519, Egypt
e-mail: amnegm85@yahoo.com; amnegm@zu.edu.eg

A. M. Negm (ed.), *Groundwater in the Nile Delta*,
Hdb Env Chem (2019) 73: 187–210, DOI 10.1007/698_2017_75,

Keywords Hydraulic conductivity, Nile Delta aquifer, Permeability, Porosity, Water salinity

Contents

1 Introduction

Well logs are used widely in the exploration of mineral and hydrocarbon resources because they provide detailed and reliable information about the geometrical and petrophysical characteristics of the geological structures [1]. They are applicable also for the investigation of shallow formations, for instance, in water prospecting and solving environmental and engineering geophysical problems [2]. In hydrogeophysics, the main target of well log analysts is to estimate the layer thickness, water saturation, groundwater salinity, effective porosity, clay content, and aquifer hydraulic conductivity accurately as possible.

Well logs can be basically used to distinguish the change of hydraulic conductivity along a well or between neighboring wells. All sorts of borehole-logging suites utilized as a part of hydrocarbon investigation can be used in groundwater prospecting. Gamma ray and SP logs are used for identifying lithology and determining the layer thickness [3]. Water saturation zones and invasion profiles surrounding the borehole could be detected by using high-resolution resistivity tools. Because freshwater has higher resistivity than salt water, traditional resistivity tests are reasonably used for groundwater investigation. While shallow resistivity devices determine the apparent resistivity of the zone invaded by mud, the deep resistivity tools measure the original formation resistivity. The resistivity readings should be corrected to estimate the true resistivity of groundwater formation which is significant for computing the aquifer water saturation [3]. Many researchers used geophysical well logs to calculate the aquifer petrophysical parameters; among them are Winslow and Kister [4], Paillet and Reese [5], Sloto et al. [6], Szabó [7, 8], Szabó and Dobróka [9], Szabó and Dobróka [10], and Szabó et al. [11]. As the Nile Delta aquifer is so important freshwater source for the highly populated area in Egypt, determining the spatial change in its petrophysical, hydrogeological, and hydrogeochemical characteristics is of great importance. This work aims to estimate the aquifer parameters as clay volume, porosity, clay thickness, permeability, TDS, and hydraulic conductivity for Bilqas and Mit Ghamr Formations.

2 Study Area

The study area is located at the central part of the middle Nile Delta (Fig. 1a). The geology of the Nile Delta has been extensively discussed by several researchers over several decades (among them, [12–20]). According to the mentioned studies,

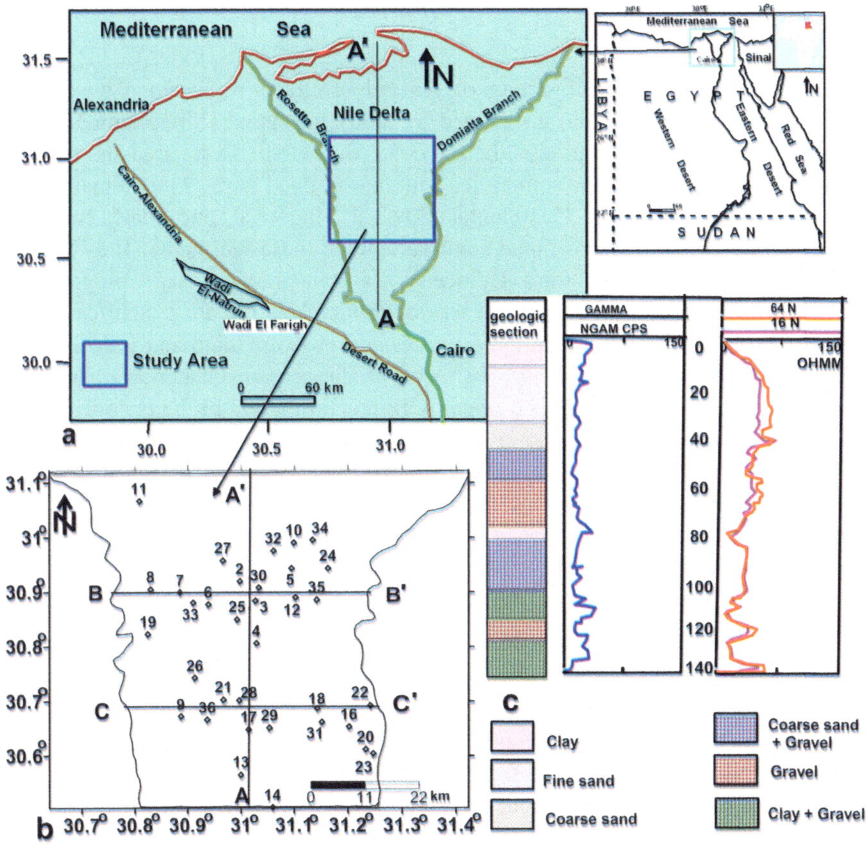

Fig. 1 Location map of the study area (**a**), boreholes location map (**b**) and representative geophysical log of well 9 (**c**). AA′ is the location of the hydrogeological cross-section shown in Fig. 2

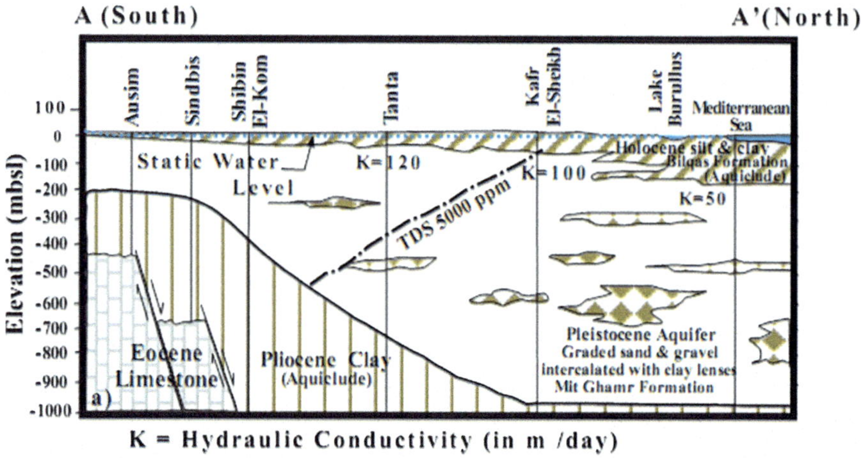

Fig. 2 The hydrogeological cross section from south to north in the Nile Delta shows the framework groundwater regime

the general hydrogeological setting of the Nile Delta is represented by clayey formation (Bilqas Formation) underlined by sandy formation (Mit Ghamr Formation) (Fig. 2). Bilqas Formation is the top cover of the Nile Delta area and mainly consists of silts and clays and sometimes with fine-grained sands. Plant remains and peat deposits are frequent. Continental, fluviatile, lagoonal, and beach environments are the probable depositional environments of these sediments. They deposited in the advent of the third Holocene sea transgression stage progressing practically from the north and northeast directions [21]. The main aquifer in the Nile Delta is Mit Ghamr Formation which is composed of sand and gravels with thin clay intercalations. It is assumed to be of Pliocene to Quaternary age. The whole sequence of Mit Ghamr is capped by Bilqas Formation of the Holocene age. Sediments of Mit Ghamr Formation were probably deposited under shallow marine to fluvial conditions [21]. On both sides of the present Delta, these deposits form a series of gravel terraces at various heights [22].

3 Methodology

Petrophysical data derived from 34 well logs with depths ranging between 80 and 140 m (Fig. 1b, c). Logging parameters collected included are gamma ray and resistivity logs. These logs were selected to estimate the spatial variability of the formation lithology, porosity, permeability, and water salinity. These logs were also used to estimate the hydraulic conductivity of the aquifer.

3.1 Shale Volume Estimation

Gamma ray method is considered as the best indicator method to calculate and identify the shale volume [23]. The Gamma ray readings after correction for borehole effects may be expressed as a linear function of shale content by the following equation [24]:

$$I_{GR} = (GR_c - GR_{min})/(GR_{max} - GR_{min}) \quad (1)$$

where:

I_{GR} is the gamma ray index.
GR_c is the corrected value of gamma ray at the interest intervals.
GR_{min} is the minimum value of gamma ray opposite clean intervals.
GR_{max} is the maximum value of gamma ray opposite shale intervals.

The volume of shale (V_{sh}) can be calculated from the gamma ray index by using the following equation [25]:

$$V_{sh} = 0.083\left[2^{3.7IGR} - 1\right] \quad \text{Larionov equation} \quad (2)$$

3.2 Formation Water Resistivity Determination

It is the water, uncontaminated by drilling mud that saturates the porous formation rock. The resistivity of the formation water (R_w) is important interpretation since it is required for the calculation of saturation water from basic resistivity logs. Archie determined experimentally that the water saturation of clean formation could be expressed in the term of its true resistivity as:

R_t is true resistivity which can be computed from long (LLD) and short (LLS) by using the following formula:

$$\text{If LLD} > \text{LLS} \quad R_t = 1.7\,\text{LLD} - 0.7\,\text{LLS} \quad (3)$$

$$\text{If LLD} < \text{LLS} \quad R_t = 2.4\,\text{LLD} - 1.4\,\text{LLS} \quad (4)$$

(Schlumberger log interpretation principle 1972)

$$R_o/R_t = R_{mf}/R_w \quad (5)$$

$$R_w = (R_{mf}/R_t)/R_o \quad (6)$$

Then

$$R_{\mathrm{w}} = (R_{\mathrm{mf}} \times R_{\mathrm{t}})/\mathrm{LLS} \tag{7}$$

R_{mf} is the resistivity of the mud filtrate.

R_{o} is the flushed zone resistivity which equals (LLS) short normal resistivity (R16″).

The formation water resistivity (R_{w}) for the Nile Delta aquifer is computed using Archie Eq. (6) after correcting the LLD readings.

3.3 Formation Factor Determination

The formation factor (F) is generally defined as the ratio of rock resistivity to water resistivity of a fully saturated rock. It is calculated as follows Archie [26]:

$$F = R_{\mathrm{t}}/R_{\mathrm{w}} \tag{8}$$

where R_{t} is the true resistivity in Ω m and R_{w} is formation water resistivity.

3.4 Porosity Determination

It is generally agreed that an empirical relationship exists between formation factor and porosity. The formation water factor is a function of porosity and also of pore structure and pore size distribution. Archie introduce this relationship in the following equation:

$$F = \frac{a}{\Phi^{m}} \tag{9}$$

where:

a is the pore geometry coefficient, dimensionless, generally, ranges from 0.6 to 2.

Φ is porosity in percent.

m is cementation factor, dimensionless, ranges from 1 to 3.

3.5 Permeability Determination

Permeability is one of the most important and least predictable fluid transport characteristics of materials. Permeability must be known to understand several natural phenomena, such as the basin-scale hydrogeological circulation [27], fault dynamics [28], the safety of waste repositories [29], and several other subsurface hydrology problems [30]. The used methods for calculating the permeability from the well logs are discussed as follows:

3.5.1 Clean Sand

Intrinsic permeability (K) is evaluated using the following equation:

$$K = 1.828 \times 10^5 \left(P^{1.1}\right) \tag{10}$$

where:

K is the intrinsic permeability in millidarcy [31].
P is a porosity factor which represented by the following equation.

$$P = \frac{\Phi^{m+2}}{(1-\Phi)^2} \tag{11}$$

3.5.2 Permeability of Clay and Clayey Sand

Based on Revil and Cathles model (1999) (Fig. 3), The classical Kozeny-Carman relationship was improved by using electric parameters which separate pore throat from total porosity and hydraulic radius. The permeability of clean sand is

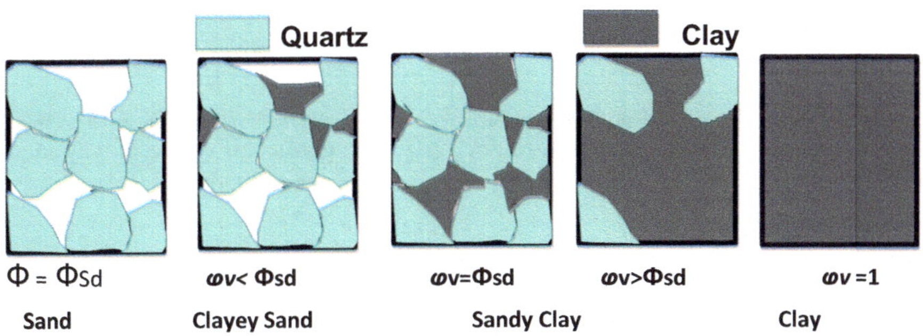

Fig. 3 Sand-shale mixture for various shale contents. The shale content increases from the *left* to the *right*. For a clay-free sand (first end-member), the noncompacted porosity is equal to Φsd. For a clayey sand, the porosity decreases because of the presence of clay particles in the pore space. This decrease continues until the critical point where all the pore space of a clean sand is occupied by clay particles, i.e., when the shale content is equal to the porosity of a clean sand. After this point the rock is sandy shale, and an increase in shale content is only possible through replacement of quartz grains by clay particles, and the porosity increases with the shale content. The second end-member is pure shale with no quartz grains and with a porosity equal to Φsh (*Note*: this figure and its description are taken from [30])

estimated as a function of the grain diameter, the porosity, and the electrical cementation exponent. The permeability of pure shale is derived in a similar way but is strongly dependent on clay mineralogy.

Clay

The shale end-member permeability k_{Sh} is related to the shale end-member porosity; Φ_{Sh} is represented by the following equation:

$$k_{\mathrm{Sh}} = K_0 \left(\Phi_{\mathrm{Sh}}/\Phi_0\right)^{3\mathrm{msh}} \tag{12}$$

where k_0 and Φ_0 are the permeability and porosity in a reference state, respectively (we take $\Phi_0 = 0.50$). Using Eq. (12), m and k_0 are calculated for each end-member clay mineralogy (kaolinite, illite, and smectite) as in (Fig. 3). The most dominant end-member clay mineral in the Nile Delta is smectite where its percent reaches more than 67% [32].

Sand-Clay Mixture

In sand-shale sediments, porosity does not depend on the shale ratio. Pure sand and shale have porosities higher than that sand-shale mixture. Porosity is decreased when pores between sand grains are filled by clay or when sand is scattered in shale. Marion et al. [33] discussed how porosity of sand and shale mixtures could be determined as follows:

$$\Phi = \Phi_{\mathrm{sd}} - \omega v (1 - \Phi_{\mathrm{sh}}) \quad \omega v \leq \Phi_{\mathrm{sd}} \tag{13}$$

$$\Phi = \omega v \cdot \Phi_{\mathrm{sh}} \quad \omega v \geq \Phi_{\mathrm{sd}} \tag{14}$$

where ωv is the clay volume fraction and Φ_{sd} and Φ_{sh} are the porosity of the clean sand and pure shale end-members, respectively. The permeability of a clayey sand (K_Φ) is related to the permeability of a clean sand (K_{sd} and Φ_{sd}):

$$K\Phi = K_{\mathrm{sd}} \left(\frac{\Phi}{\Phi_{\mathrm{sd}}}\right)^{3\mathrm{mcs}} \tag{15}$$

where mcs is the cementation exponent. As fine-grained clays filling the pores prevent fluid flow, mcs, therefore, depend mainly on the shale ratio, particularly near the boundary between the clayey sand and sandy shale domains. Because the flow blockage is severe near this boundary, it is intuitively expected that mcs is not just a positive function of the shale ratio yet can achieve values substantially higher than 2. mcs was expanded as a power function of ωv:

$$\mathrm{mcs} = \mathrm{m}_{\mathrm{cs}}^{0} + \mathrm{m}_{\mathrm{cs}}^{1} \cdot \omega v + \mathrm{O}\left(\omega^2 v\right) \tag{16}$$

where mcs = msd is the cementation exponent of the clean sand and is in the range 1.5–2.0. Combining Eqs. (13) and (16), the permeability of clayey sand is calculated as follows:

$$K = k_{sd}\left[1 - \omega v\left(\frac{1 - \Phi_{sh}}{\Phi_{sd}}\right)\right]^{3mcs} \qquad \omega v < \Phi_{sh} \tag{17}$$

Clayey Sand-Sandy Clay Boundary

In this case, $\omega v = \Phi$sd and permeability are calculated using this formula:

$$K = k\mathrm{sd}(\Phi_{sh})^{3mcs} \tag{18}$$

where

$$m_{cs} = m^{0}{}_{cs} + m^{1}{}_{cs}\ \Phi_{sd}$$

Sandy-Clay Domain

Sandy-clay permeability can be obtained by taking into account the flow blockage by the sand grains, where the permeability in the sandy shale domain is given by:

$$K = k_{sh}(\omega v)^{msd} \qquad \omega v > \Phi_{sd} \tag{19}$$

3.6 Formation Water Salinity Determination

Assessment of groundwater quality from well logs usually centers on values for total dissolved solids (TDS) and specific conductance (C_w).

The relationship between the two terms is

$$R_w\ (\Omega\ \mathrm{m}) = 10{,}000/C_w\ (\mu\ \mathrm{mhos/cm}) \tag{20}$$

where:

C_w is conductivity in micromhos/cm.
R_w is formation water resistivity.

Then the total dissolved salt (TDS) in (ppm) of the formation water is given by:

$$\mathrm{TDS} = 0.64 \times C_w \tag{21}$$

3.7 Estimation of Hydraulic Conductivity

The proportionality constant, particularly for the water flow through sediment connected pores, is known as the hydraulic conductivity; penetrability is a part of this and is just a character of the permeable media, not the liquid. Utilizing the intrinsic permeability, hydraulic conductivity can be put in term of the permeability. Given the value of hydraulic conductivity for a subsurface system, the permeability can be calculated as follows:

$$K = ki\left(\frac{\rho g}{\mu}\right) \tag{22}$$

where K is the permeability (m^2), $ki$is the hydraulic conductivity (m/s), ρ is the density (g/cm^3), and μ is the kinematic viscosity of the fluid phase (g/cm/s). g is the acceleration due to gravity (cm/s). Intrinsic permeability is often expressed in square centimeters or in darcys. Intrinsic permeability is regularly communicated in square centimeters or in darcys. 9.87×10^{-9} cm^2 equals one darcy.

4 Result and Discussion

4.1 Spatial Horizontal Distribution

4.1.1 Bilqas Formation

Figure 4 shows the horizontal spatial distribution of the petrophysical parameters of Bilqas Formation (upper clay layer). The thickness of this layer increases from 3 m at well 20 in the southwestern part to 31 m at well 34 in the northeastern direction (Fig. 4a). However, the calculated average values of shale volume increase from 54% at well 19 in the southwest to 97% at well 22 in the northeast with the average value 69.83% (Fig. 4b). The calculated average values of porosity increases from 21% at well 3 in the central parts to 55% at well 10 in northern parts of the study area with average value 24.71% (Fig. 4c). Average values of the calculated permeability increases from 16×10^{-9} mD at well 6 to 78×10^{-9} mD at well 5 in the northeastern direction of the study area with average value 30.71×10^{-9} mD (Fig. 4d). TDS of the water that contained in the clay of Bilqas Formation is an indication of the evaporation process and the chemical composition of the recharged water to the main aquifer (Mit Ghamr Formation). The minimum value of the calculated average TDS (200 mg/l) is recognized at well 23 in the southeastern parts and increases to north and northwestern direction (well 11, 1,450 mg/l). Well 33 (TDS = 1,600 mg/l) and well 5 (TDS = 1,560 mg/l) are forming a local higher TDS values which might be related to local contamination (Fig. 4e).

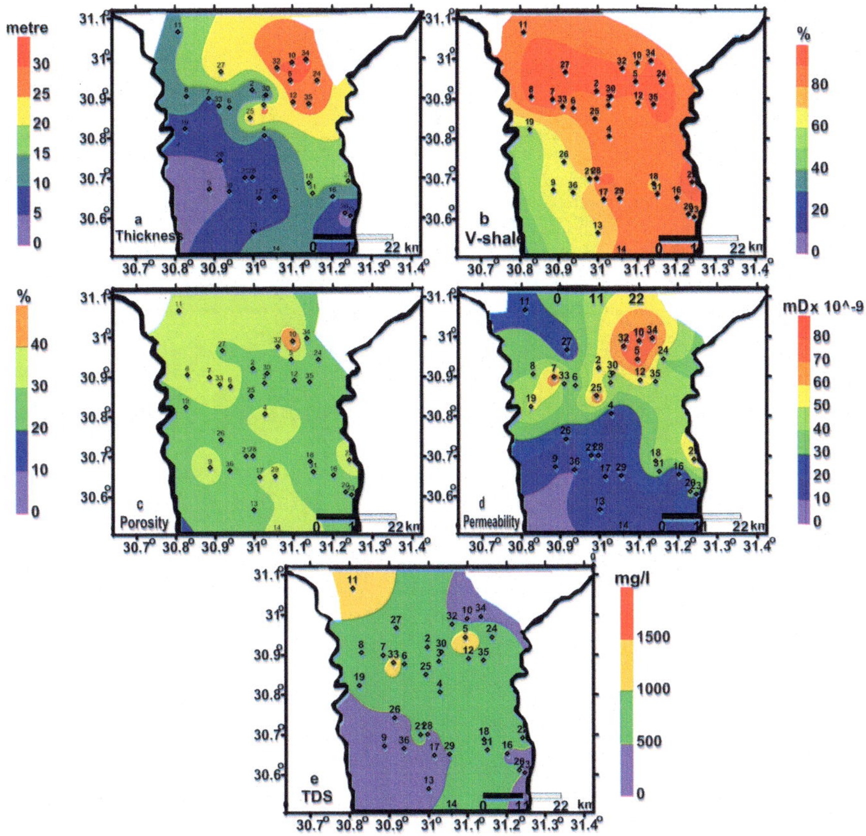

Fig. 4 Spatial distributions maps of average values of different petrophysical parameters for Bilqas formation

4.1.2 Mit Ghamr Formation

As Mit Ghamr Formation is the main aquifer in the Nile Delta, therefore, studying the spatial change in its petrophysical properties is quite important. The lowest average values of shale volume are 4.5%, 5%, and 7.1% at wells 17, 28, and 29 in the southern part, respectively, and increase to 22% at well 30 in the northern direction (Fig. 5a). Generally, the average value of shale volume increases from southern part toward northeastern parts of the study area. The minimum average value of porosity is 19% at well 2 in the central part and increases to 39% at well 10 and 11 in the northeastern and northwestern directions of the study, respectively (Fig. 5b). As a general trend, porosity increases from the south to the northwest. The minimum counted permeability value as shown in Fig. 5c is 0.1×10^{-2} mD at the location of well 35 in the northeast, and the maximum value (8.7×10^{-2} mD) is

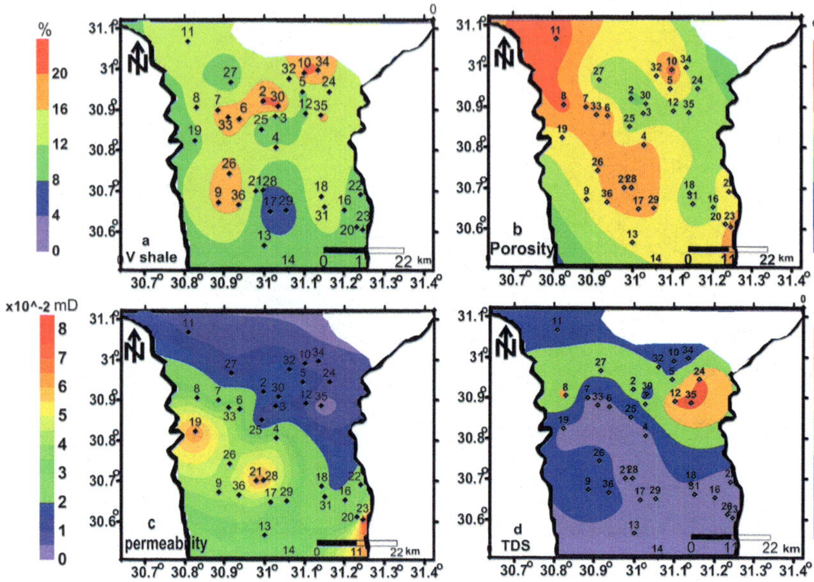

Fig. 5 Spatial distributions maps of the calculated average values of different petrophysical parameters for Mit Ghamr formation

encountered to the northwestern direction where well 11 is located. As a general trend, the average values of permeability increase from southern, eastern, and northeastern parts to the northwestern direction of the study area. In comparison, porosity and permeability have mostly the same decreasing and increasing trends. The calculated TDS values range from 220 mg/l at the location of well 28 in the central part to 2,100 mg/l at the location of well 35 in the northeastern direction (Fig. 5d). Generally, the calculated average TDS of Mit Ghamr Formation increase to the northeastern direction. The complete geochemical investigation should be done to find out whether the higher TDS in the northeastern part is related to the effect of seawater intrusion or not.

4.2 *Vertical Spatial Distribution*

The vertical distributions of the deduced petrophysical parameters along three cross sections at different locations and directions (Fig. 1b) are shown in Figs. 6, 7, 8, 9, 10, and 11. The main purpose of these cross sections is to clarify the two-dimensional images for the subsurface distributions of the petrophysical parameters. The cross section AA′ extends from south to north and includes wells 27, 2, 30, 3, 25, 4, 28, 21, 17, 13, and 14 (Figs. 6 and 7). The cross section BB′ extends east-west in the

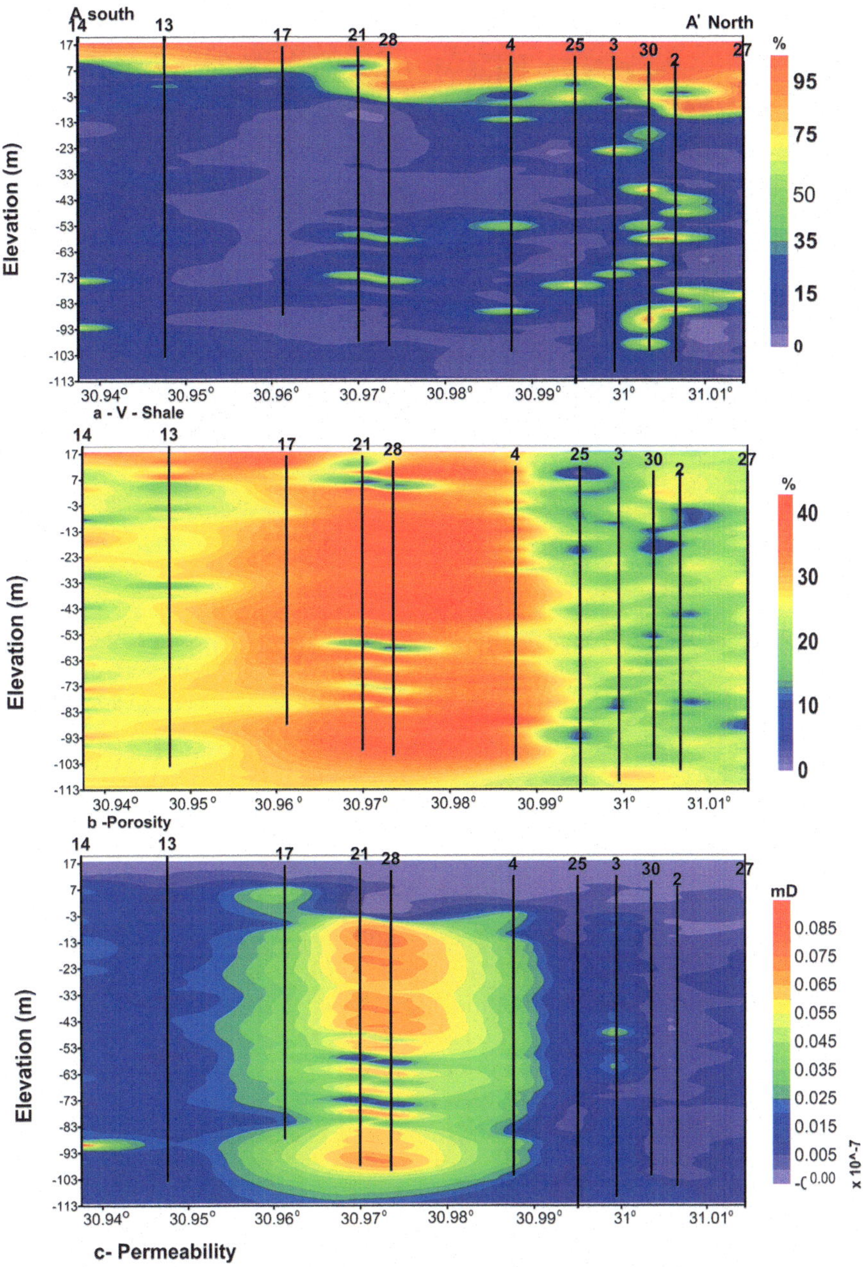

Fig. 6 Vertical distributions of the calculated values of different petrophysical parameters along the cross section AA′

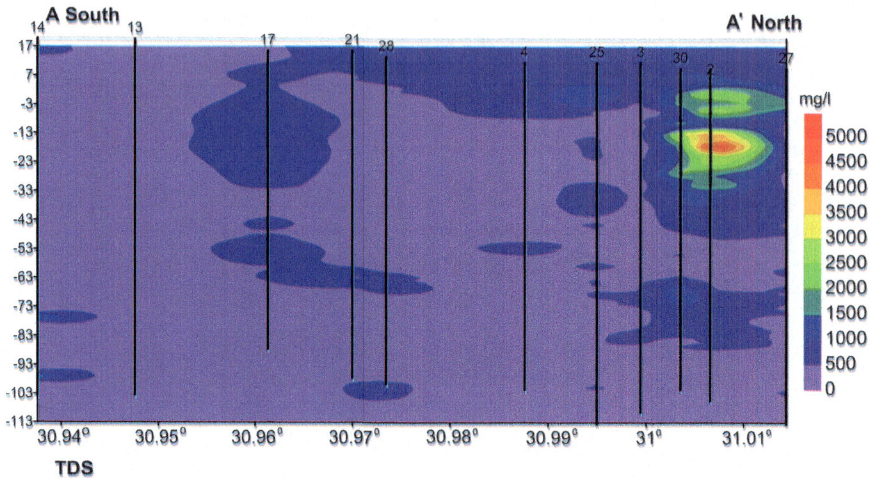

Fig. 7 Vertical distributions of the calculated values of different groundwater salinity along the cross section AA′

northern part of the study area and includes wells 35, 12, 30, 3, 2, 6, 33, 7, and 8 (Figs. 8 and 9). Cross section CC′ also extends east-west in the southern part of the study area and includes wells 20, 22, 16, 31, 18, 29, 17, 28, 21, 36, and 9 (Figs. 10 and 11).

4.2.1 The Clay Layer and Clay Content

The clay layer (V-shale is more than 75%) occupies the uppermost layer of the study area where its thickness increases gradually from the location of well 14 at the south to location of well 27 at the north and as lenses in wells 25, 30, 2, and 27 (cross section AA′, Fig. 6a), from location of well 8 at the west to location of well 27 at the eastern part (cross section BB′, Fig. 8a) and from well 9 at the south to location of well 22 at the eastern part (cross section CC′, Fig. 10a). The layers with V-shale ranging from 50 to 75% extends from well 14 at south to well 27 at north and as lenses in wells 25, 30, 2, 27, 3, 4, 21, 28, and 14 (cross section AA′, Fig. 6a), from location of well 8 at the west to location of well 27 at the eastern part and as lenses in wells 30, 2, 6, 33, 7, and 8 (cross section BB′, Fig. 8a) and from location of well 9 at the west to the location of well 20 at the eastern part and as lenses in wells 31, 18, 36, 9, 21, 28, and 22 (cross section CC′, Fig. 10a). As shown in the three cross sections, the sandy layer is interbedded with lenses of sandy clay and clayey sand.

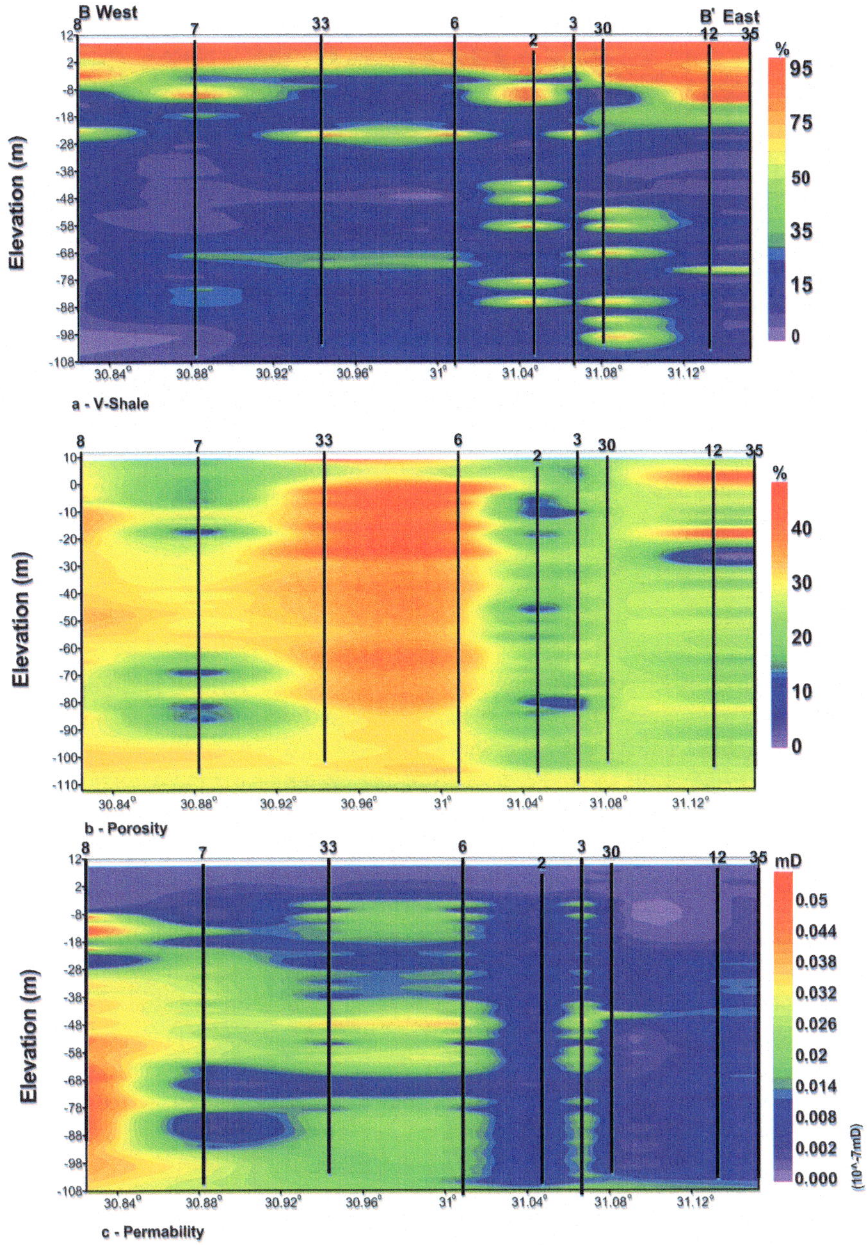

Fig. 8 Vertical distributions of the calculated values of different petrophysical parameters along the cross section BB′

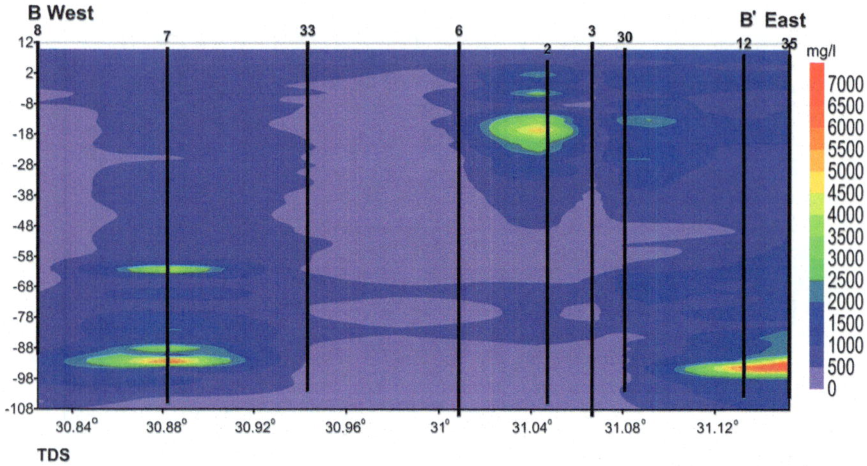

Fig. 9 Vertical distributions of the calculated values of different groundwater salinity along the cross section BB′

4.2.2 Porosity

Porosity vertical distributions showed variable ranges in vertical and horizontal directions. Low values (0–20%) are recognized as lenses at different depths in wells, 25, 2, 3, 30, and 27 at the northern part (cross section AA′, Fig. 6b), well 7 at western part and wells 2, 3, 12, and 35 at the eastern part (cross section BB′, Fig. 8b), and wells 9, 36, 21, and 28 at western part and wells 22 and 20 at the eastern part (cross section CC′, Fig. 10b). Porosity with values ranging from 20 to 30% occupies most sections of wells 25, 3, 2, and 27 at the north and as lenses in wells 13 and 14 in southern parts (cross section AA′, Fig. 6b), most wells at the eastern part and in wells 7 and 8 at western part (cross section BB′, Fig. 8b) and most wells of the cross section CC′ (Fig. 10b). The porosity values higher than 30% were represented in wells 17, 21, 28, and 4 which occupies the central part of the study area and also found in the upper parts of wells 13 and 14 and extend toward south direction (cross section AA′, Fig. 6b), wells 33 and 6 at the central part and as lenses in well 7 and 8 at the western part (cross section BB′, Fig. 8b) and most of the wells of the cross section CC′ (Fig. 10b).

4.2.3 Permeability

Permeability vertical distributions showed variable ranges in vertical and horizontal directions. Very low values (>0.005 × 10^{-7} mD) are recognized at the upper most layer of the study area (upper clay layer of Bilqas Formation) which

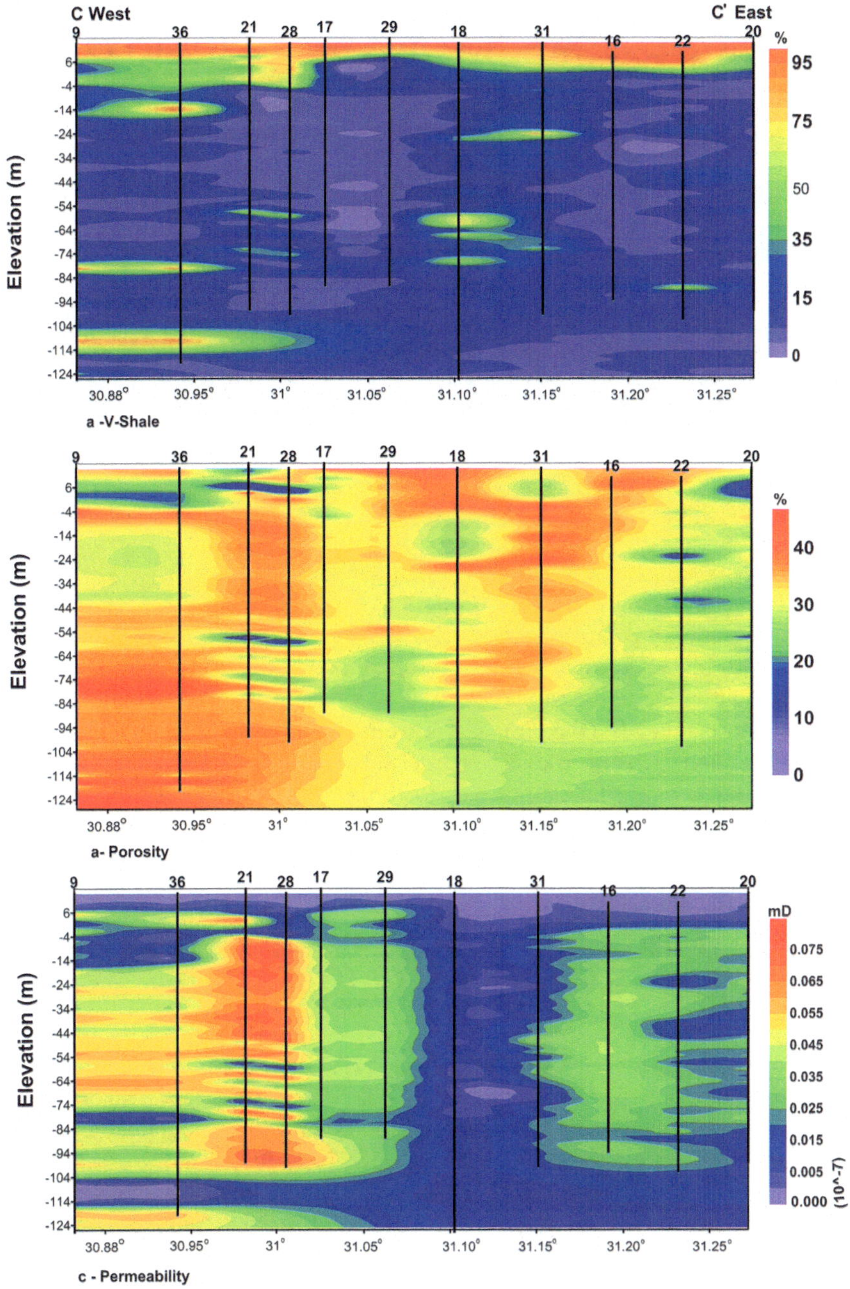

Fig. 10 Vertical distributions of the calculated values of different petrophysical parameters along the cross section CC′

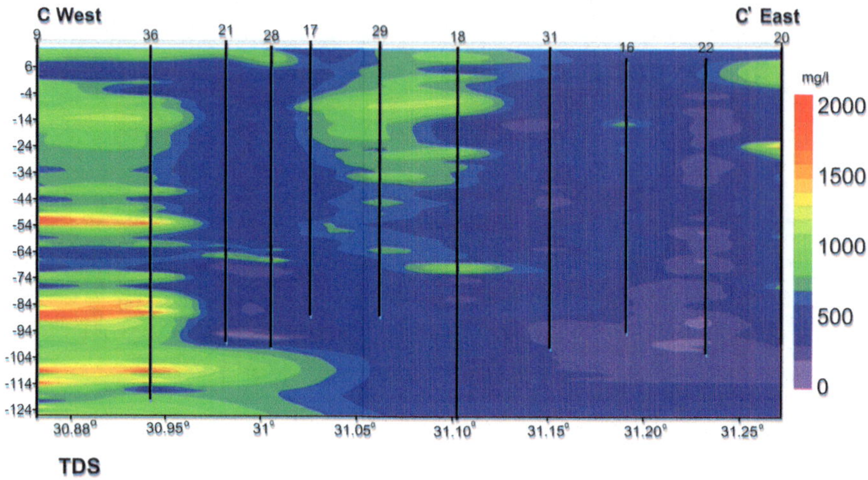

Fig. 11 Vertical distributions of the calculated values of different groundwater salinity along the cross section CC′

represented in the three cross sections (Figs. 6c, 8c, and 10c). Low permeability values of Mit Ghamr Formation ranging from 0.005×10^{-7} mD to 0.01×10^{-4} mD are shown in the northern part of cross section AA′ (Fig. 6c), in the eastern part of cross section BB′ (Fig. 8c), and in wells 18 and as lenses in wells 22, 31, 36, and 9 (section CC′, Fig. 10c). The permeability values generally increase to its highest values toward, the central part (wells 21 and 28) in cross section AA′ (Fig. 6c), the western direction (wells 7, 8, 33, 6) in cross section BB′ (Fig. 8c), and the eastern and western parts especially wells 21 and 28 in cross section CC′ (Fig. 10c).

4.2.4 Salinity

The salinity values (in term of the calculated TDS) distribution showed variable ranges in vertical and horizontal directions. TDS with values less than 500 mg/l occupy wells 13 and 14 and most parts of the other wells at depths more than 20 m (cross section AA′, Fig. 7), wells 8, 33, 6, 2, 3, 30, 2, and 30 as lenses at certain depths (cross section BB′, Fig. 9), and wells 22 and 31 and as lenses in wells 36, 21, 28, 17, 18, and 16 (cross section CC′, Fig. 11). TDS values ranging from 500 to 1,000 mg/l exhibit as lenses in wells 17, 21, 29, 2, 4, 25, 30, and 27 (cross section A-A′, Fig. 7), wells 7 and as lenses in wells 2, 30, 12, and 35 at the eastern part (cross section BB′, Fig. 9), and wells 18, 29, 26, and 9 at the west and well 20 at the east (cross section CC′, Fig. 11). The highest TDS values

(>1,000 mg/l) are represented in well 2 at the northern part of the study area (cross section A-A′, Fig. 7) and as lenses in wells 7, 2, 12, and 35 (cross section CC′, Fig. 11).

4.3 Hydraulic Conductivity

According to Eq. (22), the average of the calculated hydraulic conductivity of Mit Ghamr Formation (Fig. 12) ranges from 5.082×10^{-10} cm/s at well 35 in the northeast and increase towards the middle and the western parts of the area. The maximum value of the hydraulic conductivity is represented at well 19 (3.35×10^{-8}) in the western direction and well 23 at the southeast (3.96×10^{-8}).

Vertical distributions of the hydraulic conductivity showed variable ranges in vertical and horizontal directions. Very low values ($<2 \times 10^{-9}$ cm/s) are recognized mainly at the uppermost layer (Bilqas Formation) as shown in Fig. 13a–c. Low hydraulic conductivity values are also recognized in the northern and eastern parts of the cross sections AA′ and BB′, respectively. The hydraulic conductivity values increase toward the middle parts of the area between wells 4 and 7 (2.13×10^{-8} to 1.27×10^{-8} cm/s), and the highest values (1.93×10^{-8} cm/s) are recorded in well 8 to the west and the area between wells 9 and 17. Such horizontal and vertical change in the hydraulic conductivity could be related to the depositional system of the Nile Delta sediments.

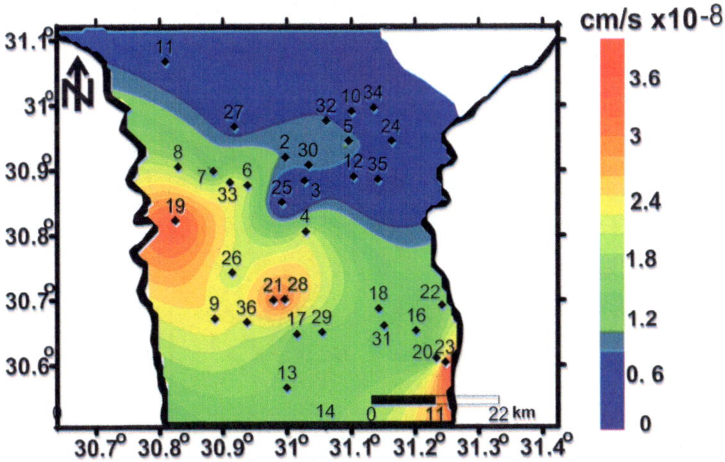

Fig. 12 Spatial distributions of the calculated average values of the hydraulic conductivity of Mit Ghamr formation

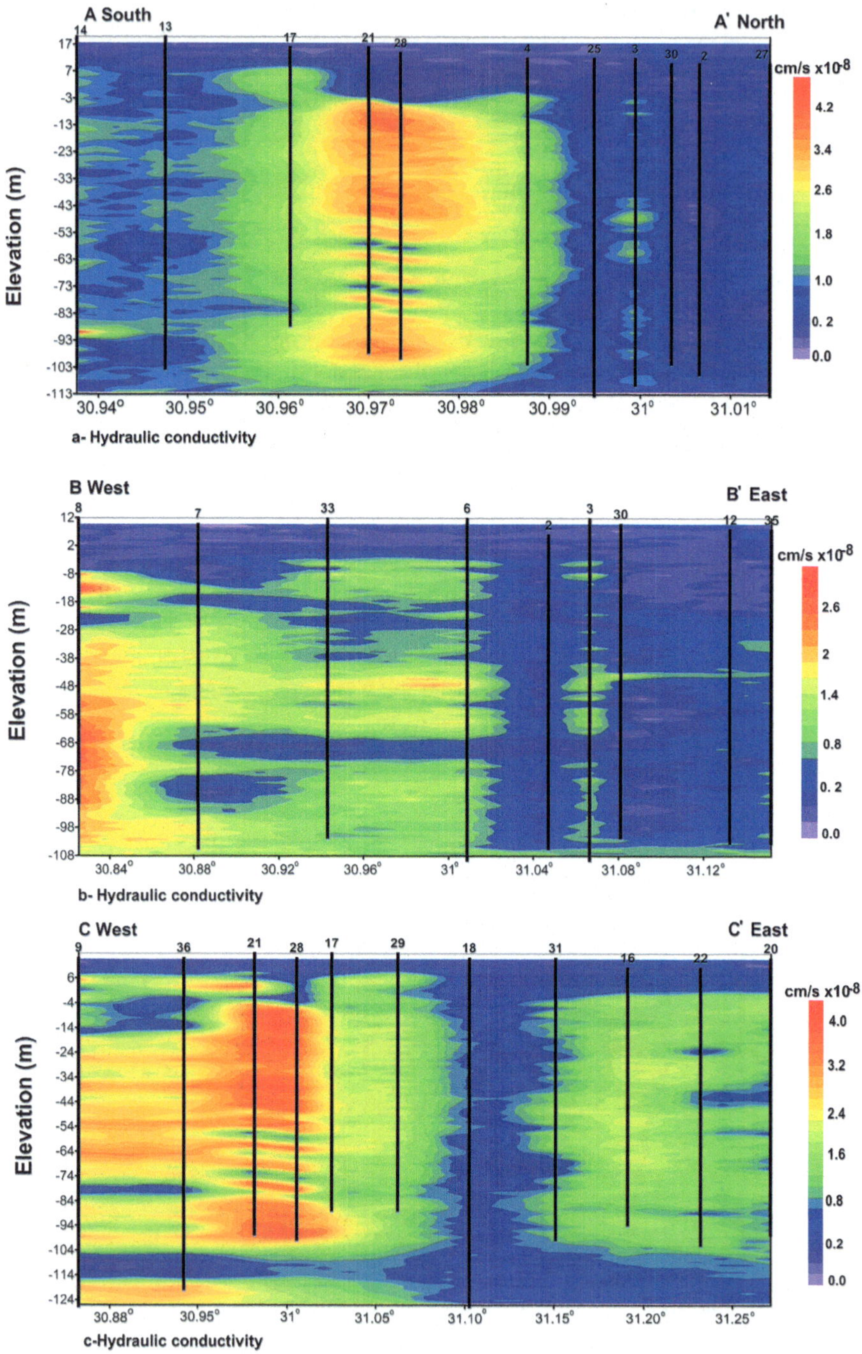

Fig. 13 Vertical distributions of the calculated values of the aquifer hydraulic conductivity along the cross sections AA′, BB′, and CC′

5 Conclusions

Electrical resistivity and gamma ray logs were used for 34 wells to identify the petrophysical, chemical, and hydrogeological characteristics of the Nile Delta aquifer due to its significant importance. The Nile Delta reservoir consists of two formations. The upper is the Bilqas Formation, which is of a clay nature and the lower formation is of a sandy nature and called Mit Ghamr Formation. According to the calculations used, the thickness of Bilqas Formation ranges from three meters in the southwest direction to 31 m in the northeast direction. The shale content ranges from 54 to 97%. The porosity ranges from 21 to 55% with an average of 24.71%. This layer has low permeability values ranged from 16×10^{-9} to 78×10^{-9} mD with an average of 30.71×10^{-9} mD. The calculations showed that the hydraulic conductivity of this layer is weak and less than 2×10^{-9} cm/s. The water salinity of this layer ranges from 200 mg to 1,600 mg/l. The increase in the thickness of this layer, the increase in porosity, the decrease in permeability and hydraulic conductivity, as well as the increase in the water salinity are to the north and northeast directions.

Mit Ghamr is the main aquifer in the region where its minimum value of the shale content is 4.5% and the maximum value is 22%. Lenses from the clay are scattered in different places and increasing in intensity in the northeastern direction of the region. Porosity ranges from a minimum of 19% to a maximum of 39%. The permeability recorded high values in this formation where it ranged from 0.1×10^{-2} to 8.7×10^{-2} mD. The calculations also showed that the hydraulic conductivity values for this formation ranged from 5.082×10^{-10} to 2.134×10^{-8} cm/s. In this layer, the increase in the shale content, the increase in porosity, decrease in the permeability and hydraulic conductivity, as well as the increase in salinity are to the north and northeast.

Acknowledgments The authors are grateful to Tanta University for the financial support offered by the project number "TU-01-12-03" during the course of this paper.

References

1. Serra O (1984) Fundamentals of well-log interpretation. Elsevier, Amsterdam
2. Tselentis GA (1985) The processing of geophysical well logs by microcomputers applied to the solution of hydrogeological problems. J Hydrol 80:215–236
3. Rubin Y, Hubbard SS (2005) Hydrogeophysics. Water Science and Technology Library Series 50, Springer, Dordrecht, Berlin
4. Winslow AG, Kister LR (1956) Saline-Water Resources of Texas: U.S. Geological Survey Water-Supply Paper 1365, 105 p
5. Paillet FL, Reese RS (2000) Integrating borehole logs and aquifer tests in aquifer characterization. Ground Water 38(5):713–725

6. Sloto RA, Goode DJ, Frasch SM (2002) Interpretation of borehole geophysical logs, aquifer-isolation tests, and water quality, supply wells 1 and 2, Willow Grove Naval Air Station/Joint Reserve Base, Horsham Township, Montgomery County, Pennsylvania Water-Resources Investigations Report 2001–4264
7. Szabó NP (2013) Shale volume estimation based on the factor analysis of well-logging data. Acta Geophys 59(5):935–953
8. Szabó NP (2015) Hydraulic conductivity explored by factor analysis of borehole geophysical data. Hydrgeol J 23(5):869–882. doi:10.1007/s10040-015-1235-4
9. Szabó NP, Dobróka M (2013) Float-encoded genetic algorithm used for the inversion processing of well-logging data. In: Michalski A (ed) Global optimization: theory, developments and applications. Mathematics research developments, computational mathematics and analysis series, Nova Science Publishers Inc, New York
10. Szabó NP, Dobroka M (2013) Extending the application of a shale volume estimation formula derived from factor analysis of wireline logging data. Math Geosci 45(7):837–850. doi:10.1007/s11004-013-9449-2
11. Szabó NP, Kormos K, Dobroka M (2015) Evaluation of hydraulic conductivity in shallow groundwater formations: a comparative study of the Csókás' and Kozeny–Carman model. Acta Geodaetica et Geophysica Hungarica. doi:10.1007/s40328-015-0105-9
12. Attia MI (1954) Deposits in the Nile valley and the Delta "Geological survey Egypt," Cairo, vol 12, pp 147–165
13. Hurst HE (1952) Long-term storage capacity of reservoirs. Trans Am Soc Civil Eng 116:770–808
14. Kashef AI (1981) The Nile-One River and nine countries. J Hydrol 53:53–71
15. Kashef AI (1981) Technical and ecological impacts of the High Aswan Dam. J Hydrol 53:73–84
16. Said R (1962) The geology of Egypt. Elsevier, Amsterdam, The Netherlands
17. Said R (1993) The Nile River: geology, hydrology, and utilization. Pergamon, New York
18. Sestini G (1989) Nile Delta; a review of depositional environments and geological history. In: Whateley MKG, Pickering KT (eds) Geological Society Special Publication No. 40, pp 99–127
19. Shahin M (1985) Hydrology of the Nile Basin. El Sevier Science Publishers B V, 575 p
20. Shata AA, El-Fayoumy IF (1970) Remarks on the hydrogeology of the Nile Delta. Proceedings of the Bucharest symposium. Deltas
21. Salem ZE (2009) Natural and human impacts on the groundwater under an Egyptian village, central Nile Delta – a case study of Mehallet Menouf. In: 13th international water technology conference (IWTC, 13), vol 3, March 12–15, 2009, Hurghada, Egypt, pp 1397–1414
22. Sandford KS, Arkell WJ (1939) Paleolithic man and the Nile Faiyum divide Chcago. Univ. Oriental Inst. Pub., I., pp 1–77
23. Poupon A, Gaymard R (1970) The evaluation of clay content from logs. Trans., SPWLA 11th Annu. Logging Symp., Pap. G
24. Dresser A (1982) Well logging and interpretation techniques; the course for home study. Dresser Atlas Division, Dresser Industries, Houston, TX, 350 p
25. Dresser A (1979) Well log interpretation charts. Dresser Atlas Division, Dresser Industries, Houston, TX
26. Archie GE (1942) The electrical resistivity log as an aid in determining some reservoir characteristics. Trans AIME 146:54–62
27. Person M, Raffensperger JP, Ge S, Garven G (1996) Basin-scale hydrogeologic modeling. Rev Geophys 34:61–87
28. Wintsch RP, Christoffersen R, Kronenberg AK (1995) Fluid-rock reaction weakening of fault zones. J Geophys Res 100:13021–13032
29. Moore DE, Morrow CA, Byerlee JD (1982) Use of swelling clays to reduce permeability and its potential application to nuclear waste repository sealing. Geophys Res Lett 9:1009–1012
30. Revil A, Cathles LM (1999) Permeability of shaly sands. Water Resour Res 35(3):651–662. http://onlinelibrary.wiley.com/doi/10.1029/98WR02700/full

31. Jorgensen DG (1980) Relationships between basic soil engineering equations and basic ground-water flow equations: U.S. Geological Survey Water-Supply Paper 2064, 40 p
32. Abu Khatita AM (2011) Assessment of soil and sediment contamination in the middle Nile Delta area (Egypt)-geo-environmental study using combined sedimentological, geophysical and geochemical methods
33. Marion D, Nur A, Yin H, Han D (1992) Compressional velocity and porosity in sand-clay mixtures. Geophysics 57:554–563

Part IV
Groundwater Investigations and Aquifer Characterization

Resistivity Characterization of Aquifer in Coastal Semiarid Areas: An Approach for Hydrogeological Evaluation

Mohamed Attwa and Halim Ali

Abstract In coastal and semiarid regions, the scientific interest lies in imaging the saltwater intrusion and delineating freshwater aquifer zones, respectively. Direct current resistivity (DCR) and induced polarization (IP) geophysical methods are commonly used to assess hydraulic characteristics of the aquifer. Particularly, the main reason for hydrogeophysical application of both DCR and IP is that the electrical characteristics of aquifers depend mainly on the geometry of the pore space and the porosity controlling the soil and rock effective transport properties. For preliminary hydrogeological investigations, these methods are applied at a wide range of field and laboratory scales. Accordingly, the vulnerable zone to the saltwater intrusion and/or contamination can be characterized by high accuracy. Furthermore, empirical and semiempirical relationships are widely used to predict the aquifer petrophysical characteristics, e.g., hydraulic conductivity, using the inversion results of such electrical methods. Equally, conventional and nonconventional DCR inversion algorithms are developed to reduce the nonuniqueness problem of actual resistivity interpretation and, consequently, to obtain more meaningful models than previously reported. As case histories, this chapter demonstrates the efficiency of DCR method for hydrogeological assessment in Nile Delta, Egypt, emphasizing on technical constraints to achieve sustainable development in coastal and semiarid areas.

Keywords DC inversion, DC resistivity, Hydraulic conductivity, Hydrogeophysics

---

M. Attwa (✉)
Geology Department, Faculty of Science, Zagazig University, Zagazig, Egypt
e-mail: attwa_m2@yahoo.com

H. Ali
National Authority of Remote Sensing and Space Sciences (NARSS), Cairo, Egypt

A. M. Negm (ed.), *Groundwater in the Nile Delta*,
Hdb Env Chem (2019) 73: 213–234, DOI 10.1007/698_2017_210,

Contents

1 Introduction

Geoelectrical methods are increasingly popular in detecting and characterizing aquifers as the resistivity of sediments and rocks depends mainly on their water and material contents (e.g., [1–3]). One of such electrical methods is direct current resistivity (DCR) technique. In DCR method, the potential difference is measured via injecting DC into the ground with electrodes at the ground surface. The DCR method has received a great attention because of its potential applications in the field of hydrogeology and saltwater intrusion tomography [4, 5]. In the arid and semiarid region, the DCR method is popularly used for groundwater exploration and aquifer mapping (e.g., [6, 7]). This can be attributed to the rapid advancement in the development of digital modeling solutions and electronic technology of hydrogeological problems (e.g., [8]). Besides the efficiency of such method for lithological characterization, the DCR technique has led to many studies on hydrogeological parameter estimation of the aquifer (e.g., [9]). On the other hand, in the coastal areas, the DCR imaging is suitable for monitoring saltwater intrusion and groundwater contamination as this method quickly investigates a large site without needing for well drilling (e.g., [10]).

Induced polarization (IP) technique is one of the geoelectrical methods based on studying the low-frequency polarization processes taking place in porous media. Generally, the IP method extends from the classical DCR method. Basically, IP measures the earth's capacity to store an electric charge over time. Accordingly, after the injected current is shut off, the voltage decay curve is measured; the higher the IP, the longer over time the stored charge. It is well known that IP is a considerable and useful method for mineral exploration. On the other hand, the real and imaginary parts of conductivity are measured over a range of frequencies in spectral induced polarization (SIP). The resistivity amplitude and the phase shift between the input current and measured voltage can be recasted into a complex conductivity. Recently, the SIP method has shown its potential to predict the petrophysical parameters in saturated conditions at the field and laboratory scales (e.g., [11, 12]). Usually, DCR method is the preferred, but the resistivity data interpretation suffers from a major flaw: the inability to discriminate between surface and bulk conductivity. This causes unrealistic interpretation in hydrogeophysics. Such main problem can be resolved by using SIP method (e.g., [11, 13, 14]).

In Egypt, the River Nile is the lifeline, where the main water source is surface water from the River. With increasing populations, urbanizations and land reclamation projects in Egypt, the surface water from the River Nile is not sufficient. The groundwater is the second important water resources. Furthermore, the saltwater has intruded from the northern and eastern parts of Egypt along the Mediterranean Sea and Suez Canal, respectively. Additionally, the overpumping of groundwater at the northern parts causes vertical upwelling of saltwater from deeper parts causing degradation to groundwater quality. It can also be noticed that a large part of agricultural areas in the northern parts has been, consequently, affected by saltwater intrusion. Accordingly, the groundwater assessment and management have become extremely important.

This chapter attempts to guide the reader to some of the important sources of information on the use of DCR and IP methods for groundwater assessment. In the first section of this chapter, fundamental of DCR and IP methods will be explained. The linkage between the data of geoelectrical and hydraulic characteristics at the field and laboratory scales will be discussed. The geoelectrical data inversion techniques will be briefly covered in the second section of this chapter. Practically, three cases studied from Nile Delta are presented in the third part. Here, a saltwater intrusion and groundwater resource characterization, i.e., hydrogeophysical characterization, in coastal and semiarid areas, respectively, are reported.

2 DC/IP Resistivity Methods

2.1 Fundamentals

Basically, DCR investigations are based on the principle of applying electric current to the earth through two electrodes and measuring the potential difference between two or more other electrodes. The distance between the electrodes and the measured potential difference is used to make interpretations of subsurface conditions. The geological formation resistivity is a physical property depending on the electric current flow in the formation. Subsurface layers resistivity is controlled by several geological parameters such as rock texture, mineralization, and fluid conductivity contained within the rock. In general, the resistivity increases with grain size and also when the rock is fine-grained and compact. Other factors that affected the rock resistivity include clay content, groundwater salinity, cation exchange capacity, pore water temperature, dissolved salt concentrations, and contaminants (e.g., [15]). Accordingly, the resistivity surveys are the best suited for clay or saline zone delineation. Table 1 shows the resistivity value range of common rocks, soils, and chemicals (e.g., [16]). The soils and sedimentary rocks have a wide variation according to their resistivity. The unconsolidated sediments have lower resistivity values than rocks.

The rock conduction is by way of pore fluids, i.e., electrolytes, and so is dependent on pore fluid presence and nature. Most of earth materials obey the empirical relationship known as Archies law (Eq. 1):

Table 1 Resistivity values of common earth's materials and chemicals [16]

Materials	Resistivity (Ω m)
Igneous and metamorphic rocks	
Granite	5×10^3–10^6
Basalt	10^3–10^6
Marble	10^2–2.5×10^8
Quartzite	10^2–2×10^8
Sedimentary rocks	
Sandstone	8–4×10^3
Shale	20–2×10^3
Limestone	50–4×10^2
Soils and water	
Clay	1–100
Alluvium	10–800
Groundwater (fresh)	10–100
Saltwater	<1

$$F = \rho_{\mathrm{w}}/\rho = a\varphi^{-m} \quad (1)$$

ρ_{w} is the water resistivity, ρ is the formation resistivity, a is an empirical constancy (typically 1 for unconsolidated sediments), m is an empirical constant (typically 2 for unconsolidated sediments), φ is the effective porosity, and F is the formation factor related to the volume and tortuosity of the pore space.

2.2 Physical Base

The DCR method is based on the principle that the earth material acts as a resistor in a circuit. After inducing an electrical current into the ground, the ability of that material is measured to resist the current by recording the electrical potential to achieve information on the resistivity structure in the ground. Generally, the current flows in radial form out from the source of current, and the equipotential surfaces forming half spheres run perpendicular to the current flow (Fig. 1).

In the common situation with both a current source and a current sink, the current flow lines and the equipotential surfaces become more complex. Electric field is created in the earth by means of low-frequency AC generators and/or DC batteries. When the electric current is sent by means of two grounded electrode, called current electrodes "A" and "B" placed at two selected points, electrons start to flow through the earth's subsurface from the negative to the positive terminals of the current source. The potential differences are measured by another two potential electrodes called "M" and "N" (see Fig. 2). If the ground is heterogeneous, the resulting resistivity obtained using Eq. (2) for any of the different types of configurations is called apparent resistivity (ρ_{a}) that constitutes a weighted average of the true resistivity.

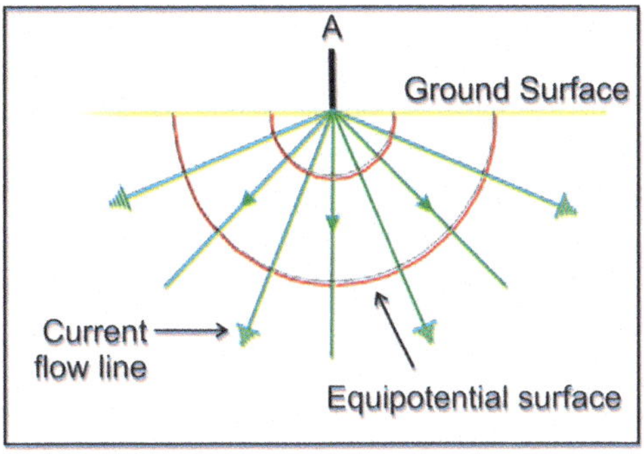

Fig. 1 Simplified current flow lines and equipotential surfaces arising from a single current source

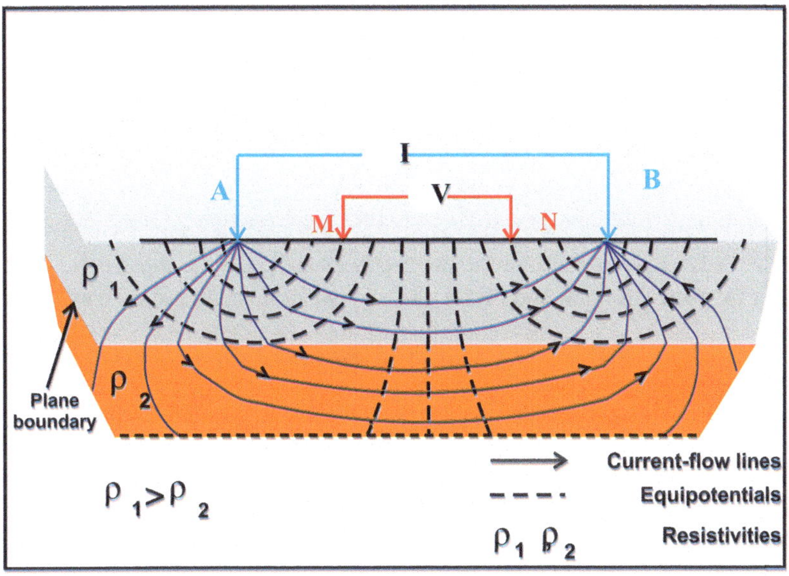

Fig. 2 Resistivity measurement principles with a four-electrode configuration

Geoelectrical data are commonly expressed as apparent resistivities:

$$\rho_a = K(^{\Delta V}/_{I}), \tag{2}$$

where

$$K = 2\pi\left(\frac{1}{r_{AM}} - \frac{1}{r_{AN}} - \frac{1}{r_{BM}} + \frac{1}{r_{BN}}\right)$$

where K is the geometrical factor having the dimension of length (m), which depends on the arrangement of the four electrodes.

The IP method is considered as an extension to the well-known geoelectric technique. It is usually assumed that IP and DCR measurements have been measured simultaneously. The IP method depends mainly on a small amount of electric charge storage when a current is passed through the rock, to be released and measured when the current is switched off. Generally speaking, the IP measurements became more established in the last 20 years.

In time domain IP, the ratio of overvoltage (V_p) to observed voltage (V_0) is measured, and this ratio is known as chargeability (M), which is expressed in millivolts per volt or percent:

$$M = \frac{V_p}{V_0} \text{ and } M = \frac{1}{V_0}\int_{t1}^{t2} V_{p(t)}dt = \frac{A}{V_0} \tag{3}$$

In frequency-domain IP, apparent resistivity at two or more AC frequencies is measured. Accordingly, the IP effect is well described by the so-called frequency effect FE (unitless):

$$\mathrm{FE} = \frac{\rho_0 - \rho_\infty}{\rho_\infty}, \tag{4}$$

where ρ_0 is the apparent resistivity at low frequency and ρ_∞ is the apparent resistivity at a higher frequency. Both M and FE are easily interrelated and they can be expressed as

$$M = \frac{\mathrm{FE}}{1 + \mathrm{FE}} = \frac{\rho_0 - \rho_\infty}{\rho_0} \tag{5}$$

In the spectral induced polarization (SIP) method, the resistivity/conductivity magnitude and phase shift are measured over a wide range of frequencies (0.625 Hz to 10 kHz).

In the SIP method, the fundamental quantities of interest are the electrical conductivity magnitude $|\sigma(\omega)|$ and phase $\varphi(\omega)$ (or the equivalent real and imaginary parts of conductivity) of complex conductivity as a function of frequency as

$$|\sigma(\omega)| = \left((\sigma')^2 + (\sigma'')^2\right)^{0.5} \tag{6}$$

$$\sigma' = |\sigma(\omega)|\cos\varphi \text{ and } \sigma'' = |\sigma(\omega)|\sin\varphi, \tag{7}$$

where ω is the angular frequency and σ' and σ'' are the real and imaginary parts of the electrical conductivity, respectively.

SIP method is based on the change of electrochemical and electronic conductivity in the pore spaces of the rocks and soil [17]. Recent reviews of the method, including a discussion of the underlying mechanisms, are given by [18, 19]. Figure 3 shows three polarization mechanisms associated with the mineral grain polarization coated by an electrical double layer (EDL) coating grain surface [21]. They include (1) the

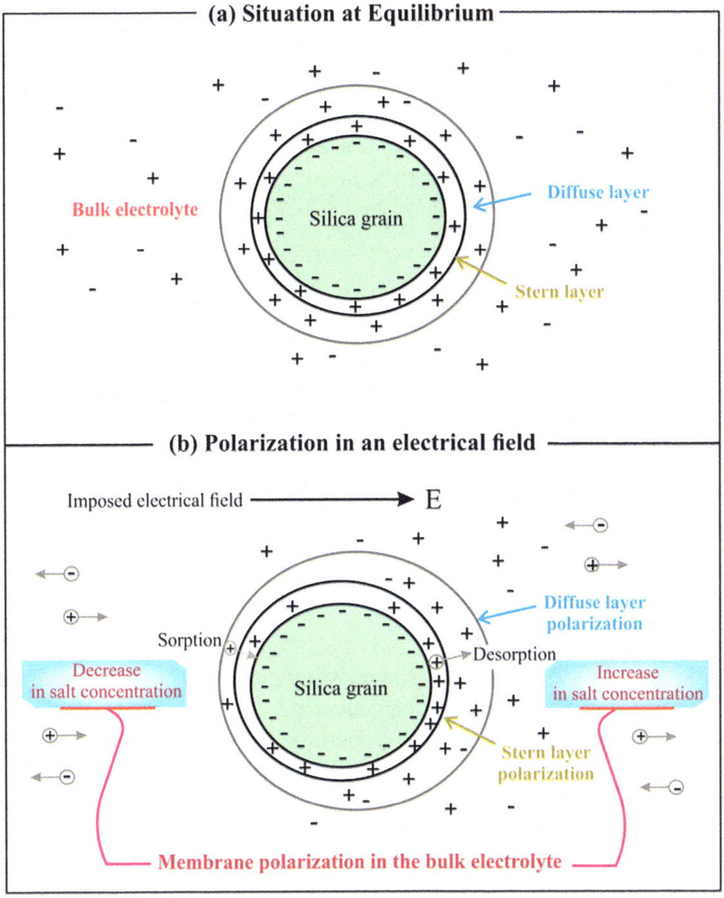

Fig. 3 A single grain polarization. (**a**) Electrical double layer (Stern plus diffuse layers) at equilibrium. (**b**) Polarization mechanisms of the grain under an electrical field *E* (modified after [20]).

diffuse layer deformation at a very high frequency >10 kHz see the so-called Debye–Falkenhagen effect; Falkenhagen [22]), (2) the Stern layer polarization [23, 24], and (3) the diffusion or membrane polarization, which is related to the difference in the transference number for the cations and anions during their migration in the porous material [25, 26].

2.3 *Electrode Arrays and Field Techniques*

There are common electrode configurations (i.e., arrangements) for placing the potentials and current electrodes for field surveying. These electrode arrays for geologic investigation are Schlumberger (SC), Wenner (MN), and dipole-dipole (DD) and pole-dipole (PD) array (Fig. 4). The choice of such arrays for a field survey depends on:

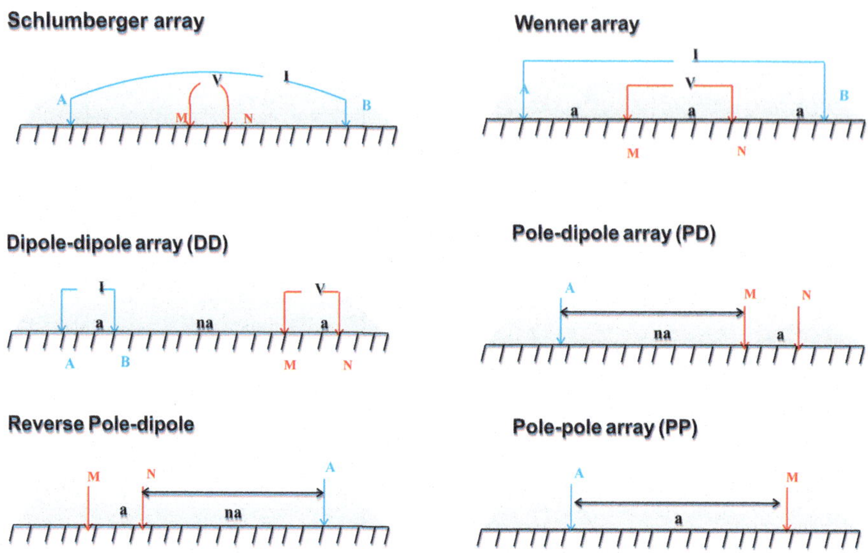

Fig. 4 Some of the most commonly used electrode arrays. The letters A and B designate the current electrodes, and the letters M and N denote the potential electrodes

1. The structure types to be mapped, the resistivity instrument sensitivity, the noise level, the signal strength, and the data coverage
2. The array sensitivity to vertical and horizontal changes in the subsurface resistivity (e.g., [27])
3. The depth of investigation (DOI) [28]
4. The array resolution for horizontal and dipping layers and sensitivity to surface heterogeneous [29]

The geoelectrical data acquisition can be carried out using three field techniques: (1) single electrical sounding (SES), (2) continuous electrical sounding (CES), and (3) pulled array continuous electrical sounding (PACES). The SES is applied to determine the resistivity variation with depth. SES is typically carried out using Schlumberger array (Fig. 4), since the ES curve can be interpreted using a 1D horizontally layered earth model. In the CES, electrical resistivity data along a profile is obtained (Fig. 5). The CES measurement's advantage is their high lateral and vertical resolution along the profile. Accordingly, CES combines the profiling and sounding surveys. On the contrary to SES, a large number of electrode arrays can be used along with a straight line. The CES data are then used for the forward and inverse modeling programs that produce the resistivity depth sections (i.e., 2D section). In the PACES, a small caterpillar drags a set (i.e., tail) of electrodes along with the processing electronics. Eight electrode arrays can be acquired simultaneously permitting for electrical sounding data to be measured continuously along the profile (Fig. 6). Further, digital data measurements can be

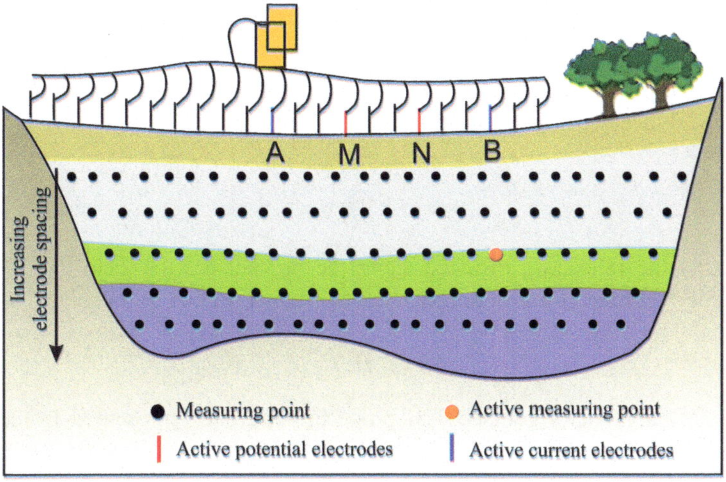

Fig. 5 A schematic model of the CES technique and data density using Wenner array

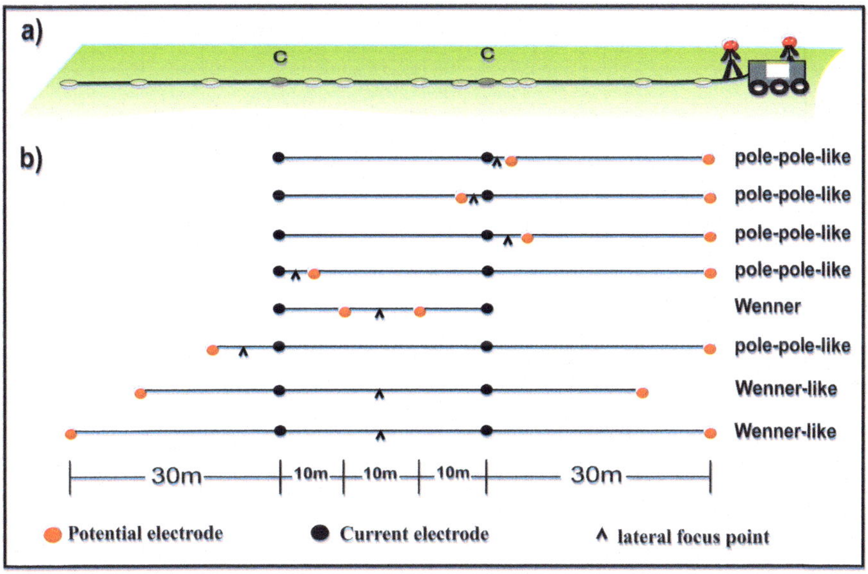

Fig. 6 A schematic model of the PACES system (**a**) showing the pulled electrodes by a small caterpillar using eight-electrode configurations (**b**)

carried out online by equipment on the caterpillar. The system in PACES has the ability to determine the inadequate galvanic contact at the potential current electrodes.

2.4 Hydraulic Conductivity Prediction

For clay-free porous media, i.e., uniformly sized particles, the relaxation time constant, τ, of IP decay curve can be given by [30, 31]

$$\tau = \frac{L^2}{D}, \tag{8}$$

where L is the relaxation distance along the applied electric field and D is the coefficient of diffusion. Accordingly, the pore radius, R, can be calculated by

$$R = (D\tau)^{0.5}, \tag{9}$$

The earth material's geoelectrical properties may be represented by complex resistivity (ρ^*):

$$\rho^* = \rho' + i\rho'', \tag{10}$$

where ρ' and ρ'' are the real and the imaginary resistivity parts, respectively, of ρ^* and φ is the phase shift of ρ^* that can be also given as follows:

$$\varphi = \arctan(\rho''/\rho') \tag{11}$$

Generally speaking, the ρ' (i.e., DC resistivity) has been commonly used to estimate hydraulic conductivity (K) value for unconsolidated aquifers. Positive or negative linear log-log relationships between ρ' and K were explained by [27]

$$K = a(\rho')^b \tag{12}$$

The correlation of K with ρ' is directly proportional on a large scale for sandy-clayey aquifer [11]. On a local scale, sometimes, such correlation is inversely proportional [32].

Many literatures [11, 33] demonstrating a power law dependence of ρ' on specific surface area (S_{por}) [34] indicated that the φ was particularly constant over a wide range of frequency and presented an empirical relationship to estimate K for a single frequency value, around 1 Hz. The proposed model, "Börner model," to calculate K (m/s) for a constant phase angle (CPA) is

$$K = \frac{a}{F\left(S_{\text{por}}\right)^c}, \tag{13}$$

where exponents a and c are adjustable parameters and F is the formation factor.

The exponent "a" is equal to 0.00475 [35] and 1 [34] for consolidated and unconsolidated sediments, respectively. The exponent $a = 0.00475$ was recommended by [36] for unconsolidated sediments. Further, the exponent "c" was formulated in the range of $2.8 < c < 4.6$ by [34] regarding the method of K measurements. The F can be calculated using [34]

$$F = \frac{\sigma_{\mathrm{w}}}{\sigma' - \sigma''/1} \tag{14}$$

where σ_{w} is the conductivity of pore fluid. For unconsolidated sandy aquifer, S_{por} can be calculated [18] as

$$S_{\mathrm{por}} = \left(\frac{\sigma''}{0.0101}\right)^{1/0.916}, \tag{15}$$

where σ'' in mS/m and Spor is in $\mu\mathrm{m}^{-1}$.

Slater and Lesmes [37] suggested another approach to calculate K by using the CPA model. The suggested model is based on a relation between the ρ'' with the grain diameter regarding the smallest 10% portion (i.e., d10) of the soil sample. Additionally, this method shows an inverse linear relation between σ'' and K around 1 Hz (Slater model):

$$K = \frac{m}{(\sigma'')^n} \tag{16}$$

where σ'' is given in µS/m. m and n are adjustable parameters with $m = (2.0 \pm 0.3)$ 10–4; $n = 1.1 \pm 0.2$, and K is in m/s.

It is generally accepted that τ(tau) is controlled by the diffusion process [20] in the electrochemical double layer (EDL). A power law relationship between some relaxation time constants and permeability (k) has been reported in published data sets:

$$k = a\tau^b, \tag{17}$$

where a and b are formation-specific parameters.

A power law relationship between the mean relaxation time (τm) and S_{por} was established by [38]. In case of a weak relation between τ and permeability (k) and/or K, this relation cannot be applied. Furthermore, [18] derived an empirical power law relationship to predict k as a function of conductivity σ'', chargeability M and τm, which are derived from DD:

$$k = a \cdot (\rho')^b \cdot M^c \cdot (\tau_m)^d. \tag{18}$$

The four empirical parameters a, b, c, and d are determined by a multivariate regression analysis.

Based on the Debye decomposition (DD) of IP spectra, [11] proposed a power law relationship in the form of

$$K = A(\rho')B(t)C, \tag{19}$$

where the A coefficient and the B and C exponents are determined by a regression analysis of the logarithmic quantities τ and K. This methodology was adopted for the estimation of the K when there is a general trend of increasing or decreasing both individual ρ' and τ with K.

2.5 Geoelectrical Data Inversion

Many researchers discussed the geoelectrical data interpretation techniques in order to solve a wide range of complex inverse problems (e.g., [29]). The main goal of geoelectrical interpretation is to define the electrical property distributions inside the earth from a set of measurements, which can be mathematically modeled from physical laws, e.g., Poisson and Laplace's equations. Figure 7 illustrates the interpretation process of geoelectrical data, which depends mainly on solving the forward and inverse problems. In the forward problem, the link between the model parameters and the model response (2D and 3D resistivity models) are solved by applying finite-difference or finite-element techniques (e.g., [39]). In both techniques, the resistivity distribution is approximated using a mesh of cells or individual elements with constant resistivity value.

The current commonly inversion approaches depend on mainly derivative-based inversion (DBI) methods [40]. In such conventional inversion schemes, the 2D forward calculations are constructed of small cells to construct 2D model. In general, the parameterization jointly with a 2D regularization usually represents a smooth and/or blurred image of the measured subsurface structures, which is known as the smearing [41]. Consequently, new parameterizations were developed for subsurface layers extraction with sharp boundaries [42]. Recently, model in the form of some interfaces can be used to represent the subsurface layers more consistent with the actual geological condition named as "structure-based (SB) model" [43]. For example, Fig. 8 shows the inverted ERTs using DBI and nonconventional with SB model approaches compared with the drilled borehole data at a new development area (Tenth of Ramadan City), Egypt, to image near-surface lateral heterogeneity. It is clear that the control point number (i.e., turquoise color lines) and the node point number (pink dots over white circles) representing the subsurface layers were suggested by the borehole data and the visual inspection of the 2D-ERT using DBI method.

The genetic algorithim (GA) and stochastic techniques belong to the same family of global optimization (GO) algorithms. The main difference between GO and the

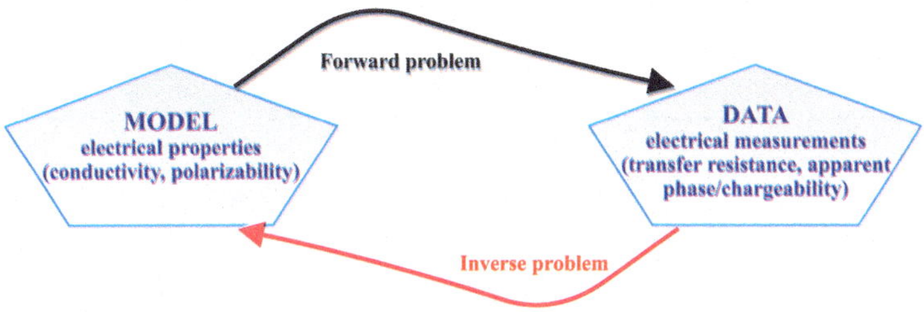

Fig. 7 DCR interpretation process in the form of forward and inverse problems

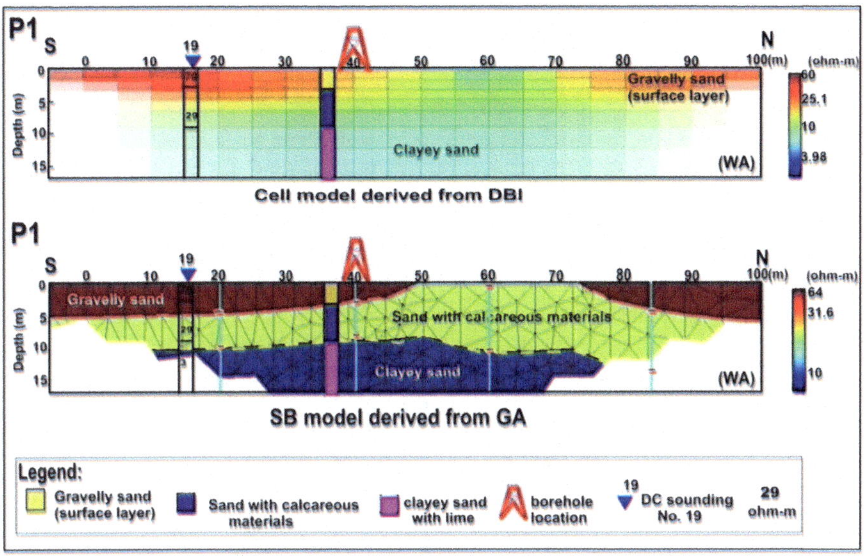

Fig. 8 2D ERTs obtained from Wenner alpha (WA) data using (*top*) DBI method and (*bottom*) nonconventional inversion method produced by GA and structure-based (SB) model. Note that (*bottom*) all model nodes (*pink* dots over *white* circle) along the control points (turquoise color) defining the layer interfaces were derived from the borehole information and the preliminary DBI results

conventional DBI is the way of data use. In the DBI methods, the differences between observed and calculated data are converted to the parameter discrepancies applying an operator called generalized inverse at each iteration step. In the GO methods, the data are not used in model update and creation. The parameter update way depends on the global minimization type. Generally speaking, the GA generates model updates by the biological process simulation and can verify a huge distinct and complicated model number. Figure 9 shows an example of DCR sounding inversion using GA, and the inversion results were calibrated with the borehole data in Atfih area, Eastern Desert (ED), Egypt.

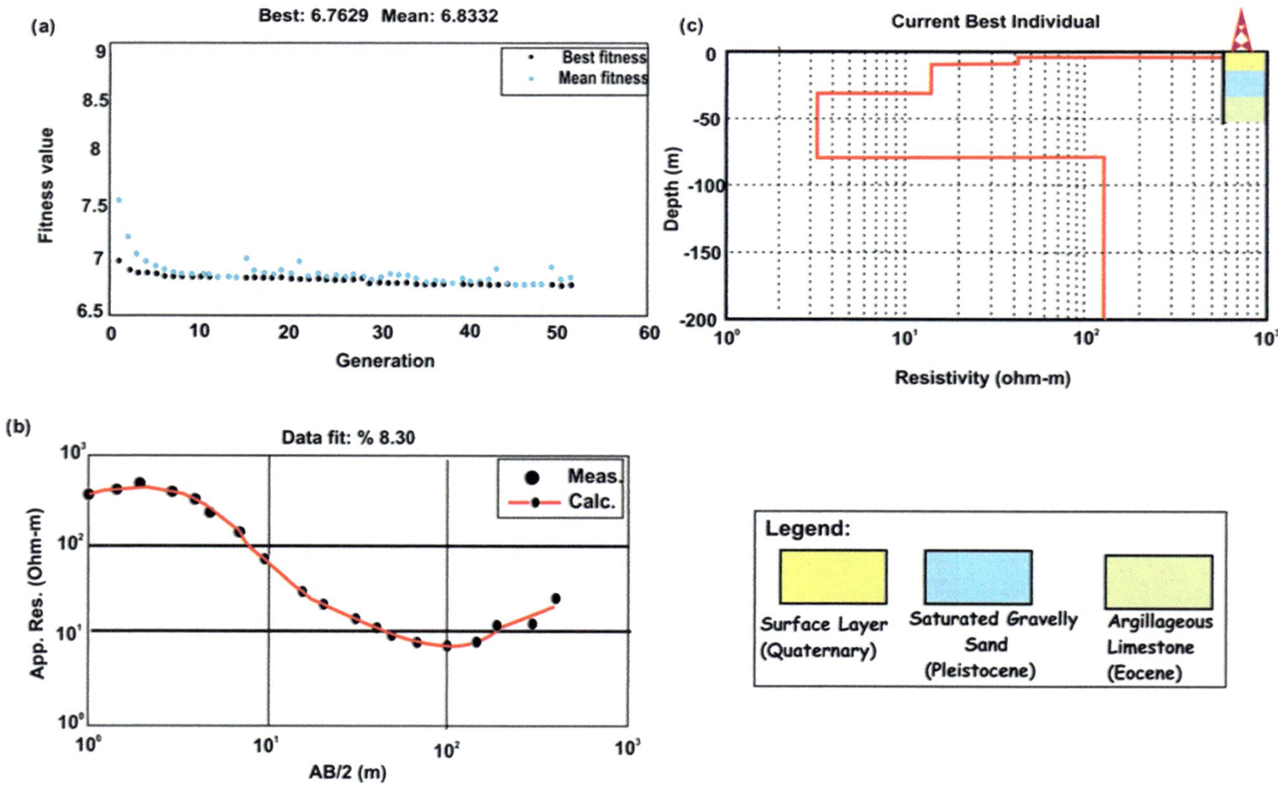

Fig. 9 The interpreted resistivity layers constrained with the geological data using a genetic algorithm (GA); (**a**) variations of mean and best misfit versus generation; (**b**) a comparison between the measured and calculated data at the best misfit value; (**c**) a comparison between the model response and borehole information

3 Applications to the Nile Delta

In the northern part of East Nile Delta (coastal area), Egypt, fresh groundwater salinization is a noticeable aspect of in groundwater quality salinization. Notably, saltwater intrusion has affected irrigation and domestic wells. Furthermore, human activity affects the salt transport. Accordingly, the main goal of the study was to delineate the saltwater intrusion due to the intensive pumping in the transition zone of eastern Nile Delta aquifer. In the course of this study, the 1D model generation using a hybrid genetic algorithm (GA) is applied and tested using borehole information. The constructed geoelectrical cross sections emphasize and delineate the extension of saltwater intrusion (Figs. 10 and 11). Cleary, it is found that due to excessive pumping from shallow wells over the last decades, the subsurface resistivity and TDS distributions can change rapidly within a short distance. Additionally, the results show that despite the dominance of brackish and saltwater at the northern part of the area, perched low conductive lenses are observed reflecting a low level of groundwater salinization. To obtain a link between water salinity and aquifer resistivity, an empirical relationship is derived to predicate the salinity variations at different depths [4]. In comparison with the measured TDS, the predicted salinity map appears realistic. These results demonstrate the important role of the integration between resistivity and salinity measurements for mapping the groundwater salinization with depth and call a further research to plan and manage the area's water resources.

Furthermore, the aquifer salinity can be predicted using the aquifer resistivity values applying a polynomial relationship, inverse first order, of both true aquifer resistivities and the measured TDS values [4]. Figure 12 shows the polynomial relationship to predict the TDS values at the northern part of East Nile Delta using 12 data points of DCR soundings and TDS values close to the observed boreholes. Notably, the TDS prediction proves the reliability of the inverted DCR data uses to predict TDS clues with various depth using the polynomial equation over the northern coastal area of East Nile Delta.

The East Nile Delta, Egypt (Fig. 13), is characterized by different geological and hydrogeological complexity conditions. Geomorphologically, the East Nile Delta decreases in elevation northward, i.e., relatively low topography (~4 m). On the other hand, the southern parts have high features with steeper slopes. Attwa et al. [2] show the efficiency of DCR soundings to identify the geometry and nature of aquifers (Fig. 14). Regionally, the inversion results of DCR soundings represent the water-bearing formations and the main structures in the form of folds and faults. Figure 14 indicates two aquifers; main Quaternary aquifer overlies the deep saline Tertiary. Further, the Quaternary groundwater-bearing formations become the main aquifer in the central part of East Nile Delta. On the other hand, Miocene aquifer represents the main aquifer in the southern portion.

Hydrogeologically, the aquifer hydraulic conductivity (K) can be predicted using the power law relation between hydraulic conductivity and aquifer resistivity (Eq. 12).

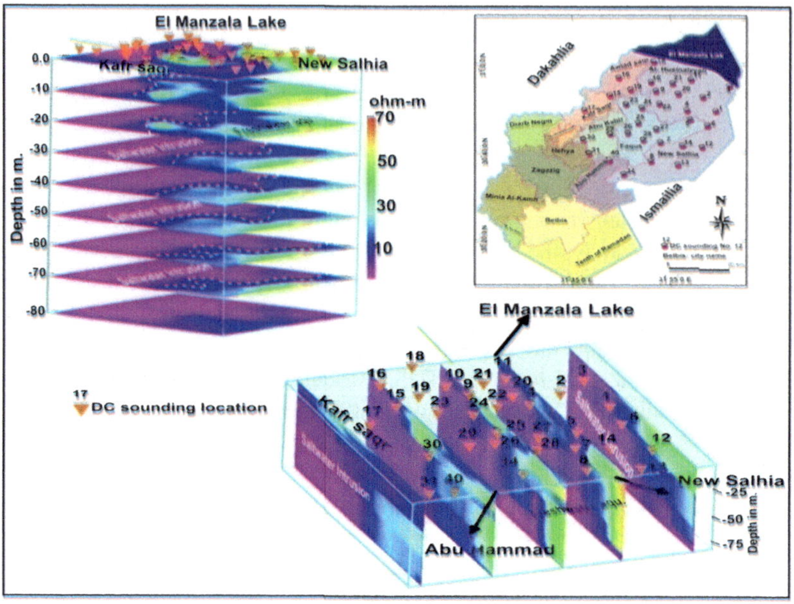

Fig. 10 3D visualization resistivity model represents the resistivity distributions at the northern part of the East Nile Delta, Egypt [4]

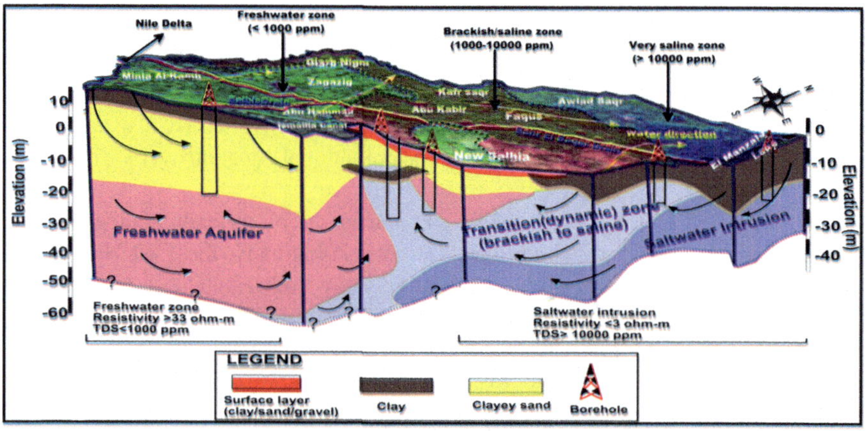

Fig. 11 3D schematic model shows the freshwater/saltwater conditions at a coastal area, East Nile Delta, Egypt [4]

Figure 15 shows log-log plot of aquifer resistivity of Quaternary and Miocene aquifers versus the observed *K*. It is clear that there is a general trend of increasing aquifer

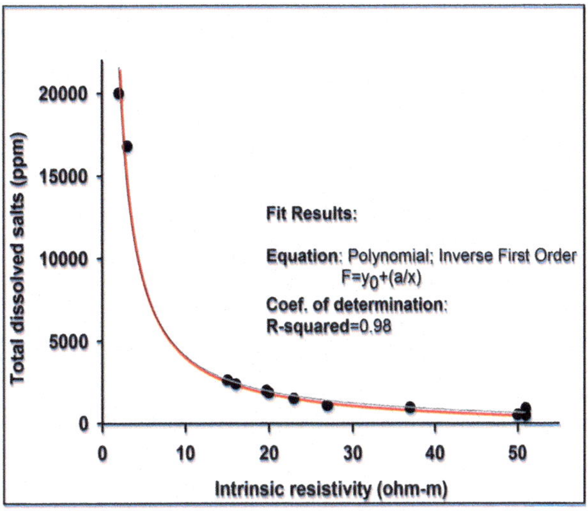

Fig. 12 Correlation between true resistivity derived from DCR sounding inversion and TDS of some collected water samples at the northern coastal part of East Nile Delta [4]

resistivity with increasing *K*. Accordingly, the *K* estimation from individual aquifer resistivity appears to be applicable.

4 Conclusions and Recommendations

The chapter shows that the application of DCR method with available boreholes, an integrative approach, is valuable in hydrogeological conditions assessment in East Nile Delta, Egypt. Based on the present studies from East Nile Delta, the results proved (1) the efficiency of the DCR method for hydrogeological evaluation in costal semiarid areas and (2) the validity of the DCR method to predict the TDS and *K* using the empirical relationships for this area. Furthermore, the joint use of the conventional and nonconventional inversion techniques is recommended for hydrogeological evaluation in East Nile Delta. In the ongoing research, the computational speed of GA requires to be improved applying parallel computation. Further research including pumping tests is required to verify the prediction of hydraulic conductivity using the geoelectrical empirical relationships in East Nile Delta. Additionally the derived salinity equation modification is desirable considering the clay content ratio and grain size analysis. Therefore, it is worthwhile to mention that the DCR method has a great importance for hydrogeological assessment to the decision-makers.

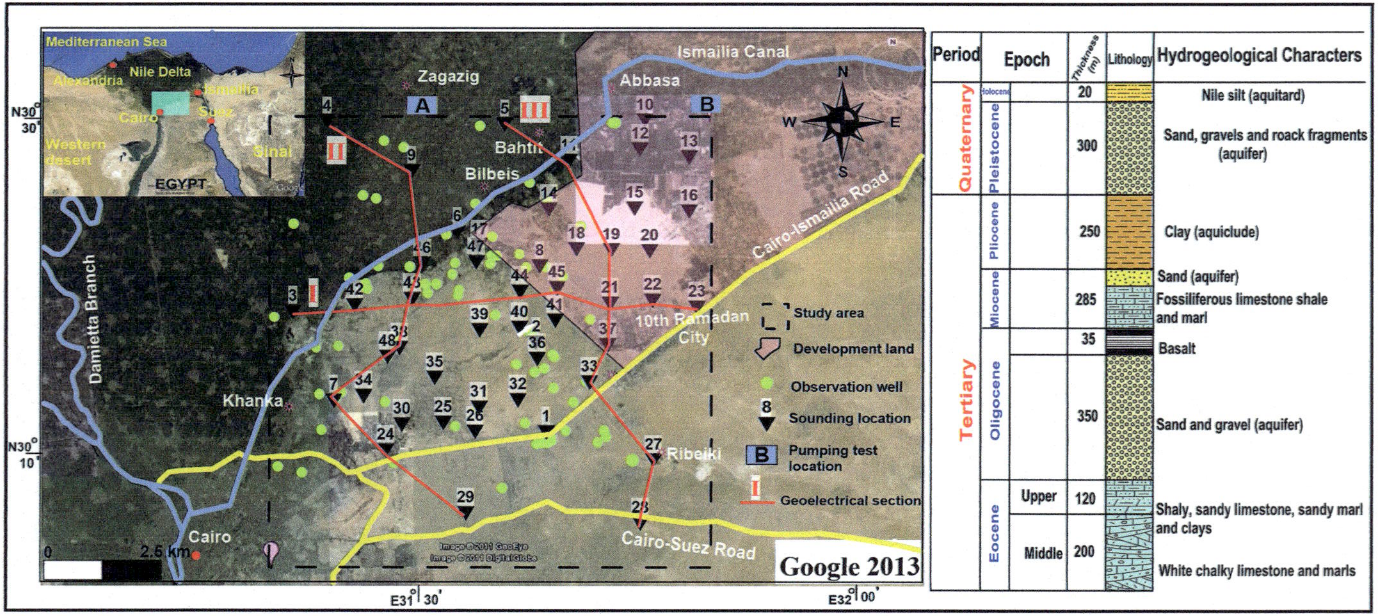

Fig. 13 Location map of the East Nile Delta showing the DCR soundings, geoelectrical cross section, and observed boreholes (*left panel*). Compiles stratigraphic section (*right panel*) of the East Nile Delta

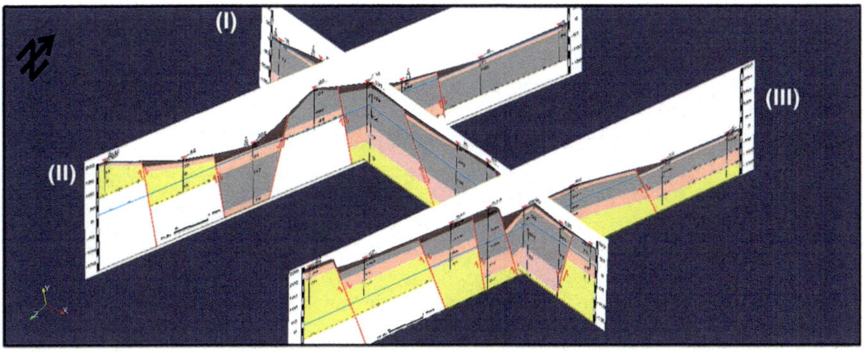

Fig. 14 Geoelectrical cross sections in the form of 3D fence diagram at East Nile Delta (for location, see Fig. 11, *left*) [2]

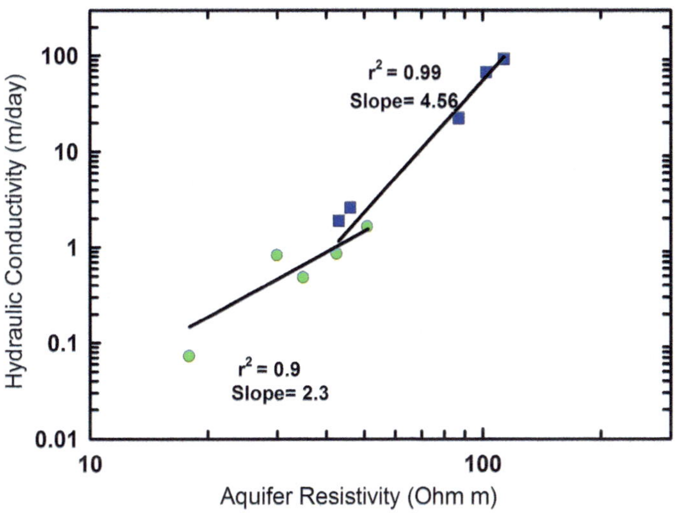

Fig. 15 The aquifer resistivity vs. the hydraulic conductivity (*K*) of ten DCR soundings and observed boreholes. *Blue* solid squares indicate Quaternary aquifer, while the Tertiary aquifer is represented by *green* solid circles [2]

References

1. Steuer A, Siemon B, Auken E (2009) A comparison of helicopter borne electromagnetics in frequency and time domain at the Cuxhaven valley in Northern Germany. Appl Geophys Sci 67:194–205
2. Attwa M, Basokur A, Akca I (2014) Hydraulic conductivity estimation using direct current (DC) sounding data: a case study in East Nile Delta Egypt. Hydrgeol J 22:1163–1178. https://doi.org/10.1007/s10040–014–1107-3

3. Ronczka M, Hellman K, Günther T, Wisén R, Dahlin T (2017) Electric resistivity and seismic refraction tomography: a challenging joint underwater survey at Äspö Hard Rock Laboratory. Solid Earth Sci 8:671–682
4. Attwa M, Gemail KS, Eleraki M (2016) Use of salinity and resistivity measurements to study the coastal aquifer salinization in a semi-arid region: a case study in northeast Nile Delta Egypt. Environ Earth Sci 75:784. https://doi.org/10.1007/s12665-016-5585-6
5. Goebela M, Adam P, Rosemary K (2017) Resistivity imaging reveals complex pattern of saltwater intrusion along Monterey coast. Hydrol Sci 551:746–755
6. Farid A, Khalid P, Jadoon KZ, Jouini MS (2014) The depositional setting of the late quaternary sedimentary fill in southern Bannu basin northwest Himalayan fold and thrust belt Pakistan environ. Monit Assess Sci 8:6587–6604
7. Muhammad M, Khalid P (2017) Hydrogeophysical investigations for assessing the groundwater potential in part of the Peshawar basin Pakistan. Arab J Sci Eng Sci 42:327–337
8. Olatunji S, Musa A (2014) Estimation of aquifer hydraulic characteristics from surface geoelectrical methods: case study of the Rima basin North Western Nigeria. Arab Sci Eng 39:5475–5487
9. Binley A, Hubbard S, Huisman J, Revil A, Robinson D, Singha K, Slater L (2015) The emergence of hydrogeophysics for improved understanding of subsurface processes over multiple scales. Water Resour Res Sci 51:3837–3866
10. Park S, Yi M-J, Kim J-H, Shin SW (2015) Electrical resistivity imaging (ERI) monitoring for groundwater contamination in an uncontrolled landfill South Korea. J Appl Geophys 135:1–7
11. Attwa M, Günther T (2013) Spectral induced polarization measurements for predicting the hydraulic conductivity in sandy aquifers. Hydrol Earth Syst Sci 17:4079–4094
12. Revil A, Binley A, Mejus L, Kessouri P (2015) Predicting permeability from the characteristic relaxation time and intrinsic formation factor of complex conductivity spectra. Water Resour Res 8:6672–6700
13. Ahmed AS, Jardani A, Revil A, Dupont JP (2016) Specific storage and hydraulic conductivity tomography through the joint inversion of hydraulic heads and self-potential data. Adv Water Resour 89:80–90
14. Maineult A, Jougnot D, Revil A (2017) Variations of petrophysical properties and spectral induced polarization in response to drainage and imbibition: a study on a correlated random tube network. Geophys J Int. https://doi.org/10.1093/gji/ggx474
15. Shevnin V, Rodríguez OD, Mousatov A, Hernández DF, Martínez HZ, Ryjov A (2006) Estimation of soil petrophysical parameters from resistivity data: their application for oil contaminated sites characterization. Geophys J Int 45(3):179–193
16. Aizebeokhai AP (2009) Geoelectrical resistivity imaging in environmental studies. In: Yanful EK (ed) Appropriate technologies for environmental protection in the developing world. Springer, Dordrecht, pp 297–305
17. Knoedel K, Lange G, Voigt H (2005) Direct resistivity methods. In: Environmental geology: handbook of field methods and case studies. Springer, New York, pp 205–237
18. Weller A, Slater L, Nordsiek S, Ntarlagiannis D (2010) On the estimation of specific surface per unit pore volume from induced polarization: a robust empirical relation fits multiple datasets. Geophysics 4:105–112
19. Revil A, Kessouri P, Torres-Verdín C (2014) Electrical conductivity induced polarization and permeability of the Fontainebleau sandstone. Geophysics 5:301–318. https://doi.org/10.1190/geo2014-0036.1
20. Revil A, Florsch N, Camerlynck C (2014) Spectral induced polarization porosimetry. Geophysics 198:1016–1033
21. Wang M, Revil A (2010) Electrochemical charge of silica surface at high ionic strength in narrow channels. Journal of Coll Interface Sci 343:381–386
22. Falkenhagen H (1934) Electrolytes. Oxford University, Oxford
23. Schwarz G (1962) A theory of the low-frequency dielectric dispersion of colloidal particles in electrolyte solution. J Phys Chem 66:2636–2642

24. Leroy P, Revil A, Kemna A, Cosenza P, Ghorbani A (2008) Spectralinduced polarization of water-saturated packs of glass beads. J Colloid Interface Sci 321:103–117
25. Marshall DJ, Madden TR (1959) Induced polarization a study of its causes. Geophysics 24:790–816
26. Bücker M, Hördt A (2013) Analytical modelling of membrane polarization with explicit parametrization of pore radii and the electrical double layer. Geophysics 194:804–813
27. Attwa M (2012) Field application "Data acquisition processing and inversion". In: Attwa M (ed) Electrical methods: practical guide for resistivity imaging and hydrogeophysics. LAP LAMBERT Academic Publishing GmbH and Co KG, Saarbrücken, Germany, pp 70–98
28. Li Y, Oldenburg DW (1992) Approximate inverse mappings in DC resistivity problems. Geophys J Int 109:343–362
29. Martorana R, Lombardo L, Messina N, Luzio D (2013) Integrated geophysical survey for 3D modeling of a coastal aquifer polluted by seawater. Near Surf Geophys 11. https://doi.org/10.3997/1873–0604.2013006
30. Vinegar HJ, Waxman MH (1984) Induced polarization of shaly sands. Geophysics 49:1267–1287
31. Vinegar HJ, Waxman MH (1987) In-situ method for determining pore size distribution capillary pressure and permeability. US Patent Sci 4:644–283
32. Mazac O, Cislerova M, Kelly WE, Landa I, Venhodova D (1990) Determination of hydraulic conductivities by surface geoelectrical methods. In: Ward SH (ed) Geotechnical and environmental geophysics, vol 2. Society of Exploration Geophysicists, Tulsa, OK, pp 125–131
33. Slater L (2007) Near surface electrical characterization of hydraulic conductivity: from petrophysical properties to aquifer geometries – a review. Surv Geophys 28:169–197
34. Börner FD, Schopper W, Weller A (1996) Evaluation of transport and storage properties in the soils and groundwater zone from induced polarization measurements. Geophysics 44:583–601
35. Pape H, Riepe L, Schopper JR (1987) Theory of self-similar network structures in sedimentary and igneous rocks and their investigation with microscopical and physical methods. Microscopy 148:121–147
36. Hördt A, Druiventak A, Blaschek R, Binot F, Kemna A, Kreye P, Zisser N (2009) Case histories of hydraulic conductivity estimation with induced polarization at the field scale. Near Surf Geophys 7:529–545
37. Slater L, Lesmes D (2002) Electric-hydraulic relationships observed for unconsolidated sediments. Water Resour Res 10:1213. https://doi.org/10.1029/2001WR00107.2002
38. Zisser N, Kemna A, Nover G (2010) Relationship between low-frequency electrical properties and hydraulic permeability of low-permeability sandstones. Geophysics 3:131–141
39. Silvester PP, Ferrari RL (1990) Finite elements for electrical engineers. 2nd edn. Cambridge University Press, Cambridge, UK. 516 p
40. Sasaki Y (1994) 3-D resistivity inversion using the finite-element method. Geophysics 59 (11):1839–1848. https://doi.org/10.1190/1.1443571
41. Szalai S, Koppan A, Szokoli K, Szarka L (2013) Geoelectric imaging properties of traditional arrays and of the optimized Stummer configuration. Near Surf Geophys 11. https://doi.org/10.3997/1873-0604.2012058
42. Blaschek R, Hördt A, Kemna A (2008) A new sensitivity-controlled focusing regularization scheme for the inversion of induced polarization data based on the minimum gradient support. Geophysics 73(2):F45–F54
43. Akca I, Basokur AT (2010) Extraction of structure-based geoelectric models by hybrid genetic algorithms. Geophysics 75(1):F15–F22

Hydrogeophysical Investigations at El-Nubariya-Wadi El-Natrun Area, West Nile Delta, Egypt

Ismael M. Ibraheem and Gad El-Qady

Abstract Establishing new communities is one of the strategic plans of the Egyptian government during the last decades. One of the targeted localities for such plan is the West Nile Delta area where it is considered as a promising area for creating new settlements. The area under investigation is located to the west of Nile Delta on both sides of the Cairo-Alexandria desert road, between latitudes 30° 17′ and 30° 42′ N and longitudes 30° 00′ and 30° 30′ E. It covers an area of about 1,825 km^2. To help such national plan, the present study aims to delineate the hydrogeological regime of the study area through qualitative and quantitative interpretation of the available electrical and electromagnetic data.

The geoelectric resistivity survey includes 93 vertical electrical soundings (VES) carried out using Schlumberger array with AB/2 ranging from 1.5 m up to 700 m in successive steps. The time domain electromagnetic (TEM) survey was executed near selected VESes where 26 TEM stations with loop spreading 50 × 50 m were measured.

The quantitative interpretation was conducted through one-dimensional resistivity inversion for both VES and TEM data using the software IX1D from Interpex Ltd. Ten cross sections (from VES data) and three cross sections (from TEM data) were constructed and well correlated with the available boreholes. The cross sections reveal that the depth to the main aquifer ranges from 6 m at the northern part near El-Nubariya city to about 90 m at the southern part. On the other hand, the aquifer system in the area is divided into Pleistocene and Pliocene aquifers. The Pleistocene aquifer which is the shallower aquifer is consisted of gravelly to clayey sand deposits.

I.M. Ibraheem
Institute of Geophysics and Meteorology, University of Cologne, Cologne, Germany
e-mail: ismail.geo@gmail.com; ismael.ibraheem@geo.uni-koeln.de

G. El-Qady (✉)
National Research Institute of Astronomy and Geophysics, Cairo, Egypt
e-mail: gadosan@yahoo.com

A. M. Negm (ed.), *Groundwater in the Nile Delta*,
Hdb Env Chem (2019) 73: 235–272, DOI 10.1007/698_2017_154,

The Pliocene aquifer, deeper, is the main aquifer and composed of sand to gravelly sand deposits where in some places, it can be divided into two aquifers. Based on the inferred structural map, it is worth to mention that Wadi El-Natrun and its lakes are structurally controlled by faulting systems trending NW direction.

Keywords El-Nubariya, Groundwater, TEM, VES, Wadi El-Natrun, West Nile Delta

Contents

1 Introduction

Limitations of water resources are the main problem to achieve the main target of constructing new settlements and land reclamation projects for sustainable development in Egypt. Great attention is paid by different Egyptian authorities for that purpose to overcome the overpopulation crisis and construct new agricultural areas. In this respect, priorities are given to the West Nile Delta area, which is considered as a promising region due to its distinct location, mild weather, easy accessibility, and the availability of water supplies. Accordingly, new desert settlements are established, e.g., South El-Tahrir, El-Sadat city, El-Nubariya, and El-Bustan pilot areas, in addition to many villages and farms.

An adequate supply of freshwater is one of the prerequisites for every type of developmental programs. On the other hand, water conservation is very important because usable water is not an unlimited resource. As water supplies dwindle, water shortages in many areas of the world constitute a major problem for both agriculture and nations. Agriculture consumes more than 80% of the water used annually in several countries, where the water is the most important need in farming, and its deficiency commonly limits plant growth.

Groundwater is the main source for domestic, industrial, and agriculture uses in most of the newly reclaimed areas in the west Nile Delta region [1]. Geophysics, especially geoelectric techniques, has been successfully used to detect the freshwater/saltwater interface in coastal aquifers [2]. Resistivity surveys are often used to search for groundwater in both porous and fissured media (e.g., [3, 4]). The methods provide detailed information about the geometry, source, and amount of contamination [5]. For these purposes, the present research is planned in order to address and evaluate the hydrogeological regime in the area of the West Nile Delta using geoelectrical and electromagnetic methods.

2 Location of the Study Area

The present study covers an area of about 1,825 km^2 of the West Nile Delta region. Wadi El-Natrun depression has nearly occupied the southwestern part of the area. The main cities in the study area are El-Nubariya, Wadi El-Natrun, and El-Sadat city. It is accessible from Cairo, Alexandria, northwestern coastal zone, and central part of the Nile Delta by good roads and railway system. The area under investigation (Fig. 1) is located to the west of Nile Delta on both sides of the Cairo-Alexandria desert road, between latitudes 30° 17′ and 30° 42′ N and longitudes 30° 00′ and 30° 30′ E.

3 Geomorphological and Geological Setting

The study area represents a wide portion of the western fringe of the Nile Delta where the landscape features of the area under study are connected to the Nile Delta and its western fringe. The Nile Delta is a flat area where the land surface slopes gently northwards. The western Nile Delta fringe is generally characterized by slightly undulating topography and higher land compared with that of cultivated part of Nile Delta. The topographic contour map (Fig. 2) is extracted from GTOPO30 DEM (digital elevation model developed by the USGS EROS Data Center) with a grid spacing of 30 arc sec (approximately 1 km). It shows that topography of the study area ranges between −23 and 60 m.

Generally, the geomorphologic studies of the area west of the Nile Delta were addressed by many authors as Sandford and Arkell [6], Philip [7], Said [8], El Fayoumi [9], Shata and El Fayoumi [10], Abu El-Izz [11], Sanad [12], Attia [13], El Shazly et al. [14], El Ghazawi [15], Abdel Baki [16], Embaby [17], Gomaa [18], and Embaby [19]. In the light of previous literature, the west Nile Delta area can be classified into the following geomorphologic units from the southwest to northeast (Fig. 3): the highlands, pediment and piedmont slopes, the lowlands, the old alluvial plain, the young alluvial plain, and the eolian sand and sand dunes.

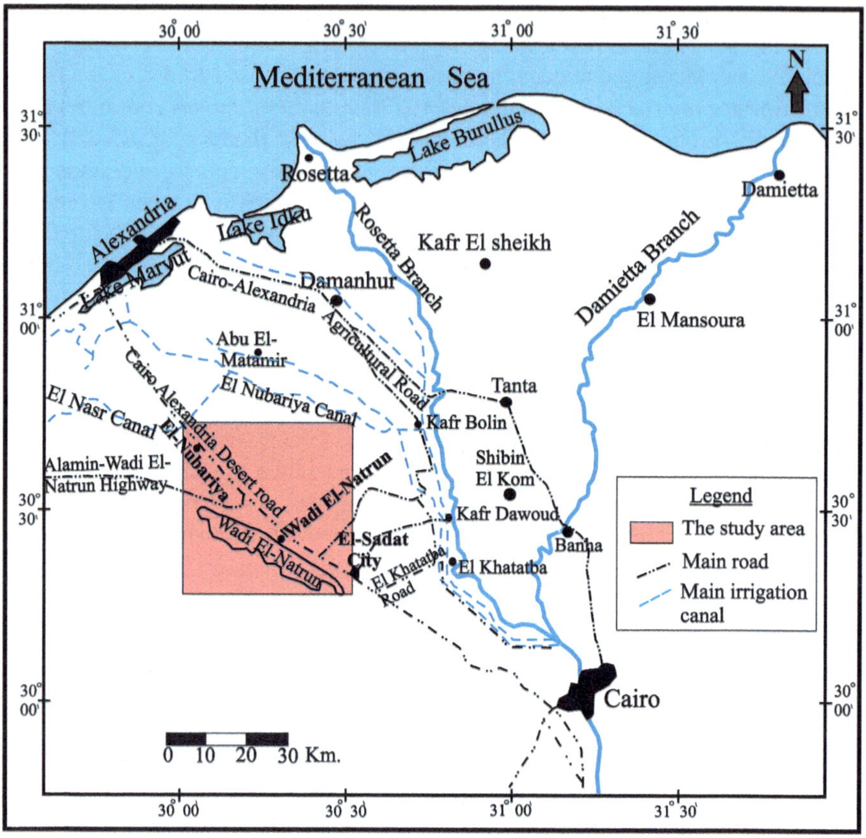

Fig. 1 Location map of the study area

The geology of the west Nile Delta is discussed by several authors such as Sandford and Arkell [6], Philips [7], Shata [20–23], [24], El Fayoumy [9], Shata et al. [25], Shata et al. [26], Sanad [12], Attia [13], El Ghazawi [15], and Shedid [27]. The exposed rocks in the area (Fig. 4) belong to Cenozoic. The Oligocene outcrops are composed of sand, sandstone, gravel, and basalt. They are locally characterizing the southwest corner of the Nile Delta. The Oligocene basalt sheets are the only exposed volcanic rocks in the West Nile Delta area, particularly in the southeastern portion. Neogene sediments are generally dominating the southern and western parts at El Ralat, Wadi El-Natrun, and Wadi El-Farigh depressions as well as at El-Washika, Dahr El-Tashasha, and Gebel El-Hadid ridges. They are composed of sand and sandstone with clay and limestone intercalations.

The Quaternary sediments are mainly clastic with essential sand facies and occasional gravel and clay intercalations. Sand sheets and sand dunes are detected

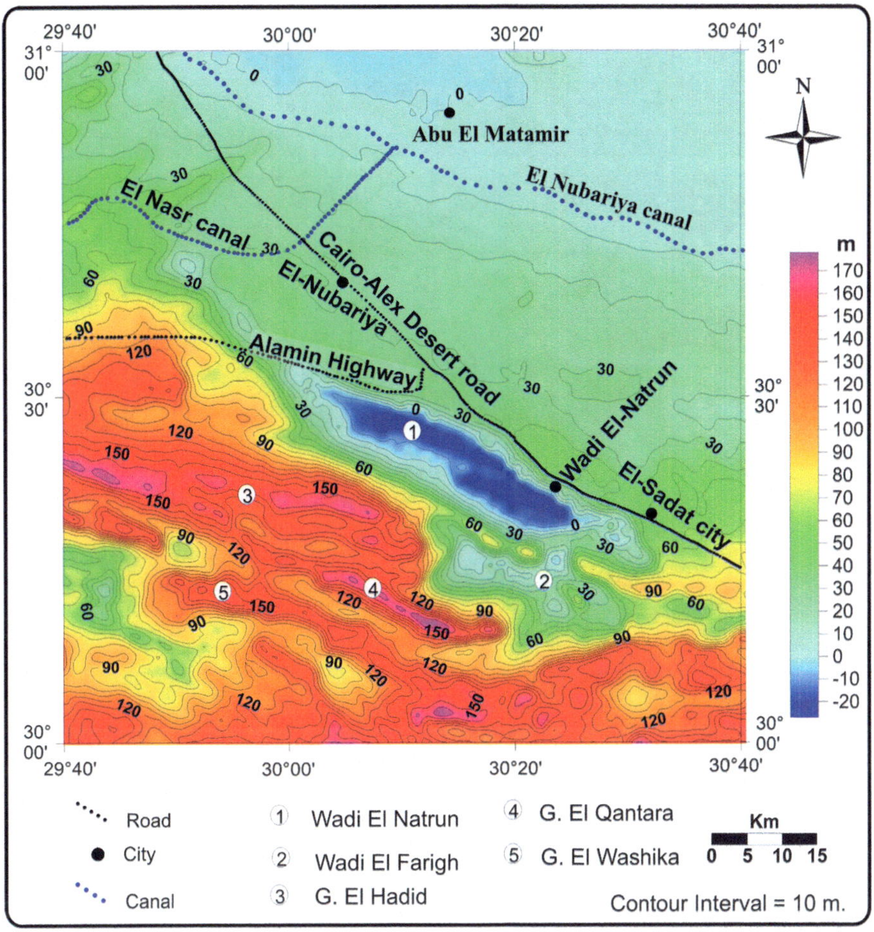

Fig. 2 Topographic contour map of the study area (Extracted from GTOPO30 DEM (digital elevation model))

toward the north. In the subsurface, the sedimentary rocks overlying the basement complex have a thickness of about 4,000 m, as recorded from Sahara Wadi El-Natrun test well (Fig. 5). The sedimentary rock succession starts from the base by Triassic rocks resting on the basement rocks and ends at the top with recent deposits belonging to the Quaternary. The surface geological structure is relatively simple, but in the subsurface, many complexities are detected according to the geophysical surveys and deep drilling. The subsurface succession is characterized by the occurrence of many unconformity surfaces and considerable faulting and folding structures.

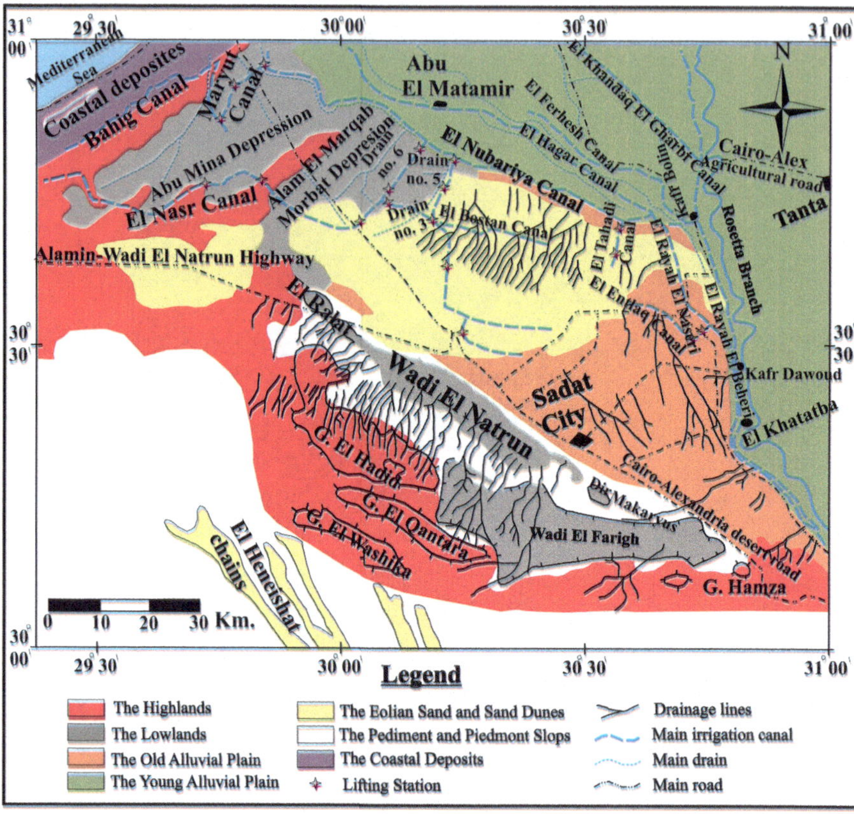

Fig. 3 Detailed geomorphologic map characterized by landforms of the study area and its environs (Reproduced after [19])

4 Structure and Tectonics

The study area is located within the unstable shelf of the northern parts of the Western Desert, so it is affected by two regional structural elements: the taphro-geosynclinal trough of the Nile Delta due to east and the northward dipping homoclinal slope of the Western Desert [29]. The area under investigation is mainly affected by folds, faults, unconformities, and basaltic intrusions. These structural elements are the most important factors influencing the groundwater conditions in the area. Many authors studied the faulting system in the area west of Nile Delta, and among them were Sigaev [30], Shata et al. [24], El Fayoumy [9], Idris [31], Sanad [12], El Shazly et al. [14, 32], El Ghazawi [15], and El Sabagh [33]. The main faults in the area can be distinguished into two groups according to their trends (Fig. 6).

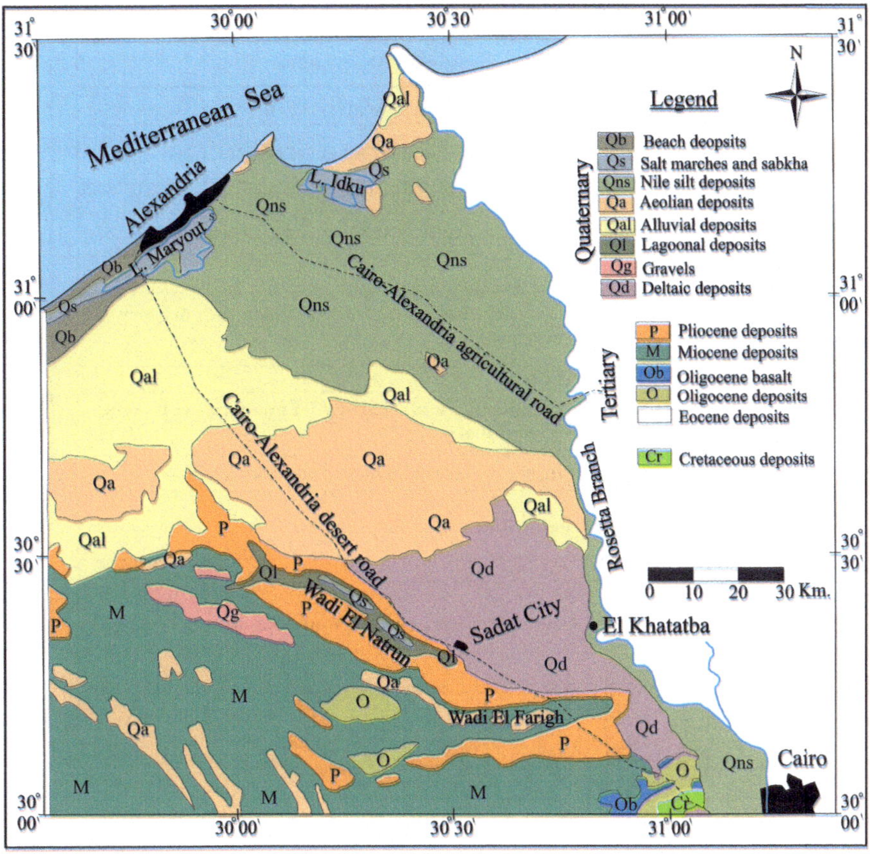

Fig. 4 Geological map of the area west of Nile Delta (Reproduced after [28])

1. The NW-SE trending faults: These faults have high displacement.
   (a) A fault (Fii) runs approximately parallel to the western edge of the Nile Delta [24].
   (b) A fault (Fiii) runs mostly tangential to the lakes of Wadi El-Natrun with a maximum displacement of about 200 m [31].
   (c) A fault is transversing the area between Wadi El-Natrun and Gabel El-Hadid with a maximum displacement of about 120 m [12].
   (d) El Marbat fault (F3) with downthrown of about 50 m due east [33].
   (e) El-Nubariya fault (F4) running parallel to El Marbat fault with its downthrown side eastward [15].

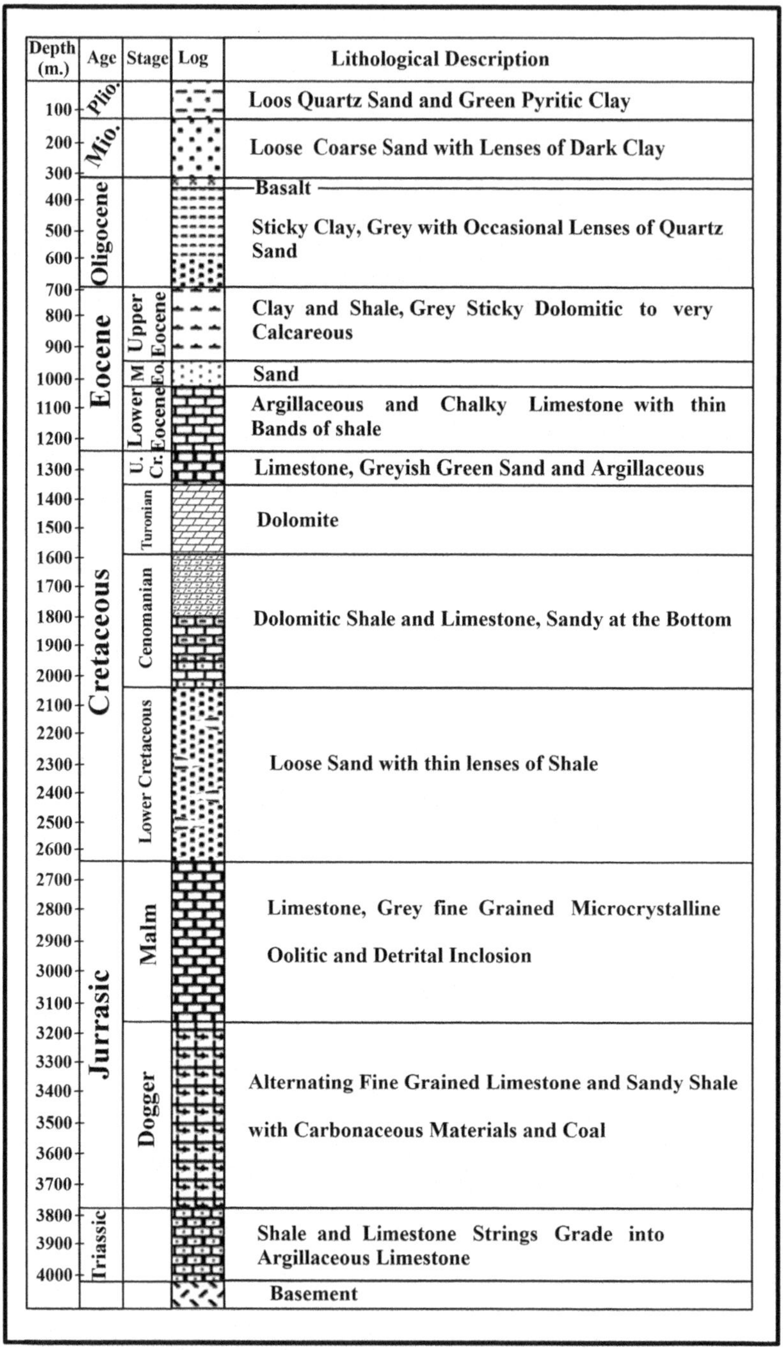

Fig. 5 Sahara Wadi El-Natrun well, Western Desert, Egypt [9]

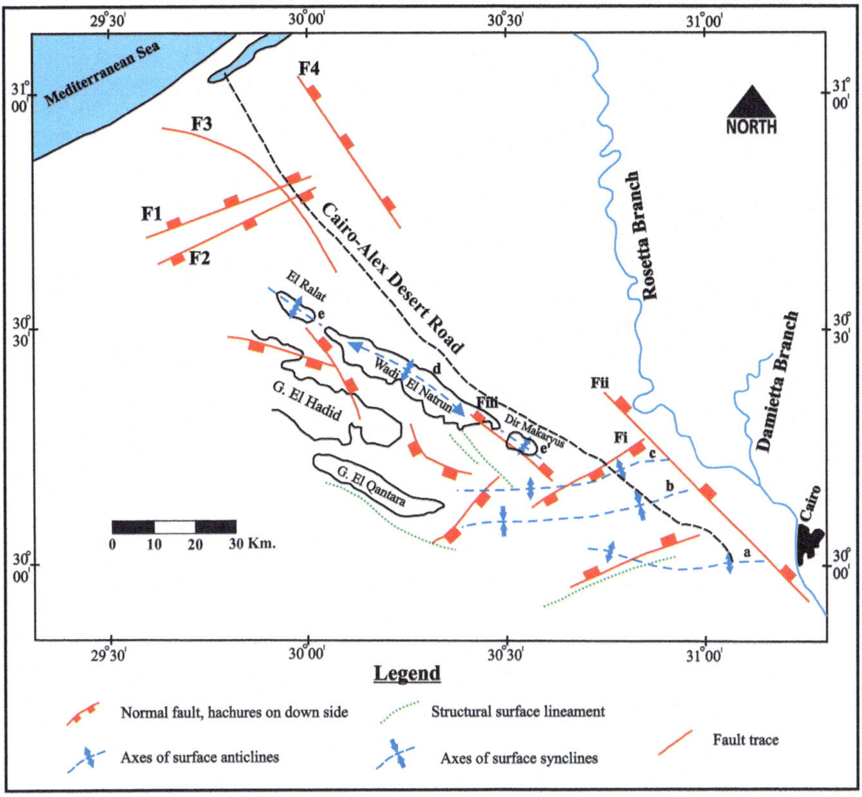

Fig. 6 Structural map of the area west of Nile Delta (Compiled after [15, 24])

2. The NE-SW trending faults:

   (a) The Rossa fault affects the area between Wadi El-Natrun and Abu Roash and has a length of about 100 km and vertical displacement of about 1 km, with its downthrown side to the north [9].
   (b) A fault (Fi) lies to the southeast of Wadi El-Natrun and extends to the edge of Rosetta branch.
   (c) Alam Shaltot fault (F1) to the north of El-Ralat depression with its downthrown side toward the north.
   (d) A fault (F2) lies to the south of Alam Shaltot fault with a downthrown to the south.

5 Geological History

The following paleogeological and paleoclimatic events are established from the Triassic to Recent in the area west of Nile Delta:

- During the *Triassic* time, a period of uplift and erosion was recorded where continental conditions prevailed which explain the absence of the Triassic deposits in the area.
- The main sea transgression occurred in the study area during the *Jurassic* producing a great section of marine deposits composed of shaley limestone and dolomites.
- Erosion and humid conditions were prevailed during the *Lower Cretaceous* leading to the deposition of a section of Nubian sandstone strata under continental conditions, while during the *Upper Cretaceous* period, the study area was subjected to subsidence associated with sea transgression resulting in the deposition of limestone, dolomite, and shale.
- The marine transgression was still prevailing during the *Eocene* times where thick limestone beds were deposited. Also, another regression phase had started during the *Late Eocene*. Consequently, shallow marine facies (calcareous shale and clay) are characterizing this period.
- The *Oligocene* period is characterized by main terrestrial conditions and the occurrence of a long period of rising and erosion. However, a restricted shallow marine environment is expected in Early Oligocene times, and at the end of Oligocene, volcanic activities dominated.
- The *Miocene* period is characterized by semiarid climatic conditions and the deposition of fluviomarine beds (Moghra Formation) with a gradual subsidence after a long period of rising, erosion, and volcanic activity. Moreover, the surface developed into different morphotectonic features such as Wadi El-Farigh anticline ridge. At the end of Miocene, the initial formation of the River Nile took place (Eonile). Also, epeirogenic movements occurred toward the western part of the Mediterranean Sea during the Miocene times. This was followed by a sea regression, and consequently the lowering of the water level of the Nile producing the Miocene deltaic plains.
- A local sea ingression into the Nile Delta and Wadi El-Natrun took place during the Early *Pliocene*. Consequently, a large marine gulf was developed occupying this area resulting in the deposition of a thick succession of pyritic clay forming aquiclude beds. The gulf was characterized from the south by freshwater tributaries, followed by the deposition of shallow marine limy facies in the Late Pliocene.
- In Early *Pleistocene* times, the prevalence of wet climatic conditions and alluvium filled Abu Mina and El Marbat depressions with gravels and sand brought from the Nile as well as from the tableland. This was followed by lowering of the Nile water level and formation of the old deltaic gravelly plains. The drainage system, particularly on the highlands, was created as a result of the extensive wet climatic conditions.

- During the *Holocene*, landscape took most of its present shape, which is represented by the sand dunes and sand sheets in addition to the formation of lakes in Wadi El-Natrun. The prevalence of the present aridity led to the development of crust layer. The Nile water deposited the young deltaic sediments in floodplains where these soils constitute the most fertile land of Egypt.

6 Hydrogeological Aspects

The western Nile Delta aquifers have been the subject of several studies where many investigations have been made to develop the western Nile Delta area; among them are El Fayoumy [9], Abdel Baki [16], Gomaa [18], Sallouma and Gomaa [34], Abd El-Wahab [35], Ahmed [36], Embaby [19], Ibraheem [37], and Ibraheem et al. [38]. The study area generally lies in an arid belt dominating the West Nile Delta area where a long hot summer and short warm winter characterize its climate with low rainfall and high evaporation intensities. The surface water system (see Fig. 3) in the area under investigation and its vicinities is mainly the following: Rosetta branch, El Rayah El Beheri, El Rayah El Naseri, the artificial canals and drains, salt lakes, and springs. This system is subjected to evaporation, rainfall, and infiltration to the groundwater aquifers. The salt lakes get their water from two sources, the springs found in the bottom of Wadi El-Natrun and the seepage into lakes from their sides as they are located on lowland. The groundwater in the study area is mainly controlled by the geological conditions including lithology, topography, and structures. The main water-bearing formations belong to Quaternary (Recent and Pleistocene) and Neogene (Pliocene and Miocene) (Fig. 7).

6.1 *Recent Aquifer*

It is detected at Wadi El-Natrun depression and mainly composed of sandy deposits with calcareous intercalations of eolian fillings. This aquifer is considered partial feeder for the salt lakes of Wadi El-Natrun.

6.2 *Pleistocene Aquifer*

Pleistocene aquifer occupies the major portion of the study area where it is composed of Nile sands and gravels intercalated with thin clay lenses. The depth to groundwater of this aquifer is close to the ground near El-Nubariya canal and increases to the south.

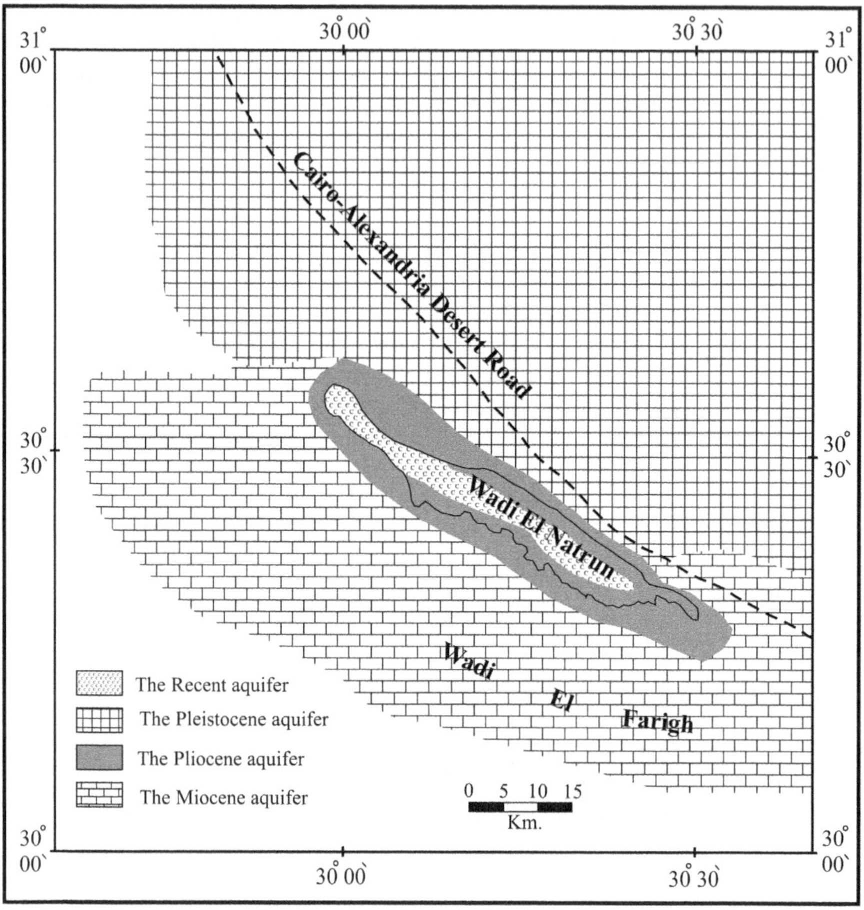

Fig. 7 Distribution of shallow aquifers in the study area and its vicinities (Compiled after [16])

6.3 *Pliocene Aquifer*

This aquifer dominates the area of Wadi El-Natrun and El-Ralat depressions. It is mainly composed of clay facies with interbeds of water-bearing sandy layers. The Pliocene aquifer sediments can be divided into two horizons [39]. The upper one is composed of loose sand and sandstone having a thickness of about 15 m where the groundwater is existing under semi-confined condition due to the presence of sandy clay layers at its top. Meanwhile, the lower one is composed of sandy facies and has a thickness varying from 1 to 10 m. The groundwater is found under confined condition due to the presence of a thick clay bed between the two horizons.

6.4 *Miocene Aquifer*

The Miocene aquifer (Moghra Formation aquifer) covers a wide portion in Wadi El-Farigh and west of Wadi El-Natrun to the south and southwest of the study area. It is mainly composed of coarse sands and sandy clay deposits, which rest on the Oligocene basaltic sheets.

7 Background Geophysical Literature

The study area and its surroundings have been subjected to some previous studies; Abdel Baki [16] classified the aquifers in the area west of Nile Delta including the study area into four items: Recent aquifer, Pleistocene aquifer, Pliocene aquifer, and Miocene aquifer. Gomaa [18] compared the hydrogeological and hydrogeochemical properties of the different aquifers west of Nile Delta and found that all the reservoirs are recharged from the Nile Delta Quaternary reservoir. Diab et al. [40] carried out a geoelectric survey (15 VESes with AB/2 ranging from 1 to 300 m) in the north of El-Hamra Lake in the main Wadi El-Natrun depression. The study aimed to interpret the phenomena of the flow up of the freshwater spring in the very saltwater of El-Hamra Lake. They found that the area is affected by inferred horstal structural feature extending into the lake. This horst faulting system brings permeable water-bearing formation against the impervious clay bed holding up the freshwater along the fault plane in the form of a faulting spring. Depending on the hydrochemical analysis of different water samples collected from the spring and the surrounding wells, they declared that the freshwater of El-Hamra spring is feeded from the Miocene artisan aquifer. Abdel Salam [41] found that the aquifers in Wadi El-Natrun area are recharged from the surface water system, El Rayah El Naseri, El Rayah El Beheri, El-Nubariya canal, and El Nasr canal, besides the recharge from the Miocene reservoir of Wadi El-Farigh area. He also confirmed the hydraulic connection between the different aquifers in Wadi El-Natrun area. Using 43 vertical electrical soundings, Abd El-Gawad and Ammar [42] indicated that the Pliocene aquifer is the main aquifer in Wadi El-Natrun area where the water-bearing formation consists of sand layers with clay intercalations and the groundwater exists under confined to semi-confined conditions. This aquifer has a maximum thickness that reaches 200 m which increases to the north and east. The depth to this aquifer ranges between 2 and 24 m. Mansour [43] carried out 35 vertical electrical soundings in the area north of El-Nubariya city and found that the main aquifer is the Pleistocene aquifer where its thickness varies from 14 to 58 m which increases in the east and southwest direction. This aquifer is composed mainly of sand and in some localities of clayey sand or silty sand. The groundwater exists under unconfined or semi-confined conditions. He also concluded that the salinity of the groundwater varies from 600 ppm in the east and southeast and 3,900 ppm (brackish type) in the south. Massoud et al. [44] executed VES and TEM data measurements at 24 stations along three profiles in El-Sadat city. They mentioned that there are two aquifer systems in the area. The shallow Pleistocene aquifer consists of sand and gravel saturated with freshwater and exhibits large thickness varying from 171 to 218 m and depth range of 6–36 m. The deep Pliocene aquifer is composed of clay and sand and shows low resistivity values. Sharaky et al. [1] studied the hydrochemical characteristics of groundwater of western Nile Delta aquifers. They found that the salinity content varies greatly from 430 mg/L (freshwater) to 24,407 mg/L (saline water) with an average of 2,705 mg/L (brackish water). Freshwater is mainly concentrated in the central-eastern part, close to Rosetta branch, indicating that the aquifer is mainly recharged by the infiltration from irrigation network and excess of irrigation water through the upper clay layer. The salinity increases gradually north- and westwards due to seawater intrusion and mixing with the Miocene aquifer, respectively.

8 Hydrogeological Mapping Using VES and TEM Techniques

Electrical resistivity and electromagnetic methods are well known in exploration geophysics and commonly applied in near-surface applications. Electrical resistivity is one of the most useful techniques applied to groundwater studies because the resistivity of rock is very sensitive to its water saturation. Similarly, EM techniques have been particularly succeeded in groundwater exploration and in the mapping of groundwater contamination and saltwater/freshwater interfaces [45–47].

The geoelectric and electromagnetic field surveys give knowledge about the subsurface and structural geology of the studied area; an identification of geometry and extent of the aquifer, some information about the groundwater quality, and a good identification about the lithology of the subsurface formations. Therefore, money, effort, and time can be saved through the reduction of the number of drilled boreholes [48]. Also, a geoelectrical method is considered to be the most promising and most suitable method for groundwater prospecting [49–51]. Although the apparent resistivity values could be considered a direct criterion for the presence or absence of groundwater, general information about water salinity could be also detected, since these values are inversely proportional to the salinity factor [52].

Among the geoelectrical methods, the electromagnetic method (EM) became commonly used for groundwater studies for several reasons.

1. The recent sophistication of instruments worldwide.
2. The field data acquisition is quicker compared to DC methods.
3. The method does not need large spreading like resistivity method and does not need a large number of people for field work.
4. The method does not need direct contact with the earth because we use a loop of wire on the surface of the earth, so if we have different conditions for working by resistivity, we can easily use the electromagnetic method.

Electromagnetic methods have been extensively developed and adapted over the past three decades for the lateral and vertical mapping of resistivity variations [53, 54]. The depth of investigation of an electromagnetic field depends upon its frequency and the electrical conductivity of the medium, through which it is propagating. The depth of penetration thus increases as both the frequency of the electromagnetic field and the conductivity of the ground decrease.

8.1 Electrical Resistivity of Rocks

Resistivity is an extremely variable parameter not only from formation to formation but even within a particular formation. There is no general correlation of the lithology with resistivity [55]. However, the resistivity depends mainly on the mineral composition, the water content, the salinity of the water and its distribution in the pores, the porosity of rocks, the texture and structure of rocks, and the presence of shale and clays [56–58]. The resistivity of common rocks, soil materials, and waters is shown in Fig. 8.

However, there are overlaps in the resistivity values of the different types of rocks and soils. This overlap due to the resistivity is a function of the quality and amount of water, porosity, and fractures in these rocks. The resistivity of the freshwater varies from less than 10 to 100 Ω m depending on the salinity of this water. Freshwater has much lower ion concentrations and lower conductivity. The resistivity of the seawater is less than 1 Ω m due to the relatively high salt content. This makes resistivity method an ideal technique for mapping the saline and freshwater interface in coastal areas.

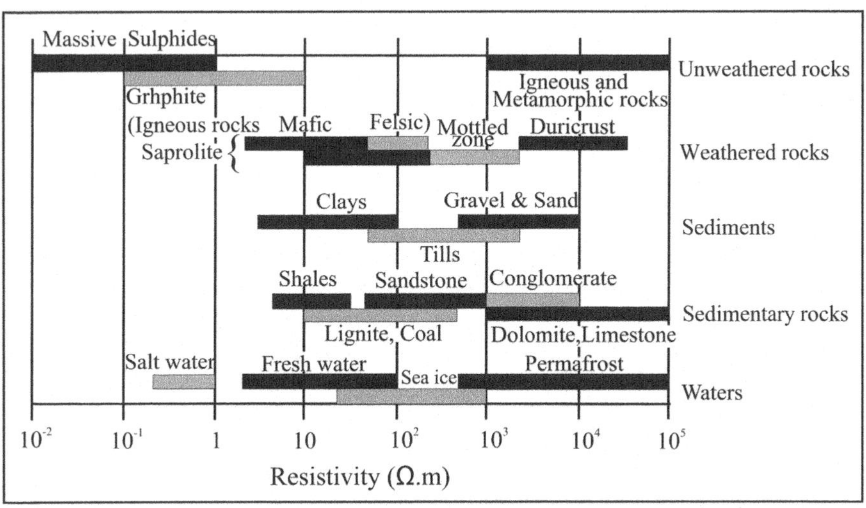

Fig. 8 Electrical resistivity of different rocks and materials [59]

8.2 Data Collection and Processing

The geophysical field data used in the present work represented by a total of 93 VESes and 26 TEM stations were collected through several campaigns to investigate aspects of groundwater resources in the survey area. Figure 9 presents the methodological flowchart to find the groundwater aquifer. The VES field measurements (Fig. 10) were carried out using direct current resistivity meter called McOhm manufactured by the OYO Corporation, using Schlumberger array with AB/2 ranging from 1.5 to 700 m. TEM measurements (Fig. 11) were made very close to the VES's sites where a simple coincident loop configuration, in which the same loop transmits and receives signals, was employed [60]. The TEM instrumentation used for the survey is the SIROTEM MK-3 Conductivity meter, and the conventional 50 $\times$ 50 TEM confirmation was applied. In all sites, the measurements were repeated up to five times. The best signal-to-noise data sets were chosen for further processing and interpretation. The major components of SIROTEM MK3 are contained in a robust, portable console unit as shown in Fig. 12.

One-dimensional resistivity inversion was carried out using the IX1D software from Interpex Ltd., which in this case solves the inversion using a ridge regression method described by Inman [61]. IX1D is a 1D direct current (DC) resistivity, induced polarization (IP), magnetotelluric (MT), and electromagnetic sounding inversion program. It supports most DC resistivity arrays, including Wenner, Schlumberger, dipole-dipole, pole-dipole, and pole-pole arrays. The RMS error is calculated by summing the squares of the difference in the log of the apparent resistivities. This result is divided by the number of data points, and the square root is taken. The antilog of this result is taken, and 1.0 is subtracted from it. Then it is multiplied by 100 to convert to percent. This method of calculating the error is used because of the large dynamic range that can be present in these data. This ensures that high apparent resistivity values do not dominate the calculated error, leaving large errors in the low apparent resistivity values [62].

8.3 Interpretation of VES Data

At each VES station, the calculated apparent resistivities are plotted against AB/2 – spacing on bi-logarithmic paper to construct the field curve for each station. The field curves were interpreted both qualitatively and quantitatively. The qualitative interpretation of the sounding curves involves the classification for the type of these curves and their aerial distribution in these areas. So, the qualitative interpretation can give an idea about the subsurface layers and their aerial distribution which will help during the quantitative interpretation.

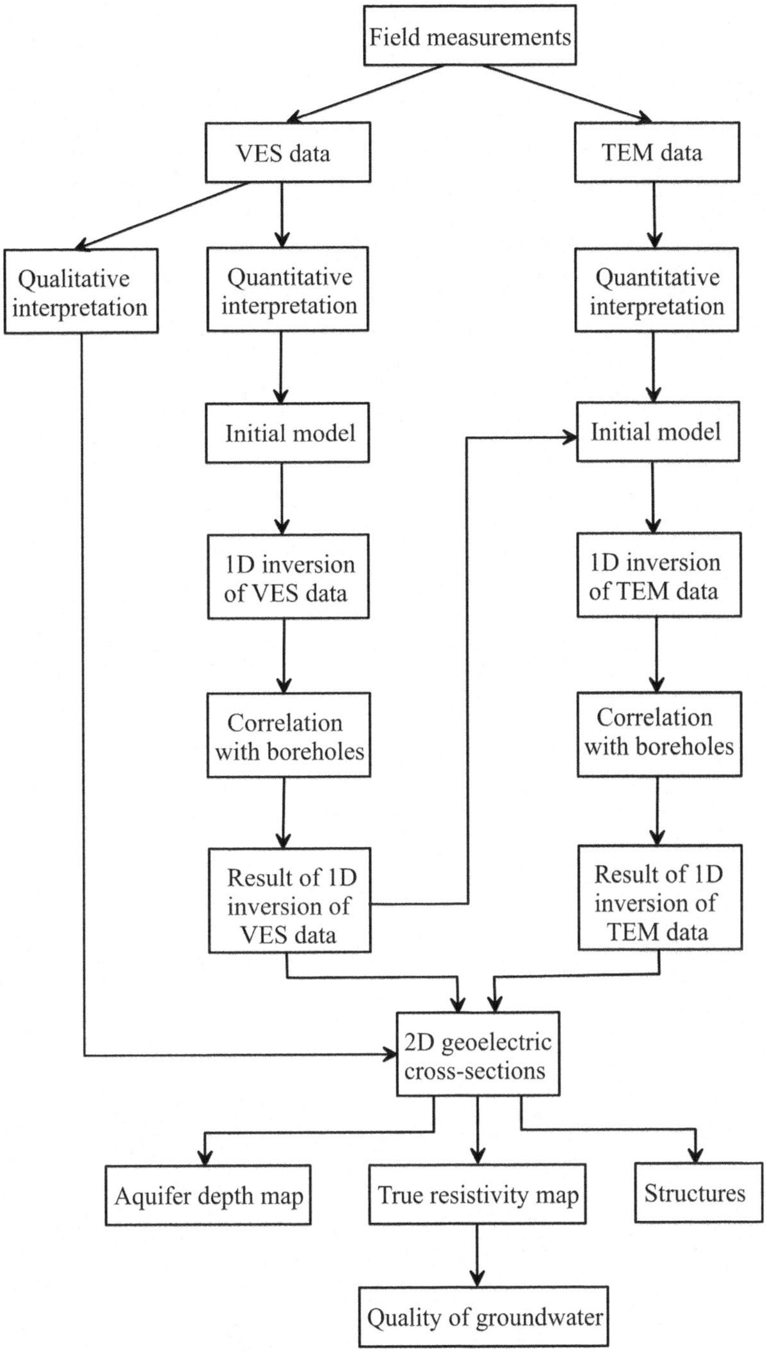

Fig. 9 Methodological flowchart to identify the groundwater aquifer using geophysical investigations

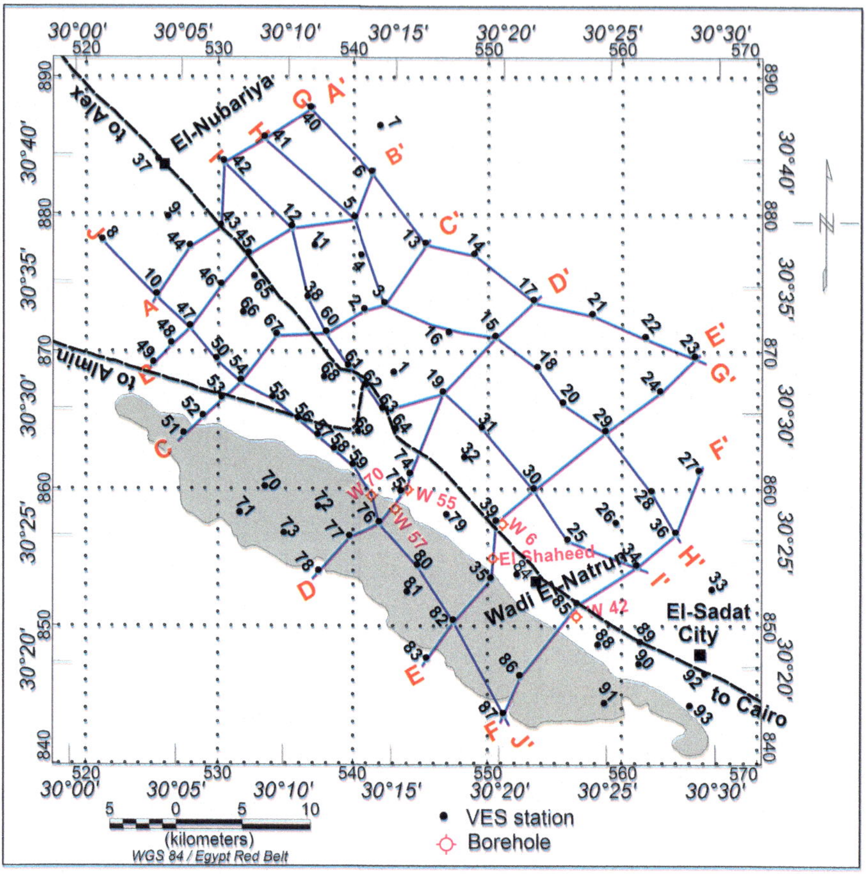

Fig. 10 Location map of the VES stations, boreholes, and selected profiles [38]

8.3.1 Qualitative Interpretation

Qualitative interpretation of the electrical resistivity data is used to give a general view of the lateral and vertical variations of the electrical properties and to shed some light on the geological and hydrogeological conditions of the area under investigation. For the general shape of sounding curves, it has to mention that the beginning of the field curves that represent the small AB/2 values reflects the surface or near-surface variations, which is, in many cases, rapid and may confuse the main objective of this interpretation. Therefore, the resistivity behavior at large AB/2 values of the field curves is the prime importance in judging the homogeneity or inhomogeneity of the subsurface layers.

The aerial distribution of the field curve types was constructed as shown in Fig. 13. We can notice that the curve-type QHK is predominant in the western part of the study area, whereas the curve-type HK characterizes the northern part of the

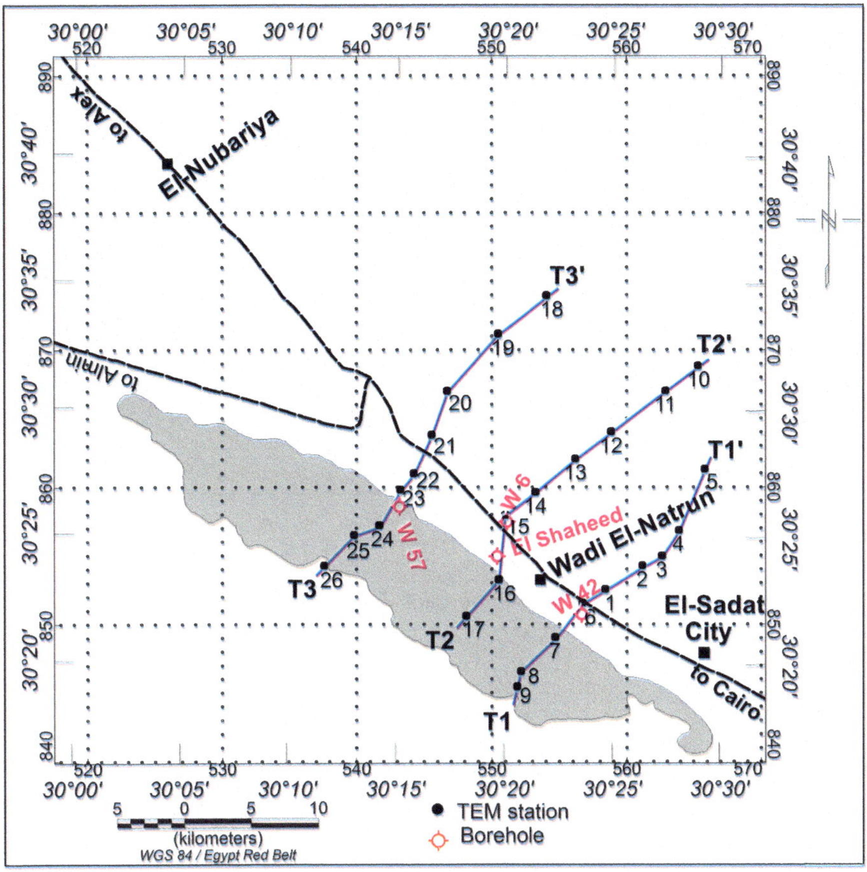

Fig. 11 Location map for TEM stations and boreholes in the study area [38]

area. Also, KQ and HKQ curve types dominate the eastern portion of the area. Added, the southern part of the survey area can be constituted by the QH curve type. Finally, KHK curve type characterizes the central part of the investigated area. As a result of curve-type inspection, we can conclude that the area under investigation may be affected by a group of normal faults that divided the study area into five main divisions, which are the northern, eastern, western, southern, and central divisions.

Iso-apparent resistivity contour maps are also constructed for different electrode spacing to illustrate the resistivity distribution at successive levels penetrated by the artificial electric current. Each map is prepared for a given electrode separation AB/2, to illustrate the geological conditions prevailing within a horizon approximately parallel to the ground surface. Iso-apparent resistivity contour maps can outline fault zones characterized by anomalies of considerable aerial extension along a given direction and having maximum horizontal electric resistivity gradient. These maps also reveal the zones of relatively low electric resistivity values, which may indicate areas of groundwater accumulations.

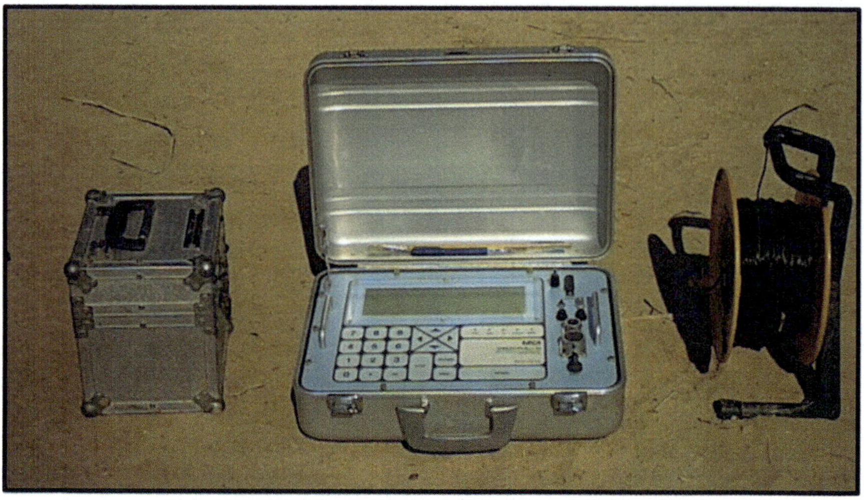

Fig. 12 The SIROTEM MK3 TEM conductivity meter

Five iso-apparent resistivity contour maps (Fig. 14) are selected and constructed for the electrode separations (AB/2: 1.5, 10, 30, 100, and 400 m) to cover most of the succession penetrated by the electric current. From the study of those maps, we can conclude the following remarks:

1. Apparent resistivity contour map of the surface layer shows variable resistivity values due to lithology heterogeneities and dryness and wetness conditions of the silt, sand, and gravel that cover the surface of the study area.
2. The conspicuous lateral variations of the iso-apparent resistivities indicate comparable lateral changes in the encountered types of rocks.
3. The comparison between the five iso-apparent resistivity contour maps reveals that the resistivity values generally begin with high values for the smaller depth of penetration and then decrease gradually by increasing the penetration depth.
4. The study of the iso-apparent resistivity contour maps shows that the eastern region is characterized by high values and may be represented by sand and gravels. Besides, it could be deduced that the areas located in the northern and western parts (of moderated resistivity values) are saturated with shallow groundwater. Added, the low resistivity values at the area around the lakes of Wadi El-Natrun may be affected by the saltwater that comes from the lakes, or it may be represented by clay deposits.

8.3.2 Quantitative Interpretation

The objective of 1D modeling of an apparent resistivity curve for certain DC sounding array is to arrive, given ρ_a, at the best estimate of the 1D true resistivity

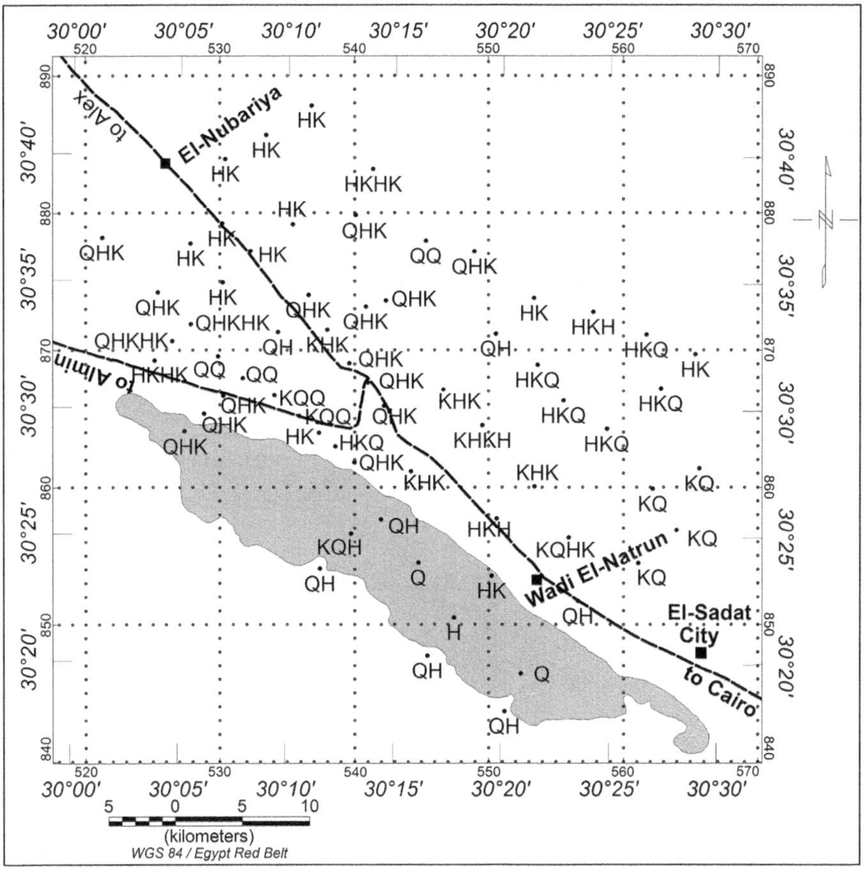

Fig. 13 Distribution of the VES curve types in the study area [38]

profile versus depth ρ_z under the measurement location. This can be done either through direct transformation or through iterative, curve fitting approaches or procedures that are of a least square nature [63]. The real meaning of quantitative interpretation of the vertical sounding curves is to determine the true resistivities and thicknesses of the different formations investigated at each station along the studied sections. One-dimensional resistivity inversion was carried out in the present study using the software IX1D.

The resistivity values of the VESes were constrained using the drilled boreholes nearby and were correlated with the lithology of those boreholes. Figure 15 is a good example of the correlation procedure where the VES 39 is correlated with the borehole W6. The geoelectric section shows the slight difference from the lithological section. During the quantitative interpretation, all available data were used to construct the initial model. These data are including the geological and hydrogeological information and the drilled boreholes in the area. Figures 16 and

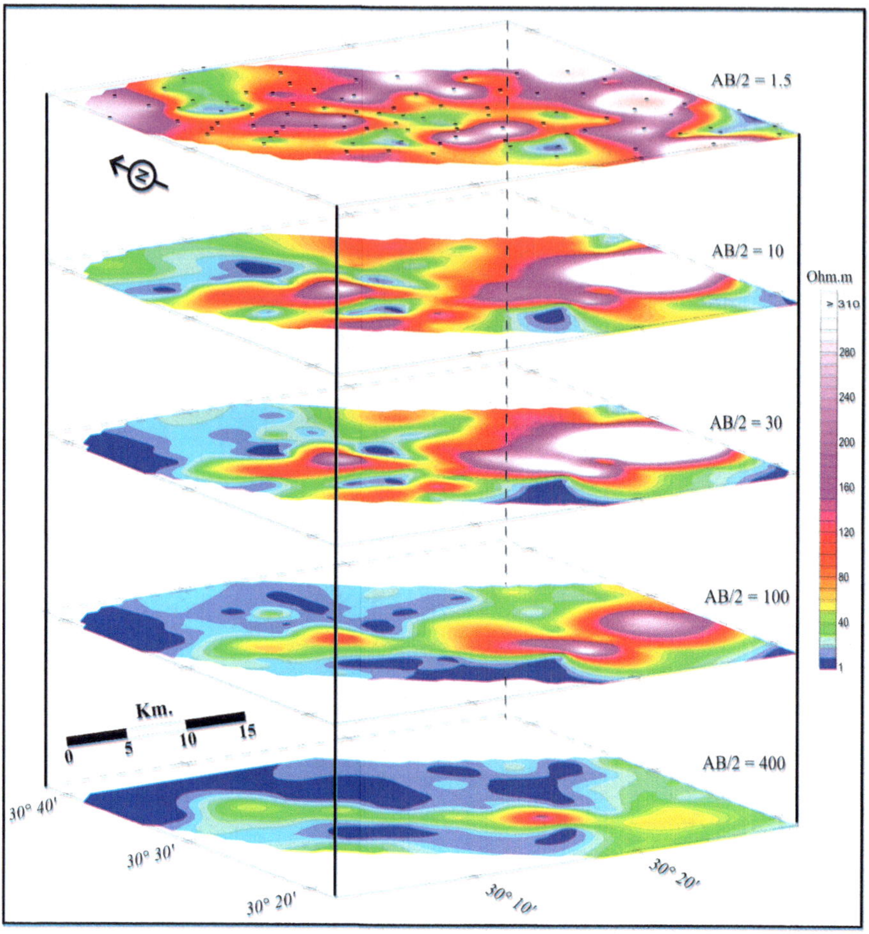

Fig. 14 Iso-apparent resistivity maps at AB/2 = 1.5, 10, 30, 100, and 400 m [38]

17 show examples of the result of applying the 1D inversion on the measured VES data in the study area. The VESes carried out in this area were used to construct ten geoelectric cross sections (See Fig. 10 for the cross sections locations) for the location of the sections. Four geoelectric cross sections are selected and shown in Figs. 18, 19, 20, and 21.

The inspection of the geoelectric cross sections shows that the area of study is divided into five geoelectric units as the following:

- The first geoelectric unit is characterized by relatively very low to very high electric resistivity values ranged between 4 Ω m at VES 35 and 4,031 Ω m at VES 30. The great differences in resistivity of this layer reflect the inhomogeneity of lithology, which corresponds to sand, silt, clay, gravel, and gravelly sand. The low resistivity values are recorded in the area of salt marshes where a layer of salt covers the surface.

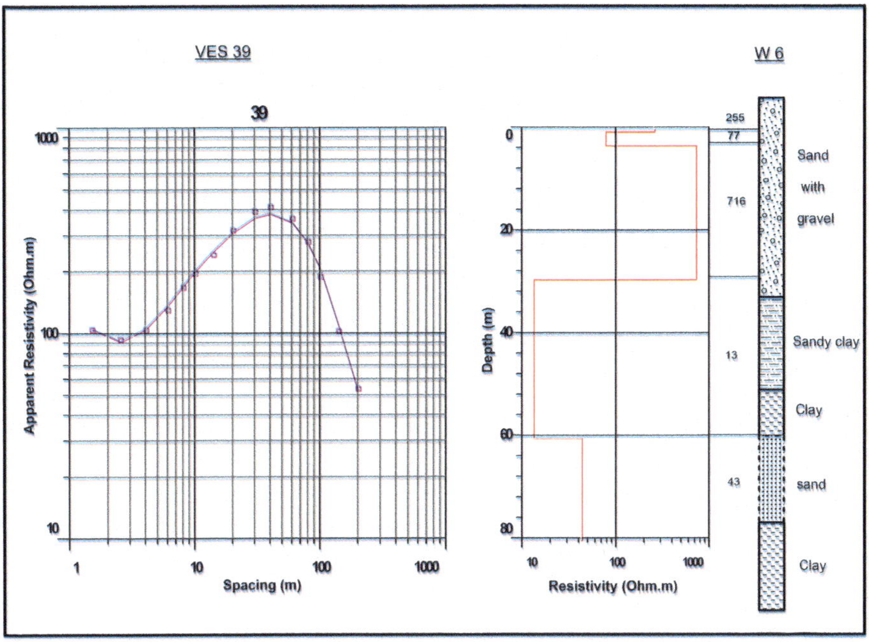

Fig. 15 Lithology log of well 6 and its corresponding geoelectric section [38]

- The second geoelectric layer shows relatively low resistivity values ranged between 2 and 15 Ω m, which corresponds to clay deposits. The thickness of this layer is ranging between 1 m at VES 40 and 44 m at VES 78. A lens of clayey sand, sand, and gravely sand that corresponds to low to high resistivity values 14–421 Ω m is found above this clay layer. This lens appears in cross sections C–C′ and D–D′.
- Relatively moderate electric resistivity values ranged between 15 Ω m (VES 12) and 75 Ω m (VES 53) characterize the third geoelectric unit constituting the main aquifer in the area. The thickness of this unit is ranged between 13 m (VES 6) and 83 m (VES 53) which corresponds to sand deposits.
- The fourth geoelectric unit is characterized by relatively low electric resistivity values ranged between 1 Ω m (VES 5 and VES 35) and 20 Ω m (VES 54 and VES 29) which corresponds to clay nature deposits.
- Uncompleted two geoelectric layers appear only in geoelectric cross section B–B′. The first layer is recorded beneath VESes 49, 48, 47, 5, and 6. It has relatively high resistivity values ranging from 147 to 275 Ω m, which reflects gravelly sand deposits. The thickness of this layer varies from 14 m beneath VES 6 and 32 m under VES 48, whereas the second geoelectric layer is determined below VESes 49, 48, 47, and 6. This layer has relatively low resistivity values varying from 1 to 11 Ω m which leads to clay nature deposits.

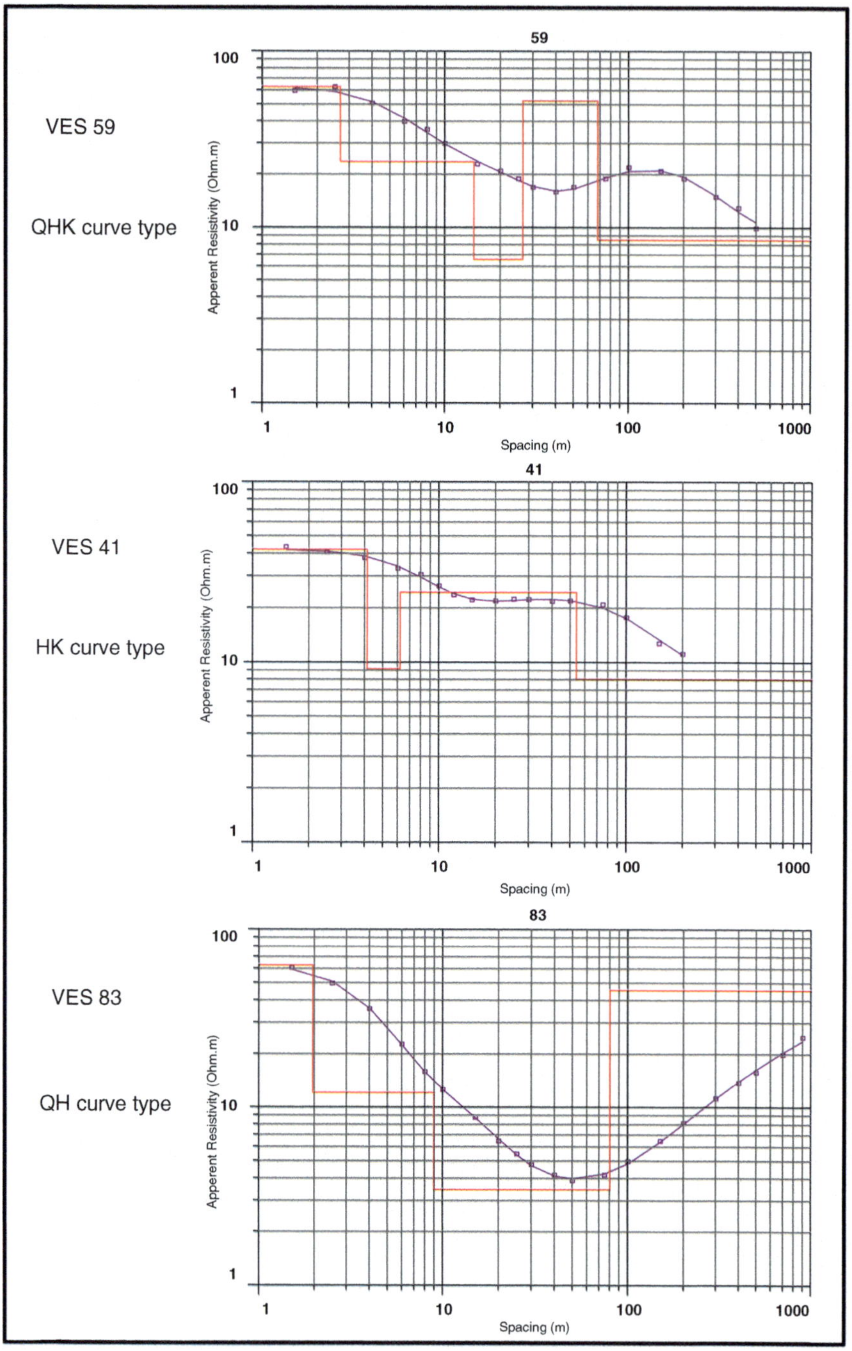

Fig. 16 Examples of the result of the 1D inversion of VESes 41, 59, and 83

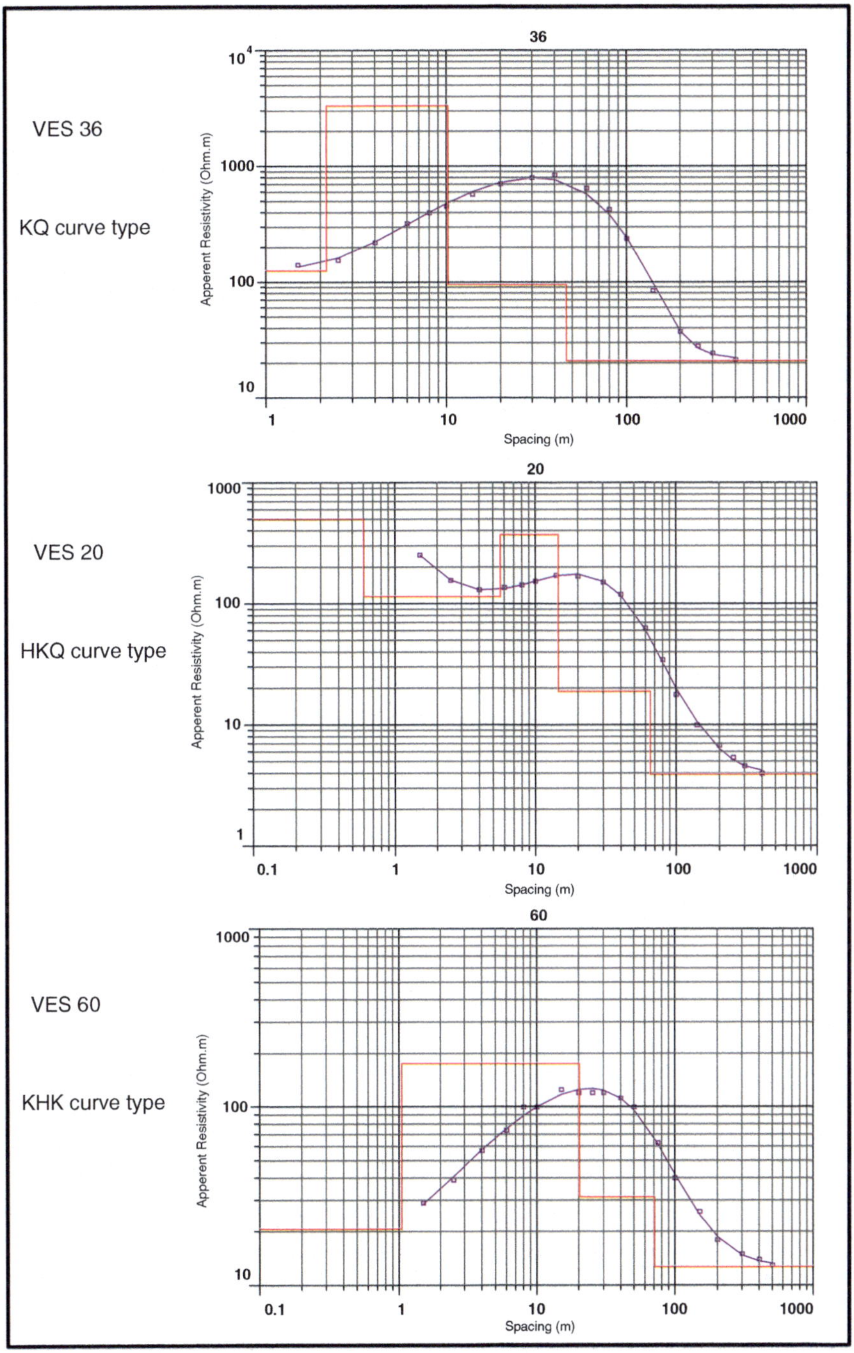

Fig. 17 Examples of the result of the 1D inversion of VESes 20, 36, and 60

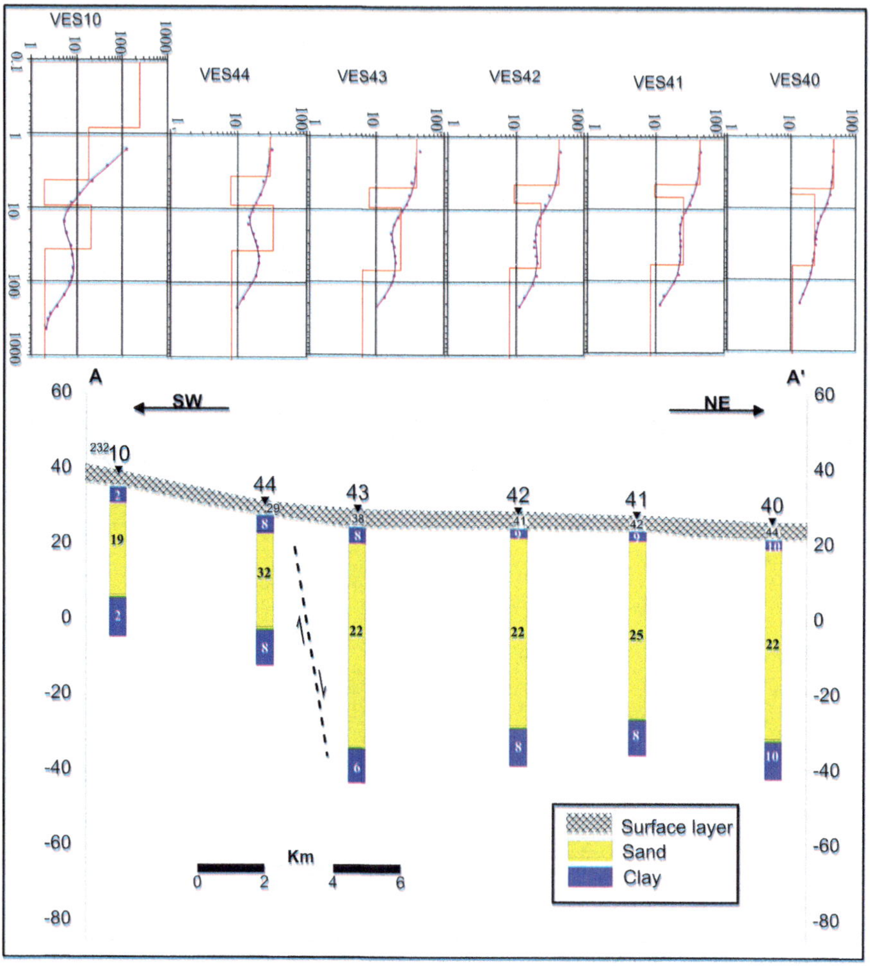

Fig. 18 Interpreted geoelectric cross section A–A′

8.4 Interpretation of TEM Data

Quantitative interpretation of TEM data has been carried out automatically using "Interpex IX1D software 2008." The models of the number of layers that were carried out from VESes insert as initial models to calculate the final models of TEM data. Figure 22 represents an example of the quantitative interpretation of TEM curves. The resistivity values of the TEM data were constrained using the drilled boreholes nearby and were correlated with the lithology of these boreholes. Figure 23 is an example of the correlation procedure where the station TEM 24 is correlated with the borehole W57. During the quantitative interpretation, all available data were used to construct the initial model. These data include the geological

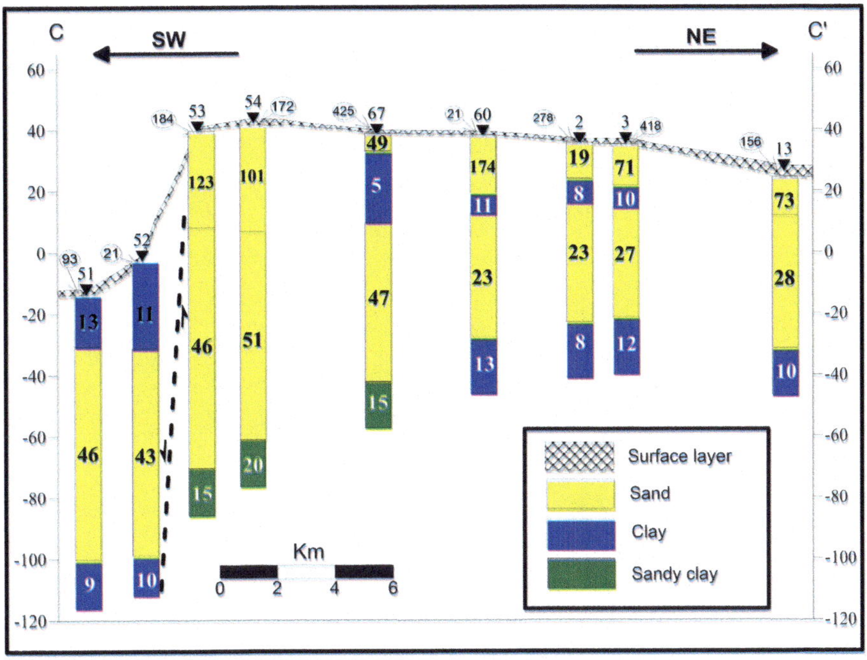

Fig. 19 Interpreted geoelectric cross section C–C′

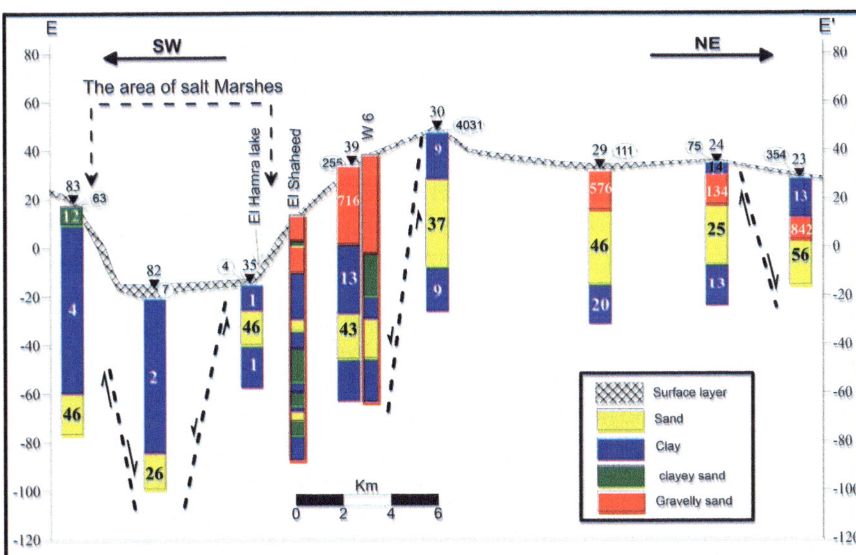

Fig. 20 Interpreted geoelectric cross section E–E′

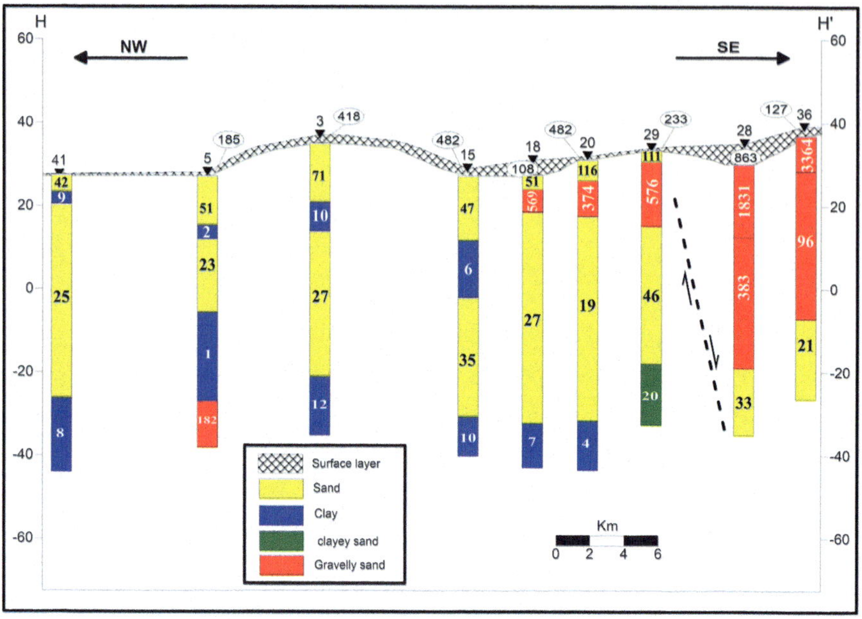

Fig. 21 Interpreted geoelectric cross section H–H′

and hydrogeological information, the geoelectric resistivity results, and the drilled boreholes in the area. The TEM stations that were carried out in this area were used to construct three geoelectric cross sections (Figs. 24, 25, and 26). The inspection of these geoelectric cross section shows that the study area can be differentiated into five geoelectric units according to their resistivity values:

- The first layer is a thin layer with resistivity range between 1 Ω m at (TEMes 8, 9, and 25) and 324 Ω m at (TEM 26). The great difference in the resistivity of this layer reflects the inhomogeneity of the lithology, which corresponds to sands, silts, clay, or gravels. The thickness of this layer ranges between 0.5 and 2.8 m.
- The second layer has relatively moderate to high resistivity values ranging between 57 and 265 Ω m and represents gravelly sand layer. The thickness of this layer ranges between 3.5 m at TEM 23 and 29 m at TEM 3.
- The third geoelectric layer has relatively low resistivity values ranging between 1 Ω m at TEM 17 and 27 Ω m at TEM 21 and is composed of clayey sand to clay deposits. It may represent the Pleistocene aquifer. In cross section T3–T3′ (Fig. 26), this layer can be divided into three sublayers: clay, clayey sand, and clay deposits.
- The fourth geoelectric layer is represented by relatively moderate resistivity values which varied from 15 to 59 Ω m. These values reflect the presence of sand and gravelly sand deposits that constitute the main Pliocene aquifer in the investigated area. The thickness varies from 9 to 57 m.

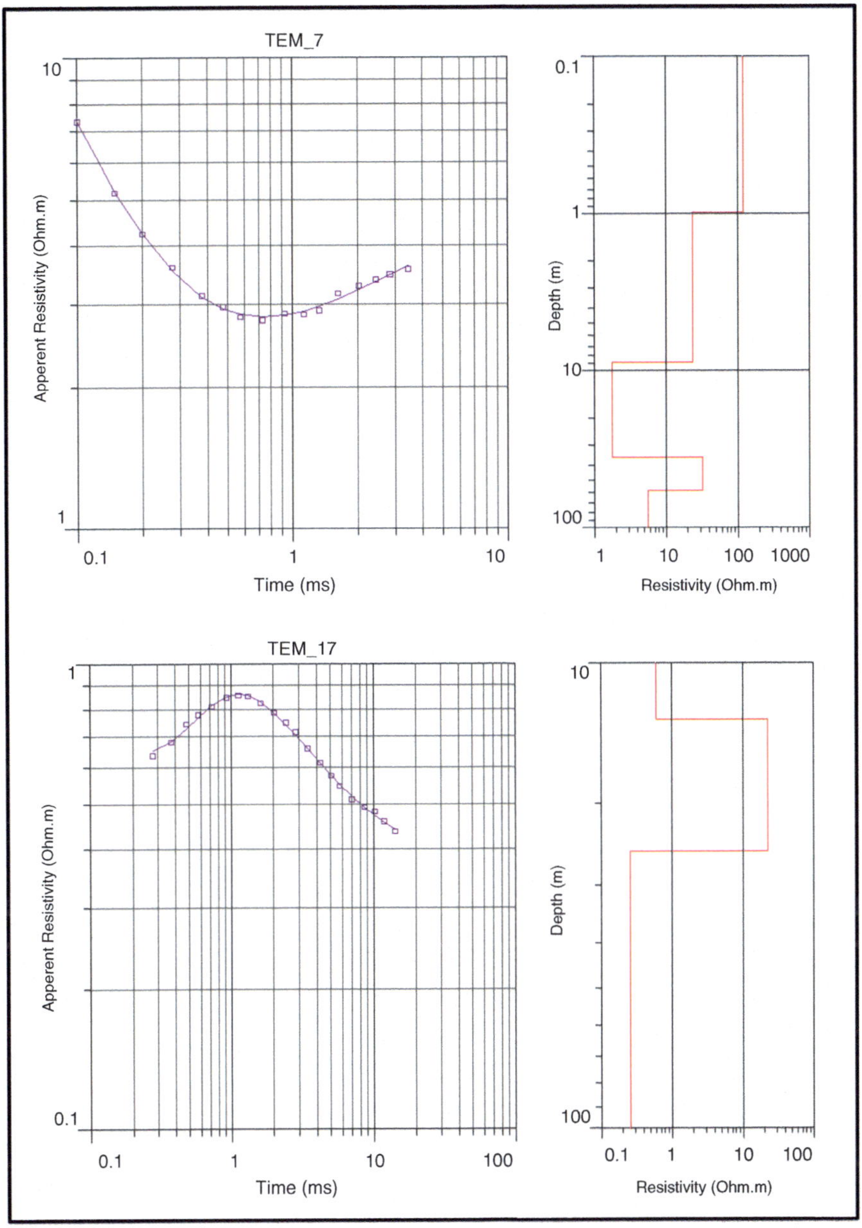

Fig. 22 TEM sounding curves and their interpretation at stations 7 and 17

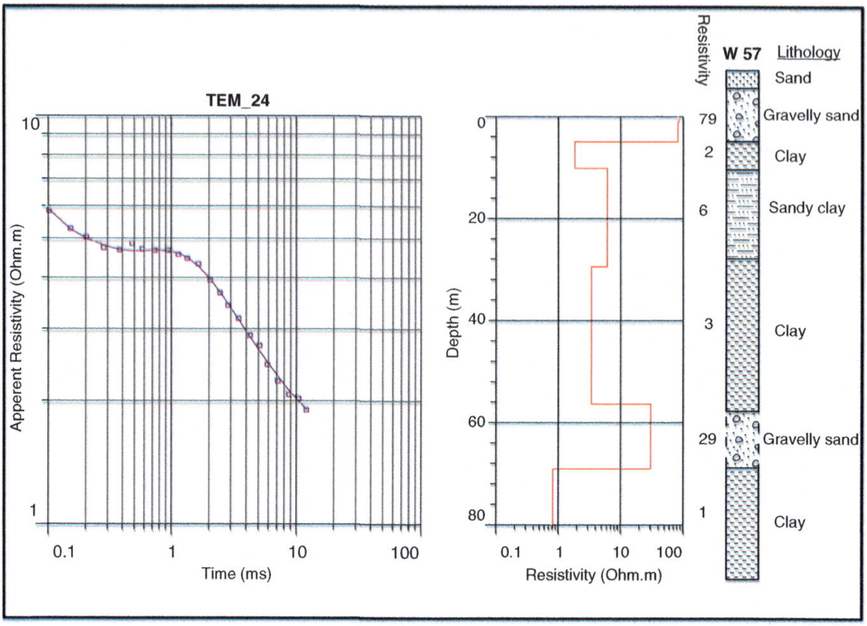

Fig. 23 Lithology of well 57 and its corresponding geoelectric section

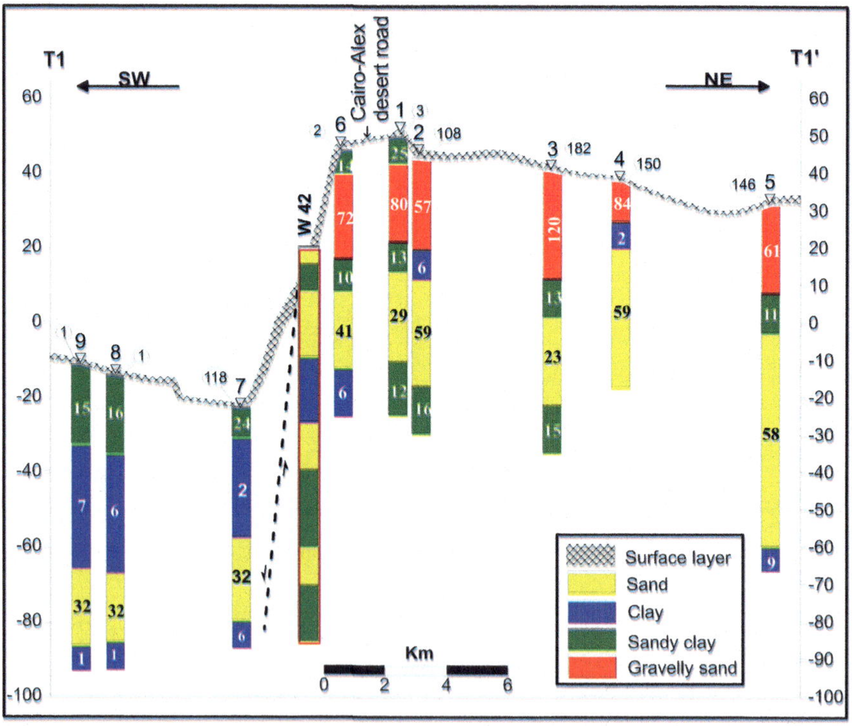

Fig. 24 Interpreted geoelectric cross section T1–T1′

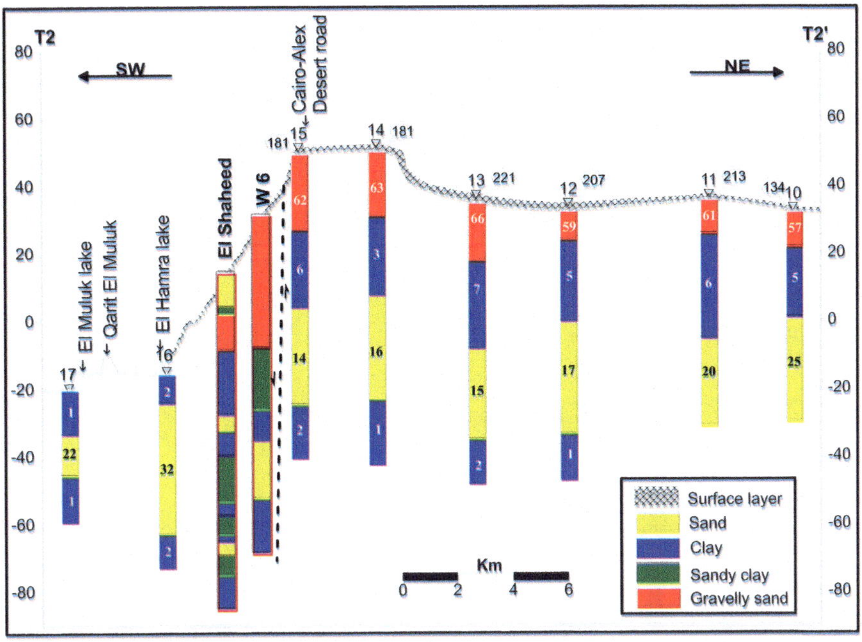

Fig. 25 Interpreted geoelectric cross section T2–T2′

– The fifth geoelectric layer has relatively low resistivity values ranging between 1 and 9 Ω m and is made up of clay. This layer is well correlated with boreholes w6, w42, and w57.

9 Results and Discussions

The quantitative interpretation was performed through one-dimensional resistivity inversion for both VES and TEM data using the software IX1D from Interpex Ltd. As a result, ten cross sections (from VES) and three cross sections (from TEM) were constructed and well correlated with the available boreholes. Two maps for the depth to the main aquifer (Fig. 27) and its true resistivity (Fig. 28) are established depending on the results of the quantitative interpretation of the VES and TEM data. The main aquifer is determined at depth range varying from 6 near El-Nubariya city to about 90 m at the southern parts of the investigated area where it increases to the south and southeast directions. The true resistivity of the aquifer ranges from 15 to 70 Ω m. It decreases in the north and northwest direction, whereas it increases in the central, southern, and northeastern parts. The highest value of the true resistivity is recorded near Wadi El-Natrun city. A low value of resistivity can be seen in the area inside of Wadi El-Natrun valley due to the influence of the

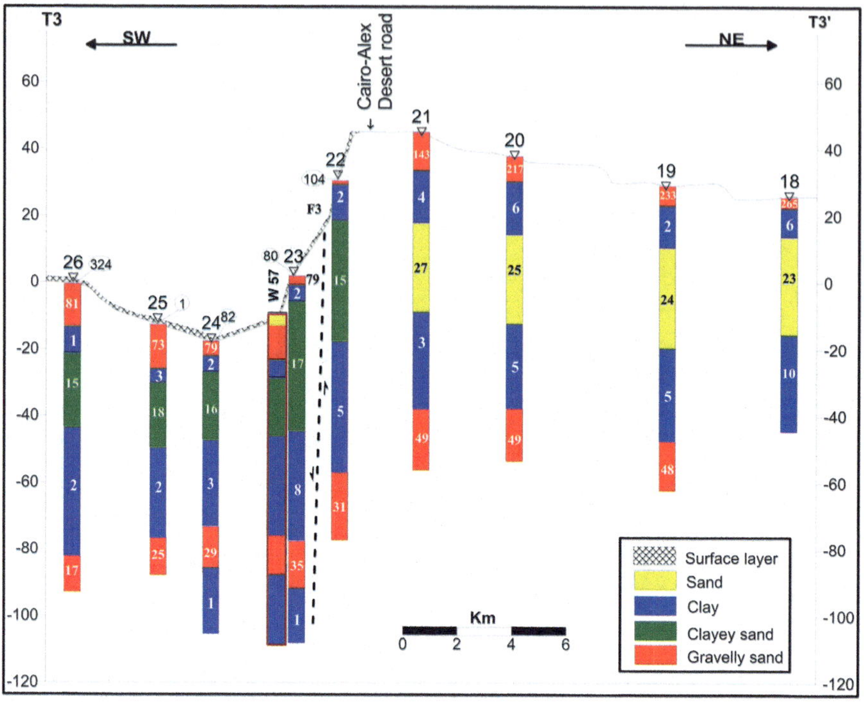

Fig. 26 Interpreted geoelectric cross section T3–T3′

saltwater of its lakes. This aquifer is composed of sand or gravelly sand. The aquifer system in the area is divided into Pleistocene and Pliocene aquifers. The Pleistocene aquifer is the shallower aquifer and composed of gravelly to clayey sand deposits. The Pliocene aquifer is the main aquifer where it consists of sand to gravelly sand deposits. In some localities, the Pliocene aquifer can be divided into two aquifers.

The quality of the groundwater can be determined depending on the results of the geoelectric exploration represented by the true resistivity map (Fig. 28). A brackish groundwater can be detected at shallow depths at the northern and northeastern parts of the survey area, whereas relatively freshwater can be determined at the southern and southeastern parts around Wadi El-Natrun city at deep depths. This information should be taken into consideration during the process of drilling new boreholes. The optimum area for drilling is the northeast of Wadi El-Natrun city.

The inferred faults from the geoelectric sections were traced and collected (Fig. 28) to construct a structure map. This map shows the location and the throw of the interpreted faults in the study area. A major fault passes parallel to the northeastern edge of Wadi El-Natrun with a southwest throw. Also another fault can be detected parallel to the southwestern edge of the valley with a throw to the northeast direction. Three faults pass nearly perpendicular to Wadi El-Natrun axis with throw almost to the east which can be seen on the structure map. It is worth to

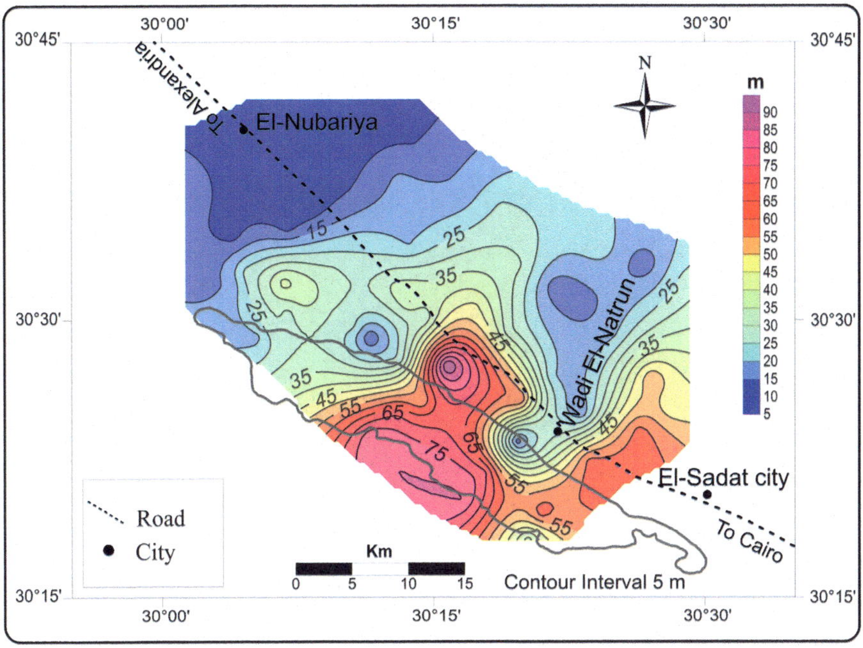

Fig. 27 Depth map to the main aquifer in the study area [28]

mention that Wadi El-Natrun depression and its lakes are structurally controlled by set of parallel faults trending NW direction.

Saltwater of the lakes of Wadi El-Natrun causes a serious problem in the area. That is why it is worthy to have a detailed investigation to delineate the extension of the brackish water zone. Also, to the north of the study area, especially near El-Nubariya city, a shallow clay layer at depth from 1.5 to 5 m causes a trapping for the surface water; this causes a very severe problem for the crops in the area.

The results showed that TEM technique is more effective than DC resistivity technique in conductive areas where it depicts more details of subsurface layers that could not be detected by VES data, especially in the deep depths. Accordingly, it is highly recommended to apply joint inversion for the available data sets, especially VES and TEM data. Also, it is important to mention the necessity of performing 2D inversion to confirm the fault structures inferred from 1D inversion of VES and TEM data. This will be conducted in the future work of the authors.

10 Conclusions and Recommendations

The main goal of this chapter is to determine the groundwater aquifers in the West Nile Delta area. From the precise inspection of the results of the VES and TEM data interpretation, it could be concluded that the aquifer system in the area extending

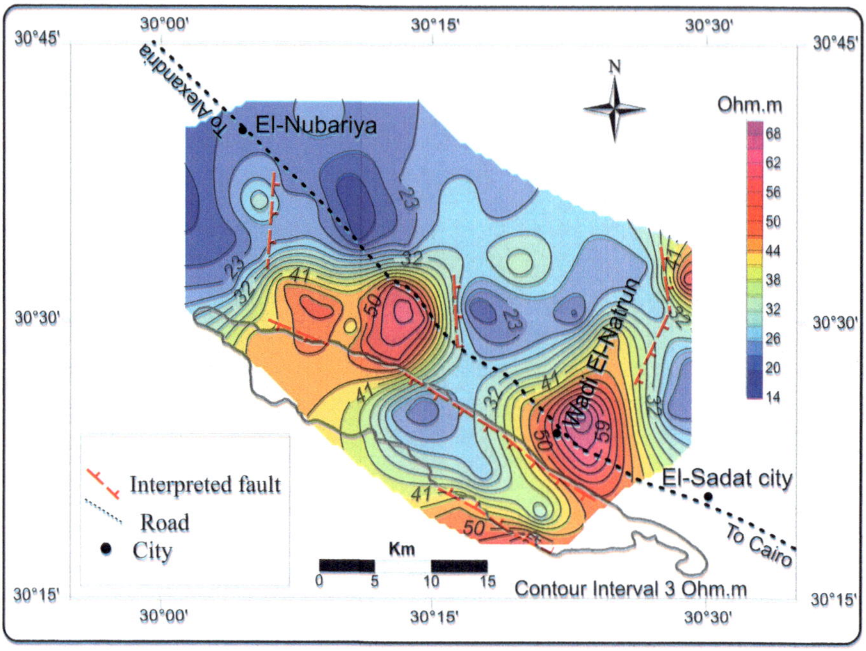

Fig. 28 True resistivity map of the main aquifer in the study area [38]

from Wadi El-Natrun city to El-Nubariya city along the Alexandria-Cairo desert road is divided into Pleistocene and Pliocene aquifers. The Pleistocene aquifer is the shallower aquifer and composed of gravelly to clayey sand deposits. The Pliocene aquifer is the main aquifer where it consists of sand to gravelly sand deposits. This aquifer can be detected at depth range between 6 m at the north part near El-Nubariya city to a depth of 90 m at the southern parts. The groundwater of the Pliocene aquifer can be described as a brackish groundwater at shallow depths at the northern and northeastern parts of the survey area, whereas it can be considered relatively freshwater at deep depths at the southern and southeastern parts around Wadi El-Natrun city. Based on the structural map inferred from the 2D geoelectric cross sections, it is worth to mention that Wadi El-Natrun depression is structurally controlled by a set of parallel faults trending NW.

We found that DC resistivity technique is very appropriate in determining the subsurface layers at shallow depths with a high resolution. On the other hand, TEM technique is more effective than DC resistivity technique in detecting more details of subsurface layers that could not be detected by VES method, especially in the deep depths. Accordingly, it is highly recommended to apply TEM and DC resistivity techniques together in groundwater studies.

In case of drilling new boreholes for drinking or agricultural purposes, we recommend the area northeast of Wadi El-Natrun city. Also, the saltwater of the lakes of Wadi El-Natrun causes a serious problem in the area. That is why it is worthy to have a detailed investigation to delineate the extension of the brackish water zone. We also recommend installing a covered drainage system to overcome the problem of trapping surface water near El-Nubariya city.

References

1. Sharaky AM, El Hasanein AS, Atta SA, Khallaf KM (2017) Nile and groundwater interaction in the western Nile Delta, Egypt. In: Negm AM (ed) The Nile Delta. Handbook of Environmental Chemistry, vol 55. Springer, Cham, pp 33–62. https://doi.org/10.1007/698_2016_127
2. Barker RD (1980) Application of geophysics in groundwater investigations. Water Surv 84:489–492
3. Adegbola RBSO, Oseni ST, Sovi KF, Oyedele LA (2010) Subsurface characterization and its environmental implications using the electrical resistivity survey: case with LASU Foundation Programme Campus Badagry, Lagos State, Nigeria. Nat Sci 8(8):146–151
4. Tarabees E, El-Qady G (2016) Seawater intrusion modeling in Rashid area of Nile Delta (Egypt) via the inversion of DC resistivity data. Am J Clim Chang 5:147–156. https://doi.org/10.4236/ajcc.2016.52014
5. Kelly W (1976) Geoelectric sounding for delineating groundwater contamination. Ground Water 14:6–10
6. Sandford KS, Arkell WJ (1939) Paleolithic man and Nile Valley in lower Egypt. Univ Chicago Oriental Inst Publ 36:1–105
7. Philip G (1955) Geology of the Pleistocene sediments of the Mediterranean Coast, west of Abu Qir. PhD thesis, Faculty of Science, Cairo University, Cairo
8. Said R (1962) The geology of Egypt. Elsevier, Amsterdam and New York, p 377
9. El Fayoumy IF (1964) Geology of groundwater supplies in Wadi El-Natrun area. MSc thesis, Faculty of Science, Cairo University, Cairo, p 109
10. Shata AA, El Fayoumi IF (1967) Geomorphological and morphopedological aspects of the region west of the Nile Delta with special reference to Wadi El-Natrun area. Bull Inst Desert d'Egypte 13(1):1–38
11. Abu El-Izz MS (1971) Landforms of Egypt. American University Press, Cairo, p 281
12. Sanad S (1973) Geology of the area between Wadi El-Natrun and the Moghra depression. PhD thesis, Faculty of Science, Assuit University, Assuit, p 184
13. Attia SH (1975) Pedology and soil genesis of the Quaternary deposits in the region west of Nile Delta (north and east of Wadi El-Natrun). PhD thesis, Faculty of Science, Ain Shams University, Cairo, p 358
14. El Shazly EM, Abdel-Hady M, El Ghawaby H, El Kassas K, Khawasik SM, El Shazly MM, Sanad S (1975) Geologic interpretation of Landsat satellite image for west Nile Delta area, Egypt. Remote Sensing Center, Academy of Scientific Research and Technology, Cairo, p 38
15. El Ghazawi MM (1982) Geological studies of the Quaternary-Neogene aquifers in the area northwest Nile Delta. MSc thesis, Faculty of Science, El Azhar University, Cairo, p 170
16. Abdel Baki AA (1983) Hydrogeological and hydrogeochemical studies on the area west of Rosetta Branch and south of El Nasr Canal. PhD thesis, Faculty of Science, Ain Shams University, Cairo, p 156
17. Embaby AAA (1995) Geomorphology and hydrogeology of El Khatatba district, west Nile Delta, Egypt. MSc thesis, Faculty of Science, Mansoura University, Mansoura, p 275
18. Gomaa MA (1995) Comparative hydrogeological and hydrogeochemical study on some aquifers, west of Nile Delta, Egypt. PhD thesis, Faculty of Science, Ain Shams University, Cairo, p 236

19. Embaby AAA (2003) Environmental evaluation for geomorphological situation in relation to the water and soil resources of the region north of the Sadat City, west Nile Delta, Egypt. PhD thesis, Faculty of Science, Mansoura University, Mansoura, p 383
20. Shata AA (1953) New light on the structural development of the Western Desert of Egypt. Bull Inst Desert d' Egypte 3(1):101–106
21. Shata AA (1955) An introductory note on the geology of northern portion of the Western Desert of Egypt. Bull Inst Desert d' Egypte 5(3):96–106
22. Shata AA (1959) Geological problems related to the groundwater supply of some desert areas of Egypt. Bull Soc Geogr d' Egypte 32:247–262
23. Shata AA (1961) The geology of groundwater supplies in some arable lands in the desert of Egypt (internal report). Desert Institute, Cairo
24. Shata AA, Pavlov M, Saad FK (1962) Preliminary report on the geology, hydrogeology, and groundwater hydrology of the Wadi El-Natrun and adjacent area (internal report), part 1. Desert Institute, Cairo
25. Shata AA, El Fayoumi IF, Tamer M (1970) Geomorphology, geology, hydrology and soils of Wadi E-Natrun – Maryot agriculture project. Desert Institute, Cairo
26. Shata AA, El Shazly MM, Attia SH, Aboul-Fetouh M (1978) The geology of the Quaternary deposits and their natural relation to soil formations in the fringes west of the Nile Delta, Egypt. Desert Institute Bull ARE 28(1):43–77
27. Shedid AG (1989) Geological and hydrogeological studies of the Sadat area and its vicinities. MSc thesis, Faculty of Science, El Menoufia University, Shibin El Kom, p 157
28. CONCO (1987) Geological map of Egypt 1: 500,000. In: Klitzsch E, List FK, Pöhlmann G (eds) Map sheets NH36NW and NH35NE. CONCO with cooperation of the Egyptian General Petroleum Corporation, Cairo
29. Shata AA, El Fayoumi IF (1970) Remarks on the regional geological structure of the Nile Delta. In: Proceedings of Bucharest symposium on Delta, 6–14 May 1969
30. Sigaev NA (1959) The main tectonic features of Egypt (internal report). Geological Survey of Egypt, Cairo, pp 1–26
31. Idris H (1970) Groundwater investigation in Wadi El-Natrun. In: Proceedings of IAEA symposium, Beirout, p 26
32. El Shazly EM, Abdel Hadi MA, El Shazly MM, Sand S, El Ghazawi MM, Abdel Mogheeth SM (1978) Subsurface geology and geochemistry of Pliocene-Quaternary aquifers northwest of the Nile Delta area. Remote Sensing Center, Academy of Scientific Research and Technology, Cairo, p 256
33. El Sabagh AA (1992) Impact of land reclamation projects on groundwater condition in the area north west of the Delta. MSc thesis, Faculty of Science, Cairo University, Cairo, p 126
34. Sallouma MK, Gomaa MA (1997) Groundwater quality in the Miocene aquifer east and west of Nile Delta and in the north of the Western Desert, Egypt. Ain Shams Sci Bull 35:47–72
35. Abd El-Wahab S (1999) Hydrogeological and isotope assessment of groundwater in Wadi El-Natrun and Sadat City, Egypt. PhD thesis, Faculty of Science, Ain Shams University, Cairo, p 237
36. Ahmed SA (1999) Hydrogeological and isotope assessment of groundwater in Wadi El-Natrun and Sadat City, Egypt. MSc thesis, Faculty of Science, Ain Shams University, Cairo, p 237
37. Ibraheem IM (2009) Geophysical potential field studies for developmental purposes at El-Nubariya – Wadi El-Natrun area, West Nile Delta, Egypt. PhD thesis, Faculty of Science, Mansoura University, Mansoura, p 193
38. Ibraheem IM, El-Qady GM, ElGalladi A (2016) Hydrogeophysical and structural investigation using VES and TDEM data: a case study at El-Nubariya–Wadi El-Natrun area, west Nile Delta, Egypt. NRIAG J Astron Geophys 5(1):198–215
39. Omara SM, Sanad S (1975) Rock stratigraphy and structural features of the area between Wadi El-Natrun and the Moghra depression, Western Desert, Egypt. Jeol Jb B 16:45–73

40. Diab MS, Hassaneen AG, El Shayeb HM, Mesbah MA, Khalil MA (1996) Geoelectrical and hydrogeological studies in Wadi El-Natrun depression. In: E.G.S. proceedings of 14th annual meeting, Cairo, pp 75–94
41. Abdel Salam AGM (1997) Groundwater evaluation and morphopedological studies for the area of Wadi El-Natrun, Egypt. MSc thesis, Faculty of Science, Mansoura University, Mansoura, p 148
42. Abd El-Gawad AMS, Ammar AI (2004) The contribution of geoelectrical investigations in delineating the shallow aquifer in Wadi El-Natrun area, Western Desert, Egypt. EGS J 2 (1):37–50
43. Mansour KKA (2005) Geophysical evaluation for groundwater aquifers at El Beheira governorate, west Delta, Egypt. MSc thesis, Faculty of Science, Zagazig University, Banha, p 152
44. Massoud U, Kenawy AA, Ragab EA, Abbas AM, El-Kosery HM (2014) Characterization of the groundwater aquifers at El-Sadat City by joint inversion of VES and TEM data. NRIAG J Astron Geophys 3:137–149
45. Ernstson K, Kirsch R (2009) Geoelectrical methods. In: Kirsch R (ed) Groundwater geophysics: a tool for hydrogeology2nd edn. Springer, Berlin, pp 85–117
46. Milsom J (2003) Field geophysics.3rd edn. Wiley, New York, p 232
47. Seidel K, Lange G (2007) Geophysics (direct current resistivity methods). In: Knödel K, Lange G, Voigt H-J (eds) Environmental geology: handbook of field methods and case studies. Springer, Berlin, pp 205–238
48. Reynolds JM (1997) An introduction to applied and environmental geophysics. Wiley, Chichester, p 796
49. Keller GV (1967) Application of resistivity methods in mineral and groundwater exploration program. Mining and groundwater geophysics, vol 26. Geological Survey of Canada, Ottawa, pp 51–66
50. Keller GV, Frischknecht FC (1966) Electrical methods in geophysical prospecting. Pergamon Press, Oxford, p 517
51. Lowrie M (2007) Fundamentals of geophysics. Cambridge University Press, New York, p 381
52. Zohdy AAR, Eaton GP, Mabey DR (1974) Application of surface geophysical to groundwater investigations. In: Techniques of water-resources investigations of the United States Geological Survey, vol 2, p 116
53. Kaufmann AA, Keller GV (1983) Frequency and transient sounding; methods in geochemistry and geophysics, vol 16. Elsevier, Amsterdam, p 685
54. Nabighian MN, Macanae JC (1991) Time-domain electromagnetic prospecting methods. In: Nabighian MN (ed) Electromagnetic methods in applied geophysics, vol 2. Society of Exploration Geophysicists, Tulsa, pp 427–514
55. Sharma PV (1986) Geophysical methods in geology. Elsevier, Amsterdam, p 442
56. Loke MH (2004) Tutorial: 2-D and 3-D electrical imaging surveys. Unpublished course notes, p 128. http://www.geoelectrical.com/downloads.php
57. Parasnis DS (1997) Principles of applied geophysics. Chapman & Hall, London, p 429
58. Yungul SH (1996) Electrical methods in geophysical exploration of deep sedimentary basins. Chapman & Hall, London, p 208
59. Palacky GJ (1987) Resistivity characteristics of geologic targets. In: Nabighian MN (ed) Electromagnetic methods in applied geophysics theory, vol 1. Society of Exploration Geophysicists, Tulsa, pp 53–129
60. Fitterman DV, Stewart MT (1986) Transient electromagnetic sounding for groundwater. Geophysics 51(4):995–1005
61. Inman JR (1975) Resistivity inversion with ridge regression. Geophysics 40:798–817
62. Interpex Ltd (2008) IX1D v3 manual, 1-D DC resistivity, IP, MT and EM inversion software. Golden. http://www.interpex.com
63. Dobrin MB, Savit CH (1988) Introduction to geophysical prospecting.4th edn. McGraw-Hill, New York, p 867

Part V
Groundwater Contamination and Degradation

Salinization and Origin of the Coastal Shallow Groundwater Aquifer, Northwestern Nile Delta, Egypt

Abbas M. Sharaky, Adel S. El Hassanein, Samir A. Atta, and Karema M.A. Khallaf

Abstract Serious salinization problem is affecting the northern Nile Delta basin and its groundwater aquifers. The hydrochemistry of major ions (K^+, Na^+, Mg^+, Ca^{2+}, Cl^-, SO_4^{2-}, HCO_3^-, CO_3^{2-}) and the trace elements (Fe, Mn, Zn, Pb, Cd, Cr, Cu, Ni) have been used to constrain the hydrochemical characteristics, source, and salinization processes of the shallow coastal aquifer, northwestern Nile Delta. Twenty groundwater wells, varying in depth from 17 to 40 m, had been examined and sampled to carry out the physicochemical parameters and chemical compositions of the groundwater and to obtain additional information on the possible contamination with major elements, trace elements (heavy metals), and nutrients (NH_4^+, NO_3^-, PO_4^{2-}).

The hydrochemical data indicated that the groundwater of the coastal aquifer, northwestern Nile Delta, is meteoric in origin and is mixed with marine water. The coastal plain aquifer is recognized to be at high risk of increasing salinization. The salinity of the groundwater as a total dissolved solid (TDS) ranges from 1,288 to 4,907 mg/l with an average of 3,155 mg/l. The electric conductivity (EC) of the groundwater ranges from 1,900 to 9,790 μS/cm with an average of 4,620 μS/cm. It is directly related to TDS and the geographical position of each well. The groundwater is slightly alkaline with pH ranges from 7.01 to 8.2. The high salinity, pH, and EC values support the conclusion of seawater intrusion.

The nutrient content such as nitrates is higher than the standard values, which mainly resulted from rural sources. The concentrations of the trace elements are lower than the standard values except for iron, manganese, and nickel. This groundwater can be used for crop irrigation but must be treated before using for drinking.

A.M. Sharaky (✉) and A.S. El Hassanein
Department of Natural Resources, Institute of African Research and Studies, Cairo University, Giza, Egypt
e-mail: sharaky@geologist.com; adelelhassanin@yahoo.com

S.A. Atta and K.M.A. Khallaf
Ministry of Water Resources and Irrigation, Giza, Egypt
e-mail: karimakhallaf@yahoo.com

A. M. Negm (ed.), *Groundwater in the Nile Delta*,
Hdb Env Chem (2019) 73: 275–306, DOI 10.1007/698_2017_183,

Keywords Coastal aquifer, Groundwater, Nile Delta, Salinization, Water quality

Contents

1 Introduction

There are several processes causing groundwater salinization. They include seawater intrusion due to intense aquifer exploitation [1–6], interactions with deep saline paleowaters, and water-rock interaction [7–12]. Other sources cause salinization are the dissolution of evaporites [13–16], evaporation of freshwater, and pollution by untreated wastewater [3, 17]. Seawater intrusion and construction of the Aswan High Dam on the Nile River in 1971 may affect the water quality in the coastal aquifer.

The Nile Delta aquifer is the principal groundwater source in Egypt. The annual rate of groundwater withdrawal from the Nile Valley and Delta aquifers increased from 5.5 billion m^3 in 2000 [18] to 6.13 billion m^3 [19] and to more than 7 billion m^3 in 2016 [20]. All these represent 12% of the annual renewable freshwater and 9% of the annual consumed water in Egypt. The Nile Delta aquifer is mainly recharged by infiltration from excess irrigation water and direct seepage from the Nile River, irrigated channels, and insufficient rainfall that infiltrates through the upper clay layer [21].

Overpumping is the most severe issue of the groundwater mainly in the Nile Delta aquifers that is followed by salinization through saltwater intrusion. The Nile River, the primary source of water in Egypt (55.5 billion m^3, 97% of the annual renewable freshwater), is insufficient to cover all the Egyptian demand of water of about 80 billion m^3 in 2016 according to the Egyptian Ministry of Water Resources and Irrigation [20]. There is a water deficit of about 23 billion m^3. Therefore, groundwater could provide a suitable source in addition to the Nile River.

The coastal aquifer, northwestern Nile Delta, is characterized by the presence of brackish water that endangers, often irreversibly, the future of water resources in the area. The overpumping of the groundwater in the coastal aquifer has severely degraded water quality due to seawater intrusions.

In general, the source of salinity in many coastal hydrogeologic systems can be identified to be modern seawater intrusion into aquifers; such a single source is not the case in many systems. Ancient marine intrusion, wind-driven sea spray and marine aerosols deposited at the topsoil, effects of mobilized salts stored in the unsaturated zone, evaporites, and local pollution may further contribute to salinization of groundwater [22, 23].

The main objectives of this study are:

1. Determination whether the brackish water in the coastal aquifer is attributed to seawater intrusion or it is just a paleo-formation water
2. Chemical analysis of pollutants and determination of the source of pollution
3. Hydrochemical characterizations of the groundwater
4. Evaluation of the groundwater suitability for different purposes

2 Local Geology and Hydrogeology

2.1 Description of the Study Area

The study area is located in the northwestern Nile Delta (see Fig. 1), just west of Alexandria, and extends up to 50 km to the west. It is bounded from the north by the Mediterranean coastline and from the south by approximately the latitude $30^{\circ}47'$ N, forming a parallel zone to the shoreline. The width of the study area ranges from 15 to 30 km, covering about 1,000 km^2.

2.2 Geology

The coastal zone has received the attention of a number of researchers such as El Fayoumy [24], El Shazly et al. [25], Gindi and Abd-Alla [26], Landis et al. [27], Leaven [21], Nagy [28], Masria et al. [29], Philip [30], Salim [31], Sanad [32], Shata [33], Swanberg et al. [34], Tarabees and El Qady [35], and Zaghloul et al. [36, 37].

Geology of the Mediterranean coastal area consists primarily of various Tertiary and Quaternary sedimentary deposits [25, 34]. The Quaternary beach deposits cover the study area [26]. The coastal zone, the northwestern zone of the Nile Delta is characterized by a series of parallel Pleistocene oolitic limestone ridges [38]. They are mainly composed of white oolotic limestone separated by shallow depressions (see Fig. 2). The width of the ridges decreases westward from Alexandria city. They are marine in origin [40]. Six distinct ridges have been recognized.

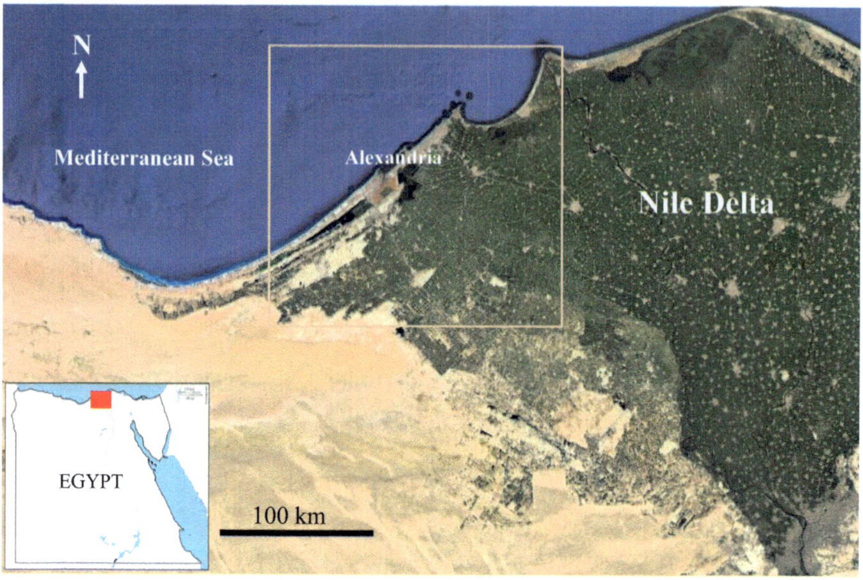

Fig. 1 Western Nile Delta from Google Earth on September 9, 2017. The square represents the chemical distribution maps in this chapter

They are Harbour Island, Costal, Abu Sir, the Gebel Maruit, Khashm El-Eish, and Alam El-Khadem ridges.

In the Nile Delta, the strata of hydrological importance belong mostly to the Quaternary and to the Tertiary. Of these strata, the deltaic deposits that belong to the Pleistocene constitute the bulk of the principal aquifer [41].

In the study area, water-bearing formation and strata of the hydrogeologic interest are found both in the Quaternary and in the Pliocene [25]. These strata have different lithological forms that include:

1. Alluvial beach sediments that are characterized by yellow color and fine to very fine texture. The sediments are limy in composition. The fresh groundwater floats on top of saltwater. The thickness of the freshwater zone is being controlled by its height above sea level and the quality of rechargeable water that enters sand dunes through direct infiltrate of rainfall or from surface runoff.
2. Alluvial deposits that are derived from the Miocene rocks, which include soil composed of almost homogenous calcareous loam and concretionary gypsum.
3. Alexandria Formation that forms the coastal beach of the Mediterranean Sea. It is composed of detrital limestone that is associated with calcareous clay soil. These members are characterized by fine, medium to coarse texture and mostly of light to gray colors. One of these varieties is composed of gypsum and clays (calcareous gypsum with soluble salts) forming islands to the south of the Mediterranean coast by about 10–12 km at Burg El Arab. These formations form the bedrock of Alexandria City. In many places, the detrital limestones

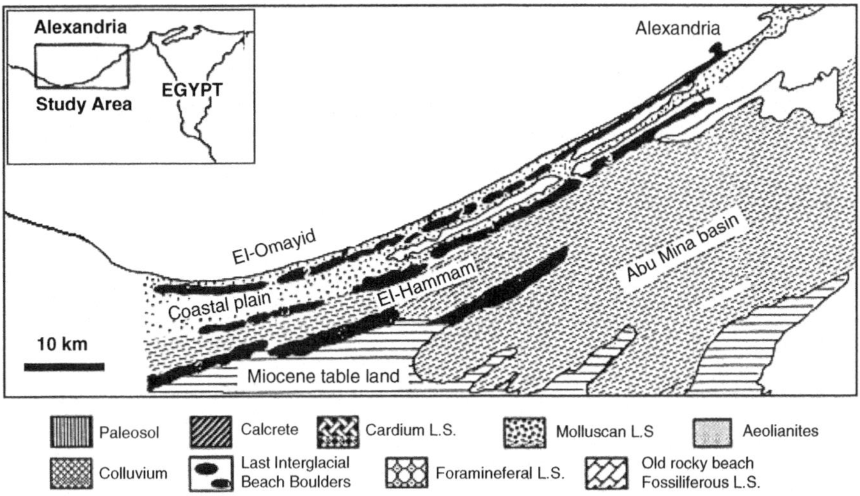

Fig. 2 Parallel elongated ridge along the Mediterranean coast separated by depressions [39]

possess indurated crust giving a brown compact rock with dark-gray surface tarnish (thickness 0–60 m).

4. The Miocene-Pliocene rocks cover brown compact sandy limestone with some gypsiferous limestones on top of small hills exposed to these small areas.

2.3 Hydrogeology

The Nile Delta aquifer consists of deltaic deposits ranging in average from 200 to 300 m thick. It is dominated by unconsolidated coarse sands and gravels with some clay lenses. The Nile Delta aquifer rests on a thick Pliocene clay layer [42]. The study area is classified as a semiarid zone. Winter precipitation varies from a year to another with an annual average range from 100 to 200 mm at the Mediterranean Sea [43].

In general, the coastal aquifers may be classified into four types [28]: (1) unconfined (phreatic), (2) confined, (3) leaky aquifers (semi-pervious confining layers), and (4) layered aquifers (several aquifers separated by impervious layers). The saturated thickness of the aquifer ranges from 40 to 60 m. The lithology ranges from sand-clay to sand-gravel with increasing depth [44]. The aquifer mostly contains brackish water that has been recharged annually by local rainfall and the Nile seepage water from Nasr and Maryut canals. The main direction of the groundwater flow in the principal aquifer is to the north toward the Mediterranean Sea. Farid [45] indicated that the brackish water in the Nile Delta is marine in origin. Seawater intrusion takes place and extends inland especially after the construction of the High Dam and the increase of demand for surface and groundwater [28].

In northern coastal areas of Egypt, groundwater aquifers are vulnerable to social and economic development. Recently, water demand for drinking, agriculture, tourism, and industry has increased.

Agriculture consumes the most significant amount of the available water in Egypt, with its share exceeding 85% of the total demand for water [20]. Water demand has been increased in Egypt during the last few decades due to fast population growth from 26 to 93.4 million in 1968 and 2017, respectively [46], new land reclamation, and urban and industrial development. The cultivated lands increased by at least 30% from 1984 to 2017 (see Fig. 3).

3 Material and Methods

Groundwater samples were collected from 20 water wells of different depths covering the coastal plain zone, northwestern Nile Delta (see Fig. 4). The wells vary in depths from 17 to 40 m. The samples were collected in two-liter polyethylene bottles in 2003. They were analyzed for chemical analyses of major cations and anions following the standard methods for the examination of water as described by the American Public Health Association [47], as well as the trace elements and nutrients (ammonia, nitrate, and phosphate). The chemical analyses were carried out in the Research Institute for Groundwater (RIGW) laboratories. The measurements of physiochemical parameters (pH, EC) were carried out in the field using portable instruments. The EC was also measured in the lab using an electric conductivity meter model YSI 32, USA.

The pH was determined using pH meter model WTW-Multilane pH, Germany. The concentrations of Na^+ and K^+ were measured using flame photometer model 410, England. Ca^{2+}, Mg^{2+}, Cl^-, CO_3^{2-}, and HCO_3^- were determined using chemical titration following the standard methods of the American Public Health Association

Fig. 3 Extension of cultivated lands (*green*) in the northwestern zone of the Nile Delta, from 1984 to 2017 (Google Earth)

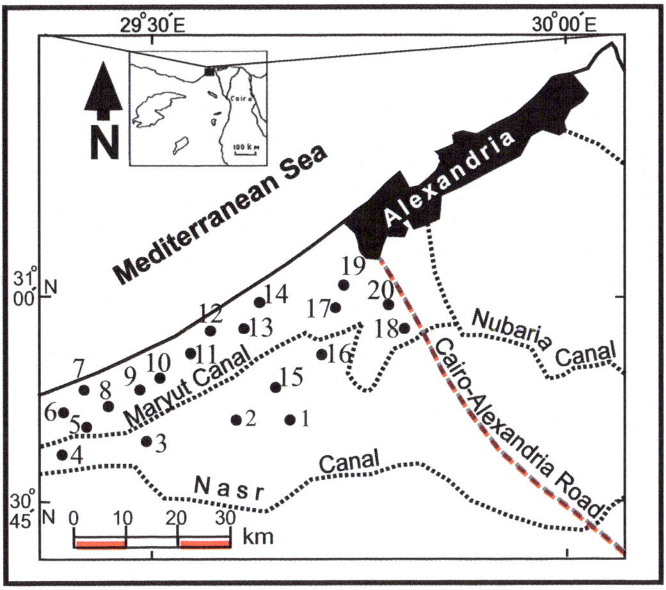

Fig. 4 Locations of the investigated wells in the coastal aquifer

[47]. SO_4^{2-} was measured by turbid metric method spectrophotometer model Spectronic 21D Milton Roy, England. NO_3^- was analyzed using cadmium reduction method, NH_4^+ using ammonia selective electrode method, and PO_4^{2-} using plasma method. The concentrations of trace elements (Fe, Mn, Zn, Pb, Cd, Cr, Cu, Ni) were performed using atomic absorption flame spectrometric technique (GBC 908 AA). The chemical data were displayed in graphical forms of the trilinear Piper [48] and Sulin [49] diagrams to make a visual comparison for the composition of the different water wells.

4 Results and Discussion

4.1 Hydrogen Ion Activity (pH)

The pH of the natural water is controlled by various possible reactions involving the major ions, other nonionic solutes, solids, and gases associated with a solution [50]. Streams in arid regions show pH values between 7 and 8 [51], while rainfall water has values from 4 to 8 [52]. Seawater near the surface normally shows a pH of 8.1 to 8.3. Soil water and playa-lake water in deserts may have pHs of 9 or even higher [51]. The pH values of the coastal aquifer range from 7.01 to 8.20 with an average of 7.73 (see Table 1). They are slightly higher than those of the surface water of the Nile River in Egypt, which shows an average pH of 7.5 at

Table 1 Hydrochemical composition of groundwater from the coastal aquifer, northwestern Nile Delta, Egypt

Well	Depth,		EC µS/	Anions, mg/l					Cations, mg/l					TDS	Well	Trace elements, ppm							
no.	m	pH	cm	Cl^-	SO_4^{2-}	HCO_3^-	NO_3^-	PO_4^{2-}	Na^+	K^+	Mg^+	Ca^{2+}	NH_4^+	mg/l	no.	Fe	Mn	Zn	Pb	Cd	Cr	Cu	Ni
1	27.5	7.99	6,500	1,420	1,416	79	8	0.15	1,426	20	26	150	0.04	4,545	1	9.3	07.0	3.1	003.0	003.0	042.0	1.0	1.0
2	27.5	7.04	5,080	1,136	1,090	183	2	0.30	915	31	128	150	0.04	3,635	2	1.0	06.0	2.1	003.0	0023.0	041.0	05.0	06.0
3	30.0	7.89	2,160	266	528	305	3	0.40	322	10	85	50	0.04	1,569	3	1.0	3.0	6.0	004.0	0025.0	044.0	06.0	2.0
4	37.0	7.03	5,530	1,136	1,243	183	11	0.50	869	14	167	210	0.03	3,834	4	5.4	1.0	9.2	0035.0	0024.0	045.0	06.0	1.0
5	30.0	8.10	5,960	1,349	1,128	201	5	0.16	1,035	16	146	160	0.04	4,040	5	7.0	05.0	9.0	003.0	003.0	043.0	1.0	1.0
6	35.0	7.86	4,040	888	1,152	275	7	0.20	713	16	182	150	0.04	3,383	6	3.0	06.0	8.2	002.0	003.0	043.0	1.0	09.0
7	30.0	7.08	5,400	1,136	1,129	183	7	0.15	888	32	154	190	0.02	3,719	7	6.0	1.0	4.2	003.0	0025.0	044.0	1.0	3.0
8	27.5	8.20	1,900	178	456	263	4	0.50	230	15	61	80	0.07	1,288	8	7.0	1.0	1	004.0	003.0	044.0	1.0	07.0
9	27.5	8.04	9,790	2,840	792	214	6	0.30	869	14	92	80	0.08	4,907	9	8.2	1.0	8.0	004.0	0025.0	043.0	08.0	06.0
10	17.5	7.14	5,140	1,172	1,085	226	7	0.40	888	31	143	180	0.1	3,733	10	3.0	06.0	4.2	003.0	0024.0	042.0	07.0	2.0
11	27.5	7.42	3,700	444	1,032	244	6	0.20	460	31	131	144	0.1	2,492	11	3.1	1.0	8.0	0035.0	0021.0	043.0	1.0	3.0
12	17.0	7.56	4,380	834	931	329	12	0.20	681	31	170	94	0.1	3,082	12	1.0	06.0	2.0	004.0	003.0	043.0	1.0	1.0
13	30.0	7.77	3,600	888	389	226	13	0.15	483	14	101	150	0.1	2,264	13	1.0	03.0	04.0	004.0	003.0	043.0	1.0	1.0
14	40.0	8.10	3,600	586	787	305	4	0.10	869	14	164	120	0.07	2,849	14	15.0	05.0	1.0	004.0	0025.0	033.0	12.0	15.0
15	30.0	7.78	2,040	302	614	232	6	0.13	345	23	88	60	0.09	1,670	15	2.0	01.0	15.0	003.0	003.0	001.0	1.0	1.0
16	30.0	8.10	2,100	320	408	275	4	0.20	294	23	102	30	0.1	1,456	16	15.0	04.0	16.0	004.0	0025.0	031.0	1.0	1.0
17	30.0	7.01	5,410	1,136	1,229	183	4	0.20	888	31	154	190	0.1	3,815	17	3.0	1.0	1.0	004.0	0025.0	045.0	07.0	09.0
18	40.0	8.20	6,150	1,598	336	592	17	0.30	1,150	10	109	61	0.1	3,873	18	25.0	05.0	1.0	0045.0	0026.0	044.0	06.0	05.0
19	30.0	8.20	5,600	1,243	576	519	15	0.20	869	14	160	158	0.09	3,554	19	35.0	2.0	09.0	003.0	0027.0	034.0	05.0	06.0
20	32.0	8.15	5,220	1,420	192	580	6	0.15	1,024	8	77	81	0.08	3,388	20	4.0	1.0	1.0	002.0	0023.0	035.0	1.0	08.0
Max	40.0	8.20	9,790	2,840	1,243	592	17	0.30	1,426	32	170	210	0.1	6,530	Max.	4.500	0.300	2.900	0.005	0.005	0.045	0.120	0.300
Min	17.0	7.03	1,900	178	192	79	2	0.10	230	8	26	50	0.07	765	Min.	0.100	0.010	0.040	0.002	0.002	0.001	0.050	0.050
Average	29.8	7.73	4,665	1,015	816	280	7.4	0.24	761	19.9	122.0	124	0.072	3199.8	Mean	0.705	0.088	0.886	0.003	0.003	0.039	0.085	0.122

Aswan-Cairo [53] and 7.7 at Aswan-Sohag [54]. The pH shows indirect correlation with Ca^+ ions [$R^2 = 0.42$, (see Fig. 5)] due to releasing of HCO_3^- ions from the Pleistocene carbonate rocks.

4.2 *Electric Conductivity (EC)*

Electric conductivity (EC) measures water's ability to conduct an electric current and is directly related to the total dissolved salts (ions) in the water. The studied well water samples show electrical conductivities in the range of 1,900–9,790 μS/cm, with an average of 4,665 μS/cm (see Table 1). They show a reasonable spatial variability in EC. The EC values are much higher than those of the surface Nile water 289 μS/cm at Aswan-Sohag [54].

4.3 *Salinity Content*

The hydrochemical data of the coastal aquifer indicated that the TDS varies from 1,288 to 4,907 mg/l with an average of 3,155 mg/l (see Table 1). The TDS was calculated from the sum of the concentrations of the major ions in the groundwater. The coastal aquifer is considered as a shallow aquifer because the water table ranges from 7 to 25 m [44]. The sources of water recharge are the local winter rainfall (200 mm/year) and water seepage from Nile water. The salinity content varies spatially all over the study area (see Fig. 6).

The variation of TDS is mainly affected by recharge rate from rainwater and seepage water from canals, exploitation rate, aquifer-bearing strata and the geographic position of wells. The wells containing water of more than 2,000 mg/l are mainly due to the aquifer-bearing strata. In the central area, the groundwater varies

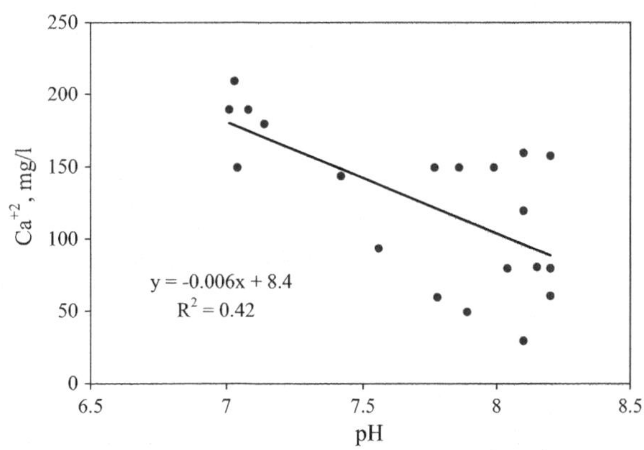

Fig. 5 The relationships between Ca^{+2} ions and pH values of the groundwater in the coastal aquifer, northwestern Nile Delta

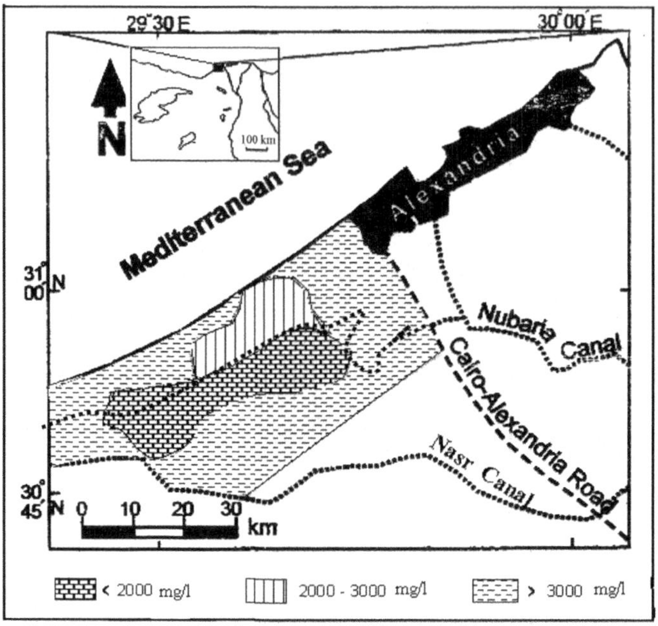

Fig. 6 Distribution of salinity content in the coastal aquifer

from freshwater (<1,000 mg/l) to brackish (2,000 mg/l) due to seepage from Maryut canal that reduces the high saline water of the aquifer. This conclusion is conformable with the results of El Shazly et al. [25] and Ministry of Irrigation, Desert Irrigation Department [55]. The TDS are not apparently affected by changes in water depths (see Fig. 7a) due to the shallowness of the wells.

4.4 Major Anions

4.4.1 Chloride (Cl^-)

Chloride content in the well water samples shows a wide range of concentration, ranging between 178 and 2,840 mg/l with an average of 1,015. A few samples have chloride concentrations less than 500 mg/l, but most of the water wells have relatively high chloride concentrations that are greater than 1,000 mg/l (see Fig. 8), which are mostly attributed to well intakes from seawater intrusion and local sources such as seepage from Maryut canal. Like TDS, low chloride content ranging from 200 to 500 mg/l occurs in the central area around the Maryut canal that recharges the groundwater and dilutes the water aquifer. A significant correlation exists between the chloride content and salinity (see Fig. 7b), indicating that the major part of the salinity of the groundwater is due to halite. The majority of groundwater samples

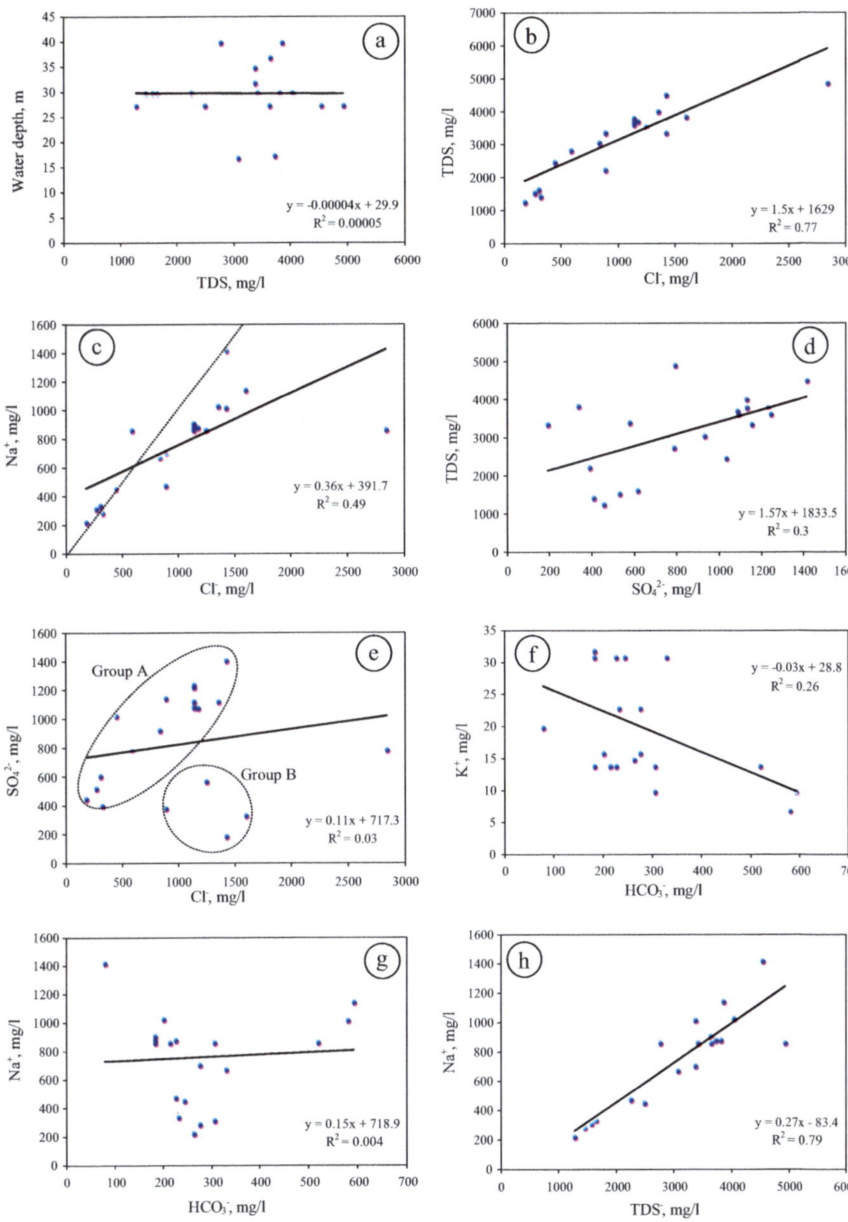

Fig. 7 X-Y diagrams of water depth, TDS, Cl^-, SO_4^{2-}, HCO_3^-, K^+, and Na^+

have Na/Cl ratios that are slightly lower than unity, indicating that some sodium is lost through reverse ion exchange. This effect is more pronounced in the groundwaters that have high Na-Cl content, in which Na/Cl ratio are close to 3:4. There is a

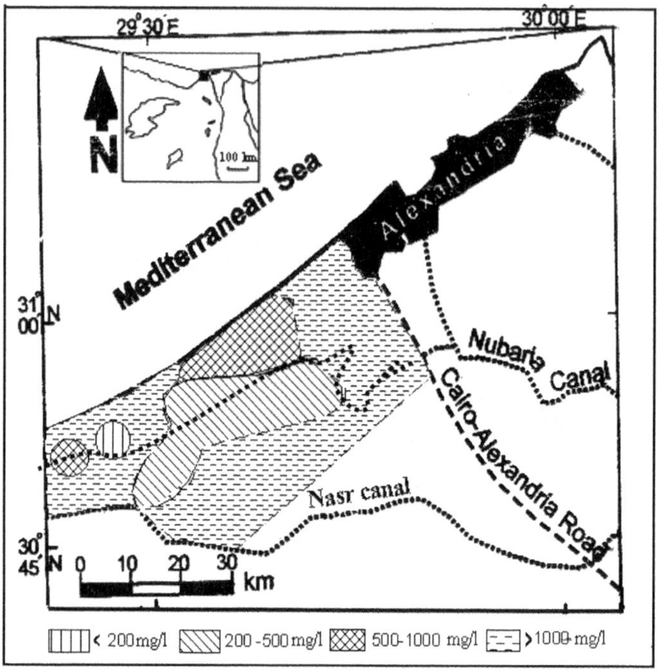

Fig. 8 The distribution of chloride content in the coastal aquifer

linear positive correlation between Cl^- and Na^+ concentrations ($R^2 = 0.45$, see Fig. 7c). A few samples show Na/Cl ratios that are slightly higher than unity, indicating an additional source for Na^+.

4.4.2 Sulfate (SO_4^{2-})

Sulfate content varies from 192 to 1,416 mg/l with an average of 824 mg/l. Most of the water wells (southern and northwestern areas) are characterized by high sulfate content (>1,000 mg/l, see Fig. 9). In the eastern area, the concentration of sulfate content is less than 600 mg/l due to seepage from Maryut canal and rainwater. Most of the central area (around the Maryut canal) has sulfate content ranging from 600 to 1,000 mg/l due to seepage from Maryut canal, rainwater, and seepage from excess irrigation water. Unlike chlorine, sulfate concentrations in groundwater have a relatively less direct relationship with TDS (see Fig. 7d). The plot of Cl^- versus SO_4^{2-} (see Fig. 7e) illustrates bimodal relationships. One group, about 75% of the water wells, has good correlation suggesting that their accumulation in the groundwater is a result of a common surface process (evapo-concentration or dissolution of secondary anhydrite). The other small group has a lack of correlation between sulfate and chloride ions, suggesting another source of sulfate resulted from the water-rock interaction of gypsum associated with the Pleistocene sediments.

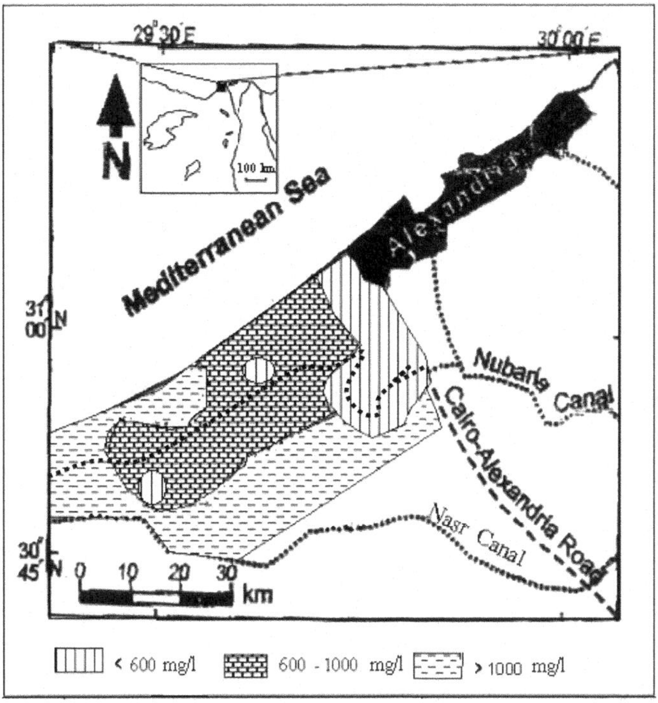

Fig. 9 Distribution of sulfate content in the coastal aquifer

4.4.3 Carbonate (CO_3^{2-}) and Bicarbonate (HCO_3^-)

Carbonate ions (CO_3^{2-}) are not detected in the groundwater, while bicarbonates range from 79 to 592 mg/l with an average of 285 mg/l. The bicarbonates are most likely contributed to the groundwater by the dissolution of the carbonate fraction of the Pleistocene sediments. Another source could be the dissolution that occurs rapidly on initial infiltration because of soil-generated CO_2, which dissociates to carbonic acid, which influences the ability to infiltrate groundwater to dissolve calcium carbonate [56]. The bicarbonates (HCO_3^-) have lack correlation to K (see Fig. 7f) and Na (see Fig. 7g). However, they show relatively more negative correlation with K than Na, suggesting seawater intrusion. It is observed that Na is excess over K, indicating that NaCl is the major source of Na, in addition to the greater resistance of K minerals to weathering.

4.5 Major Cations

4.5.1 Sodium (Na^+)

Sodium is one of the most abundant elements on earth. Its salts are highly soluble in water. The igneous rocks are composed to the extent of 60% feldspars that are aluminum silicate of alkaline earth metals. Albite is the pure sodium feldspar ($NaAlSi_3O_8$); white anorthite is the pure calcium feldspar ($CaAl_2Si_2O_8$). Albite and anorthite are decomposed more easily than potassium feldspar. Albite dissolves under the influence of CO_2 to form a solution that is characterized by the solutes HCO_3^-, Na^+, H_4SiO_4, and H^+ [57].

$$NaAlSi_3O_8 + CO_2 + H_2O = Na^+ + HCO_3^- + H_2SO_4 + Al_2Si_2O_5\,(OH)_4\ \text{(kaolinite)}$$

Feldspathiod is also an important source of Na. Evaporites such as halite (NaCl) are the most important source of Na. The sodium content in groundwater is mainly dependent on rock type of the aquifer.

The groundwater in the coastal aquifer is characterized by a relatively high concentration of sodium ions (Na^+) that change from 230 to 1,426 mg/l with an average of 761 mg/l (see Table 1). Most of the aquifer has Na^+ ranging from 500 to 1,000 mg/l (see Fig. 10), suggesting seawater intrusion. The groundwater containing Na^+ higher than 1,000 mg/l is restricted to the eastern part of the aquifer (see Fig. 10) close to Maryut Lake and other two wells. In the central area, the sodium content is less than 500 mg/l due to seepage from Maryut canal that reduces the high saline water of the aquifer.

Like chloride, sodium content has a significant direct correlation with TDS ($R^2 = 0.79$, see Fig. 7h), indicating that the major part of the salinity of the groundwater is due to NaCl.

The K^+ content in the coastal aquifer varies from 8 to 32 mg/l with an average of 19.9 mg/l (see Table 1). It is higher than that of the surface Nile water (1.4 mg/l, [53]) and the average of the world's major rivers (1.4 mg/l; from [58]).

4.5.2 Potassium (K^+)

The concentration of potassium in the coastal aquifer changes from 7 to 32 mg/l with an average of 21 mg/l. The central part is relatively higher in K content than the west of the coastal zone (see Fig. 11). The increasing of K in most of the coastal wells is most probably due to rural contamination (fertilizers).

The relatively high potassium content in the groundwater of the coastal aquifer is mainly due to the invasion of the seawater. The observed excess of Na^+ over K^+ is caused principally by the inflow of NaCl brines derived from seawater. Calcium and magnesium concentrations show a weak direct correlation (see Fig. 12a) mainly due to carbonate weathering more than silicates. The majority of the groundwater samples have a concentration of Ca^{2+} much higher than that of Mg^{2+}. This may be explained by the lack of ferromagnesian and abundance of carbonate minerals that

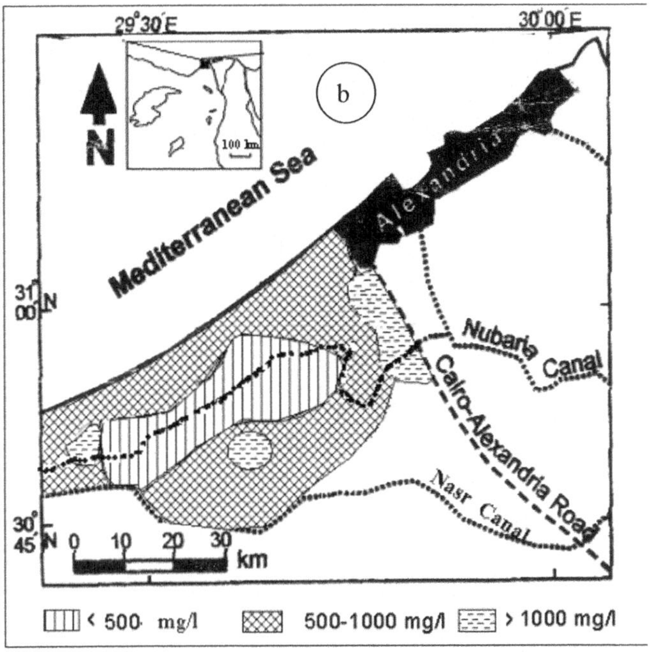

Fig. 10 Distribution of sodium content in the coastal aquifer

are present within the rocks of the area, by ion exchange processes, and by the precipitation of calcite. Similar to Ca^{2+}, Mg^{2+} shows weak direct correlation with TDS (see Fig. 12b), indicating that the salinity in the groundwater is mainly attributed to NaCl.

4.5.3 Magnesium (Mg^{2+})

Magnesium is common in natural water as Mg^{2+}. Along with calcium, magnesium is a primary contributor to water hardness. The solubility of magnesium carbonate like that of calcium carbonate is increased in the presence of CO_2. The presence of sodium ions in solution may also help easily the dissolution of magnesium ions from rocks [59].

In the coastal aquifer, the magnesium content ranges between 26 and 182 mg/l with an average of 123 mg/l. Spatial distribution of magnesium ions shows that most of the groundwater has more than 100 mg/l of Mg^{2+}, indicating that magnesium ions are mostly marine in origin. The central area shows less magnesium concentration due to seepage from irrigation channels (see Fig. 13). Mg^{2+} shows weak direct correlation with TDS ($R^2 = 0.034$, see Fig. 12b), indicating that the salinity in the groundwater is mainly attributed to NaCl.

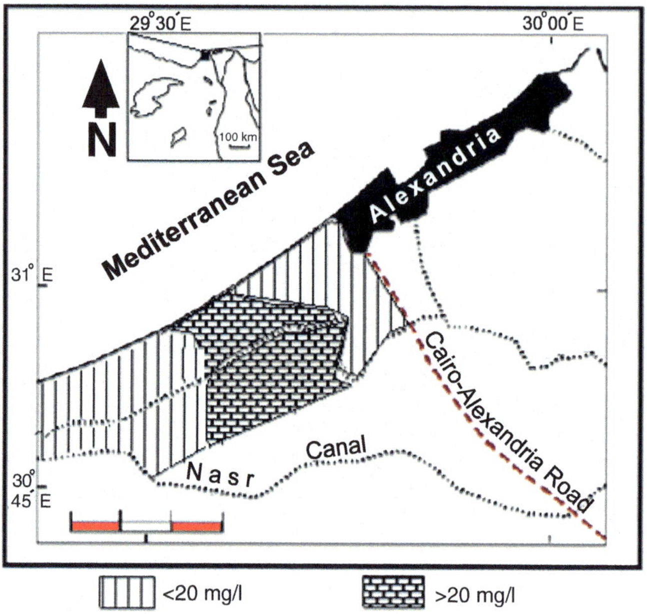

Fig. 11 The distribution of potasium content in the coastal aquifer

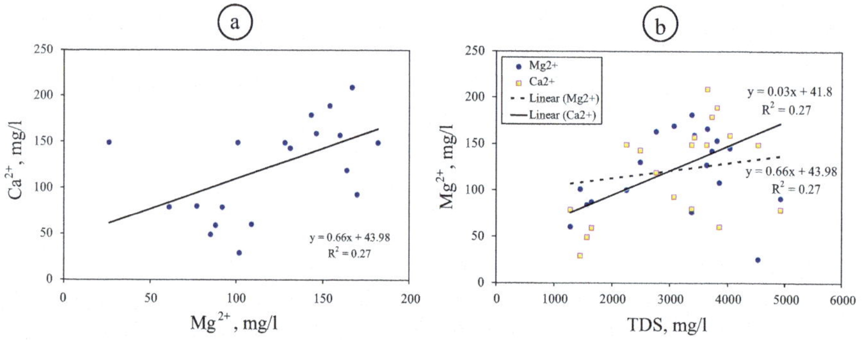

Fig. 12 X-Y diagrams of Ca^{2+}, Mg^{2+}, and TDS

4.5.4 Calcium (Ca^{2+})

Calcium is present in all water as Ca^{2+} ions and is readily dissolved from rocks enriched in calcium minerals such as carbonates (limestone) and sulfates (gypsum). The salts of calcium together with those of magnesium are responsible for the hardness of water.

In the present study, calcium ions range between 30 and 210 mg/l with an average of 124 mg/l. The majority values range from 100 to 200 mg/l. The concentration of

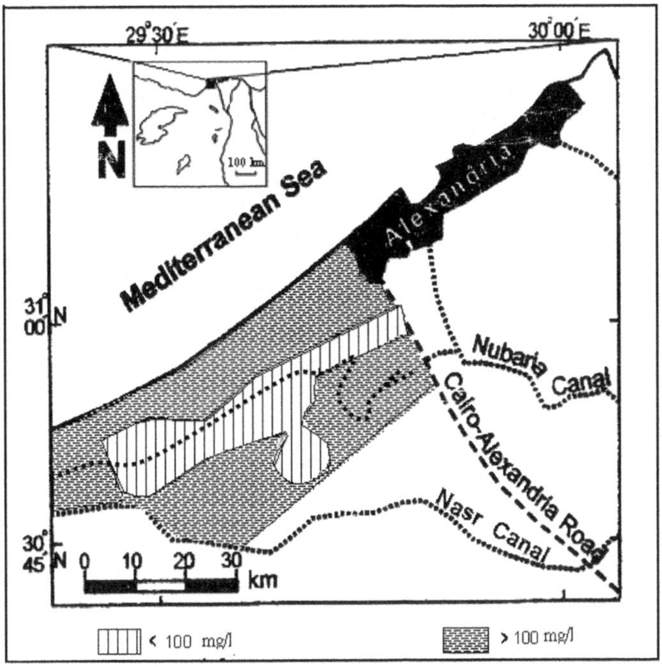

Fig. 13 Distribution of magnesium content in the coastal aquifer

calcium is low in the central part (see Fig. 14) due to seepage of water from the irrigation channel. Similar to Mg^{2+}, the Ca^{2+} shows weak direct correlation with TDS ($R^2 = 0.274$, see Fig. 12b), indicating that the salinity in the groundwater is mainly attributed to NaCl. The concentration of calcium in coastal zone area is relatively higher than that in the western Nile Delta aquifer (98 mg/l, [60]). The presence of calcium content is related to the type of water-bearing strata in which limestone is responsible for enriching the groundwater with calcium ions.

4.6 *Nutrients*

The term "nutrients" refers broadly to those chemical elements that are essential to life, but more specifically to nitrogen and phosphorus in water pollution content. Plants and animals are made up mostly of compounds of nitrogen (N) and phosphorus (P). These are the primary macronutrients. Many other elements are necessary for growth, but in tiny amounts, and they are classified as micronutrients [61].

Nitrogen gets into the water in inorganic and organic forms. The primary inorganic forms of N are ammonia (NH_3), ammonium (NH_4^+), nitrate (NO_3^-), nitrite (NO_2^-), nitrous oxide (N_2O), nitric oxide (NO), and inorganic nitrogen gas (N_2). Ammonia and ammonium salts are found in small quantities in rainwater. Organic

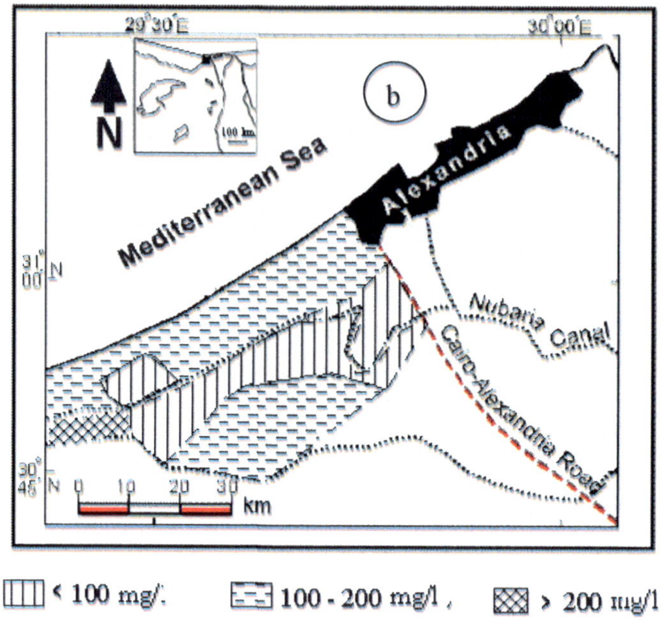

Fig. 14 The distribution of calcium content in the coastal aquifer

nitrogen (organic-N) is found in proteins, amino acids, urea, living or dead organisms (i.e., algae and bacteria), and decaying plant material [62].

4.6.1 Ammonium (NH_4^+)

Ammonium (NH_4^+) is discharged into water bodies by some industrial processes (e.g., ammonia-based pulp and paper production) and also as a component of municipal or community waste. The ammonium content in the coastal zone varies between 0.02 and 0.0 mg/l (see Table 2). According to the World Health Organization [63], the threshold odor concentration of ammonia at alkaline pH is approximately 1.5 mg/l, and a taste threshold of 35 mg/l has been proposed for the ammonium cation. There is no published maximum contaminant level (MCL) for ammonia that is "very toxic to fish and aquatic life. Ammonia concentrations of 0.06 mg/l can cause gill damage in fish, and 0.2 mg/l is lethal to trout. Concentrations in excess of 0.1 mg/l suggest domestic or agricultural sources of waste" [64].

4.6.2 Nitrate (NO_3^-)

Nitrogen is a decaying organic matter; such compounds are decomposed bacterially to gas ammonia (NH_3), most of which is oxidized by soil bacteria to nitrate (NO_3^-), the

Table 2 Nutrient content (mg/l) in the groundwater the coast aquifer

Sample no.	NH_4^+	PO_4^{2-}	NO_3^-
1	0.05	3.0	3.0
2	0.02	3.0	3.0
3	0.04	2.0	4.0
4	0.03	2.0	2.0
5	0.02	2.5	4.0
6	0.02	0.16	6.0
7	0.05	0.18	5.0
8	0.02	3.0	2.0
9	0.04	3.0	2.5
10	0.03	2.0	2.0
11	0.02	2.0	4.0
12	0.02	2.5	3.4
13	0.05	0.16	2.4
14	0.02	0.11	7.0
15	0.04	0.12	4.0
16	0.03	2.0	2.5
17	0.02	2.2	3.0
18	0.02	3.0	4.0
19	0.06	3.0	2.5
20	0.03	3.0	4.1

form of nitrogen that is most readily utilized by plants [65]. Nitrate in water is undetectable without testing because it is colorless, odorless, and tasteless [66].

Large amounts of nitrates (fertilizers), owing to their solubility, are washed into streams, rivers, lakes, and groundwater. Nitrate is the most stable oxidized form of combined nitrogen in most environmental media. Most nitrogenous materials in natural water tend to be converted to nitrates that may occur naturally in mineral deposits (generally sodium or potassium nitrate), soils, seawater, freshwater systems, the atmosphere, and biota. Nitrate minerals include soda niter ($NaNO_3$), nitrocalcite ($Ca(NO_3)_2 + nH_2O$), and nitroglauberite ($6NaNO_3{\cdot}2Na_2SO_4 + 3H_2O$) [67]. Lakes and other static water bodies usually have less than 1.0 mg/l of nitrate/nitrogen. Levels of nitrate/nitrogen in groundwater may range up to 20 mg/l or more, with higher levels characteristically occurring in shallow aquifers.

According to the World Health Organization [63], the recommended maximum limit for drinking water is 10 mg/l (NO_3^--N). Water with a higher concentration of nitrate represents a significant health risk. According to FAO [68], the guidelines for nitrogen in irrigation water can be classified as follows:

(a) Water with less than 5 mg/l NO_3^--N (about 23 mg/l as NO_3^-) is very good water for irrigation and can be used without any restrictions.
(b) Water with nitrate (NO_3^--N) 5–30 mg/l (23–135 mg/l NO_3^-) can be used for irrigation but with slight to moderate restriction.
(c) Water with more than 30 mg/l NO_3^--N (>135 mg/l as NO_3^-) is bad water.

The nitrate content in the coastal zone area is very low (see Table 2) and varies between 2 and 6 mg/l with an average of 3.2 mg/l (see Fig. 15). Nitrates (NO_3^-) are more abundant than ammonia ions (NH_4^+). All the groundwater is suitable for irrigation because nitrate is less than the permissible limits of guidelines.

4.6.3 Phosphate (PO_4^{2-})

Phosphates are chemical compounds containing phosphorus that is a nonmetallic element necessary for life. It is found in rock as inorganic phosphates.

During the natural process of weathering, the rocks gradually release the phosphorus as phosphate ions that are soluble in water. Phosphates exist in three forms: orthophosphate, metaphosphate (or polyphosphate), and organically bound phosphate. These forms of phosphate occur in living and decaying plants and animal remains, as free ions or weakly chemically bounded in aqueous systems, chemically bounded to sediments and soils, or as mineralized compounds in soil, rocks, and sediments [61].

Phosphate rock is enriched in phosphate minerals, most commonly those of the apatite family with the general formula [$Ca_5(PO_4,CO_3)3(F,OH,Cl)$]. About 90% of phosphate mined is used to produce chemical fertilizers. Phosphate deposits can be classified into three types:

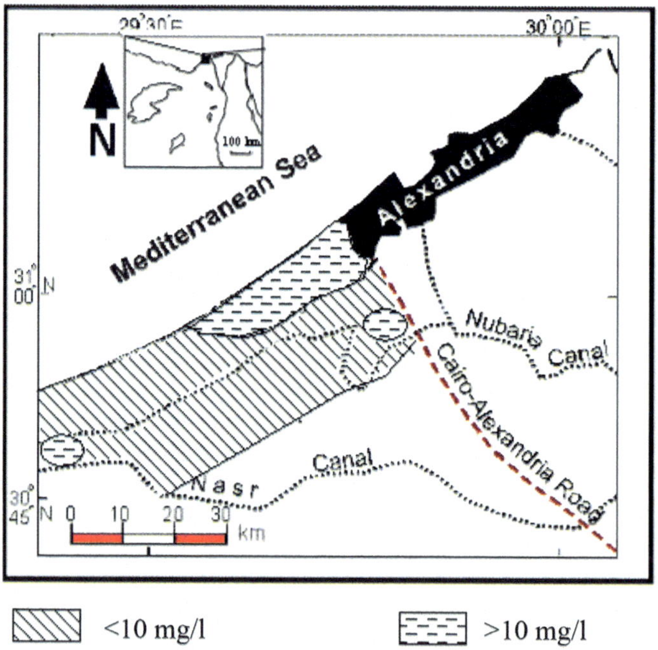

Fig. 15 Distribution of nitrate content in the coastal aquifer

1. Marine sedimentary phosphorites, where the phosphorus was chemically and biologically precipitated as apatite group minerals in regions of reducing condition of upwelling cold, nutrient-rich, and shallow marine water. It forms 76% of world phosphate. The commercial phosphate rock should contain at least 20% P_2O_5, e.g., phosphates in North Africa.
2. Igneous phosphate, where the phosphorus (apatite) is a primary component of alkali carbonatite complexes such as Palabora mine, South Africa. This type forms 23% of world phosphate.
3. Island phosphate (guano accumulations): Island phosphates have concentrated excreted as guano on tropical islands. This type of phosphate forms 1% of world phosphates. A minor amount of guano occurs in some coastal cavities of Mozambique.

Sulfuric acid is used in the conversion of the phosphate rock to form superphosphate fertilizer. The organic phosphorus in drinking water was recommended by the National Academy of Sciences and the National Academy of Engineering [69] to be not >0.1 mg/l.

The concentration of phosphate ions in the groundwater of the coastal aquifer varies between 0.1 and 3.0 mg/l (see Table 2). The aquifer is polluted by phosphate due to the composition of the aquifer rocks and decomposition process and overuse of fertilizes. The water is unsuitable for drinking water, but it can be used for irrigation if the other solutes are suitable.

4.7 *Origin of the Groundwater*

According to Sulin's classification system, 95% of the groundwater samples are located in the lower quadrant of Sulin's diagram (see Fig. 16).

They are characterized by Na-SO_4 genetic water type, indicating meteoric water that is mixed with seawater. The groundwater of well No. 20 is located in Na-HCO_3, indicating rainwater type that is not mainly affected by seawater. It is most probably recent rainwater in an area where flushing process partially occurs and recharges the groundwater. The groundwater of well No. 13 is located in Cl-Mg water type, indicating recent marine water in Genesis. A Piper plot indicates that the groundwaters vary from chloride-rich to sulfate-rich waters in the anions triangle, while they are all located in the alkali (Na + K)-rich field (see Fig. 17). The diamond-shaped field indicates that all the groundwater samples are located in the field of Na-K and Cl-SO_4 water type, supporting seawater intrusion to the freshwater in the coastal aquifer.

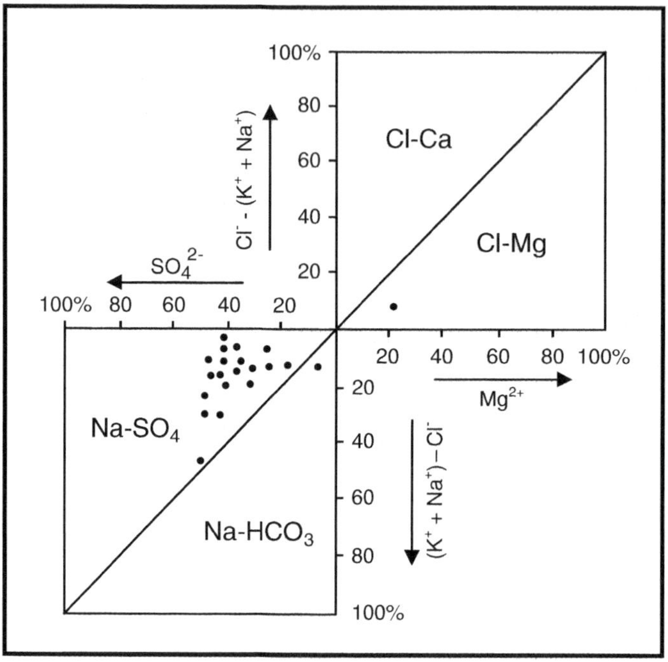

Fig. 16 Sulin diagram showing genetic types of groundwater in the coastal aquifer, northwestern Nile Delta

4.8 Trace Elements

The immediate source of trace elements "metals" in the groundwater is considered to be secondary concentrations. Chemical analysis results of the trace elements (see Table 3) indicate that most of the groundwaters are seriously contaminated with iron and partially with zinc and nickel (see Fig. 18). The iron contaminant is most probably produced from iron-oxide coating layers of quartz sand grains of the Pleistocene sediments. The other trace elements (Zn, Pb, Cd, Cr, Cu) are much lower than the maximum contaminant level (MCL), at the highest allowable concentration of a contaminant in drinking water supplied by the US Environmental Protection Agency [70].

Based on salinity and chemistry of the groundwater, the coastal aquifer water is unsuitable for drinking due to its high salinity, nitrates, phosphates, and metal content of iron, manganese, and nickel.

Table 3 Concentration of trace elements in the groundwaters from the costal aquifer, northwestern Nile Delta, Egypt, and drinking water standards (MCL)

		Field parameters		Anions, mg/l					Cations, mg/l							Trace elements, mg/l							
Well no.	Depth, m	pH	EC µS/cm	Cl^-	SO_4^{2-}	HCO_3^-	NO_3^-	PO_4^{2-}	Na^+	K^+	Mg^+	Ca^{2+}	NH_4^+	TDS mg/l	Well no.	Fe	Mn	Zn	Pb	Cd	Cr	Cu	Ni
1	27.5	7.99	6,500	1,420	1,416	79	8	0.15	1,426	20	26	150	0.04	4,545	1	90.3	07.0	30.1	003.0	003.0	042.0	10.0	10.0
2	27.5	7.04	5,080	1,136	1,090	183	2	0.30	915	31	128	150	0.04	3,635	2	10.0	06.0	20.1	003.0	0023.0	041.0	05.0	06.0
3	30.0	7.89	2,160	266	528	305	3	0.40	322	10	85	50	0.04	1,569	3	10.0	30.0	60.0	004.0	0025.0	044.0	06.0	20.0
4	37.0	7.03	5,530	1,136	1,243	183	11	0.50	869	14	167	210	0.03	3,834	4	50.4	10.0	90.2	0035.0	0024.0	045.0	06.0	10.0
5	30.0	8.10	5,960	1,349	1,128	201	5	0.16	1,035	16	146	160	0.04	4,040	5	70.0	05.0	90.0	003.0	003.0	043.0	10.0	10.0
6	35.0	7.86	4,040	888	1,152	275	7	0.20	713	16	182	150	0.04	3,383	6	30.0	06.0	80.2	002.0	003.0	043.0	10.0	09.0
7	30.0	7.08	5,400	1,136	1,129	183	7	0.15	888	32	154	190	0.02	3,719	7	60.0	10.0	40.2	003.0	0025.0	044.0	10.0	30.0
8	27.5	8.20	1,900	178	456	263	4	0.50	230	15	61	80	0.07	1,288	8	70.0	10.0	00.1	004.0	003.0	044.0	10.0	07.0
9	27.5	8.04	9,790	2,840	792	214	6	0.30	869	14	92	80	0.08	4,907	9	80.2	10.0	80.0	004.0	0025.0	043.0	08.0	06.0
10	17.5	7.14	5,140	1,172	1,085	226	7	0.40	888	31	143	180	0.1	3,733	10	30.0	06.0	40.2	003.0	0024.0	042.0	07.0	20.0
11	27.5	7.42	3,700	444	1,032	244	6	0.20	460	31	131	144	0.1	2,492	11	30.1	10.0	80.0	0035.0	0021.0	043.0	10.0	30.0
12	17.0	7.56	4,380	834	931	329	12	0.20	681	31	170	94	0.1	3,082	12	10.0	06.0	20.0	004.0	003.0	043.0	10.0	10.0
13	30.0	7.77	3,600	888	389	226	13	0.15	483	14	101	150	0.1	2,264	13	10.0	03.0	04.0	004.0	003.0	043.0	10.0	10.0
14	40.0	8.10	3,600	586	787	305	4	0.10	869	14	164	120	0.07	2,849	14	15.0	05.0	10.0	004.0	0025.0	033.0	12.0	15.0
15	30.0	7.78	2,040	302	614	232	6	0.13	345	23	88	60	0.09	1,670	15	20.0	01.0	15.0	003.0	003.0	001.0	10.0	10.0
16	30.0	8.10	2,100	320	408	275	4	0.20	294	23	102	30	0.1	1,456	16	15.0	04.0	16.0	004.0	0025.0	031.0	10.0	10.0
17	30.0	7.01	5,410	1,136	1,229	183	4	0.20	888	31	154	190	0.1	3,815	17	30.0	10.0	10.0	004.0	0025.0	045.0	07.0	09.0
18	40.0	8.20	6,150	1,598	336	592	17	0.30	1,150	10	109	61	0.1	3,873	18	25.0	05.0	10.0	0045.0	0026.0	044.0	06.0	05.0
19	30.0	8.20	5,600	1,243	576	519	15	0.20	869	14	160	158	0.09	3,554	19	35.0	20.0	09.0	003.0	0027.0	034.0	05.0	06.0
20	32.0	8.15	5,220	1,420	192	580	6	0.15	1,024	8	77	81	0.08	3,388	20	40.0	10.0	10.0	002.0	0023.0	035.0	10.0	08.0
Max	40.0	8.20	9,790	2,840	1,243	592	17	0.30	1,426	32	170	210	0.1	6,530	Maximum	4.500	0.300	2.900	0.005	0.005	0.045	0.120	0.300
Min	17.0	7.03	1,900	178	192	79	2	0.10	230	8	26	50	0.07	765	Minimum	0.100	0.010	0.040	0.002	0.002	0.001	0.050	0.050
Mean	29.8	7.73	4,665	1,015	816	280	7.4	0.24	761	19.9	122.0	124	0.072	3,200	Average	0.705	0.088	0.886	0.003	0.003	0.039	0.085	0.122
															MCL[a]	0.30	0.05	1.50	0.05	0.01	0.05	1.00	0.10

[a]Primary standards set a limit, called the maximum contaminant level (MCL), on the highest allowable concentrations of a contaminant in drinking water, which are set by the US Environmental Protection Agency [70]. Ni standard is from Fetter [71]

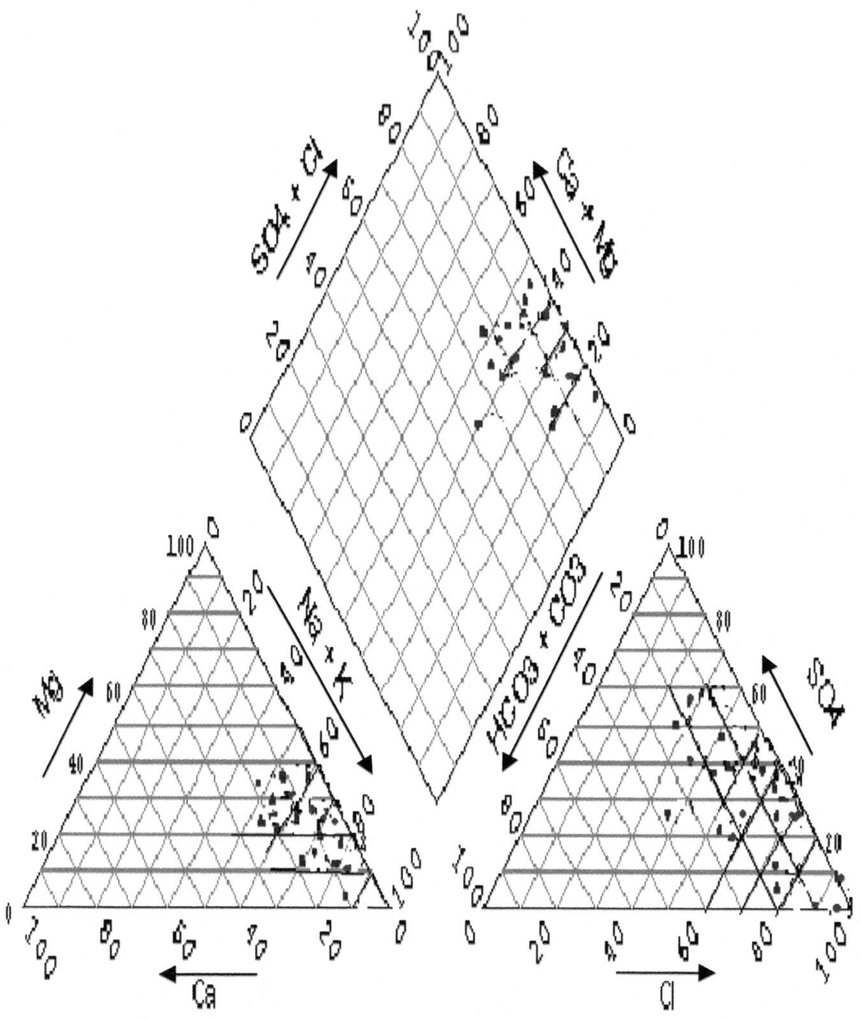

Fig. 17 Piper diagram of major ion chemistry of groundwater samples in the coastal aquifer, northwestern Nile Delta, showing the various groundwater types

4.9 Water Evaluation and Suitability

The evaluation of groundwater for different purposes is very necessary, especially when taking into consideration the expansion of reclamation areas that depend mainly on the groundwater as a principal source. The following section deals with the suitability of groundwater for different uses of a human. It can be classified as follows:

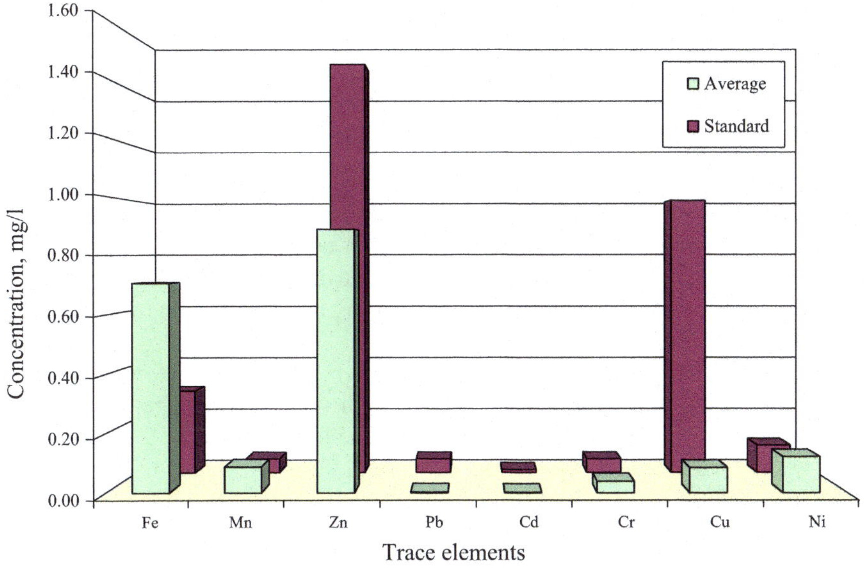

Fig. 18 Quality of the groundwater in the coastal aquifer relative to the acceptable standard concentrations of trace elements in drinking water established by USEPA

4.9.1 Suitability of Water for Drinking

Water for drinking and domestic uses should be colorless, odorless, clear, and free from excessive dissolved solids as well as harm. The TDS content is more than the permissible limits of WHO standards, WHO [72]. It is unsuitable for drinking. On the other hand, the concentration of trace element in the groundwater of coastal the zone is generally within the permissible limits for drinking except iron and manganese. Their concentrations are higher than permissible limits due to the presence of gypsiferous limestone at the top of small hills [25, 55]. This water should be treated prior to using it by reverse osmosis method, aeration, sedimentation, filtration, pH adjustment, ozone or potassium permanganate oxidation with precipitation/filtration, or other treatments.

4.9.2 Suitability of Groundwater for Irrigation

The agriculture expansion largely depends on the groundwater resources and its quality. The present section deals with the suitability of the examined groundwater from wells for the irrigation purposes and also the impact of groundwater quality on the soil texture and salt tolerance of crops to recommend the measures to avoid any harmful effect.

Classification of Groundwater Quality According to TDS

In the present work, the total dissolved salts (TDS in mg/l) is considered as a standard for evaluating the water for irrigation and other purposes, the US Salinity Laboratory [73] classification according to salt concentration is discussed as follows. The groundwater samples are divided into two categories:

1. Very high saline water category ranges in TDS between 1,500 and 3,000 mg/l. This type of water is not suitable for irrigation under ordinary conditions. The soil must be permeable; drainage must be adequate; and very salt-tolerant crops should be selected. The groundwater samples represented this category are 3, 8, 15, and 16.
2. Excessively high saline water includes >3,000 mg/l TDS. Therefore, this water is undesirable for irrigation. Plants of high salt tolerance must be chosen, and frequent leashing must be done to overcome the high salinity. The groundwater samples represented this class are 1, 2, 4, 5, 6, 7, 9, 10, 11, 12, 13, 14, 17, 18, 19, and 20.

Sodium Adsorption Ratio (SAR)

Excessive amount of sodium ion affects the soil permeability and soil structure and creates toxic condition for plants [74]. High sodium concentration leads to development of alkaline soils that are difficult to take into agricultural production because of the low infiltration capacity [75]. The classification of groundwater with respect to SAR is a simple method of evaluating the danger of high sodium water in sodium-adsorption ratio that is calculated by Richard [73].

$$SAR = Na^{+}/\{(Ca^{2+} + Mg^{2+})/2\}^{½}$$

where Na^{+}, Mg^{2+}, and Ca^{2+} are expressed in meq/l of the respective ions.

In the coastal zone, most of the samples are located in S1 and S2 classes indicating low to medium sodium content (see Table 4). This water can be used for irrigation. Also, two water samples have high SAR values, which can be used for irrigation of sandy soils. Two water samples have very high SAR, which cannot be used for irrigation.

Residual Sodium Carbonates (RSC)

Concentration of bicarbonate ions in irrigation water is considered as an important factor, with regard to precipitation of calcium and magnesium in the soil, which leads to an increase in the sodium hazard. The residual sodium carbonate (RSC) was calculated by the following formula:

Table 4 The classification of groundwater according to SAR ratio [73]

Class	SAR	Sodium content	Usage	Well no.
S 1	0–10	Low	Can be used for all soils	3, 6, 8, 11, 12, 13, 14, 15, 16
S 2	10–18	Medium	Preferably used for coarse texture, soils of good permeability	2, 4, 5, 7, 10, 17, 19
S 3	18–26	High	Can produce harmful effects and good soil management is essential	18, 20
S 4	26–100	Very high	Not satisfactory for irrigation purposes	1, 9

$$RSC = \left(CO_3{}^{2-} + HCO_3{}^{-}\right) - \left(Ca^{2+} + Mg^{2+}\right)$$

where cations and anions are expressed in meq/l.

A negative RSC indicates that sodium buildup is unlikely since sufficient calcium and magnesium are in excess of what can be precipitated as carbonate. A positive RSC indicates that sodium buildup in the soil is possible. By applying RSC equation, the results for all samples showed negative values, indicating a good class.

Quality of Groundwater According to Chloride Content

The chloride content is very important for the suitability of groundwater for irrigation purposes. The chloride ion compared to other ions is very strongly absorbed by plants. Most plants are very sensitive to chloride content. In this connection as the chloride ions are toxic, they are not accepted, and certain plants have the ability to accumulate chlorides even from water to a low chloride content that has no harmful effect. In the present work, the chloride content is distributed to four classes:

Class 1: Water has chloride content from 0 to 200 mg/l. It is represented by well No. 8.

Class 2: Water has chloride content from 200 to 500 mg/l. It is represented by well Nos. 3, 15, and 16.

Class 3: Water has chloride content ranges between 500 and 1,000 mg/l. It is represented by well Nos. 6, 12, 13, and 14.

Class 4: Water has chloride content >1,000 mg/l. It is represented by well Nos. 1, 2, 4, 5, 7, 9, 10, 11, 17, 18, 19, and 20.

Quality of Groundwater According to Trace Elements

The concentrations of trace elements in the groundwater of coastal zone area are generally within the permissible limits for irrigation except for iron, manganese, and nickel. To remove iron from the groundwater in the study area, aeration, sedimentation, filtration, pH adjustment, and ozone or potassium permanganate oxidation with precipitation/filtration can be used separately or in combination.

5 Conclusions

The hydrochemical data indicated that the groundwater of the coastal aquifer, northwestern Nile Delta, is meteoric in origin (Nile River and local rainfall), which is mixed with marine water due to seawater intrusion. The coastal aquifer is recognized to be at high risk of increasing salinization. The groundwaters of the shallow coastal aquifer, northern Nile Delta, are brackish (3,155 mg/l). The salinity of the groundwater as a total dissolved solids (TDS) ranges from 1,288 to 4,907 mg/l. A few samples show Na/Cl ratios that are slightly higher than unity, indicating an additional source for Na^+ that is mostly silicate weathering. The groundwater is slightly alkaline with pH range from 7.01 to 8.2. The electric conductivity (EC) of the groundwater ranges from 1,900 to 9,790 μS/cm with an average of 4,620 μS/cm. It is mostly related to the geographical position of each well.

The primary sources of excess salinization of the groundwater are (1) seawater intrusions that provided mostly NaCl and some Mg and (2) aquifer rocks that contain carbonates and gypsum, which provide mostly Ca^{2+}, HCO_3^-, K^+, some Na^+, and SO_4^{2-}. Chloride (Cl^-) and sulfate (SO_4^{2-}) are the significant anions, while sodium (Na^+) and magnesium (Mg^{2+}) are the primary cations. The concentration of the examined significant ions is higher than the maximum contaminant level (MCL), on the highest allowable concentrations of a contaminant in drinking water, which are set by the US Environmental Protection Agency. The nutrient content such as nitrates is also higher than the standard values, which is mainly produced from rural sources.

The brackish groundwaters of the shallow coastal aquifer, northern Nile Delta, are slightly alkaline (pH = 7.73) and partially polluted with nitrates (7.4 mg/l) and phosphates (0.24 mg/l). The groundwaters also contain an excess of some metals (Fe, Mn, Ni) that are higher than the standard levels. Fortunately, the poisonous metals such as Cd, Cr, and Pb are lower than the maximum contaminant levels. As a result of the physiochemical characters of the groundwater, the coastal aquifer (brackish water) is unsuitable for drinking purposes, but it may be used safely for irrigation of some suitable crops.

6 Recommendations

Salinization of the coastal aquifer in the northwestern zone of the Nile Delta is the result of concomitant processes related to both seawater intrusion and water-rock interaction, in addition to overpumping from the groundwater aquifer that can damage water quality through acceleration of seawater intrusion. The recommendation is to allow the groundwater users to balance supply and demand and to meet the future needs of farms, cities, and the environment. The balance between using surface water from channels derived from the Nile and local groundwater is required to prevent the groundwater from further degradation.

As demand for water increases, water managers and planners need to look widely for ways to improve water management and augment water supplies such as rainfall harvesting, water use efficiency, water recycling, and desalinated water for domestic and industrial purposes.

The demand is hardening as a result of new governmental development projects in conjunction with the private expansion in both agriculture and urbanization. There is a definite need for detailed monitoring of the groundwater parameters to ensure sustainable development of water resources.

References

1. Andersen MS, Nyvang V, Jakobsen R, Postma D (2005) Geochemical processes and solute transport at the seawater/freshwater interface of a sandy aquifer. Geochim Cosmochim Acta 69:3979–3994
2. Bianchini G, Pennisi M, Cioni R, Muti A, Cerbai N, Kloppmann W (2005) Hydrochemistry of the high-boron groundwaters of the Cornia aquifer (Tuscany, Italy). Geothermics 34:297–319
3. Cary L, Casanova J, Gaaloul N, Guerrot C (2013) Combining boron isotopes and carbamazepine to trace sewage in salinized groundwater: a case study in Cap Bon, Tunisia. Appl Geochem 34:126–139
4. Cary L, Petelet-Giraud E, Bertrand B, Kloppmann W, Aquilina L (2015) Origins and processes of groundwater salinization in the urban coastal aquifers of Recife (Pernambuco, Brazil): a multi-isotope approach. Sci Total Environ 530–531:411–429
5. de Montety V, Radakovitch O, Vallet-Coulomb C, Blavoux B, Hermitte D, Valles V (2008) Origin of groundwater salinity and hydrogeochemical processes in a confined coastal aquifer: case of the Rhone delta (Southern France). Appl Geochem 23:2337–2349
6. Werner AD, Bakker M, Post VEA, Vandenbohede A, Lu C, Ataie-Ashtiani B et al (2013) Seawater intrusion processes, investigation and management: recent advances and future challenges. Adv Water Resour 51:3–26
7. Aquilina L, Landes AA-L, Ayraud-Vergnaud V, Labasque T, Roques C, Davy P et al (2013) Evidence for a saline component at shallow depth in the crystalline Armorican basement (W France). Procedia Earth Planet Sci 7:19–22
8. Armandine Les Landes A, Aquilina L, Davy P, Vergnaud-Ayraud V, Le Carlier C (2015) Timescales of regional circulation of saline fluids in continental crystalline rock aquifers (Armorican Massif, western France). Hydrol Earth Syst Sci 19:1413–1426. https://www.hydrol-earth-syst-sci.net/19/1413/2015/
9. Duriez A, Marlin C, Dotsika E, Massault M, Noret A, Morel JL (2008) Geochemical evidence of seawater intrusion into a coastal geothermal field of central Greece: example of the Thermopylae system. Environ Geol 54:551–564
10. Han D, Kohfahl C, Song X, Xiao G, Yang J (2011) Geochemical and isotopic evidence for palaeoseawater intrusion into the south coast aquifer of Laizhou Bay, China. Appl Geochem 26:863–883
11. Khaska M, La Salle CLG, Lancelot J, Mohamad A, Verdoux P, Noret A (2013) Origin of groundwater salinity (current seawater vs. saline deep water) in a coastal karst aquifer based on Sr and Cl isotopes. Case study of the La Clape massif (southern France). Appl Geochem 37:212–227

12. Sola F, Vallejos A, Daniele L, Pulido-Bosch A (2014) Identification of a Holocene aquifer-lagoon system using hydrogeochemical data. Quatern Res 82:121–131
13. Cendón DI, Ayora C, Pueyo JJ, Taberner C, Blanc-Valleron MM (2008) The chemical and hydrological evolution of the Mulhouse potash basin (France): are "marine" ancient evaporites always representative of synchronous seawater chemistry? Chem Geol 252:109–124
14. Lucas Y, Schmitt AD, Chabaux F, Clément A, Fritz B, Elsass P (2010) Geochemical tracing and hydrogeochemical modelling of water–rock interactions during salinization of alluvial groundwater (Upper Rhine Valley, France). Appl Geochem 25:1644–1663
15. Merchán D, Auqué LF, Acero P, Gimeno MJ, Causapé J (2015) Geochemical processes controlling water salinization in an irrigated basin in Spain: identification of natural and anthropogenic influence. Sci Total Environ 502:330–343
16. Mongelli G, Monni S, Oggiano G, Paternoster M, Sinisi R (2013) Tracing groundwater salinization processes in coastal aquifers: a hydrogeochemical and isotopic approach in the Na-Cl brackish waters of northwestern Sardinia, Italy. Hydrol Earth Syst Sci 17:2917–2928
17. Forcada GE, Evangelista MI (2008) Contributions of boron isotopes to understanding the hydrogeochemistry of the coastal detritic aquifer of Castellón Plain, Spain. Hydrogeol J 16:547–557
18. Allam MN, Allam GI (2007) Water resources in Egypt: future challenges and opportunities. Water Int 32:205–218
19. Abdel-Galil I, Hegazi AM, El-Demerdashe S, Affifi MY, El-Shourbagy M, Metwall A (2004) National report for combating desertification, Desert Research Center, Egypt, 63 p
20. MWRI (Ministry of Water Resources and Irrigation, Egypt) (2016) Water scarcity in Egypt: the urgent need for regional cooperation among the Nile Basin Countries, 5 p. http://www.mfa.gov.eg/SiteCollectionDocuments/Egypt%20Water%20Resources%20Paper_2014.pdf
21. Leaven MT (1991) Hydrogeological study of the Nile Delta and adjacent desert areas, Egypt, with emphasis on hydrochemistry and isotope hydrology, thesis, Free University, Amsterdam, also published by RIGW/IWACO as Technical note
22. Cartwright I, Weaver TR, Fulton S, Nichol C, Reid M, Cheng X (2004) Hydrogeochemical and isotopic constraints on the origins of dryland salinity, Murray Basin, VIC, Australia. Appl Geochem 1233–1254
23. Cruz JV, Silva MO (2000) Groundwater salinization in Pico Island, Azores, Portugal: origin and mechanisms. Environ Geol 39:1181–1189
24. El Fayoumy IF (1964) Geology of groundwater supplies in Wadi El Natrun area, M.Sc. Thesis, Fac. Sci., Cairo Univ., Cairo, Egypt, 109 p
25. El Shazly EM, Abdel Hady MA, El Ghawaby IA, El Kassas SM, El Khawasik MM, Sanad S (1975) Geological interpretation of Landsat satellite images for west Nile Delta area, Remote Sensing Center, Academy of Scientific Research and Technology, Cairo, Egypt, 86 p
26. Gindi A, Abd-Alla MA (2000) Stable isotopes and microfacies of the middle Miocene Marmarica formation, north Western Desert, Egypt. Egypt J Geol 44:109–125
27. Landis GP, Swanberg CA, Morgan P, Boulos FK, El-Serif AA, El-Sayed AA, Basta NZ, Melek YS (1995) Reconnaissance hydrogen isotope geochemistry of Egyptian groundwaters. Ann Geol Surv Egypt 20:655–673
28. Nagy HM (2001) Hydrological and environmental characteristics of wetlands in Egypt. Lowland Technol Int 3:41–56
29. Masria A, Negm A, Iskander M, Saavedra O (2014) Coastal zone issues: a case study (Egypt). Procedia Eng 70:1102–1111
30. Philip G (1955) Geology of the Pleistocene sediments of the Mediterranean coast, west of Abu Qir, Ph.D. Thesis, Fac. Sci., Cairo Univ. Egypt
31. Salim MG (2012) Selection of groundwater sites in Egypt, using geographic information systems, for desalination by solar energy in order to reduce greenhouse gases. J Adv Res 3:3–11
32. Sanad S (1973) Geology of the area between Wadi El Natrun and the Moghra Depression Ph. D. Thesis Fac. Sci, Assuit Univ, Assuit, Egypt
33. Shata A (1962) The geology, origin and age of groundwater supplies in some desert areas of the U.R.E. Bulletin de l'Institut du Desert a Egypt 12:61–120

34. Swanberg CA, Morgan P, Boulos FK (1984) Geochemistry of the groundwaters of Egypt. Ann Geol Surv Egypt 14:127–150
35. Tarabees E, El-Qady G (2016) Sea water intrusion modeling in Rashid area of Nile Delta (Egypt) via the inversion of DC resistivity data. Am J Clim Chang 5:147–156. https://doi.org/10.4236/ajcc.2016.52014
36. Zaghloul ZM, Taha AA, Hegab OA, El-Fawal FM (1979) The Plio-Pliostocene Nile Delta subenvironments, stratigraphic section and genetic class. Ann Geol Surv Egypt 9:282–291
37. Zaghloul ZM, Abdallah AM, Serag El-Din H, Hefny K (1984) Groundwater pollution in the Nile Delta area. Egypt J Geol 28:131–140
38. Sharaky AM (1990) Geomorphological studies on sand dunes and ridges is some African deserts, M.S. Thesis, Dept. Natural Resources, Institute of African Research and Studies, Cairo University, Egypt, 175 p
39. El-Asmar HM, Wood P (2000) Quaternary shoreline development: the northwestern coast of Egypt. Quat Sci Rev 19:1137–1149
40. Shukri NM, Philip G, Said R (1955) The geology of the Mediterranean coast between Rosetta and Bardia, part 2: pleistocene sediments: geomorphology and microfacies. Bull Inst Egypt 37:395–427
41. Shata A, Hefny K (1995) Strategies for planning and management of groundwater in the Nile Valley and Nile Delta in Egypt, Working Paper Series No. 31-1, Strategic Research Program (SRP), NWRC-MPWWR, Cairo
42. Sherif MM (1999) The Nile Delta aquifer, chapter 17. In: Bear J, Cheng AHD, Sorek S, Ouazar D, Herrera I (eds) Sea-water intrusion in coastal aquifers: concepts, methods and practices, book series theory and application of transport in porous media, vol 14. Kluwer Academic Publishers, Dordrecht, The Netherlands, pp 559–590
43. NWRP (2005) National Water Resources Plan, Facing the Challenge, Ministry of Water Resources and Irrigation, Integrated Water Resources Management Plan for 2017
44. RIGW and IWACO (1991) Monitoring and control groundwater pollution groundwater quality survey, in the Nile Delta and adjacent desert areas, V. I and II. Cairo
45. Farid MS (1980) Nile Delta groundwater study M.Sc. Thesis, Fac. Engin., Cairo Univ., Cairo
46. PRB (2017) World Population Data Sheet. www.worldpopdata.org
47. American Public Health Association (1989) Standard methods for the examination of water and waste water, New York
48. Piper AM (1944) A graphic procedure in the geochemical interpretation of water analyses. Trans Am Geophys Union 25914–25923
49. Sulin VA (1948) Condition of formation, principals of classification and constituents of natural waters, particularly water of petroleum accumulation. Moscow, Leningrad, Acad Sci USSR
50. Brownlow AH (1996) Geochemistry, 2nd edn. Prentice, Englewood Cliffs, NJ, 582 p
51. Krauskopf KB, Bird DK (1995) Introduction to geochemistry, 3rd edn. McGraw-Hill, Inc., New York, 647 p
52. Baas Becking LGM, Kaplan IR, Moore D (1960) Limits of the natural environment in terms of pH and oxidation-reduction potentials. J Geol 68(3):243–284
53. Ismail SS, Ramadan A (1995) Characterization of Nile and drinking water quality by chemical and cluster analysis. Sci Total Environ 173/174:69–81
54. Komy ZR, El-Samahy AA (1995) Dissolved ions of trace and major elements and in suspended sediments in the Nile, Egypt. Chem Ecol 11:25–37
55. Ministry of Irrigation, Desert Irrigation Department (1976) Groundwater resources of the north western coastal zone, groundwater series in the Arab Republic of Egypt Number 5, part 1, pp 8–11
56. Dyke L (1999) Regional groundwater and stream chemistry survey, Oak Ridges Moraine, ON. In: Current research 1999-E, geological survey of Canada, pp 111–121
57. Stumm W, Morgan JJ (1996) Aquatic chemistry, 3rd edn. Wiley, New York
58. Faure G (1991) Principles and applications of inorganic geochemistry. Macmilian Publ. Co., New York, 626 p

59. Hem JD (1970) Study and interpretation of the chemical characteristics of natural water. Second Education a Geological Survey Water Supply, Washington, 140 p
60. Sharaky AM, El Hasanein AS, Atta SA, Khallaf K (2017) Nile and groundwater interaction in the Western Nile Delta, Egypt. In: Negm AM (ed) The Nile Delta, volume 55, the handbook of environmental chemistry, Springer International Publishing AG, pp 33–62
61. Smith VH, Tilman GD, Nekola JC (1999) Eutrophication: impacts of excess nutrient inputs on freshwater, marine and terrestrial ecosystems. Environ Pollut 3:179–196
62. Wall W (2013) Nitrogen in waters: forms and concerns, nitrogen in Minnesota surface waters. Minn Pollut Contr Agency A2:1–22
63. World Health Organization (WHO) (2017) Guidelines for drinking-water quality, 4th edn. Geneva, Switzerland, 541 p
64. Brian Oram PG (2017) water quality terms glossary. http://www.water-research.net/ndex.php/glossary
65. Gill R (1989) Chemical fundamentals of geology. Unwin Hyman Ltd., London, 291 p
66. Brian Oram PG (2017) Nitrates and nitrites in drinking water and surface waters. http://www.water-research.net/index.php/nitrate
67. Gale HS (1912) Nitrate deposits, Department of the Interior United States Geological Survey, Smith GO, Director, Bulletin, p 523
68. FAO (1985) Water quality for agriculture, FAO irrigation and drainage paper No. 29, Rome
69. NAS and NAE (National Academy of Science and National Academy of Engineering) (1972) Water quality criteria, report prepared by Committee of Water Quality Criteria at Request of U.S. Environmental Protection Agency, Washington, DC, 594 p
70. USEPA (2005) Drinking water standards. http://www.epa.com/
71. Fetter CW (1994) Applied hydrogeology, 3rd edn. Prentice Hall, Englewood Cliffs, NJ, 691 p
72. WHO (1998) World Health Organization, Copper. Geneva, World Health Organization, International Programme on Chemical Safety (Environmental Health Criteria, monograph No. 200)
73. Richard LA (1954) Diagnosis and improvement of saline and alkali soils, Agric. Handbook No. 60, M.S. Dept. Agric., Washington, pp 69–82
74. Prasad MSVKV, Siva Praveena G, Prasada Rao PVV (2015) Assessment of groundwater suitability for irrigation purpose: a case study of Narsapur-Mogalthur mandals, West Godavari district, Andhra Pradesh, India. IOSR J Environ Sci Toxicol Food Technol 9(3):7–11. ver. III
75. Nag SK, Suchetana B (2016) Groundwater quality and its suitability for irrigation and domestic purposes: a study in Rajnagar Block, Birbhum District, West Bengal, India. J Earth Sci Clim Change 7:337. https://doi.org/10.4172/2157-7617.1000337

Soil Aquifer Treatment System Design Equation for Organic Micropollutant Removal

Ahmed M. Abdel Sattar, Hossein Bonakdari, Abdelazim Negm, Bahram Gharabaghi, and Mohamed Elhakeem

Abstract Rapid population growth and mass migration from rural to urban centers have contributed to a new era of water sacristy, and a significant drop in per capita freshwater availability, resulting in the reuse of wastewater emerging as a viable alternative. The reuse of wastewater after treatment using the soil aquifer treatment (SAT) has recently gained popularity due to low operating/maintenance cost of the method. However, the presence of organic micropollutants (OMPs) may present a health risk if the SAT is not adequately designed to ensure required attenuation of the OMPs. An important aspect of the design of the SAT system is the large degree of natural variability in the OMP concentrations/loads in the wastewater and the uncertainty associated with the current methods for calculation of the removal efficiency of

A.M. Abdel Sattar (✉)
Department of Irrigation and Hydraulics, Faculty of Engineering, Cairo University, Giza, Egypt

German University in Cairo, New Cairo, Egypt
e-mail: ahmoudy77@yahoo.com

H. Bonakdari
Department of Civil Engineering, Razi University, Kermanshah, Iran

A. Negm
Department of Water and Water Structures Engineering, Faculty of Engineering, Zagazig University, Zagazig, Egypt
e-mail: amnegm@zu.edu.eg

B. Gharabaghi
School of Engineering, University of Guelph, Guelph, ON, Canada

M. Elhakeem
University of Tennessee, Knoxville, TN, USA

Abu Dhabi University, Abu Dhabi, United Arab Emirates
e-mail: mohamed.elhakeem@adu.ac.ae

A. M. Negm (ed.), *Groundwater in the Nile Delta*,
Hdb Env Chem (2019) 73: 307–326, DOI 10.1007/698_2017_136,

the SAT for the OMPs. This study presents a novel model for more accurate prediction of the removal efficiency of the SAT system for the OMPs and the fate of the OMPs trapped within the vadose zone. A large data set is compiled covering a broad range of aquifer conditions, and the SAT system parameters, including hydraulic loading rate and dry/wet ratio. This study suggests that removal of OMPs in SAT systems is most affected by biodegradation rate and soil saturated hydraulic conductivity, in addition to dry to wet ratio. This conclusion is reached by the application of the developed prediction model using data sets from the case study SAT systems in Egypt.

Keywords Extreme learning machine (ELM), Fivefold cross-validation, Monte Carlo simulation (MCS), Organic micropollutants (OMPs), Soil aquifer treatment (SAT)

Contents

1 Introduction

SAT system is considered attractive unconventional water resources for Egypt, which is suffering water sacristy. The usage of SAT can provide treatment for the wastewater and recharge in groundwater aquifers. While guidelines are available for the use of SAT system in Egypt for removal of nitrogen and organic matter, no guidelines are available for the SAT removal potential of organic micropollutants. This chapter discusses this issue and provides a prediction model for the OMP removal in SAT systems and is based on the author's work on soil aquifer treatment system and analysis models [1–21].

2 Soil Aquifer Treatment System

The increasing development of urbanization and population growth has been caused by water resource pollution control and proper recharge of soil aquifer. There is the possibility of organic micropollutants (OMPs) into the aquifer during groundwater aquifer recharge. Organic micropollutants consist of toxic minerals of endocrine disruptors, pharmaceutically active compounds, and personal care product. The environmental and health risks are increased by entering of the organic micropollutants into an aquifer soil. Therefore, the control and elimination of organic micropollutants from soil aquifer treatment system are of paramount importance to protect the aquifer from pollution. Because of the importance of control and removal of organic micropollutants from water resources, many studies have been conducted to remove these compounds. Drillia et al. [22] in an experimental study removed six different pharmaceutical compounds from various soils for municipal wastewater. Xu et al. [23] using the degradation and adsorption removed the personal care products and pharmaceuticals in agricultural soils in an experimental study. Maeng et al. [24] examined the OMP removal during bank penetration and feeding and recovery of aquifers. Time and place of transfer were provided to remove the active pharmaceutical ingredients. Yu et al. [25] studied three different soil types to remove the five types of pharmaceuticals and personal care products (PPCPs). Also, they studied the seasonal changes of organic micropollutant compounds in sewage treatment plant (STP). Personal care products and endocrine-disrupting were removed from the California wastewater.

Soil aquifer treatment (SAT) is a perfect option to artificial recharge of the groundwater aquifer. On the other hand, there is likely the penetration of the organic micropollutants (OMPs) into the soil aquifer during soil aquifer treatment as well. Hence, OMP removal in SAT system is of considerable importance. In Fig. 1, a schematic plan of soil aquifer treatment system at a sewage treatment station has been shown.

Lin et al. [26] eliminated the heavy metals for soil aquifer treatment in a wastewater treatment plant. Also, Fox et al. [27] in an experimental study analyzed the organic carbon content in the soil aquifer treatment for the five different types of

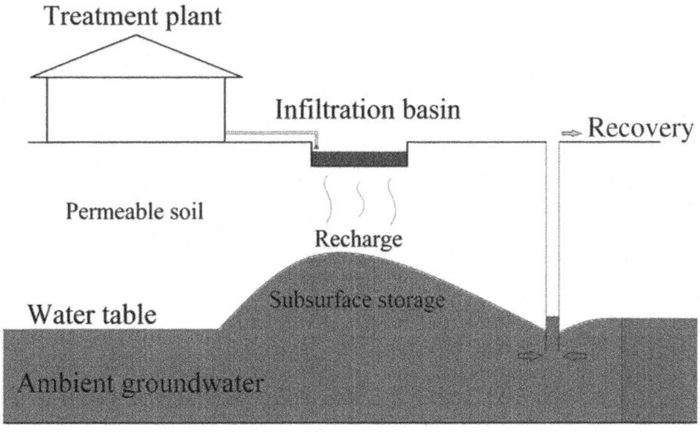

Fig. 1 Schematic plan of soil aquifer treatment system at a sewage treatment plant

soil. Amy and Drewes [28] investigated the removal/transformation mechanism of organic matter from wastewater stream through the soil aquifer treatment for wastewater treatment plant. Then, Sharma et al. [29] analyzed the removal of organic materials in wastewater during soil aquifer treatment system. Their results showed that the redox conditions, input flow rate, and residence time are sufficient for the removal of organic waste. Xu et al. [23] studied the characteristics and behavior of dissolved organic matter on the soil aquifer treatment. It was also showed that during soil aquifer treatment, 70% of the organic material is removed. Also, Caballero [30] studied the SAT system as a pretreatment process to remove organic micropollutant compounds from the wastewater stream. Sharma et al. [29] analyzed the effects of horizontal roughing filtration, coagulation, and sedimentation during pretreatment operations on the mechanism of soil aquifer treatment. It was showed that sedimentation and coagulation lead to less head loss and reduce the clogging effects. Abel et al. [31] examined the effects of temperature and redox conditions on the reduction of organic matter of wastewater on the soil columns and SAT systems. Abel et al. [31] studied the organic matter reduction, pharmaceutical compounds, and nitrogen in the soil aquifer treatment process in a laboratory model. Onesios-Barry et al. [32] conducted an experimental study of the SAT system in the removal of pharmaceuticals and personal care products (PPCPs) for different concentrations in wastewater treatment plants. Suzuki et al. [33] studied the mechanism of the organic matter removal and disinfection by-product formation potential in the higher layer of the soil aquifer treatment system in an experimental study.

3 Simulations of SAT Organic Micropollutant Removal

3.1 Model Setup

To capture the change in characteristics of OMPs during infiltration in the vadose zone, Sattar [13] chose a 2D vertical section in the soil beneath the SAT pond. Study domain dimensions were taken (10 m $\times$ 30 m). The upper boundary was selected to be a variable flux boundary of width 5 m, to simulate the intermittent water infiltration from a spreading basin, and lies in the center of the domain width. The lower boundary is chosen as free drainage to allow flow passage below the study domain. The OMP attenuation was simulated throughout 90 days from the day of application of wastewater in the ponds. This time was considered sufficient for OMP plume to infiltrate through soil layers and gets attenuated.

3.2 Attenuation of OMPs in SAT System

Using the average values of the SAT system parameters presented in Sattar [11], HYDRUS simulations were carried to model the fate of OMPs in SAT system under

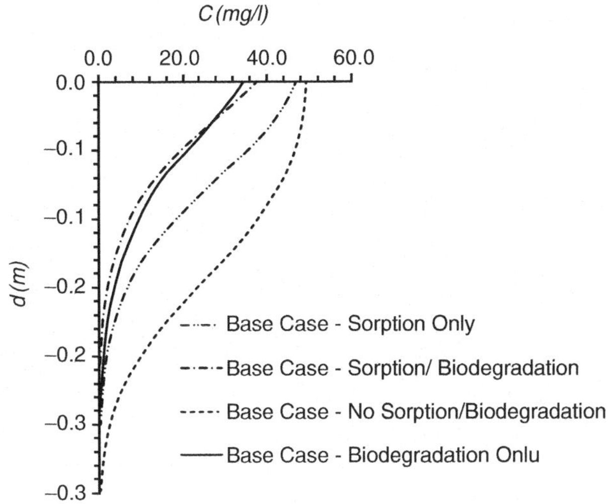

Fig. 2 Variation of OMPs along depth

effects of sorption and biodegradation, both individually and combined. Figure 2 shows the variation of OMP concentration along the depth of vadose zone captured along the vertical centerline of the study domain. It was clear that the biodegradation process was more efficient in reducing the contaminant concentration than the sorption process although the plume sizes for both simulations were almost the same after 90 days. Moreover, it was noted that the OMP concentration at the topsoil layer was higher, in case of considering both sorption and biodegradation than in case of considering biodegradation only. The reason for this was attributed to the effect of sorption which distributes the contaminant mass between the sorbed phase and the liquid phase soon after the contaminant injection which explains the lower concentration after day 86. On the other side, biodegradation process is only active for the contaminant mass existing in the liquid phase which means that the mass that is sorbed onto the soil is not available for biodegradation. This leads to higher biodegradation rates in the case of biodegradation only than the case of combined sorption and biodegradation.

4 Extreme Learning Machine

Various studies have been conducted concerning to the use of artificial intelligence in performance evaluation of SAT system. Recently, Sattar [11] predicted the organic micropollutant removal during soil aquifer treatment system by gene expression programming model.

One of the most popular technique in data mining field is feedforward neural network (FFNN) which is trained by a gradient-descent algorithm such as back-propagation (BP). The disadvantages of FFNN-BP which lead to the low performance of this technique are imprecise learning rate and slow rate of convergence and presence of local minima. Therefore, Huang et al. [34] introduced a new algorithm for FFNN training based on single hidden layer feedforward neural networks (SLFNs), namely, extreme learning machine (ELM). The ELM required adjusting only activation function type and a number of hidden layers, while there are several user-defined parameters such as adjustment of hidden layer biases during execution of the algorithm and input weights. ELM compared to other learning algorithms such as BP in the learning process perform very fast and present appropriate performance in extended processing generation function. In this study, the removal mechanism of organic micropollutants (OMPs) in soil aquifer treatment (SAT) using extreme learning machine (ELM) method is modeled.

4.1 Architecture of ELM

Huang et al. [34] presented a new algorithm in terms of extreme learning machine learning to train the single-layer feedforward neural network (SLFFNN) as it does not require iterative tuning and has the ability to achieve global minima. The use of ELM in training the SLFFNN leads to a significant reduction in training time compared to the algorithms based on gradient-descent. Also, the use of active ELM as training algorithm does not require additional parameters such as stopping criterion and learning rate. Experimental observations by Huang et al. [34] showed that the ELM has an excellent ability in universal approximation and useful generalization. In an SLFFNN with random hidden nodes, at first, the input data set and real actual output of the model [(X), (Y)] are determined. Subsequently, the number of hidden nodes [K] and the type of activation function [$g(\cdot)$] are determined. Then, its weight and bias values are presented in random order [(W), (b)]. Then, the hidden layers' matrix [H] is determined, and then the weight of output as analytic is calculated [β]. Input variables can be defined as the following matrix:

$$X = \begin{bmatrix} X_{11} & \cdots & X_{1j} \\ \vdots & \cdots & \vdots \\ X_{n1} & \cdots & X_{nj} \end{bmatrix}_{n \times j} \tag{1}$$

where n and j are the numbers of samples and variables, respectively. Also, the actual output is defined as follows:

$$Y = [Y_1 \ \cdots \ Y_n]^T \tag{2}$$

In the following, a positive integer value for hidden nodes (K) and a differentiable function for $g(\cdot)$ should be defined. Thus, an input weight matrix W is created randomly to make the connection between the input and hidden nodes:

$$W = \begin{bmatrix} W_{11} & \cdots & W_{1k} \\ \vdots & \cdots & \vdots \\ W_{j1} & \cdots & W_{jk} \end{bmatrix}_{j \times k} \tag{3}$$

Hidden layer matrix H by multiplying the input matrix X in the weight matrix W is calculated as follows:

$$H = XW \tag{4}$$

Hidden layer active matrix H with the function $g(\cdot)$ leads to the hidden layer output matrix:

$$H = g(H) \tag{5}$$

Output matrix of the hidden layer, H_{out}, and an input vector $\hat{Y}$ by output layer weight β are connected. ELM network output is calculated as follows:

$$\hat{Y} = H_{\text{out}}\beta \tag{6}$$

and

$$\left\| H_{\text{out}}\widehat{\beta} - Y \right\| = \min_{\beta} \| H_{\text{out}}\beta - Y \| \tag{7}$$

The minimum norm to solve the least squares is calculated as follows:

$$\beta = H_{\text{out}}^{+} Y \tag{8}$$

where H_{out}^{+} calculated as follows is the Moore-Penrose generalized inverse of H_{out}:

$$H_{\text{out}}^{+} = \left(H^T H\right)^{-1} H^T \tag{9}$$

4.2 Performance Evaluation Criteria

In this study, to investigate the accuracy of numerical models, statistical indices root-mean-square error (RMSE), mean absolute percentage error (MARE), correlation coefficient (R), BIAS, scatter index (SI), and ρ are used as follows:

$$\text{RMSE} = \sqrt{\frac{1}{n'} \sum_{i=1}^{n'} \left(T_{(\text{Predicted})_i} - T_{(\text{Observed})_i} \right)^2} \tag{10}$$

$$\text{MARE} = \frac{1}{n'} \sum_{i=1}^{n'} \left(\frac{\left| T_{(\text{Predicted})_i} - T_{(\text{Observed})_i} \right|}{T_{(\text{Observed})_i}} \right) \tag{11}$$

$$\text{BIAS} = \frac{1}{n'} \sum_{i=1}^{n'} \left(T_{(\text{Predicted})_i} - T_{(\text{Observed})_i} \right) \tag{12}$$

$$R = \frac{\sum_{i=1}^{n'} \left(T_{(\text{Observed})i} - \bar{T}_{(\text{Observed})} \right) \left(T_{(\text{Predicted})i} - \bar{T}_{(\text{Predicted})} \right)}{\sqrt{\sum_{i=1}^{n'} \left(T_{(\text{Observed})i} - T_{(\text{Observed})} \right)^2 \sum_{i=1}^{n'} \left(T_{(\text{Predicted})i} - \bar{T}_{(\text{Predicted})} \right)^2}} \tag{13}$$

$$\text{SI} = \frac{\text{RMSE}}{\bar{T}_{(\text{Observed})}} \tag{14}$$

$$\rho = \frac{\text{SI}}{1 + R} \tag{15}$$

where $T_{(\text{Observed})i}$ is the observed values, $T_{(\text{Predicted})i}$ the predicted values by the numerical model, $\bar{T}_{(\text{Observed})i}$ the mean of observed values, and n' the number of observed data.

5 ELM Prediction of SAT Organic Micropollutant Removal

Table 1 shows the parameters controlling the operation of a SAT system and the removal of OMPs, while Table 2 illustrates the ranking of parameters according to their contribution on the removal of OMPs in a SAT system [35]. It is observed that three parameters had the highest ranking: first-order biodegradation rate, saturated hydraulic conductivity, and dry to wet ratio. These high influential parameters have been chosen as predictors for developing prediction models in Sattar [11] and in this study. Using the 50,000 data sets simulated by Sattar [11], 25,000 (50%) were used to develop the models, 12,500 (25%) were used to test the models, and 12,500 (25%) were used to validate the developed models.

To develop an OMP attenuation prediction model, the OMP plume mass, normalized mass ratio, and zero concentration depth are addressed. The mass stored in the contaminant plume can be calculated as:

$$\text{Plume Mass} = V \times \sum_{i=1}^{N} C_i \cdot \theta_i \tag{16}$$

where N = the total number of 2D FE mesh nodes in the study domain, i = the index of the node number, C_i = contaminant concentration at node i, θ_i = unsaturated moisture content at node i, and V = soil volume of each node.

Table 1 Parameters controlling the fate and transport of OMPs in SAT system [11]

Parameter classes	Parameter	Min.	Max.	Avg.
Soil hydraulic	Residual water content, θ_r (cm^3/cm^3)	0.024	0.066	0.045
	Saturated water content, θ_s (cm^3/cm^3)	0.364	0.448	0.406
	Saturated hydraulic conductivity, K_s (cm/day)	20	800	410
	Van Genuchten parameter, α (1/cm)	0.0059	0.0514	0.0287
	Van Genuchten parameter, n	1.34	3.54	2.44
OMP transport and fate	Freundlich coefficient, K_f ((μg kg^{-1}/μg l^{-1}) η)	0.1	500	250.1
	Freundlich exponent, n	0.40	1.6	1
	First-order biodegradation rate, μ_1 (day^{-1})	0.01	0.85	0.43
	Soil bulk density, ρ_b (gm/cm^3)	1.35	1.85	1.6
	Longitudinal dispersivity, α_L (cm)	30	300	165
	Ratio long./trans. dispersivity, α_T/α_L	0.05	0.1667	0.1084
	Molecular diffusion, D_d (cm^2/day)	0.9	1.7	1.3
Operation	HLR/K_s	0.05	0.12	0.085
	HLR (cm/day)	1	95	48
	Initial concentration, C_{in} (μg/l)	0.01	100	50.0
	Dry to wet ratio, DWR	1	7	4

Table 2 Ranking parameters controlling the fate and transport of OMPs in SAT system according to their contribution in system output uncertainty [11]

Parameters classes	Parameter	% importance
0% concentration depth (depth under SAT pond where OMPs are completely attenuated)	First-order biodegradation rate, μ_1 (day^{-1})	48.80
	Saturated hydraulic conductivity, K_s (cm/day)	36.50
	Dry to wet ratio, DWR	6.90
	HLR/K_s	4.30
Plume mass (OMP mass stored in contaminant plume)	First-order biodegradation rate, μ_1 (day^{-1})	66.39
	Initial concentration C_{in} (μg/l)	13.33
	Saturated hydraulic conductivity, K_s (cm/day)	12.05
	Dry to wet ratio, DWR	5.25
Plume normalized mass ratio (ratio of plume mass and cumulative total injected mass)	First-order biodegradation rate, μ_1 (day^{-1})	94.57
	Dry to wet ratio, DWR	5.25

To assess the SAT system OMP removal efficiency, the normalized plume mass has to be calculated, where a system with high removal efficiency would yield smaller fractions of the normalized mass. The normalized plume mass can be calculated as the ratio between the plume mass to the cumulative total injected mass, with time.

Table 3 The quantities needed for modeling the OMP removal by ELM model

Parameter	Maximum	Minimum	Average
Ks (cm/day)	800	20	410
μ_1 (day^{-1})	0.85	0.01	0.43
HLR (cm/day)	95	1	48
C_0 (μg/l)	100	0.01	50.0
DWR	7	1	4

To estimate the parameters of plume mass, normalized mass ratio and concentration depth of 0% of the five parameters are presented in Table 3, as ELM model inputs are used. Therefore, a total of 15 different models of ELM are introduced. It should be noted that the Monte Carlo simulation (MCS) to determine the uncertainty of plume mass, normalized mass ratio, and concentration depth of 0% with 1,000 realizations to generate random inputs of ELM models are used. Sattar [11] stated that the plume mass for organic micropollutants in the soil, 90 days after the SAT operation, is calculated as follows:

$$\text{Plume mass} = 6.14K_s^{1.1}C_0\mu_1^{-1}\text{DWR}^{1/2} \tag{17}$$

Here, K_s is saturated hydraulic conductivity, C_0 the amount of concentration, μ_1 the rate of first-order biodegradation, and DWR the intermittent application of wastewater in the soil aquifer treatment system. Also, to calculate the normalized mass ratio, the following equation was used:

$$\text{Mass ratio} = 0.0055_0\mu_1^{-1}\text{DWR}^{1/5} \tag{18}$$

The minimum depth under a SAT system required to remove more than 98% contamination to the depth of 0% concentration is defined as follows:

$$Y_{\text{zero}} = 0.17e^{\text{DWR}^{0.5}K_s^{-1}}\mu_1^{-1/2}\text{HLR}^{1/2}\text{DWR}^{-3/8}K_s^{1/4} \tag{19}$$

where HLR is the hydraulic loading rate.

5.1 Plume Mass

Plume mass parameter is calculated by Eq. (16). The results of ELM models 1–5 for this parameter with GEP model provided by Sattar [11] were compared. In Table 4, different statistical indices for ELM models 1–5 and Sattar [11] model in the prediction of plume mass parameter are arranged. Also, scattering plots of the models for plume mass are depicted in Fig. 3. The highest correlation coefficient value for ELM 2 and ELM 5 models is calculated. The lowest of R value for ELM 4 equal to 0.964 is computed. The SI and ρ values for ELM 4 have been predicted, 0.811 and 0.413, respectively. ELM 5 among all ELM models has the highest correlation coefficient value and the least amount of errors. For this model, the

Table 4 Statistical indices of ELM models 1–5 and [11] to predict the plume mass parameter

	Plume mass					
	RMSE	MARE	*R*	BIAS	SI	ρ
[11]	731,788.8	38.274	0.933	6,603.806	1.073	0.555
ELM 1	527,939.2	6.164	0.968	132,269.1	0.774	0.393
ELM 2	514,951.5	5.625	0.970	116,093.1	0.755	0.383
ELM 3	527,269.9	5.634	0.968	118,727.1	0.773	0.393
ELM 4	553,288	5.272	0.964	−120,723	0.811	0.413
ELM 5	508,715.7	5.473	0.970	−113,364	0.746	0.379

BIAS value is predicted which is − 113,364. However, the MARE and the correlation coefficient values for the Sattar [11] model have been calculated, 38.274 and 0.933, respectively.

5.2 *Mass Ratio*

In the following, the results of ELM models 1–5 to predict the mass ratio parameter are evaluated. In Table 5, statistical index values to predict the mass ratio by ELM models [11] are shown. Also, scatter plots for the model presented in Fig. 4 are visible. Based on the results of ELM, the highest RMSE value has been predicted for ELM 1 (RMSE = 0.019). For this model, the *R* statistical index is calculated to equal to 0.974, while ELM 3 has the highest amount of correlation ($R = 0.977$) and the lowest MARE value(MARE = 1.661). For ELM 3, BIAS and ρ parameter are calculated, 0.0036 and 0.239, respectively. In contrast, GEP model introduced by Sattar [11] has less correlation ($R = 0.823$). Hence, RMSE and MARE values for the model [11] have been calculated, 0.521 and 341.867, respectively. Therefore, ELM models to predict the mass ratio parameter have an acceptable accuracy.

5.3 *Zero Concentration Depth*

Also, the accuracy of ELM models 1–5 in the modeling of Y_{zero} parameter is examined (see Fig. 5). In Table 6, statistical indices calculated for ELM models and GEP model proposed by Sattar [11] are shown. ELM 1 between the ELM models has the least accuracy. For this model, RMSE and ρ values are calculated, 1.479 and 0.145, respectively. However, the value of correlation coefficient for ELM 1 is estimated, 0.959. Also, ELM 5 predicts the Y_{zero} parameter more accurately compared to other ELM models. The parameters MARE, BIAS, and ρ values for this model are predicted, 0.237, 0.00106, and 0.141, respectively. However, the accuracy of the model [11] to predict Y_{zero} parameter is less than ELM models. In other words, the correlation coefficient value for the model is

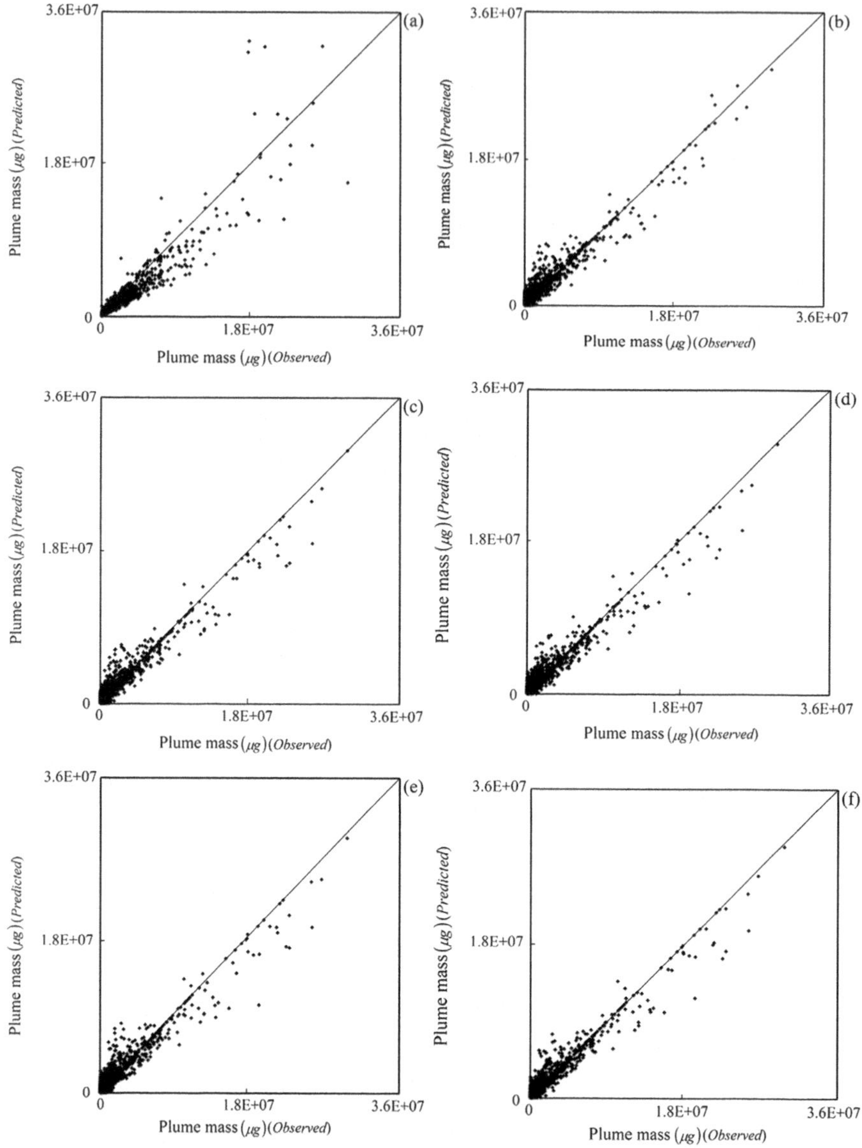

Fig. 3 Scatter plot for prediction of plume mass parameter (**a**) Sattar [11–13] (**b**) ELM 1 (**c**) ELM 2 (**d**) ELM 3 (**e**) ELM 4 (**f**) ELM 5

0.944 [11]. Therefore, based on the analysis of simulation results, ELM model estimates parameters plume mass, mass ratio, and Y_{zero} with reasonable accuracy.

To analyze the results of ELM, the parameter discrepancy ratio (DR) as the ratio of modeled values to measured values is introduced ($DR = T_{(\text{Predicted})}/T_{(\text{Observed})}$) [11]. The proximity of discrepancy ratio to 1 represents the proximity of predicted

Table 5 Statistical indices of ELM models 1–5 and [11] to predict mass ratio parameter

	Mass ratio					
	RMSE	MARE	R	BIAS	SI	ρ
[11]	0.521	341.867	0.823	0.213	13.821	78.084
ELM 1	0.019	5.848	0.974	0.0040	0.494	0.250
ELM 2	0.018	1.979	0.976	0.0039	0.488	0.247
ELM 3	0.018	1.661	0.977	0.0036	0.472	0.239
ELM 4	0.018	1.714	0.976	0.0038	0.480	0.243
ELM 5	0.018	1.807	0.977	0.0037	0.471	0.238

values to measured results. In Table 7, the values DR_{max}, DR_{min}, and DR_{ave} are the maximum, minimum, and mean discrepancy ratio. For plume mass parameter, the lowest DR_{ave} for ELM 4 is calculated, while the average of discrepancy ratio for the model has been computed, 39.175 [11]. For ELM 3, the mass ratio parameter has the lowest DR_{ave} value ($DR_{ave} = 2.508$). The DR_{max} and DR_{min} values for this table model have been estimated, 335.291 and 0.0004, respectively. As can be seen, the DR_{ave} for the proposed GEP model [11] has been obtained, 342.861. The lowest DR_{ave} value to ELM models in prediction of Y_{zero} parameter for ELM 5 is obtained ($DR_{ave} = 1.060$). The DR_{max} and DR_{min} values for the model ELM 5 have been calculated, 65.876 and 0.022, respectively. For model DR_{max}, DR_{min} and DR_{ave} values are obtained, 109.803, 0.545 and 1.250, respectively [11]. Based on the analysis results of discrepancy ratio parameter, results predicted by the model's ELM compared with GEP model introduced by Sattar [11] are closer to the measured values.

6 SAT Site Selection in Egypt

SAT systems would be an attractive unconventional water resource in Egypt, which is true, especially in rural communities. Egyptian researchers are keeping this in mind. Recently, RIGW [36] published recommended characteristics of groundwater aquifer for a successful and efficient SAT operation and removal of organic matter. These included an infiltration rate of more than 0.25 m/day, minimum depth to groundwater of 5 m, high values of porosity, and saturated zone transmissivity, and most importantly, the aquifer should not be flowing into the Nile River. Recently, El Arabi et al. [37] have provided guidelines for the selection of potential SAT sites in Egypt. The primary target for these guidelines was to ensure adequate removal of nitrogen and biochemical oxygen demand (BOD) from treated wastewater. However, the OMP removal criterion has not been considered in these guidelines despite the ecological and health risks imposed by their presence in Egyptian soils and native groundwater. Figure 6 shows the potential locations for construction of SAT systems in Egypt [38]. It was found that the best sites existed on the Western fringes of the Nile delta, W1 to W6, as shown in Fig. 6. SAT

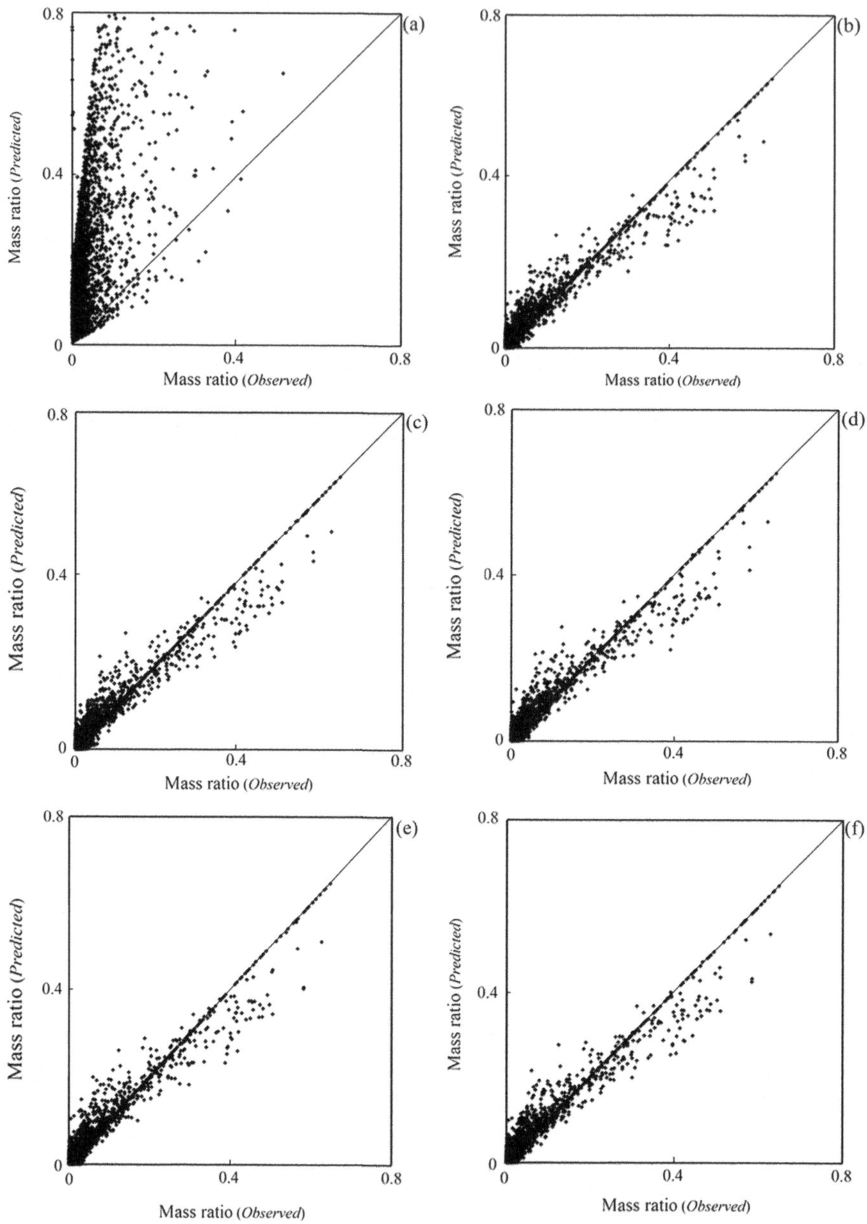

Fig. 4 Scatter plots for prediction of mass ratio parameter (**a**) Sattar [11–13] (**b**) ELM 1 (**c**) ELM 2 (**d**) ELM 3 (**e**) ELM 4 (**f**) ELM 5

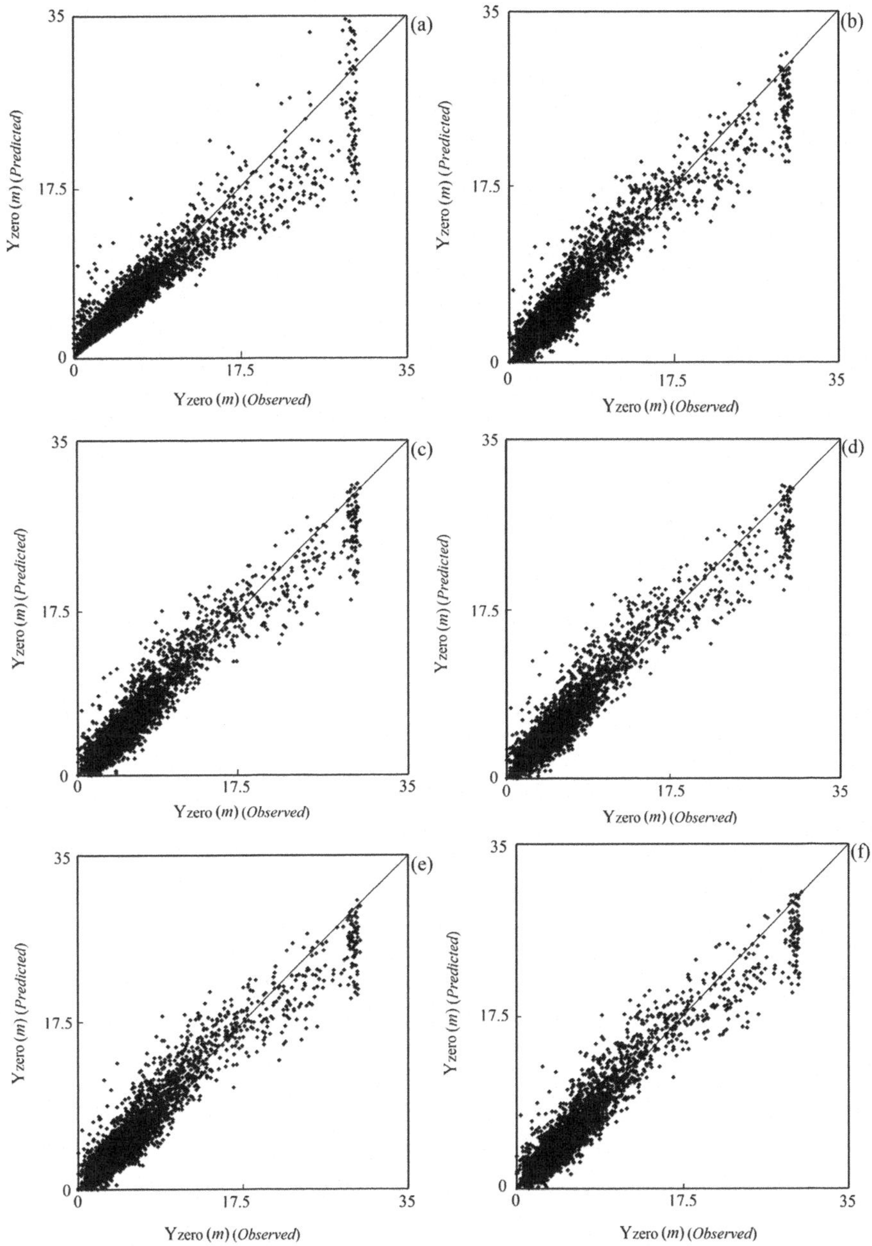

Fig. 5 Scatter plot for prediction of Y_{zero} parameter (**a**) Sattar [11–13] (**b**) ELM 1 (**c**) ELM 2 (**d**) ELM 3 (**e**) ELM 4 (**f**) ELM 5

Table 6 Statistical indices of ELM models 1–5 and [11] to predict the parameter Y_{zero}

	Y_{zero}					
	RMSE	MARE	R	BIAS	SI	ρ
[11]	1.748	0.325	0.944	0.16620	0.335	0.172
ELM 1	1.479	0.252	0.959	0.00115	0.283	0.145
ELM 2	1.457	0.243	0.961	0.00109	0.279	0.142
ELM 3	1.454	0.250	0.961	0.00053	0.278	0.142
ELM 4	1.477	0.245	0.959	0.00079	0.283	0.144
ELM 5	1.447	0.237	0.961	0.00106	0.277	0.141

Table 7 DR_{max}, DR_{min}, and DR_{ave} values for ELM models 1–5 and [11]

	Plume mass			Mass ratio			Y_{zero}		
	DR_{max}	DR_{min}	DR_{ave}	DR_{max}	DR_{min}	DR_{ave}	DR_{max}	DR_{min}	DR_{ave}
[11]	163,392.1	0.368	39.175	1,403,391	0.5618	342.861	109.803	0.545	1.250
ELM 1	2,108.627	0.0018	7.0281	483.6609	0.0037	2.765	104.356	0.032	1.073
ELM 2	4,098.162	0.003	6.490	1,045.368	0.0053	2.829	82.452	0.008	1.068
ELM 3	4,014.867	5.61E-05	6.495	335.291	0.0004	2.508	114.419	0.004	1.073
ELM 4	1,469.444	0.0005	6.133	537.4445	0.0017	2.575	63.755	0.003	1.062
ELM 5	2,311.012	0.006	6.335	822.4562	0.0010	2.652	65.876	0.022	1.060

systems in these areas would improve the quality of groundwater regarding reducing the salinity and would help treat more than 350 million m^3/year of wastewater produced from nearby treatment plants, I–V.

For the potential SAT sites in Egypt (as shown in Fig. 6), the average depth to groundwater is contained in Table 8 [36, 39, 40]. Table 8 presents the results of Sattar model study [11], and the model developed in this study for the zero concentration depth, i.e., the depth at which the concentration of OMPs reaches zero. The best locations for potential SAT system with the highest efficiency in removal of OMPs are Wadi El Natrun, Sadat City, and Alexandria, respectively, with an average removal efficiency of 90%. On the other hand, Rashed and Abu Rawash had the shallowest groundwater table disabling vadose soil to completely attenuate OMPs during wastewater infiltration, making these sites less favorable for SAT system construction. With the availability of detailed hydrogeological investigations for the potential SAT site, the uncertainty in model predictions [41–45], for OMP removal, can be significantly reduced, and the influence of SAT operational aspects can be studied.

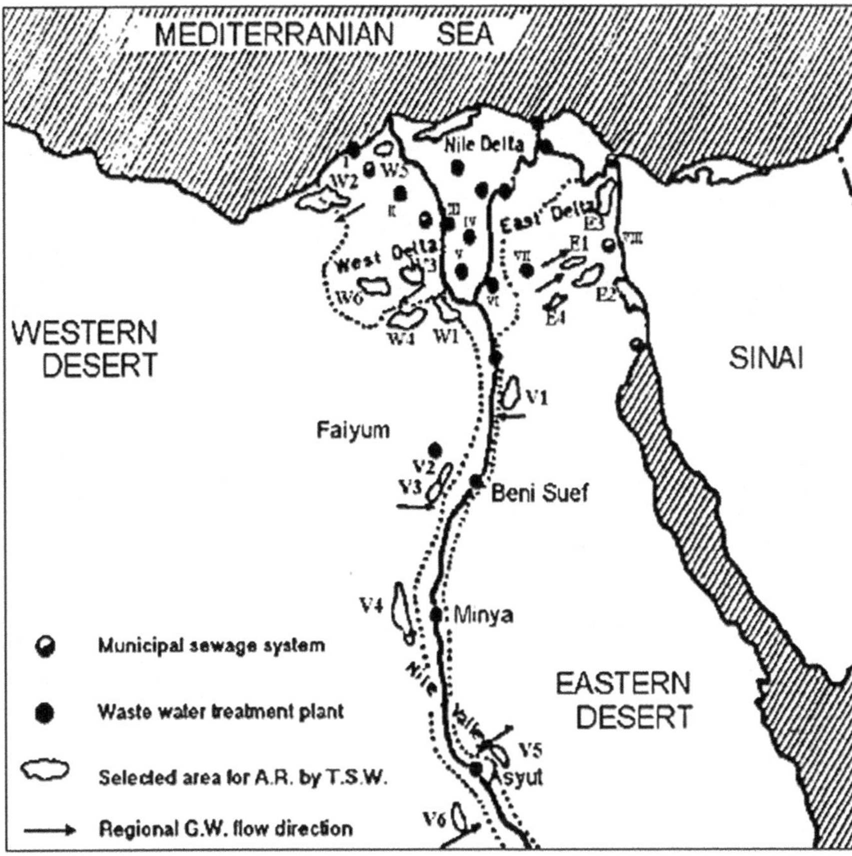

Fig. 6 Potential locations for construction of SAT systems in Egypt [38]

Table 8 Potential SAT system locations in Egyptian Western delta fringes and corresponding OMP removal

SAT location	Depth to groundwater table (m)	Removal of OMP (% from studied range)
Abu Rawash	4–7	20–60
Alexandria	3–14	10–90
Sadat City	15–25	95–100
Wadi EL Natrun	0–30	0–100
Rashed	1–5	5–50

7 Conclusions and Recommendations

One of the most important methods to remove organic micropollutants (OMPs) at water and wastewater treatment plants is using the soil aquifer treatment (SAT). In this study, the mechanism of organic micropollutant removal using extreme learning machine (ELM) method was evaluated. Therefore, five different ELMs for each of the parameters' plume mass, mass ratio, and depth of concentration of 0% (Y_{zero}) were defined. Also, the results of extreme learning machine with results of gene expression programming model provided by [11] were compared. Analysis of numerical model results indicated the acceptable accuracy of ELM models in prediction of the pollutant removal efficiency during the SAT operation. Moreover, the model prediction accuracy of the optimum ELM model is assessed using the statistical performance parameters, including MARE, correlation coefficient, and scatter index, which were found to be 5.473, 0.970, and 0.746, respectively. Also, the discrepancy ratio for the parameter Y_{zero} by the best ELM model was calculated to be 1.060.

References

1. Sattar AM, Dickerson JR, Chaudhry MH (2009) Wavelet-Galerkin solution to the water hammer equations. J Hydraul Eng 135(4):283–295
2. Foda A, Sattar A (2013) Morphological changes in river Nile at Bani-Sweif for probable flood flow releases. In: Proceedings of the international conference on fluvial hydraulics, RIVER FLOW 2014, Switzerland
3. Sattar AM (2013) Using gene expression programming to determine the impact of minerals on erosion resistance of selected cohesive Egyptian soils. Experimental and computational solutions of hydraulic problems, part of the series geoplanet: earth and planetary sciences. Springer, Berlin, pp 375–387
4. Sattar AM (2014) Predicting morphological changes ds new Naga-Hammadi barrage for extreme Nile flood flows: a Monte Carlo analysis. J Adv Res 5(1):97–107
5. Sattar AM (2014) Gene expression models for prediction of dam breach parameters. J Hydroinf 16(3):550–571
6. Sattar AM (2014) Gene expression models for the prediction of longitudinal dispersion coefficients in transitional and turbulent pipe flow. J Pipeline Syst Eng Pract ASCE 5 (1):04013011
7. El-Hakeem M, Sattar AM (2015) An entrainment model for non-uniform sediment. Earth Surf Process Landf 4(9):1216–1226. https://doi.org/10.1002/esp.3715.
8. Sattar AM, Gharabaghi B (2015) Gene expression models for prediction of longitudinal dispersion coefficient in streams. J Hydrol 524:587–596
9. Najafzadeh M, Sattar AM (2015) Neuro-fuzzy GMDH approach to predict longitudinal dispersion in water networks. Water Resour Manag 29:2205–2219. https://doi.org/10.1007/s11269-015-0936-8
10. Sattar AM, Gharabaghi B, McBean E (2016) Prediction of timing of watermain failure using gene expression models. Water Resour Manag 30(5):1635–1651
11. Sattar AM (2016) Prediction of organic micropollutant removal in soil aquifer treatment system using GEP. J Hydrol Eng ASCE 21(9):04016027. https://doi.org/10.1061/(ASCE)HE.1943-5584.0001372
12. Sattar AM (2016) A probabilistic projection of the transient flow equations with random system parameters and internal boundary conditions. J Hydraul Res 54(3):342–359. https://doi.org/10.1080/00221686.2016.1140682
13. Sattar AM (2016) Closure to "gene expression models for the prediction of longitudinal dispersion coefficients in transitional and turbulent pipe flow" by Ahmed M. A. Sattar. J Pipeline Syst Eng Pract 7(4):07016002. https://doi.org/10.1061/(ASCE)PS.1949-1204.0000254

14. Thompson J, Sattar AM, Gharabaghi B, Richard W (2016) Event based total suspended sediment particle size distribution model. J Hydrol 536:236–246. https://doi.org/10.1016/j.jhydrol.2016.02.056
15. Sabouri F, Gharabaghi B, Sattar AM, Thompson AM (2016) Event-based stormwater management pond runoff temperature model. J Hydrol 540:306–316. https://doi.org/10.1016/j.jhydrol.2016.06.017
16. Atieh M, Taylor G, Sattar AM, Gharabaghi B (2017) Prediction of flow duration curves for ungauged basins. J Hydrol 545:383–394
17. Gharabaghi B, Sattar AM (2017) Empirical models for longitudinal dispersion coefficient in natural streams. J Hydrol. https://doi.org/10.1016/j.jhydrol.2017.01.022
18. Sattar AM, Gharabaghi B, Sabouri F, Thompson AM (2017) Urban stormwater thermal gene expression models for protection of sensitive receiving streams. Hydrol Process 31 (13):2330–2348. https://doi.org/10.1002/hyp.11170
19. El-Hakeem M, Sattar AM (2017) Explicit solution for the specific flow depths in partially filled pipes. J Pipeline Syst Eng Pract 8(4):06017004. https://doi.org/10.1061/(ASCE)PS.1949-1204.0000283
20. Sattar AM, Ertuğrul ÖF, Gharabaghi B, McBean EA (2017) Extreme learning machine model for water network management. Neural Comput Applic, 1–13. https://doi.org/10.1007/s00521-017-2987-7DO
21. Sattar AM, Beltagy M (2016) Stochastic solution to the water hammer equations using polynomial chaos expansion with random boundary and initial conditions. J Hydraul Eng 143(2):04016078. https://doi.org/10.1061/(ASCE)HY.1943-7900.0001227
22. Drillia P, Stamatelatou K, Lyberatos G (2005) Fate and mobility of pharmaceuticals in solid matrices. Chemosphere 60(8):1034–1044
23. Xu J, Wu L, Chang AC (2009) Degradation and adsorption of selected pharmaceuticals and personal care products (PPCPs) in agricultural soils. Chemosphere 77(10):1299–1305
24. Maeng SK, Sharma SK, Amy G (2011) Framework for assessment of organic micropollutant removals during managed aquifer recharge and recovery. Riverbank filtration for water security in desert countries. Springer, Dordrecht, pp 137–149
25. Yu Y, Liu Y, Wu L (2013) Sorption and degradation of pharmaceuticals and personal care products (PPCPs) in soils. Environ Sci Pollut Res 20(6):4261–4267
26. Lin C, Shacahr Y, Banin A (2004) Heavy metal retention and partitioning in a large-scale soil-aquifer treatment (SAT) system used for wastewater reclamation. Chemosphere 57 (9):1047–1058
27. Fox P, Aboshanp W, Alsamadi B (2005) Analysis of soils to demonstrate sustained organic carbon removal during soil aquifer treatment. J Environ Qual 34(1):156–163
28. Amy G, Drewes J (2007) Soil aquifer treatment (SAT) as a natural and sustainable wastewater reclamation/reuse technology: fate of wastewater effluent organic matter (EfOM) and trace organic compounds. Environ Monit Assess 129:19–26
29. Sharma SK, Harun CM, Amy G (2008) Framework for assessment of performance of soil aquifer treatment systems. Water Sci Technol 57:941–946
30. Caballero MH (2010) Soil aquifer treatment as pre-treatment for organic micropollutant removal during membrane filtration. MSc thesis, MWI 2019-5, UNESCO-IHE, Delft
31. Abel CDT, Sharma SK, Malolo YN, Maeng SK, Kennedy MD, Amy GL (2012) Attenuation of bulk organic matter, nutrients (N and P), and pathogen indicators during soil passage: effect of temperature and redox conditions in simulated soil aquifer treatment (SAT). Water Air Soil Pollut 223:5205–5220
32. Onesios-Barry KM, Berry D, Proescher JB, Sivakumar IA, Bouwer EJ (2014) Removal of pharmaceuticals and personal care products during water recycling: microbial community structure and effects of substrate concentration. Appl Environ Microbiol 80(8):2440–2450
33. Suzuki R, Kameda I, Takabe Y, Nishimura F, Itoh S (2015) Removal of dissolved organic matter and disinfection by-products formation potential in the upper layer during soil aquifer treatment. J Water Environ Technol 13(2):107–118

34. Huang GB, Zhu QY, Siew CK (2006) Extreme learning machine: theory and applications. Neurocomputing 70(1):489–501
35. Abokifa A (2013) Modeling soil aquifer treatment processes during managed aquifer recharge considering soil heterogeneity and parameter uncertainty. MSc thesis, Department of Irrigation and Hydraulics, Cairo University, Giza
36. Regional Institute for Groundwater Modeling (RIGW) (2006) Evaluation of different uses for the groundwater in Sadat city. Technical Report No. 3, Research Institute for Groundwater
37. El Arabi N, Mohamed A, Dawoud A (2012) Groundwater aquifer recharge with treated wastewater in Egypt: technical, environmental, economical and regulatory considerations. Desalin Water Treat 47:266–278
38. Fadlelmawal A, Halem M, Vissers M (1999) Preliminary plans for artificial recharge with reclaimed wastewater in Egypt. In: Proceedings, 9th biennial symposium on the artificial recharge of groundwater, Tempe
39. Abu Zeid K (2009) Assessment of groundwater potential in Alexandria governorate. Report submitted to sixth framework programme, CEDARE International
40. Ahmed MA, Abdel Samie SG, El-Maghrabi HM (2011) Recharge and contamination sources of shallow and deep groundwater of pleistocene aquifer in El-Sadat industrial city: isotope and hydrochemical approaches. Environ Earth Sci 62:751–768
41. Safadoust A, Amiri Khaboushan E, Mahboubi AA, Gharabaghi B, Mosaddeghi MR, Ahrens B, Hassanpour Y (2016) Comparison of three models describing bromide transport as affected by different soil structure types. Arch Agron Soil Sci 62(5):674–687. https://doi.org/10.1080/03650340.2015.1074184
42. Gharabaghi B, Safadoust A, Mahboubi AA, Mosaddeghi MR, Unc A, Ahrens B, Sayyad G (2015) Temperature effect on the transport of bromide and E. coli NAR in saturated soils. J Hydrol 522:418–427. https://doi.org/10.1016/j.jhydrol.2015.01.003
43. Safadoost A, Yousefi G, Mahboubi A, Gharabaghi B, Mosaddeghi M, Yousefi G, Safadoust A, Mahboubi AA, Gharabaghi B, Mosaddeghi MR, Ahrens B, Shirani H (2014) Bromide and lithium transport in soils under long-term cultivation of alfalfa and wheat. Agric Ecosyst Environ 188:221–228. https://doi.org/10.1016/j.agee.2014.02.031
44. Safadoust A, Mahboubi AA, Mossaddeghi MR, Gharabaghi B, Unce A, Voroney P, Heydari A (2012) Effect of regenerated soil structure on unsaturated transport of Escherichia coli and bromide. J Hydrol 430–431:80–90. https://doi.org/10.1016/j.jhydrol.2012.02.003
45. Safadoust A, Mahboubi AA, Gharabaghi B, Mosaddeghic MR, Voroney P, Unce A, Khodakaramian G (2012) Significance of physical weathering of two-texturally different soils for the saturated transport of E. coli and bromide. J Environ Manag 107:147–158. https://doi.org/10.1016/j.jenvman.2012.04.007

Part VI
Saltwater Intrusion

Investigation of Saltwater Intrusion in Coastal Aquifers

Ismail Abd-Elaty, Hany F. Abd-Elhamid, and Abdelazim M. Negm

Abstract Groundwater is considered the main source of water in many coastal areas. The increase of water demands increases the abstraction from aquifers which has resulted in lowering water tables and caused saltwater intrusion. Coastal aquifers lie within some of the most intensively exploited areas of the world. Saltwater intrusion is one of the main causes of groundwater quality degradation and a major challenge in the management of groundwater resources in coastal regions. Saltwater intrusion causes an increase of salt concentration in groundwater which places limitations on its uses. Excessive pumping always leads to a dramatic increase in saltwater intrusion. In coastal aquifers, the hydraulic gradient exists towards the sea which leads to flow of the excess freshwater to the sea. Seawater intrusion is a special category of groundwater contamination that threatens the health and possibly lives of the people living in coastal areas. The problems of saltwater intrusion into groundwater had become a considerable concern in many countries particularly in coastal areas. Seawater intrusion leads to the depletion of groundwater resources and should be prevented or controlled to protect water resources in coastal regions. The intrusion of saltwater in coastal aquifers has been investigated by several methods including geophysical methods, geochemical methods, experimental studies and mathematical models. This chapter presents a brief history of saltwater intrusion in coastal aquifers. The Nile Delta aquifer is one of the largest underground freshwater reservoirs in the world that attacked by saltwater intrusion. Large amounts of freshwater were damaged by salinization. Extensive studies were carried out to investigate saltwater intrusion in Nile Delta aquifer using numerical and field studies. Most of these investigations revealed that the seawater intrusion in the Nile Delta aquifer has extended to a distance of more

I. Abd-Elaty (✉), H.F. Abd-Elhamid, and A.M. Negm
Water and Water Structures Engineering Department, Faculty of Engineering, Zagazig University, Zagazig, Egypt
e-mail: Eng_abdelaty2006@yahoo.com; Eng_abdelaty@zu.edu.eg; Hany_farhat2003@yahoo.com; Amnegm@zu.edu.eg; Amnegm85@yahoo.com

A. M. Negm (ed.), *Groundwater in the Nile Delta*, 329
Hdb Env Chem (2019) 73: 329–354, DOI 10.1007/698_2017_190,

than 100 km from the Mediterranean coast. The effect of climatic changes including the rise in the sea level has a significant effect on the position of the transition zone, and the groundwater quality would deteriorate in large areas of the Nile Delta aquifer.

Keywords Coastal aquifers, Egypt, Groundwater, Modelling, Nile Delta, Saltwater intrusion

Contents

1 Introduction

Groundwater is considered the main sources of water in many coastal areas. The increase of water demands increases the abstraction from aquifers which has resulted in lowering water tables and caused saltwater intrusion. Saltwater intrusion is one of the main causes of groundwater quality degradation and a major challenge in the management of groundwater resources in coastal regions. Saltwater intrusion causes an increase of salt concentration in groundwater which places limitations on its uses. Seawater intrusion is a special category of groundwater contamination that threatens the health and possibly lives of many people living in coastal areas. Therefore, efficient control of seawater intrusion is very important to protect groundwater resources from depletion.

The problems of saltwater intrusion into groundwater have become a considerable concern in many countries with coastal areas. This chapter presents a brief history of saltwater intrusion in coastal aquifers and also presents the traditional methods that are applied to investigate saltwater intrusion in coastal aquifers around the world. New methods to control saltwater intrusion in coastal aquifers are presented and discussed in details; also the advantages and disadvantages of each method were highlighted. Finally, investigation and control of saltwater intrusion in Egypt, especially in the Nile Delta aquifer, are discussed. The possibility of applying new methods to control saltwater intrusion in Egypt is presented.

2 Saltwater Intrusion Problem Identification

Coastal zones are among the most densely populated areas in the world with over 70% of the world's population. These regions face serious hydrological problems such as scarcity of freshwater, contamination of groundwater and seawater intrusion. The growing of global population and raising standards of living have increased water demands and pumping from aquifers. Excessive pumping has led to a dramatic increase in saltwater intrusion problems. In coastal aquifers, a hydraulic gradient exists towards the sea which leads to flow of the excess freshwater to the sea. Owing to the presence of seawater in the aquifer formation under the sea bottom, a zone of contact is formed between the lighter freshwater flowing to the sea and the heavier, underlying seawater (Fig. 1) [1]. Freshwater and seawater are miscible fluids, and therefore the zone of contact between them takes the form of a transition zone caused by hydrodynamic dispersion. Across this zone, the density of the mixed water varies from that of freshwater to that of seawater. Under certain conditions, the width of this zone is relatively small, when compared with the thickness of the aquifer so that the boundary can be considered as a sharp interface that separates the two regions occupied by the two fluids, in this case, the two fluids are assumed to be immiscible. However, if the transition zone is wide, this assumption becomes invalid [1, 2].

Under natural undisturbed conditions in a coastal aquifer, freshwater flows to the sea. By excessive pumping from a coastal aquifer, the water table or piezometric surface could be lowered to the extent that the piezometric head in the freshwater body becomes less than in the adjacent seawater wedge and the interface could start to advance inland until a new equilibrium is reached. This phenomenon is called seawater intrusion. As the interface advances, the transition zone widens. When the interface reaches the inland pumping well, the well becomes contaminated. When

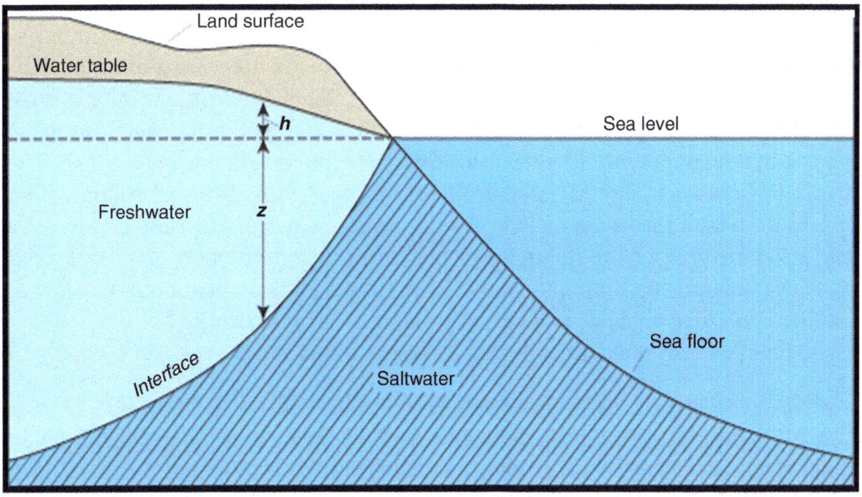

Fig. 1 Saltwater/freshwater interfaces

the pumping takes place in a well located above the interface, upconing towards the pumping well could occur. Unless the rate of pumping is carefully controlled, seawater will eventually enter the pumping wells, and the wells may be abandoned. Seawater intrusion leads to the depletion of groundwater resources and should be prevented or controlled to protect water resources in coastal regions [1].

3 Types and Causes of Saltwater Intrusion

Coastal aquifers lie within some of the most intensively exploited areas of the world. If current levels of population growth and industrial development are not controlled shortly, "the amount of groundwater use will increase dramatically, to the point that the control of seawater intrusion becomes a major challenge to future water resources management engineers. Saltwater intrusion is a serious problem in the coastal regions all over the world. It may occur due to human activities and by natural events such as climate change and sea level rise" [1, 2].

The main causes of saltwater intrusion include [3]:

1. Overabstraction of the aquifers
2. Seasonal changes in natural groundwater flow
3. Tidal effects
4. Barometric pressure
5. Seismic waves
6. Dispersion
7. Climate change – global warming and associated sea level rise

Overabstraction is considered one of the main causes of saltwater intrusion as shown in Fig. 2. Some of the previous causes of saltwater intrusion are periodic (e.g. seasonal changes in natural groundwater flow), some have short-term implications (tidal effects and barometric pressure), and others have long-term implications (climate changes and artificial influences). Salinization is one of the most widespread forms of groundwater contamination in the world. Saltwater intrusion poses a major limitation to the utilization of groundwater resources [1]. The intrusion of seawater should be controlled to protect groundwater resources from depletion. To take measures to control and prevent seawater intrusion and to get a clear understanding of the relationship between the consumption of groundwater and seawater intrusion, in-depth study either from the theoretical point of view or from the numerical analysis and practical point of view is required.

The types of seawater intrusion can be summarized as:

1. Lateral intrusion of seawater into an aquifer adjacent to the sea
2. Downwards seepage from a saline surface waterway that overlies an aquifer
3. Upconing of deep saline seawater that is present within an aquifer
4. Alteration to natural barriers between seawater and an aquifer
5. Increasing groundwater abstraction

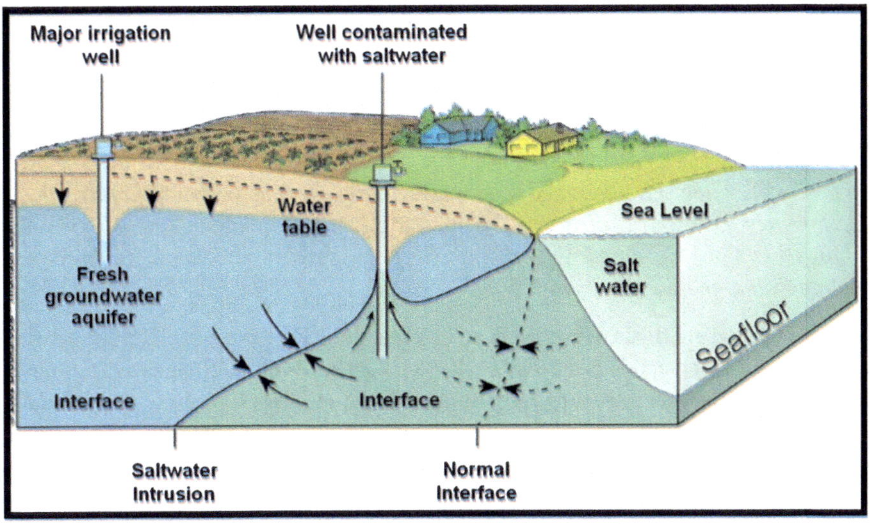

Fig. 2 Overabstraction and saltwater intrusion (https://sweetscienceyspirit.wordpress.com/2014/11/25/over-extraction-of-groundwater-and-its-consequences/)

4 Methods of Investigating Saltwater Intrusion

The analysis of salinization process in coastal regions requires information on both geological and hydrochemical characteristics of aquifers. The intrusion of saltwater in coastal aquifers has been investigated by several methods including:

1. Geophysical methods
2. Geochemical methods
3. Experimental studies
4. Mathematical models

These approaches attempted to ascertain the position of freshwater/saltwater interface and predict changes in water levels and salinity. A brief description of the methods that have been used is presented in the following sections [1].

4.1 Geophysical Investigations

Saltwater intrusion can be distinguished by the movement of water with high total dissolved solids (TDS) content into freshwater. The degree of salinity varies in the transition zone between freshwater and saline groundwater. The contour map of TDS is one measure of salinity, and the location of the contour of TDS = 1,000 mg/l marks the area under the influence of seawater intrusion. Geophysical methods measure the spatial distribution of physical properties of the earth, such as bulk conductivity of seismic velocity. Several geophysical methods can be used for hydrogeologic investigations such as [1, 2]:

1. *Surface geophysical methods*: electrical methods, such as DC resistivity, frequency domain electromagnetic methods, airborne EM, loop-loop EM, time-domain electromagnetic sounding and very low-frequency EM. Seismic methods, such as seismic refraction and seismic reflection. Ground penetrating radar
2. *Borehole methods*: electric logs, radiometric logs and integrated use of borehole logs
3. *Integrated geophysical surveys*

Geophysical methods were used to investigate saltwater into different coastal aquifers. Fitterman and Deszcz-Pan [4] used the airborne electromagnetic technique for mapping saltwater intrusion in Everglades National Park, Florida. Ekwurzel et al. [5] carried out a hydrogeologic investigation using isotopes to determine the source of salinization in the Souss-Massa Basin, Morocco. Sherif et al. [6] presented field investigations and measurements that revealed the nature and extent of the seawater intrusion in the coastal aquifer of Wadi Ham, UAE. They presented a 3D geological and true earth resistively modelling for the aquifer. They used the results of the earth resistivity imaging surveys and chemical analyses of collected water samples to obtain an empirical relationship between the inferred earth resistivity and the amount of total dissolved solids.

4.2 Geochemical Investigations

The distinction of different salinization mechanisms is crucial to the evolution of the origin, pathways, rates and future salinization of coastal aquifers. Several geochemical criteria can be used to identify the origin of salinity in coastal aquifers [1, 2]:

1. Cl concentration: the Cl concentration of 200 mg/l in groundwater is used as an index of saltwater intrusion.
2. Cl/Br ratios: the bromide ion can be considered as a good indicator of saltwater intrusion. The concentration of bromide in freshwater is less than 0.01 mg/l.
3. Na/Cl ratios.
4. Ca/Mg, Ca/(HCO_3 + SO_4) ratios.
5. O and H isotopes.
6. Boron isotopes.

Geochemical methods were used to investigate saltwater intrusion. Edet and Okereke [7] examined the extent of seawater intrusion using vertical electrical sounding and hydrochemical data into shallow aquifers beneath the coastal plains of Southeastern Nigeria. Allen et al. [8] presented a multidisciplinary study involving hydrochemical sampling and borehole geophysics to study the nature and occurrence of saline groundwater in the Gulf Islands, British Columbia, Canada.

Sherif et al. [9] conducted a 2D earth resistivity survey in Wadi Ham, UAE, in the area between Fujairah and Kalba to delineate the seawater intrusion. They used monitoring wells to measure the horizontal and vertical variations in water salinity and thus to improve the interpretation of earth resistivity imaging data. They used the results of vertical electrical soundings and chemical analyses of collected water samples to obtain an empirical relationship between the inferred earth resistivity and the amount of total dissolved solids. This relationship was used along with the true resistivity sections resulting from the inversion of 2D resistivity data to identify three zones of water-bearing formation (fresh, brackish and saltwater zones) [1].

4.3 Experimental Studies

Some researchers presented experimental studies to investigate saltwater intrusion. Koch and Starke [10] presented an experimental study using a Plexiglass for carrying out 2D tracer experiment of macrodispersion in density-dependent flow within a stochastic realization of a heterogeneous media. Van Meir et al. [11] developed in situ laboratories where an integrated set of tools were developed for characterizing the saltwater intrusion process in a hard rock aquifer with an appreciable level of heterogeneity. The final goal of the geophysical and hydraulic investigations was to derive 3D porosity and permeability models as input for the analysis of specific saltwater intrusion experiments [1]. Chang and Clement [12] used a combination of laboratory experiments and numerical simulations to study the impacts of changes in groundwater flux, on saltwater intrusion process. Abd-Elaty [13] investigated the effect of climate change and sea level rise on saltwater intrusion using a combination of experimental and numerical models. The experimental model was carried out using flow tank model, but SEAWAT code was used for numerical simulation.

4.4 Mathematical Models

Mathematical models help to understand the relevant processes that cause saltwater intrusion in coastal aquifers [2]. A large number of analytical and numerical models have been used to predict the location and movement of the saltwater/freshwater interface. The numerical models can be categorized as sharp-interface models or diffusive (dispersive) interface models. The first attempt to model seawater intrusion was made by Ghyben [14] and Herzberg [15]. This simple model is known as Ghyben-Herzberg model which assumes that saltwater and freshwater are immiscible and separated by a sharp interface. Henry [16] presented numerical solutions of steady saltwater intrusion based on the assumption of the sharp interface. This assumption treats the interface between saltwater and freshwater as a sharp and well-defined interface. However, in practice, the interface is not sharp, and the saltwater merges gradually with the freshwater by the process of mechanical dispersion. The width of the dispersion zone depends on the characteristics of the aquifer and the movement of water particles [1, 17]. The assumption of the sharp interface can be applied in some field conditions to obtain a first approximation to the freshwater pattern if the transition zone is relatively narrow. Otherwise, the dispersion process should be taken into account, and the entrainment of saltwater by the moving freshwater is necessary to describe the phenomena [1, 18, 19].

Extensive research is being carried out in many parts of the world with the objectives of understanding the mechanism of seawater intrusion and improving the methods to control it in order to protect groundwater resources in coastal aquifers. The developed numerical models are based on different concepts and can be categorized as [1]:

1. Sharp-interface or diffusive interface (density-dependent) models
2. Saturated or unsaturated flow models
3. Transient or steady flow models

4.4.1 Sharp-Interface or Diffusive Interface Models

Some mathematical models have been developed to simulate saltwater intrusion based on sharp-interface assumption. Recently, researchers have been focusing more on diffusive interface (density-dependent) models which are more realistic [1].

Sharp-Interface Models

Modelling of seawater intrusion in coastal aquifers is not a new subject. The initial model was developed independently by Ghyben [14] and Herzberg [15]. This simple model is known as Ghyben-Herzberg model and is based on the hydrostatic equilibrium between freshwater and saline water as shown in Fig. 3 and is described by Ghyben-Herzberg equation as

$$h_s = \left(\frac{\rho_f}{\rho_s - \rho_f}\right) h_f \qquad (1)$$

where ρ_s is the saltwater density, ρ_f is the freshwater density, h_s is the depth of interface below mean sea level, h_f is the height of the potentiometric surface above the mean sea level.

According to Ghyben-Herzberg equation, if the saltwater density is 1,025 kg/m^3 and freshwater density is 1,000 kg/m^3, this gives $h_s = 40h_f$.

Some mathematical and numerical models were developed to simulate saltwater intrusion based on sharp-interface assumption. Sbai et al. [20] used the finite element method for the prediction of saltwater intrusion under steady-state and transient conditions based on a sharp-interface model assuming the freshwater and saltwater to be immiscible. Bower et al. [21] presented an analytical model for saltwater upconing in a leaky confined aquifer. The model assumed the existence of

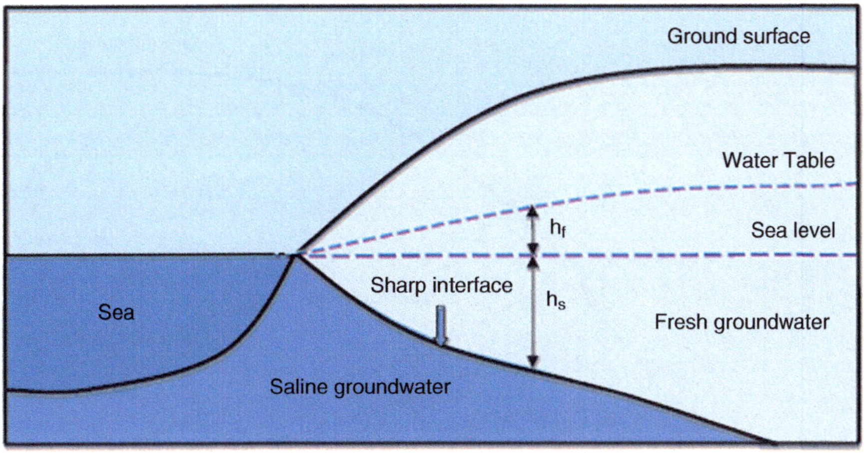

Fig. 3 Hydrostatic equilibrium between freshwater and saline water

a sharp interface between freshwater and saltwater. Aharmouch and Larabi [22] presented a numerical model, based on the sharp-interface assumption, to predict the location, shape and extent of the interface that occurs in coastal aquifer due to groundwater pumping. El Fleet and Baird [23] developed a 2D two-layer sharp-interface model to predict the impact of overabstraction from the coastal aquifer on saltwater intrusion and applied the model to the coastal aquifer in Tripoli, Libya.

Diffusive Interface Models

Saltwater and fresh groundwater are separated by either a sharp interface or a transition zone. Hydrodynamic dispersion and diffusion tend to mix the two fluids forming a transition zone at the interface as shown in Fig. 4. The transition zone depends on the extent of seawater intrusion and aquifer properties. This transition zone is a result of hydrodynamic dispersion of the dissolved matter and could be wide or narrow according to the aquifer depth. Steady flow minimizes the transition zone thickness, whereas pumping, recharge, tides and other excitations increase its thickness. The thickness of transition zones could range from a few meters to several kilometres in overpumped aquifers. The thickness of a transition zone depends on [1, 24]: structure of the aquifer, extraction from the aquifer and the variability of recharge, tides and climate change.

A number of mathematical models have been developed to simulate saltwater intrusion based on diffusive interface. Cheng et al. [25] developed a 2D finite element model for density-dependent flow and solute transport through saturated and unsaturated soils. Sakr [26] presented a finite element model to simulate density-dependent solute transport. He used the model to investigate the limitation of the sharp-interface approach in coastal aquifers for both steady and unsteady states.

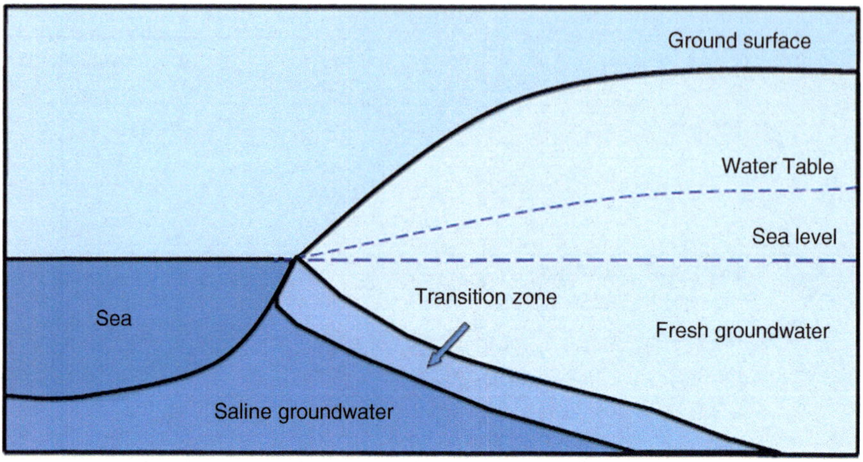

Fig. 4 Location of transition zone

A numerical model to study density-dependent groundwater flow and solute transport in unsaturated soil was developed by Jung et al. [27]. Gingerich and Voss [28] expanded a cross-sectional 2D model developed by Voss [29] to 3D. The developed model showed the effect of variable-density flow and solute transport on groundwater flow, freshwater/saltwater zone and salinity distribution. Abarca et al. [30] studied the effect of aquifer bottom topography on seawater intrusion, using a three-dimensional density-dependent groundwater flow and transport model. Lecca et al. [31] used a 3D density-dependent flow and contaminant transport code to improve the understanding of groundwater degradation mechanism within the complex hydrodynamic system in the coastal plain, Muravera, Italy. Doulgeris and Zissis [32] developed a 3D finite element model for the simulation of saltwater intrusion in coastal confined aquifers by considering the development of a transition zone between freshwater and seawater and thus density-dependent flow and transport. Shalabey et al. [33] presented a numerical model of advective and dispersive saltwater transport below a partially penetrating pumping well. The numerical solution included a solution of the coupled flow and solute transport equations.

4.4.2 Steady-State and Transient Flow Models

Numerous models have been presented to study coupled fluid flow and solute transport in porous media. Several researchers assumed steady-state solution for simplicity and avoided the transient simulation because of high computational requirements. Bixio et al. [34] developed a fully coupled steady-state flow and solute transport finite element model for the prediction of the maximum extent of saltwater intrusion in a vertical section in the Venice aquifer, Italy. Liu et al. [35] presented a 2D finite volume model to simulate saltwater intrusion in coastal aquifers, the flow was considered transient, and the variability of fluid density was taken into account. Bear et al. [36] developed a 3D model using the finite element framework. The model took into account the development of a transition zone and variation of fluid density with time.

4.4.3 Saturated and Unsaturated Flow Models

Various models were developed to simulate solute transport in saturated soil. However, few models were developed to simulate solute transport in the unsaturated zone. Teatini et al. [37] developed a finite element model for saturated flow and transport to simulate saltwater intrusion in shallow confined aquifers. Cau et al. [38] developed a 3D model supported by GIS to investigate the causes of saltwater intrusion, using coupled flow and solute transport in saturated porous media. The model was used to predict the movement of the freshwater-saltwater mixing zone under different future scenarios [1]. Paniconi et al. [39] developed a numerical model that treats density-dependent flow and miscible salt transport. The developed model was used to investigate the occurrence of seawater intrusion in the Korba plain in northern Tunisia.

4.4.4 Computer Codes Used to Simulate Saltwater Intrusion

A number of computer codes have been used to model coupled flow and solute transport and saltwater intrusion in coastal aquifers. The common computer codes used to investigate saltwater intrusion in coastal aquifers are presented below.

A 2D and 3D finite element code is called SUTRA. It can simulate density-dependent groundwater flow with energy transport or solute transport [29]. This code has been widely used to simulate variable-density groundwater flow. Narayan et al. [40] used SUTRA to define the current and potential extent of saltwater intrusion in the Burdekin delta aquifer under various pumping and recharge conditions. Ojeda et al. [41] modelled the saline interface of the Yucatan peninsula aquifer using SUTRA. The results showed that the interface position is very sensitive to head changes. The SEWAT which is a combination of MODFLOW and MT3DMS is designed to simulate 3D variable-density groundwater flow and solute transport [42]. This code was used by some researchers to simulate different cases of saltwater intrusion. Tiruneh and Motz [43] investigated the effect of sea level rise on the freshwater-saltwater interface, using SEWAT. Lakfifi et al. [44] applied SEWAT to study groundwater flow and saltwater intrusion in the coastal aquifer of Chaouia in Morocco. Qahman and Larabi [45] used SEAWAT code for simulating the spatial and temporal evolution of hydraulic heads and solute concentrations of groundwater in the Gaza aquifer (Palestine).

On the other hand, CODESA-3D is a 3D finite element model that can simulate saltwater interface in saturated and variably saturated porous media by solving the convective-dispersive transport equation. Barrocu et al. [46] used CODESA 3D with GIS to study the migration of contaminant and the impact of land management using different groundwater abstraction schemes and artificial recharge in Sardinia, Italy. Paniconi et al. [39] used CODESA-3D to investigate the occurrence of seawater intrusion in the Korba plain in northern Tunisia. Also, SWIFT is a 3D code to simulate groundwater flow, heat transfer and brine and radio nuclide transport in porous and fractured media [47, 48]. Ma et al. [49] used SWIFT to simulate saltwater upconing and its effect on the salinity of the pumped water.

A 3D code to simulate density-dependent groundwater flow and solute transport was developed and *MOCDENS3D* [50]. Oude Essink [51] used the code to model the movement of freshwater, brackish and saline groundwater in coastal aquifers including advection and dispersion. Barreto et al. [52] used a finite difference model called SHARP to simulate groundwater flow in aquifers and established a relationship between the pumping flow, the location of the wells, the number of the wells, the drop of freshwater heads and the movement of the interface. Zhou et al. [53] used a quasi-three-dimensional finite element to simulate the spatial and temporal distribution of groundwater levels in the three-aquifer system. Mahesha [54] used a quasi-three-dimensional finite element to examine the efficiency of a battery of injection wells in controlling saltwater intrusion. Scholze et al. [55] used SHEMAT code to simulate saltwater intrusion in the coastal aquifers of Metro Cebu. Some other codes have been used to simulate saltwater intrusion in coastal aquifers [e.g. FEFLOW, HST3D, FAST-C (2D/3D) and MOCDENSE] [3].

Recently, some computer codes have been used to model coupled flow and solute transport and saltwater intrusion in coastal aquifers. Abd-Elaty et al. [56, 57] used MODFLOW and SEAWAT to study the possible impacts of climatic change on seawater intrusion in Nile Delta aquifer using SEAWAT and concluded that the aquifer would be subjected to additional seawater intrusion. Abd-Elhamid et al. [58] presented a numerical simulation of seawater intrusion in Gaza aquifer using coupled transient finite element model considering different scenarios of sea level rise and overpumping using two models SEWAT and 2D-FEST.

5 Saltwater Intrusion in Egypt's Coastal Aquifer

Groundwater represents the main source of freshwater in many countries around the globe. This is typically true in arid and semiarid regions, where rainfall is scarce, random and insignificant and surface water bodies are limited and sometimes absent. Because of the high population density near shorelines and the tendency of people to live in and develop the coastal regions, coastal aquifers are exposed to extensive groundwater pumping. Egypt is located in the arid region and suffering from water scarcity (Fig. 5). Egypt has a large hydrogeologic potential with many aquifers widely distributed throughout the country. Egypt is facing increasing water demanded due to rapid population growth, urbanization and higher standards of living and by an agricultural policy which emphasizes expanded production to feed the growing population.

With the growing population, Egypt currently became a water-stressed country, and water is becoming a limiting factor for development. There are three dependent

Fig. 5 Location map of Egypt (http://www.123rf.com/stock-photo/egypt_map.html)

sources of water in Egypt: Nile water, groundwater and drainage water. The average annual rainfall is around 150 mm. Due to the scarcity and randomness, rainfall does not contribute to the water resources budget of the country. The water of the Nile River represents the only renewable water resource in the country. The Egyptian share from the Nile water, according to the international agreements, is limited to 55.5×10^9 m^3 per year since 1959 when the population was about 25 million. The per capita share of the renewable water is estimated at 650 m^3/year. Currently, about 5.5 billion m^3/year of groundwater is pumped from the different aquifers, 85% of which originates from the Nile water and the rest is fossil water. About 5 billion m^3/year of agricultural drainage water is being reused, while about 12 billion m^3 of low-quality drainage water is expelled to the Mediterranean Sea and the northern lakes [56, 57].

5.1 *Major Aquifers in Egypt*

The major groundwater system in Egypt consists of several aquifers as shown in Fig. 6. These aquifers are (El-Arabi [59]):

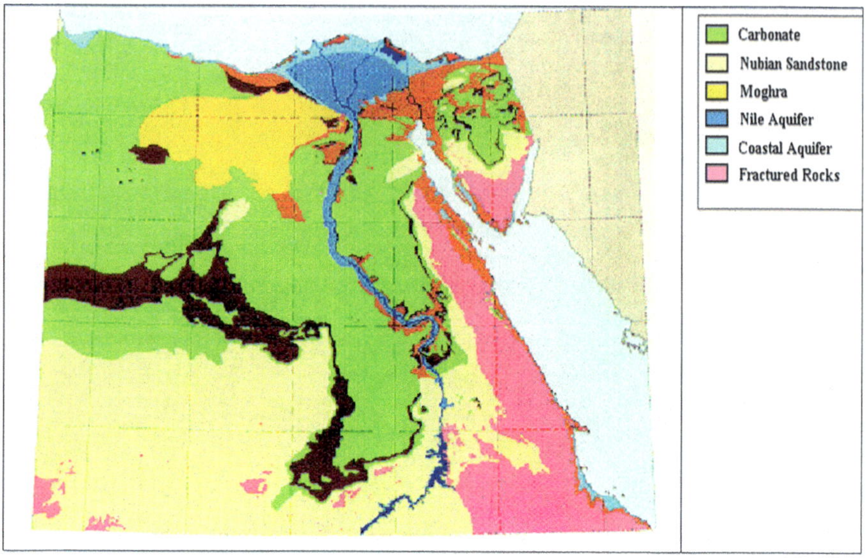

Fig. 6 The major aquifer system in Egypt [60]

1. *The Nile aquifer system*, assigned to the Quaternary and Late Tertiary, occupies the Nile floodplain and desert fringes.
2. *The Nubian Sandstone Aquifer System*, assigned to the Paleozoic-Mesozoic, occupies a large area in the Western Desert and parts of the Eastern Desert and Sinai.
3. *The Moghra aquifer system*, assigned to the Lower Miocene, occupies mainly the western edge of the Delta.
4. *The coastal aquifer systems*, assigned to the Quaternary and Late Tertiary, occupy the northwestern and eastern coasts.
5. *The karstified carbonate aquifer system*, assigned to the Eocene and to the Upper Cretaceous, predominates essentially in the north and middle parts of the Western Desert.
6. *The fissured and weathered hard rock aquifer system*, assigned to the Precambrian, predominates in the Eastern Desert and Sinai.

5.2 The Nile Delta Aquifer

The Nile Delta represents the most fertile land in Egypt. However, less than two-thirds of its area is under cultivation. The uncultivated lands are found along the sea to the north and in the vicinity of Delta fringes close to the desert. Irrigation is mostly practiced by surface water through an intensive network of canals branching out from the Nile River. The growth of population in the Nile Delta region along with the increase in agricultural and industrial activities has imposed an increasing demand for freshwater. This increase in demand in the Delta is covered by intensive pumping of fresh groundwater, causing subsequent lowering of the piezometric head and upsetting the dynamic balance between freshwater and saline water bodies in the aquifer. Like any coastal aquifer, an extensive saltwater body has intruded the Nile Delta aquifer forming the major constraint against aquifer exploitation.

The groundwater in the Nile Delta aquifer fills a vast underground bowl situated between Cairo and the Mediterranean Sea. If it were not for the presence of a saline body of seawater along the bottom of this bowl, the Nile Delta aquifer could be easily exploited to the best interests of Egypt. Seepage from the Nile River, canals and the surplus of irrigation water percolate downward, representing the main source of aquifer recharge. It may also be recharged, nominally, by northward flow from the Nile Valley aquifer. On the other hand, the aquifer loses some of its freshwater to the Mediterranean Sea and the drainage system in the northern part of the Delta.

The Nile Delta aquifer, Fig. 7, is among the largest underground freshwater reservoirs in the world [61]. Previous investigations revealed that the seawater intrusion in the Nile Delta aquifer has extended to a distance of more than 100 km from the Mediterranean coast as shown in Fig. 8. This has left only a small triangular zone of freshwater, extending about 40 km north of the apex of the Delta. The risks of the migration of the seawater and possible upconing of lower quality groundwater in the deeper zones have constraint the freshwater volumes that can be safely withdrawn.

The Nile Delta aquifer system forms an immense and complex groundwater system. It is a leaky Pleistocene aquifer system overlain by a semi-pervious Holocene aquitard (clay cap) and underlain by an impermeable Miocene aquiclude. The clay cap (aquitard), being the main source through which aquifer recharge occurs, is of great importance. Based on the thickness and vertical permeability of the upper confining clay layer, the aquifer system is identified as confined, semi-confined or phreatic. On the contrary, the aquiclude layer has no hydrologic importance, except that it forms an impermeable bottom and defines the aquifer geometry. The bulk of the Nile Delta aquifer consists of deltaic deposits 200–300 m thick in average. It is dominated by unconsolidated coarse sand and gravel with occasional clay lenses. The top boundary of the deltaic deposits, which acts as a cap for the aquifer, is a semi-pervious clay and silt layer. The thickness of the upper clay cap increases uniformly towards the Mediterranean Sea from about 5 m near Cairo and reaching about 70 m near the shoreline. The clay cap is intermeshing with the aquifer near the shore.

The Nile Delta with its fringes area expands over an area of about 22,000 km^2. It lies between latitudes 30° 25′ and 31° 30′ North and longitudes 29° 50′ and 30° 15′ East. The Nile Delta has its apex about 20 km north of Cairo. The length of its base at the Mediterranean Sea is about 245 km. The length of its right branch (Damietta) is about 240 km and that of the left branch (Rosetta) is about 235 km (see Fig. 8). Water flowing through the Damietta and Rosetta branches accounts for less than half of the original total northward flow across the delta plain that is now captured by artificial waterways comprising over 10,000 km of canals, usually only 2–3 km

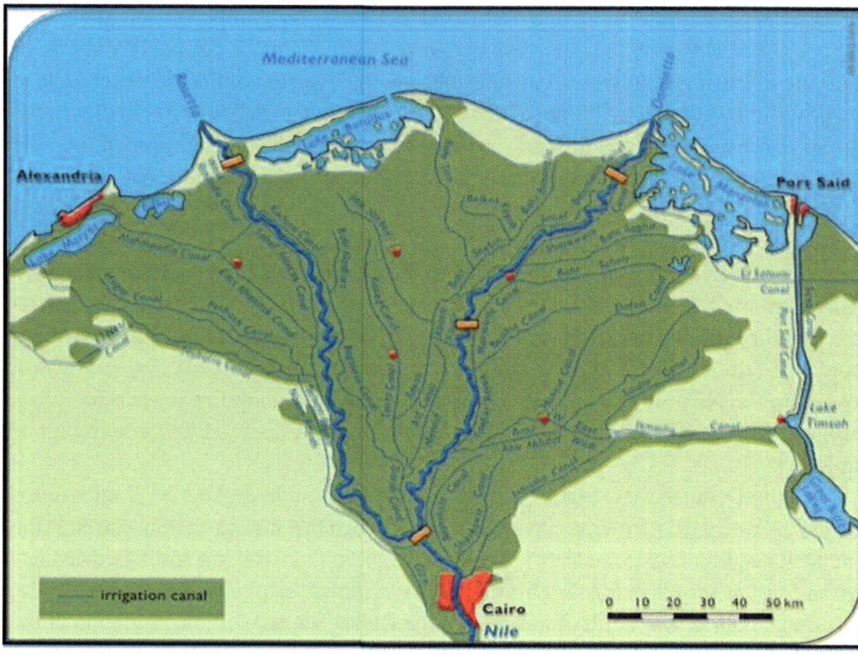

Fig. 7 Location map of the Nile Delta aquifer, Egypt [62]

apart [64]. This huge surface water network interacts with the groundwater system of the Nile Delta.

Horizontal and vertical dimensions of the Nile Delta aquifer were obtained using deep oil borings and data from test bore holes [65]. The thickness of the Pleistocene aquifer increases towards the Mediterranean Sea and decreases southward, almost vanishing near El Manawat and partly separating the Nile Delta aquifer from the Nile Valley aquifer. South of Tanta in the transversal direction, the bottom boundary forms a concave shape. The aquifer thickness decreases eastward and southward with a maximum depth in the middle of the Delta. North of Tanta, the maximum thickness of the aquifer shifts towards the east. The clay cap takes different profiles along the shore of the Mediterranean Sea. It is divided into two layers. The first layer is the upper clay cap which acts as an aquitard, about 20 m thick, with low permeability. The second layer is the lower clayey sand layer, a few metres thick, with a higher permeability. The thickness of the upper clay cap (aquitard) and the clayey sand layer are well defined at many locations in the Delta area. Their thicknesses are mostly less than 20 m and 15 m, respectively, except in the vicinity of the shoreline. No clay cap exists at the Delta fringes.

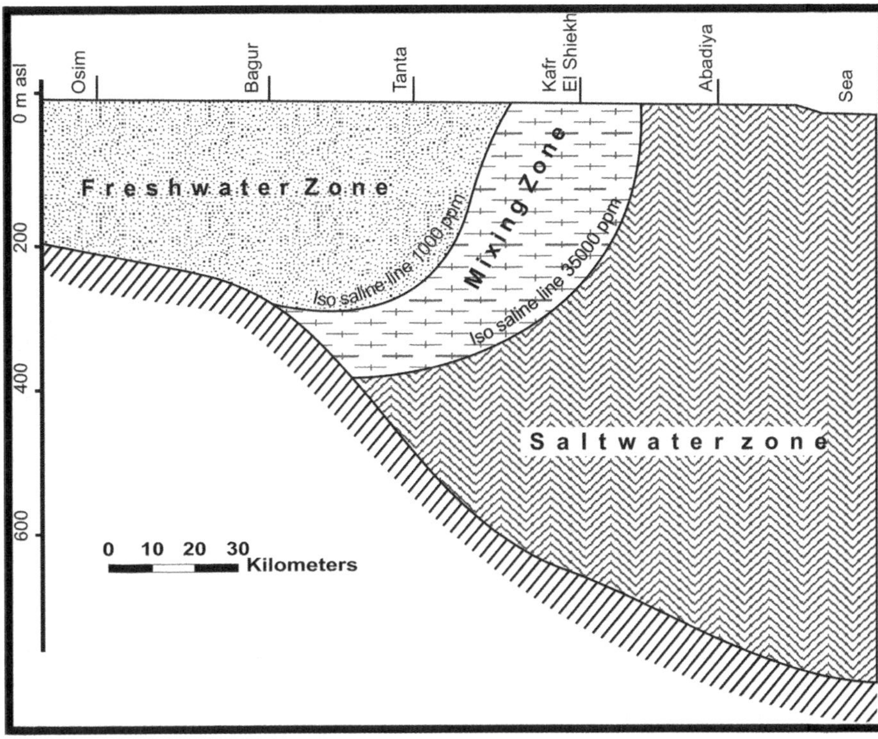

Fig. 8 A cross-section in the Delta region from north to south [63]

5.3 *Investigation of Saltwater Intrusion in Egypt*

The Nile Delta aquifer in Egypt is one of these aquifers which subject to sever seawater intrusion from the Mediterranean Sea. The Nile Delta aquifer is among the largest underground freshwater reservoirs in the world [66]. Water resources in the Nile Delta region are threatened with a number of factors including overpopulation, overpumping, land subsidence, coastal erosion and contamination due to anthropogenic activities and salinization due to seawater intrusion. All these factors are responsible for making the water resources at the Nile Delta at risk and the management of these resources a necessity.

Investigations of the Nile Delta aquifer have shown that saltwater intrusion has been extended to a distance more than 100 km from the Mediterranean coast. The alluvial graded sandy aquifer underlying the Delta contains a huge amount of water; the amounts that can be safely withdrawn are governed by their present quality or potential quality deterioration by seawater intrusion, saltwater upconing and/or general contamination. A number of studies were conducted to simulate seawater intrusion in the Nile Delta aquifer using different numerical techniques. Examples of these studies include among others [67–73].

Sherif and Al-Rashed [74] studied groundwater quality in the Northern parts of the Delta. The study approved that the aquifer has been deteriorated considerably due to the excessive pumping over the last few decades and any additional pumping should be practiced in the Middle Delta and pumping from the Eastern and Western parts should be reduced. Sherif and Hamza [75] developed a finite element model (2D-FED) to simulate the effect of pumping brackish water from the transition zone to restore the balance between freshwater and saline water in the coastal aquifer. They considered dispersion and variable-density flow in the modelling processes. They proposed different alternative solutions to use the abstracted saline water for some purposes such as cooling, desalting, injection into deep wells, irrigation or disposal to the sea. El-Arabi [76] used Visual MODFLOW to study the environmental impacts of new settlements on the groundwater. The model was used to simulate the groundwater behaviour and the migration of the pollution plum under the initiated industrial and agriculture activities. Morsy [77] used Visual MODFLOW and solute transport model MT3D to evaluate the future potential quantitative and qualitative impacts of the proposed national water policy on the aquifer system of the Nile Delta. Hendy [78] used Visual MODFLOW to study groundwater management at the north of Sharkia. The study indicated that increasing the abstraction rate will lead to more saltwater intrusion. Sherif et al. [79] discussed the concept of the equivalent freshwater head in successive horizontal simulations of seawater intrusion in Nile Delta using FEFLOW (a 3D finite element variable-density model). Their results demonstrated that the location of the transition zone moves towards land side as moving down with depth.

Abd-Elaty [13] carried out a numerical and experimental study for impacts of climate change and sea level rise on the Nile Delta aquifer using MODFLOW and SEAWAT. The study proved that increasing seawater level increased the intrusion of saline water in the aquifer. Abd-Elaty et al. [56] studied the impact of climatic change and sea level rise on saltwater intrusion in Nile Delta aquifer using SEAWAT using VISUAL MODFLOW. The results showed that increasing sea level, decreasing surface water and increasing extraction rate have a significant effect on groundwater level, but the combination of these scenarios will have a very dangerous effect on water resources in Nile Delta aquifer. Also, Abd-Elaty et al. [57] used 3D model (SEAWAT) to study seawater intrusion in Nile Delta aquifer considering different scenarios including the increase of sea level, decrease of the surface water system, increase of the extraction rate and combination of the three scenarios. The results showed that these changes could affect groundwater level and loss large quantity of freshwater in the Nile Delta aquifer due to saltwater intrusion.

Recently, a limited number of studies were conducted for the environmental management of groundwater resources in the Nile Delta region. They mainly considered the impact of sea level rise. Mohamed [80] studied the global and local factors affecting the rise of sea level and its impact on coastal regions and specifically deltaic zones. He assessed the impact of such factors on northern coasts of Egypt extending from Alexandria to El-Arish for the years 2020 and 2050. He used the artificial neural network (ANN) to estimate the change in seawater level from available data of tide gauges and weather stations in the vicinity of the northern

Egyptian coasts. Nassar [81] conducted a numerical and experimental study to assess the environmental impacts of abstracting brackish water for desalination purposes and recharging the brine water through deep wells to control the seawater intrusion into the Nile Delta aquifer. A number of design charts for abstraction and recharge wells including discharge and recharge rates and locations and depths of pumping and injection wells were presented. Saleh [82] developed a numerical model using MODFLOW and MT3D to quantitatively and qualitatively assess the possible impacts of the policies related to groundwater extraction and development activities in the Nile Delta region. Sefelnasr and Sherif [73] studied the impacts of seawater rise on seawater intrusion in the Nile Delta aquifer, Egypt; they used FEFLOW, a 3D finite element variable-density model. Negm et al. [83] used MODFLOW to evaluate the impact of irrigation canals covering on groundwater in the Nile Delta through comparing the cases of covering and non-covering at Abu Kebir city, Sharkia, Egypt. The results proved that the covering process has an insignificant effect on groundwater levels. The differences in groundwater level were 0.04 and 0.03 m as an average value for the cases of maximum and minimum water levels consecutively.

Abd-Elhamid et al. [84] presented a coupled transient finite element model for simulation of fluid flow and solute transport in saturated and unsaturated soils (2D-FEST) and employed to study seawater intrusion in the Nile Delta aquifer. The developed model was used to investigate the problem considering the possible impacts of climate change. The results were compared with SEAWAT code, and good agreement was obtained. Figures 9, 10 and 11 show a real view of groundwater levels in Nile Delta aquifer; horizontal and vertical distributions of TDS in Nile Delta aquifer were considered. Three scenarios were studied, rise in sea level, a decline of the piezometric head at the land side due to excessive pumping and the combination of sea level rise and decline of the piezometric head at the land side. The results showed that the rise in the sea level has a significant effect on the position of the transition zone. The third scenario represents the worst case under which the groundwater quality would deteriorate in large areas of the Nile Delta aquifer.

6 Summary and Conclusion

Most of the research that had been carried out to simulate saltwater intrusion in the Nile Delta aquifer were reviewed in this chapter. This chapter discussed some interesting topics including groundwater contamination, saltwater intrusion types and causes and investigation and modelling of saltwater intrusion. It is concluded that seawater intrusion represents a severe risk to groundwater resources in coastal aquifers. It raises the water salinity to levels exceeding the acceptable drinking water standards. Salinization of groundwater is considered one of the main pollutants in coastal aquifers. The Nile Delta aquifer is among the largest underground freshwater reservoirs in the world. Previous investigations revealed that the seawater intrusion

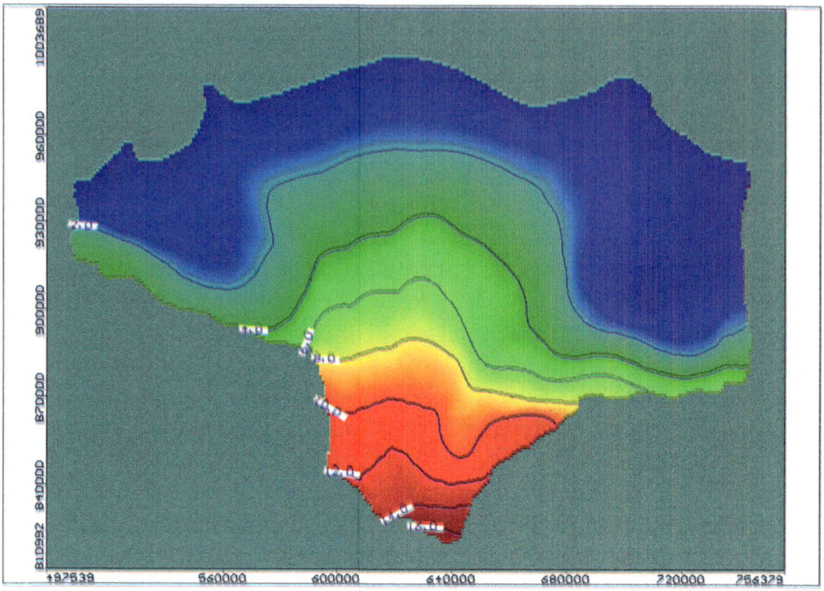

Fig. 9 Areal view of groundwater levels in Nile Delta aquifer

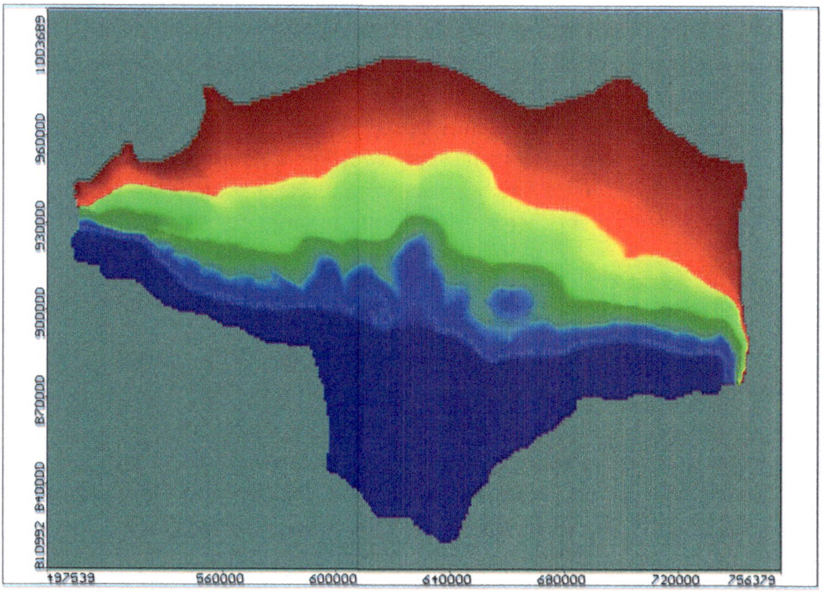

Fig. 10 Horizontal distribution of TDS in Nile Delta aquifer

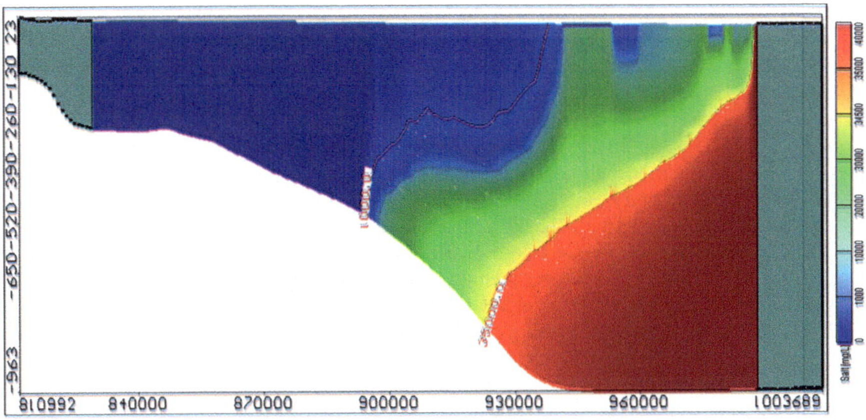

Fig. 11 Vertical distribution of TDS in Nile Delta aquifer

in the Nile Delta aquifer has extended to a distance of more than 100 km from the Mediterranean coast.

7 Recommendations

Several methods are available to investigate the control of saltwater intrusion into the coastal aquifers. These methods should be tested using the numerical simulation to decide its technical feasibility to apply to control the saltwater intrusion to the Nile Delta aquifer. Also, a comprehensive feasibility studies should be conducted before implementing any of the technically feasible methods.

References

1. Abd-Elhamid HF (2010) A simulation-optimization model to study the control of seawater intrusion in coastal aquifers. Ph.D. Thesis, College of Engineering, Mathematics and Physical Sciences, University of Exeter
2. Bear J (1979) Hydraulics of groundwater. McGraw-Hill, New York
3. Bear J, Cheng, AH, Sorek S, Quazar D, Herrera I (1999) Seawater intrusion in coastal aquifers, concepts, methods and practices. Kluwer Academic, Dordrecht. ISBN 0-7923-5573-3
4. Fitterman DV, Deszcz-Pan M (2001) Saltwater intrusion in Everglades National Park, Florida measured by airborn electromagnetic surveys. In: Proceeding of the 1st international conference and workshop on saltwater intrusion and coastal aquifers, monitoring, modelling, and management
5. Ekwurzel B, Moran JE, Hudson GB, Bissani M, Blake R, Krimissa M, Mosleh N, Marah H, Safsaf N, Hsissou Y, Bouchaou M (2001) An isotopic investigation of salinity and water sources in the Souss-Massa Basin, Morocco. In: Proceeding of the 1st international conference and workshop on saltwater intrusion and coastal aquifers, monitoring, modelling, and management

6. Sherif M, Kacimov A, Ebraheem A, AlMulla M (2010) Three-dimensional mapping of seawater intrusion using geophysical methods. In: Proceedings of the world environmental and water resources congress 2010
7. Edet AE, Okereke CS (2001) A regional study of saltwater intrusion in south eastern Nigeria based on the analysis of geoelectrical and hydrochemical data. Environ Hydrol 40:1278–1289
8. Allen DM, Mastuo G, Suchy M, Abbey DG (2001) A multidisciplinary approach to studying the nature and occurrence of saline groundwater in the gulf islands, British Columbia, Canada. In: Proceeding of the 1st international conference and workshop on saltwater intrusion and coastal aquifers, monitoring, modelling, and management
9. Sherif M, El Mahmoudi A, Garamoon H, Kacimov A, Akram S, Ebraheem A and Shetty A (2005) Geoelectrical and hydrogeochemical studies for delineating seawater intrusion in the outlet of Wadi Ham, UAE, Environtal Geology
10. Koch M, Starke B (2001) Experimental and numerical investigation of macrodispersion of density-dependent flow and transport in stochastically heterogeneous media. In: Proceeding of the 1st international conference and workshop on saltwater intrusion and coastal aquifers, monitoring, modelling, and management
11. Van Meir N, Jaeggi D, Herfort M, Low S, Lods G, Pezard P, Gouze P (2004) Development of a test site for the investigation of saltwater intrusion process in a karstified limestone aquifer (Campos Mallorca, Spain). In: Proceeding of the 18th salt water intrusion meeting, Cartagena
12. Chang SW, Clement TP (2012) Experimental and numerical investigation of saltwater intrusion dynamics in flux-controlled groundwater systems. Water Resour Res J 48:1–10
13. Abd-Elaty IM (2014) Numerical and experimental study for simulating climatic changes effects on Nile Delta aquifer. Ph.D. Thesis, Zagazig University
14. Ghyben BW (1889) Nota in verband met de voorgenomen put boring Nabij Amsterdam. Tydscrift Van Het Koninkyky Institute Van Ingenieurs, The Hahue, p 21
15. Herzberg A (1901) Die wasserversorgung einnger nordsecbader. Journal Gasbeleuchtung U. Wasservesurg, Jahrg 44, Munich, pp 815–819
16. Henry HR (1959) Sat intrusion into fresh-water aquifer. J Geophys Res 64:1911–1919
17. Cooper HH (1959) A hypothesis concerning the dynamic balance of fresh water and saltwater in a coastal aquifer. J Geophys Res 64:461–467
18. Kohout FA (1960) Cyclic flow of fresh water in the Biscayne aquifer of southeastern Florida. J Geophys Res 65:2133–2141
19. Kohout FA (1960) Flow pattern of fresh water and saltwater in the Biscayne aquifer of the Miami, Florida. Int Assoc Sci Hydrol Publ 52:440–448
20. Sbai MA, Larabi A, De Smedt F (1998) Modelling saltwater intrusion by a 3-D sharp interface finite element model. Trans Ecol Environ 17:1743–3541
21. Bower JW, Motez LH, Durden DW (1999) Analytical solution for the critical conditions of saltwater upconing in a leaky artesian aquifer. J Hydrol 221:43–54
22. Aharmouch A, Larabi A (2001) Numerical modelling of saltwater interface upconing in coastal aquifers. In: Proceeding of the 1st international conference and workshop on saltwater intrusion and coastal aquifers, monitoring, modelling, and management
23. El Fleet M, Baird J (2001) The development and application of groundwater models to simulate the behaviour of groundwater resources in the Tripoli aquifer, Libya. In: Proceeding of the 1st international conference and workshop on saltwater intrusion and coastal aquifers, monitoring, modelling, and management
24. Todd DK (1974) Salt-water intrusion and its control. Water technology/resources. J Am Water Works Assoc 66(3):180–187
25. Cheng J, Strobl RO, Yeh G, Lin H, Choi W (1998) Modelling of 2D density-dependant flow and transport in the subsurface. J Hydrol Eng 3(4):248–257
26. Sakr SA (1999) Validity of a sharp-interface model in a confined coastal aquifer. Hydrol J 7:155–160

27. Jung B, Kim J, Chang H (2002) Finite element modeling of density-dependent groundwater flow and solute transport in the unsaturated layered coastal aquifer system. In: Proceeding of the 17th salt water intrusion meeting, Delft
28. Gingerich SB, Voss CI (2002) Three-dimensional variable-density flow simulation of coastal aquifer in southern Oahu, Hawaii, USA. In: Proceeding of the 17th salt water intrusion meeting, Delft
29. Voss CI (1984) SUTRA, a finite element simulation for saturated-unsaturated fluid density-dependant groundwater flow with energy transport or chemically reactive single species solute transport. U S Geol Surv Water Resour Invest Rep 84:4369:409
30. Abarca E, Carrera J, Voss CI, Sanchez-Vila X (2002) Effect of aquifer bottom morphology on seawater intrusion. In: Proceeding of the 17th salt water intrusion meeting, Delft
31. Lecca G, Cau P, Ardau F (2004) Modelling study of the seawater intrusion in the muravera coastal plain (SE Sardinia, Italy). In: Proceeding of the 18th salt water intrusion meeting, Cartagena
32. Doulgeris C, Zissis T (2005) Simulation of seawater intrusion in three-dimensional confined aquifers. Geophys Res 7:02780
33. Shalabey ME, Kashyap D, Sharma A (2006) Numerical model of saltwater intrusion transport toward a pumping well. J Hydrol Eng (ASCI) 11(4):306–318
34. Bixio A, Putti M, Tosi L, Carbognin L, Gambolati G (1998) Finite element modelling of saltwater in the Venice aquifer system. Trans Ecol Environ 17:1743–3541
35. Liu F, Turner I, Anh V (2001) A finite volume unstructured mesh method for modelling saltwater intrusion into aquifer systems. In: Proceeding of the 1st international conference and workshop on saltwater intrusion and coastal aquifers, monitoring, modelling, and management
36. Bear J, Zhou Q, Bensabat J (2001) Three dimensional simulation of seawater intrusion in heterogeneous aquifers, with application to the coastal aquifer of Israel. In: Proceeding of the 1st international conference and workshop on saltwater intrusion and coastal aquifers, monitoring, modelling, and management
37. Teatini P, Gambolati G, Gonella M, Brunone M, Ferrante M, Marconi S (2001) Modelling seawater intrusion in the protorecanati, Italy. In: Proceeding of the 1st international conference and workshop on saltwater intrusion and coastal aquifers, monitoring, modelling, and management
38. Cau P, Lecca G, Muscas L, Barrocu G, Uras G (2002) Seawater intrusion in the plain of Oristano (Sardinia, Italy). In: Proceeding of the 17th salt water intrusion meeting, Delft
39. Paniconi C, Khlaifi I, Lecca G, Giacomelli A, Tarhouni J (2001) A modelling study of seawater intrusion in the Korba plain, Tunisia. Phys Chem Earth B 26(4):345–351
40. Narayan KA, Schleeberger C, Charlesworth PB, Bristow KL (2002) Modelling saltwater intrusion in the lower Burdekin Delta north Queensland, Australia. In: Denver Annual Meeting, The Geological Society of America (GSA)
41. Ojeda CG, Gallardo P, Hita LG, Arteaga LM (2004) Saline interface of the Yucatan peninsula aquifer. In: Proceeding of the 18th salt water intrusion meeting, Cartagena
42. Zheng C, Wang PP (1999) MT3DMS, a modular three-dimensional multi-species transport model for advection, dispersion and chemical reactions of contamination and user's guide. US. AERDCC, Report SERDP-99-1
43. Tiruneh ND, Motz LH (2004) Three dimensional modelling of saltwater intrusion coupled with the impact of climate changes and pumping. In: Proceeding of World Water Congress
44. Lakfifi L, Larabi A, Bzioui M, Benbiba M, Lahmouri A (2004) Regional model for seawater intrusion in the chaouia coastal aquifer (Morocco). In: Proceeding of the 18th salt water intrusion meeting, Cartagena
45. Qahman K, Larabi A (2004) Three dimensional numerical models of seawater intrusion in Gaza aquifer, Palatine. In: Proceeding of the 18th salt water intrusion meeting, Cartagena
46. Barrocu G, Cau P, Muscas L, Soddu S, Uras G (2004) Predicting groundwater salinity changes in the coastal aquifer of Arborea (central-western Sardinia). In: Proceeding of the 18th salt water intrusion meeting, Cartagena
47. Ward DS, Benegar J (1998) Data input guide for SWIFT-98, version 2.57. HIS Geotrans, Inc., Sterling, VA

48. Ward DS (1991) Data input for SWIFT/386, version 2.5. Geotrans Technical Report, Sterling, VA
49. Ma Tain S, Sophocleous M, Yu Y, Buddemeier RW (1997) Modelling saltwater upconing in a freshwater aquifer south-central Kansas. J Hydrol 201:120–137
50. Oude Essink GH (1998) MOC3D adapted to simulate 3-D density-dependant groundwater flow. In: Proceeding of MODFLOW-98 conference, pp 291–303, Golden, CO
51. Oude Essink GH (2001) Saltwater intrusion in 3D large-scale aquifers: a Dutch case. Phys Chem Earth B 26:337–344
52. Barreto J, Nuno MA, Haie N (2001) Modelling of rural usage of groundwater subject to saltwater intrusion. In: Proceeding of the 1st international conference and workshop on saltwater intrusion and coastal aquifers, monitoring, modelling, and management, (Morocco)
53. Zhou X, Chen M, Liang C (2003) Optimal schemes of groundwater exploitation for prevention of seawater intrusion in the Leizhou Peninsula in southern China. Environ Geol 43:978–985
54. Mahesha A (1996) Control of seawater intrusion through injection-extraction well system. J Irrig Drain Eng (ASCE) 122:314–317
55. Scholze O, Hillmer G, Schneider W (2002) Protection of the groundwater resources of Metropolis CEBU (Philippines) in consideration of saltwater intrusion into the coastal aquifer. In: Proceeding of the 17th salt water intrusion meeting, Delft
56. Abd-Elaty IM, Abd-Elhamid HF, Fahmy MR, Abdelaal GM (2014) Investigation of some potential parameters and its impacts on saltwater intrusion in Nile Delta aquifer. J Eng Sci 42 (4):931–955
57. Abd-Elaty IM, Abd-Elhamid HF, Fahmy MR, Abdelaal GM (2014) Study of impact climate change and other on groundwater system in Nile Delta aquifer. Egypt J Eng Sci Technol 17 (4):2061–2079
58. Abd-Elhamid HF, Javadi AA, Qahman K (2015) Impact of over-pumping and sea level rise on seawater intrusion in Gaza aquifer. J Water Clim Chang 6(4):891–902
59. El-Arabi N (2012) Environmental management of groundwater in Egypt via artificial recharge extending the practice to soil aquifer treatment (SAT). Int J Environ Sustain 1(3):66–82. ISSN 1927-9566
60. RIGW (Research Institute for Groundwater) (1999) Environmental management of groundwater resources in Egypt. Final report
61. Sherif MM, Singh VP, Amer AM (1990) A note on saltwater intrusion in coastal aquifers. Water Resour Manag 4:123–134
62. MWRI (2013) Adaptation to climate change in the Nile Delta through integrated coastal zone management. Ministry of Water Resources and Irrigation
63. RIGW (Research Institute for Groundwater) (1980) Project of safe yield study for groundwater aquifers in the Nile Delta and Upper Egypt. Part 1. Ministry of Irrigation, Academy of Science Research and Technology, and Organization of Atomic Energy, Egypt
64. Kashef A (1976) Control of saltwater intrusion by recharge wells. J Irrig Drain Div 102:445–456
65. Farid MS (1980) Nile Delta groundwater study. M.Sc. Thesis, Faculty of Engineering, Cairo University
66. Sherif MM, Singh VP (1999) Effects of climate change on sea water intrusion in coastal aquifers. Hydrol Process J 13:1277–1287
67. Farid MS (1985) Management of groundwater system in the Nile Delta. Ph.D. Thesis, Cairo University
68. Sherif MM, Singh VP, Amer AM (1988) A two-dimensional finite element model for dispersion (2D-FED) in coastal aquifers. J Hydrol 103:11–36
69. Darwish MM (1994) Effect of probable hydrological changes on the Nile Delta aquifer system. Ph.D. Thesis, Faculty of Engineering, Cairo University, Cairo
70. Amer A, Sherif MM (1996) An integrated study for seawater intrusion in the Nile Delta aquifer. Working Paper for SRP. NWRC-MPWWR, Cairo

71. Amer A, Farid MS (1981) About sea water intrusion phenomenon in the Nile Delta aquifer. In: Proceedings of the international workshop on management of the Nile Delta groundwater aquifer. CU/Mit, Cairo
72. Sherif MM, Singh VP (1997) Groundwater development and sustainability in the Nile Delta aquifer. Final Report, Binational Fulbright Commission, Egypt
73. Sefelnasr A, Sherif MM (2014) Impacts of seawater rise on seawater intrusion in the Nile Delta aquifer, Egypt. J Groundwater 52(2):264–276
74. Sherif MM, Al-Rashed MF (2001) Vertical and horizontal simulation of seawater intrusion in the Nile Delta Aquifer. In: Proceeding of the 1st international conference and workshop on saltwater intrusion and coastal aquifers, monitoring, modelling, and management
75. Sherif MM, Hamza KI (2001) Mitigation of seawater intrusion by pumping brackish water. J Transp Porous Media 43(1):29–44
76. El-Arabi M (2007) Environmental impact of new settlements in groundwater in a region in the Nile Delta. M.Sc. Thesis, Faculty of Engineering, Zagazig University
77. Morsy WS (2009) Environmental management to groundwater resources for Nile Delta region. Ph.D. Thesis, Faculty of Engineering, Cairo University
78. Hendy M (2012) Study groundwater management at the north of Sharkia Directorate's Zagazig University. M.Sc. Thesis, Faculty of Engineering, Zagazig University
79. Sherif MM, Sefelnasr A, Javad A (2012) Incorporating the concept of equivalent freshwater head in successive horizontal simulations of seawater intrusion in the Nile Delta aquifer. Egypt J. Hydrol 465:186–198
80. Mohamed MA (2004) Seawater level variation and its estimation along the Northern Egyptian coasts using Artificial Neural Network model. Ph.D. Thesis, Alexandria University
81. Nassar MK (2004) Design and management of discharge and recharge well field for coastal desalination plants. M.Sc. Thesis, Cairo University
82. Saleh WS (2009) Environmental management of groundwater resources in the Nile Delta region. Ph.D. Thesis, Cairo University
83. Abdel Azim S, Negm AM, Abd-elhamid HF, Abdelaal GM, Nassar M (2015) The impact of irrigation canals covering on groundwater in the Nile Delta, a case study: Abu Kebier city, Sharkia, Egypt. Int Water Technol J 5(3):222–234
84. Abd-Elhamid HF, Javadi AA, Abdelaty IM, Sherif MM (2016) Simulation of seawater intrusion in the Nile Delta aquifer under the conditions of climate change. J Hydrol Res 47(5):1–14

Control of Saltwater Intrusion in Coastal Aquifers

Hany F. Abd-Elhamid, Ismail Abd-Elaty, and Abdelazim M. Negm

Abstract Seawater intrusion occurs in many coastal and deltaic areas around the world. When saltwater travels inland to production wells, underground water supplies become useless. Intrusion of saltwater is the most common contamination occurrence in coastal aquifers. A number of several methods have been used to control seawater intrusion to protect groundwater reserves in coastal aquifers. Extensive research has been carried out to investigate saltwater intrusion in coastal aquifers. Although some research has been done to investigate saltwater intrusion, however, only a limited amount of work has concentrated on the control of saltwater intrusion to protect groundwater resources in coastal areas which represent the most densely populated areas in the world, where 70% of the world's population live. The coastal aquifers' management requires careful planning of withdrawal strategies for control of saltwater intrusion. Therefore, efficient control of seawater intrusion is very important to protect groundwater resources from depletion. New methods to control saltwater intrusion in coastal aquifers are presented and discussed in details; also the advantages and disadvantages of each method were highlighted. Finally, control of saltwater intrusion in Egypt, especially in the Nile Delta aquifer, is discussed. The possibility of applying new methods to control saltwater intrusion in Egypt is presented.

Keywords Coastal aquifers, Control, Egypt, Modeling, Nile Delta, Saltwater intrusion

Contents

H.F. Abd-Elhamid (✉), I. Abd-Elaty, and A.M. Negm
Water and Water Structures Engineering Department, Faculty of Engineering, Zagazig University, Zagazig, Egypt
e-mail: hany_farhat2003@yahoo.com; eng_abdelaty2006@yahoo.com; amnegm@zu.edu.eg; amnegm85@yahoo.com

A. M. Negm (ed.), *Groundwater in the Nile Delta*,
Hdb Env Chem (2019) 73: 355–384, DOI 10.1007/698_2017_138,

1 Introduction

Seawater intrusion is often a major constraint to optimal use of fresh groundwater from coastal aquifers. Excessive groundwater abstraction to meet growing demands from an increasing coastal population and the expected rise in mean sea level from global warming will cause seawater to encroach farther inland and threaten the available groundwater supply. Seawater intrusion occurs in many coastal and deltaic areas around the world. When saltwater travels inland to production wells, underground water supplies become useless. In agriculture, this could cause soil salinity problems resulting to poor crop yields and the substitution of more salt-tolerant crops over indigenous crops [1].

The contaminated aquifer is sometimes abandoned resulting in the loss of a precious groundwater resource. Thus, present-day and future water supply engineers and managers face the challenges of optimal exploitation of fresh groundwater and the control of seawater intrusion [2]. There are different types of pollutants that can be found in groundwater, such as nitrate, heavy metals, and saltwater. Intrusion of saltwater is the most common contamination occurrence in coastal aquifers [3]. Intrusion of saltwater occurs when saltwater displaces freshwater in an aquifer. The phenomenon can occur in deep aquifers with the advance of saline waters of geologic origin, in shallow aquifers from surface waste discharge, and in coastal aquifers from the invasion of seawater [4].

New methods to control saltwater intrusion in coastal aquifers are presented and discussed in details; also the advantages and disadvantages of each method were highlighted. Finally, investigation and control of saltwater intrusion in Egypt, especially in the Nile Delta aquifer, are discussed. The possibility of applying new methods to control saltwater intrusion in Egypt is presented.

2 Saltwater Intrusion Control Methods

A number of methods had been adopted to control seawater intrusion to protect groundwater reserves in coastal aquifers. Todd [5] presented various methods of preventing seawater from contaminating groundwater sources including:

1. Reduction of pumping rates
2. Relocation of pumping wells
3. Use of subsurface barriers
4. Natural recharge
5. Artificial recharge
6. Abstraction of saline water
7. Combination techniques

A number of numerical models had been developed and used to help in understanding the relevant process that causes saltwater intrusion in coastal aquifers and identify suitable methods of control. Extensive research has been carried out to investigate saltwater intrusion in coastal aquifers. However, only limited amount of research has been directed to study the control of saltwater intrusion. The following sections present the traditional methods that have been used to control saltwater intrusion in different locations. Also, the advantages and disadvantages of each one are discussed [6].

2.1 Reduction of Pumping Rates

The natural balance between freshwater and saltwater in coastal aquifers is disturbed by abstraction and other human activities that lower groundwater levels, reduce the amount of fresh groundwater flowing to the sea, and ultimately cause saltwater to intrude coastal aquifer. Increasing pumping rate is considered the main cause of saltwater intrusion along the coasts. Other parameters that cause saltwater intrusion have smaller impact in comparison with pumping. Population growth in coastal areas has increased water demand which has in turn resulted in increase in abstraction from aquifers. The reduction of pumping rates aims to achieve the sustainable yield and use other water resources to supply adequate water demand. This can be achieved by a number of measures including:

1. Increase in public awareness of the necessity of water to save water
2. Reduction of losses from the water transportation and distribution systems
3. Reduction of water requirement in irrigation by changing the crop pattern and using new methods for irrigation such as drip irrigation, canal lining, etc.
4. Recycling of water for industrial uses, after appropriate treatment
5. Reuse of treated wastewater in cooling and irrigation and recharge of groundwater
6. Desalination of seawater

A number of models have been developed to control saltwater intrusion by reducing pumping rates from the aquifer or using optimization models to optimize the abstraction and control the intrusion of saline water. Zhou et al. [7] used a quasi-three-dimensional finite element model to simulate the spatial and temporal distribution of groundwater levels. The objective of the model was to maximize the total groundwater pumping from the confined aquifer and control saltwater intrusion by relocating the wells. Amaziane et al. [8] coupled the boundary element method and a genetic algorithm for the optimization of pumping rates to prevent saltwater intrusion. Qahman and Larabi [9] investigated the problem of extensive saltwater intrusion in Gaza aquifer, Palestine, using the SEWAT code. Different scenarios were considered to predict the extension of saltwater intrusion with different pumping rates over the time.

Advantages Reduction of abstraction rates and use of other water resources help to increase the volume of freshwater which retards the intrusion of saltwater.

Disadvantages Control of saltwater intrusion using this method has some limitations. The control of abstraction rates cannot be fully achieved especially from private stakeholders. Also the alternatives proposed for facing the shortage of water and the increased demands may be very costly, especially when freshwater is not available and requires transportation. Desalination of seawater as an alternative has some limitations; it is still expensive, is a source of pollution, and requires a large area of land. The treated wastewater is also costly. Furthermore, this can only be a temporary solution because it does not prevent saltwater intrusion but it attempts to reduce it; but with the population growth and increasing demands, the problem will not be controlled [6].

2.2 Relocation of Pumping Wells

Changing the location of pumping wells by moving the wells to more inland positions aims to raise the groundwater level and maintain the groundwater storage. This is because in the inland direction, the thickness of the freshwater lens increases and the risk of upconing of saltwater decreases accordingly. Hong et al. [10] developed an optimal pumping model to evaluate optimal groundwater withdrawal and the optimal location of pumping wells in steady-state condition while minimizing adverse effects such as water quality in the pumping well, drawdown, saltwater intrusion, and upconing. The study involved experimental verification of the optimal pumping model to develop sustainable water resources in the coastal areas. Ojeda et al. [11] modeled the saline interface of the Yucatan Peninsula aquifer using SUTRA assuming an equivalent porous medium with laminar flow and two-dimensional flow. The models showed that the interface position is very sensitive to head changes, and a simple groundwater abstraction scheme was presented which was based on the well depth and the distance from the coast. Sherif and Hamza [12] used two models to simulate the problem of saltwater intrusion in the Nile Delta aquifer in Egypt in the vertical and horizontal directions. The two models were 2D-FED and SUTRA. The 2D-FED model was employed to simulate the current condition and predict the effect of the seawater level rise in the Mediterranean Sea under the condition of global warming. SUTRA was used to define the best location of additional groundwater pumping wells from the Nile Delta aquifer and to assess the effect of various pumping scenarios on the intrusion process.

Advantages Moving pumping wells further inland helps to decrease the occurrence of upconing of saltwater [6].

Disadvantages Control of saltwater intrusion by this method is costly and may face some obstructions such as buildings or the size of the aquifer may not allow such movement. It is also a temporary solution and does not prevent the intrusion of saline water into aquifers.

2.3 Subsurface Barriers

This method involves establishment of a subsurface barrier to reduce the permeability of the aquifer to prevent the inflow of seawater into the basin. Construction of barriers could be achieved using sheet piling, cement grout, or chemical grout. Figure 1 shows a sketch of a subsurface barrier to control saltwater intrusion.

Barsi [13] developed two methods for optimal design of subsurface barriers to control seawater intrusion through the development of implicit and explicit simulation-optimization models. The main objective was to find the optimal design of subsurface barrier to minimize the total construction cost through the selection of the width and location of the barrier. James et al. [14] developed a process for selectively plugging permeable strata with microbial biofilm. These biofilm barriers can aid the prevention of saltwater intrusion by reducing the subsurface hydraulic

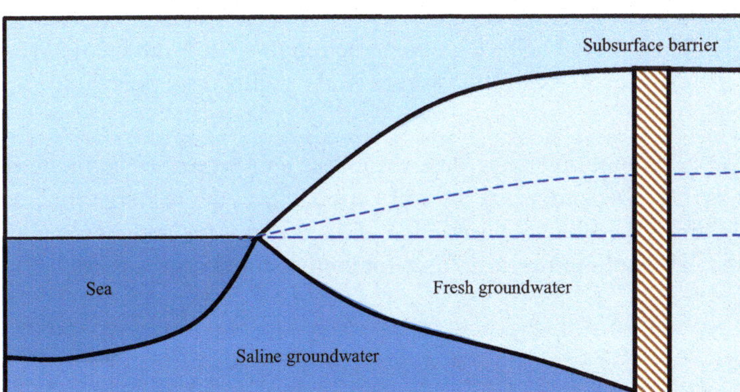

Fig. 1 Sketch of a subsurface barrier

conductivity. The main advantage offered by the biofilm barrier technology is that biofilm barrier construction can be achieved without excavation and therefore will be economic. Harne et al. [15] presented a 2-D subsurface transport model of saltwater considering the soil to be homogenous and isotropic under the influence of constant seepage velocity. They used the finite difference method to solve the transport equation. The model examined the efficiency of subsurface barrier to control saltwater intrusion.

Advantages Using subsurface barriers helps to reduce the intrusion of saline water.

Disadvantages Control of saltwater intrusion using subsurface barriers could be costly in terms of construction, operation, maintenance, and monitoring. It is also not efficient for deep aquifers [6].

2.4 Natural Recharge

This method aims to feed aquifers with additional surface water by constructing dams and weirs to prevent the runoff from flowing to the sea. This method can also be used for flood protection. The retained water infiltrates into the soil and increases the volume of groundwater storage. Figure 2 shows a sketch of the natural recharge process.

This method could be efficient for unconfined aquifers, but it could take a long time to recharge the aquifer depending on its properties. Ru et al. [16] used a quasi-three-dimensional model to simulate the movement of the interface between sea-water and freshwater and evaluate the effect of constructing a dam to protect groundwater resources. The function of the subsurface dam was to collect the rainfall water and recharge the aquifer to increase the groundwater storage and retard saltwater intrusion. Bajjali [17] applied geostatistical techniques of GPI, IDW with GIS, to examine the effect of recharge dam on groundwater quality. The infiltrated water below the dam increases the aquifer storage of freshwater and pushes the saline water toward the sea.

Advantages This method helps to prevent the runoff to flow directly to the sea and uses it to increase the groundwater storage in the aquifer and prevent the intrusion of saline water.

Disadvantages Natural recharge depends on the soil properties and requires high permeability soil. Depending on the soil permeability, the recharge process could take a long time. The cost of construction of dams and weirs and their maintenance is very high. This solution is unsuitable for confined and deep aquifers.

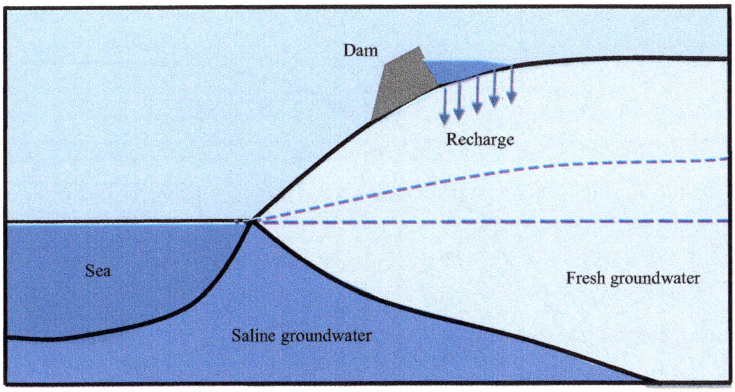

Fig. 2 Sketch of the natural recharge process

2.5 *Artificial Recharge*

Artificial recharge aims to increase the groundwater levels, using surface spread for unconfined aquifers and recharge wells for confined aquifers. The potential sources of water for injection may be from surface water, pumped groundwater, treated wastewater, desalinated seawater, or desalted brackish water. Surface water can be taken from rivers or canals through pipelines. Figure 3 shows a sketch of a recharge well.

A number of researchers used this method to control saltwater intrusion in coastal aquifers. Narayan et al. [18] used SUTRA to define the current and potential extent of saltwater intrusion in the Burdekin Delta aquifer under various pumping and recharge conditions. The results addressed the effects of seasonal variation in pumping rate and artificial and natural recharge rates on the dynamic of saltwater intrusion. Papadopoulou et al. [19] developed a 3-D finite element-finite difference groundwater flow simulation model. The extension of the saltwater front along the coastal zone was calculated only hydraulically because his model did not consider diffusion due to different densities of saltwater and freshwater. Artificial recharge was presented using different scenarios including different well locations and different injection rates. The source of water used for recharge was the effluent of the wastewater treatment plant of the industrial zone after tertiary treatment.

Mahesha [20] studied the effect of battery of injection wells on seawater intrusion in confined coastal aquifers. He used a quasi-three-dimensional areal finite element model considering a sharp interface. He studied various conditions by changing the well spacing and intensity and duration of freshwater injection. He concluded that spacing between wells, injection rate, and duration of injection control the repulsion of the saline wedge. Mahesha [21] presented steady-state solutions for the movement of the freshwater/seawater interface due to a series of injection wells in a confined aquifer using a sharp interface finite element model. The model was used to perform parametric studies on the effect of location of the series of injection wells, spacing of

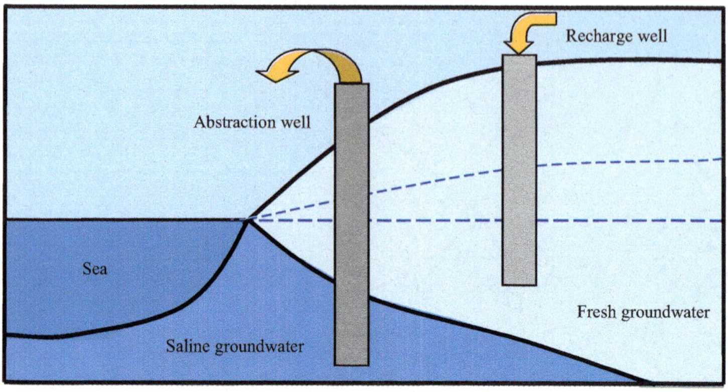

Fig. 3 Sketch of recharge and abstraction wells

the wells, and freshwater injection rate on the seawater intrusion. It was found that reduction of seawater intrusion (of up to 60–90%) could be achieved through proper selection of injection rate and spacing between the wells. Also, he studied the effect of double series of injection wells and compared it with the single series. He found that a double series performs slightly better than single series. Also, it was found that the staggered system of wells in the series is slightly better than the straight well system for long spacing. Vandenbohede et al. [22] presented a simulation model to study sustainable water management using artificial recharge of freshwater in the dunes of the western Belgian coastal plain by two recharge ponds. The recharged water was produced from secondary-treated wastewater effluent by the combination of ultrafiltration and reverse osmosis.

Advantages Artificial recharge helps to increase the groundwater storage in the aquifer and prevent the intrusion of saline water.

Disadvantages The artificial recharge has been discussed in many papers as a solution to control saltwater intrusion. However, some of these models ignored the source of freshwater especially in the areas that suffer from scarcity of water. On the other hand, the cost of freshwater in cases where it is not available in these areas and transported from other places has not been taken into consideration. Using desalinated seawater is costly and could be a source of pollution. The treated wastewater was used in many areas for recharge to control saltwater intrusion. This technique is often costly and ineffective in the areas where excessive groundwater pumping occurs [23], [18]. Treated wastewater affects the aquifer properties and requires at least 1 year before abstraction from the aquifer. At present there is a growing opposition against artificial recharge by infiltration at the land surface, because it occupies a large area.

2.6 Abstraction of Saline Water

This method of control aims to reduce the volume of saltwater by extracting brackish water from the aquifer and returning to the sea. Figure 3 shows a sketch of an abstraction well. Johnson and Spery [24] presented different methods to control saltwater intrusion in different states in the USA. In California they extracted saline water and desalinated it using RO treatment process. The treated water was blended with untreated groundwater to produce water suitable for domestic delivery. In Los Angeles, injection wells were used to protect the coastal aquifers from saltwater intrusion. Potable and highly treated wastewater was injected into the wells. Maimone and Fitzgerald [25] applied three-dimensional groundwater flow models, dual-phase sharp interface intrusion model, radial upconing model, and single-phase contaminant transport model to develop a coastal aquifer management plan. Two techniques were used: the development of new well locations further inland and use of RO treatment for desalinating brackish water and using it for domestic purposes. Sherif and Kacimov [26] suggested pumping brackish water, encountered between the freshwater and saline water bodies, to reduce the extension of seawater intrusion. They used SUTRA to examine different pumping scenarios in the vertical view, and the equiconcentration lines and velocity vectors were identified for the different cases. They concluded that seawater intrusion problems could be controlled through proper pumping of saline groundwater from the coastal zone.

Kacimov et al. [27] investigated seawater intrusion in coastal unconfined aquifers of Oman, experimentally, analytically, and numerically. Water table elevation, capillary fringe, moisture distribution, and EC were observed and measured in a pilot site Al-Hail, Oman. Laboratory measurements of conductivity and capillary rise were conducted in repacked columns in the laboratory. Evaporation from a shallow horizontal water table to a dry soil surface was modeled by HYDRUS2D. An analytical Dupuit-Forchheimer model was developed for the planar part of the catchment with explicit expressions for the water table, sharp interface, and stored volume of freshwater. SUTRA code is used to study a variable density flow in a leaky aquifer with line sinks modeling freshwater withdrawal and evaporation. Both analytical and numerical models proved that pumping of saline groundwater from coastal aquifers would mitigate the migration of seawater into the aquifer and would contribute to the enhancement of the groundwater quality that is consistent with the findings of Sherif and Hamza [12].

Advantages Abstraction of saline water decreases the volume of saline water in the aquifer and protects pumping wells from upconing.

Disadvantages The main problem in abstraction of brackish water is the disposal of the saline water. Many researchers have attempted to solve this problem, but it is still a subject of current research. Brackish water is suitable only for certain types of crops, and using it in cooling may cause corrosion to the systems. The disposal of brackish water into the sea could affect the marine life in these areas, fishing, and tourism activities. Another problem is that increasing abstraction of saline water could, in some cases, increase the intrusion of saltwater [6].

2.7 Combination Techniques

Combination between two or more of the above methods can help to eliminate the disadvantages of these methods and give better control of saltwater intrusion. Zhou et al. [7] and Hong et al. [10] used a combination of reduction in pumping rates and relocation of pumping wells to control saltwater intrusion. Maimone and Fitzgerald [25] used the development of new well locations further inland and use of RO treatment for desalinating brackish water and using it for domestic consumption to reduce the abstraction from the aquifer. Narayan et al. [18], Paniconi et al. [28], and Barrocu et al. [29] used reduction of pumping rates and recharge of freshwater to the aquifer. Johnson et al. [30] used a combination of three techniques: subsurface barriers, reduction of abstraction rates, and recharge of freshwater. Fitzgerald et al. [31] used change of well locations, reduction of pumping rates, and recharge of freshwater.

The combination of freshwater's injection and extraction of saline water can reduce the volume of saltwater and increase the volume of freshwater. Only few models have been developed for this technique. Mahesha [32] studied the control of seawater intrusion by a series of abstraction wells for saline water alone and also in combination with freshwater injection wells in a confined aquifer using a vertically integrated two-dimensional sharp interface model under steady-state conditions. It was found that the combination of injection wells with extraction wells produced excellent results to the individual cases for larger well spacing and small rates of injection. Rastogi et al. [33] developed a two-dimensional steady-state numerical model to study seawater intrusion problem involved. The model was used to investigate the efficiency of seawater control measures involving two scenarios: (1) freshwater recharge and (2) combination of freshwater recharge and saltwater discharge wells. The study found that depth, location, and head of recharge wells are important parameters that can control saltwater intrusion. The model confirmed that combined recharge and discharge system is more effective in controlling seawater intrusion.

Tanapol et al. [34] studied the effectiveness of controlling methods in unconfined aquifers by using scaled-down physical models. The investigated methods included freshwater injection, saltwater extraction, and subsurface barrier. They concluded that freshwater injection is more favorable than the saltwater extraction and subsurface barrier method. A freshwater injection rate of about 10% of the usage rate can effectively push the interface toward the shoreline and maintain the water quality in the vicinity of pumping wells. Acostaa and Donadoa [35] conducted a laboratory scale simulation of seawater hydraulic barriers in confined coastal aquifers considering the effects of stratification. The best hydraulic barrier performance was observed in the extreme point of wedge, where 17.8% of intrusion reduction, in homogeneous media, and 78.9% in stratified media were observed.

The highest reduction of seawater intrusion was achieved with the highest injection rate. Stratification affected the performance of the hydraulic barrier. Smaller injection rates were necessary to reduce seawater intrusion in stratified medium.

3 New Methodologies to Control Saltwater Intrusion

Coupling of flow and solute transport models with optimization techniques to address important groundwater quality management problems has started early 1980s. The motivation for the development of simulation-optimization (S/O) approach for the control of saltwater intrusion is the enormous cost of groundwater remediation. The S/O approach has been shown to be capable of reducing the remediation costs of contaminated land in several real-world applications [36]. The S/O approach can be used efficiently in groundwater remediation system design and in other groundwater quality management problems including saltwater intrusion control. The S/O approach is interesting because it can both account for the complex behavior of a groundwater flow system and identify the best management strategy to achieve a given set of prescribed constraints [37].

3.1 Simulation-Optimization Models

Simulation models have been applied for the management of groundwater resources seeking for optimal management strategy by *trial-and-error*. This has proved to be time-consuming and laborious. In addition, the results obtained may not be optimal. The main reason for this is the inability of this approach to consider important physical and operational restrictions. To accommodate these restrictions, coupling of the simulation model with a management model is the generally adopted procedure. For groundwater systems, combined simulation and management models may adequately predict the behavior of the system and provide the best solution for problems such as planning of long-term water supply or preventing seawater intrusion. A relatively small number of studies have concentrated on the control of saltwater intrusion linking simulation with optimization models to effectively determine the optimal solution for control of seawater intrusion [37].

3.1.1 Simulation Models

Modeling of fluid flow and solute transport for a site involves development of a mathematical model of the site system being studied and to use this model to predict the value of hydraulic heads and salt concentrations at points of interest at different times. Numerical simulation of seawater intrusion, assuming that mixing occurs at the transition zone between seawater and freshwater, involves the solution of the partial differential equations representing the conservation of mass for the variable density fluid (flow equations) and for the dissolved solute (transport equation). The numerical solution of coupled fluid flow and solute transport is based on solving the governing equations with the boundary and initial conditions by the iterative solution scheme. Recently, a number of simulation codes were developed and capable of simulating saltwater intrusion such as SEAWAT and SUTRA [37].

3.1.2 Optimization Models

The optimization problem can be solved through trial-and-error adjustment or through a formal optimization technique. Numerical simulation models can be used to examine a limited number of design options by trial-and-error. Due to the simplicity of the trial-and-error method, it is widely used. However, testing and checking hundreds to thousands of trial solutions is tedious and cannot guarantee that the optimal solution has been identified. On the other hand, the optimization technique can be used to search for the optimal solution in a wide search space of design variables and, equally important, to prove whether a particular management scenario or remedial alternative is feasible in terms of meeting the management objective and satisfying all the constraints [37].

Optimization model can be defined in terms of an objective function and a set of constraints. Groundwater management problem includes two sets of variables: decision variables and state variables. Decision variables are the variables that are used to define different alternative decisions, such as pumping or injection rate of a well, well location, and well depth. Decision variables can be managed in the calculation process to identify the optimal management policy or strategy. State variables are the variables that describe the flow and transport conditions of an aquifer, such as head and salt concentrations, which are the dependent variables in flow and transport equations. In a coupled S/O model, simulation components update the state variables, and optimization components determine optimal values of decision variables [36]. Optimization models are widely used in groundwater planning and management. In recent years, a number of simulation models have been combined with optimization techniques to address groundwater management problems. The complex behavior of the groundwater system and identification of the best management strategy under consideration of the management objectives and constraints can be achieved easily by the combined simulation and optimization modeling processes. In the last two decades, genetic algorithms (GAs) have received increased attention from academic and industrial communities for dealing with a wide range of optimization problems.

3.2 Genetic Algorithm (GA)

GA is an optimization technique based on the process of biological evolution. GA was first introduced by Holland [38] and followed by Goldberg [39], Davis [40], Michalewicz [41], Mckinney and Lin [42], and Haupt and Haupt [43] among others. GAs are a family of combinatorial methods that search for solutions of complex problems using an analogy between optimization and natural selection. GA mimics biological evolution based on the Darwinist theory of survival of the fittest, where the strongest offspring in a generation are more likely to survive and reproduce. The GA method starts with randomly generating an initial set of solutions, called the initial population. Each member of this population represents a possible solution of the problem, encoded as a chromosome. A chromosome is a string of symbols, usually a binary bit string. The population of chromosomes evolves through a cycle that involves selection, crossover, and mutation. Each cycle is referred to as a generation. After many generations, the population will contain chromosomes that represent near optimal solutions to the problem. GA was applied to a wide variety of problems in engineering including groundwater management, water resources, and seawater intrusion. An overview to the application of GA in these fields is presented in the following sections [36].

GA has been applied to a number of groundwater management problems. A number of researchers incorporated groundwater simulation models with GA to solve groundwater management problems such as maximizing extraction from an aquifer, minimizing the cost of water supply, and minimizing the cost of aquifer remediation and pump-and-treat. Among the researchers who applied GA to groundwater management problems are Mckinn and Lin [44], Rogers and Dowla [45], El Harrouni et al. [46], and Aly and Peralta [47]. Recently, GA has been used to identify cost-effective solutions for pump-and-treat groundwater remediation design. In the field of water resource management, GA has been applied to a variety of problems such as calibration of rainfall-runoff models and pipe network systems and operation of reservoir systems. A number of researches applied GA for calibration of rainfall-runoff models (e.g., [48–50]). A large number of applications in pipe network optimization using GA were presented by Murphy et al. [51] and Farmani et al. [52, 53].

On the other hand, development and application of optimization techniques in association with saltwater intrusion considering density-dependent flow are relatively recent and very few. Das and Datta [54, 55] presented a multi-objective management model. The objectives were to maximize pumping from the freshwater zone and minimize pumping from the saline zone in order to control saltwater intrusion. The nonlinear finite difference of the density-dependent miscible flow and salt transport model for seawater intrusion in coastal aquifers was embedded within constraints of the management model. Das and Datta [56] developed a nonlinear optimization method for minimizing total sustainable yield from specific location of the aquifer while satisfying salinity constraints. Gurdu et al. [57] developed a management model to maximize the total pumping rate from wells considering saltwater intrusion into the aquifer. They used SUTRA code for groundwater simulation and the general algebraic modeling system code to execute the optimization model.

Gordon et al. [58] developed a model for optimal management of a regional aquifer under salinization. The source of salinization was from irrigation water percolating over some part of the aquifer, influx of saline water from faults in the aquifer bottom, and inflow from laterally adjacent saline water bodies. The objectives of management were to maximize the total amount of water pumped for use and to minimize the total amount of salt extracted with the water. The simulation model used a finite element formulation for the flow and a streamline upwind Petrov-Galerkin formulation for the transport. The model computed the gradient of the state variables (heads and concentrations) with respect to the decision variables (pumping rates at wells). The gradients were then used in a Bundle-Trust non-smooth optimization procedure to achieve an improved solution. Rao and Rao [59] presented S/O approach to control saltwater intrusion through a series of extraction wells. They used SEAWAT code to simulate saltwater intrusion and the simulated annealing algorithm for solving the optimization problem. Two objective functions were considered: first to maximize groundwater pumpage for beneficial purposes and second to minimize the pumping in extraction wells which were used to extract saline water and dispose to the sea.

Javadi et al. [60] presented the development and application of a simulation-optimization model to control seawater intrusion in coastal aquifers using different management scenarios: abstraction of brackish water, recharge of freshwater, and combination of abstraction and recharge. The model was based on the integration of a genetic algorithm optimization technique and a coupled transient density-dependent finite element model. The developed model was applied to analyze the control of seawater intrusion in a hypothetical confined coastal aquifer. The efficiencies of the three management scenarios are examined and compared. The results showed that combination of abstraction and recharge wells is significantly better than using abstraction wells or recharge wells alone as it gives the least cost and least salt concentration in the aquifer. The results from the study would be useful in designing the system of abstraction/recharge wells to control seawater intrusion in coastal aquifers and can be applied in areas where there is a risk of seawater intrusion.

3.3 *Examples of New Methods to Control Saltwater Intrusion*

The new methodologies used to control saltwater intrusion in different locations of the world are present in this section.

3.3.1 Abstraction of Saline Water and Recharge Using Surface Ponds

This method includes a combination of using surface ponds to recharge the aquifer and abstraction of saline water near the shoreline. Javadia et al. [61] presented a new method for optimal control of seawater intrusion. The method was based on a combination of abstraction of saline water near shoreline and recharge of aquifer using surface ponds as shown in Fig. 4. The source of water for the surface pond could be treated wastewater or excess of desalinated brackish water (if any), etc. The variable density flow and solute transport model, SUTRA, is integrated with a genetic algorithm optimization tool in order to investigate the efficacy of different

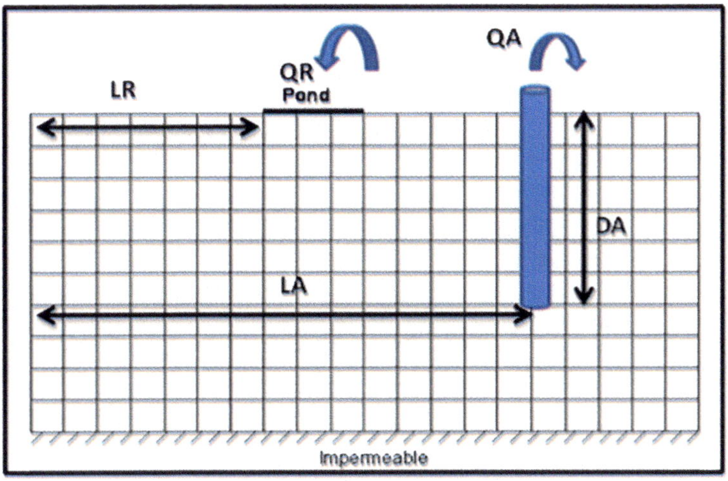

Fig. 4 Schematic sketch for abstraction of saline water and recharge using surface ponds

scenarios of the seawater intrusion control in an unconfined costal aquifer. The locations of the pond and the abstraction well in relation to the shoreline, depth of abstraction well, and the rates of abstraction and recharge are considered as the main decision variables of the optimization model, which aims to minimize the costs of construction and operation of the abstraction wells and recharge ponds as well as the salt concentrations in the aquifer. The results indicated that the combined system is better than using abstraction alone in controlling seawater intrusion as it gave the least cost and least salinity in the aquifer.

3.3.2 Abstraction, Desalination, and Recharge (ADR)

A new methodology ADR (abstraction, desalination, and recharge) to control saltwater intrusion in coastal aquifers was presented by [6]. Figure 5 shows a sketch of the new methodology (ADR). ADR methodology consists of three steps: abstraction of brackish water from the saline zone using abstraction wells, desalination of the abstracted brackish, and recharge of the treated water using recharge wells into the aquifers. This method is a combination of two techniques: abstraction of saline water and recharge of freshwater in addition to desalination of abstracted water and treatment to be ready for recharge or domestic use. It has the advantages of the following three steps:

1. Abstraction of brackish water to reduce the saline water volume in the aquifer and reduce the intrusion of saltwater.
2. Desalination of abstracted brackish water using RO treatment process to produce freshwater from brackish water for recharge. This step is very important to produce freshwater in the areas where freshwater is scarce. The advantages of RO as a desalination method are discussed in details in the next section.

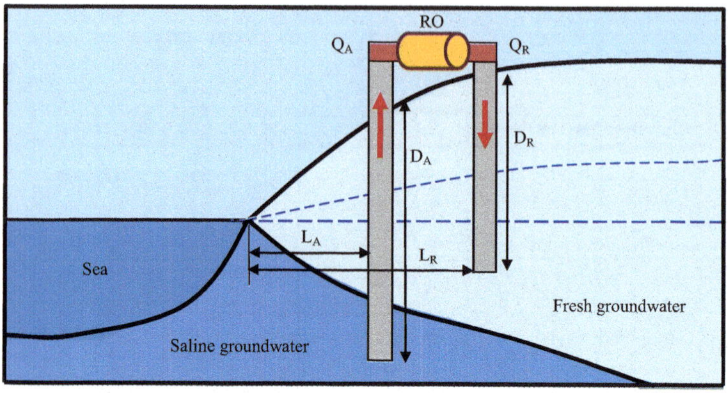

Fig. 5 Schematic sketch for abstraction, desalination, and recharge (ADR)

3. Recharge of treated water to increase the fresh groundwater volume to prevent the intrusion of saltwater.

The combination of abstraction and recharge techniques is considered one of the most efficient methods to control saltwater intrusion. It increases the volume of fresh groundwater and decreases the volume of saltwater. This method is capable of retarding saltwater intrusion. It also has lower energy consumption, lower cost, and lower environmental impact. ADR can be used to increase the water resources in coastal regions by increasing the abstraction of saline water and desalination. The excess reclaimed water can be used directly for different purposes or injected into the aquifer to increase the groundwater storage [6].

Abd-Elhamid and Javadi [62] presented a cost-effective method to control seawater intrusion in coastal aquifers. The methodology (ADR) includes abstraction of saline water and recharge to the aquifer after desalination. A coupled transient density-dependent finite element model was developed by for simulation of fluid flow and solute transport and used to simulate seawater intrusion. The simulation model was integrated with an optimization model to examine three scenarios to control seawater intrusion including abstraction, recharge, and a combination system, ADR. A comparison between the combined system (ADR) and the individual abstraction or recharge system was made in terms of total cost and total salt concentration in the aquifer and the amount of repulsion of seawater achieved. The results showed "that the proposed ADR system performs significantly better than using abstraction or recharge wells alone as it gives the least cost and least salt concentration in the aquifer. ADR is considered an effective tool to control seawater intrusion and can be applied in areas where there is a risk of seawater intrusion" [62].

3.3.3 Treatment, Recharge, Abstraction, and Desalination (TRAD)

This new methodology to control saltwater intrusion was based on a combination of abstraction and recharge techniques. The methodology can help to solve saltwater intrusion problem and take into consideration different parameters such as physical constraints, practical aspects, economical aspects, and environmental impacts. The proposed methodology, (*TRAD*), includes a combination of *T*reatment of wastewater and *R*echarge to the aquifer, *A*bstraction of brackish water from the aquifer, and *D*esalination. The abstracted saline water will be desalinated and used to meet a part of the demand for water or used as a source of freshwater to recharge the aquifer to control saltwater intrusion. Figure 6 shows a sketch of the new methodology (*TRAD*).

The main benefits of the (*TRAD*) methodology are to return the mixing zone (the interface between fresh groundwater and saline water) into the original status and to reach the dynamic balance between fresh- and saline groundwater through two processes of (A) abstraction of brackish groundwater to reduce the volume of saline water and (R) recharge of the aquifer by treated wastewater to increase the volume of fresh groundwater. The abstraction and recharge processes help to push saline water toward the sea. The desalination of brackish water (D) aims to produce freshwater from the brackish water and use it for different purposes such as domestic, irrigation, and industry. Desalination of seawater has a lot of problems such as high cost, high pollution (mainly carbon emission), large area of land that is required for plants, and disposal of the brine. "Desalinating brackish water is an efficient alternative to seawater desalination, because the salinity of brackish water is less than one-third of that of seawater. Therefore, brackish water can be desalinated at a significantly lower cost than sea water" [63]. Several methods can be used for desalination, but RO has been selected as a method for desalting brackish water,

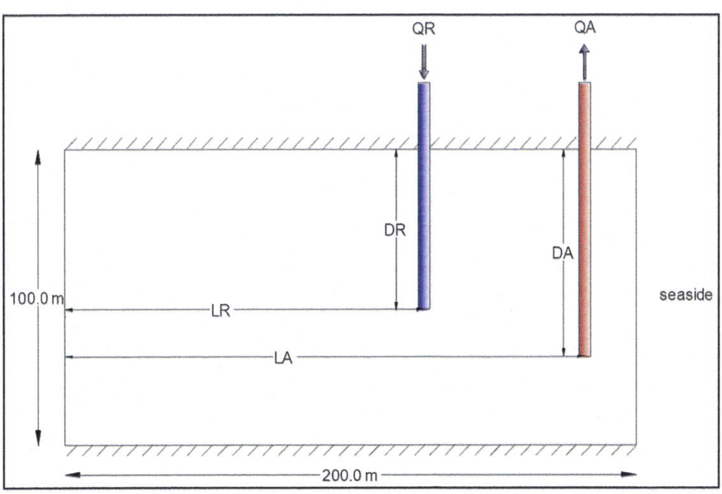

Fig. 6 Schematic sketch for TRAD method

because it has many advantages; it requires simpler equipment, low energy, and low cost. The treatment of waste water (T) aims to overcome the scarcity of water in these areas and also to prevent the disposal of wastewater into seas and rivers. Overall, the methodology has lower cost and lower environmental impacts. It can be used to increase the water resources by increasing the abstraction of brackish water and desalination. The excess treated wastewater may be used directly for different purposes. The abstraction rates of the main aquifer can be increased to provide more freshwater and prevent loss of water that flows to the sea.

Javadia et al. [64] presented the results of an investigation on the efficiencies of different management scenarios for controlling saltwater intrusion using a simulation-optimization approach. A new methodology was proposed to control SWI in coastal aquifers. The proposed method was based on a combination of abstraction of saline water near shoreline, desalination of the abstracted water for domestic consumption, and recharge of the aquifer by deep injection of the treated wastewater to ensure the sustainability of the aquifer. The efficiency of the proposed method was investigated in terms of water quality and capital and maintenance costs in comparison with other scenarios of groundwater management. A multi-objective genetic algorithm-based evolutionary optimization model was integrated with the numerical simulation model to search for optimal solution of each scenario of SWI control. The main objective was to minimize both the total cost of management process and the total salinity in aquifer. The results indicated that the TRAD method is efficient in controlling SWI as it offers the least cost and least salinity in the aquifer.

4 Control of Saltwater Intrusion in Egypt

"Saltwater intrusion in a coastal aquifer is a highly complex and nonlinear process. The management of coastal aquifers requires careful planning of withdrawal strategies for control and remediation of saltwater intrusion." A number of researches have been carried out to simulate saltwater intrusion in the Nile Delta aquifer. However, a limited number of studies were conducted to control saltwater intrusion into the aquifer and protect the groundwater storage. Sherif and Singh [65] examined different scenarios for an additional pumping of 2.3 billion m^3/year from the Nile Delta aquifer. It was concluded that additional pumping practices should be located in the middle Delta to decelerate and minimize the intrusion of saline water. Gaamea [66] examined the use of scavenger wells to control saltwater intrusion in the Nile Delta aquifer. He used SUTRA code to generalize and simplify the use of scavenger wells as individual wells or in the shape of the field. Nasar [67] presented the environmental impact assessment of abstracting brackish water for desalination and recharging the brine water through deep wells to control the seawater intrusion into the Nile Delta aquifer. Designed charts for abstraction and recharge wells including discharge and recharge rates and locations and depths of pumping and injection wells were presented.

Abd-Elhamid [68] used SEAWAT code to study groundwater flow and seawater intrusion in the Eastern Nile Delta (END) aquifer. The distribution of saltwater intrusion in the END aquifer is shown in Fig. 7a which is considered the base case. A vertical cross section is taken in the middle of the aquifer from the top to the bottom. The vertical cross section (Fig. 7b) showed that the intrusion length of equiconcentration line 35 reached 75.75 km from shoreline. However, equiconcentration line 1 intruded to a distance of 90.25 km from shoreline. Different scenarios of climate change were considered including sea level rise, increasing abstraction, decreasing recharge, and combination of these scenarios. The results showed that decreasing recharge has a significant effect on seawater intrusion. However, the combinations of these scenarios resulted in harm intrusion and lose large quantity of groundwater. Also, the soil salinity increased which decreased the agricultural production. The results of different scenarios are presented in Table 1.

Different scenarios were employed to control seawater intrusion in the END aquifer including decreasing abstraction, increasing recharge, abstraction of brackish water, and combination of these scenarios. The results of the different scenarios of seawater intrusion control are shown in Fig. 8a–d and summarized in Table 1.

4.1 Application of New Methods to Control Saltwater Intrusion in Egypt

Groundwater is an important source for drinking, agriculture, and industry in Egypt. But pollution leads to lose large quantity of water and increase the gap between available and demand. Saltwater intrusion in the Nile Delta aquifer threatens billions of cubic meters of freshwater that are available for different purposes. According to the literature a limited number of studies have been carried out to control saltwater intrusion in the Nile Delta aquifer. The control of saltwater intrusion is very important to protect the available water resources and increase the national income through using this water in the development in agriculture and industry. The key to control saltwater intrusion is to maintain a proper balance between water being pumped from the aquifer and water recharged to the aquifer. Not all the solutions are economically feasible because they are considered long-term solutions, and the time to reach the state of dynamic equilibrium between freshwater and brackish water may takes tens of years [69].

From the literature, a number of control methods can be applied in Egypt. The selection of the method is very important and should consider environmental, social, and economical issues. The most suitable methods that can be applied to control saltwater intrusion in the Nile Delta are:

1. Optimization of abstraction rates
2. Pump and treat (P&T)
3. Recharge and abstraction (R&A)

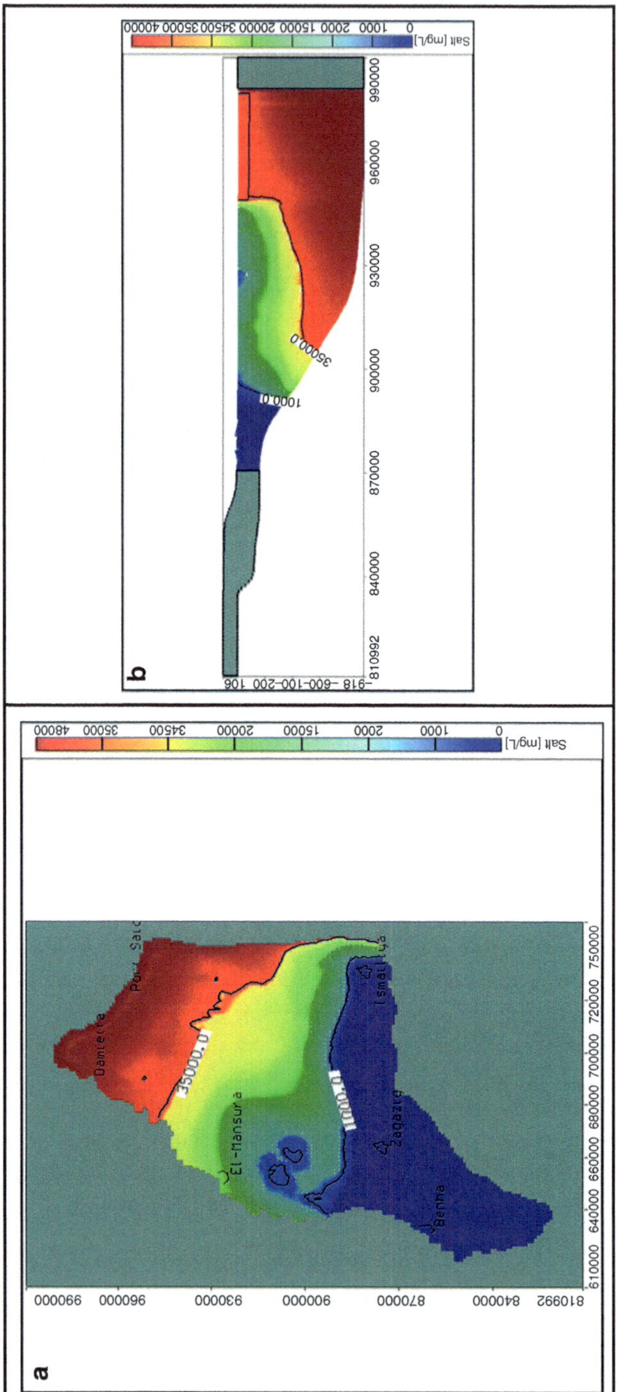

Fig. 7 TDS distribution in the END aquifer for base case (**a**) areal and (**b**) vertical distribution

Table 1 Results of different scenarios to investigate and control SWI in the Eastern Nile Delta aquifer

Case	Scenario	Scenario description	Intrusion length (km)	
			Equiconcentration line 35	Equiconcentration line 1
Base case		Current situation	75.75	90.25
Different scenarios of investigating SWI due to climate change	1	SLR 100 cm	77.00	90.75
	2	Increasing abstraction rate 100%	79.60	91.50
	3	Decreasing recharge 100%	85.25	95.40
	4	Combination of 1, 2, and 3	99.40	110.25
		SLR 100 cm		
		Increasing abstraction rate 100%		
		Decreasing recharge 100%		
Different scenarios of controlling SWI	1	Decreasing abstraction 50%	69.95	88.40
	2	Increasing recharge 50%	70.75	88.00
	3	Abstracting brackish water 50%	61.00	87.80
	4	Combination of 1, 2, and 3	59.80	85.90
		Decreasing abstraction 50%		
		Increasing recharge 50%		
		Abstracting brackish water 50%		

These three techniques have been applied in different locations of the world (e.g., Gaza, Florida, USA, India). These three methods are capable of preventing the intrusion of saltwater into the Nile Delta aquifer, but the cost of applying these techniques may be high and may have some environmental and social impacts. The use of linked simulation-optimization models to assess these techniques should be

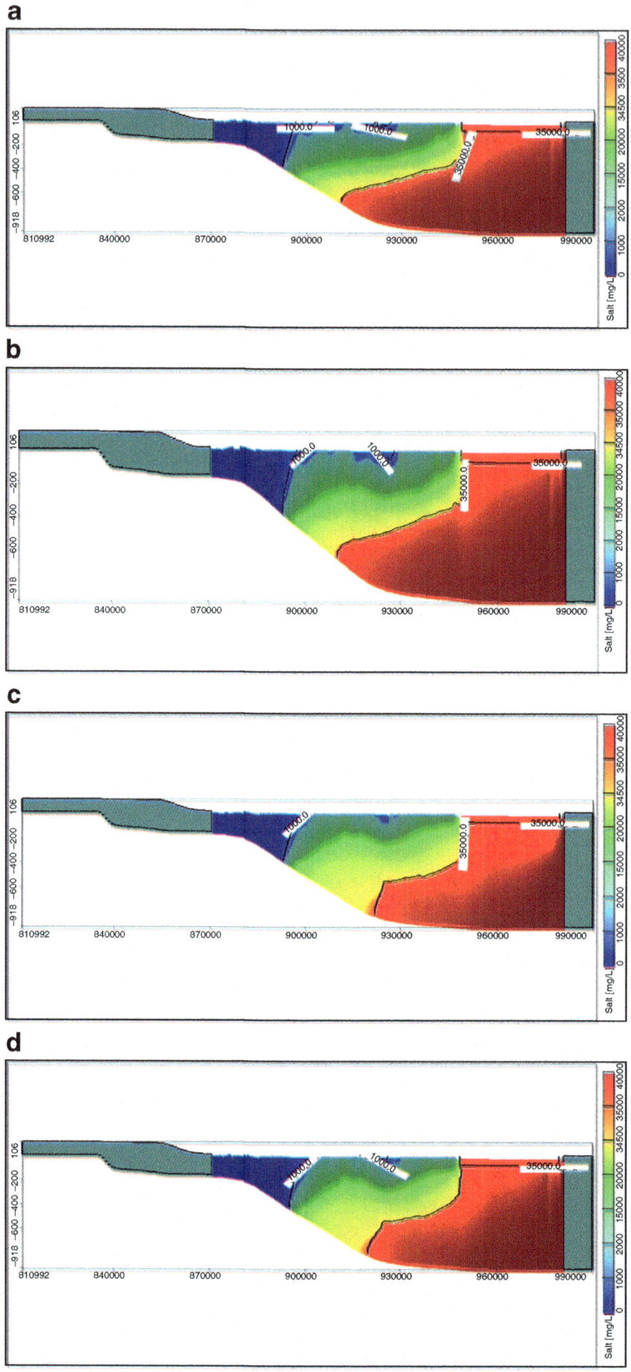

Fig. 8 Vertical distribution of TDS in the END aquifer due to different scenarios of SWI control (**a**) Scenario 1, (**b**) Scenario 2, (**c**) Scenario 3, and (**d**) Scenario 4

considered. S/O models can help to optimize the abstraction from the aquifer and minimize the intrusion of saltwater. Pump and treat (P&T) can help to remediate the aquifer and provide a new source of freshwater. Recharge and abstraction (R&A) technique can help to prevent the intrusion of saltwater by keeping a proper balance between water being pumped from the aquifer and water recharged to the aquifer considering social, economical, and environmental issues.

4.1.1 Optimization of Abstraction Rates

Increasing pumping rate is considered the main cause of saltwater intrusion along the coasts because the natural balance between freshwater and saltwater in coastal aquifers is disturbed by abstraction which lowers groundwater levels, reduces the amount of fresh groundwater flowing to the sea, and ultimately causes saltwater to intrude coastal aquifer. The traditional method of reducing abstraction rates to control the intrusion of saline water was applied in different areas using simulation models (e.g., Sherif and Singh [65] , Scholze et al. [70], and Zhou et al. [7]).

But linked simulation-optimization models were recently applied to maximize the abstraction and control the intrusion of saline water. Bhattacharjya and Datta [71] used an artificial neural network as a simulator for flow and solute transport in coastal aquifer to predict salt concentrations of the pumped water. A linked simulation-optimization model was formulated to link the trained ANN with a GA-based optimization model to solve saltwater management problems. The objective of the management model was to maximize the permissible optimal abstraction of groundwater from the coastal aquifer for beneficial use while maintaining salt concentration in the pumped water under specific permissible limits. Qahman and Larabi [9] investigated the problem of extensive saltwater intrusion in Gaza aquifer, Palestine, using the SEWAT code linked with an optimization model. Different scenarios were considered to predict the extension of saltwater intrusion with different pumping rates over the time. Dokou et al. [72] presented a simulation-optimization model to manage saltwater intrusion in two unconfined coastal aquifers in Crete, Greece. The optimization model seeks to maximize groundwater withdrawal rates while maintaining the saltwater intrusion. A combination of a groundwater flow model (MODFLOW) with simple algorithm (GWM) was employed. The results showed that under the current pumping strategies, the saltwater intrusion front will continue to move inland, posing a serious threat to the groundwater quality of these regions. In both case studies, significant reductions in pumping are required to retract the saltwater intrusion front closer to the shoreline.

For Egypt, this technique should be considered with the other methods of controlling saltwater intrusion because using any control method with increasing abstraction rate is ineffective. Application of optimization techniques linked to simulation models to assess the efficiency of this method in controlling saltwater intrusion can help to avoid the limitations of the traditional method of reduction of abstraction rates and use of other water resources. The optimal management of abstraction from the Nile Delta aquifer has become necessary to protect the aquifer from depletion.

4.1.2 Pump and Treat (P&T)

Optimization models are widely used in groundwater remediation. Recently, a number of researchers incorporated groundwater simulation models with GA to minimize the cost of aquifer remediation and pump-and-treat. Among the researchers who applied GA to groundwater management problems are Rogers and Dowla [45] and Aly and Peralta [47]. Mckinn and Lin [44] applied a GA to groundwater resource management and pump-and-treat system design. Johnson and Rogers [73] used GA and neural network to select the optimal well locations and pumping rates in a remediation design problem. Zheng and Wang [74] applied GA to pump-and-treat system design optimization under field conditions. Recently, GA has been used to identify cost-effective solutions for pump-and-treat groundwater remediation design. Some researchers have applied GA to solve groundwater pollution remediation problems (e.g., [75–77]).

For Egypt, this technique can be applied for remediation of aquifers contaminated by saltwater intrusion and/or other types of contaminants. The polluted water can be pumped, treated, and used for different proposes. GA can be linked with simulation models to minimize the cost of aquifer remediation and pump-and-treat. This technique can provide a new source of freshwater from polluted water to face the increased demands.

4.1.3 Recharge and Abstraction (R&A)

Combination techniques can help to eliminate the limitations of using individual methods and give better control of saltwater intrusion. Combination of recharge of freshwater and extraction of saline water can reduce the volume of saltwater and increase the volume of freshwater which help to control saltwater intrusion. A number of models have been used to assess the control saltwater intrusion using R&A (e.g., [33–35, 78]). Recently, linked S/O models were applied to maximize the abstraction and minimize the recharge to control the intrusion of saline (e.g., [59, 60]).

Three new forms of the recharge and abstraction (R&A) have been used to assess the control of saltwater intrusion using linked simulation-optimization models:

(a) Abstraction of saline water and recharge using surface ponds
(b) Abstraction, desalination, and recharge (ADR)
(c) Treatment, recharge, abstraction, and desalination (TRAD)

The developed models to assess the efficiency of these methods to control SWI were discussed above and the possibility of application to Egypt is discussed below.

Application of Abstraction of Saline Water and Recharge Using Surface Ponds to Egypt

This method is based on a combination of abstraction of saline water near shoreline and recharge freshwater to the aquifer using surface ponds. This method was presented by Javadia et al. [61]. The study presented the integration between a simulation model SUTRA and a genetic algorithm optimization tool in order to investigate the efficacy of abstraction of saline water and recharge using surface ponds. The results indicated that the combined system is better than using abstraction alone in controlling seawater intrusion as it gave the least cost and least salinity in the aquifer. For Egypt, this technique can be used to control saltwater intrusion in locations where freshwater for recharge is available. The source of freshwater could be treated wastewater or excess desalinated brackish water (if any). Also, storm water can be collected and used for recharging the aquifer. The abstracted saline water could be desalinated and used for different purposes.

Application of Abstraction, Desalination, and Recharge (ADR) to Egypt

The methodology ADR (abstraction, desalination, and recharge) includes abstraction of saline water and recharge to the aquifer after desalination. Abd-Elhamid and Javadi [62] presented the integration between coupled transient density-dependent finite element model and an optimization model to examine the efficiency of ADR to control saltwater intrusion. A comparison between the combined system (ADR) and the individual abstraction or recharge system was made in terms of total cost and total salt concentration in the aquifer and the amount of repulsion of seawater achieved. The results showed that ADR system performs significantly better than using abstraction or recharge wells alone as it gives the least cost and least salt concentration in the aquifer. For Egypt, this technique can be used to control saltwater intrusion in locations where the width of the transition zone is small to reduce the cost of transporting the desalinated brackish water and use for recharge.

Application of Treatment, Recharge, Abstraction, and Desalination (TRAD) to Egypt

These methods is based on a combination of abstraction of saline water near shoreline, desalination of the abstracted water for different purposes, and recharging the aquifer by deep injection of the treated wastewater. Javadia et al. [64] presented the integration between SUTRA code and a genetic algorithm optimization tool in order to investigate the efficacy of (TRAD) method to control saltwater intrusion. The results indicated that the TRAD method is efficient in controlling saltwater intrusion in coastal aquifer as it offers the least cost and least salinity in the aquifer. For Egypt, the application of this technique to control saltwater intrusion could be useful as the width of the transition zone reaches some kilometers. This technique does not depend on the distance between recharge wells and abstraction wells because the treated wastewater is used for recharge and the abstracted saline water is desalinated and used for different uses.

For Egypt, these three techniques can be applied and tested using linked simulation-optimization models to assess the efficiency of these methods in controlling saltwater intrusion. Optimization of abstraction rates from the Nile Delta aquifer should be implemented to protect the aquifer from saltwater intrusion. Remediation of the aquifer from different types of pollutants and saltwater intrusion using pump and treat method could be useful. Combination of recharge of freshwater and abstraction (R&A) of saline water can be applied to reduce the volume of saltwater and increase the volume of freshwater which help to control saltwater intrusion. These methods overcome two main problems in Egypt:

1. *The source of freshwater for recharges* can be addressed using treated wastewater or collect the storm water in a drainage system and inject to the aquifer will retard the intrusion of saline water and protect the coastal area from flash flood such as in the case of Alexandria city in 2015.
2. *The disposal of abstracted saline water* can be addressed using desalination and use the desalinated water for different purposes or for recharge. The desalination problems including energy and brine disposal can be addressed using renewable energy such as solar energy with new techniques of desalination such as reverse osmoses (RO) or forward osmosis (FA). Brine can be used to produce salt or used for certain types of crops or fish.

Protection of groundwater resources in Egypt has become an essential issue due to the increased gap between demands and available resources. The optimal management of the Nile Delta aquifer has become essential to protect the aquifer from depletion under the new circumstances such as:

1. Population growth
2. Climate change and sea level rise
3. New reclamation projects
4. Reducing recharge due to new structures on the Nile River and associated increase in abstraction rates from the aquifer.

5 Summary, Conclusion, and Recommendation

This chapter gave a background of different control methods that can be applied in different locations where saltwater intrusion is found. Excessive groundwater abstraction to meet growing demands due to increasing population and the expected rise in mean sea level due to global warming will increase seawater and threaten the available groundwater supply. The methods used to control saltwater intrusion in coastal aquifers were presented in this chapter. Also, the new methods that could be applied to control saltwater intrusion in coastal aquifers in Egypt, especially in the

Nile Delta aquifer, were presented. Saltwater intrusion in costal aquifer must be controlled to protect freshwater. It is recommended to secure the needed financial support to apply the suitable technique to control saltwater intrusion in the coastal aquifer (Nile Delta aquifer) to protect groundwater in these aquifers. Also, monitoring of the performance of implemented techniques is very important.

References

1. Roger LJ (2010) Effects of subsurface physical barrier and artificial recharge on seawater intrusion in coastal aquifers, PhD thesis, Kagoshima University, Japan
2. Bear J, Cheng AHD (1999) Introduction. In: Seawater intrusion in coastal aquifers concepts, methods and practices, Bear J, Cheng AHD, Sorek S, Ouazar D, Herrera I (eds) Kluwer Academic Publishers, Dordrecht, The Netherlands, pp 1–8
3. Charbeneau R (2000) Groundwater hydraulics and pollutant transport. Prentice Hall, Upper Saddle River, NJ
4. Todd D (1980) Groundwater hydrology, 2nd edn. Wiley, New York
5. Todd DK (1974) "Salt-water intrusion and its control", *Water technology/resources*. J Am Water Works Assoc 66(3):180–187
6. Abd-Elhamid HF, Javadi AA (2008) Mathematical models to control saltwater intrusion in coastal aquifer. Proceeding of GeoCongress, New Orleans, LA
7. Zhou X, Chen M, Liang C (2003) Optimal schemes of groundwater exploitation for prevention of seawater intrusion in the Leizhou Peninsula in southern China. Environ Geol 43:978–985
8. Amaziane B, Cheng A H-D, Naji A, Ouazar D (2004) Stochastic optimal pumping under no-intrusion constraints in coastal aquifers. Proceeding of the 18th salt water intrusion meeting, Cartagena, Spain
9. Qahman K, Larabi A (2004) Three dimensional numerical models of seawater intrusion in Gaza aquifer, Palatine. In: Proceeding of the 18th Salt Water Intrusion Meeting, Cartagena (Spain)
10. Hong S, Park N Bhopanam N, Han S (2004) Verification of optimal model of coastal pumping with sand-tank experiment. Proceeding of the 18th salt water intrusion meeting, Cartagena, Spain
11. Ojeda CG, Gallardo P, Hita LG, Arteaga LM (2004) Saline interface of the Yucatan peninsula aquifer. In: Proceeding of the 18th Salt Water Intrusion Meeting, Cartagena (Spain)
12. Sherif MM, Hamza KI (2001) Mitigation of seawater intrusion by pumping brackish water. J Transport Porous Media 43(1):29–44
13. Barsi MH (2001) Two new methods for optimal design of subsurface barrier to control seawater intrusion, PhD thesis, Department of Civil and Geological Engineering, University of Manitoba, Winnipeg, Manitoba, Canada
14. James G, Hiebert R, Warwood B, Cunningham A (2001) Subsurface biofilm for controlling saltwater intrusion. Proceeding of the 1st international conference and workshop on saltwater intrusion and coastal aquifers, monitoring, modelling, and management, Morocco
15. Harne S, Chaube UC, Sharma S, Sharma P, Parkhya S (2006) Mathematical modelling of salt water transport and its control in groundwater. Nat Sci 4:32–39
16. Ru Y, Jinno K, Hosokawa T, Nakagawa K (2001) Study on effect of subsurface dam in coastal seawater intrusion. Proceeding of the 1st international conference and workshop on saltwater intrusion and coastal aquifers, monitoring, modelling, and management, Morocco
17. Bajjali W (2005) Model the effect of four artificial recharge dams on the quality of groundwater using geostatistical methods in GIS environment, Oman. J Spat Hydrol 5(2):1–15

18. Narayan KA, Schleeberger C, Charlesworth PB, Bristow KI (2006) Effects of groundwater pumping on saltwater intrusion in the lower Burdekin Delta, North Queensland, CSIRO Land and Water, Davies Laboratory, Townsville, Australia
19. Papadopoulou MP, Karatzas GP, Koukadaki MA, Trichakis Y (2005) Modelling the saltwater intrusion phenomenon in coastal aquifers – a case study in the industrial zone of Herakleio in Crete. Global NEST J 7(2):197–203
20. Mahesha A (1996) Transient effect of battery of injection wells on seawater intrusion. J Hydraul Eng ASCE 122:266–271
21. Mahesha A (1996) Steady-state effect of freshwater injection on seawater intrusion. J Irrigat Drai Eng ASCE 122:149–154
22. Vandenbohede A, Lebbe L, Houtte EV (2008) Artificial recharge of fresh water in the Belgian Coastal Dunes. In: Proceeding of 20th SWIM, Naples, Florida, USA
23. Narayan KA, Schleeberger C, Charlesworth PB, Bristow KL (2002) Modelling saltwater intrusion in the lower Burdekin Delta North Queensland, Australia. In: Denver Annual Meeting, The Geological Society of America (GSA)
24. Johnson T, Spery M (2001) Direct potable use of intruded seawater from west coast basin aquifers, Los Angeles Country, California. Proceeding of the 1st international conference and workshop on saltwater intrusion and coastal aquifers, monitoring, modelling, and management, Morocco
25. Maimone M, Fitzgerald R (2001) Effective modelling of coastal aquifers systems. Proceeding of the 1st international conference and workshop on saltwater intrusion and coastal aquifers, monitoring, modelling, and management, Morocco
26. Sherif M, Kacimov A (2008) Pumping of brackish and saline water in coastal aquifers: an effective tool for alleviation of seawater intrusion. Proceeding of 20th SWIM, Naples, FL, USA
27. Kacimov AR, Sherif MM, Perret JS, Al-Mushikhi A (2009) Control of sea-water intrusion by salt-water pumping: coast of Oman. Hydrgeol J 17:541–558
28. Paniconi C, Khlaifi I, Lecca G, Giacomelli A, Tarhouni J (2001) A modelling study of seawater intrusion in the Korba plain, Tunisia. Elsevier Phys Chem Earth (B) 26(4):345–351
29. Barrocu G, Cau P, Muscas L, Soddu S, Uras G (2004) Predicting groundwater salinity changes in the coastal aquifer of Arborea (central-western Sardinia). In: Proceeding of the 18th Salt Water Intrusion Meeting, Cartagena (Spain)
30. Johnson T, Reichard E, Land M, Crawford S (2001) Monitoring, modelling, and managing saltwater intrusion, central and west coast groundwater basins, Los Angeles Country, California. Proceeding of the 1st international conference and workshop on saltwater intrusion and coastal aquifers, monitoring, modelling, and management, Morocco
31. Fitzgerald R, Riordan P, Harley B (2001) An integrated set of codes to support a variety of coastal aquifer modelling approaches, first international conference on saltwater intrusion and coastal aquifers monitoring, modeling, and management. Essaouira, Morocco
32. Mahesha A (1996c) Control of seawater intrusion through injection-extraction well system. J Irrigat Drain Eng ASCE 122:314–317
33. Rastogi AK, Choi GW, Ukarande SK (2004) Diffused interface model to prevent ingress of seawater in multi-layer coastal aquifers. J Spec Hydrol 4(2):1–31
34. Tanapol S, Chaowarin W, Decho P, Kittitep F (2012) Physical model simulations of seawater intrusion in unconfined aquifer. Songklanakarin J Sci Technol 34(6):679–687
35. Acostaa A, Donadoa LD (2015) Laboratory scale simulation of hydraulic barriers to seawater intrusion in confined coastal aquifers considering the effects of stratification, 7th groundwater symposium of the international association for hydro-environment engineering and research. Proc Environ Sci 25:36–43
36. Zheng C, Bennett GD (2002) Applied contaminant transport modelling, 2nd edn. Wiley, New York
37. Abd-Elhamid HF (2010) A simulation-optimization model to study the control of seawater intrusion in coastal aquifers, PhD thesis, University of Exeter, UK

38. Holland JH (1975) Adaptation in natural and artificial systems. University of Michigan Press, Ann Arbor
39. Goldberg DE (1989) Genetic algorithms in search, optimization and machine learning. Addison-Wesley-Longman, Reading, MA
40. Davis L (1991) Handbook of genetic algorithm. Van Nostrand Reinhold, New York
41. Michalewicz Z (1992) Genetic algorithm+data structures=evolutionary programs. Springer, New York
42. Mckinney DC, Lin MD (1995) Approximate mixed-integer nonlinear programming methods for optimal aquifer remediation design. Water Resour Res 31(3):731–740
43. Haupt R, Haupt SE (1998) Practical genetic algorithms. Wiley, Reading, NY
44. Mckinn DC, Lin MD (1994) Genetic algorithm solution of groundwater management models. Water Resour Res 30(6):1897–1906
45. Rogers LL, Dowla FU (1994) Optimization of groundwater remediation using artificial neural networks with parallel solute transport modeling. Water Resour Res 30(2):457–481
46. El Harrouni K, Ouazar D, Walters GA, Cheng AH-D (1998) Groundwater optimization and parameter estimation by genetic algorithm and dual reciprocity boundary element method. Eng Anal Boundary Manag 124:129–139
47. Aly AH, Peralta RC (1999) Optimal design of aquifer cleanup systems under uncertainty using a neural network and a genetic algorithm. Water Resour Res 35(8):2523–2532
48. Wang QJ (1987) The genetic algorithm and its application to calibration of conceptual rainfall-runoff models. Water Resour Res 27(9):2467–2471
49. Franchini M (1996) Use of a genetic algorithm combined with a local search method for the automatic calibration of conceptual rainfall-runoff models. J Hydrol Sci 41(1):21–40
50. Mulligan AE, Brown LC (1998) Genetic algorithm for calibrating water quality models. J Environ Eng 124(3):202–211
51. Murphy LJ, Simpson AR, Dandy GC (1993) Design of a pipe network using genetic algorithms. Water 20:40–42
52. Farmani R, Walters GA, Savic DA (2005) Trade-off between total cost and reliability for Anytown water distribution network. J Water Resour Plan Manag 131(3):161–171
53. Farmani R, Savic DA, Walters GA (2005) Evolutionary multi-objective optimization in water distribution network design. J Eng Optim 37(2):167–185
54. Das A, Datta B (1999) Development of multiobjective management models for coastal aquifers. J Water Resour Plan Manag 125:76–87
55. Das A, Datta B (1999) Development models for sustainable use of coastal aquifers. Irrigat Drain Eng 125:112–121
56. Das A, Datta B (2000) Optimization based solution of density dependent seawater intrusion in coastal aquifers. J Hydrol Eng 5(1):82–89
57. Gurdu F, Motz L H, Yurtal R (2001) Simulation of seawater intrusion in the Goksu Delta at Silifke, Turkey. Proceeding of the 1st international conference and workshop on saltwater intrusion and coastal aquifers, monitoring, modelling, and management, Morocco
58. Gordon E, Shamir U, Bensabat J (2001) Optimal extraction of water from regional aquifer under salinization. J Water Resour Plan Manag 127(2):71–77
59. Rao KLN, Rao PB (2004) Salinization of coastal fresh water aquifers of Andhra Pradesh, India. Proceeding of the 18th salt water intrusion meeting, Cartagena, Spain
60. Javadi AA, Abd-Elhamid HF, Farmani R (2011) A simulation-optimization model to control seawater intrusion in coastal aquifers using abstraction/recharge wells. Int J Numer Anal Methods Geomech. https://doi.org/10.1002/nag.1068
61. Javadia AA, Hussaina M, Sherif M (2013) Optimal control of seawater intrusion in coastal aquifers, international conference on computational mechanics (CM13), 25–27 March 2013, Durham, UK
62. Abd-Elhamid HF, Javadi AA (2011) A cost-effective method to control seawater intrusion in coastal aquifers. J Water Resour Manag 25:2755–2780

63. Abd-Elhamid HF, Javadi AA (2008) An investigation into control of saltwater intrusion considering the effects of climate change and sea level rise. Proceeding of 20th SWIM, June 23–27, 2008, Naples, FL, USA
64. Javadia AA, Hussaina M, Sherif M, Farmani R (2015) Multi-objective optimization of different management scenarios to control seawater intrusion in coastal aquifers. Water Resour Manage 29:1843–1857
65. Sherif MM, Singh VP (1999) Effects of climate change on sea water intrusion in coastal aquifers. Hydrol Process J 13:1277–1287
66. Gaamea OMA (2000) Behavior of the transition zone in the Nile Delta aquifer under different pumping schemes, PhD thesis, Cairo University, Egypt
67. Nasar MKK (2004) Design and management of discharge and recharge well field for coastal desalination plants. MSc Thesis, Cairo University, Egypt
68. Abd-Elhamid HF (2016) Investigation and control of seawater intrusion in the eastern Nile Delta aquifer considering climate change. Water Sci Technol. https://doi.org/10.2166/ws.2016.129
69. Bear J, Cheng AH, Sorek S, Quazar D, Herrera I (1999) Seawater intrusion in coastal aquifers, concepts, methods and practices. Kluwer Academic Publisher, Dordrecht, The Netherlands. ISBN 0-7923-5573-3
70. Scholze O, Hillmer G, Schneider W (2002) Protection of the groundwater resources of Metropolis CEBU (Philippines) in consideration of saltwater intrusion into the coastal aquifer. In: Proceeding of the 17th Salt Water Intrusion Meeting, Delft, The Netherlands
71. Bhattacharjya RK, Datta B (2004) An ANN-GA approach for solving saltwater intrusion management problem in coastal aquifers. Proceeding of the 18th salt water intrusion meeting, Cartagena, Spain
72. Dokou Z, Dettoraki M, Karatzas G, Varouchakis1 E, Pappa A (2016) Utilizing successive linearization optimization to control the saltwater intrusion phenomenon in unconfined coastal aquifers in Crete, Greece. J Environ Model Assess. https://doi.org/10.1007/s10666-016-9529-z
73. Johnson VM, Rogers LL (1995) Location analysis in ground-water remediation using neural networks. Ground Water 33(5):749–758
74. Zheng C, Wang PP (2001) A field demonstration of a simulation-optimization approach for remediation system design, Ground Water
75. Chan Hilton AB, Culver TB (2000) Constraint handling for genetic algorithms in optimal remediation design. J Water Resour Plan Manag 126(3):128–137
76. Maskey S, Jonoski A, Solomatine DP (2002) Ground water remediation strategy using global optimization algorithms. J Water Resour Plan Manag 431–439
77. Chan Hilton AB, Culver TB (2005) Groundwater remediation design under uncertainty using genetic algorithms. J Water Resour Plan Manag 131(1):25–34
78. Georgpoulou E, Kotronarou A, Koussis A, Restrrepo PJ, Gomez-Gotor A, Jimenez J, Rodiguez J (2001) A methodology to investigate brackish groundwater desalination coupled with aquifer recharge by treated wastewater as an alternative strategy for water supply in Mediterranean areas. Desalination 136:307–315. Elsevier

Part VII
Delineation of Groundwater Flow and Seawater Intrusion

Use of One-Dimensional Subsurface Temperature Profiles to Characterize the Groundwater Flow System in the Northwestern Part of the Nile Delta, Egypt

Zenhom El-Said Salem and Osman M. Osman

Abstract The temperature-depth profiles of 47 boreholes were investigated to characterize the groundwater flow system in the northwestern part of the Nile Delta, Egypt (hereafter referred to as the study area). A vertical subsurface thermal system was used for the investigation.

The groundwater was recharged in the reclaimed area located in the southwestern direction of the study area, where high subsurface temperatures were recorded. The discharge regions were located in the old agricultural lands (northern and northeastern areas) and were characterized by low subsurface temperatures compared with those in the recharge area. This abnormal thermal system was attributed to contrasts in the surface air temperature, where higher values were recorded in the recharge area (Wadi El-Natrun station; annual average temperature 23.15°C) and lower values were detected in the old agricultural lands (Damanhour station; annual average temperature 20.37°C). Regardless of this unusual framework, the geothermal gradient was low in the recharge territory (average 0.0198°C/m) and high in the discharge area (average 0.0343°C/m).

The effect of irrigation canals on the thermal system was detected from the constructed vertical two-dimensional (2D) cross-sections, where the gaining streams were underlined by a warm zone, while the losing streams were underlined by a cooler zone. Vertical groundwater fluxes in the study area were assessed from a comparison between measured and simulated one-dimensional (1D) temperature

---

Z. E. Salem (✉)
Geology Department, Faculty of Science, Tanta University, Tanta, Egypt
e-mail: Zenhomsalem@yahoo.com; zenhoum.eisa@science.tanta.edu.eg

O. M. Osman
Geology Department, Faculty of Science, Damanhour University, Damanhur, Egypt

A. M. Negm (ed.), *Groundwater in the Nile Delta*,
Hdb Env Chem (2019) 73: 387–424, DOI 10.1007/698_2018_248,

profiles. All temperature profiles related to wells in the Wadi El-Natrun station area were of the recharge type, with groundwater flux (*U*) values ranging from 0.3 to 1.5 m/year, except for well 21, which was the discharge type (*U* range −0.1 to −0.5 m/year). The temperature profiles related to Damanhour station were of the discharge type (*U* range −0.1 to −1.5 m/year), except for wells 22, 47, and 39, which were the recharge type (*U* range 0.1 to 0.5 m/year).

We concluded that the geothermal gradient is superior to the thermal system for following the groundwater stream framework in regions such as those we examined.

Keywords Northwestern Nile Delta, Groundwater flow system, Subsurface thermal regime, Landuse change, One dimensional thermal modeling

Contents

1 Introduction

It is known that water table data alone are not sufficient for calibrating groundwater flow models, and assessments of groundwater flux, as well as information on solute development and heat flow, are required for model adjustment [1]. Subsurface heat exchange occurs everywhere on Earth, at higher rates in some areas than in others. Geothermal heat is considered to be a good tracer for various geologic processes; the best example of these processes is the groundwater flow system [2], in which temperature profiles can be utilized to gauge the recharge and discharge rates of groundwater, the effects of surface warming, the groundwater-surface water relationship, the hydraulic conductivity of the stream base layer, and basin penetrability [1].

Stallman [3] was one of the early specialists who investigated the potential for utilizing subsurface temperature to estimate groundwater velocity and aquifer hydraulic conductivity. The pioneering hypothetical work of Suzuki [4], Stallman [5], and Bredehoeft and Papadopulos [6] demonstrated that the subsurface temperature could be utilized to analyze one-dimensional (1D) heat transfer in order to estimate groundwater velocity. The augmentation of this kind of investigation to assess groundwater upward and downward movement rates, interaction with surface water, and flow in fractures, as well as the impact of surface warming, is considered here in relation to the temperature-depth profile.

The utilization of subsurface temperature at the basin scale was spearheaded by Cartwright [7], who utilized the Bredehoeft and Papadopulos [6] model to estimate the velocity of groundwater upward movement in the Illinois Basin. Domenico and Palciauskas [8] introduced a 2D analytical solution regarding heat transfer in a vertical cross-section of a basin. Subsequent utilization of analytical solutions in regard to sedimentary basins was essentially made by Japanese investigators, including Sakura [9], Inagaki and Taniguchi [10], Taniguchi [11, 12], and Dapaah-Siakwan and Kayane [13]. Numerous specialists became interested in the investigation of numerical solutions for coupled groundwater and heat transfer models in two and three dimensions. The best monograph on heat transfer in a groundwater basin was published in 1989 and distributed by the American Geophysical Union [14].

Taniguchi et al. [15] reported that surface temperatures in Tokyo had increased by 0.025°C/year in the past 100 years, and they computed vertical groundwater flux by comparing their temperature profiles with those measured by Dapaah-Siakwan and Kayane [13]. Ferguson and Woodbury [16] further developed the model crafted by Taniguchi et al. [15] by utilizing a numerical model to include measured ground surface temperature (GST) at the upper boundary of the model. Different scientists in Japan explored the subjective impact of surface warming on temperature profiles, e.g., Sakura et al. [17]. Salem et al. [18, 19] inferred that current temperature profiles measured in the Nagaoka region were influenced by surface warming to a depth of 60 m. Related work explored the impact of land use on subsurface temperatures [20–23]. Investigations the consequences of surface warming on groundwater are limited to date; however, future investigations will be valuable in determining the potential impact of worldwide warming on water resources [24].

Changes in climatic conditions, particularly surface warming, influence the subsurface temperature, and have been recorded as inversions in subsurface temperature profiles [25]. Human activities affect the thermal conditions of the shallow subsurface under urbanized areas and create what is called subsurface urban heat islands. These islands have been detected in various regions around the world; for example, underneath Osaka, Japan [26]; Albuquerque, New Mexico, United States [27]; Winnipeg, Canada [28]; urban areas in England [29, 30]; and some urban areas in Germany [31, 32].

All these investigations dealing with the relationship between the subsurface temperature and the groundwater flow system assume that the temperature profile in a well is illustrative of the temperature in the relevant aquifer. In open and screened boreholes, differences in water levels of a regional scale flow system that interface with a well create vertical flow inside the borehole [33]. These differences in hydraulic heads are attributed to nearby flow conditions: downward flow in recharge zones and upward flow in discharge zones. The resulting vertical flow within the borehole significantly disturbs the temperature profile [34, 35].

Gravity-driven groundwater flow is associated with a forced convective heat flow and produces cooling and a low geothermal gradient under groundwater recharge regions, an undisturbed thermal system and gradient underneath horizontal groundwater flow domains, and higher temperatures and gradients underneath groundwater discharge zones [15]. Such a framework could be regarded as a typical subsurface heat framework caused by the groundwater stream. In the area of the present study, an abnormal thermal regime was recognized, where the recharge area had a higher groundwater temperature than that of the discharge area. Accordingly, the purpose of this study was to explain this thermal system and use it to clarify the regional groundwater flow system in the area.

Studies dealing with the relation between subsurface temperature and groundwater flow systems in Egypt have been performed recently. Salem and Osman [36] and Salem and El Bayumy [37] modeled the vertical 2D groundwater flow and heat transport to estimate vertical groundwater flow velocities in the northwestern part of the Nile Delta (present study area) and the east Wadi El-Natrun region, respectively. They concluded that permeability was the controlling parameter for the groundwater flow system. Salem et al. [38] and Salem [39] used temperature-depth profiles to depict the groundwater stream framework in the middle Nile Delta and at Wadi El-Assiuty in the Nile Valley, respectively. Also, borehole temperatures were used to show the surface water-groundwater interaction [40]. Salem [41] used the subsurface thermal regime to investigate the paleo-groundwater flow system in Al Kufra basin, Libya, as a case study for an arid area.

One of the principal national concerns of Egyptian governments during the past five decades has been the reclaiming of Egyptian deserts, especially the desert located in the west Nile Delta. Numerous horticultural land improvement projects have been undertaken in the west Nile Delta, depending on the surface water and groundwater. Since these improvement practices have an impact on groundwater quantity, quality, and flow system, the present study relied on the usage of borehole temperatures to trace groundwater stream lines and to assess groundwater vertical flux rates in the study region.

2 Study Area

The study area is located in the northwestern part of the Nile Delta (Fig. 1a, b) and is bounded by latitudes 30° 30′ and 31° 00′ N and longitudes 30° 00′ and 30° 30′ E. In this area, the eastern part of the El-Nubaria canal is categorized as belonging to old agricultural land (young alluvial plain). The western El-Nubaria canal area belongs to the old alluvial plain and is one of the desert areas recently reclaimed for agricultural purposes (see Fig. 2). Many canals and drains were recently constructed for irrigation purposes (see Fig. 3). These waterways play important roles in groundwater recharge and discharge, water level changes, flow direction, and water quality. A connection between these surface water streams and the shallow groundwater is assumed because these canals and drains were constructed through the Quaternary sediments of the present aquifer. The principal irrigation canals in the investigated area are the El-Nubaria, Ferhash, El-Hager El-Nasr, El-Bustan, and El-Bustan extension canals.

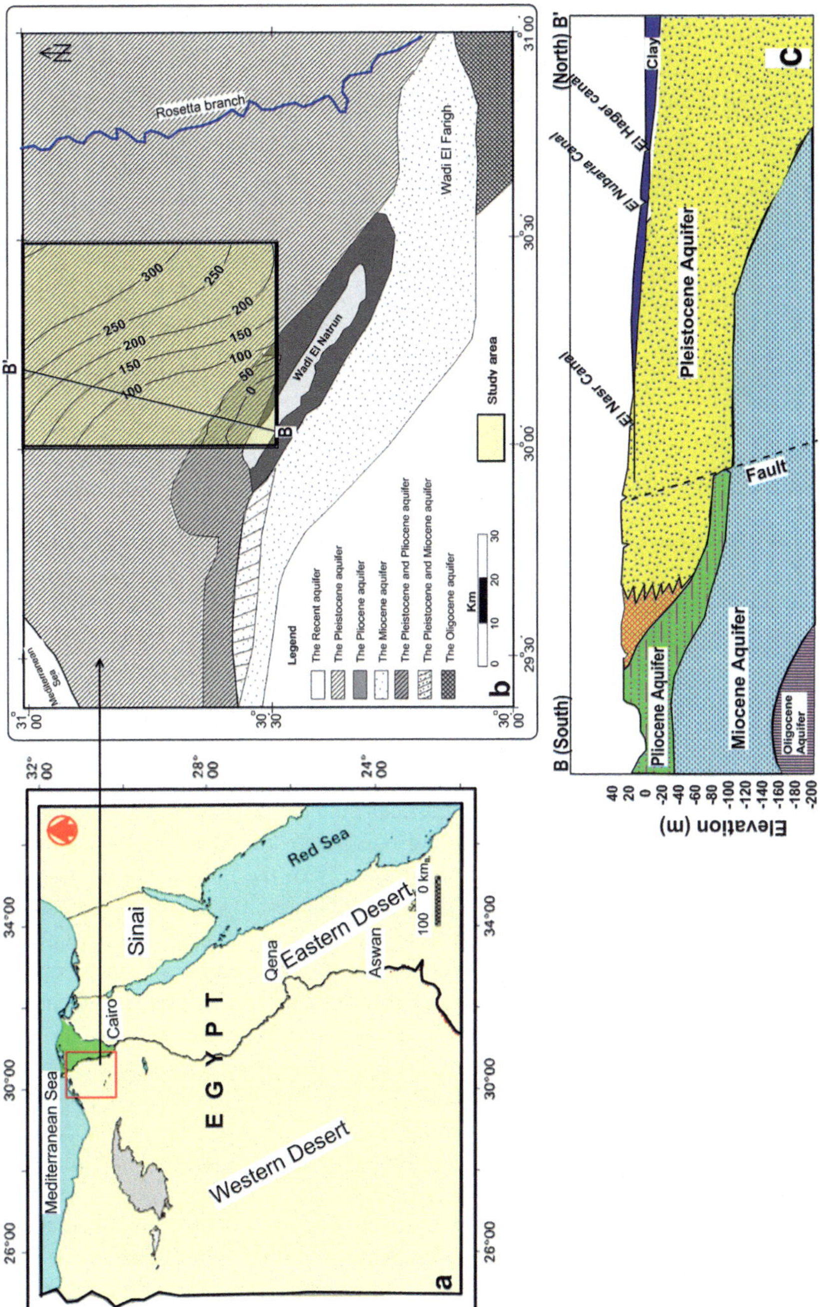

Fig. 1 Location maps of the study area (**a**, **b**). Hydrogeological map of the western Nile Delta and aquifer thickness in the study area (**b**) and southwest-northeast hydrogeological cross-section of the study area (**c**)

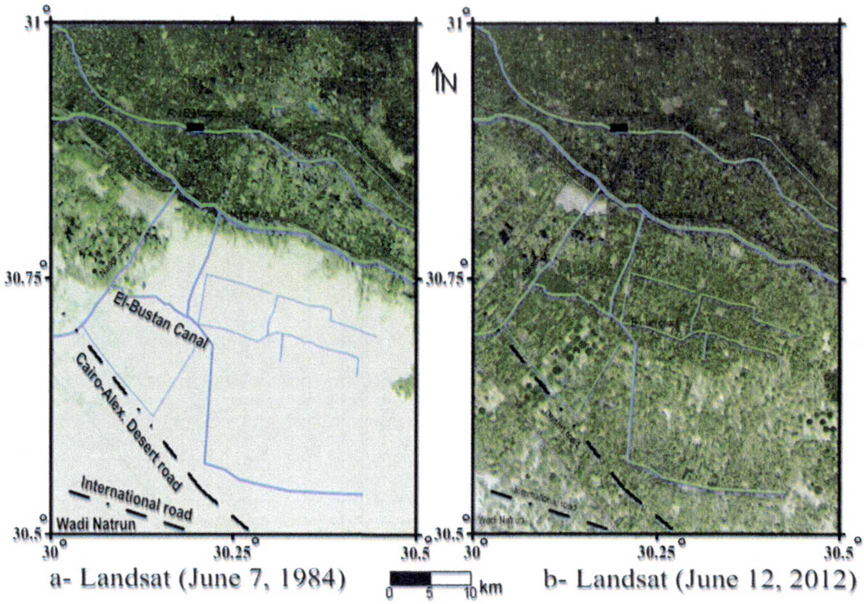

Fig. 2 Maps indicate land use change where the desert land shown in (**a**) (1984) was completely reclaimed by agricultural activities in 2012 (**b**). Modified from [42]

3 Hydrogeology

The western Nile Delta area has been the subject of previous intensive hydrogeological, hydrogeochemical, and geophysical studies. Among these studies is the work of Saad [43], who stated that the groundwater flow direction in the area is from northeast to southwest, and the transmissivity of the Pleistocene aquifer is 0.615–0.897 m^2/min. El Fayoumy [44] reported that Quaternary sediments represent the soil cover and constitute the main aquifer, which is mostly clastic with dominant sand facies. He added that the subsurface sedimentary section in the western Delta is about 4,000 m thick, and rests upon the basement complex, with a complicated structure system and many unconformity surfaces. The Desert Institute, in an internal report [45], stated that the groundwater level in the entire area increased by about 2.5–3 m from March 1972 to March 1973. Before the reclamation projects began, the groundwater seeped from the El-Nubaria canal to the west, but since the construction of the pumping stations, the groundwater has been flowing towards the El-Nubaria canal, where it acts as a drain.

Mabrouk [46] used an electrical resistivity technique to determine the successive layers and their thicknesses between Wadi El-Natrun and the El-Nasr canal area. He noted that the geologic succession consisted of three main layers – a surface layer of aeolian sand, loam, and sandy gravel – and two subsurface layers, consisting of deltaic deposits (thickness 90–347 m) and an impervious clay bed. El Ghazawi [47]

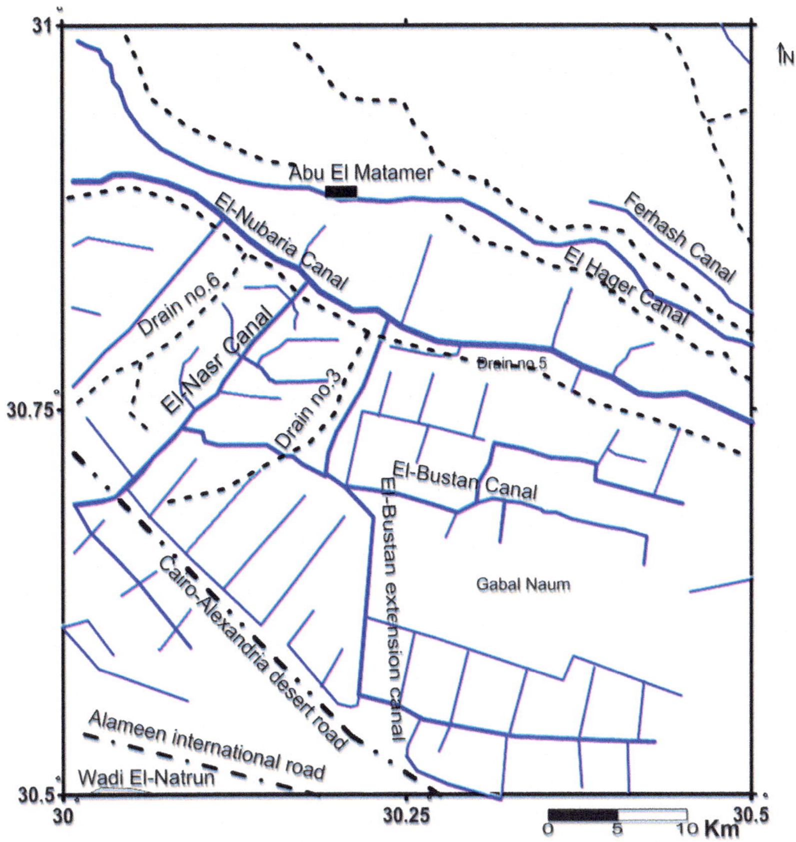

Fig. 3 Surface water irrigation system in the study area

proved that the El Galaa and El-Nubaria faults were the most important factors influencing the groundwater conditions in the area. Abdel Baki [48] stated that the recharge in the area arose from two sources, the groundwater inflow and the water infiltrated from irrigation, and he found that the groundwater depth varied widely, from a few meters close to the Delta to about 60 m near the Cairo-Alexandria desert road. Abdel Baki [48], Ismail [49], and Gomaa [50] noted that the water-bearing formations in the western Nile Delta were Recent, Pleistocene, Pliocene, Miocene, and Oligocene aquifers. Gomaa [50] showed that all the western Nile Delta reservoirs were recharged from the Nile Delta Quaternary reservoir.

Abdel Baki [48] and Diab and Dahab [51] stated that the Rosetta branch of the Nile and the El Rayah, El Behiri, and El Rayah El Nasseri canals cut through sands, and these authors noted that these canals have a direct connection with the Quaternary aquifer and play an important role in the water regime of the area. El Sheikh [52] mentioned that the general groundwater flow north of Wadi El-Natrun is from the south and southeast to the direction of the Wadi El-Natrun depression. He added that the Nile Delta reservoir is the most important recharging source for the southern part of the area, as well as being an important source of seepage from the surface water system. Embaby [53] noted that if irrigation systems and management were to be improved, soil surface and groundwater salinization would be reduced. He suggested installing an artificial drainage system to avoid the problem of waterlogging. The groundwater in the Quaternary aquifer generally exists in a free water table condition. The depth of the fresh water body, which lies under a saline water body, decreases toward the west of the area.

Based on a geoelectric resistivity technique, Nawar [54] found that the depth to the fresh water-bearing zone increased from the southwest towards the northeast in the Nile Delta area. The resistivity of the fresh water bearing-zone also increased from the southwest towards the northeast. He added that the general groundwater flow was from the southern parts (recharge area) to the northern and northeastern (discharge) parts of the Nile Delta area. Fouad and Zhi [55] stated that because of the direct hydraulic contact between surface water bodies and the Nile Delta aquifer, any change in the flow and water table would affect the groundwater levels and the regional water balance, owing to surface water/groundwater interaction and the complex flow system. They found that the highest water levels in the northwestern parts of the area were due to the thinness of the aquifer. The total amount of groundwater taken for irrigation is larger than the groundwater recharge, and this factor, will, in the long term, undermine the capacity and sustainability of the aquifer, and will lead to saltwater intrusion.

Abdel-Raouf and Abdel-Galil [56] have stated that the thickness of the Pleistocene aquifer is 200 m in the central parts and increases toward the north, reaching 350 m, while the thickness decreases toward the south, thinning out in the Wadi El-Natrun area. The saltwater intrusion increases toward the northern and southern portions owing to seawater intrusion and the shallowness of the Miocene aquifer. El Arabi and Morsy [57] have stated that the regional flow direction in the Nile Delta area is from south to north. In areas with high recharge from local reclamation water conveyors in old reclamation projects that use surface water, groundwater mounds exist owing to the downward seepage of excess irrigation water and leakage from irrigation.

Salem et al. [58] investigated the hydrogeochemical changes accompanying reclamation processes in the small Abu Mina basin, which is part of the western Nile Delta region. These authors compared the groundwater data in pre-reclamation (1974) and post-reclamation (2008) periods, and found that the groundwater seemed to have been subjected to many changes: a rise in water level, modification of the flow system, improvement of water quality, and the addition of new salts through dissolution processes.

Salem and El Bayumy [59] studied the groundwater aquifer in the area east of Wadi El-Natrun and they stated that

> The studied groundwater flows from the east and north directions where the surface water sources are located to the west and south directions. Wadi El-Natrun is a natural discharge area for the groundwater and the hydraulic gradient increases gradually from east to west which could be related to topographic and fault effects. Water table change between 1960

and 2010 showed a drawdown of groundwater levels reached about 25 m mainly in the southern part of the region. Seepage from the irrigation channel and excess uses of surface irrigation water at the northern part of the area lead to water level rise about 5 m to the east and 10 m to the west. In the southern part, to avoid wells drying and aquifer depletion, it is recommended to regulate the pumping rate of the current wells and construct surface channel network to minimize groundwater pumping.

As mentioned by Salem and Osman [36, 42], the groundwater in the investigated region is contained in sediments of the Quaternary (Recent and Pleistocene), Neogene (Pliocene and Miocene), and Oligocene ages (Fig. 1b, c). The main aquifer in the area is the Pleistocene aquifer, where its thickness increases to the east and northeast, reaching 300 m in the northeastern part (see Fig. 1a). The sediments of the Holocene age are located in the western part of the area and are composed of recently formed aeolian sand conglomerations (4- to 10-m-thick) with intercalations of calcareous deposits (Fig. 4b). The Pleistocene aquifer includes Nile gravels and sands with thin clay intercalations (Fig. 1b, c). The underlying sediments of this aquifer are clay-rich Pliocene and Miocene sediments.

The Quaternary aquifer is unconfined in the western part of the El-Nubaria Canal, while it is semi-confined in the eastern part, owing to the effect of the clay cover of the young alluvial plain where the old cultivated land is located (Fig. 1c). The main groundwater recharge source of the present aquifer is seepage from the constructed canal and drain network [48]. According to Salem and Osman [36], the water table changes from 2 mbsl in the northern parts (Abu El Matameer zone) up to a level of 27 mbsl in the southwestern direction (north of Wadi El-Natrun) (Fig. 4). The groundwater mainly flows from the southwestern corner to the northern and northeastern directions. Local groundwater flow was noted in the southern part. An extra groundwater flow to Wadi El-Natrun in the southwestern course was also discerned.

Salem and Osman [60] studied the freshwater and saltwater interaction in the study area using a geoelectrical survey, and they stated that "two zones of groundwater quality were delineated: the slightly freshwater zone in the southern part, with resistivity range of 15–90 Ω m, and the brackish water to saltwater zone, with a very low resistivity of <2 Ω m in the northwestern parts. In addition to tracing the freshwater-seawater contact zone, these authors detected three geoelectric layers. The surface layer was composed of sand, clay, and silt. Its resistivity ranged from 5 to 512 Ω, and the thickness varied from 1 to 25 m. The aquifer layer was composed of sand with intercalations of clay, with resistivity ranging from 15 to 90 Ω and thickness ranging from 25 to 120 m. The clay layer resistivity ranged from 2 to 15 Ω and thickness ranged from 2 to 69 m".

Abdelhameed et al. [61] investigated the sedimentological characteristics of the Quaternary aquifer in the study area and they stated that "the majority of the samples are medium to coarse clayey sand, moderately sorted and coarse skewed. This may support the multi-directional depositional currents. Wind action influences skin of the sediments. The samples lie within the field of river processes. Mineralogical analysis by X-ray diffractometry revealed that smectite is the most abundant clay mineral, followed by kaolinite, whereas illite is the less abundant clay mineral. The essential carbonate minerals are calcite and dolomite whereas non-carbonate minerals include quartz, feldspar, and hematite."

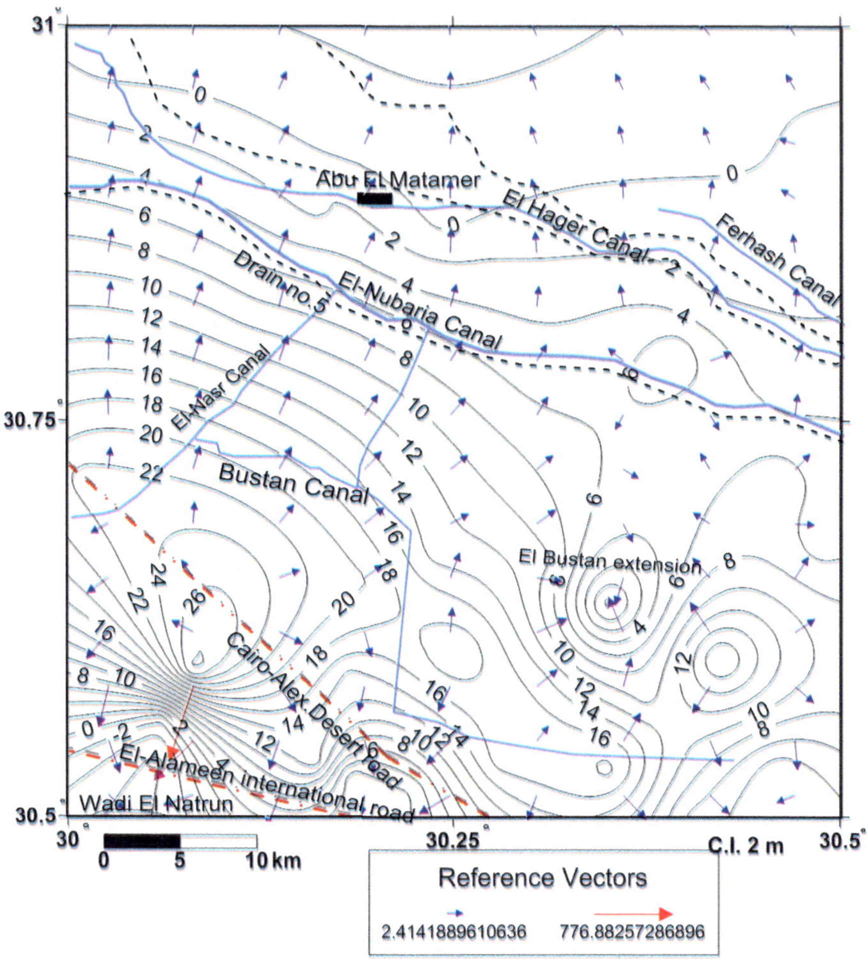

Fig. 4 Spatial distribution map of hydraulic heads in the study area. Modified from [42]

4 Methodology

A flow-chart illustrating the methodology of the present work is shown in Fig. 5. During 2010, the temperature was measured in 47 boreholes, at 5 to 10 m depth intervals, (Fig. 6), using thermistor thermometers, which have an accuracy of 0.1°C. Well screens (average length 20 m) were placed at the bases of the wells. The borehole depths ranged from 22 to 200 m (usually 80–120 m) and their average diameter was 20 cm. The wells used were encased, and were drilled at least 1 year before the temperature logging was performed. It is assumed that groundwater temperature in the boreholes is the same as the temperature of the aquifer around the well, where a physical temperature balance occurs between the well water

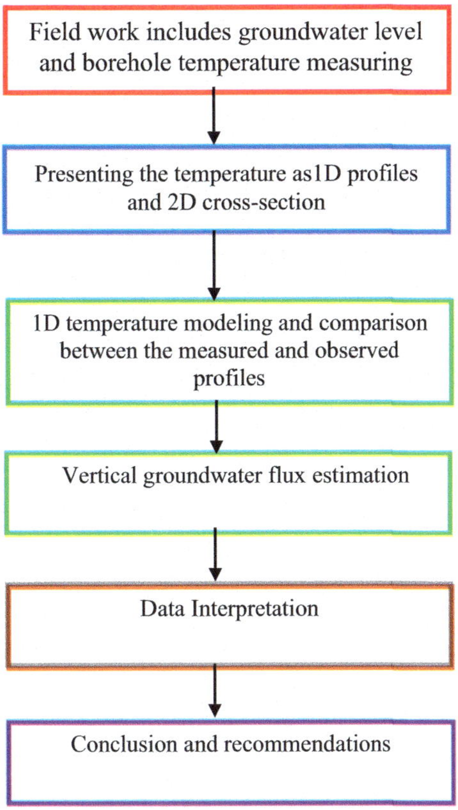

Fig. 5 Flow-chart illustrating the working steps of the current study. *1D* one-dimensional

temperature and the formation temperature. The measured borehole temperatures were plotted, and were classified as the Damanhour group and the Wadi El-Natrun group, and then 2D cross-sections were created parallel or perpendicular to the flow direction. Then 1D modeling was performed according to the methods mentioned by Carslaw and Jaeger [62] and Taniguchi et al. [15, 23]. The groundwater vertical velocity was then estimated and spatially plotted on the map of the area. The data used in this work were used by Salem and Osman [36] to model the subsurface thermal regime in the study area in two dimensions.

5 Climate

The climate in Egypt is dry, characterized by a hot summer and a short warm winter. Damanhour and Wadi El-Natrun are the two accessible stations in the climate investigation range, and their geographical positions are shown in Fig. 6. Changes in the mean annual air temperature with time are presented in Fig. 7. The annual average air temperatures at Wadi El-Natrun station and Damanhour station are

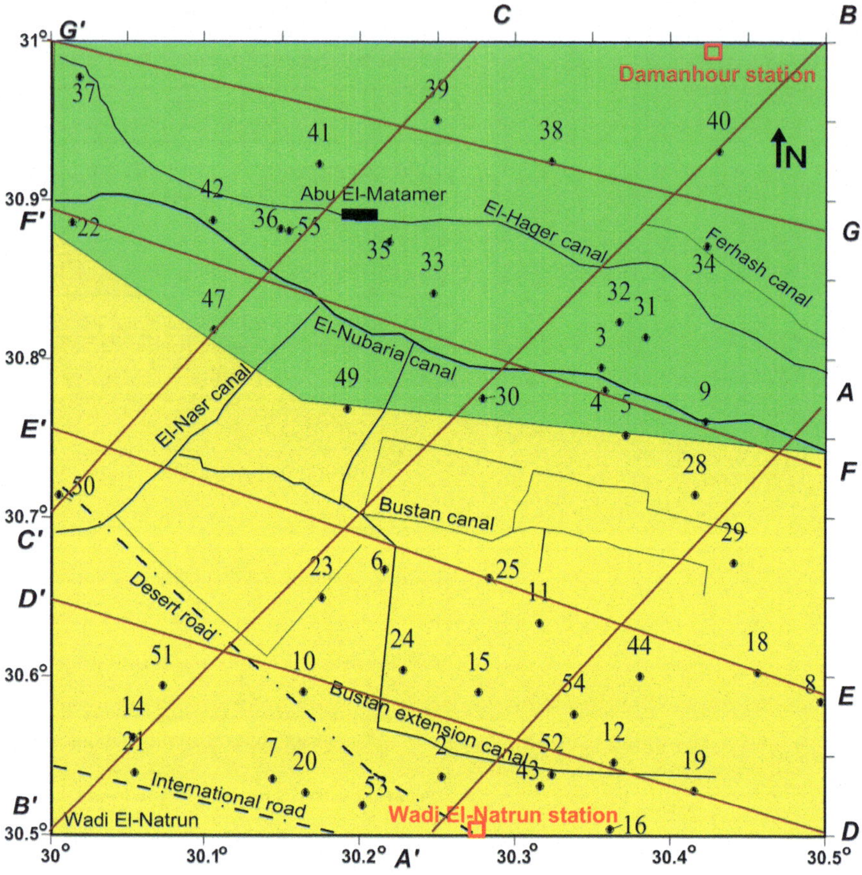

Fig. 6 Borehole location map. The yellow color background represents the recently reclaimed desert land and the green color represents the old cultivated land. Wadi El-Natrun and Damanhour stations are plotted. The brown lines indicate the locations of the hydraulic heads and temperature cross-sections

23.15°C and 20.37°C, respectively. During the 14 years before we carried out our study, the annual recorded air temperatures had a warming pattern at the Damanhour station and a cooling pattern at the Wadi El-Natrun station. The warming pattern detected at Damanhour station could be attributed to a general worldwide environmental change. The cooling pattern at Wadi El-Natrun station could be related to the effect of agricultural reclamation in the desert part of the western Nile Delta during the past few decades [36, 37, 59]. The estimated mean annual surface air temperature changes are 0.044°C/year in the Damanhour area and 0.03°C/year in the Wadi El-Natrun area.

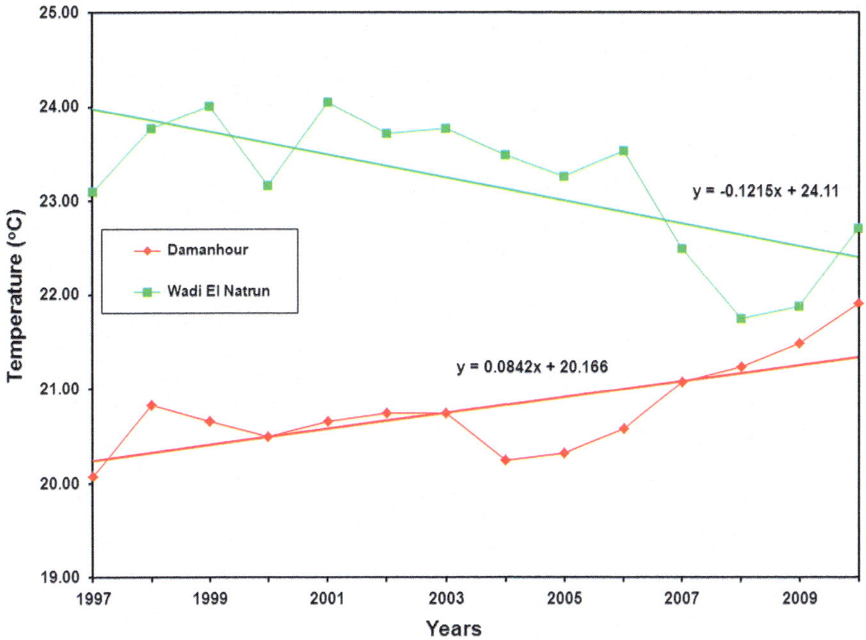

Fig. 7 Surface air temperature changes in the study area. Modified from [36, 37]

6 Results

6.1 One-Dimensional Temperature Profiles

The measured borehole temperature profiles were classified into two groups (Figs. 8 and 9) according to the locations of the wells (Fig. 6) and the groundwater flow system (Fig. 4), The wells in the first group are located in the reclaimed western desert area (El-Nubaria canal) where the recharge area is located. The wells in the second group are in the old developed terrain of the eastern El-Nubaria trench, where the groundwater mostly discharges. In general, the measured temperatures in the second group were lower than those in the first group. In contrast to the second group of wells, the first group had a lower geothermal gradient, ranging from 0.0145 (well 29) to 0.0326°C/m (well 21), with an average value of 0.0198°C/m. The geothermal gradient in the second group ranged from 0.016 (well 47) to 0.049°C/m (well 38), with an average of 0.0343°C/m.

Groundwater flow causes a forced convective heat flow that brings down temperatures underneath the main groundwater recharge region, with an undisturbed thermal pattern underneath regions of horizontal groundwater flow, and warming underneath the groundwater discharge regions. Accordingly, the present information points to an abnormal temperature framework, where the recharge area profiles had higher temperatures (first group) than those in the discharge zone (second group).

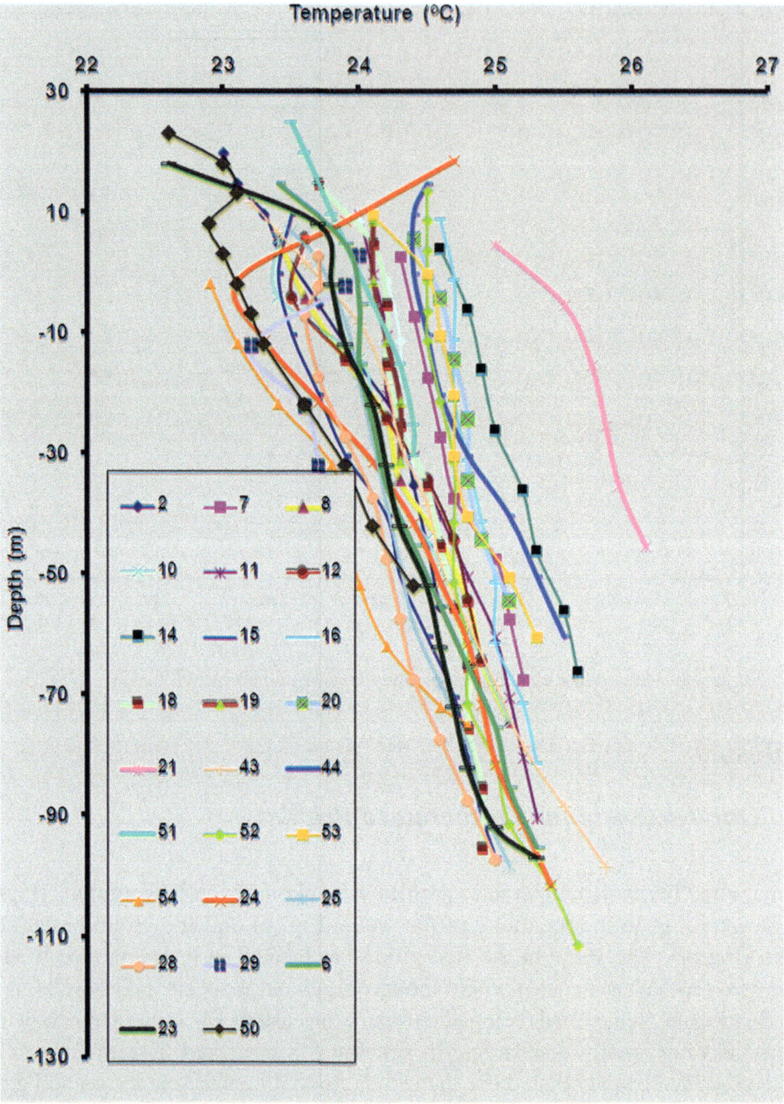

Fig. 8 Temperature depth profiles representing the subsurface temperatures in the newly reclaimed desert land

This unusual subsurface thermal framework could be identified by the differences in the average annual air temperatures, with the temperature in the recently reclaimed desert land being higher (23.15°C at Wadi El-Natrun station) than that in the old horticultural land (20.37°C at Damanhour station). The geothermal gradient is considered to be the principal thermal parameter for distinguishing between the recharge and the discharge zones, and the present geothermal gradient of the recharge zone was lower than that of the discharge zone, as mentioned before.

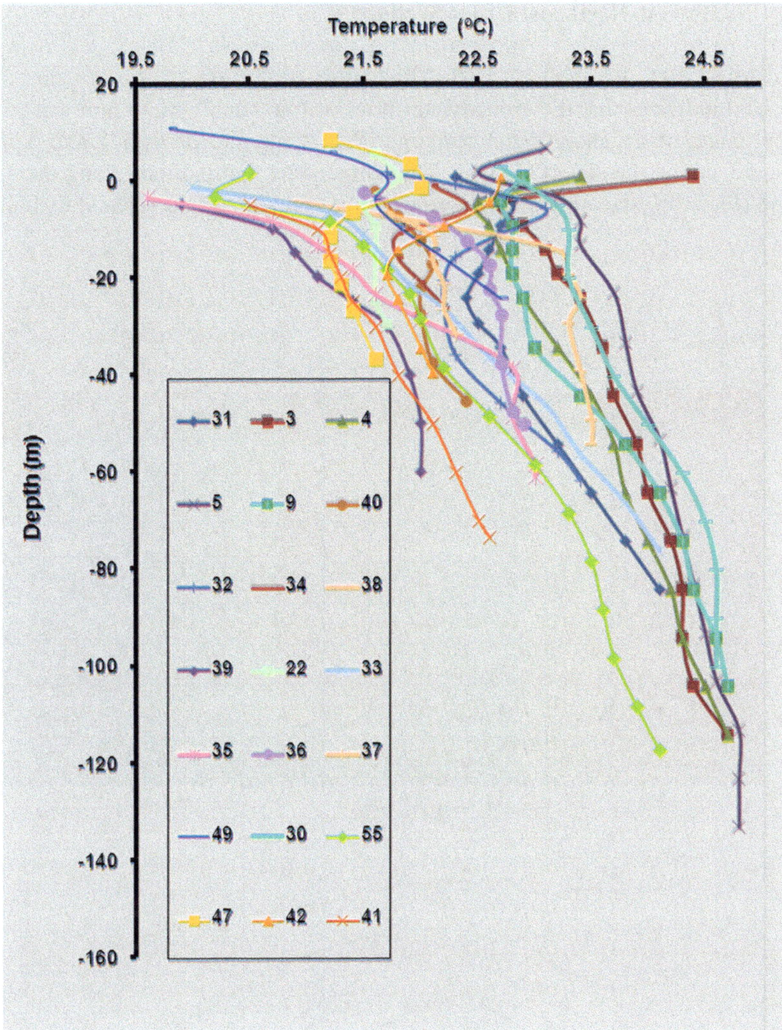

Fig. 9 Temperature depth profiles representing the subsurface temperatures in the old agricultural lands

6.2 *Vertical Two-Dimensional Subsurface Temperature Distribution*

Seven vertical 2D subsurface temperature distributions were selected (Fig. 6). Three cross-sections were matched with the groundwater flow direction (southwest-northeast), while four cross-sections were perpendicular or parallel to the groundwater flow direction (southeast-northwest).

6.2.1 Southwest-Northeast Cross-Sections

The vertical 2D distribution of the hydraulic heads along cross-section AA' (Fig. 10a) indicates that the groundwater flow is from southwest to northeast, with the existence of a general local recharge area at the site of well 1 and a local discharge area at the site of well 11. The temperature distributions along the same section (Fig. 10b) show high temperatures in the recharge area compared with those

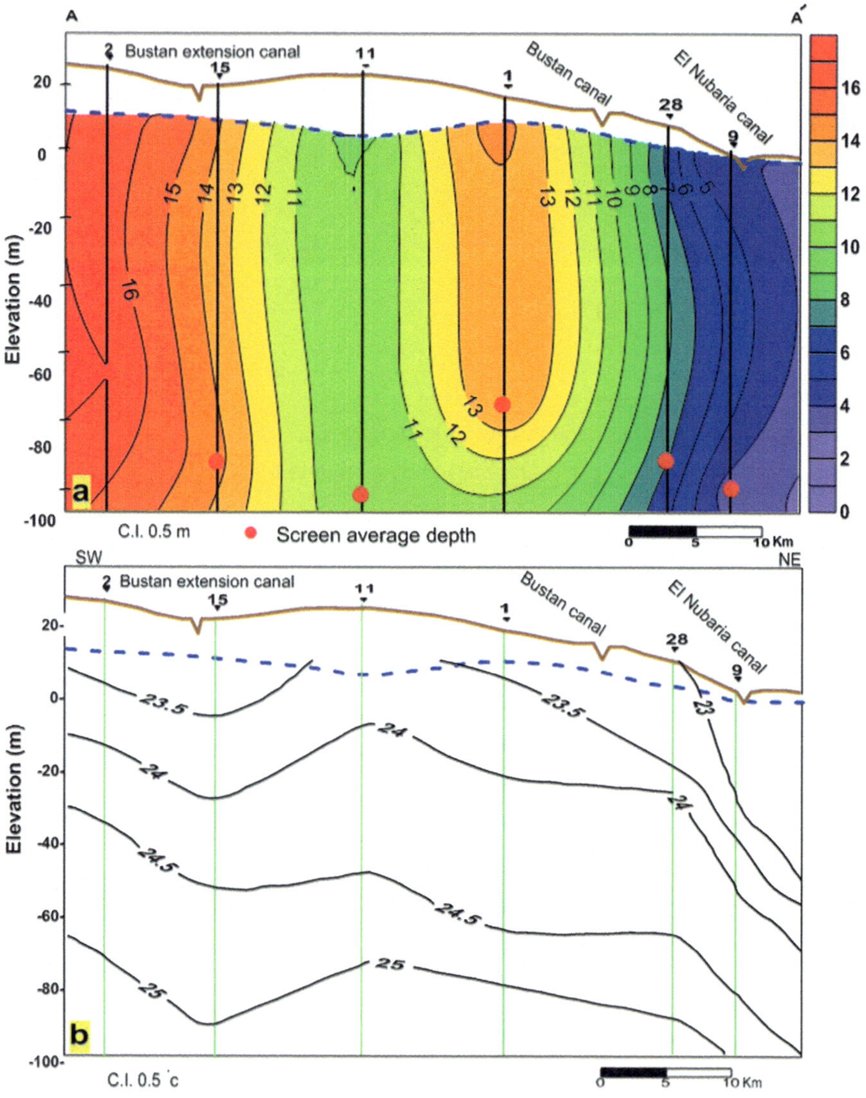

Fig. 10 Vertical 2D hydraulic heads (**a**) and subsurface temperature (**b**) distributions along the cross-section AA'. Temperature values are in °C and the vertical lines indicate the boreholes

in the discharge area. High temperatures were encountered in the local discharge area at the site of well 11, while low temperatures were noted in the local recharge area at the site of well 1.

As shown in Fig. 11a, there are two main flow systems; one towards the southwest to the Wadi El-Natrun direction, and the other trending to the northeastern

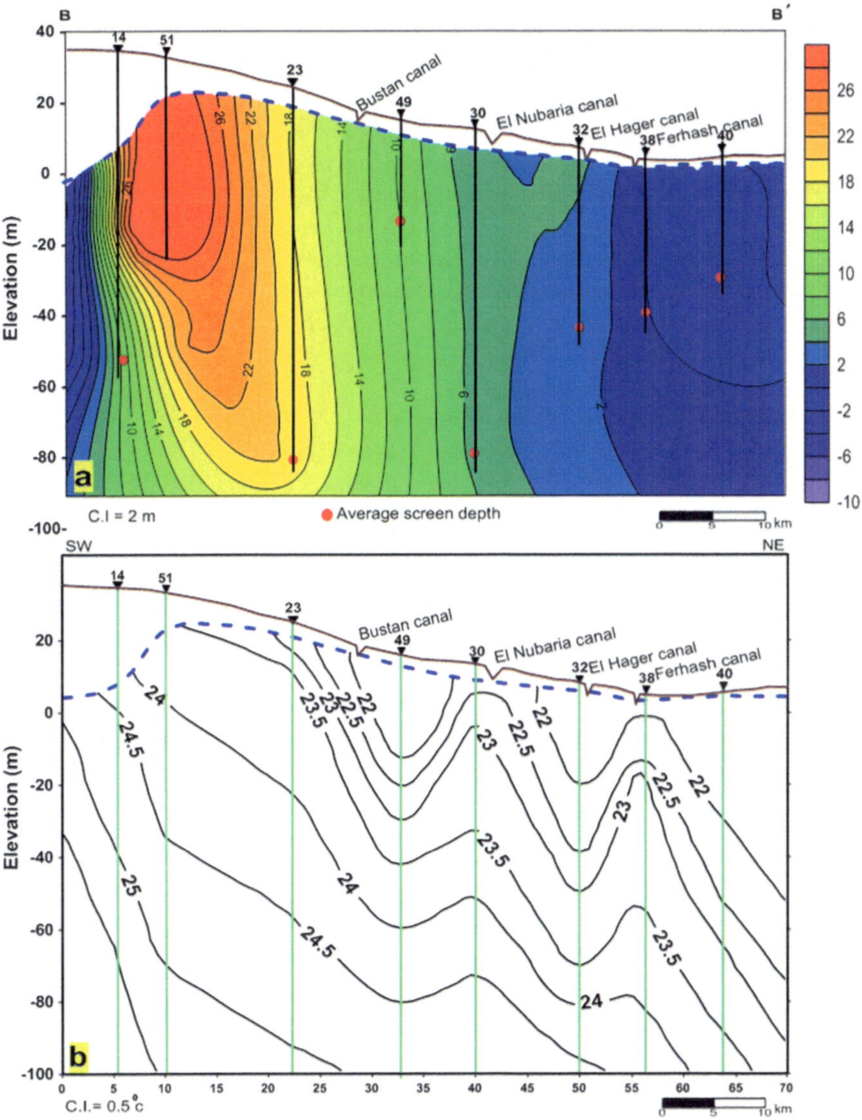

Fig. 11 Vertical 2D hydraulic heads (**a**) and subsurface temperature (**b**) distributions along the cross-section BB'. Modified from [36]. Temperature values are in °C and the vertical lines indicate the boreholes

direction, which has two local recharge areas at well sites 49 and 32 and two local discharge areas at well sites 30 and 38. Figure 11b shows the vertical 2D temperature distribution along the cross-section BB. The water discharged to the Wadi El-Natrun direction has a higher temperature than that of the recharge area. Isotherms are deeper along the regional flow system from the southwest to the northeast. Isotherms show higher values in the local discharge areas under the El-Nubaria and Ferhash canals and lower values in the local recharge areas under the El-Bustan and El-Hager canals. This means that, in these locations, the El-Nubaria and Ferhash canals are gaining streams, while the El-Bustan and El-Hager canals are losing streams.

Along the CC' cross-section, the hydraulic heads are high in the southwestern direction, owing to the higher topography, and they are decreased in the northeast, where the lowlands are located (Fig. 12b). Only one flow system from southwest to northeast is noted. Higher isothermal line values are recognized in the recharge area, while horizontal isothermal lines are noted in the northeastern direction between well sites 41 and 39 (Fig. 12b). Such horizontal isotherms indicate almost no flow conditions, a condition which may be caused by seawater intrusion. However, the El-Nubaria canal at this location acts as a losing stream, and the cooling effect of the seeped water from this canal is recorded in the area of well site 47.

6.2.2 Southeast-Northwest Cross-Sections

The cross-section DD' (Fig. 13a) shows the vertical 2D distribution of the hydraulic heads. Hydraulic heads are high in the southwestern direction and decrease to the southeast toward the lowlands. The vertical 2D temperature distributions along the cross-section DD' (Fig. 13b) show a cooling effect on the subsurface temperature owing to groundwater seepage from the El-Bustan extension and El-Nasr canals. Hydraulic heads are high in the northwest of the cross-section EE' owing to the location of the main recharge area and water seepage from the El-Nasr canal (Fig. 14a). The water table elevation decreases in a southeastern direction with a high local value at well site 44. Low isothermal line values were recorded in the northwest in the proposed recharge area, whereas higher values were seen in the southeast, owing to groundwater discharge (Fig. 14b).

The vertical 2D distribution of the hydraulic heads along the FF' cross-section (Fig. 15a) shows higher values of the hydraulic heads at well site 47 and a decrease toward well site 5. The higher water table might be related to the effect of the El-Nubaria canal and irrigation processes. The Subsurface temperature has low values under the area of higher hydraulic heads (Fig. 15b) owing to the recharge effect. Higher temperature isotherms are encountered in the southeastern parts, where the discharge process occurs (at well sites 5 and 28).

A local discharge area was noted in the cross-section GG' between well sites 39 and 38, and higher potentiometric surface values were noted to the east, while low

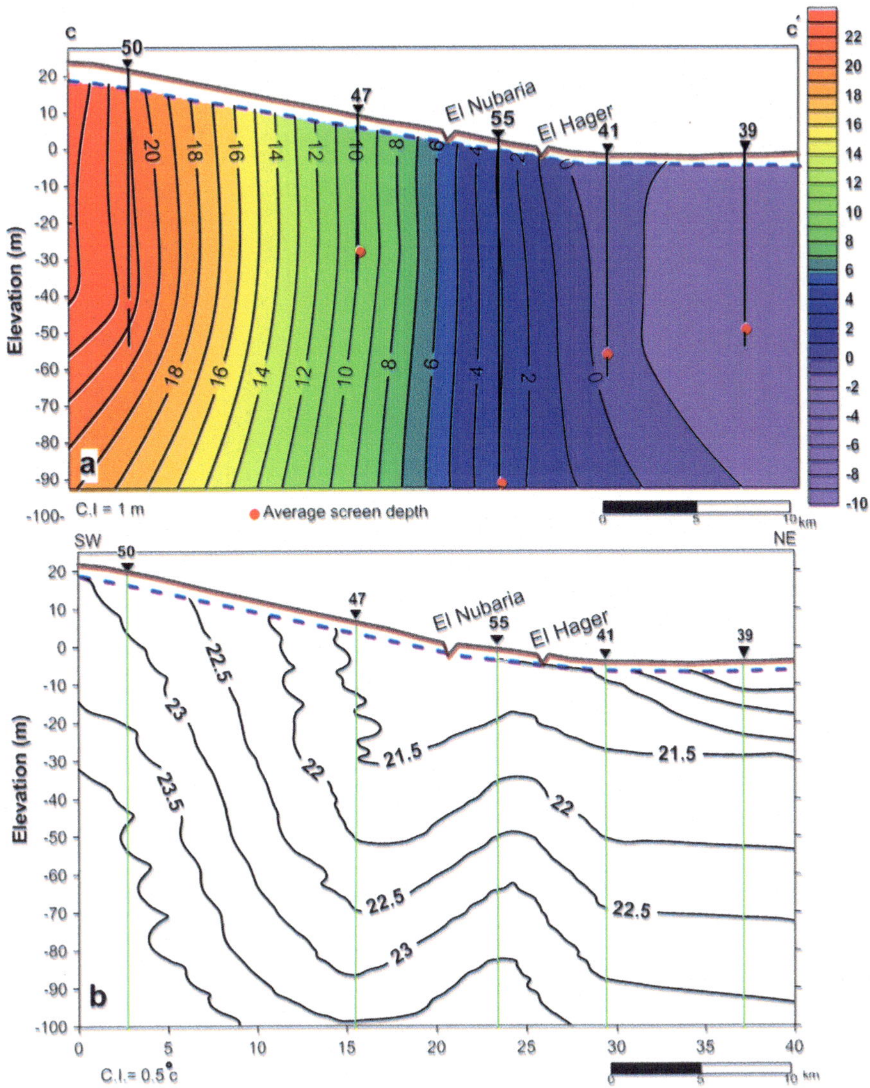

Fig. 12 Vertical 2D hydraulic heads (**a**) and subsurface temperature (**b**) distributions along the cross-section CC'. Temperature values are in °C and the vertical lines indicate the boreholes

values were recorded to the west (Fig. 16a). In general, subsurface temperatures had lower values where the hydraulic heads had higher values (well site 34) and the subsurface temperatures had higher values where the hydraulic heads had low values (well site 37) (Fig. 16b). Local cooling and warming effects, owing to local recharge and discharge at well sites 41 and 39, and well site 34, respectively, were noted. The

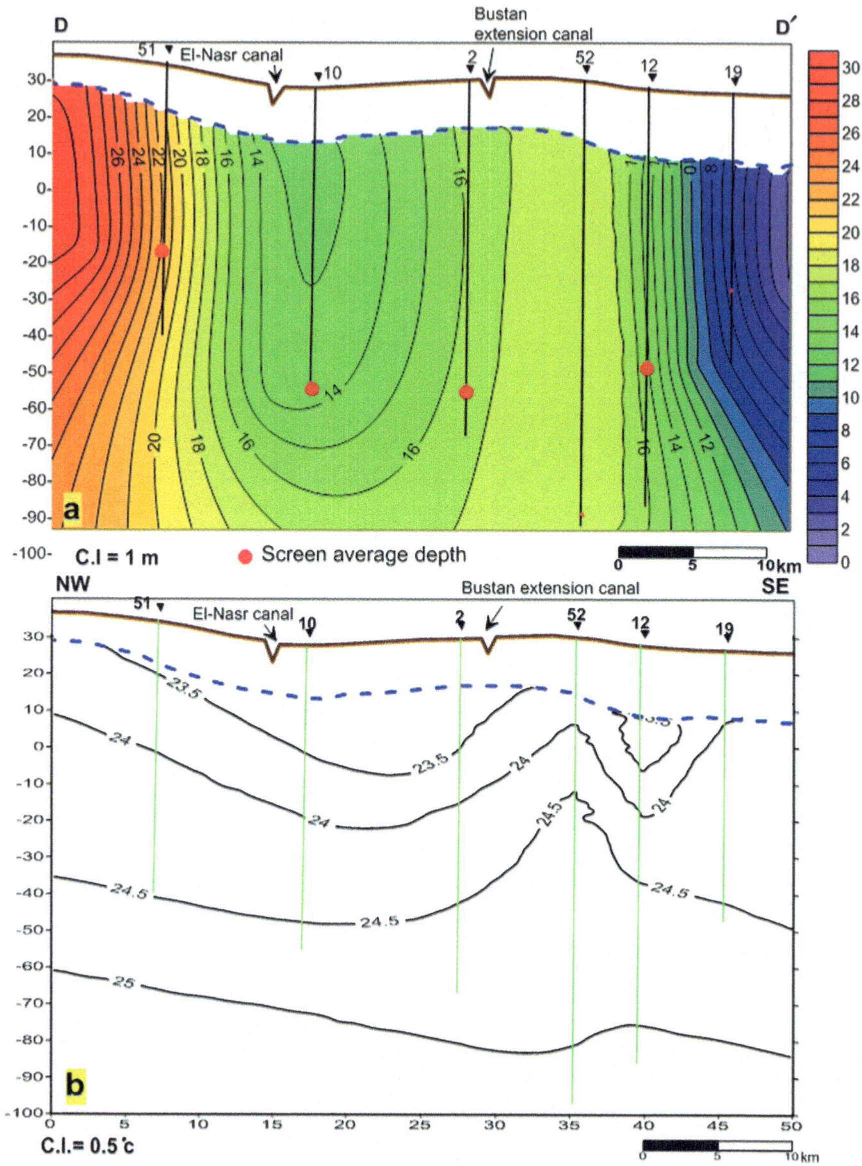

Fig. 13 Vertical 2D hydraulic heads (**a**) and subsurface temperature (**b**) distributions along the cross-section DD’. Temperature values are in °C and the vertical lines indicate the boreholes

abrupt change in subsurface temperatures between well sites 34 and 38 could be related to a fault effect [37].

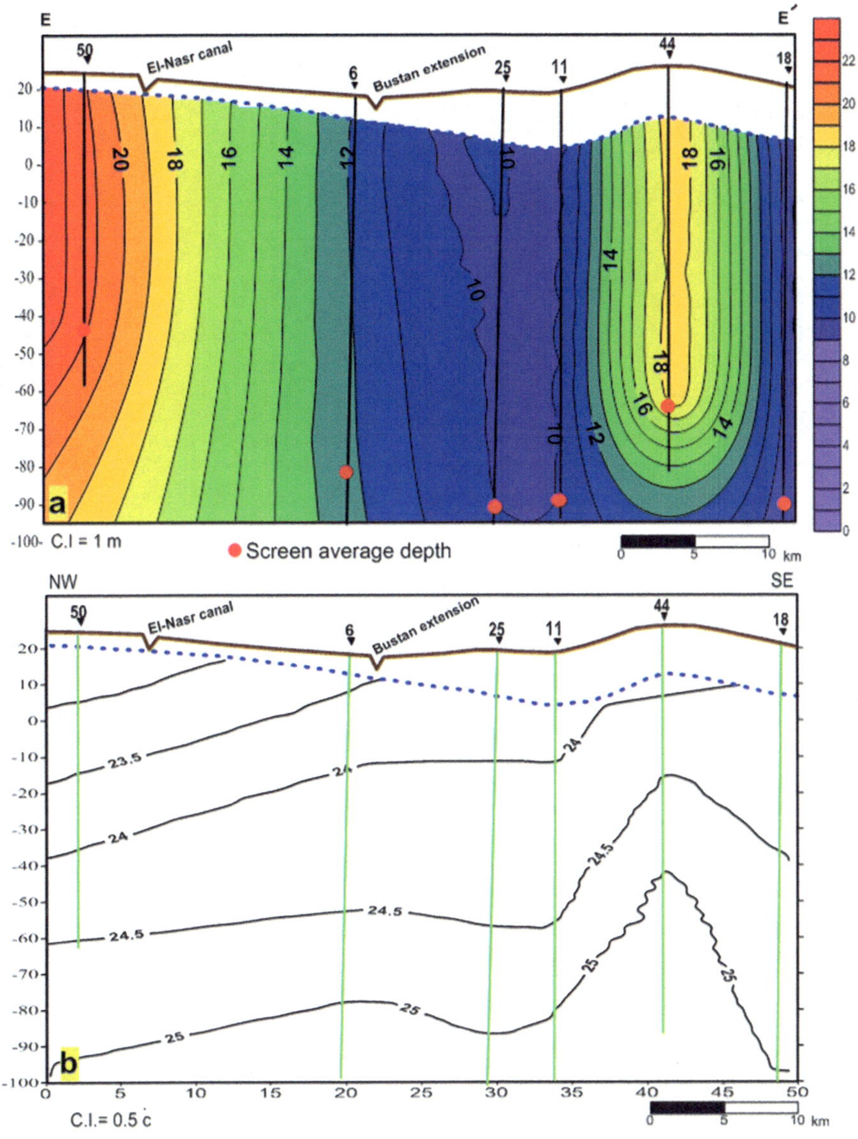

Fig. 14 Vertical 2D hydraulic heads (**a**) and subsurface temperature (**b**) distributions along the cross-section EE'. Temperature values are in °C and the vertical lines indicate the boreholes

6.2.3 Model Calculations

To analyze the temperature-depth profiles with heat advection owing to groundwater flow, subsurface temperatures have been calculated using different vertical

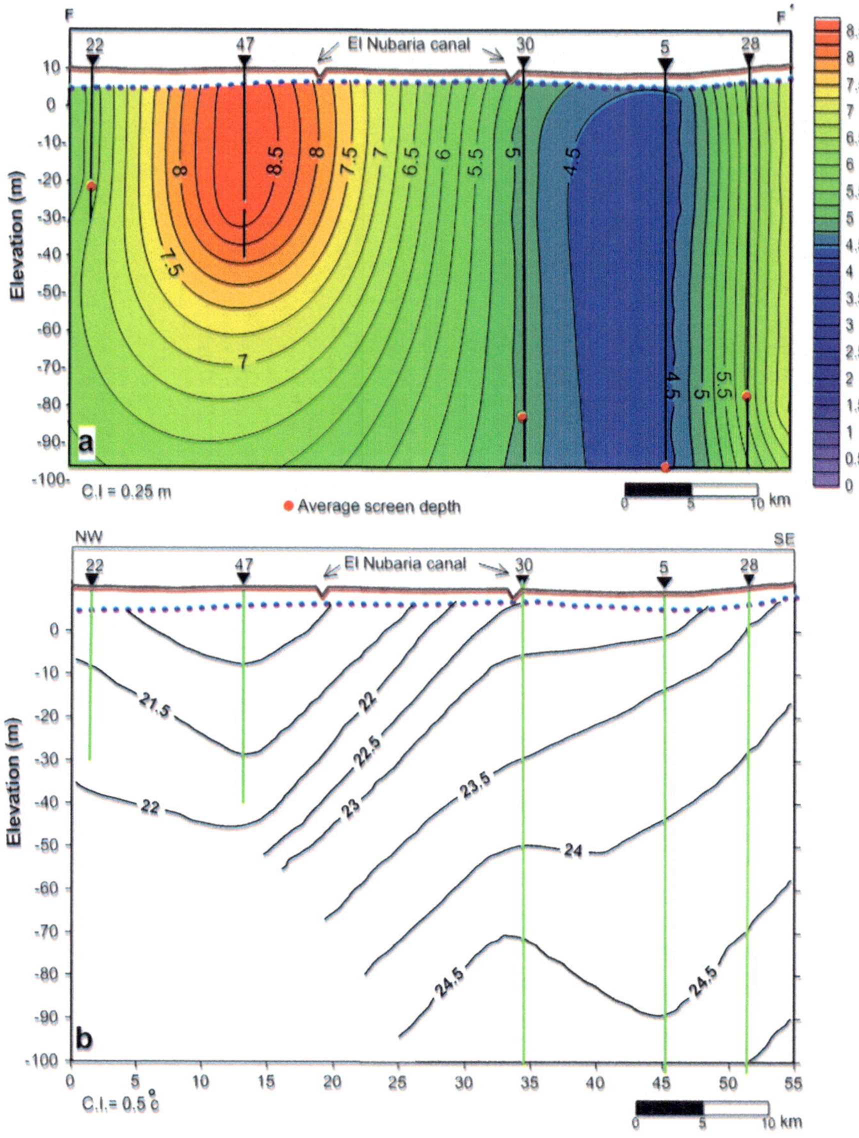

Fig. 15 Vertical 2D hydraulic heads (**a**) and subsurface temperature (**b**) distributions along the cross-section FF'. Temperature values are in °C and the vertical lines indicate the boreholes

groundwater fluxes. Carslaw and Jaeger [62] calculated the temperature using a 1D heat conduction–advection equation under the condition of a linear increase in surface temperature, as shown in Eq. (1):

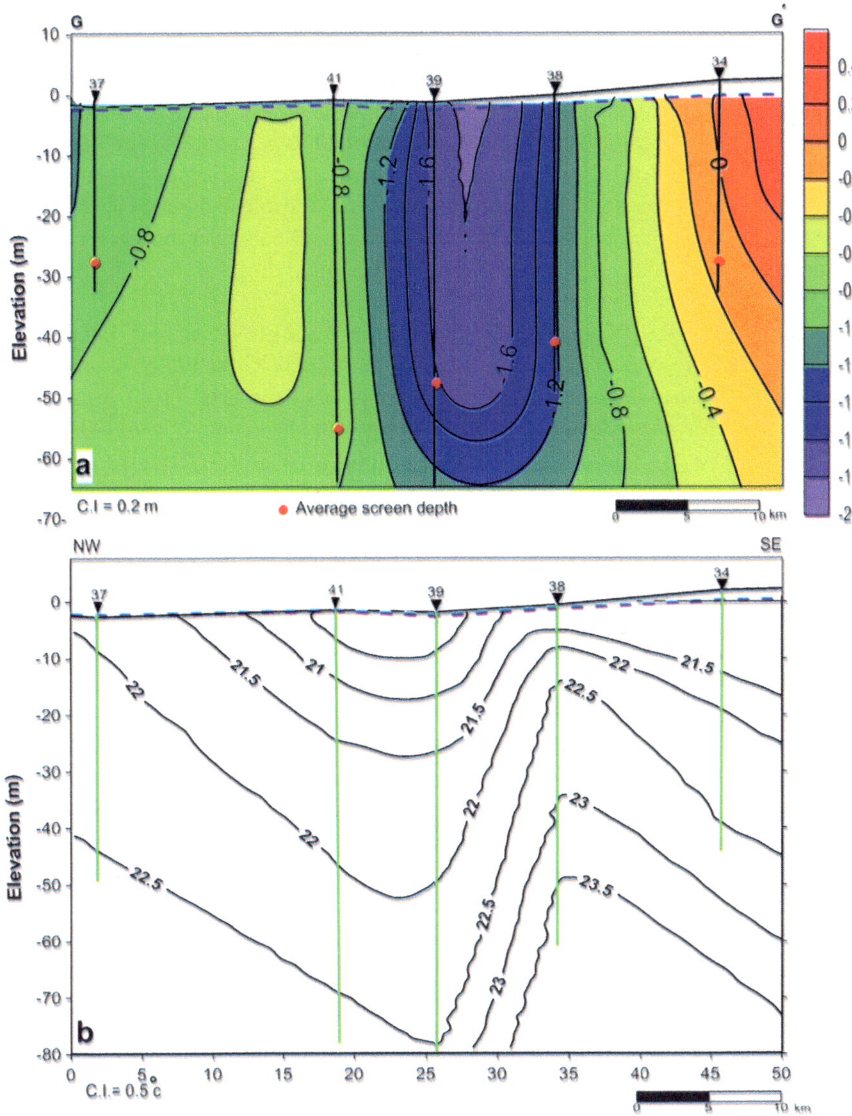

Fig. 16 Vertical 2D hydraulic heads (**a**) and subsurface temperature (**b**) distributions along the cross-section GG'. Temperature values are in °C and the vertical lines indicate the boreholes

$$T(z,t) = T_o + T_G(z - Ut) + \{(b + T_G U)/2U\} \times \left[(z + Ut)e^{Uz/\alpha}\mathrm{erfc}\left\{(z + Ut)/2(\alpha t)^{1/2}\right\} + (Ut - z)\mathrm{erfc}\left\{(z - Ut)/2(\alpha t)^{1/2}\right\}\right] \quad (1)$$

where b is the rate of the surface temperature increase, t is the time after the semi-equilibrium condition [15, 23], T_o is the surface temperature, T_G is the general geothermal gradient, and $U = \upsilon c_o\rho_o/c\rho$, in which υ is the vertical groundwater flux, $c_o\rho_o$ is the heat capacity of the water, and $c\rho$ is the heat capacity of the aquifer, α is the thermal diffusivity of the aquifer, and erfc is the complementary error function. The modeling is limited to semi-infinite layers with only vertical conduction and convection, and vertical groundwater flux is assumed to be constant with depth.

Land use has an effect on surface air temperature. Therefore, we modeled borehole temperatures under reclaimed desert land and old cultivated land under different surface air temperatures and geothermal gradients. The boreholes in the reclaimed desert belong to Wadi El-Natrun station, and those of the old cultivated land belong to Damanhour station.

T_G values were 0.0198 and 0.0343°C/m for the Wadi El-Natrun and Damanhour stations, respectively, and T_o values were 24.0 and 20.37°C for these two stations, respectively. For the whole area, the thermal diffusivity (α) was 6×10^{-7} m^2/s. For a t value of 50 years, the b values for Wadi El-Natrun and Damanhour were −0.03 and 0.044, respectively. Therefore, the modeled profiles of Wadi El-Natrun show a near surface cooling pattern, while those of Damanhor exhibit a near surface warming pattern. Temperature-depth profiles were computed using the above equation for different vertical groundwater flux values (U), according to Taniguchi et al. [63].

Evaluation of the vertical groundwater flux using the measured temperature profiles was done in relation to the nearest meteorological station. Wadi El-Natrun station climatic data were used for modeling the temperature profiles measured in the newly reclaimed desert, and those of Damanhour station were used for the profiles measured in the old agricultural land. A comparison between the modeled and observed profiles is shown in Figs. 17a–d and 18a–d.

There were 26 temperature profiles related to the Wadi El-Natrun station. All these profiles were of the recharge type, except for one profile of the discharge type (well 21) (Fig. 17a–d). The recharge (downward groundwater flux) rates varied among recharge wells (Fig. 18a–d). The rates were 1.0–1.5 m/year in wells 11, 15, 24, 23, 25, 29, 50, 52, and 54; 0.5–1 m/year in wells 2, 6, 7, 8, 10, 12, 16, 18, 19, 20, 28, 43, 51, and 53; and 0.3–0.5 m/year in wells 44 and 14. The discharge (upward groundwater flux) rate in well 21 was −0.1 to −0.5 m/year (Fig. 19). It was noted that most of the wells showed cross-cutting of the modeled profiles in their shallower zones (Fig. 17a–d). This feature may be related to irrigation with groundwater, which is warmer than the surface water. The maximum recharge rate of the groundwater was 1.5 m/day, owing to water seepage from the irrigation canals, as the wells were mostly located in the area between the branches of the El-Bustan Canal (Fig. 19), where the land surface is covered by accumulations of sand deposits [64]. The minimum recharge value of the wells located in the reclaimed desert was 0.3 m/day, which was attributed to local geology or to a thin aquifer, as seen with well 14, which is located close to Wadi El-Natrun, where the Quaternary aquifer has its minimum thickness and rests on Pliocene clay-rich sediments.

In contrast to the profiles of the Wadi El Natrun station, the 21 temperature profiles of Damanhour station were mostly of the discharge type (Fig. 18a–d). The discharge rates varied among the wells. The rates were −0.1 to −1.5 m/year in wells 3, 4, 5, 9, 31, and 38; −0.5 to −1.0 m/year in wells 33, 36, and 37; and −0.1

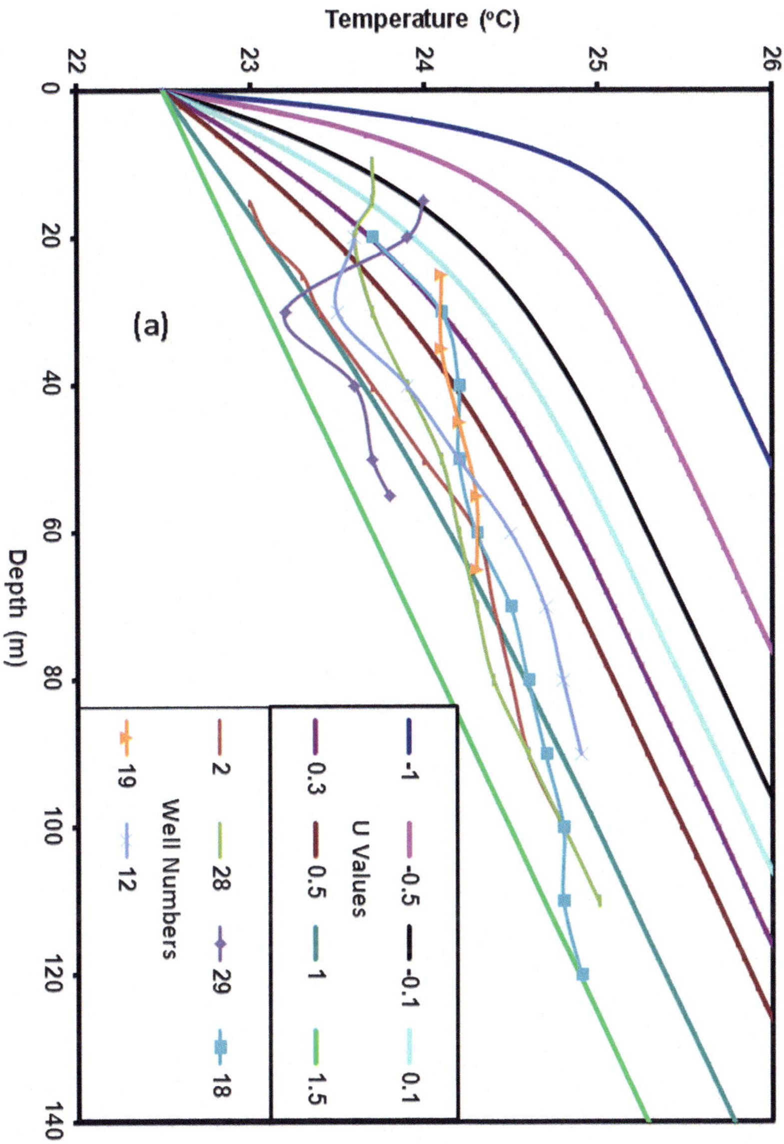

Fig. 17 Comparison between the modeled 1D and the measured temperature-depth profiles for the newly reclaimed desert land. The modeled profiles show cooling trend due to present decrease in the surface air temperature at Wadi El-Natrun station as a result of agricultural reclamation activities

to −0.5 m/year in wells 32, 34, 35, 40, 42, and 49 (Fig. 19). A lower discharge rate of these profiles was recognized in the deeper zones and the rates increased to higher values in the shallower zones. As the discharge area is located near the Mediterranean Sea coast, such a vertical change in the discharge rate might be attributed to a low-salinity shallow water zone which lies over the seawater intrusion-related

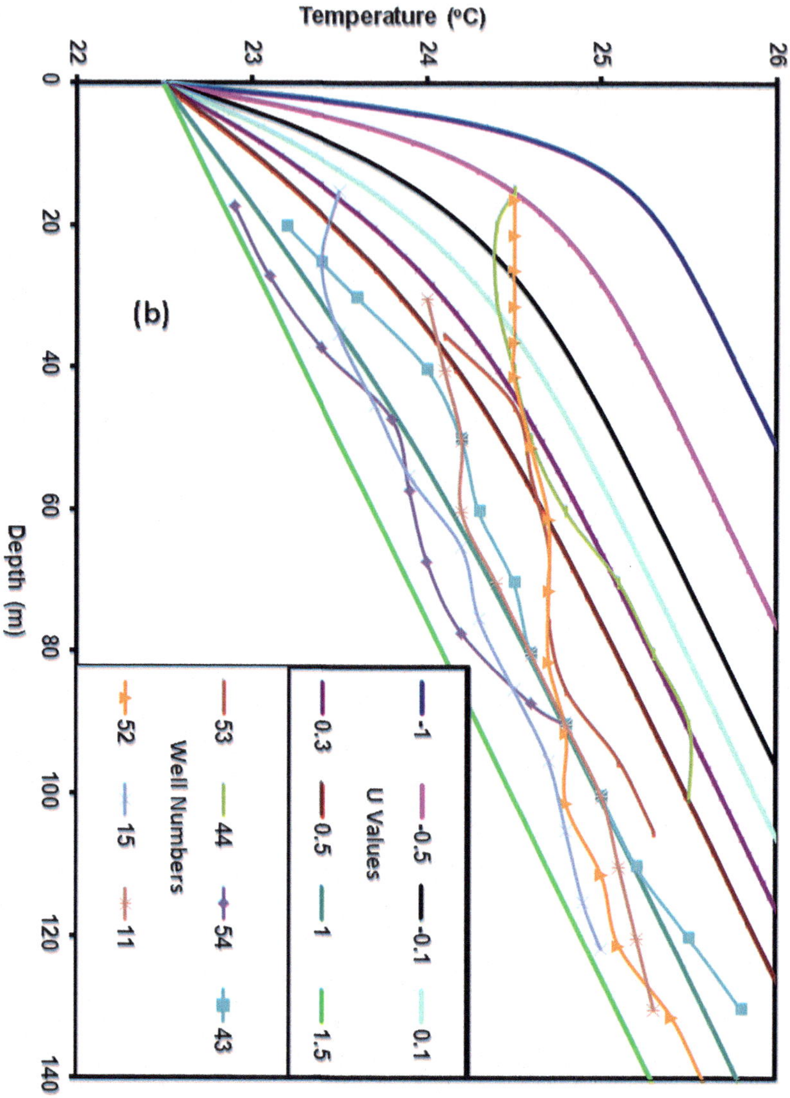

Fig. 17 (continued)

groundwater zone. These findings are consistent with the findings of geochemical and geophysical studies of the groundwater aquifer in the study area carried out by Osman [65]. Of note, the change from recharge type in the deeper zone for well 55 (up to 0.3 m/year) to a discharge type in the shallower zone (up to −0.5 m/year) could also be related to seawater intrusion. The maximum discharge rate (−1.5 m/year) was recognized around the southern part of the El-Nubaria Canal close to the region of contact between the old and the new agricultural lands (Fig. 19). The latter discharge process would make the southern part of El-Nubaria canal a gaining stream. The wells that had the

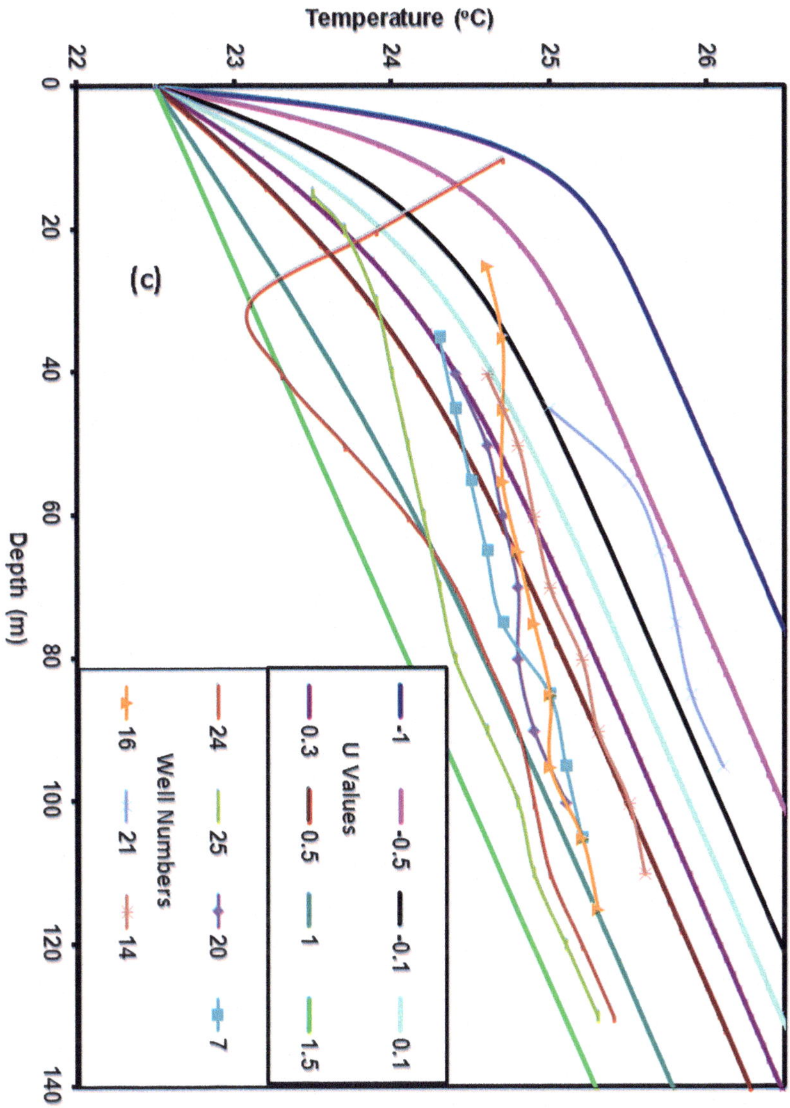

Fig. 17 (continued)

minimum discharge rate (−0.1 m/day) are mostly located to the east of the El-Nubaria Canal (Fig. 19). Therefore, the reduction in the discharge rate of these wells could be related to eastward seepage from the El-Nubaria Canal. The temperature profiles in wells 22, 47, and 39 indicate that the recharge occurs either from irrigation processes or water seepage from the irrigation canals, with a variable rate ranging from 0.1 to 0.5 m/year. Well number 41 is considered to have an intermediate-type profile.

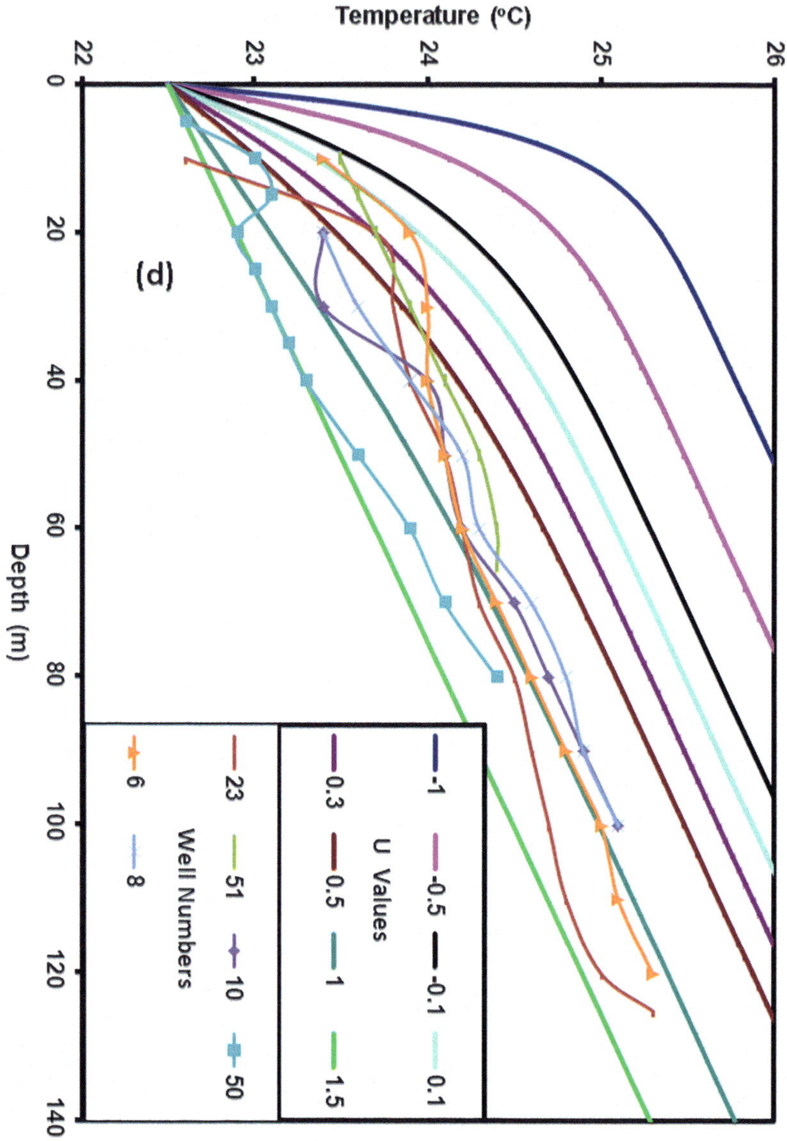

Fig. 17 (continued)

7 Discussion and Conclusions

In regard to the relationship between subsurface temperature distribution and the groundwater stream framework, we consider the area we investigated to show an abnormal thermal system. We analyzed land use and the groundwater stream framework, on the one hand, and the subsurface thermal regime (in light of 1D

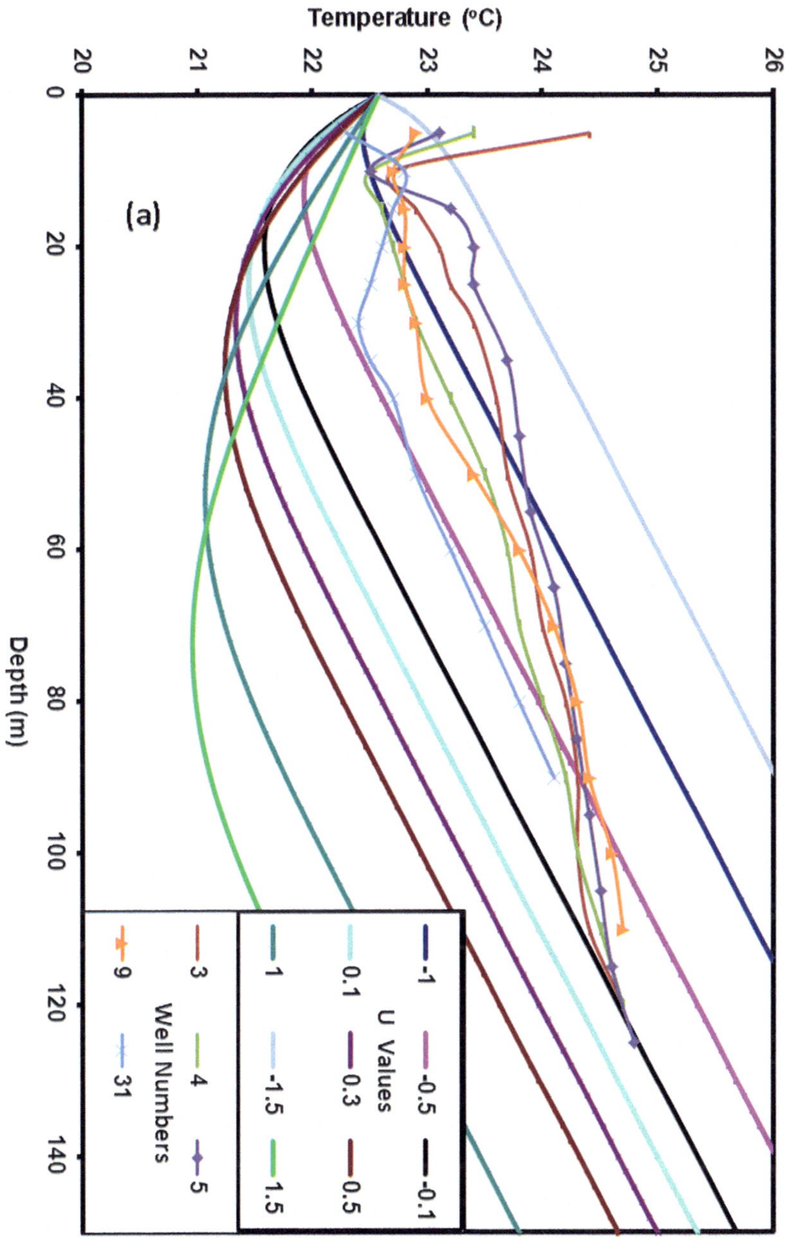

Fig. 18 Comparison between the modeled 1D and the measured temperature-depth profiles for the old agricultural land. The modeled profiles show warming trend due to recent increase in the surface air temperature as shown in Damanhour station

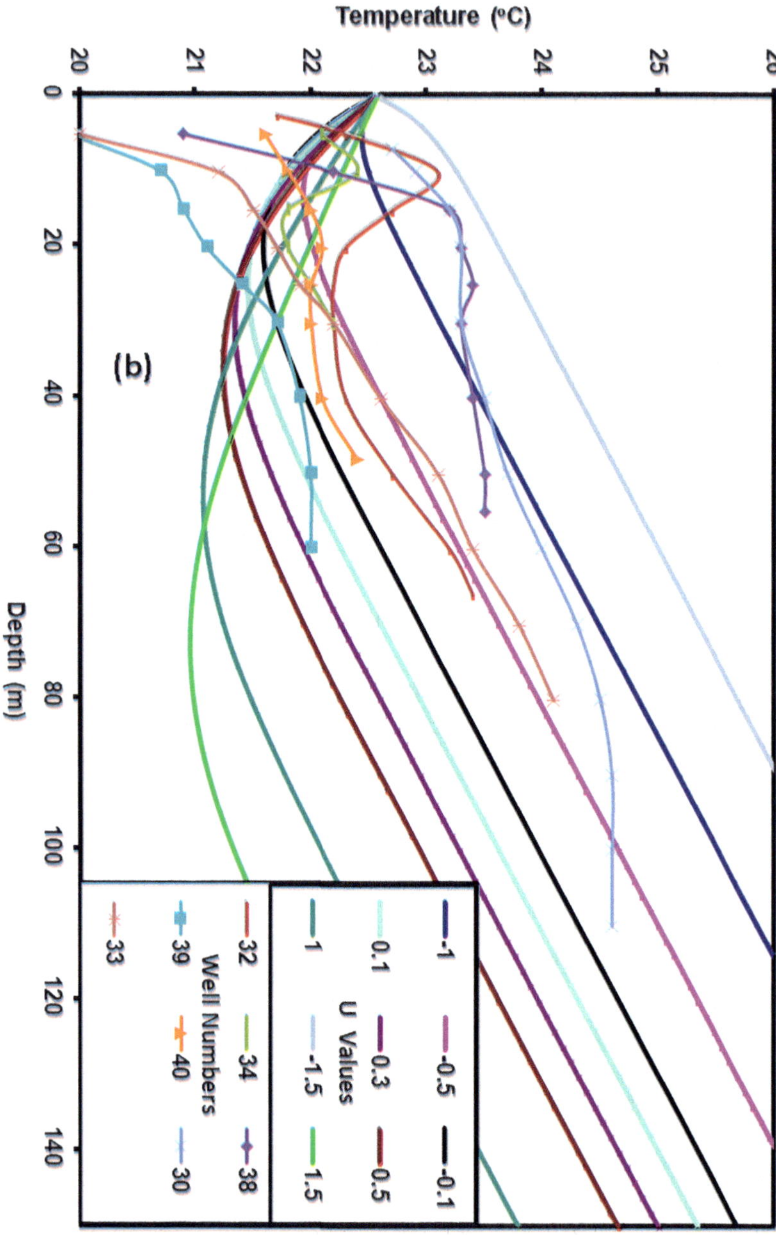

Fig. 18 (continued)

profiles, and vertical 2D thermal framework), on the other hand.. The subsurface temperatures in the reclaimed desert areas, where the recharge territory was found, were higher than those in the discharge region in the old agrarian land. This

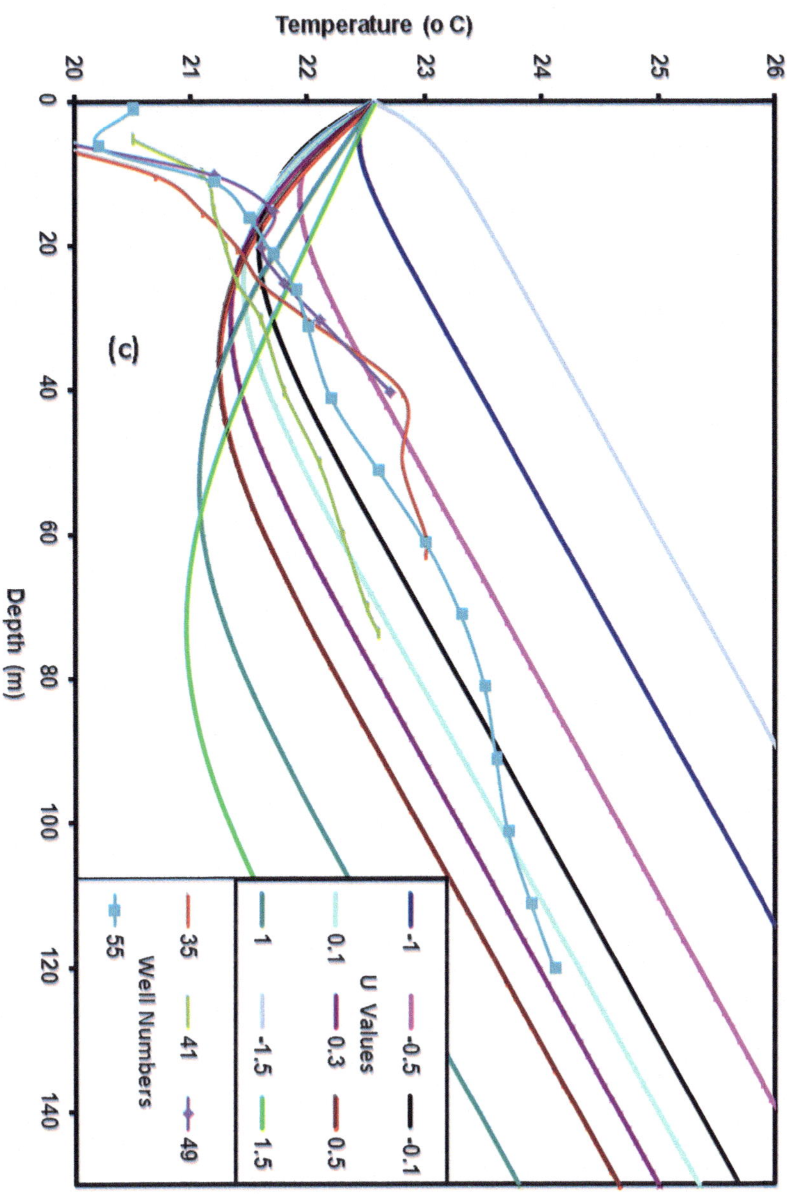

Fig. 18 (continued)

abnormal thermal system was ascribed to the surface air temperature, where the reclaimed deserts have a higher average annual temperature (23.15°C at Wadi El-Natrun station) than that experienced in the old agricultural terrains (20.37°C at Damanhour station). In contrast to this abnormal thermal regime, the geothermal

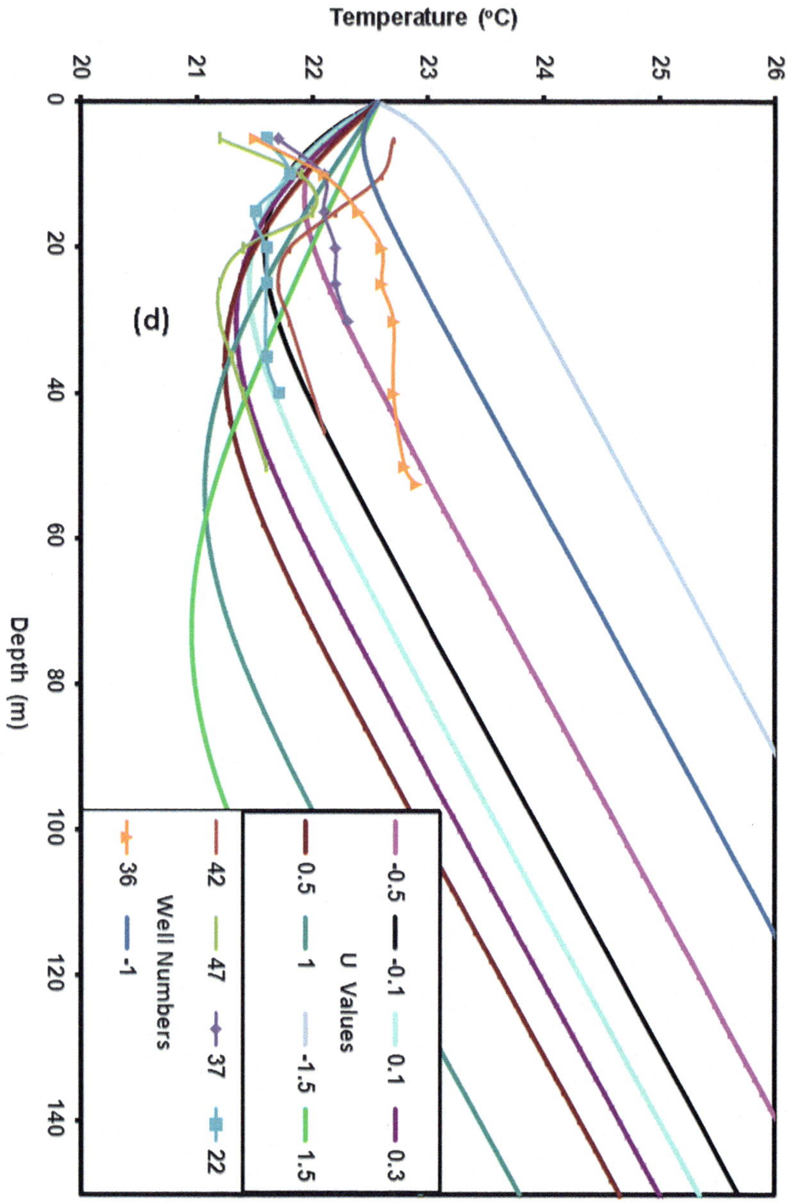

Fig. 18 (continued)

gradient showed normal values; in the recharge areas it had low values (0.0145–0.0326°C/m; average, 0.0198°C/m), while in the discharge areas higher values were encountered (0.016–0.049°C/m; average, 0.0343°C/m). For this reason, the geothermal gradient is superior to the thermal system for following the groundwater stream framework in regions such as those we examined.

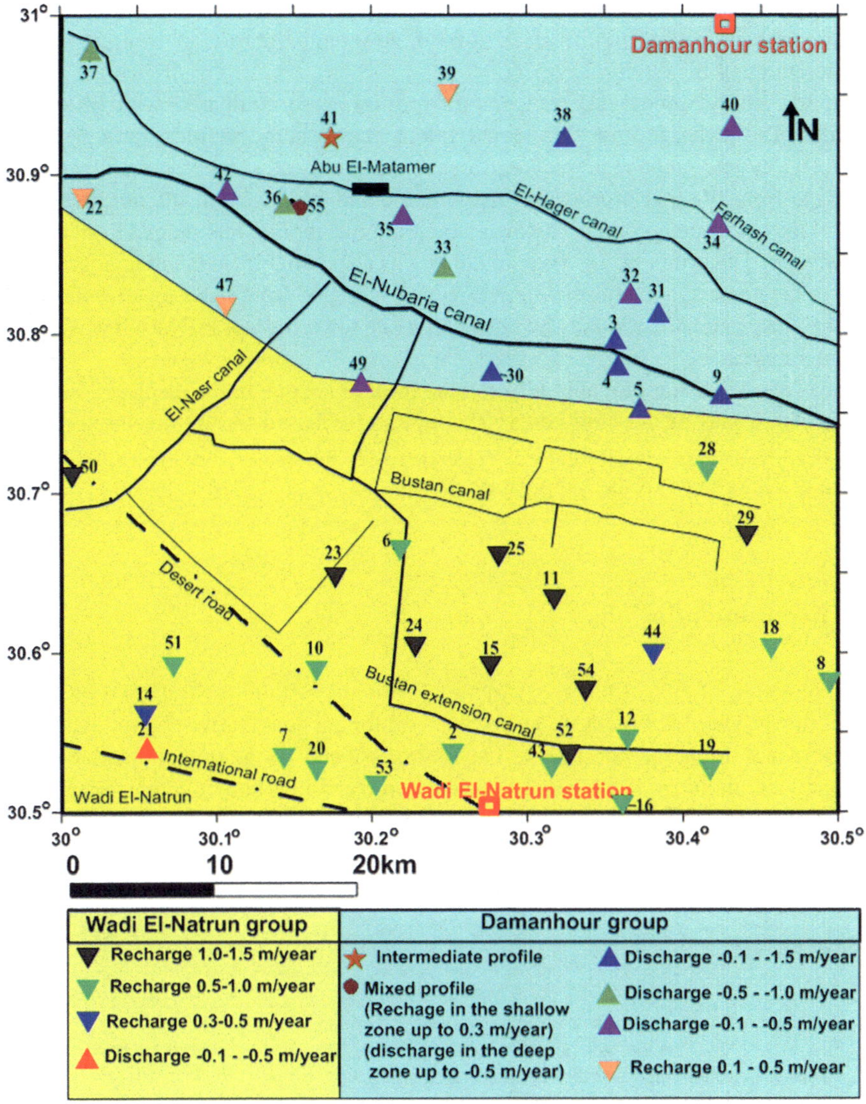

Fig. 19 Spatial distribution of the groundwater flux rates based on the data derived from Figs. 17 and 18

The connection between groundwater and surface water is obviously recognized from the subsurface temperature distribution. A warm zone is formed under a gaining stream because of upward movement of the groundwater, and a cool zone exists around a losing stream because of leakage descending from the surface water. The El-Nubaria canal is a gaining stream in its southern part, but a losing stream in its northern part. The El-Hager canal is viewed as a losing stream along its entire course. Water intrusion from the El-Bustan canal influences the groundwater temperatures, creating a subsurface zone of low temperatures, in contrast to the temperatures of the environs.

In this study, model 1D temperature profiles were compared with observed profiles. The recharge flow flux varied among the recharge profiles, ranging from 0.3 to 1.5 m/year in the main recharge area to the south and southwestern directions. Seepage from the irrigation canals (El-Nubaria and El-Hager) in the northwestern part of the study area produced a downward groundwater flux that ranged from 0.1 to 0.5 m/year. The groundwater discharge flux was also variable; in the main discharge area to the north and the northeastern direction it ranged from −0.1 to −1.5 m/year. On the other hand, the groundwater discharge rate toward Wadi El-Natrun was −0.1 to −0.5 m/year.

Seawater intrusion probably affected the thermal regime in the discharge area to the northern part of the study area. This opinion is based on the occurrence of a colder temperature zone in the northern parts of the area and a downward reduction in the discharge rate of the temperature profiles.

8 Recommendations

Determination of subsurface temperatures is a good method for tracing a groundwater flow system; this method can also be used for the quantitative determination of recharge and discharge rates, as well as for assessments of the spatial circulation of groundwater in an aquifer. Therefore, we recommend that borehole temperature be used widely in Egypt to help in solving some complicated groundwater problems.

References

1. Anderson MP (2005) Heat as a groundwater tracer. Ground Water 43:951–968
2. Saar MO (2011) Review: geothermal heat as a tracer of large-scale groundwater flow and as a means to determine permeability fields. Hydrogeol J 19:31–52
3. Stallman RW (1963) Methods of collecting and interpreting ground-water data: U.S. Geological Survey Water-Supply Paper 1544–H, pp 36–46
4. Suzuki S (1960) Percolation measurements based on heat flow through soil with special reference to paddy fields. J Geophys Res 65:2883–2885. https://doi.org/10.1029/JZ065i009p02883
5. Stallman RW (1965) Steady one-dimensional fluid flow in a semi-infinite porous medium with sinusoidal surface temperature. J Geophys Res 70:2821. https://doi.org/10.1029/JZ070i012p0282
6. Bredehoeft JD, Papadopulos IS (1965) Rates of vertical groundwater movement estimated from the Earth's thermal profile. Water Resour Res 1:325–328
7. Cartwright K (1970) Groundwater discharge in the Illinois Basin as suggested by temperature anomalies. Water Resour Res 6:912–918
8. Domenico PA, Palciauskas VV (1973) Theoretical analysis of forced convective heat transfer in regional groundwater flow. Geol Soc Am Bull 84:3803–3814

9. Sakura Y (1993) Groundwater flow estimated from temperatures in the Yonezawa Basin, Northeast Japan. In: Tracers in hydrology. IAHS Publication 215. IAHS, Oxfordshire, pp 161–170
10. Inagaki N, Taniguchi M (1994) Estimations of hydraulic conductivity and groundwater flow systems by using groundwater temperature in Nara Basin, Japan. J Jpn Assoc Hydrol Sci 24:171–182
11. Taniguchi M (1993) Evaluation of vertical groundwater fluxes and thermal properties of aquifers based on transient temperature-depth profiles. Water Resour Res 29:2021–2026
12. Taniguchi M (1994) Estimated recharge rates from groundwater temperatures in the Nara Basin, Japan. Hydrogeol J 2:7–14
13. Dapaah-Siakwan S, Kayane I (1995) Estimation of vertical water and heat fluxes in the semi-confined aquifers in Tokyo metropolitan area, Japan. Hydrol Process 9:143–160
14. Beck AE, Garven G, Stegena L (1989) Hydrogeological regimes and their subsurface thermal effects. Geophysical monograph 47. American Geophysical Union, Washington
15. Taniguchi M, Shimada J, Tanaka T, Kayane I, Sakura Y, Shimano Y, Dapaah-Siakwan S, Kawashima S (1999) Disturbances of temperature – depth profiles due to surface climate-change and subsurface water flow: (1) an effect of linear increase in surface temperature caused by global warming and urbanization in the Tokyo metropolitan area, Japan. Water Resour Res 35:1507–1517
16. Ferguson G, Woodbury AD (2005) The effects of climatic variability on estimates of recharge from temperature profiles. Ground Water 43:837–842
17. Sakura Y, Uchida Y, Taniguchi M, Kayane I, Anderson MP (2000) Change of subsurface temperature caused by climatic change in Japan. In: Proceedings of the 30th international IAH congress. Groundwater: past achievements and future challenges, Cape Town, South Africa. AA Balkema, Rotterdam
18. Salem ZE, Sakura Y, Mohamed Aslam MA (2004) The use of temperature, stable isotopes and water quality to determine the pattern and spatial extent of groundwater flow: Nagaoka area, Japan. Hydrogeol J 12:563–575
19. Salem ZE, Taniguchi M, Sakura Y (2004) Use of temperature profiles and stable isotopes to trace flow lines: Nagaoka area, Japan. Ground Water 42:83–91
20. Heath RC (1964) Seasonal temperature fluctuations in surficial sand near Albany, NY professional paper 475-D. USGS, Washington
21. Pluhowski EJ, Kantrowitz IH (1963) Influence of land surface conditions on ground water temperatures in southwestern Suffolk County, Long Island, NY professional paper 475-B. USGS, Washington
22. Taniguchi M, Williamson DR, Peck AJ (1997) Changes in surface and subsurface temperatures after clearing forest in Western Australia. In: Taniguchi M (ed) Subsurface hydrological responses to land cover and land use changes. Kluwer Academic, Boston, pp 139–151
23. Taniguchi M, Williamson DR, Peck AJ (1999b) Disturbances of temperature – depth profiles due to surface climate-change and subsurface water flow: (2) an effect of step increase in surface temperature caused by forest clearing in southwest of Western Australia. Water Resour Res 35:1519–1529
24. Sophocleous M (2004) Climate change: why should water professionals care? Ground Water 42:637
25. Jessop AM (1990) Thermal geophysics, developments in solid earth, geophysics. Elsevier, Amsterdam, p 306
26. Taniguchi M, Uemura T (2005) Effects of urbanization and groundwater flow on the subsurface temperature in Osaka, Japan. Phys Earth Planet Int 152:305–313
27. Reiter M (2006) Vadose zone temperature measurements at a site in the northern Albuquerque Basin indicate ground-surface warming due to urbanization. Environ Eng Geosci 12:353–360
28. Ferguson G, Woodbury AD (2007) Urban heat island in the subsurface. Geophys Res Lett 34:4. https://doi.org/10.1029/2007GL032324

29. Banks D, Gandy CJ, Younger PL, Withers J, Underwood C (2009) Anthropogenic thermo-geological 'anomaly' in Gateshead, Tyne and Wear, UK. Q J Eng Geol Hydrogeol 42:307–312
30. Westaway R, Scotney PM, Younger PL, Boyce AJ (2015) Subsurface absorption of anthropogenic warming of the land surface: the case of the world's largest brickworks (Stewartby, Bedfordshire, UK). Sci Total Environ 508:585–603
31. Menberg K, Bayer P, Zosseder K, Rumohr S, Blum P (2013) Subsurface urban heat islands in German cities. Sci Total Environ 442:123–133
32. Menberg K, Blum P, Schaffitel A, Bayer P (2013) Long-term evolution of anthropogenic heat fluxes into a subsurface urban heat island. Environ Sci Technol 47:9747–9755
33. Paillet FL (1998) Flow modeling and permeability estimation using borehole flow logs in heterogeneous fractured formations. Water Resour Res 34:997–1010
34. Borner F, Berthold S (2009) Vertical flows in groundwater monitoring wells. Groundwater geophysics. Springer, Berlin, pp 367–389. https://doi.org/10.1007/978-3-540-88405-7
35. Chatelier M, Ruelleu S, Bour O, Porel G, Delay F (2011) Combined fluid temperature and flow logging for the characterization of hydraulic structure in a fractured karst aquifer. J Hydrol 400:377–386
36. Salem ZE, Osman MO (2016) Shallow subsurface temperature in the environs of El-Nubaria canal, northwestern Nile Delta of Egypt: implications for monitoring groundwater flow system. Environ Earth Sci 75:1241. https://doi.org/10.1007/s12665-016-6046-y
37. Salem ZE, El Bayumy DA (2016) Use of the subsurface thermal regime as a groundwater-flow tracer in the semi-arid western Nile Delta, Egypt. Hydrogeol J 24(4):1001. https://doi.org/10.1007/s10040-016-1377-z
38. Salem ZE, Gaame OM, Hassan TM (2008) Using temperature logs and hydrochemistry as indicators for seawater intrusion and flow lines of groundwater in Quaternary aquifer, Nile Delta, Egypt. In: Proceedings of the fifth international symposium on geophysics (ISG 5), 27–29 Nov, pp 25–38
39. Salem ZE (2009) Hydraulic head, subsurface temperature and water quality as reasons for deciphering the groundwater resources and flow pattern in Wadi El-Assuity, Egypt. Sedimentol Egypt 17:199–216
40. Salem ZE (2009) Natural and human impacts on the groundwater under an Egyptian village, central Nile Delta – a case study of Mehallet Menouf. In: Proceedings of the thirteenth international water technology conference (IWTC 13), Hurghada, 12–15 Mar, vol 3, pp 1397–1414
41. Salem ZE (2016) Subsurface thermal regime to delineate the paleo-groundwater flow system in an arid area, Al Kufra, Libya. NRIAG J Astron Geophys 5(2):451–462. https://doi.org/10.1016/j.nrjag.2016.10.002
42. Salem ZE, Osman OM (2017) Use of major ions to evaluate the hydrogeochemistry of groundwater influenced by reclamation and seawater intrusion, West Nile Delta, Egypt. Environ Sci Pollut Res 24:3675–3704. https://doi.org/10.1007/s11356-016-8056-4
43. Saad K (1962) Groundwater studies of Wadi El-Natrun and its vicinities. Desert Institute, Cairo, p 61
44. El Fayoumy IF (1964) Geology of groundwater supplies in Wadi El-Natrun area. MSc thesis, Faculty of Science, Cairo University, Cairo
45. Desert Research Institute (Cairo) (1974) Report on the regional hydrogeological studies of the West El Nubaria area (300,000 Feddans reclamation project), internal report, Cairo, Egypt. p 125
46. Mabrouk MA (1978) Electrical prospecting on the groundwater in the area west of Cairo-Alexandria Desert Road (between Wadi El-Natron and El-Nasr Canal). MSc thesis, Faculty of Science, Ain Shams University, Cairo
47. El Ghazawi MM (1982) Geological studies of the Quaternary-Neogene aquifers in the area northwest Nile Delta. MSc thesis, Al-Azhar University, Cairo

48. Abdel Baki AA (1983) Hydrogeological and hydrogeochemical studies in the area west of Rosetta branch and south of El Nasr canal. PhD thesis, Faculty of Science, Ain Shams University
49. Ismail YL (1994) Hydrological studies on the ground water reservoir in some localities under reclamation in west Al-Nubariya province. MSc thesis, Faculty of Science, Al-Azhar University, Cairo
50. Gomaa MAA (1995) Comparative hydrogeological and hydrogeochemical study on some aquifer, west of Nile Delta, Egypt. PhD thesis, Faculty of Science, Ain Shams University, Cairo
51. Diab MSh, Dahab KA (1992) New results on salt water intrusion in the Nile Delta Egypt (abstract). In: Thirty years of international cooperation, the geology of Egypt and related science, Cairo, pp 56–57
52. El Sheikh AS (2000) Hydrogeology of the area north and west of Wadi El-Natrun. MSc thesis, Faculty of Science, Minufiya University, Menofia Governorate
53. Embaby AA (2003) Environment evaluation for geomorphological situation in relation to the water and soil resources of the region north of the Sadat city, West Nile Delta, Egypt. PhD thesis, Faculty of Science, Damietta, Mansoura University, Cairo
54. Nawar A (2008) Hydrogeology and hydrogeophysics studies on eastern El Behira governorate, Egypt. M.Sc. Thesis. Faculty of Science Alexandria University, Egypt
55. Fouad R, Zhi C (2010) Regional groundwater flow modeling in western Nile Delta, Egypt. World Rural Observations 2(2):37–42. http://www.sciencepub.net/rural
56. Abdel-Raouf O, Abdel-Galil K (2013) Conjunctive use of DC resistivity method and hydrochemical analysis for groundwater potentiality of Wadi El Natrun area, Egypt. Int J Water Resour Arid Environ 2:110–119
57. El Arabi NE, Morsy WS (2013) Applying integrated ground- and surface-water management (case study: Nubaryia Basin, West Delta, Egypt). J Am Sci 9:43–53
58. Salem ZE, Atwia MG, El-Horiny MM (2015) Hydrogeochemical analysis and evaluation of groundwater in the reclaimed small basin of Abu Mina, Egypt. Hydrogeol J 23:1781–1797
59. Salem ZE, El Bayumy DA (2016) Hydrogeological, petrophysical and hydrogeochemical characteristics of the groundwater aquifers east of Wadi El-Natrun, Egypt. NRIAG J Astron Geophys 5:106–123
60. Salem ZE, Osman OM (2017) Use of geoelectrical resistivity to delineate the seawater intrusion in the northwestern part of the Nile Delta, Egypt. The handbook of environmental chemistry. Springer, Berlin, Heidelberg
61. Abdelhameed AT, Salem ZE, Osman OM (2017) Sedimentological characteristics of the quaternary groundwater aquifer, northwestern Nile Delta, Egypt. The handbook of environmental chemistry. Springer, Berlin, Heidelberg
62. Carslaw HS, Jaeger JC (1959) Conduction of heat in solids, 2nd ed. Oxford University Press, Oxford, p 510
63. Taniguchi M, Shimano Y, Kayane I (1989) Groundwater flow analysis using the temperature in the upland areas at the western foot of Aso volcanoes. J Jpn Assoc Hydrol Sci 19:171–179
64. El Shazly EM, Abdel-Hady MA, El-Ghawaby MA, El Kassas IA, Khawasil SM, El Shazly MM, Sanad S (1975) Geological interpretation of Landsat Satellite Images for west Nile Delta area. Egypt. Remote Sensing Research project, Academy of Scientific Research Technology, Egypt
65. Osman OM (2014) Hydrogeological and geoenviromental studies of El Behira Governorate, Egypt. PhD Dissertation. Tanta University, Egypt

Use of Geoelectrical Resistivity to Delineate the Seawater Intrusion in the Northwestern Part of the Nile Delta, Egypt

Zenhom E. Salem and Osman M. Osman

Abstract Mapping of the boundaries between freshwater and saltwater was helpful in surface resistivity surveys because of the high electric conductivity of saltwater relative to freshwater. A total of 30 electrical soundings were measured to configurate the seawater intrusion. Accordingly, two zones of groundwater quality were delineated: the slightly freshwater zone in the southern part, with resistivity range of 15–90 Ω m, and the brackish water to saltwater zone, with a very low resistivity of <2 Ω m in the northwestern parts. In addition to tracing the freshwater-seawater contact zone, three geoelectric layers were detected. The surface layer composed of sand, clay, and silt. Its resistivity ranges from 5 to 512 Ω, and the thickness varies from 1 to 25 m. The aquifer layer is composed of sand with intercalations of clay with resistivity ranging from 15 to 90 Ω m and thickness from 25 to 120 m. The clay layer resistivity ranges from 2 to 15 Ω m and thickness from 2 to 69 m.

Keywords Geoelectrical resistivity, Northwestern Nile Delta, Quaternary aquifer, Seawater intrusion

Contents

Z.E. Salem (✉)
Geology Department, Faculty of Science, Tanta University, Tanta, Egypt
e-mail: zenhomsalem@yahoo.com

O.M. Osman
Geology Department, Faculty of Science, Damanhour University, Damanhour, Egypt

A. M. Negm (ed.), *Groundwater in the Nile Delta*,
Hdb Env Chem (2019) 73: 425–460, DOI 10.1007/698_2017_175,

1 Introduction

Coastal regions are of incredible ecological, economic, and social importance. Hence, the execution of reasonable observing and insurance activities is basic for their safeguarding and for guaranteeing future utilization of this asset [1]. Post [2] has characterized the coastal aquifers as the subsurface reciprocals of seaside zones where mainland new groundwater and seawater meet. Seaside fields are frequently polluted by saltwaters, and the procedure related to the marine water entering an aquifer is for the most part called seawater intrusion. Seawater intrusion becomes a severe problem in arid and semiarid regions where the groundwater constitutes the main freshwater resource. Mixing of only 3% seawater with freshwater in a coastal aquifer would render the freshwater resource unsuitable for human consumption [3, 4]. Seawater intrusion is regarded as a natural process that might be accelerated or retarded by external factors such as the increase or decrease in the groundwater pumping, irrigation system, recharge rates, land use, and possible seawater rise due to the impact of global warming [3]. Diverse methodologies have been received to evaluate seawater intrusion. For instance, many studies including Wilson et al. [5], Diab and Saleh [6], Sherif et al. [7–9], Diab et al. [10], Petalas and Diamantis [11], Sherif [12], Polemio et al. [13], Petalas and Lambrakis [14], Somay and Gemici [15], Salem et al. [16], Salem and El-horiny [17], Sefelnasr and Sherif [18], and Salem and Osman [19] have utilized geochemical strategies in view of modeling technique, stable isotopes, and hydrochemical data to evaluate the seawater intrusion.

Werner et al. [20] stated that the field of coastal hydrogeology, considered as a subdiscipline of hydrogeology, spans seawater intrusion, submarine groundwater discharge (SGD), beach-scale hydrology, sub-seafloor hydrogeology, and studies on geological time scales involving coastline geomorphology. Despite this, coastal aquifer hydrodynamics and seawater intrusion remain challenging to measure and quantify, commonly used models and field data are difficult to reconcile, and predictions of future coastal aquifer functioning are relatively uncertain across both regional and local (individual well) scales. The extent of seawater intrusion into the fresh groundwater is affected by subsurface geology, hydraulic gradient, rate of groundwater recharge, and groundwater pumping amounts [21]. Investigating the geometry of seawater intrusion extent, utilization of data from geological research, groundwater geochemistry, well drilling, and exploitation boreholes are used. These strategies are costly and tedious, keeping their utilization on an expansive scale. Interestingly, geophysical measurements can give a more affordable approach to enhance the information of an arrangement of boreholes [22]. Geophysical prospecting strategies can give corresponding information that empower geological correlations, even in areas where there are no information from boreholes. Indirect geophysical strategies (like electrical resistivity tomography (ERT) and VES survey) create continuous information throughout a given profile. It helps in understanding spatial relations between fresh and seawater which normally coincide in beach front aquifers [23].

Recently, geophysical methods (especially the direct current resistivity) are considered as major tools for solving the complicated hydrogeological and environmental problems especially in the coastal areas, e.g., Steeples [24], Lapenna et al. [25], Pantelis et al. [26], Chianese and Lapenna [27], and Naudet et al. [28]. Electrical resistivity method was used by many authors around the world to characterize the coastal areas; among them are Urish and Frohlich [29], Ebraheem et al. [30], Kruse et al. [31], Nowroozi et al. [32], Abdul Nassir et al. [33], Balia et al. [34], Choudhury and Saha [35], Batayneh [36], Bauer et al. [37] Khalil [38], Sherif et al. [39], Koukadaki et al. [40], Cimino et al. [41], Satriani et al. [1], and Fadili et al. [42].

In addition to surface contaminants [43], the groundwater quality in the Nile Delta is affected mainly by seawater intrusion [30, 44–47]. In the last decades, a common problem of Nile Delta sandy aquifer is the degradation of the groundwater quality due to seawater intrusion from the north and excessive pumping in relation to average natural recharge [48]. A detailed review of seawater intrusion in the Nile Delta groundwater system and the basis for assessing impacts due to climate changes and water resources development was discussed in Mabrouk et al. [49]. The encountered recent studies that used electrical resistivity to delineate the seawater-freshwater relationship in the northern part of the Nile delta are Attwa et al. [50], Mohamed [51], Salem et al. [3], and Tarabees and El-Qady [52]. The subsurface thermal regime was also used to trace the seawater intrusion in the Nile Delta [16, 53].

The purpose of this chapter is to present a detailed vertical electrical sounding survey in the proposed area of the Nile Delta aquifer to delineate and follow the seawater intrusion of the study area located in the northwestern part of the Nile Delta (Fig. 1). The interpretation includes the correlation of these similar or nearly similar parameters to illustrate the vertical, horizontal, and lateral extent of the seawater intrusion as well as the subsurface geologic structural configuration. This study supplies the decision maker with information for groundwater management strategy.

2 Description of the Study Area

The west of the Nile Delta represents one of the largest groundwater provinces in northern Egypt. Groundwater occurrence and distribution are governed by the geographic location, the sedimentary sequence, the geologic structure, and the

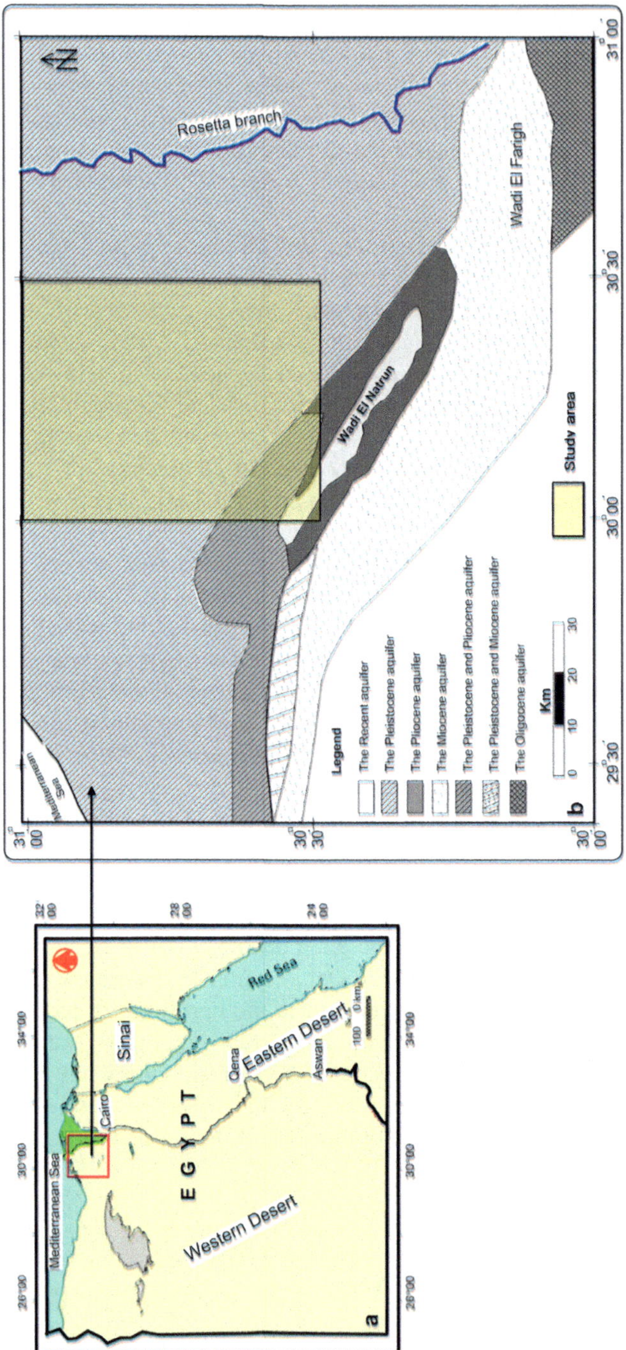

Fig. 1 Maps show the location of the study area (**a**) and describe its hydrogeology (**b**)

physiographic features. The study of groundwater is very important to detect the flow direction and saltwater intrusion as well as to ensure its quality for several purposes. The area is located to the west of the Nile Delta in northern Egypt (Fig. 1). It is bounded geographically by latitudes 30° 30′ and 31° 00′ N and longitudes 30° 00′ and 30° 30′ E. The study area is characterized by many human activities such as agricultural, industrial, and desert land reclamation. The reclamation processes are increased in the past decades, so the land uses are changed [54]. Over the years, agricultural areas have begun to move from the fertile soils of the Nile River to the east [19]. While most fields are irrigated by water from the Nile, high water tables have allowed the use of groundwater to support additional vegetable crops.

The sedimentary succession in the western Nile Delta reached up to 4 km in thickness and ranged in age from Triassic to Quaternary. The surface deposits of the West Nile Delta (Fig. 1b) are of Oligocene, Miocene, Pliocene, and Quaternary ages. The Oligocene rocks are portrayed by ferruginous sandstone and sands and sometimes gravels. These deposits are covered by some dissected basaltic sheet and located in the southern part of the West Nile Delta (Fig. 1b). Miocene deposits (Moghra Formation, Moghra aquifer) are up to 200 m in thickness and made out of shallow marine intercalations. Sediments are unconformably overlain the Miocene deposits. The Pliocene is portrayed by estuarine clayey facies at the base and fluvio-marine and shallow marine white limestones at the top in Wadi El Natrun and its districts (Fig. 1b).

The Quaternary pink deposits of sandy, thick, and pebbly sand with conglomerates cover the Pliocene deposits. The Pleistocene ruddy sand, silt, and gravels are on a very basic level scattered west of Rosetta branch and north and east of Wadi El Natrun. Windblown sand sheets and longitudinal slopes (NNW–SSE) are found in the swamp scopes of Wadi El Natrun and in the Wadi El Farigh lowlands. Lagoonal deposits are created from the trademark pools of Wadi El Natrun. Deltaic gravels and sands in the eastern part of Wadi El Natrun gradually decrease in thickness from 300 m at Rosetta branch to few meters thickness north of Wadi El Natrun [55]. The Pleistocene aquifer is the main aquifer that covers the range between the Rosetta branch and the eastern part of Wadi El Natrun. This aquifer is mostly unconfined but under semi-confined conditions at northern El Nubaria canal. The hydraulic conductivity of the Quaternary aquifer in the examined region ranges from 30 up to 77 m/day [56–58]. As indicated by Osman [59], most of the Pleistocene deposits in the investigation region are of river depositional environment with multidirectional sedimentation environment. These deposits are poorly sorted, and the mean grain estimate ranges from medium pebbles to medium sand with average size that of extremely coarse sand. The Holocene aquifer is made out of present-day sand gatherings of aeolian sandy stores with calcareous intercalations from 4 m to around 10 m thickness. It is situated in the lowland regions of Wadi El Natrun (Fig. 1b).

The surface geology of the area (Fig. 2) shows that, in the south, there are stabilized dunes and sand sheets, while Wadi deposits (sand and gravels) cover the central part around Nubaria canal. In the northern part, the surface is covered by

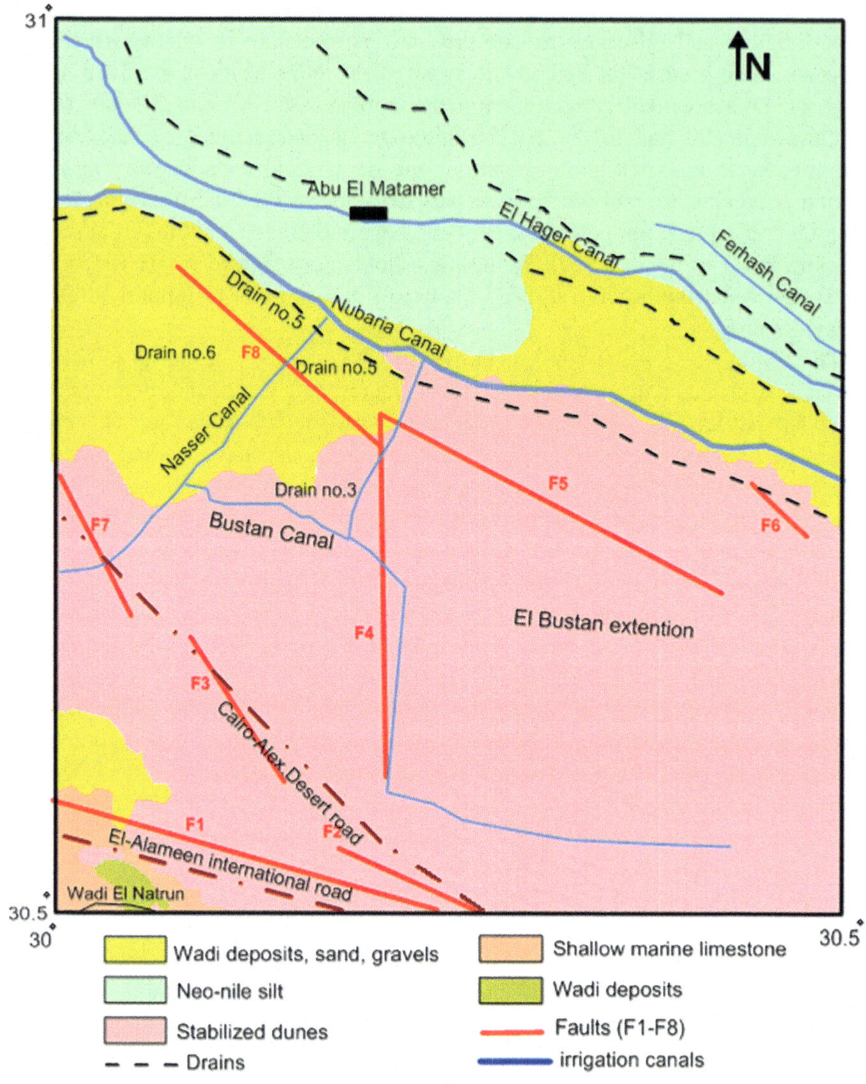

Fig. 2 Geomorphologic and structural map of the study area

neo-Nile silt and clay. The structural map of the area compiled by different authors is shown in Fig. 2. Major faults with downthrown sides due east have been delineated along the western margin of the Nile Delta. These faults have led, among other geologic features, to the facing of the highly permeable Pleistocene gravel of the Nile Delta opposite to older impermeable sediments to the west. This situation accelerates the subsurface westward flow of Nile water to the older sediments [60]. Two subsurface faults in the area were detected by El Ghazawi [61] comprising El Galala fault (F7) with downthrown to east and displacement reaches about 65 m and El Nubaria fault (F8) with downthrown also to east (Fig. 2). They are related to clysmic system during Neogene times. However, the rapid vertical and lateral facies changes of the Quaternary sediments hide this displacement. On the other hand, four probably subsurface faults (F3, F4, F5, and F6) have been detected with downthrown to the Nile Delta [62]. They greatly affect the thickness of the Pleistocene aquifer.

Wadi El Natrun anticlinal structure trends E 35° W and extends for about 60 km from El Ralat depression in the north to Beni Salama depression in the south [63]. It is a symmetrical structure affected by parallel and diagonal faults, among them F1 and F2 that led to the formation of the central main depression (Fig. 2). Unconformities are recognized in the area between the Quaternary deposits and Pliocene deposits, where Wadi El Natrun Formation (early Pliocene) is covered directly by Quaternary clastics, and El Hagif Formation is absent (late Pliocene) [61].

Saad [64] stated that the groundwater flow direction in this area is from NE to SW and the transmissivity of the Pleistocene aquifer is 0.615–0.897 m^2/min. The water level all over the area was increased by about 2.5–3 m from March 1972 to March 1973. Before reclamation projects, the groundwater seeps from El Nubaria canal to the west, but after constructing the pumping stations, the groundwater flows toward El Nubaria canal, where it acts as a drain (The Desert Institute in its internal report [65]). Abdel Baki [66] stated that the recharge in the area is from two sources: the groundwater inflow and the infiltrated water from irrigation and the groundwater depth vary widely from a few meters close to the Delta to about 60 m near the Cairo-Alexandria desert road. Recently, groundwater flow system during 2010 was constructed by Salem and Osman [19, 53]. It was found that the water table height differs from −2 m in the northern sites to more than +27 m above sea level in the southwestern direction (Fig. 3). The general groundwater flow is from south to northern and northeastern directions. The main recharge area is seen in the southern part of the territory. The groundwater likewise streams toward Wadi El Natrun in the southwestern parts. They also stated that the piezometric level difference between 1966 and 2010 revealed an increase of the water table reaching up to 30 m in the southwestern part of the area which might have occurred due to irrigation processes and seepage from the canal systems.

3 Methodology

The two basic electrical properties of the rocks are electrical resistivity and polarization. The polarization refers to a substance that exhibits a charge separation in an electric current. The resistivity of rock depends on many factors, including rock type, the conductivity of pores and nature of the fluid, and metallic content of the solid matrix. The rock resistivity is roughly equal to the resistivity of the pore fluids divided by the fraction porosity [67]. The resistivities of the common rocks and minerals are shown in Fig. 4. This work was done in three steps: data acquisition (field work), data processing, and data interpretation.

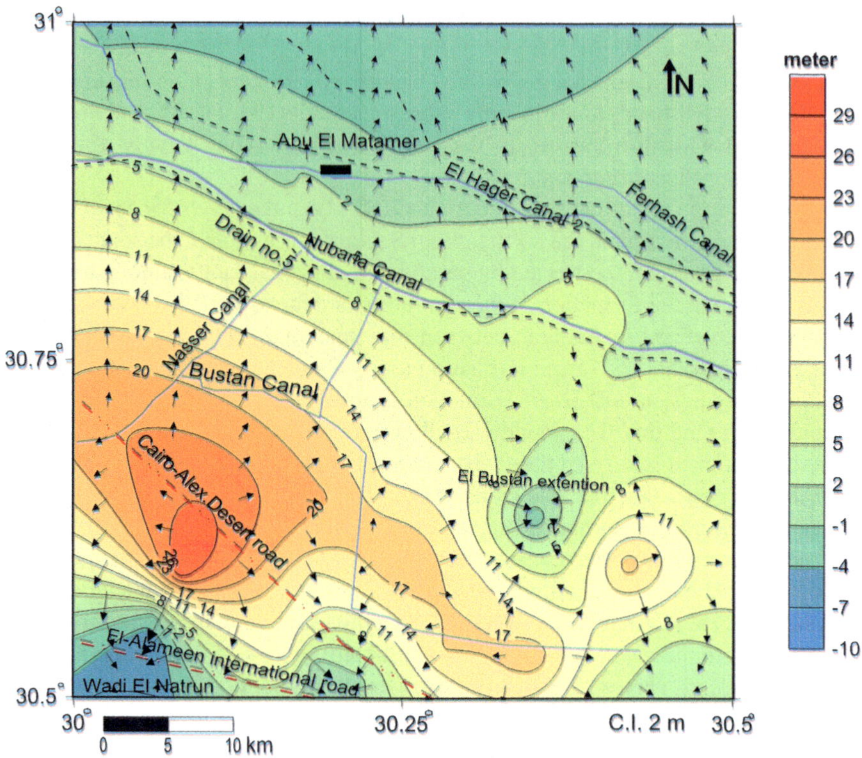

Fig. 3 Groundwater flow system in the study area (after [19])

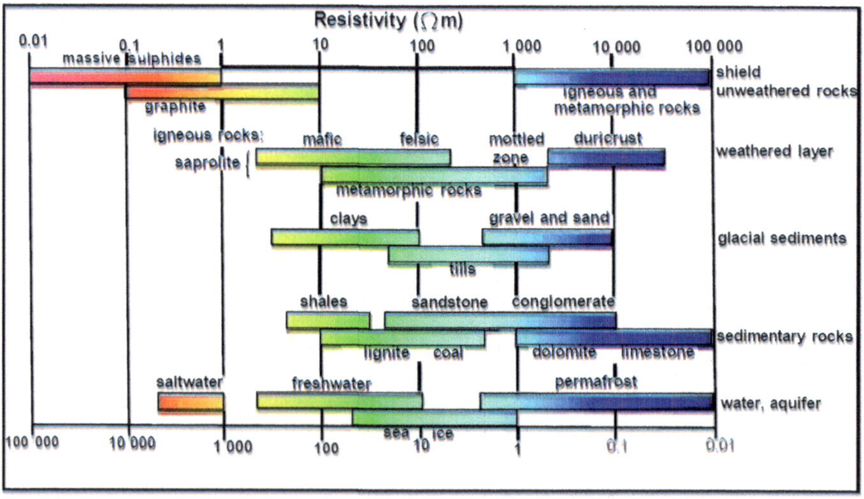

Fig. 4 Electrical properties of the common rocks and minerals [68]

3.1 Data Acquisition (Field Work)

The geoelectric survey of this work is executed by ELERIC-T SYSCAL-R2 (Fig. 5). In this work, the electrical resistivity survey was carried out in the form of vertical electrical sounding (VES), using Schlumberger 4-electrode array. Thirty-two vertical electrical soundings were measured in the study area. The VES locations were planned on topographic map for the study area (scale 1:50,000) and projected in the field by GPS instrument. The sounding points and geoelectrical cross sections were distributed to cover the area (Fig. 6). Some VES stations were located beside wells or boreholes to correlate the measured resistivity data with known geologic and lithologic successions. The current electrode spreading (AB/2) starts with 1–1,000 m, while the potential electrode spreading (MN/2) starts with 0.25–25 m to obtain a measurable potential difference. This electrode separation is sufficient to reach the required depth that fulfills the main aim of the study given the geologic and hydrogeologic information.

3.2 Data Processing

The SYSCAL R2 computes and displays the apparent resistivity automatically for the most common electrode arrays (Schlumberger and Wenner sounding and profiling, gradient, dipole-dipole, etc.) in the field, i.e., it has automatic processing software. Data interpretation can be made qualitatively and quantitatively. The qualitative interpretation gives initial information about layer numbers, resistivity, direction, and homogeneity degree. The quantitative interpretation determines

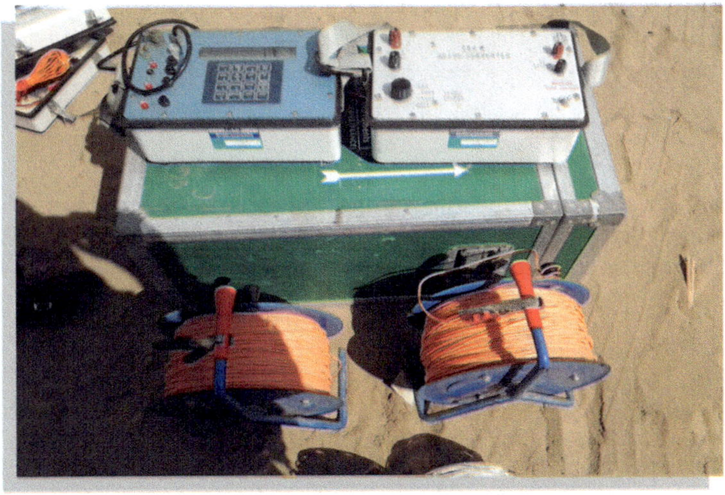

Fig. 5 The used Earth resistivity instruments (ELERIC-T, SYSCAL R2)

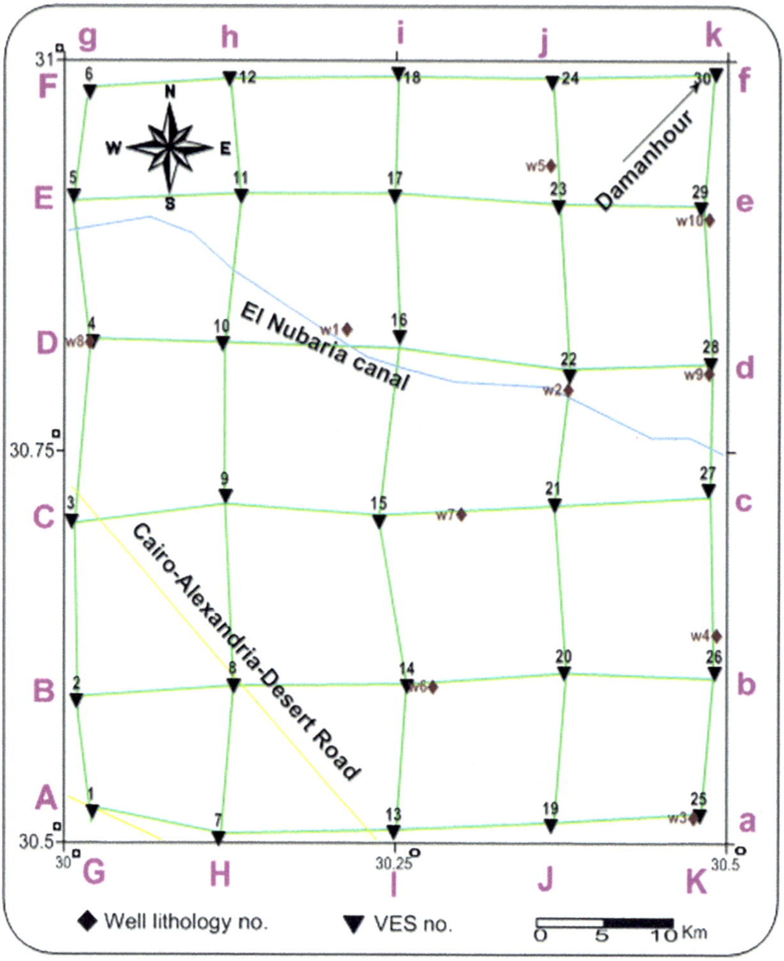

Fig. 6 Location map of the soundings (VESs) and geoelectrical cross-section distribution in the area

geoelectrical layers, true depth, thickness, and resistivity of each layer. The geophysical boundaries may cross over different lithologies and may not coincide with lithologic boundaries or geologic formation boundaries.

3.3 Data Interpretation

There are many computer programs commonly used in quantitative interpretation. In this work, a program called IPI2win, version 2.0 [69], was used. The obtained lithology (thickness, type) data from the nearby wells was used as a preliminary 1D model for regularizing the process of the fitting error minimization. An important advantage of this program is the concept of interpreting a profile, where its points are treated as a unity representing the geological structure of the survey area as a whole, rather than a set of independent objects dealt with separately.

According to geologic information and nearby wells lithology to the VES stations, a good starting model for each one is used during the inverse modeling operation by using IPI2win software. The fixing option in this program is used to fix some parameters of the model to not be free during the inverse modeling process. By this way, all the available layer thickness which obtained from some wells are introduced as a ground truth. Also, reliable geoelectric cross sections are constructed by reinterpretation of some VES according to the results obtained from interpretation of the neighbored ones. The final model which is calculated by using this program is given in the appendix. Also, the interpreted layer parameters in the form of true resistivities and thicknesses of the surface geoelectric layers of these soundings are listed in Table 1. The aquifer depth, aquifer thickness, and water table are shown in Table 2.

4 Results and Discussion

4.1 *Resistivity Spectrum*

There are many VESs carried out beside drilled wells in the area to calibrate its resistivity results with those of well lithology. These calibrations reveal a good matching of the interpreted resistivities with the lithology layers (Fig. 7) and give a good resistivity spectrum for the area. The electric resistivity of sediments depends on lithology, clay and water contents, as well as salinity [21]. Therefore, the resistivity spectrum shows a wide range of resistivity values. So, there is an importance to correlate the wells lithology with the VES values to ensure the final results. The resistivity spectrum of this work is shown in Table 3.

4.2 *Geoelectrical Cross Sections*

There are 11 cross sections at different locations and directions; they are constructed from the deduced layer parameters. These include six lateral cross sections in an east-west direction containing five VES points and five traverse cross sections in a north-south direction having six VES points. The main purpose of these cross sections is clarifying two-dimensional pictures for the subsurface layer distributions (depth, thickness, lithology, aquifer extension, and seawater intrusion limitation). These cross sections mainly consist of four layers. They are surface, sand (fine sand with intercalations of clay, medium to coarse sand), clay layers (clay and sandy clay), and brackish water to saltwater.

Table 1 The obtained true resistivities and thicknesses of the detected geoelectric layers

				Layer 1		Layer 2		Layer 3		Layer 4		Layer 5		Layer 6	
No.	Lat. (N)	Long. (E)	Elev. (m)	ρ (Ω)	h (m)	ρ (Ω)	h (m)	ρ (Ω)	h (m)	ρ (Ω)	h (m)	ρ (Ω)	h (m)	ρ (Ω)	h (m)
1	30.52	30.02	30	17	8	6	23	21	60	15					
2	30.58	30.01	37	10	4	2	10	5	25	15	80	2.6			
3	30.71	30.00	30	9.3	1.5	68	1.5	15	60	5					
4	30.82	30.02	19	10	3	15	29	5	45	2.5					
5	30.92	30.00	0	8	1.5	6	19	10	22	2.5					
6	30.98	30.02	−3	15	1.5	5	17	2.5	25	1					
7	30.50	30.12	40	30.2	1	7.2	10	15.8	13	6.2	25.3	15			
8	30.57	30.13	31	77.4	3	24	15	5							
9	30.72	30.12	25	131	1.3	66.8	1.3	25.9	18.4	15.3	59	5	37.5	1	
10	30.82	30.12	12	22.4	1.5	65	2	17	25	34	36	8			
11	30.92	30.13	−3	33.2	1	5	58	1							
12	30.99	30.13	−3	10	0.5	6	15	2.5	22	1					
13	30.51	30.25	36	17	9	28.2	100	5							
14	30.60	30.27	28	510	4	12	7	8	40	24	60	3			
15	30.71	30.24	21	109.5	3.3	20.4	90	5							
16	30.83	30.25	6	12.3	1	35	6	18	30	46	38	5.4			
17	30.92	30.25	−2	16	3.5	7	16	15	28	6	36	1			
18	31.00	30.25	−2.5	10.6	2	6	15	2.7	22	1					
19	30.52	30.40	30	37	1	23	4	307	8	21	86	4			
20	30.61	30.37	27	31.5	1	18.2	5	184	4.3	8	40	24			
21	30.72	30.37	18	73.6	1	261	1.3	10	4	26	120	8			
22	30.80	30.38	6	70.8	1.5	5.9	1	17	45	30	55	14			
23	30.91	30.38	−2	16	1.5	12	9.5	23	35	8	42	1.4			
24	30.99	30.37	−4	4.2	0.5	27.7	0.5	11	32	6	35	1			

25	30.52	30.48	32	163	1.9	25	2.5	296	17	26	47	23			
26	30.61	30.50	21	91	2	116	9	50	54.7	29					
27	30.73	30.49	10	55	4	7.9	50	24.6	65	7.2					
28	30.81	30.49	4	39.6	2.9	9.1	2.1	20.1	13	14.2	13.4	83.2	61.2	12	
29	30.91	30.48	3	10.1	1.5	142	3.3	16	31	44	44.3	7.2			
30	31.00	30.50	3	7.2	2	2.5	7	59.5	67	6.6					

Table 2 The hydrogeological parameters of the Pleistocene aquifer of the study area

No.	Water level	Water table	Aquifer depth	Brackish water depth	Aquifer thickness	Aquifer resistivity	Average surface resistivity
1	−24	6	24			18	17
2	−31.9	5.1	31.9			15	6
3	−2.8	27.2	2.8			15	40
4	−1.9	17.1	1.9			15	10
5	−1.5	−1.5	1.5				8
6	0.5	−2.5	0.5	44	44		15
7	−13.4	26.6	13.4			15	23
8	−21	10	21				50
9	−2.5	22.5	2.5	117	114.5	21	100
10	−1.7	10.3	1.7			26	43
11	−0.4	−3.4	0.4	60	59.6		33
12	−0.3	−3.3	0.3	38	37.7		10
13	−13	23	13			28	17
14	−6.7	21.3	6.7			24	260
15	−3.3	17.7	3.3			20	110
16	−1	5	1			32	25
17	−0.5	−2.5	0.5	84	83.5	15	16
18	−1	−3.5	1	40	39		10
19	−19.5	10.5	19.5			21	122
20	−24.1	2.9	24.1			24	228
21	−12	6	12			26	115
22	−1.6	4.4	1.6			24	26
23	−1.2	−3.2	1.2	88	87	23	16
24	−1	−5	1	58	57		28
25	−21.7	10.3	21.7			25	160
26	−14.5	6.5	14.5			40	70
27	−4	6	4			25	55
28	−4.9	−0.9	4.9			52	25
29	−2.7	0.3	2.7			30	76
30	−1.2	1.8	1.2			60	7

4.2.1 Cross-Section Aa

This cross section passes at the southern part of the area from east to west and includes five VES stations (1, 7, 13, 19, and 25). The geoelectric horizons (Fig. 8) are (from top to bottom):

1. The surface layer: its resistivity ranges from 16 (VES no.7) to 307 Ω m (VES no.19). It consists of sand, calcareous sand, and shale. Its thickness varies from 10 to 25 m.

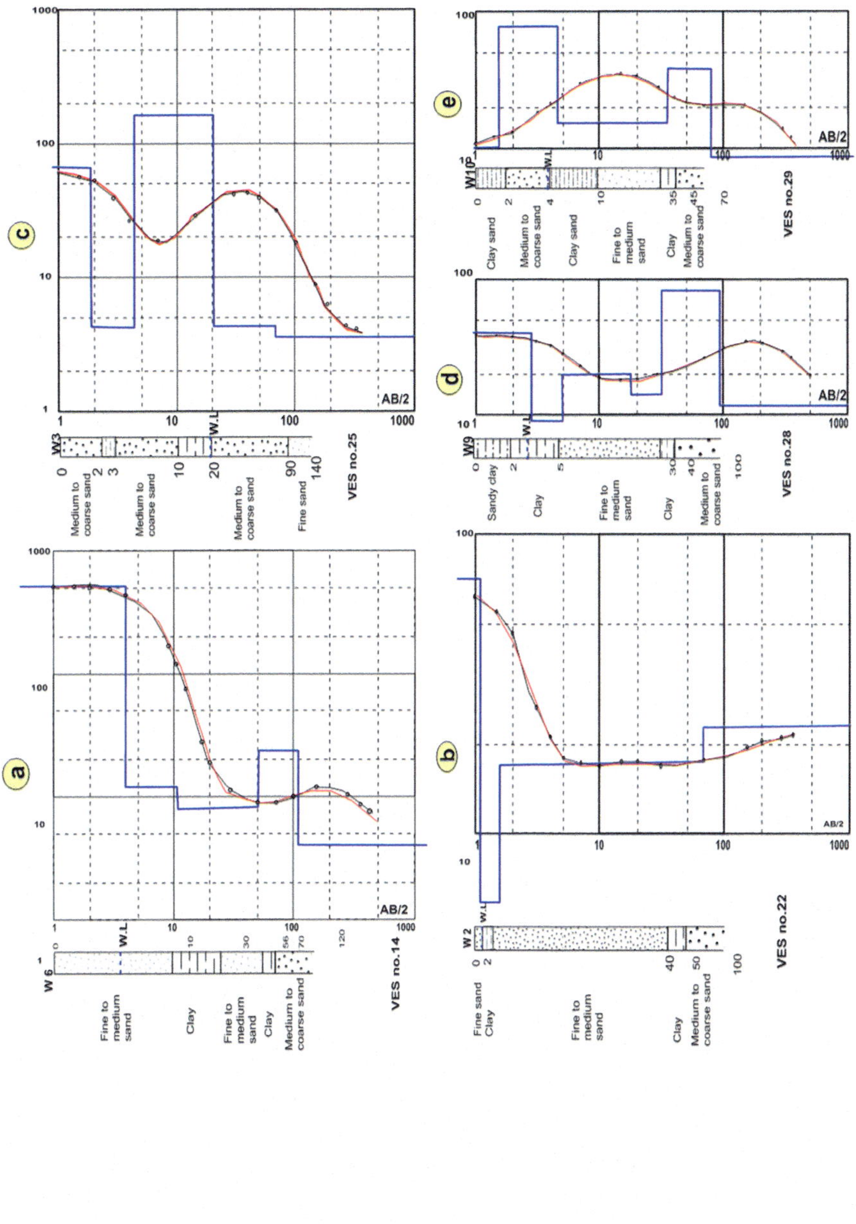

Fig. 7 Examples of the calibration between interpreted VES and lithology of the nearest wells of the study area (**a**, VES 14; **b**, VES 22; **c**, VES 25; **d**, VES 28; and **e**, VES 29)

Table 3 Resistivity spectrum results of the detected layers in the study area

No.	Phase	Lithology type	(Ω m)
1	Surface layers (dry)	Sand, clay, gravels, ... etc.	10–500
2, 3	Saturated zone	Medium to coarse sand	30–90
		Clayey sand	15–30
		Sandy clay	5–15
		Clay	2–10
4	Brackish water to saltwater zone		<2

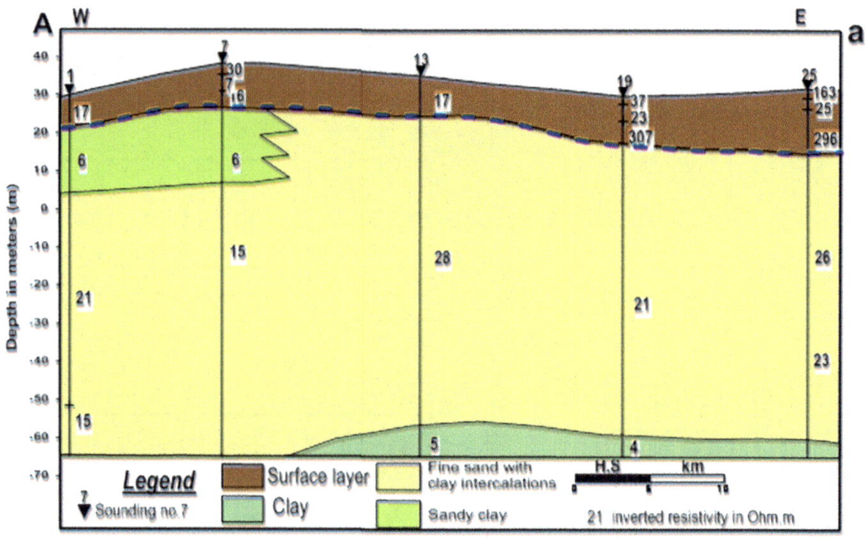

Fig. 8 Geoelectric cross section along profile Aa (W-E) in the study area

2. Aquifer layers: resistivity ranges from 15 to 28 Ω m and it consists of fine sand with intercalations of clay. The thickness varies from 60 m (VES no. 19) to 90 m (VES no. 13).
3. Clay layers: resistivity varies from 4 to 6 Ω m (VES no.19, 7, respectively). They are composed of clay at the base of VES no. 13, 19 and 25 and sandy clay between the surface and fine sand in VES no. 1 and 7.

4.2.2 Cross-Section Bb

It also extends from east to west and lies in the southern part of the area (Fig. 9). This cross section contains five VESs (2, 8, 14, 20, and 26). Interpretation of cross-

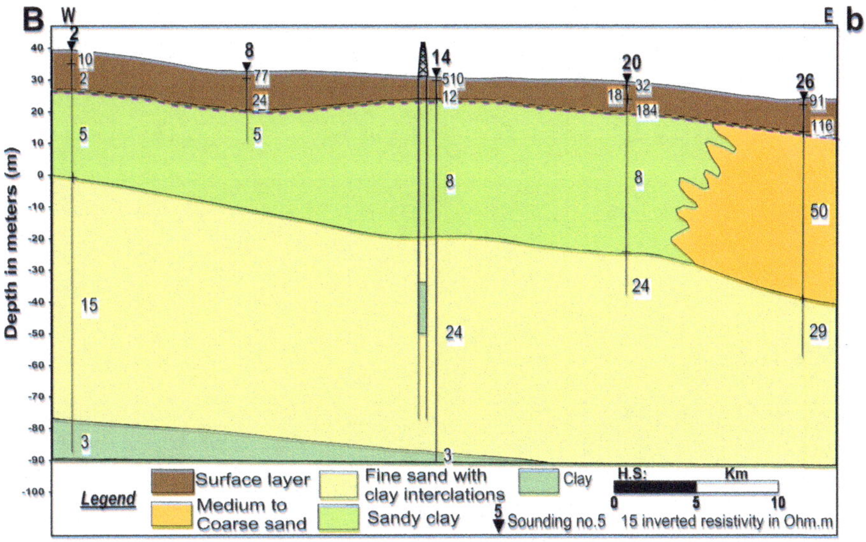

Fig. 9 Geoelectric cross section along profile Bb (W-E) in the study area

section VESs is mainly constrained by the geologic information from the drilled well close to sounding no. 14; the layers of this section are:

1. The surface layer: its resistivity ranges from 10 to 512 Ω m (VES 2 and 14) with a variable thickness ranging from 10 to18 m (VES 20 and 8).
2. Aquifer layers: resistivity ranges from 15 to 50 Ω m (VES 2 and 26), and the thickness varies from 60 m to unknown value (VES 14 and 26). These layers are composed of fine sand with clay intercalations and medium to coarse sand.
3. Clay layers: their resistivities vary from 3 to 8 Ω m (VES no. 2 and 4). They are composed of clay at the base (VES no. 2 and 14, 20) and sandy clay between surface and fine sand (VES no. 2 and 20) with thickness ranging between 25 m and 40 m (VES no. 2 and 20).

4.2.3 Cross-Section Cc

The geoelectric cross section along this profile contains five soundings; they are 3, 9, 15, 21, and 27 (Fig. 10). This section lies at the southern central part of the area. It extends from west to east and consists of the following geoelectric horizons:

1. The surface layer: Its resistivity ranges from 10 to 261 Ω m (VES 3 and 21), while the thickness varies from 2.6 to 6.3 m (VES 9 and 21).
2. The aquifer layers: With resistivity ranging from 15 to 26 Ω m (VES 3 and 21). The thickness ranged from 60 to 120 m (VES 3 and 21) and is composed of fine sand with clay intercalations.

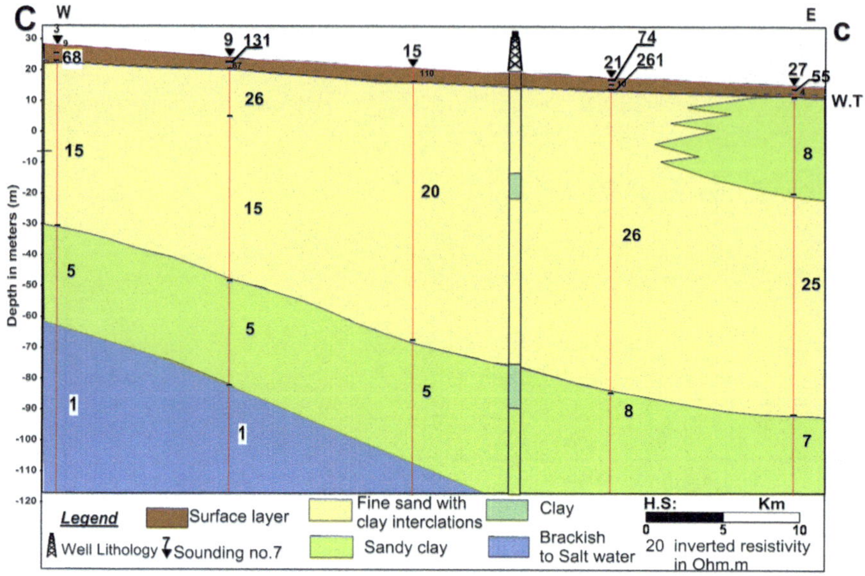

Fig. 10 Geoelectric cross section along profile Cc (W-E) in the study area

3. The clay layer: Its resistivity ranges from 5 to 9.3 Ω m (VES 3, 9, 15 to 21, 27). The thickness varies from 37.5 m (VES 9) to unknown value at the rest of the cross section.
4. The brackish water to saltwater intrusion zone: Its resistivity in the range of 1 Ω m and depth − 90 m (VES 9) and to be expected in VES no. 3 and 15 (Fig. 10).

4.2.4 Cross-Section Dd

This cross section is parallel to the previous cross section and lies in the north-central part extending from west to east. It contains five soundings; they are 4, 10, 16, 22, and 29. It includes the following geoelectric layers (Fig. 11):

1. The surface layer: its resistivity ranges from 6 to 71 Ω m (VES no. 22), and thickness varies from 2.5 to 7 m (VES 22 and 16).
2. The aquifer layers: can be differentiated into two zones:
   (a) Fine sand with clay intercalations: at the top with resistivity ranging from 15 to 20 Ω m (VES 4 and 28). The thickness varies from 25 to 30 m (VES 10 and 16).
   (b) Medium to coarse sand: at the base with resistivity ranging from 30 to 83 Ω m (VES no. 22 and 28). Its thickness varies from 36 to 61 m (VES no. 10 and 28).

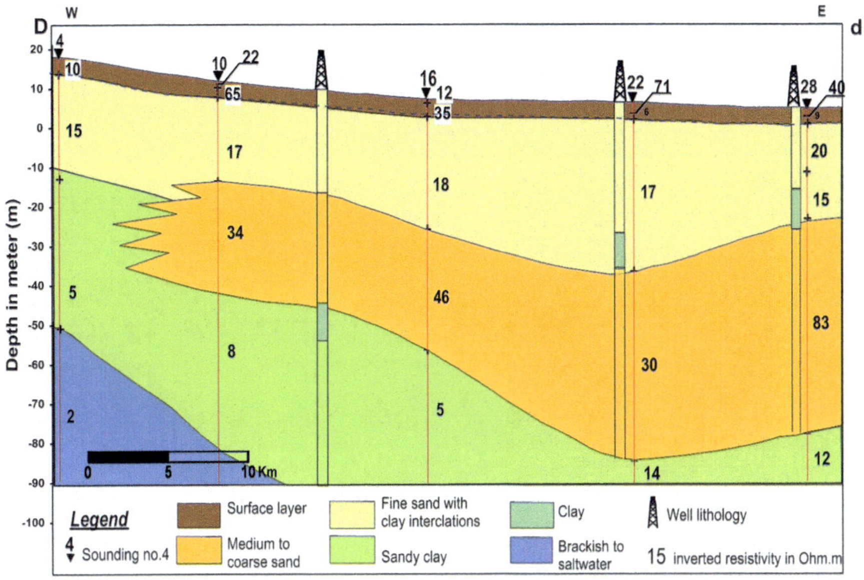

Fig. 11 Geoelectric cross section along profile Dd (W-E) in the study area

3. The clay layer: with resistivity ranging from 2 to 14 Ω m (VES 4 and 22), and thickness varies from 45 m (VES 4) to unknown value in the others. It is composed of sandy clay and clay.

4.2.5 Cross-Section Ee

This cross section includes five VES stations (5, 11, 17, 23, and 29). It extends from the west to the east and lies to the north and parallel to the previous cross sections. This cross section (Fig. 12) consists of the following layers:

1. The surface layer: with resistivity ranging from 8 to 142 Ω m (VES 5 and 29) and thickness from 1 to 3.5 m (VES 11 and 17).
2. The aquifer layers: the resistivity ranges from 15 to 44 Ω m (VES 17 and 29). The thickness varied from 30 to 75 m (VES 15 and 29) and is composed of fine sand with clay intercalations and medium to coarse sand.
3. The clay layer: its resistivity ranges from 5 to 12 Ω m (VES 11 and 23). The thickness varies from 52 to 69 m (VES 5 and 11).
4. The brackish water to saltwater intrusion zone: the resistivity is less than 2 Ω m (VES 5, 11, 17, 23). Its depth ranges from −47 to −100 m (VES 5 and 23).

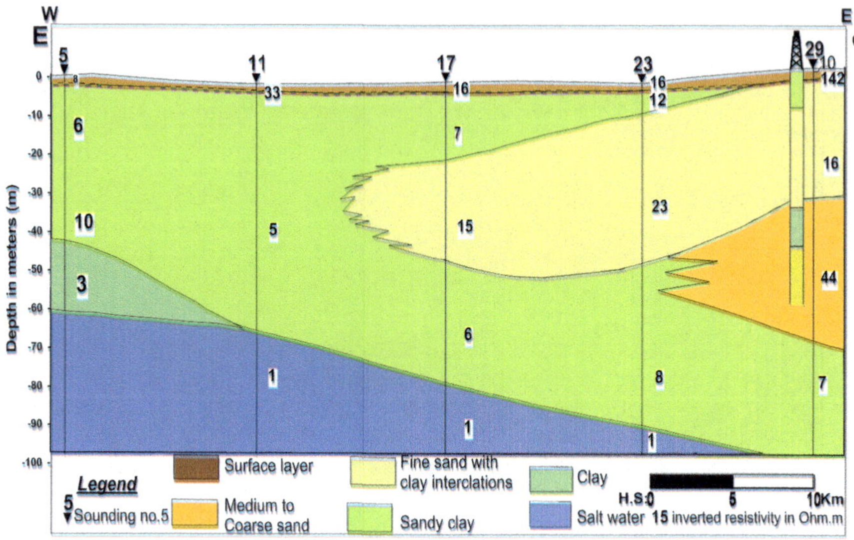

Fig. 12 Geoelectric cross section along profile Ee (W-E) in the study area

4.2.6 Cross-Section Ff

This cross section lies in the most northern part of the area and extends from west to east direction parallel to the previous cross sections. It includes five VES stations (6, 12, 18, 24, and 30). The sedimentary succession is (Fig. 13) as follows:

1. The surface layer: Its resistivity ranges from 6 to 28 Ω m (VES 12 and 24). The thickness varies from 1 to 2.5 m (VES 12 and 30).
2. The aquifer layers: Resistivity is 60 Ω m and thickness is 67 m (VES 30). They are composed of medium to coarse sand.
3. The clay layer: Its resistivity is 2 Ω m (VES 6, 12, 18), and the thickness varies from 30 to 35 m (VES 6, 12).
4. The brackish water to saltwater intrusion zone: Its resistivity is less than 2 Ω m (all cross-section VES). The depth ranges from −35 to −57 m (VES 6, 18).

4.2.7 Cross-Section Gg

This cross section lies in the western part of the area and extends in NS direction. It includes six VES stations; they are 1, 2, 3, 4, 5, and 6. The geoelectric layers (Fig. 14) are the following:

1. The surface layer: Its resistivity ranges from 2 to 68 Ω m (VES 2 and 3), and thickness from 2 to 14 m (VES 6 and 2).

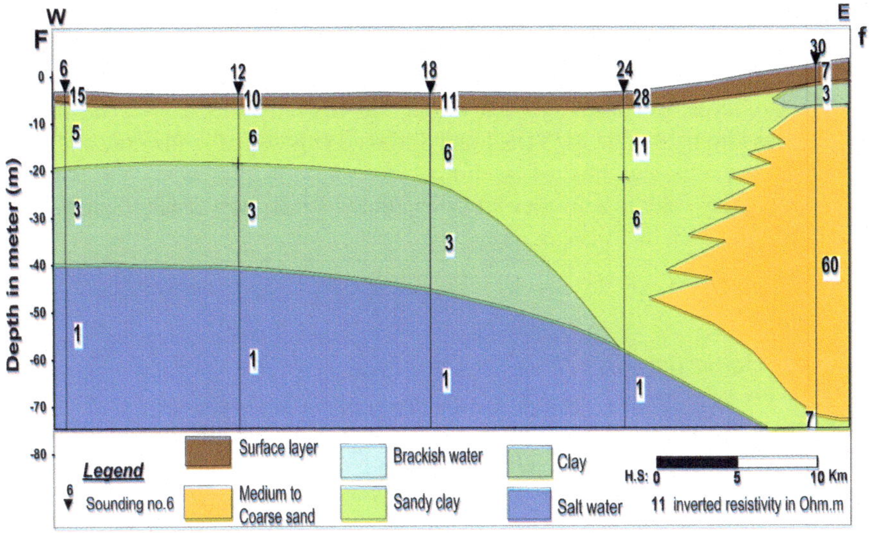

Fig. 13 Geoelectric cross section along profile Ff (W-E) in the study area

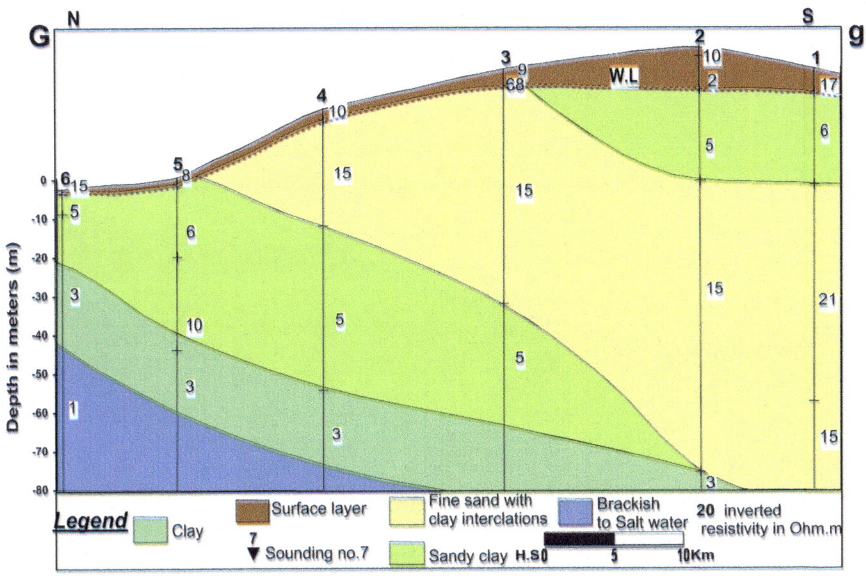

Fig. 14 Geoelectric cross section along profile Gg (N-S) in the study area

2. The aquifer layers: The resistivity ranges from 15 to 21 Ω m (VES 1, 2, 3, 4, and 1) and the thickness from 29 to 80 m (VES 4 and 2). The aquifer is composed of fine sand with clay intercalations.
3. The clay layer: Its resistivity ranges from 5 to 10 Ω m (VES 2, 3, 4, and 5) and thickness from 23 to 52 m (VES 1 and 5). It is composed of sandy clay and clay layers (at the base of the cross section).
4. The brackish water to saltwater intrusion zone: Its resistivity is less than 2 Ω m, and depth is −37 m (VES 1).

4.2.8 Cross-Section Hh

This cross section lies in the central western part of the area and extends in NS direction. It includes six VES stations; they are 7, 8, 9, 10, 11, and 12 and extend parallel to the previous one. The geoelectric layers (Fig. 15) are the following:

1. The surface layer: Its resistivity value ranges from 10 to 131 Ω m (VES 12 and 9), and the thickness ranges from 1 to 18 m (VES 12 and 8).
2. The aquifer layers: Resistivity ranges from 15 to 34 Ω m (VES 9 and 10), and the thickness varies from 61 to 77 m (VES 10 and 9). They are composed of fine sand with clay intercalations and medium to coarse sand.

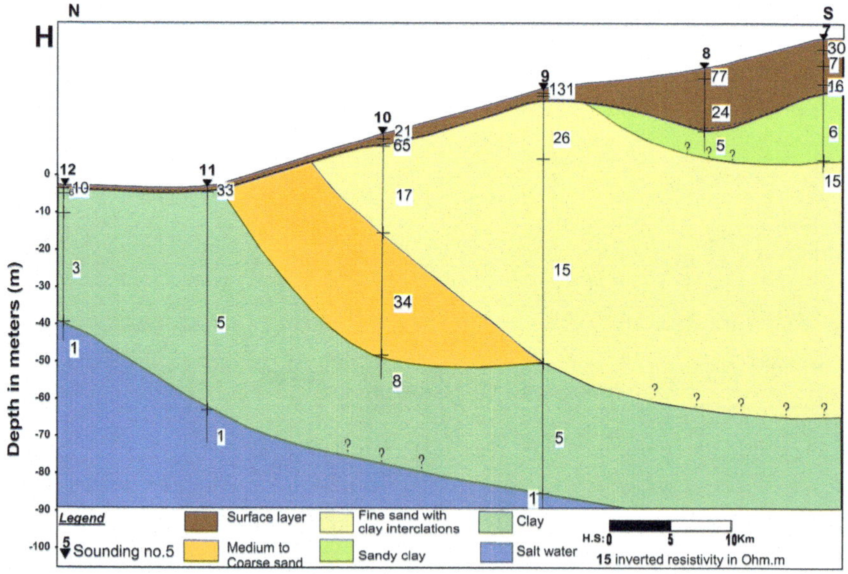

Fig. 15 Geoelectric cross section along profile Hh (N-S) in the study area

3. The clay layer: Its resistivity ranges from 2 to 8 Ω m (VES 12 and 10), and the thickness varies from 25 to 69 m (VES 7 and 11). It is composed of sandy clay layers at the top and clay layers at the base of the section.
4. The brackish water to the saltwater zone: Its resistivity is less than 2 Ω m (VES 9, 11, 12), and the depth varies from −37 to −90 m b.s.l (VES 12 and 9).

4.2.9 Cross-Section Ii

This cross section lies in the middle part of the area and extends in NS direction. It includes six stations; they are 13, 14, 15, 16, 17, and 18. The geoelectric layers of this section (Fig. 16) are the following:

1. The surface layer: Its resistivity ranges from 10 to 510 Ω m (VES 18 and 14), and the thickness varies from 2 to 14 m (VES 18 and 14).
2. The aquifer layers: The resistivity ranges from 15 to 46 Ω m (VES 17, 16), while the thickness varies from 30 to 90 m (VES 17 and 16). They are composed of fine sand with clay intercalations and medium to coarse sand (VES 16).
3. The clay layers: With resistivity ranges from 2 to 8 Ω m (VES 14 and 18) and thickness from 8 to 40 m (VES 18 and 14). They are composed of sandy clay and clay layers (base of the section).
4. The brackish water to saltwater intrusion zone: Its resistivity is less than 2 Ω m (VES 17 and18).

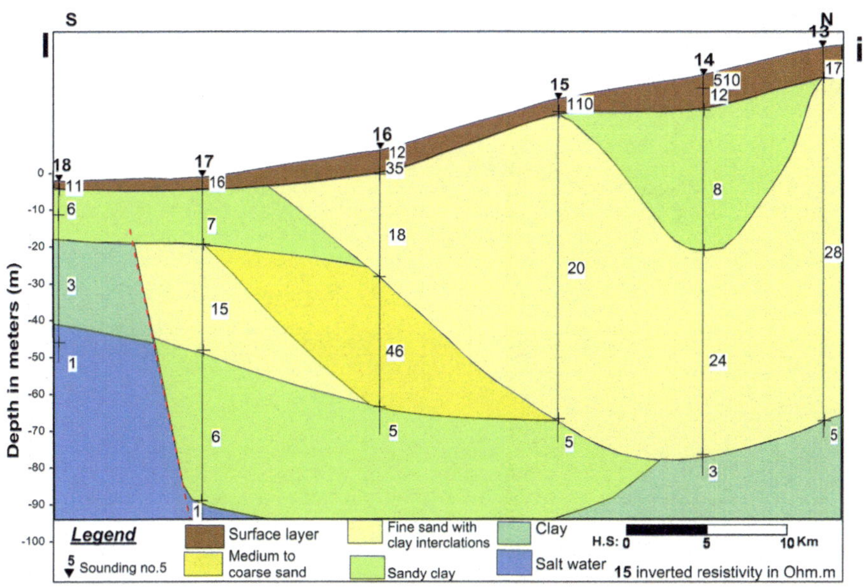

Fig. 16 Geoelectric cross section along profile Ii (S-N) in the study area

4.2.10 Cross-Section Jj

This cross section lies in the central eastern part of the area and extends in NS direction. It includes six VES stations; they are 19, 20, 21, 22, 23, and 24. The geoelectric layers of this section (Fig. 17) are the following:

1. The surface layer: Its resistivity ranges from 6 to 307 Ω m (VES 22 and 19), and the thickness ranges from 1 to 13 m (VES 24 and 19).
2. The aquifer layers: The resistivity ranges from 17 to 30 Ω m (VES 22). Its thickness ranges from 40 to 120 m (VES 23 and 21). They are composed of fine sand with clay intercalations and medium to coarse sand.
3. The clay layer: Its resistivity ranges from 4 to 14 Ω m (VES 19 and 22). The thickness of this layer ranged from 35 to 40 m (VES 24 and 20) and is composed of sandy clay and clay beds.
4. The brackish water to saltwater intrusion zone: Its resistivity is less than 2 Ω m (VES 23, 24).

4.2.11 Cross-Section Kk

This cross section lies in the eastern part of the area and extends in NS direction. It includes six VES stations; they are 25, 26, 27, 28, 29, and 30. The geoelectric layers of this section (Fig. 18) are the following:

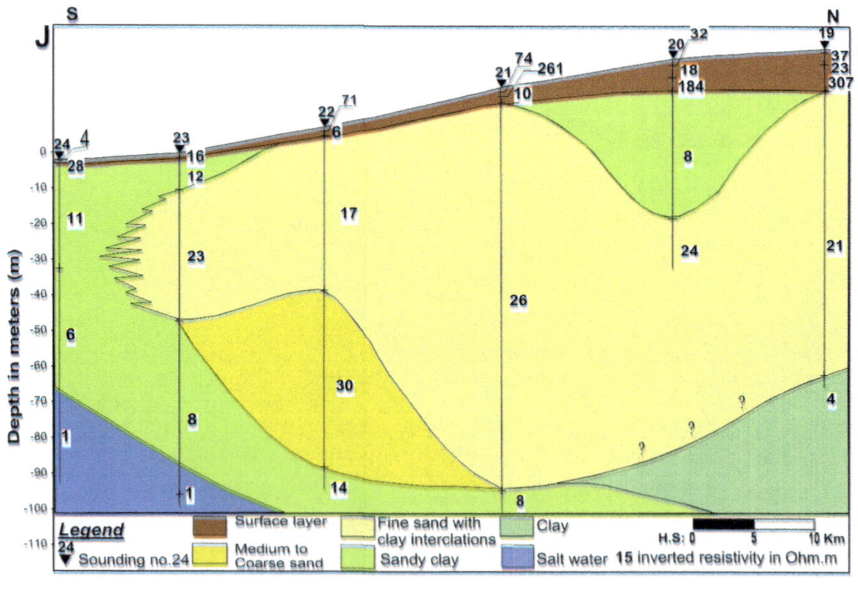

Fig. 17 Geoelectric cross section along profile Jj (S-N) in the study area

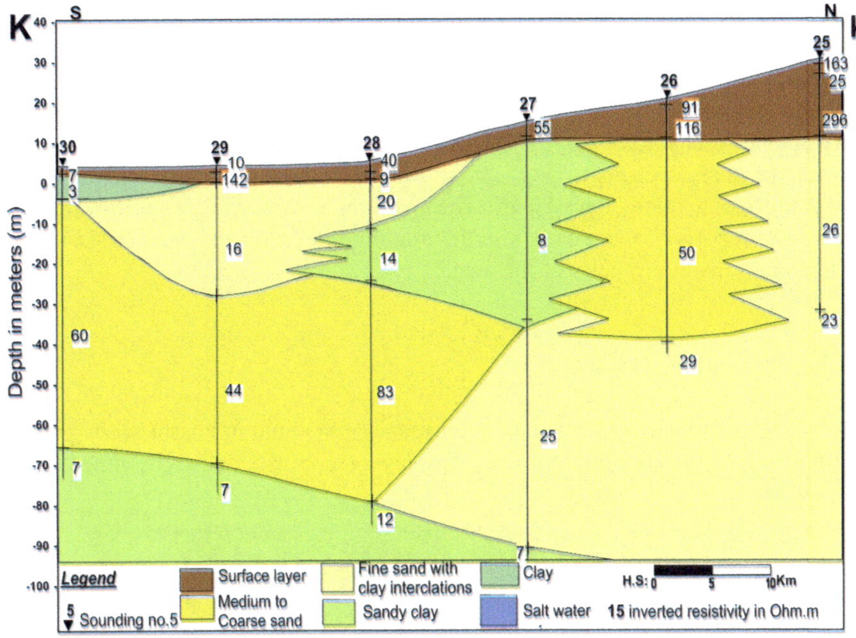

Fig. 18 Geoelectric cross section along profile Kk (S-N) in the study area

1. The surface layer: Its resistivity ranges from 7 to 296 Ω m (VES 30 and 25) with thickness varying from 2 to 21.5 m (VES 30 and 25).
2. The aquifer layers: The resistivity ranges from 16 to 83 Ω m (VES 29 and 28). The thickness varies from 65 m to unknown value (VES 27 and 25, 26). They are composed of medium to coarse sand and fine sand with clay intercalations.
3. The clay layer: Its resistivity ranges from 3 to 14 Ω m (VES 30 and 28), and the thickness varies from 2 to 50 m (VES 30 and 27). It is composed of sandy clay and clay layers.

4.3 Geoelectrical Maps

Several isoparametric maps have been constructed to illustrate the lateral variations of the geoelectrical data all over the area. The following are the description and interpretation of these maps:

4.3.1 The Surface Layer

This layer represents the recent deposits, deltaic deposits, stabilized dunes, and neo-Nile deposits. It exhibits a wide resistivity value ranging in average from 20 up to 240 Ω m. The true resistivity geoelectric map of this upper layer (Fig. 19) shows that the maximum value (240 Ω m) is toward the southeast of the area, reflecting the low salinity of sand dunes. On the other hand, the minimum value (20 Ω m) is recorded toward the northwestern parts of the area, due to the soil type (clay materials) that characterized by high salinity value. The thickness distribution map of the surface deposits (Fig. 20) shows an average thickness ranging from less than 1 m to more than 30 m. The maximum value (30 m) is toward the southwestern parts of the area, while the minimum value (1 m) is recorded in the north and northwestern parts of the area.

4.3.2 The Aquifer Layers

These layers are water saturated and composed of medium to coarse sand, gravel, and fine sand with clay intercalation. The layers exhibit a resistivity range from

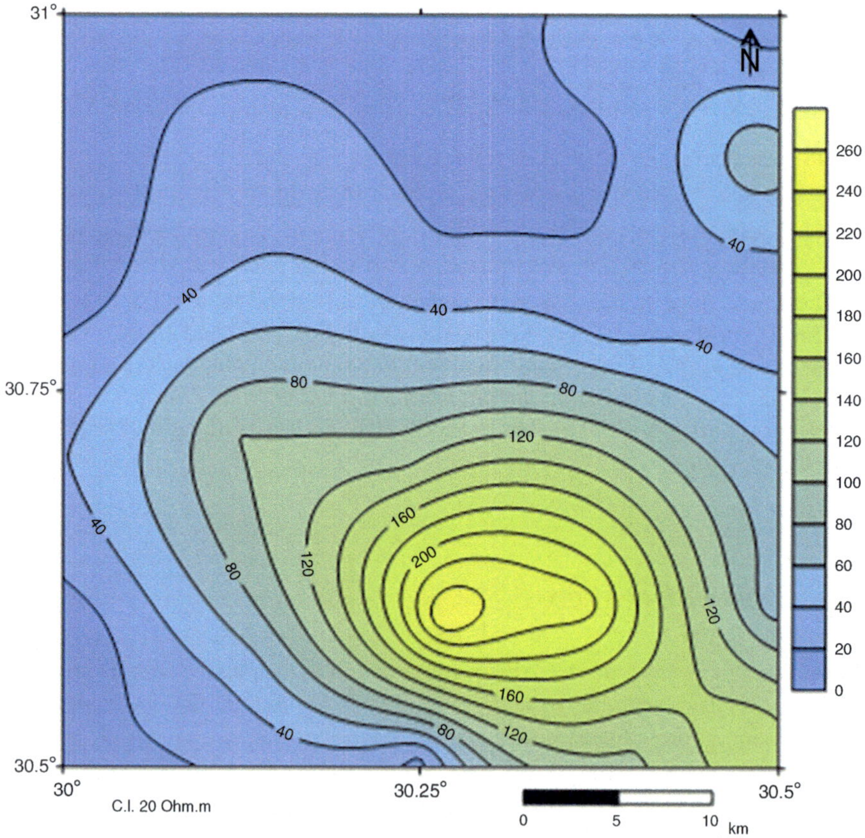

Fig. 19 The average surface layer resistivity of the area as deduced from resistivity electrical method

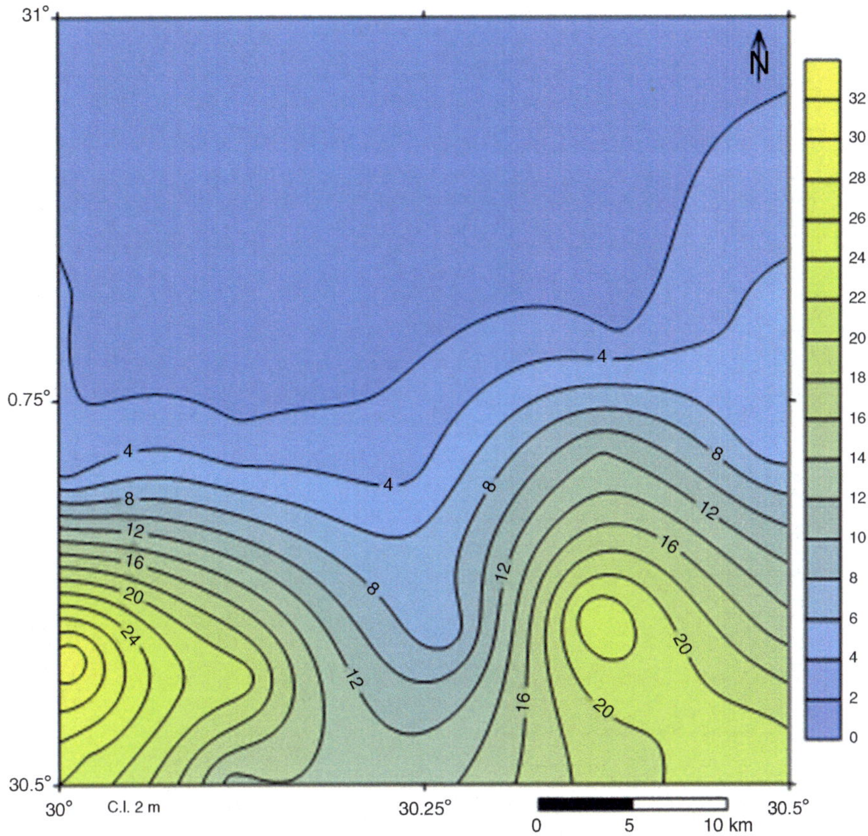

Fig. 20 The surface layer thickness of the area as deduced from resistivity electrical method

15 to 60 Ω m all over the area (Fig. 21). The minimum resistivity value is toward the northwestern parts, indicating seawater intrusion in this direction, while the maximum resistivity value is noted at the eastern parts toward the freshening water recharge from surface canals (as El Hager, Ferhash, El Nubaria). The aquifer thickness (Fig. 22) ranges from 40 up to 150 m. The minimum thickness is 40 m at the northwestern parts due to the effect of seawater intrusion. The maximum thickness (150 m) is recorded in the central part, and the aquifer thickness is increased toward the southeastern parts. It must be noted that the resistivity method applied here seldom reach more than 150 m.

4.3.3 The Seawater Intrusion Limitations

In the present study, the resistivity of the brackish water to saltwater zone is less than 2 Ω m as deduced from resistivity method. The seawater intrusion depth ranges

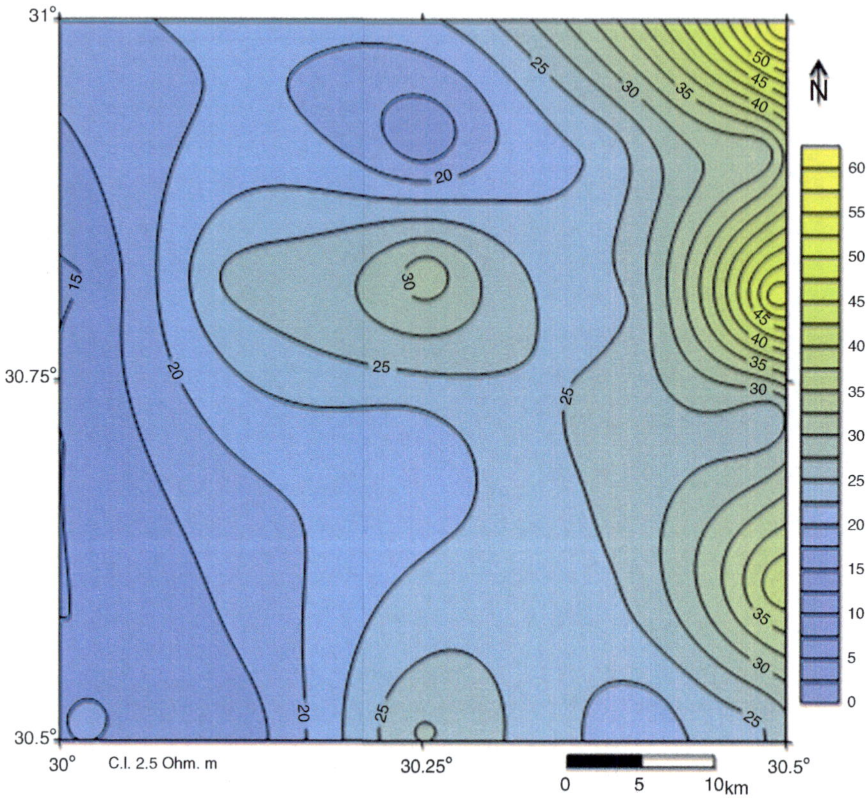

Fig. 21 The freshwater zone resistivity of the area as deduced from resistivity electrical method

from 40 to 150 m b.s.l (Fig. 23). The minimum depth reaches about 40 m in the northwestern parts of the area due to seawater intrusion. The maximum depth (150 m) is recorded toward the central parts. The applied resistivity method did not reach the seawater intrusion zone in the south and southeastern portions of the area.

4.4 3D Resistivity Model

The interpreted true resistivities and thicknesses at all sounding points are interpolated to construct the 3D model. The interpreted parameter was used to visualize the area in the form of 3D and 2D slices by using commercial Rockwork Software [70] to build solid geophysical models for the area in 3D of the dataset.

The model (Fig. 24a, b) illustrates the location of high resistivity zone in the southern and eastern directions with an average of 30–90 Ω m. Cutoff model is

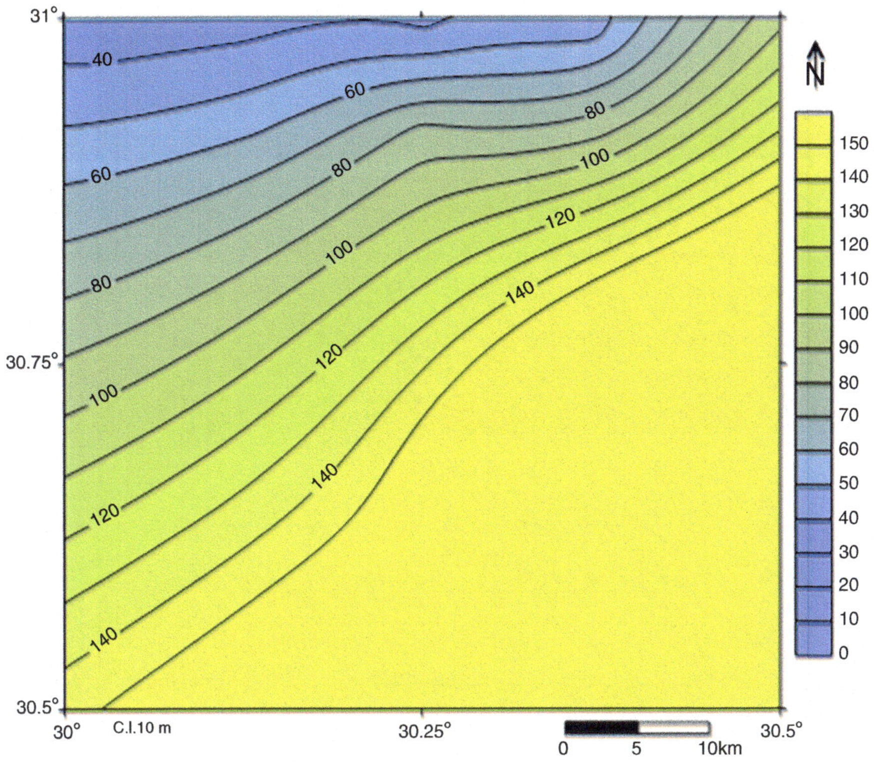

Fig. 22 The freshwater zone thickness of the area as deduced from resistivity electrical method

constructed for this resistivity zone to show only the thickness and depths of this zone which would be interpreted as the groundwater aquifer of the study area. It shows resistivity range from about 30–90 Ω at a depth range from the land surface (at the northwestern portions) and extends to nearly 30 m (at the southern portions) of the study area.

2D slices in the E-W and N-S directions of the resulted 3D model (Fig. 24c, d) were carried out to understand the lateral and vertical extension of the aquifer in the study area. It must be emphasized that the scale relations and orientations of the X, Y, and Z axes are changed for a better perspective view of the variation. The S-N slices show linear high resistivity in the south that decreases to north direction reflecting seawater intrusion trend. The east-west slices show that the main trend of high resistivity zone is toward the east direction.

The horizontal slices (Fig. 24e) reflect the high resistivity zone in the shallow depths at the southern part of the area as a result to dry surface layer. This value is decreased with depth until reaching the low resistivity zone of the brackish water. Finally, the resistivity cutoff (Fig. 24f) illustrates the aquifer of the study area in 3D direction. It shows that the great thickness is at the southeastern portions which decrease to the northwestern direction toward the seawater intrusion trend.

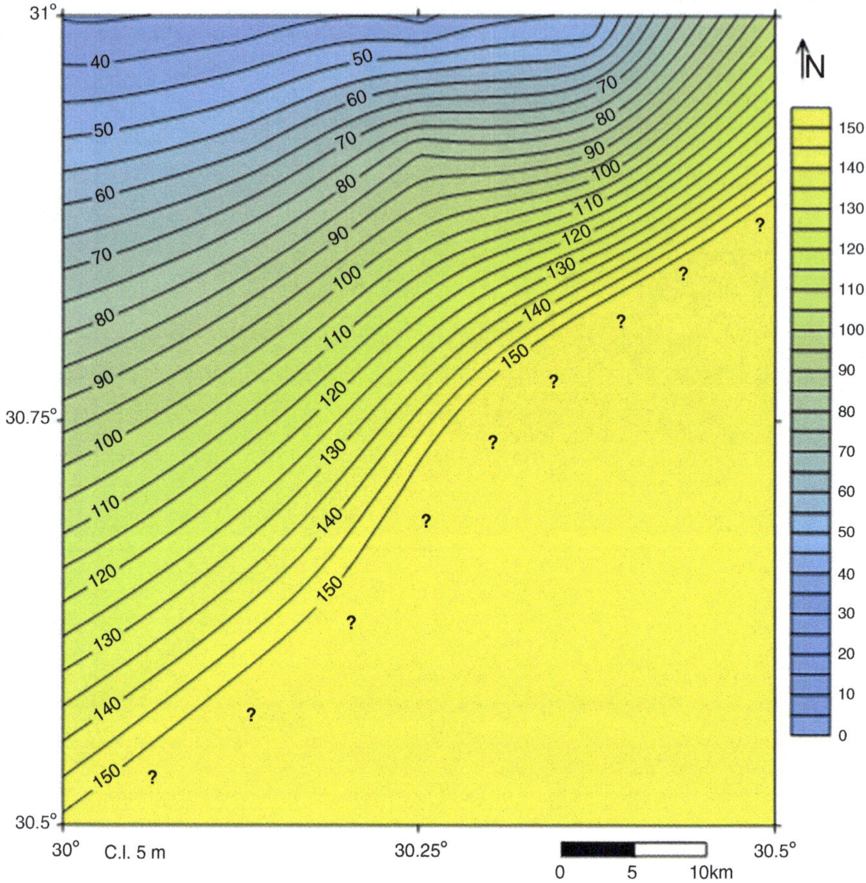

Fig. 23 The brackish water to saltwater zone depth of the area as deduced from resistivity electrical method

5 Conclusions

The resistivity data revealed four geoelectrical layers, they are:

1. The surface layer: This is composed of sand, clay, and silt. Its resistivity ranges from 5 to 512 Ω, and the thickness varies from 1 to 25 m.
2. The aquifer layer: Is composed of sand (fine sand with intercalations of clay and medium to coarse sand). Its resistivity ranges from 15 to 90 Ω m and thickness from 25 to 120 m.
3. The clay layer: Its resistivity ranges from 2 to 15 Ω m and thickness from 2 to 69 m.
4. The brackish water to saltwater intrusion zone: Its resistivity is nearly 1 Ω m, and its depth ranges from −35 in the northwest to −100 m in the south.

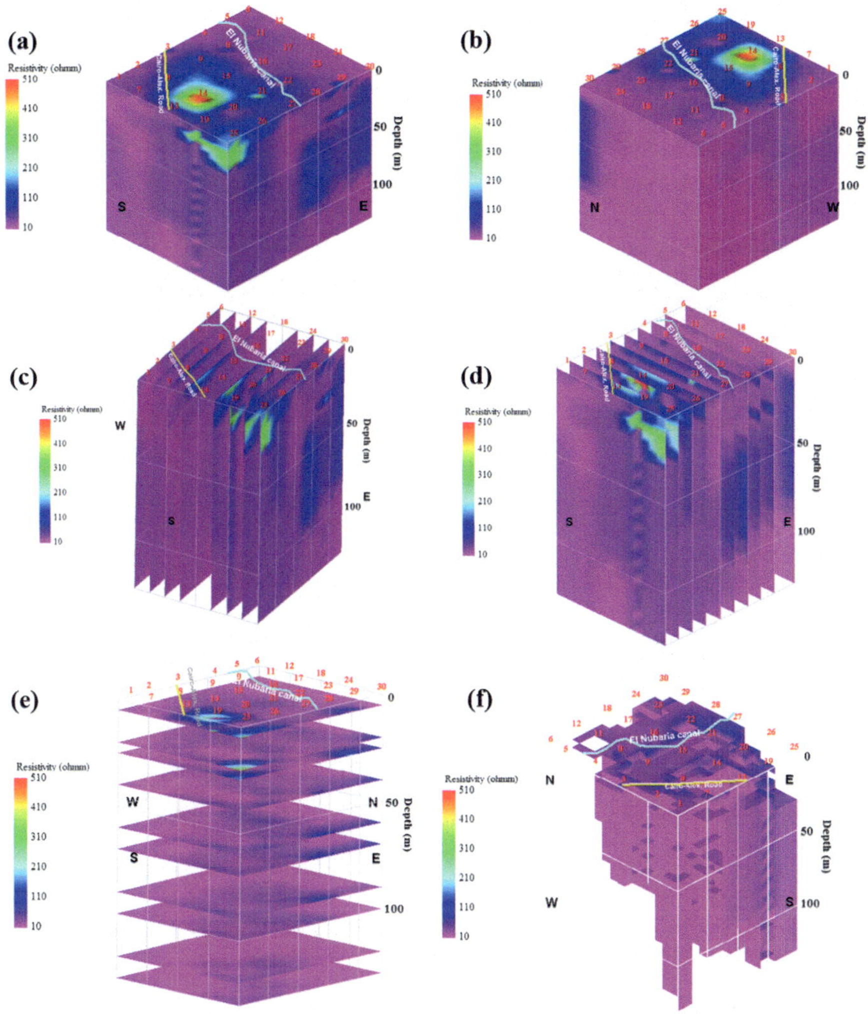

Fig. 24 3D presentation of the resistivity distribution, (**a**, **b**) resistivity solid diagram, (**c**) west-east slices, (**d**) south-north slices, (**e**) horizontal slices, and (**f**) resistivity cutoff diagram

The resistivity of the aquifer layer is increased toward the south and east directions where both the freshwater quality and thickness are increased. In contrast, the resistivity of the aquifer layer is decreased toward the northwest (the Mediterranean Sea direction) due to the effect of seawater intrusion.

6 Recommendations

Geoelectrical resistivity was successful for imaging the seawater-freshwater relationship and classifying the aquifer lithology in the target area. Therefore, the results of this work are recommended to be used by the decision makers for groundwater management planning in the study area.

Acknowledgments The authors are grateful to Tanta University for the financial support offered during the course of this research work. The authors thank the editor Prof. Dr. Abdelazim Negm for his constructive remarks.

References

1. Satriani A, Loperte A, Imbrenda V, Lapenna V (2012) Geoelectrical surveys for characterization of the coastal saltwater intrusion in metapontum forest reserve (Southern Italy). Int J Geophys. https://doi.org/10.1155/2012/238478
2. Post VEA (2005) Fresh and saline groundwater interaction in coastal aquifers: is our technology ready for the problems ahead. Hydrogeol J 13:120–123
3. Salem ZE, Al Temamy AM, Salah MK, Kassab M (2016) Origin and characteristics of brackish groundwater in Abu Madi coastal area, Northern Nile Delta, Egypt. Estuar Coast Shelf Sci 178:21–35
4. Sherif MM, Kacimov A (2007) Seawater intrusion in the coastal aquifer of Wadi Ham, UAE: a new focus on groundwater seawater interactions. Proceedings of symposium HS 1001 at IUGG 2007, Perugia, vol 312. IAHS, Wallingford, pp 315–325
5. Wilson J, Townley LR, Sa Da Costa A (1979) Mathematical development and verification of a finite element aquifer flow model AQUIFEM-1, technology adaptation program, Report No. 79-2. M.I.T., Massachusetts
6. Diab MS, Saleh MF (1982) The hydrogeochemistry of the pleistocene aquifer, Nile Delta area, Egypt. Environmental International, Alex, pp 75–85
7. Sherif MM, Singh VP, Amer AM (1988) A two-dimensional finite element model for dispersion (2D-FED) in coastal aquifer. J Hydrol 103:11–36
8. Sherif MM, Singh VP, Amer AM (1990) A note on saltwater intrusion in coastal aquifers. J Water Resour Manag 4:113–123
9. Sherif MM, Sefelnasr A, Javadi A (2012) Incorporating the concept of equivalent freshwater head in successive horizontal simulations of seawater intrusion in the Nile Delta Aquifer. Egypt. J Hydrol 464-465:186–198. https://doi.org/10.1016/j.jhydrol.2012.07.007
10. Diab MS, Dahab K, El Fakharany M (1997) Impacts of the paleohydrological conditions on the groundwater quality in the northern part of Nile Delta. The geological society of Egypt, Cairo. J Geol 4112B:779–795
11. Petalas CP, Diamantis IB (1999) Origin and distribution of saline groundwaters in the upper Miocene aquifer system, coastal Rhodope area, northeastern Greece. Hydrogeol J 7:305–316
12. Sherif MM (1999) The Nile delta aquifer in Egypt. In: Bear J et al (eds) Seawater intrusion in coastal aquifers: concepts, methods and practices. Theory and application of transport in porous media, vol 14. Kluwer Academic, Dordrecht, pp 559–590
13. Polemio M, Limoni PP, Mitolo D, Santaloia F (2002) Characterisation of the Ionian-Lucanian coastal aquifer and seawater intrusion hazard. In: Proceedings of the 17th salt water intrusion meeting, Delft, pp 422–434

14. Petalas C, Lambrakis N (2006) Simulation of intense salinization phenomena in coastal aquifers-the case of the coastal aquifers of Thrace. J Hydrol 324:51–64
15. Somay MA, Gemici U (2009) Assessment of the salinization process at the coastal area with hydrogeochemical tools and geographical information systems (GIS): Selçuk plain, Izmir, Turkey. Water Air Soil Pollut 201:55–74
16. Salem ZE, Gaame OM, Hassan TM (2008) Using temperature logs and hydrochemistry as indicators for seawater intrusion and flow lines of groundwater in the Quaternary aquifer, Nile Delta, Egypt. In: Proceeding of the 5th international symposium on geophysics, Tanta, pp 25–38
17. Salem ZE, El-Horiny MM (2014) Hydrogeochemical evaluation of calcareous eolianite aquifer with saline soil in a semiarid area. Environ Sci Pollut Res 21:8294–8314
18. Sefelnasr A, Sherif MM (2014) Impacts of seawater rise on seawater intrusion in the Nile Delta aquifer, Egypt. Ground Water 52:264–276
19. Salem ZE, Osman OM (2017) Use of major ions to evaluate the hydrogeochemistry of groundwater influenced by reclamation and seawater intrusion, West Nile Delta, Egypt. Environ Sci Pollut Res 24:3675–3704. https://doi.org/10.1007/s11356-016-8056-4
20. Werner AD, Bakker M, Post VEA, Vandenbohede A, Lu C, Ataie-Ashtiani B, Simmons CT, Barry DA (2013) Seawater intrusion processes, investigation and management: recent advances and future challenges. Adv Water Resour 51:3–26
21. Choudhury K, Saha DK, Chakraborty P (2001) Geophysical study for saline water intrusion in a coastal alluvial terrain. J Appl Geophys 46:189–200
22. Maillet GM, Rizzo E, Revil A, Vella C (2005) High resolution electrical resistivity tomography (ERT) in a transition zone environment: application for detailed internal architecture and infilling processes study of a Rhône River paleo-channel. Mar Geophys Res 26:317–328
23. Gurunadha Rao VVS, Tamma Rao G, Surinaidu L, Rajesh R, Mahesh J (2011) Geophysical and geochemical approach for seawater intrusion assessment in the Godavari Delta Basin, India. Water Air Soil Pollut 217:503–514
24. Steeples DW (2001) Engineering and environmental geophysics at the millennium. Geophysics 66:31–35
25. Lapenna V, Lorenzo P, Perrone A, Piscitelli S, Rizzo E, Sdao F (2005) 2D electrical resistivity imaging of some complex landslides in the Lucanian Apennine chain, southern Italy. Geophysics 70:11–18
26. Pantelis S, Kouli M, Vallianatos F, Vafidis A, Stavroulakis G (2007) Estimation of aquifer hydraulic parameters from surficial geophysical methods: a case study of Keritis Basin in Chania (Crete – Greece). J Hydrol 338:122–131
27. Chianese D, Lapenna V (2007) Magnetic probability tomography for environmental purposes: test measurements and field applications. J Geophys Eng 4:63–74
28. Naudet V, Lazzari M, Perrone A, Loperte A, Piscitelli S, Lapenna V (2008) Integrated geophysical and geomorphological approach to investigate the snowmelt-triggered landslide of Bosco Piccolo village (Basilicata, southern Italy). Eng Geol 98:156–167
29. Urish DW, Frohlich RK (1990) Surface electrical resistivity in coastal groundwater exploration. Geoexploration 26:267–289
30. Ebraheem AAM, Senosy MM, Dahab KA (1997) Geoelectrical and hydrogeochemical studies for delineating ground-water contamination due to salt-water intrusion in the northern part of the Nile Delta, Egypt. Ground Water 35:216–222
31. Kruse SE, Brudzinski MR, Geib TL (1998) Use of electrical and electromagnetic techniques to map seawater intrusion near the Cross-Florida Barge Canal. Environ Eng Geosci 4:331–340
32. Nowroozi AA, Horrocks SB, Henderson P (1999) Saltwater intrusion into the freshwater aquifer in the eastern shore of Virginia: a reconnaissance electrical resistivity survey. J Appl Geophys 42:1–22
33. Abdul Nassir SS, Loke MH, Lee CY, Nawawi MNM (2000) Salt-water intrusion mapping by geoelectrical imaging surveys. Geophys Prospect 48:647–661

34. Balia R, Gavaudò E, Ardau F, Ghiglieri G (2003) Geophysical approach to the environmental study of a coastal plain. Geophysics 68:1446–1459
35. Choudhury K, Saha DK (2004) Integrated geophysical and chemical study of saline water intrusion. Ground Water 42:671–677
36. Batayneh AT (2006) Use of electrical resistivity methods for detecting subsurface fresh and saline water and delineating their interfacial configuration: a case study of the eastern Dead Sea coastal aquifers, Jordan. Hydrogeol J 14:1277–1283
37. Bauer P, Supper R, Zimmermann S, Kinzelbach W (2006) Geoelectrical imaging of groundwater salinization in the Okavango Delta, Botswana. J Appl Geophys 60:126–141
38. Khalil MH (2006) Geoelectric resistivity sounding for delineating salt water intrusion in the Abu Zenima area, west Sinai. Egypt. J Geophys Eng 3:243–251
39. Sherif M, El Mahmoudi A, Garamoon H, Kacimov A (2006) Geoelectrical and hydrogeochemical studies for delineating seawater intrusion in the outlet of Wadi Ham, UAE. Environ Geol 49:536–551
40. Koukadaki MA, Karatzas GP, Papadopoulou MP, Vafidis A (2007) Identification of the saline zone in a coastal aquifer using electrical tomography data and simulation. Water Resour Manag 21:1881–1898
41. Cimino A, Cosentino C, Oieni A, Tranchina L (2008) A geophysical and geochemical approach for seawater intrusion assessment in the Acquedolci coastal aquifer (Northern Sicily). Environ Geol 55:1473–1482
42. Fadili A, Najib S, Mehdi K, Riss J, Malaurent P, Makan A (2017) Geoelectrical and hydrochemical study for the assessment of seawater intrusion evolution in coastal aquifers of Oualidia, Morocco. J Appl Geophys 146:178–187
43. Salem ZE (2009) Natural and human impacts on the groundwater under an Egyptian village, central Nile Delta: a case study of Mehallet Menouf. In: 13th international water technology conference (IWTC, 13), Hurghada, 12–15 Mar 2009, vol 3, pp 1397–1414
44. Attwa M, Basokur A, Akca I (2014) Hydraulic conductivity estimation using direct current (DC) sounding data: a case study in East Nile Delta, Egypt. Hydrogeol J 22:1163–1178. https://doi.org/10.1007/s10040-014-1107-3
45. Gemail K, El-Shishtawy AM, El-Alfy M, Ghoneim MF, El-Bary MHA (2011) Assessment of aquifer vulnerability to industrial waste water using resistivity measurements. A case study, along El-Gharbyia main drain, Nile Delta, Egypt. J Appl Geophys 75:140–150
46. Kashef A (1983) Salt water intrusion in the Nile Delta. Groundwater 21:160–167
47. Mabrouk B, Arafa S, Gemail K (2015) Water management strategy in assessing the water scarcity in Northern Western region of Nile Delta, Egypt. Geophysical research abstracts, vol 17. EGU General Assembly 2015, Vienna. EGU2015-15805, 2015
48. McGranahan G, Balk D, Anderson B (2007) The rising tide: assessing the risks of climate change and human settlements in low elevation coastal zones. Environ Urban 19:17–37
49. Mabrouk MB, Jonoski A, Solomatine D, Uhlenbrook S (2013) A review of seawater intrusion in the Nile Delta groundwater system – the basis for assessing impacts due to climate changes and water resources development. Hydrol Earth Syst Sci Discuss 10:10873–10911. https://doi.org/10.5194/hessd-10-10873-2013
50. Attwa M, Gemail KS, Eleraki M (2016) Use of salinity and resistivity measurements to study the coastal aquifer salinization in a semi-arid region: a case study in northeast Nile Delta. Egypt. Environ Earth Sci 75:784
51. Mohamed AK (2016) Application of DC resistivity method for groundwater investigation, case study at West Nile Delta, Egypt. Arab J Geosci 9:11
52. Tarabees E, El-Qady G (2016) Sea water intrusion modeling in Rashid area of Nile Delta (Egypt) via the inversion of DC resistivity data. Am J Clim Chang 5:147–156. https://doi.org/10.4236/ajcc.2016.52014
53. Salem ZE, Osman MO (2016) Shallow subsurface temperature in the environs of El-Nubaria canal, northwestern Nile Delta of Egypt: implications for monitoring groundwater flow system. Environ Earth Sci 75:1241. https://doi.org/10.1007/s12665-016-6046-y

54. Salem ZE, Atwia MG, El-Horiny MM (2015) Hydrogeochemical analysis and evaluation of groundwater in the reclaimed small basin of Abu Mina, Egypt. Hydrogeol J. https://doi.org/10.1007/s10040-015-1303-9
55. Fadlelmawla AA, Dawoud MA (2006) An approach for delineating drinking water wellhead protection areas at the Nile Delta, Egypt. J Environ Manag 79:140–149
56. El-Bayumy DA (2014) Sedimentological and hydrological studies on the area west of the Nile Delta, Egypt. MSc thesis, Tanta University
57. Salem ZE, El Bayumy DA (2016a) Hydrogeological, petrophysical and hydrogeochemical characteristics of the groundwater aquifers east of Wadi El-Natrun, Egypt. NRIAG J Astron Geophys 5:106–123
58. Salem ZE, El Bayumy DA (2016b) Use of the subsurface thermal regime as a groundwater-flow tracer in the semi-arid western Nile Delta, Egypt. Hydrogeol J 24:1001–1014. https://doi.org/10.1007/s10040-016-1377-z
59. Osman OM (2014) Hydrogeological and geoenviromental studies of El Behira Governorate, Egypt. PhD dissertation, Tanta University
60. El Shazly EM, Abdel-Hady M A, El-Ghawab MA, El Kassas IA, Khawasil SM, El Shazly MM, Sanad S (1975) Geological interpretation of Landsat Satellite Images for west Nile Delta area, Egypt. Remote Sensing Research project. Academy of Scientific Research Technology, Cairo
61. El Ghazawi MM (1982) Geological studies of the Quaternary-Neogene aquifers in the area northwest Nile Delta. MSc thesis, Al-Azhar University
62. Mabrouk MA (1978) Electrical prospecting on the groundwater in the area west of Cairo-Alexandria desert road (between Wadi El- Natron and El-Nasr Canal). MSc thesis, Faculty of Science, Ain Shams University
63. Abdel Wahab S (1999) Hydrogeological and isotope assessment of groundwater in Wadi El-Natrum and Sadat city, Egypt. MSc thesis, Faculty of Science, Ain Shams University
64. Saad K (1962) Groundwater studies of Wadi El-Natrun and its vicinities. Desert Institute, Cairo, p 61
65. Desert Research Institute (DRI) (1974) Report on the regional hydrogeological studies of the West El Nubaria area (300,000 Feddans reclamation project), internal report, Cairo, 125 p
66. Abdel Baki AA (1983) Hydrogeological and hydrochemical studies on the area west of Rosette branch and south of El-Nasr Canal. PhD thesis, Faculty of Science, Ain Shams University
67. Milsom J (2003) Field geophysics, 3rd edn. Wiley, Chichester, p 247
68. Palacky GJ (1987) Resistivity characteristics of geologic targets. In: Nabighian MN (ed) Electromagnetic methods in applied geophysics theory, vol 1. Society of Exploration Geophysicists, Tulsa, Okla, pp 53–129
69. Bobachev AA, Modin IN, Chevnin VA (2001) IPI2win v. 2.0: user's guide. Moscow State University, Geological Faculty, Department of Geophysics, Moscow, p 35
70. Rockwork (2004) Earth Science and GIS Software (Geosof), www.rockware.com. Suite 101 Golden, CO 80401 USA

Integrated Subsurface Thermal Regime and Hydrogeochemical Data to Delineate the Groundwater Flow System and Seawater Intrusion in the Middle Nile Delta, Egypt

Zenhom E. Salem, Osama M. Gaame, and Taher M. Hassan

Abstract Several aquifers around the world are situated in the coastal zones and influenced by seawater intrusion. The development of populace in coastal territories and the conjugate increment in human, farming, and industrial activities have forced an increase in the needs for freshwater. The Quaternary aquifer in the Nile Delta is among the biggest groundwater aquifers in the world. Along its northern side, the aquifer is highly affected by the Mediterranean Sea. Because of the inordinate pumping in the course of the most recent couple of decades, the groundwater quality in the northern parts of the Nile Delta has been decreased extensively. Therefore, this chapter aims to trace the groundwater flow system and seawater intrusion in the study area using the multi-tracing technique. The integration between borehole temperatures and groundwater chemistry was good to conduct the aim of this study. Borehole temperature was measured in eight boreholes, and the groundwater was sampled from the same wells but sometimes from the shallow and deep zones. Tala well located to the south of the study area indicated the recharged fresh groundwater with downward flux of 0.8 m/year. The fresh groundwater started to discharge from south Tanta City till south Kafr Elsheikh City where the calculated upward fluxes were −0.1 to −0.5, −0.35, and −0.23 m/year for Kafelarab, Nawag, and Elkarada wells. Hydrochemically, the groundwater in the area northern Kafr Elsheikh City is highly affected by seawater intrusion, and the measured temperature profiles are of discharge type, and their calculated upward fluxes were −0.6, −1.2, and −2.8 m/year for Kafr Mesaaed, Elhadady, and Motobes wells, respectively.

In comparison, temperature profile (Motobes well) affected by seawater intrusion has higher upward flux, while the freshwater recharge-type profile (Tala well) has

Z. E. Salem (✉)
Geology Department, Faculty of Science, Tanta University, Tanta, Egypt
e-mail: zenhomslem@yahoo.com

O. M. Gaame and T. M. Hassan
Research Institute of Groundwater, National Water Research Centre, Cairo, Egypt

A. M. Negm (ed.), *Groundwater in the Nile Delta*,
Hdb Env Chem (2019) 73: 461–486, DOI 10.1007/698_2018_253,

lower downward flux. Hydrochemically, the seawater intrusion highly affected the wells from ElKarada wells to Motobes wells (northern Kafr Elsheikh City) and close to the Mediterranean Sea. Two types of saline water were recognized. The shallow groundwater is highly affected by seawater intrusion (TDS around 20 g/l), and the deeper groundwater is of hypersaline characters (80 g/l). These two types of saline water could deteriorate the groundwater quality in the Nile Delta in case of unresponsible severe pumping rates.

Keywords Groundwater flow system, Hydrogeochemistry, Nile Delta, Seawater intrusion, Subsurface temperature

Contents

1 Introduction

There are three inquiries in hydrogeological research: (1) Where is the recharge area? (2) Where is the discharge area? (3) What is the pattern of the groundwater circulation in deep and shallow zones? Accordingly, the groundwater flow in vertical and horizontal directions could be classified "local" and "regional" groundwater framework (e.g., [1–3]). Local flow is restricted inside a single basin surrounded by topographic highs and vertically circulates in the shallow subsurface. Regional flow system is not mostly limited to a single basin and groundwater circulated to deeper zones. There are three techniques used to indicate the groundwater stream in a wide territory: measuring liquid potential, numerical modeling, and utilization of tracers. Six types of tracers were used for the groundwater investigation: colorimetry, photometry, mass spectrograph, electric conductivity, hydrochemistry, and temperature [4].

Seawater intrusion and submarine groundwater discharge (SGD) are commonly expected in coastal areas. Seawater intrusion (Fig. 1) is of fundamental interest for researcher and groundwater users in every coastal aquifer around the world. The seawater intrusion is a vital issue in arid and semiarid areas where the groundwater constitutes the major freshwater resources. About 3% of seawater mix with the freshwater in a coastal aquifer would render the freshwater unacceptable for human utilization. That is why the coastal aquifer groundwater should be intensively studied to keep up the dynamic freshwater-seawater interaction. The issue of salt-water/new water interface exists in Nile Delta because of overpumping for water system and seaside nature of its aquifer. Sherif et al. [5, 6], Amer and Sherif [7],

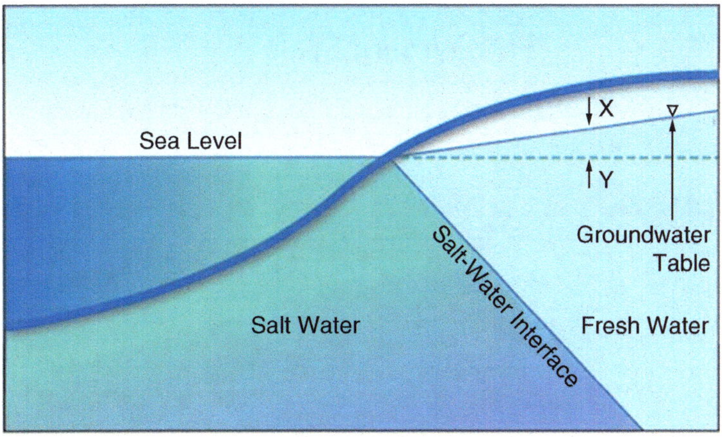

Fig. 1 Illustrates the seawater intrusion in the continental fresh groundwater

Sherif and Singh [8], Sherif [9], and Sherif and Singh [10] used different techniques to study the seawater intrusion issue in the Nile Delta aquifer; however, none of them utilized subsurface thermal regime.

The SGD considers the water output from a basin-scale hydrological cycle that represents an input into the ocean. SGD nearshore scale (Fig. 2) is the embayment or inner continental shelf scale operationally defined as extending 10 m to 10 km offshore and to depths of about 5–50 m below the seafloor, including the first confined or semi-confined portion of a submarine flow system. The shallowest submarine confining unit [11] permits primarily topographically driven the regional flow of fresh or brackish water offshore beneath the nearshore zone and sometimes entirely beneath lagoons, shallow estuaries, embayments, and barrier islands [12].

The next larger scale is that of the entire continental shelf (Fig. 3), which may include multiple confined aquifer systems extending below the first confined aquifer to depths of 500 m or more below the seafloor and to the outer continental shelf edges, submarine canyon incisions, and even the continental slope, especially during sea-level low stands. The primary process driving flow at this scale is usually geothermal convection, which produces seawater recirculation through the shelf [13–16]. Sediment compaction and associated dewatering, as well as brine-related processes, are also important in some settings [17]. It could be argued that there is actually more known about the occurrence of submarine groundwater systems at the shelf scale, including relict-reduced salinity groundwater than is known about intermediate embayment-scale submarine aquifer systems in some settings.

Cartwright [18] and Sakura [19] utilized subsurface temperature information to investigate the groundwater stream in a basin. Mostly, the typical geothermal gradient of a subsurface system is influenced by groundwater stream and changes in surface temperature. A hypothetical analysis of a distorted temperature field by geographically

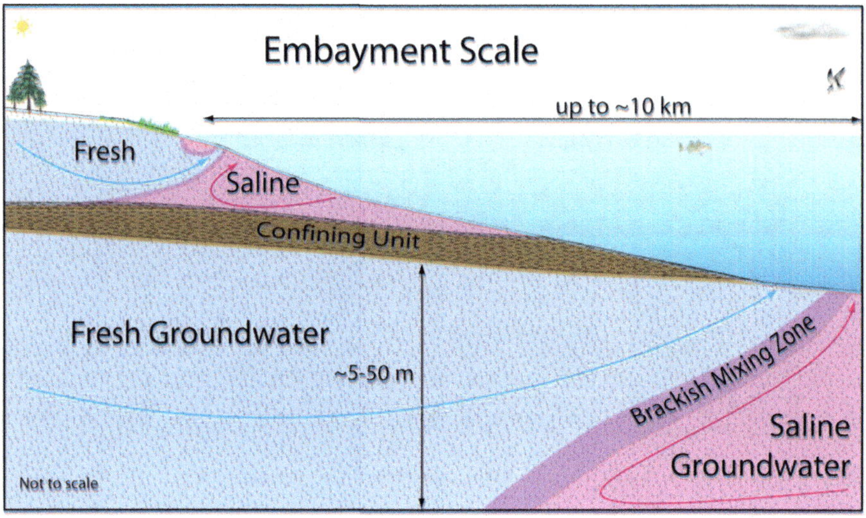

Fig. 2 Schematic diagram of the embayment or inner shelf scale of submarine groundwater flow and discharge showing submarine flow of low-salinity water in the first confined aquifer and the zone of offshore discharge beyond the edge of the submarine confining unit [11]

forced groundwater stream was introduced by Domenico and Palciauskas [20]. Their outcome demonstrates that groundwater temperature increments with depth affected by both thermal conduction and surface warming conditions. A temperature-depth profile without groundwater stream, for the most part, has a consistent slope with depth (Fig. 4a) and stratified thermal system (Fig. 4b) in the 2D cross section. At a similar height, the subsurface temperature in the recharge zone is brought down under the state of groundwater stream than under no stream condition. In the discharge zone, the temperature is higher with groundwater stream than under no stream conditions (Fig. 4c, d). The impact of groundwater flow and surface warming on subsurface temperature is delineated in Fig. 1e, f.

The chemical characteristics of the groundwater are mostly settled in the unsaturated zone. In contrast, less intense geochemical evolution occurs in the saturated zone, and changes take after dynamic changes in water quality toward regions of discharge. These procedures are time dependent and the chemical changes and, additionally, isotopic varieties might be utilized to give data on water stream pathways [22]. Utilizing multi-following strategy, Salem et al. [23, 24] prevailing to decide the connection between the groundwater and Shinano River in Nagaoka territory, Japan. In the present work, temperature and water chemistry were utilized for following the groundwater stream framework and seawater intrusion in the Nile Delta. Flow analysis is essentially in light of the idea of various leveled groundwater stream frameworks [2, 3].

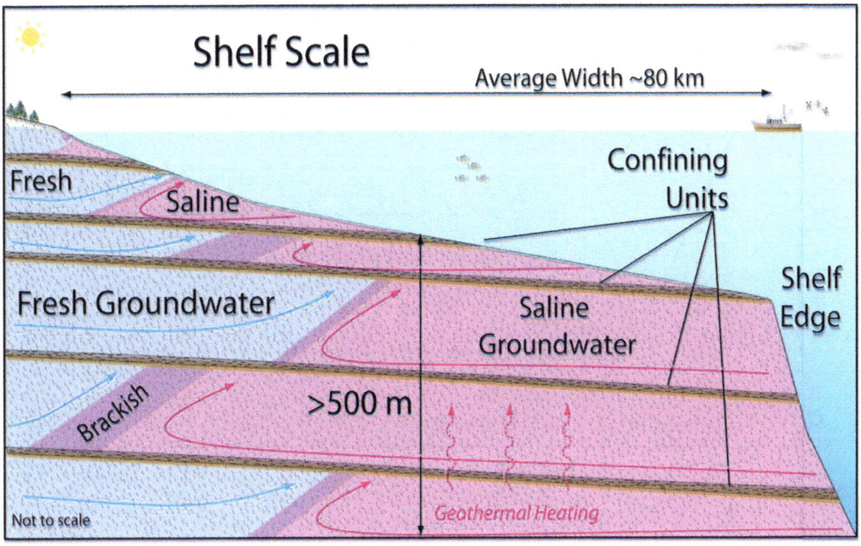

Fig. 3 Schematic diagram of the continental shelf scale of submarine groundwater phenomena showing the variable position of the fresh-saline interface in multiple confined aquifers on the shelf, the variable widths of the mixed zone at the interface, the flow of saline water inward from the exposed edges of confined aquifers, and the upward movement of saline groundwater induced by geothermal heating at depth [11]

2 Hydrogeology of the Nile Delta

In ancient times, the Nile River at Cairo formed a wide estuary that has been occupied by river deposits to frame the current fruitful delta of 250-km-wide base at the Mediterranean coast and around 160 km from Cairo (south) to sea coast (north). Seven branches of the Nile ran through the delta, but humans and nature have closed five of them [25]. The only two remaining channels are Rosetta, whose mouth is located just east of Alexandria, and Damietta, whose mouth is located at the northeast tip of the Delta. The other five mouths, which have been proven both historically and geologically to have existed [25].

The level of the lands of the delta runs between 17 m amsl at the southern limit and less than 1 m amsl at the northern parts [26]. The rice is cultivated every year in around 420,000 ha in the Nile Delta [27]. It is incorporated into summer trimming design, which additionally incorporates cotton and maize as principle summer crops in 2 or 3 years' harvest turn. As rice cultivations use a huge quantity of water, it widely recharges the groundwater contrasted with alternate crops. In the northern territories, the rice is intensively cultivated compared to the other regions (Fig. 5). These regions are required to have hydrogeological characters not the same as alternate less rice developed regions as will be mentioned later.

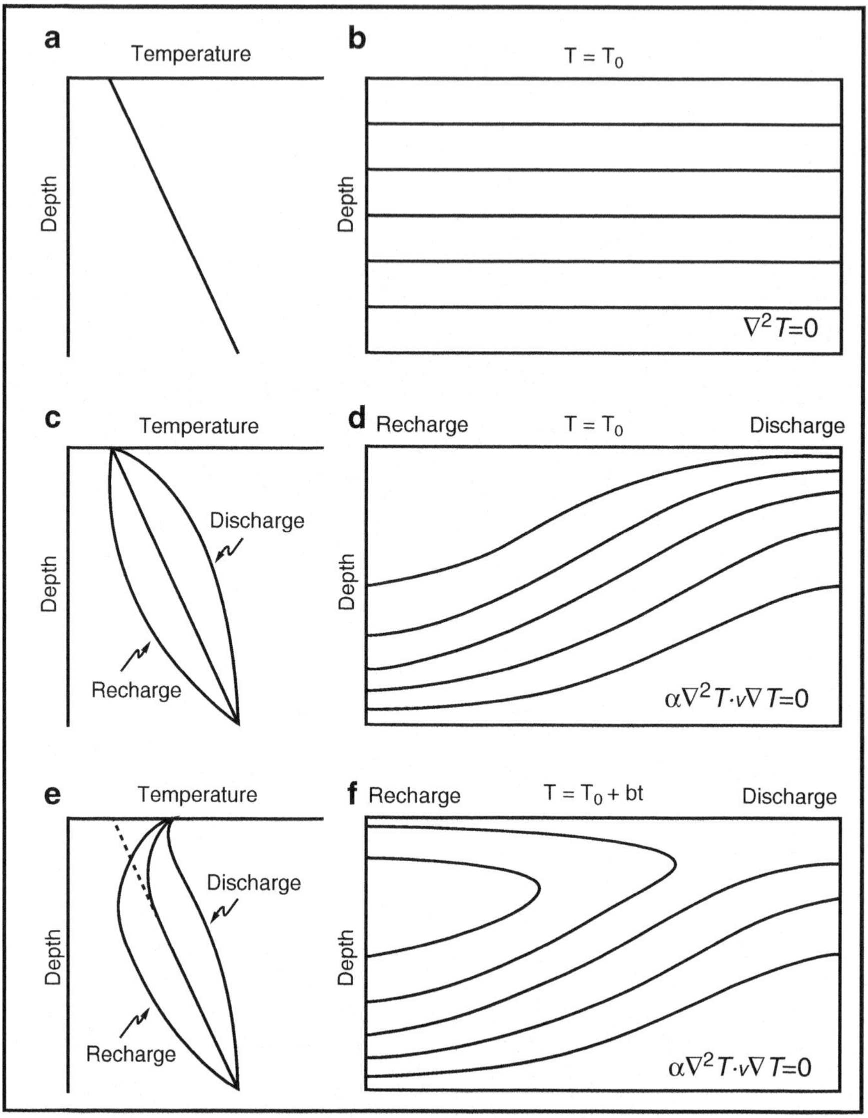

Fig. 4 Schematic diagrams of the groundwater flow system and subsurface thermal regime under the condition of (**a**) and (**b**) no groundwater flow, (**c**) and (**d**) regional groundwater flow, and (**e**) and (**f**) regional groundwater flow with surface warming. Note: T, T_o, and t are subsurface temperature, constant surface temperature, and time, respectively [21]

"The Nile Delta is one of the oldest intensely cultivated areas on earth. It is very heavily populated, with population densities up to 1,600 inhabitants per square kilometer" [25]. Current increments in the populace, together with upgraded ways of life and additionally the restricted freshwater sources in the Nile Delta, have made

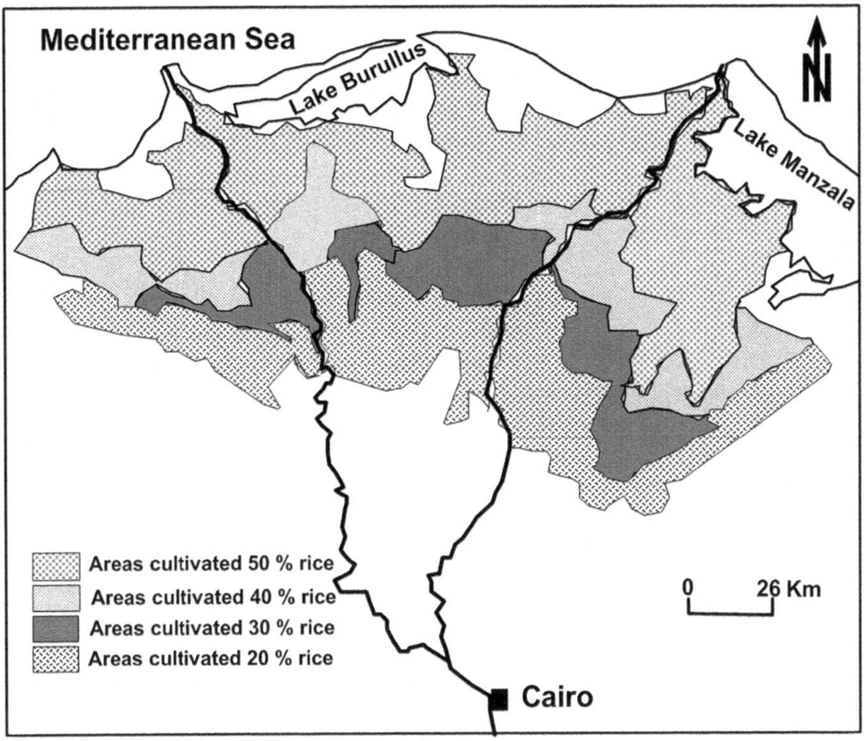

Fig. 5 Areal distribution of the rice cultivated lands in the Nile Delta (Modified after [27])

a more prominent request on water resources, requiring enhanced groundwater administration. Any new groundwater improvement in the Nile Delta should consider the potential outcomes of seawater intrusion and guarantee satisfactory control, with the counteractive action of saline intrusion being seen as perfect. Up till now, various groundwater investigations chiefly centered on seawater intrusion on the upper 100 m of the groundwater framework and expected salinities not surpassing that of Mediterranean water. There was no information on groundwater in the deeper parts of the Quaternary Nile Delta aquifer (depth up to 1,000 m). Recent studies gathered salinity estimations and found an across-the-board event of "hypersaline" groundwater: groundwater with salinities to a great extent surpassing that of seawater at depth higher than 400 m [28, 29]. This hypersaline groundwater incredibly influences the groundwater flow system and the new water capability of the aquifer. Engelen et al. [30] investigated the causes of the hypersaline groundwater and its transport system. They considered all the applicable salinization forms in the Nile Delta aquifer, over a time space of up to 2.5 million years, which is the time traverse in which the aquifer got deposited.

Several research works dealing with the hydrogeology and hydrogeochemistry of the Nile Delta aquifer were done covering the period from 1959 to 2017, among them: Zaghloul [31], El-Fayoumy [32], Shata [33], Anon [34], El-Dairy [35], El-Hefny et al. [36], Sallouma [37], Hamza et al. [38], Kotb [39], Mousa [40], Mousa [41], El-Shamy and Greish [42], Ezz and Deen [43], Sollouma and Gomaa [44], Salem et al. [29], Salem [45], Atwia [46], Shabaan [47], Frihy and Lawrence [48], El Banna [49, 50], El Banna and Frihy [51], El Banna and Frihy [52], El-Asmar and Hereher [53], Elewa et al. [54], Mabrouk et al. [55] and Shehata and El-Sabrouty [56], Salem et al. [57], Salem and El Bayumy [58, 59], Salem et al. [60], Salem and Osman [61–63], and Salem et al. [64–66].

The Nile Delta aquifer system is considered to be a leaky aquifer in the southern and middle parts and a free aquifer in the western and eastern borders, where the thickness of the top Holocene deposits reaches its minimum value. In the northern parts, the top Holocene deposits strongly retard upward discharge from the Plio-Pleistocene aquifer; therefore, the aquifer becomes of a confined type. The existing aquifer, in the study area, belongs to the main Nile Delta aquifer system. It is mainly formed of the Pleistocene graded sand and gravel, changing to fine sand intercalated with clay lenses. This aquifer is a semi-confined one. The thickness of the semi-confining layer is generally between 0 and 20 m and increases reaching a thickness of 70 m at the northern part of the Nile Delta. There are not too many studies with estimations of the hydraulic parameters for the overlying clay layer in literature. Farid [67] reported that its vertical hydraulic conductivity is at 0.0025 m/day, while Leaven [68] reported it at 0.0484 m/day. With slightly lower values, Wolf [69] reported it as 0.0011 m/day, and Arlt [70] at 0.0046 m/day. On the other hand, Sherif et al. [71] reported the vertical hydraulic conductivity to be about 0.67 m/day. Recently, Salem et al. [64] studied the petrophysical and the hydrogeological properties of this clay layer (Bilqas Formation) in the central part of the middle Nile Delta. They stated that "Bilqas Formation ranges in thickness from 3 m in the southwest direction to 31 m in the northeast direction. The shale content and porosity ranges are 54% to 97% and 21% to 55%, respectively. This layer has low values of permeability (16×10^{-9} to 78×10^{-9} mD) and hydraulic conductivity ($<2 \times 10^{-9}$ cm/s). The water salinity of this layer ranges from 200 to 1,600 mg/l."

The Nile Delta Quaternary aquifer (Mit Ghamr Formation) is the main source of groundwater in the Nile Delta area [72]. It covers the whole Nile Delta. Its thickness varies from 200 m in the southern parts to 1,000 m in the northern parts [73] (Fig. 6). It constitutes variable proportions of sands, clays, and gravels with lateral variation and variable thickness. From those sections, the hydrogeological setting of the Nile Delta classified into one water-bearing zone (Quaternary) which involves confining thin beds (clay beds) that could prevent intrusion of seawater (Fig. 6). Dahab [75] estimated the hydraulic parameters of the aquifers. Transmissivity reaches its minimum values in the southwestern area, where it ranges between 2,000 and 3,000 m^2/day. It reaches its maximum values in the middle and southeastern parts, where it ranges between 15,000 and 9,000 m^2/day. Hydraulic conductivity values range between 60 and 70 m/day in the southwestern and northern parts. In the middle and southeastern parts, it reaches its maximum values, between 100 and 180 m/day. Effective porosity over the whole area ranges between 12 and 19%, which indicates that the aquifer is mainly composed of coarse sand and gravel. Recently, Salem et al.

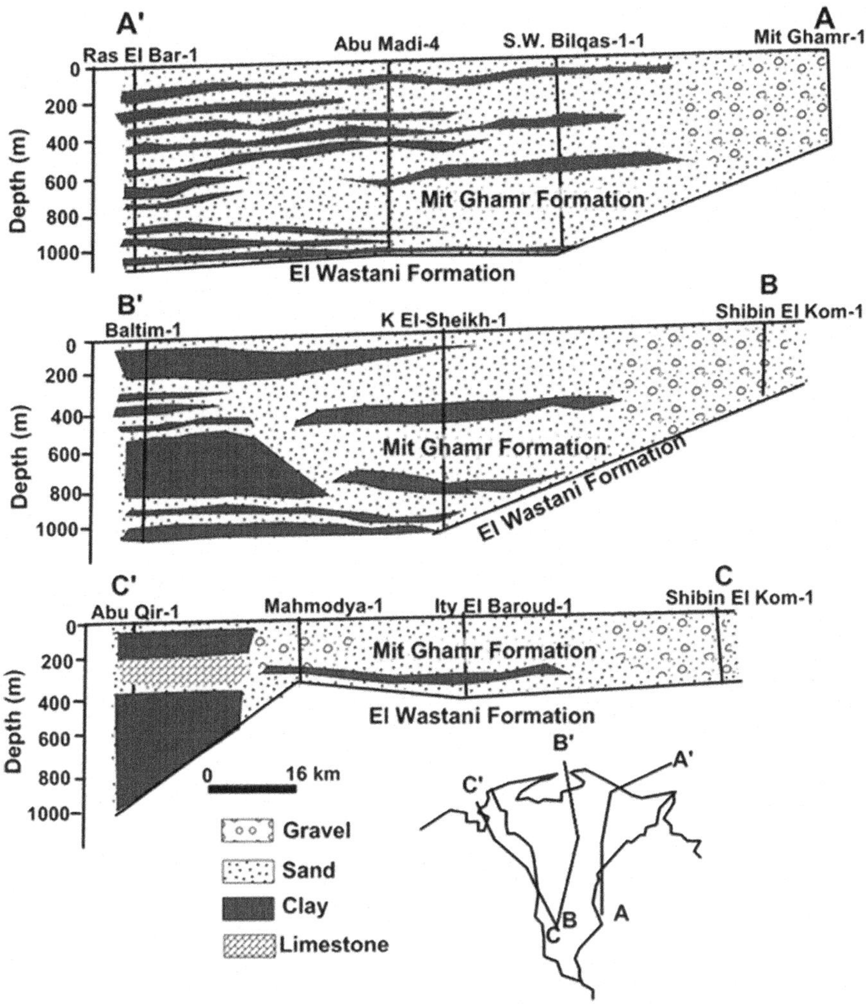

Fig. 6 Locations and details of north-south hydrogeological cross sections in the Nile Delta area [60, 74]

[64] studied the petrophysical and the hydrogeological properties of this Quaternary aquifer (Mit Ghamr Formation) in the central part of the middle Nile Delta. They stated that "in Mit Ghamr Formation, average shale content ranges from 4.5 to 22%. Numbers of scattered clay lenses are detected in different places with high intensity in the northeastern direction. Porosity ranges from 19 to 39%. High permeability values are recorded in this formation and ranged from 0.1×10^{-2} to 8.7×10^{-2} mD. The water salinity average values in this aquifer range from 220 mg to 2,100 mg/l. The calculated hydraulic conductivity values for this formation are of range 5.082×10^{-10} to 2.134×10^{-8} cm/s. In this layer, the increase in the shale content, the increase

in porosity, decrease in the permeability and hydraulic conductivity, as well as the increase in salinity, are to the northern and northeastern directions."

The depth to the groundwater table in this aquifer ranges between 1–2 m in the north, 3–4 m in the middle, and 5 m in the south. Different estimated depths to groundwater table that have been reported by RIGW [76] and Morsy [77]. Direction of groundwater movement is northward to the sea. The irrigation and drainage networks are the main sources of groundwater recharge. The Rosetta branch, the Mediterranean Sea, coastal lakes, and pumped water used for irrigation, industrial, and domestic purposes are the main discharge forms.

3 Methodology

Flowchart illustrates the methodology of this work is shown in Fig. 7. Field work of this study includes groundwater sampling and borehole temperature measurement. The measured temperature-depth profiles were classified, and the subsurface temperature was modeled using 1D modeling techniques. The modeled and measured profiles were compared to estimate the vertical groundwater flux. The collected water samples were subjected to hydrochemical analysis and then graphically presented on Piper and Stiff diagrams. The results of subsurface temperature and hydrogeochemistry were integrated to get a complete image about the seawater-freshwater interaction. After that, conclusion and recommendation were written. Subsurface thermal measurements generally are made in the observation wells that are assumed to be in equilibrated thermal state between the water and the surrounding solid material. The Borehole temperature measurement

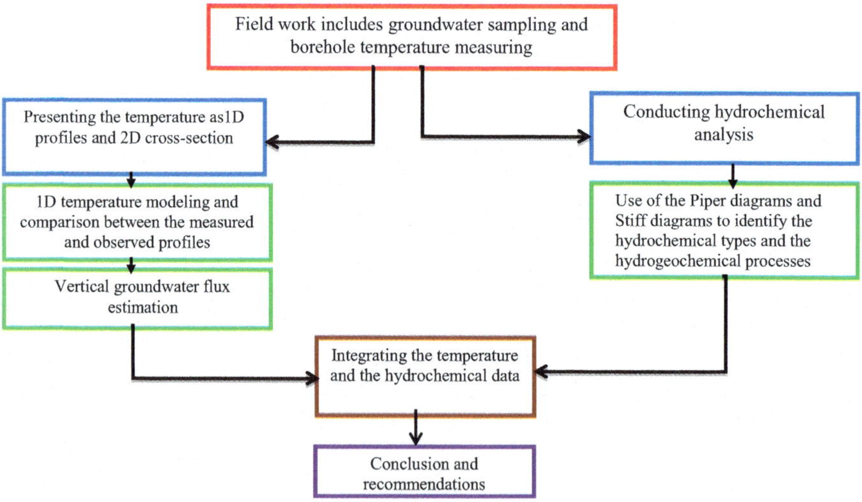

Fig. 7 Flowchart illustrates the methodology of the current study

technique was discussed in detail in Salem and El Bayumy [59]. Temperature measurements in this study were carried out in eight observation wells (Tala 2, Kafr Elarab 2, Nawag, Elkarada 2, Abu Mesaaed, Elhadady, Motobes 1, and Motobes 2) (Fig. 8). The equipment used for the measurements was a digital thermistor thermometer (resolution of 0.1°C) attached to a cable of 500 m length. Data were recorded from the water table to the bottom of the borehole.

Water samples for chemical analysis were collected from all the observation wells (Tala 1 and 2, Kafr Elarab 1 and 2, Nawag, Elkarada 1 and 2, Abu Mesaaed, Elhadady, and Motobes 1 and 2) and two pumping wells in Qutor as well as seawater. The water table, electrical conductivity, TDS, as well as pH were measured in situ. All the chemical analysis of anions and cations were made in central laboratories of National Water Research Centre, Cairo. The hydrochemical analysis procedures were discussed in detail in Salem et al. [60]. Well, location map and the listed hydrochemical analysis are shown in Fig. 8 and Table 1. This chapter is a modified and updated version of Salem et al. [29], where 1D temperature modeling was added and data presentation and interpretation were updated.

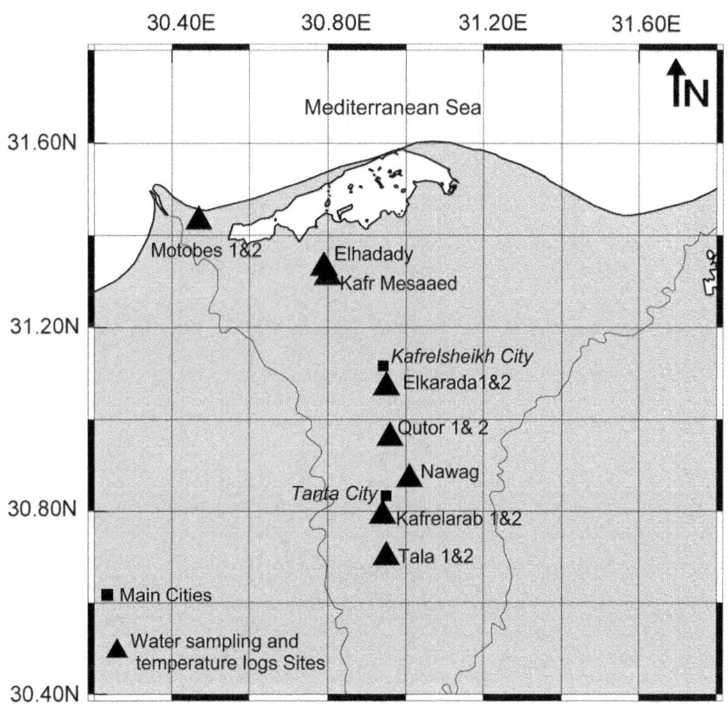

Fig. 8 Wells location map [29]

Table 1 Chemical composition of the collected water samples [29]

Physicochemical parameters		Tala (1)	Tala (2)	Kafrelarab (1)	Kafrelarab (2)	Nawag	Qutor (1)	Qutor (2)	Elkarada (1)	Elkarada (2)	Kafr Mesaaed	Elhadady	Motobes (1)	Motobes (2)	Seawater
Total depth	m	100	320	100	280	45	45	100	45	75	85	90	140	430	
Water table	m	3.9	3.9	3.5	3.5	3.0	ND	ND	1.6	0.2	−0.7	−0.5	0.0	−14.5	
pH	–	7.85	7.9	9.51	9.32	8.16	8.65	7.1	10.71	8.03	7.47	7.08	5.87	6.69	7.5
TDS	mg/l	514	916	198	280	291	531	741	1,604	1,751	6,220	8,288	19,840	80,192	38,157
EC	mmhos/cm	0.802	1.429	0.31	0.447	0.454	0.845	1.48	2.51	2.74	9.72	12.95	31.00	125.3	59.523
CO_3	mg/l	0.0	0.0	17.1	29.6	0.0	ND	ND	57.6	0.0	0.0	0.0	0.0	0.0	ND
HCO_3	mg/l	365.9	551.8	53.2	40.7	44.8	122	188	234.0	200.0	23.2	29.0	11.6	174	157.0
Total alkalinity	mg/l	365.9	551.8	70.4	70.2	44.8	ND	ND	291.6	200.0	23.2	29.0	11.6	174	157
Ca	mg/l	71.4	92.3	14.86	17.1	18.6	60.4	54.4	31.85	40.5	293	472	539	2,150	513
Mg	mg/l	33	55.9	11.32	14.87	13.21	29	71.3	26.7	23.4	153	220	672	2,260	1,417
Na	mg/l	38.6	66	30.4	47.6	46.9	60	96	500	523	1,650	1,870	4,650	22,700	11,850
K	mg/l	11.6	19.4	7.42	8.4	7.5	5	5.75	39.7	16.4	26.9	37.2	151	737	220
Cl	mg/l	45.2	78.5	33.5	52.4	64.1	160	292	630	780	3,290	3,820	6,844	43,500	20,800
NO_3	mg/l	<0.2	<0.2	<0.2	<0.2	0.91	9.4	ND	46.90	18.70	105.0	111	66	4,283	6.6
SO_4	mg/l	57.20	92.40	44.10	57	73.50	67.00	42	68.20	55.40	100.2	520	5,237	908.4	3,260

Note: *ND* not determined

4 Result and Discussion

The two most illustrative parameters were utilized to be specific, temperature and hydrochemistry. The first of these tools gives data about vertical, rising, and downward flows. The second gives data about the arrangement of the water influenced, which thus gives data about the "palaeoprocesses," the existence of salts in the medium, and the water origin.

4.1 Subsurface Thermal Regime

4.1.1 One-Dimensional Temperature Profiles

The vertical well logs demonstrate the thermal properties of the aquifers and enable separation of segments with generally contrasting advancements, as far as the presences or non-presence of warm water or streams of an alternate origin [78]. Temperature logs of groundwater additionally enabled the evaluation of the aquifer to be separated point by point, as for the presence of hot or cold water streams.

Regarding the temperature profiles of the present study (Fig. 9), they can be classified into two groups. The first group characterizes the area southern Kafr Elsheikh City, and the second group includes the wells located northern Kafr Elsheikh City. Based on the shape of the profiles, the first group includes recharge- and discharge-type profiles (Tala and Kafrelarab wells, respectively). This means in the study area, the fresh groundwater in the Nile Delta starts to flow southern Tanta City upward. The reason could be related to the existence of denser seawater in deeper zones which forced the light, fresh groundwater to flow upward. Kafr Elarab, Nawag, and Elkarada temperature logs have higher temperatures compared to Tala and Kafrelarab logs which indicate that the groundwater continues to discharge northward. The latter three profiles also have higher temperatures in the shallower zones; as these wells located in a cultivated land out of the urban effect so, such warm zones might be related to upward movement of the warm freshwater and its movement laterally northward in the shallow zones.

The second group of wells characterizes the northern part of Kafr Elsheikh City where Kafr Mesaaed, Elhadady, and Motobes wells (shallow and deep) are located. It has no significant profile type, but it seems different compared to the first group. Elhadady well and the shallow Rasheed well have warm water in their topmost parts. It also means that a discharge of the groundwater took place in the subsurface into the sea northward. For the same depth in the four wells, Kafr Mesaaed well has the lowest temperatures compared to the other three wells. This means that north Kafr Elsheikh City could be another recharge area different from the main recharge area of the Nile Delta aquifer. The later phenomenon happened due to excessive irrigation of highly cultivated rice fields in the northern part of the Nile Delta (Fig. 5). Motobes wells (shallow and deep) show different behavior in the shallow zone (Fig. 9). As the

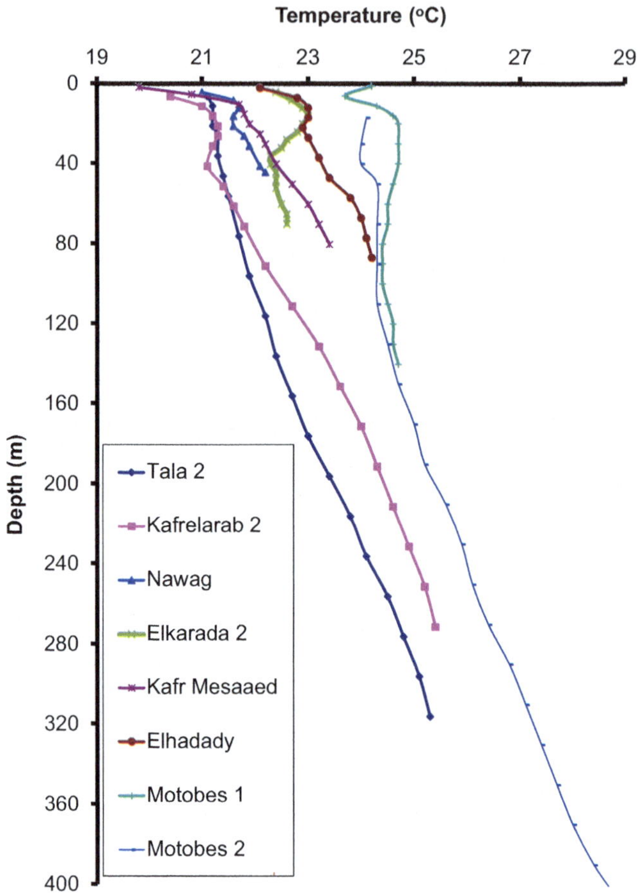

Fig. 9 The measured temperature-depth profiles [29]

two wells lie close to the shoreline, this temperature difference could be related to the presence of two active groundwater flow systems one deeper and the other shallow.

The measured temperature profiles show three important observations as criteria for seawater intrusion in the Nile Delta. These observations are (1) although the horizontal distance between Nawag and Elkarada is about 30 km, their profiles met each other at a depth of 45 m below the surface; (2) in the same way, the distance between Kafrelarab well and Motobes wells is about 85 km, but the temperature difference at deep zones is less than 1°C; and (3) the profiles of Motobes and Elhadady wells are expected to meet each other at about 90 m below surface. These obtained three observations reflect an inland encroachment of seawater which is probably reached Nawag area in the deeper zones (northern Tanta City).

4.1.2 Temperature Cross Section

The vertical two-dimensional temperature distributions can be studied from combining the vertical logs of temperature made in different piezometers over the study area to show the spatial behavior of subsurface temperature along the fresh and saline groundwater flow direction from south to north (Fig. 10). As shown from the isothermal lines, Nile Delta basin in this study could be classified as mentioned before into two main areas: the area to the south of Kafr Elsheikh City and that northern Kafr Elsheikh City. The southern area shows the regional flow in the delta basin, where the freshwater is recharged south Tala, and it discharges starting from Kafrelarab well (south Tanta City) and flowing laterally in the shallow zones till Kafr Elsheikh (Elkarada wells). The northern area shows a local flow system where the water has recharged in the area lying south of Kafr Mesaaed well, and it discharges in the northern part, where Elhadady and Motobes wells are located. The recharge source of the groundwater in the latter area is expected to be related to agriculture and human activities (Fig. 5). Seawater intrusion could also be seen from Fig. 10 where the isothermal lines are nearly horizontal in the deep zones between Motobes and Nawag.

4.1.3 Model Calculations

To investigate the one-dimensional temperature profiles affected by groundwater flow, the estimation of subsurface temperature in the investigation region has been made utilizing distinctive vertical groundwater fluxes. Carslaw and Jaeger [79] got the mathematical solution for the subsurface temperature utilizing one-dimensional heat

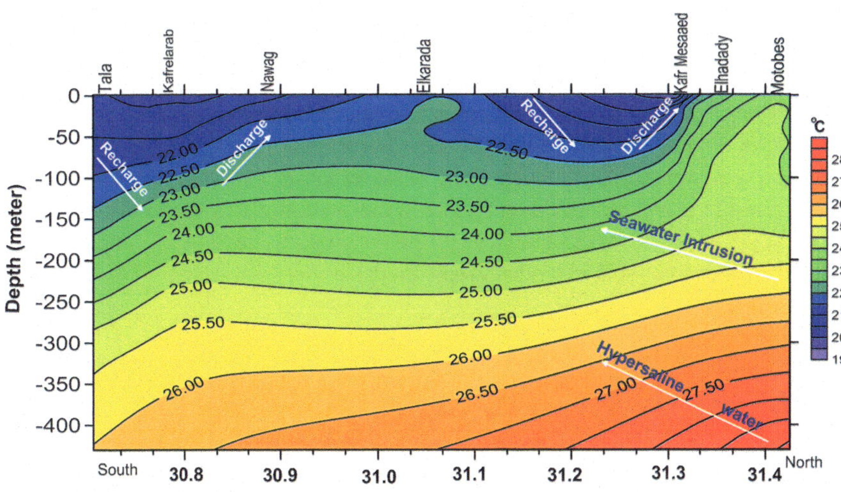

Fig. 10 South-north cross section shows the subsurface thermal regime of the study area. Groundwater flow system is also indicated by the *white arrows* [29]

conduction-advection equation under the state of linear increment in surface temperature as presented in Eq. 1:

$$T(z,t)=T_o+T_G(z-Ut)+\{(b+T_GU)/2U\} \times\left[(z+Ut)e^{Uz/\alpha}\text{erfc}\left\{(z+Ut)/2(\alpha t)^{1/2}\right\}+(Ut-z)\text{erfc}\left\{(z-Ut)/2(\alpha t)^{1/2}\right\}\right] \quad (1)$$

where b is the rate of the surface temperature increment; t is the time after semi-equilibrium condition [21]; T_o is the surface temperature; T_G is the general geothermal gradient; $U = vc_o\rho_o/c\rho$ in which v is the vertical groundwater flux, $c_o\rho_o$ is the heat capacity of the water, and $c\rho$ is the heat capacity of the aquifer; α is the heat diffusivity of the aquifer; and erfc is the correlative error function. The modeling is restricted for semi-infinite layers with just vertical conduction and convection, and vertical groundwater flux is thought to be consistent with depth.

Setting T_G is 0.19, 0.29, and 0.15°C/m for Tala and Kafrelarab; Nawag, Elkarada, and Kafr Mesaaed; and Motobes (shallow and deep) wells, respectively. T_o values is 20.4°C for all wells. The utilized warm diffusivity (α) is 6×10^{-7} m^2/s. For t meets 50 years, the utilized estimation of b is 0.05°C. In this manner, Salem and Osman [61] studied the change in the surface air temperature in the Nile Delta and detected warming trend in Damanhour weather station. Temperature depth profiles are modeled utilizing the above condition for various estimations of vertical groundwater motion (U) following Taniguchi et al. [21]. The comparisons between the modeled and the observed profiles are shown in Fig. 11a–c.

Recharge-type (downward groundwater flux) profile is only detected in Tala well (Fig. 11a) with U value equals 0.8 m/year. The other wells showed a discharge type (upward groundwater flux) with varying rates: −0.1 to −0.5, −0.35, −0.23, −0.6, −1.2, and −2.8 m/year for Kafelarab, Nawag, Elkarada, Kafr Mesaaed, Elhadady, and Motobes wells, respectively, as shown in Fig. 11a–c. These estimates indicate that fresh groundwater starts to move vertical upward southern Tanta City. The most dangerous recorded phenomenon is that seawater intrusion upward flux recorded in Motobes wells is higher than the downward recharge flux calculated in Tala well. Two observations were detected from this model. These observations include (1) the geothermal gradient decreases in the areas influenced by seawater intrusion like Motobes wells, and (2) recognized temperature profiles curvature in the shallow zones in most of the wells cannot be modeled by the current model because the groundwater flow is complicated in the shallow zones to be solved by this simple model.

4.2 *Hydrogeochemistry*

The hydrochemical compositions of the collected samples are variable (Table 1). In all groundwater samples, EC and TDS are increased with depth and from south to north. Deeper freshwater in Tala 2 and Kafrelarab 2 wells have higher EC and TDS

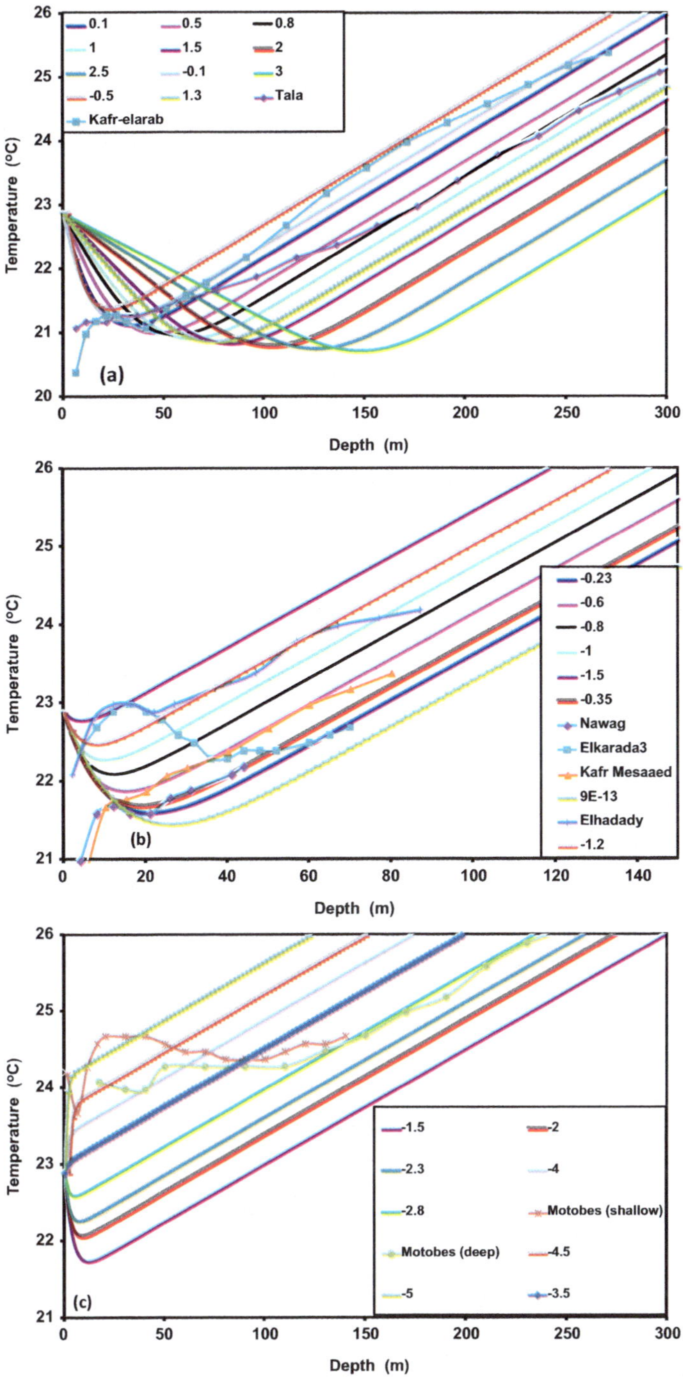

Fig. 11 Binary diagram shows the comparison between the modeled and the measured temperature profile. Groundwater flux (U) values are either positive or negative indicating the downward or the upward movement, respectively

compared to the shallow wells of the same areas. This phenomenon could be related to long time of water-rock interaction in deeper zones compared to the shallow ones. TDS increases from Nawag to Motobes wells in lateral and vertical directions. The most recognized fact is that groundwater of Motobes 1 (shallow) has TDS around 20,000 mg/l, while Motobes 2 (deep) has TDS value around 80,000 mg/l. That means the Nile Delta groundwater close to the Mediterranean Sea is affected by seawater intrusion in the shallow zones but is of brine hypersaline origin in the deeper zones. Shallow well values of pH are also variables ranging from 5.87 in Motobes 1 to 10.71 in Elkarada 1. According to Inskeep and Bloom [80] and Herman and Lorah [81], high pH values are related to calcite precipitation which occurs at pH > 8 and $P_{CO2} < 0.1$ atom. The mechanism of precipitation reaction is shown in Eq. 2:

$$Ca^{2+} + HCO_3^{-} + OH^{-} \rightarrow CaCO_3 + H_2O \tag{2}$$

This reaction could happen in Kafrelarab and Nawag which contain the very low Ca^{2+} and HCO_3^{-} concentrations in contrast to Tala and Qutor. Presences of nitrates in groundwater usually indicate an infiltration of water as a result of agricultural and human activities. Groundwater from the area south Nawag has very low nitrate concentrations which means that clay layer of Bilqas Formation and the irrigation system do not lead to an excess of nitrate to reach groundwater. The wells located northern Kafr Elsheikh City have high nitrate concentration marking a great contribution from the irrigation system where those areas are highly cultivated with rice as shown in Fig. 5.

Two main chemical types are recognized in Piper diagram (Fig. 12). The first type is the freshwater samples from Tala wells (1 and 2), and the second type is represented as seawater type including seawater sample, Kafr Mesaaed, Elhadady, and Motobes (1 and 2) wells. The later samples are shifted up from the seawater sample in the direction of seawater intrusion proving primary salinity characters. Elkarada samples (1 and 2) show a little downward shift from the seawater sample as a result of a little replacement of Ca by Na forming $NaHCO_3$-type water due to receiving calcium bicarbonate-rich water by surface infiltration. Although, Kafrelarab (1 and 2) and Nawag wells are freshwater but shifted toward seawater type. This could be related to loosing Ca and carbonates as calcite precipitation due to higher pH values. Qutor samples (1 and 2), on the other hand, are located toward seawater intrusion type as a result of ion exchange between Na of seawater with Ca and Mg from the aquifer material.

To be able to understand the interaction between seawater intrusion and freshwater in Nile Delta aquifer, ion exchange processes should be explained. Appelo and Postma [82] explained the ion exchange processes. The infiltrated freshwater is dominant by Ca^{2+} and HCO_3^{-} ions. Cation exchangers in the aquifer, therefore, have mostly Ca^{2+} ions absorbed on the surfaces. In the intruded seawater, Na^{+} and Cl^{-} are the dominant ions, and the sediment in contact with seawater will have absorbed Na^{+} for the large part. When seawater intrudes in a coastal freshwater aquifer, an exchange of cations takes place according to Eq. 3:

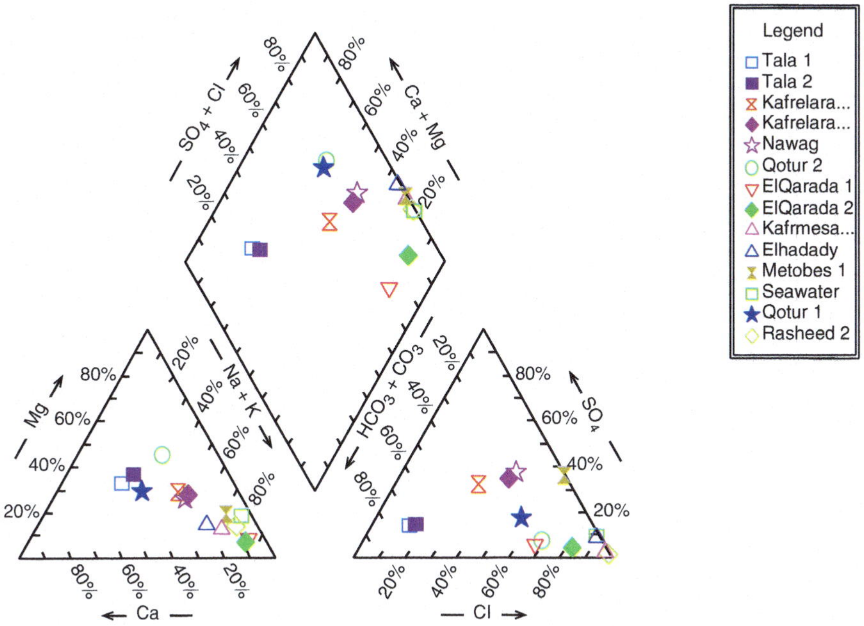

Fig. 12 Piper diagram illustrates the hydrochemical processes [29]

$$Na^{+} + 1/2\ Ca\text{-}X_2 \rightarrow Na\text{-}X + 1/2\ Ca^{+2} \quad (3)$$

where X indicates the soil exchanger. Sodium is taken by the exchanger and Ca^{2+} released. The water type then changes from NaCl into a $CaCl_2$ water type. The reverse process takes place with freshening, i.e., when freshwater flushes a salt water aquifer. Where Ca^{2+} is taken up from water in return for Na^{+} with a $NaHCO_3$ water type as a result. Water chemistry can thus indicate upcoming of seawater or, conversely, that saltwater is flushed by freshwater. These two processes could happen completely or partially.

Figure 13 showed the vertical two-dimensional distribution of water chemistry along the south-north direction using Stiff diagrams. Regardless of the ionic concentrations, the present groundwater could be classified into three main types. The first one is recorded as Ca-Mg $(HCO_3)_2$ freshwater type in Tala wells. The second type has NaCl composition and including all the rest of the wells except Qutor wells. The third hydrochemical type encountered in Qutor wells is of $CaCl_2$ in well 1 (at shallow depths) and $MgCl_2$ type in well 2 (at deeper depths). The minor details of Stiff diagrams reflect the hydrochemical processes that happened in the Quaternary aquifer along the flow path as follows:

1. Kafrelarab and Nawag waters are freshwater of NaCl type; this may happen as a result of calcite precipitation during upward movement of calcium carbonate type in Tala due to an increase of pH.

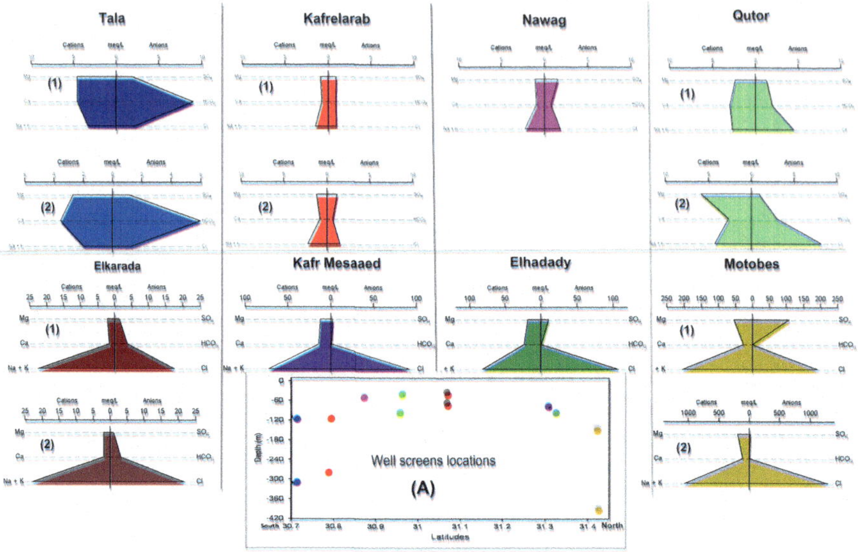

Fig. 13 Stiff diagram shows the difference in the chemical types of the collected samples. (1) and (2) indicate the shallow and the deep groundwater at each site. (**a**) is south–north cross section and illustrates the average depth of the screens where the groundwater was sampled; color of the screen dots is the same of the related stiff diagrams

2. Compared to the Motobes 1 which shows a typical seawater type, Kafr Mesaaed, Elhadady, and Motobes 2 glasses of water have a relative increase in Ca^{2+} and lower concentration of Na^+ than Cl^- indicating seawater intrusion as mentioned by Appelo and Postma [82]. In the other hand, Elkarada samples show a relative increase in HCO_3^-, and Na^+ is slightly greater than Cl^-. This could be related to freshwater infiltration from cultivated land.
3. In the same way, composition of Qutor groundwater ($MgCl_2$ type in deeper zones and $CaCl_2$ in shallow zones) could be explained as an indicator for seawater intrusion, and it might be located in the intermediate zone between the regional fresh groundwater and the intruded seawater in the Quaternary aquifer of the Nile Delta.

5 Conclusions

The integrated tracer technique between subsurface temperature and water chemistry was a good technique for tracing the groundwater flow system and seawater in the Nile Delta Quaternary aquifer. Two flow systems were detected: one is regional, and the other is local, and both of them are affected by seawater intrusion. The regional one is recharged south Tala and discharges northward in the area from Kafrelarab (south Tanta City) till ElKarada (south Kafr Elsheik City. The shape of the temperature profiles reflects a recharge type in

Tala and discharge type in Kafrelarab, and the shape of the isothermal lines indicates a deep warm water flows upward to the shallow zones. Water chemistry is supporting this idea where the recharge area is characterized by $CaHCO_3$ water type in Tala. Such water type is indicating an inland recharged freshwater. The discharge area of the system is characterized by NaCl type which is the most common water type due to calcite precipitation during upward flowing due to the change in pH. Recharge-type profile detected in Tala well has a groundwater flux (U) value equals 0.8 m/year. The other wells in this regional groundwater flow system showed a discharge type (upward groundwater flux) with varied U values: −0.1 to −0.5, −0.35, and −0.23 in Kafelarab, Nawag, and Elkarada wells, respectively. These estimates indicate that fresh groundwater starts to move vertically upward southern Tanta City.

Very low concentrations of nitrates in the groundwater of Tala and Kafrelarab and Nawag wells provide another evidence for being the water in the area south Nawag is contributed by deep circulating water. The shallow zones between Kafrelarab and Elkarada show a warming behavior as indicated from the temperature profiles. As those wells are not located in the urban areas, therefore, this thermal regime could be related to an upward and northward groundwater discharge.

The local flow system is identified from the cooling of the isothermal lines that are recognized in the area from north Kafr Elsheik till Motobes. This system is isolated from the regional one. The maximum depth of this local system is about 140 m in Kafr Mesaaed and 90 m in Elhadady. The recorded excess in nitrate concentrations indicates the agricultural and human surface activities. The temperature profiles measured in the area of this local system are of discharge value with upward flux (U) equals −0.6, −1.2, and −2.8 m/year for Kafr Mesaaed, Elhadady, and Motobes wells, respectively.

The subsurface thermal regime and the hydrogeochemical data in the study area gave an image about the spatial extend of seawater intrusion in the Nile Delta. The estimated criteria for seawater intrusion from the subsurface temperature are as follows:

1. The low-temperature difference between deep wells of Kafrelarab and Motobes wells in the deeper zones despite the long distance between them indicates nearly static seawater in the mentioned area.
2. The intersections between the temperature profiles. For example, Nawag and Elkarada profiles intersect at 45 m deep; Kafr Mesaaed and Elhadady profiles met Motobes profiles at 140 m and 90 m deep, respectively.
3. The nearly horizontal isothermal lines under most of the studied area in the deeper zones (till Nawag).
4. The low thermal gradient noticed in Motobes wells which located close to the Mediterranean Sea compared to other wells.

The abovementioned thermal criteria indicate the existence of a nearly static water body deeper in south Kafr Elsheikh City and shallower northern Kafr Elsheikh City which might be seawater intrusion.

Recognized hydrogeochemical criteria indicating seawater intrusion in the study area are as follows:

1. Groundwater in Qutor wells has both $CaCl_2$ and $MgCl_2$ water types, in the shallow and deeper zones, respectively. This indicates the existence of an intermediate zone between seawater and freshwater under Qutor area. This estimation is limited to 100 m depth which is the maximum depth of Qutor wells.
2. Seawater intrusion was also indicated from the partial ion exchange observed as a relative increase in Ca^{2+} and fewer Na^+ than Cl^- as in Kafr Mesaaed and Elhadady wells.
3. In contrast to the regional groundwater flow system of the southern part and the local flow system in the northern part, Elkarada samples show a relative increase of HCO_3^- and higher Na^+ compared to Cl^-. Such feature indicates partial ion exchange during freshwater propagation into the seawater. In another meaning, the groundwater under Elkarada is originally seawater mixed with a little of infiltrated freshwater.

Finally, the most dangerous information given from this research is that the groundwater in the Nile Delta is not affected only by seawater intrusion but also affected by hypersaline brine water inland propagation. The seawater affects the upper 200–250 m, but the hypersaline water was indicated in Motobes well which is of 420 m depth. The upward seawater intrusion flux rates in Motobes wells was 2.8 m/year which is much higher than the groundwater recharge flux at Tala well which was 0.8 m/day.

6 Recommendations

Governmental regulation of the pumping process from the Quaternary Nile Delta aquifer is urgently needed as the groundwater is not just facing the pollution from surface human activities and seawater intrusion but also suffers from the deeper hypersaline brine groundwater inland propagation. Integration between subsurface thermal and hydrogeochemical data is a good tool in recognition of the groundwater flow system and seawater intrusion.

Acknowledgment This chapter is an update and revised version of the first draft which was presented at the 5th International Symposium on Geophysics, 2008, Tanta, Egypt. Also, the authors thank the editor Prof. Dr. Abdelazim Negm and the reviewers for their constructive remarks and comments.

References

1. Freeze RA, Witherspoon PA (1967) Theoretical analysis of regional groundwater flow: effect of water-table configuration and subsurface permeability variation. Water Resour Res 3:623–634
2. Toth J (1962) A theory of groundwater motion in small drainage basins in central Alberta. J Geophysical Res 67:4375–4387

3. Toth J (1963) A theoretical analysis of groundwater flow in small drainage basins. J Geophysical Res 68:4795–4812
4. Todd DK (1959) Ground water hydrology. Wiley, New York, 336 p
5. Sherif MM, Singh VP, Amer AM (1988) A two-dimensional finite element model for dispersion (2D-FED) in coastal aquifer. J Hydrol 103:11–36
6. Sherif MM, Singh VP, Amer AM (1990) A note on saltwater intrusion in coastal aquifers. Water Resour Manag 4:123–113
7. Amer A, Sherif MM (1996) An integrated study for seawater intrusion in the Nile Delta aquifer. Working Paper for SRP, NWRC-MPWWR, Cairo, Egypt
8. Sherif MM, Singh VP (1997) Groundwater development and sustainability in the Nile Delta aquifer. Final report submitted to Binational Fulbright Commission, Egypt
9. Sherif MM (1999) The Nile Delta aquifer in Egypt, chapter (17) in seawater intrusion in coastal aquifers: concepts, methods and practices. In: Bear J, Cheng A, Sorek S, Ouazar D, Herrera A (eds) Theory and application of transport in porous media, vol 14. Kluwer Academic Publishers, Dordrecht, The Netherlands, pp 559–590
10. Sherif MM, Singh VP (2002) Effect of groundwater pumping on seawater intrusion in coastal aquifers. J Agric Sci 7:61–67
11. Bratton JF (2007) The importance of shallow confining units to submarine groundwater flow. In: Sanford W, Langevin C, Polemio M, Povinec P (eds) A new focus on groundwater-seawater interactions, vol 312. IAHS Publ, pp 28–36
12. Bratton JF, Bohlke JK, Krantz DE, Tobias CR (2009) Flow and geochemistry of groundwater beneath a back-barrier lagoon: the subterranean estuary at Chincoteague Bay, Maryland, USA. Mar Chem 92:78–113
13. Hughes JD, Vacher HL, Sanford WE (2009) Temporal response of hydraulic head, temperature, and chloride concentrations to sea-level changes, Floridan aquifer system, USA. Hydrogeol J. https://doi.org/10.1007/s10040-008-0412-0
14. Hughes JD, Vacher HL, Sanford WE (2007) Three-dimensional flow in the Florida platform: theoretical analysis of Kohout convection at its type locality. Geology 35:663–666
15. Kohout FA (1967) Ground-water flow and the geothermal regime of the Floridian plateau. Trans Gulf Coast Assoc Geol Soc 17:339–354
16. Wilson AM (2005) Fresh and saline groundwater discharge to the ocean: a regional perspective. Water Resour Res 41. https://doi.org/10.1029/2004WR003399
17. Wilson A, Ruppel C (2007) Salt tectonics and shallow subseafloor fluid convection: models of coupled fluid-heat-salt transport. Geofluids 7:377–386. https://doi.org/10.1111/j.1468-8123.2007.00191.x
18. Cartwright K (1970) Groundwater discharge in the Illinois Basin as suggested by temperature anomalies. Water Resources Res 6:912–918
19. Sakura Y (1978) Studies on groundwater circulation using temperature data. In: Studies on water balance in Japan, Kokon Syoin, Tokyo. (In Japanese)
20. Domenico PA, Palciauskas VV (1973) Theoretical analysis of forced convective heat transfer in regional ground-water flow. Geol Soc America Bull 84:3803–3813
21. Taniguchi M, Shimada J, Tanaka T, Kayane I, Sakura Y, Shimano S, Dapaah-Siakwan S, Kawashima S (1999) Disturbances of temperature–depth profiles due to surface climate-change and subsurface water flow: (1) an effect of linear increase in surface temperature caused by global warming and urbanization in Tokyo metropolitan area, Japan. Water Resour Res 35:1507–1517
22. Herczeg AL, Edmunds WM (2000) Inorganic ions as tracers. In: Cook P, Herczeg A (eds) Environmental tracers in subsurface hydrology. Kluwer, Boston, pp 31–76
23. Salem ZE, Sakura Y, Mohamed Aslam MA (2004) The use of temperature, stable isotopes and water quality to determine the pattern and spatial extent of groundwater flow: Nagaoka area, Japan. Hydrogeol J 12:563–575
24. Salem ZE, Taniguchi M, Sakura Y (2004) Use of temperature profiles and stable isotopes to trace flow lines: Nagaoka area, Japan. Ground Water 42:83–91

25. Zeidan BA (2005) The Nile Delta in a global vision. 9th international water technology conference (IWTC-IX). p 10
26. Farid MS (1985) Management of groundwater system in the Nile Delta. PhD thesis, Fac of Eng Cairo Univ, Cairo
27. El Atfy H (1999) Modified drainage system for rice growing areas: a tool for water saving. Watsave Award: Innovation Water Management Award Winning Paper. 15 p
28. Nofal ER, Amer MA, El-Didy SM, Akram MF (2015) SeaWater intrusion in Nile Delta in perspective of new configuration of the aquifer heterogeneity using the recent stratigraphy data. J Am Sci 11:567–570
29. Salem ZE, Gaame OM, Hassan TM (2008) Using temperature logs and hydrochemistry as indicators for seawater intrusion and flow lines of groundwater in the quaternary aquifer, Nile Delta, Egypt. In: Proceeding of the 5th international symposium on geophysics, Tanta, Egypt, pp 25–38
30. Engelen J, Essink GHPO, Kooi H, Bierkens MFP (2017) On the origins of hypersaline groundwater in the Nile Delta Aquifer. EGU General Assembly, Geophysical Research Abstracts, 19
31. Zaghloul MG (1959) Flow distribution through the groundwater aquifer of the Nile Delta. MSc thesis, Fac Eng, Alex Univ, Egypt, p 103
32. El-Fayoumy IF (1968) Geology of groundwater supplies in the region east of the Nile Delta. PhD thesis, Fac Sci, Cairo Univ, p 201
33. Shata AA (1978) Hydrogeology of the area between Belbeis and Ismalia includes Wadi El-tumylat. MSc thesis, Fac Agric, Zagazig Univ, Egypt, p 117
34. Anon A (1980) Groundwater studies for tenth of Ramadan city. Ministry of Irrigation, Water Research Center, RIGW, p 50
35. El-Dairy MD (1980) Hydrogeological studies on the eastern part of the Nile Delta using isotopes techniques. MSc thesis, Fac Sci, El-Azhar Univ, Egypt, p 220
36. El-Hefny K, Morsi A, Kalil JB, Farid MS, El-Ridi MR, Khater AR (1983) Groundwater study for El-Obour city area. RIGW, Giza, Egypt, p 37
37. Salluma MKM (1983) Hydrogeological and hydrochemical studies east of the Nile Delta. PhD thesis, Fac Sci, Ain Shams Univ, Cairo, Egypt, pp 166
38. Hamza MS, Aly AI, Nada AA, Awad MA (1988) Estimation of seepage from Ismailia canal using iodine 13. Isotopenpraxis 24:110–114
39. Kotb AM (1988) Geological, hydrogeological and geoelectrical studies on the eastern portion of Delta. MSc thesis, Fac Sci, Al-Azhar Univ, Cairo, Egypt, p 120
40. Mousa SA (1988) Geophysical and hydrogeological studies on the area north of El-Mokattam-Ataqa plateau and the south of El-Ismailia canal, Egypt. PhD thesis, Fac Sci, Ain Shams Univ, Cairo, Egypt
41. Mousa BM (1990) Petrology and soil genesis of the surface Quaternary deposits, east of the Nile Delta. PhD thesis, Fac Sci, Ain Shams Univ, Cairo, Egypt, p 391
42. El-Shamy IZ, Greish AH (1992) Hydrogology of west Ismailia area, Egypt. Desert Inst Bull 42:271–288
43. Ezz EL, Deen HM (1993) Sedimentological and geophysical studies of Heliopolis basin, Cairo, Ismailia desert road, and their application. MSc thesis, Fac Sci, Ain Shams Univ, Cairo, Egypt, p 228
44. Sollouma M, Gomaa MA (1997) Groundwater quality in the Miocene aquifers east and west of the Nile Delta and the north west desert, Egypt. Sci J Fac Sci Ain Shams Univ Cairo, Egypt 35:47–72
45. Salem ZE (2009) Natural and human impacts on the groundwater under an Egyptian village, central Nile Delta e a case study of Mehallet Menouf. In: 13th international water technology conference (IWTC, 13), March 12–15, Hurghada, Egypt, vol 3, pp 1397–1414
46. Atwia MG (2000) Mathematical modeling of unconsolidated sediments of Pleistocene aquifer, south of Ismailia canal area, Egypt. J Environ Res Zagazig Univ Egypt 2:191–214

47. Shabaan M (2002) Environmental impacts of development on the hydrogeology and hydrochemistry, West of Delta, Egypt. PhD Dissertation, Faculty of Science, Cairo University, Egypt, p 208
48. Frihy OE, Lawrence D (2004) Evolution of the modern Nile delta promontories: development of accretional features during shoreline retreat. Environ Geol 46:914–993
49. El Banna MM (2004) Damietta sand spit, Nile delta, Egypt. The Egyptian Sedimentology 12:269–282
50. El Banna MM (2004) Nature and human impact on Nile delta coastal sand dunes,
51. El Banna MM, Frihy OE (2008) Human-induced changes in the geomorphology of the north-eastern coast of the Nile delta, Egypt Coastal Research Institute, Alexandria 21514, Egypt p 7
52. El Banna M, Frihy O (2009) Human-induced changes in the geomorphology of the northeastern coast of the Nile Delta, Egypt. Geomorphology 107:72–78
53. El-Asmar HM, Hereher ME (2010) Change detection of the coastal zone east of the Nile Delta using remote sensing. Environ Earth Sci. https://doi.org/10.1007/s12665-010-0564-9
54. Elewa HH, Shohaib RE, Qaddah AA, Nousir AM (2012) Determining groundwater protection zones for the quaternary aquifer of northeastern Nile Delta using GIS-based vulnerability mapping. Environ Earth Sci 68:313–331
55. Mabrouk MB, Jonoski A, Solomatine D, Uhlenbrook S (2013) A review of seawater intrusion in the Nile Delta groundwater system – the basis for assessing impacts due to climate changes and water resources development. Hydrol Earth Syst Sci Discuss 10:10873–10911
56. Shehata M, El-Sabrouty MN (2014) Applications of hydrogeochemical modeling to assessment geochemical evolution of the quaternary aquifer system in Belbies area, East Nile Delta, Egypt. J Biol Earth Sci 14
57. Salem ZE, Atwia MG, El-Horiny MM (2015) Hydrogeochemical analysis and evaluation of groundwater in the reclaimed small basin of Abu Mina, Egypt. Hydrogeol J. https://doi.org/10.1007/s10040-015-1303-9
58. Salem ZE, El Bayumy DA (2016) Hydrogeological, petrophysical and hydrogeochemical characteristics of the groundwater aquifers east of Wadi El-Natrun, Egypt. NRIAG J Astron Geophys 5:106–123
59. Salem ZE, El Bayumy DA (2016) Use of the subsurface thermal regime as a groundwater-flow tracer in the semi-arid western Nile Delta, Egypt. Hydrogeol J 24:1001–1014. https://doi.org/10.1007/s10040-016-1377-z
60. Salem ZE, Al Temamy AM, Salah MK, Kassab M (2016) Origin and characteristics of brackish groundwater in Abu Madi coastal area, Northern Nile Delta, Egypt. Estuar Coast Shelf Sci 178:21–35
61. Salem ZE, Osman MO (2016) Shallow subsurface temperature in the environs of El-Nubaria canal, northwestern Nile Delta of Egypt: implications for monitoring groundwater flow system. Environ Earth Sci 75:1241. https://doi.org/10.1007/s12665-016-6046-y
62. Salem ZE, Osman OM (2017) Use of geoelectrical resistivity to delineate the seawater intrusion in the northwestern part of the Nile Delta, Egypt. In: Negm AM (ed) Groundwater in the Nile Delta, Hdb Env Chem, DOI https://doi.org/10.1007/698_2017_175, © Springer International Publishing AG
63. Salem ZE, Osman OM (2017) Use of major ions to evaluate the hydrogeochemistry of groundwater influenced by reclamation and seawater intrusion, West Nile Delta, Egypt. Environ Sci Pollut Res 24:3675–3704. https://doi.org/10.1007/s11356-016-8056-4
64. Salem ZE, Negm AM, ElNahrawy A (2017) Hydrogeophysical characteristics of the central Nile Delta aquifer. In: Negm AM (ed) Groundwater in the Nile Delta, Hdb Env Chem, Doi: https://doi.org/10.1007/698_2017_75, © Springer International Publishing AG
65. Salem ZE, Elsaiedy G, Elnahrawy A (2017) Hydrogeochemistry and quality assessment of groundwater under some central Nile Delta villages, Egypt. In: Negm AM (ed) Groundwater in the Nile Delta, Hdb Env Chem, Doi: https://doi.org/10.1007/698_2017_111,© Springer International Publishing AG
66. Salem ZE, Elsaiedy G, Elnahrawy A (2017) Assessment of the groundwater quality for drinking and irrigation purposes in the central Nile Delta region, Egypt. In: Negm AM (ed) Groundwater

in the Nile Delta, Hdb Env Chem, Doi: https://doi.org/10.1007/698_2017_137, © Springer International Publishing AG
67. Farid MSM (1980) Nile Delta groundwater study, MSc thesis, Cairo University, Egypt
68. Leaven MT (1991) Hydrogeological study of the Nile Delta and adjacent desert areas, Egypt, with emphasis on hydrochemistry and isotope hydrology, PhD thesis, Free University, Amsterdam, also published by RIGW/IWACO as Technical note TN 77.01300-91-01
69. Wolf P (1987) The problem of drainage and its solution in the Nile Valley and Nile Delta. Natural Res Develop 25:62–73
70. Arlt HD (1995) A hydrogeological study of the Nile Delta aquifer with emphasis on saltwater intrusion in the northern area, Mitteilung/Institut fur Wasserbau und Wasserwirtschaft, Technis- ¨ che Universitat Berlin, 130: 291–302
71. Sherif MM, Sefelnasr A, Javadi A (2012) Incorporating the concept of equivalent freshwater head in successive horizontal simulations of seawater intrusion in the Nile Delta Aquifer, Egypt. J Hydrol 464:186–198. https://doi.org/10.1016/j.jhydrol.2012.07.007
72. Ebraheem AM, Senosy MM, Dahab KA (1997) Geoelectrical and hydrogeochemical studies for delineating groundwater contamination due to saltwater intrusion in the northern part of the Nile Delta, Egypt. Ground Water 35(2):216–222
73. Research Institute for Groundwater (RIGW) (1992) Hydrogeological map of Nile Delta, Scale 1:500,000, 1st edn
74. Serag ElDin H (1989) Geological, hydrogeological and hydrological studies on the Nile Delta Quaternary aquifer. PhD thesis, Fac Sci, Mansoura Univ, Mansoura, Egypt
75. Dahab K (1993) Hydrogeological evaluation of the Nile Delta after High Dam construction, PhD thesis, Fac of Sci, Menoufia Univ, Egypt
76. Research Institute for Groundwater (RIGW) (2002) Nile Delta groundwater modeling report. Research Institute for Groundwater, Kanater ElKhairia, Egypt
77. Morsy WS (2009) Environmental management to groundwater resources for Nile Delta region. PhD thesis, Faculty of Engineering, Cairo University, Egypt
78. Chengjie Z (1988) A study of geothermal field and karstic leakage in Karstic area. International Association of Hydrologists 21st Congress, 10–15 October, Guilin, China, vol 2, pp 1127–1135
79. Carslaw HS, Jaeger JC (1959) Conduction of heat in solids. 2nd edn. Oxford University Press, London
80. Inskeep WP, Bloom PR (1985) An evaluation of rate equations for calcite precipitation kinetics at pCO_2 less than 0.01 atm and pH greater than 8. Geochim Publ Cosmochim Acta 49:2165–2180
81. Herman JS, Lorah MM (1988) Calcite precipitation rates in the field: measurement and prediction for a travertine-depositing stream. Geochim Cosmochim Acta 52:2347–2355
82. Appelo CAJ, Postma D (1994) Geochemistry, groundwater and pollution. AA Balkema Publishers, The Netherlands, 536 p
83. Shata A, El-Fayoumi I (1970) Remarks on the hydrogeology of the Nile Delta. In: Hydrology of deltas, UNESCO, Bucharest symp., I., Unesco, Paris UNDP "United Nation Environment Program"/UNESCO. Coastal protection studies. Final Technical Report, Paris, 1, 155 p
84. Shata AA (1966) Geological structure of the Nile Delta, Tenth Arab Engineering Conference, Jerusalem, Jordan (in Arabic)

Part VIII
Groundwater Modelling

Integrated Groundwater Modeling for Simulation of Saltwater Intrusion in the Nile Delta Aquifer, Egypt

Asaad M. Armanuos and Abdelazim Negm

Abstract Groundwater in the Nile Delta aquifer is considered one of the most important water resources of Egypt. In the last 30 years, the abstraction rates from groundwater wells in the Nile Delta increased dramatically. The Nile Delta region is considered very vulnerable to the sea level rise in the Mediterranean Sea due to climate change. The main objective of this study is to build an integrated 3D groundwater model of the Nile Delta aquifer to simulate the saltwater intrusion under different climate change scenarios using the MODFLOW and SEAWAT programs with the actual irrigation canal network in the Nile Delta region. Also, it proposed different scenarios for management and control of saltwater intrusion in the Nile Delta aquifer. Google Earth Pro software was used to estimate the bank levels and top width of the irrigation canals within the Nile Delta region. The spatial and temporal variation of groundwater recharge from rainfall in the Nile Delta aquifer was estimated by using WetSpass hydrological model. ENVI software was used to come up with land use classification based on available land cover images of the Nile Delta for 1972, 1984, 1990, 2000, and 2009. The WetSpass model was calibrated by comparing the simulated groundwater recharge with the calculated one by using the water balance equation method. The results indicated close agreement in groundwater recharge between the two methods' outputs. The WetSpass model was then applied in respect of 1970, 1980, 1990, and 2010 for the purpose of validation. The SEAWAT program was used to build an integrated groundwater model for the Nile Delta aquifer to simulate the saltwater intrusion

A.M. Armanuos (✉)
Department of Irrigation and Hydraulics Engineering, Faculty of Engineering, Tanta University, Tanta, Egypt
e-mail: asaad.matter@f-eng.tanta.edu.eg

A. Negm
Water and Water structures Engineering Department, Faculty of Engineering, Zagazig University, Zagazig, Egypt
e-mail: amnegm@zu.edu.eg; amnegm85@yahoo.com

A. M. Negm (ed.), *Groundwater in the Nile Delta*, Hdb Env Chem (2019) 73: 489–544, DOI 10.1007/698_2017_184,

where three scenarios are proposed: sea level rise, decreasing the south groundwater head due to additional groundwater pumping and combination of the two scenarios previously mentioned. The third scenario (combination between sea level rise and decreasing the groundwater head due to additional pumping) is the worst scenario. Finally, seven scenarios based on the built model results were proposed through decreasing the pumping discharges for management and control of saltwater intrusion in the Nile Delta aquifer. Reduction of the pumping discharges from all areas of the Nile Delta has a significant impact more than regional reduction and redistribution of pumping discharges in order to control the saltwater intrusion.

Keywords Integrated model, Nile Delta aquifer, Saltwater intrusion, SEAWAT

Contents

1 Introduction

Groundwater in coastal aquifers represents an important source of freshwater in many areas around the world especially in arid and semiarid areas, where the rainfall is scarce and the surface freshwater bodies are limited. In arid and semiarid regions, the development and the urbanization of coastal zones are exploiting the groundwater resources [1]. In coastal aquifers, the excessive groundwater pumping has a significant impact on a saltwater intrusion as the freshwater flux toward the sea decreases and causes the saltwater to intrude inland into the coastal aquifer [1]. The Nile Delta aquifer is one of these coastal aquifers which are vulnerable to the

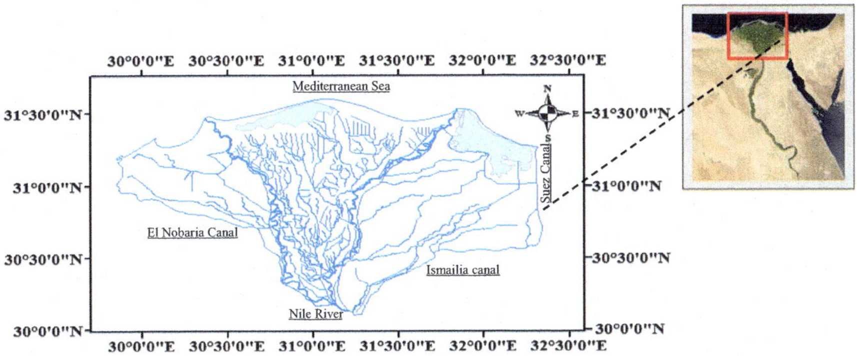

Fig. 1 Location map of the Nile Delta aquifer, Egypt

saltwater intrusion from the Mediterranean Sea. Figure 1 shows the location map of the Nile Delta aquifer, Egypt.

Over the last 30 years, the Nile Delta aquifer has subjected to severe groundwater pumping so a relatively large dispersion zone is encountered between the freshwater in the Nile Delta aquifer and the intruded saltwater from the Mediterranean Sea. Climate change also accelerates the degree of saltwater intrusion in the Nile Delta aquifer. Sea level rise has a significant impact on saltwater intrusion in the Nile Delta aquifer. Building an integrated groundwater model of the Nile Delta aquifer needs input data of a long-term groundwater recharge from rainfall and irrigation canal network for accurate representation and simulation. Integrated groundwater modeling of the Nile Delta is essential for assessing the effect of climate change (sea level rise and hydrological conditions) and the related impacts of increasing development on the management of groundwater resources of the Nile Delta aquifer to quantify the variations of groundwater level, Nile Delta water budget, and groundwater salinity conditions.

2 Saltwater Intrusion in the Nile Delta Aquifer

Anon [2] determined the sharp interface between freshwater and saltwater in the Nile Delta aquifer by using electrical logging method. The saltwater interface extended to about potentiometric contour +10.00 m. Farid [3] presented a new model by modifying AQUIFEM-1 model in the two dimensions to handle the seawater intrusion problem in the Nile Delta aquifer as a movable interface. The constructed model was applied to the Nile Delta aquifer to determine the shape and position of the interface and consequently the toe point under different withdrawal conditions. The present pumping from the Nile Delta aquifer was assumed to change to be double, triple, and zero, having a recharge of the same amount. Kashef [4] presented a method to determine the toe of the saltwater wedge across the Nile Delta and parts of its extensions in the Eastern and Western Deserts. He indicated that the saltwater intrudes deep in the Nile Delta aquifer to about 130 km from the

Mediterranean Sea approximately near El Bagour city. Sherif et al. [5] employed a two-dimensional finite element model (2D-FED) for dispersion in coastal aquifers to predict the concentration distribution and the flow pattern in the aquifer. The equi-concentration line 1,000 ppm intruded at the aquifer bottom to 106 km at the south of Tanta city and near to equipotential line +12.00 above the sea level. RIGW/IWACO [6] mentioned that the interface between saltwater and freshwater is actively moving inland into the Nile Delta aquifer and seawater intrusion advanced to the aquifer at Tanta city. Other authors believed that the flow of saltwater intrusion in the aquifer has reached the steady state. Soliman et al. [7] studied the groundwater flow and the solute transport in porous media using developed finite element model. The model was applied to Ruehen region in Germany and the eastern part of the Nile Delta aquifer to simulate the saltwater intrusion. The equi-concentration line 1,000 ppm reached Ismailia city in the Nile Delta aquifer. Sherif and Sighn [8] simulated the saltwater intrusion in the Nile Delta aquifer using SUTRA model in the aerial view considering the pumping discharges in each governorate. The equi-concentration of 1,000 ppm intruded through the aquifer at a distance of 41 km, and the equi-concentration line of 35,000 ppm intruded inland to a distance of 84 km. Sherif and Al-Rashed [9] presented a vertical simulation of the seawater intrusion by using a two-dimensional finite element model (2D-FED). The model was used to identify the dispersion zone and the flow pattern in the Nile Delta aquifer using the density-dependent approach. The equi-concentration line 35,000 ppm (seawater) intruded into the Nile Delta aquifer to a distance of 63.0 km from the seashore measured at the bottom boundary, and the equi-concentration line 1,000 ppm advanced into the aquifer to a distance of 108.0 km. There was a flux from the freshwater to the saltwater through the upper semi-previous layer with a depth of at least 22.0 km. RIGW [10] introduced the contour maps of groundwater salinity in the Nile Delta aquifer in the period between the years 1960 and 2000. RIGW concluded from the maps that the groundwater salinity in the Nile Delta aquifer was affected by the development activities. The groundwater salinity increased from 1980 to 1990, and the iso-salinity line of 1,000 ppm moved southward showing more saline water intrusion. Sherif et al. [11] used a 3D finite element variable density model (FEFLOW) to study the saltwater intrusion in the Nile Delta aquifer in a horizontal view. The concept of the equivalent freshwater head was studied in a two-dimensional real vertical simulation. The horizontal simulation was conducted with four horizontal sections located at different levels (100, 200, 300, and 400 m), assigning an appropriate pressure head to identify the depth of the horizontal section. The results showed that at a depth of 100 m, the seawater migrated a few kilometers inland, and at a depth of 200 m, the transition zone moved further inland to reach Tanta city. At a depth of 300 m, the seawater advanced to Mansoura, and at a depth of 400 m, the entire active area of the domain is occupied by seawater. Abdelaty et al. [12] used a three-dimensional model (SEAWAT) to study the seawater intrusion in the Nile Delta aquifer. The available data of salinity concentrations in different wells in the year 2008 was used according to RIGW. The results indicated that equi-concentration 35,000 ppm moved inland by a distance of 63.75 km from the shoreline in cross section in the middle of the delta, and the equi-concentration line 1,000 ppm intruded inland the aquifer and reached 93.75 km from the shoreline of the Mediterranean Sea. Abd-Elhamid et al. [13] conducted a coupled transient (2D–FEST) finite element model in order to simulate the fluid flow and the solute transport in both saturated and unsaturated zone for studying the saltwater intrusion in the Nile Delta aquifer under the impacts of climate change. The results of the model indicated that the isoline 35 intruded inland into the Nile Delta aquifer by a distance of about 64 km from the shoreline, while the isoline 1 moved inland into the aquifer at a distance of 112 km at a cross section in the central part of the Nile Delta. Nofal et al. [14] used the three-dimensional finite difference model SEAWAT to simulate the saltwater intrusion in the Nile Delta aquifer with considering the available heterogeneity data and the variation of groundwater density values in the recent drilled boreholes. The model results showed that the appreciable agreement is related to the flow fluxes and piezometric head. The saltwater intrusion in the Nile Delta aquifer was delineated by the model. Wassef and Schüttrumpf [15] built a three-dimensional finite element model in the western area of the Nile Delta by using FEFLOW software to study the saltwater intrusion under different climate change scenarios. The results showed that by the year 2100, it is expected that the interface of the seawater will reach a maximum of about 43 km according to RCP 2.6 scenario, while it is expected to reach 57 km according to RCP 8.5 scenario. The groundwater overexploitation will cause an increase in the salinity concentration to about 5,000 mg/L. Mazi et al. [16] based on the results of site-specific vulnerability assessments concluded that the Nile Delta aquifer was expected to threaten by the advance of saltwater intrusion. A 10% decline of the groundwater head will lead to advancement of seawater intrusion by about 5–8 km, and a 20% decline may lead the Nile Delta aquifer to its tipping point with a complete aquifer intrusion.

2.1 Effect of Excessive Groundwater Pumping on Saltwater Intrusion in the Nile Delta Aquifer

Sherif [17] used SUTRA model to examine the effect of pumping activities in the Nile Delta aquifer on the saltwater intrusion and determine the best sites for pumping in six different scenarios. The Nile Delta was divided into three zones according to pumping activities: the middle, eastern, and western zones. When the same amounts of water were pumped from the eastern and western Nile Delta, a considerable reduction of saltwater intrusion was founded in the middle Delta. The withdrawal from the middle Delta only reduces the saltwater intrusion in the Nile Delta with specific reduction in the eastern side. Where an additional pumping was increased in the middle Delta, the concentration line 35,000 advanced inland into the Nile Delta aquifer. Sherif [18] used the SUTRA model to simulate the saltwater intrusion in the Nile Delta aquifer on the horizontal direction to assess the effect of various pumping scenarios on the saltwater intrusion process. It is recommended that any additional pumping should be practiced in the middle of the Delta to decrease the adverse impacts and the redistribution of pumping activities, and land use for agricultural practices in the Nile Delta may help mitigate the intrusion migration. Sheriff [19] investigated the effect of land use on the seawater intrusion; additional scenarios for rice cultivation were considered in the northern and southern parts of the Nile Delta. The results showed for all scenarios that any additional pumping should be practiced from the middle of the delta and decrease in the eastern and western parts. The rice cultivation in the southern part would reduce the saltwater intrusion because the clay layer in it is more permeable and thinner, so he recommended increasing the rice cultivation in the southern area of the Nile Delta. Abdelaty et al. [12] used the (SEAWAT) model to study the effect of pumping activities on seawater intrusion in the Nile Delta aquifer. Three different scenarios of increasing pumping activities were considered to the model by 25, 50, and 100%. The results showed that the increase of pumping by 25, 50, and 100% caused the equi-concentration 35,000 ppm at a section in the middle of the Delta to intrude inland to distances of 63.75, 66.5 and 65.75 km, respectively. On the other hand, the equi-concentration 1,000 ppm moved inland into the aquifer to distances of 98.25, 101.25, and 107.75 km, respectively, at the same cross section. Kashef [4] and Anon [2] stated that the water supply and industry consume about 1.6×10^9 m^3/year from the groundwater in the Nile Delta aquifer system, and survey from the Egyptian ministry of irrigation demonstrated that the total amount of extraction due to irrigation is 0.82 km^3/h and the total volume according to irrigation and water supply and industrial equals to 2.42 km^3/h. Dahab [20] stated that the total annual abstraction from the groundwater in the Nile Delta due to irrigation and domestic pumping equals to 2.123×10^9 m^3/year at 1993. RIGW in Egypt monitored the pumping activities at each governorate in the Nile Delta. According to the extraction well inventory in 1992, the total volume of extraction was about 1.92×10^9 m^3/year. El Gharbia governorate recorded the maximum number of extraction wells that equals to 3,391 followed by Sharkia and Menoufia governorates recording 1953 and 1719, respectively [21]. RIGW also carried the abstraction well inventory for the Nile Delta region for the years 1995, 1997, and 2002. Information sheets are uploaded to calculate the amount of extraction in the year 2008. The total volume rate of extraction from the Nile Delta was 3.03×10^9 m^3/year and 4.90×10^9 m^3/year for the years 1992 and 2008, respectively [22]. Serag El Din [23] mentioned that the upward flow of groundwater to upper clay layer due to saltwater intrusion is about 3×10^3 m^3/year and the recharge to groundwater in the Nile Delta is almost three times as much as the upward loss. The total annual abstraction from the Nile Delta aquifer increased dramatically from 1.6×10^9 m^3 in the year 1981 to 2.6×10^9 m^3 in 1991 and increased slightly to 3.2×10^9 m^3 in 1997 and to 3.5×10^9 m^3 in 2003. It can be observed that a dramatic increase occurred in the total abstraction since 2003 to 2010 and arrived 4.5×10^9 m^3/year with 0.2×10^9 m^3/year increase [24]. The distribution of abstraction wells in the Nile Delta was collected by RIGW and was shown in Abdelaty et al. [12]. A huge number of extraction wells are concentrated in the central and southern parts of the Nile Delta in El Gharbia and El Monufia governorates. In the eastern part, they are concentrated in El Dakahlia, El

Sharqia, and El Qalyubia governorates and in El Beheira governorate in the western part of the Nile Delta. El-Arabi [25] mentioned that groundwater in the Nile Delta aquifer system and desert fringes is not a resource in itself as it is replenished from the Nile River by seepage from canals and deep percolation from irrigation application. About 4.6×10^9 m^3 is the total abstraction from the Nile Delta and its fringes, and 0.5×10^9 m^3 is abstracted only from the desert aquifers and the coastal areas. The total annual groundwater abstraction in the year 2015 from the Nile Delta is expected to increase considerably to reach 11.4×10^9 m^3. This increase is due to the relative importance of groundwater in the national water resources management. Abd-Elhamid et al. [13] used a coupled transient (2D–FEST) finite element model to simulate the saltwater intrusion in the Nile Delta aquifer under excessive groundwater pumping. The results showed that a 100 cm decrease in the piezometric head due to the excessive groundwater pumping will lead to additional saltwater intrusion in the Nile Delta aquifer with about 8 km. Also, the combination of sea level rise by 100 cm and piezometric head decreasing by 100 cm will lead to an additional saltwater intrusion of about 15 km.

2.2 *Effect of Sea Level Rise on Saltwater Intrusion in the Nile Delta Aquifer*

Sherif and Sighn [8] studied the effect of climate change on seawater intrusion in two coastal aquifers, Nile Delta aquifer in Egypt and Madras aquifer in India. A numerical model was used to investigate the effect of sea level rise by 0.2 and 0.5 m on the saltwater intrusion. The results indicated that a rise of 0.5 m in the sea level of Mediterranean Sea would cause additional intrusion of 9.0 km in the Nile Delta. On the other hand, Madras aquifer will cause additional intrusion of 0.4 km when the Bay of Bengal rises with the same value. It was concluded that Nile Delta aquifer is more endangered than Madras aquifer under the condition of climate change and sea level rise. Sherif [17] studied the effect of sea level rise on the saltwater intrusion in the Nile Delta aquifer in the vertical direction by using 2D-FED model. The rise of the water level of the Mediterranean Sea by 0.2 m and no change in piezometric head lead the concentration line 1,000 ppm to advance slightly into the aquifer, and the concentration 35,000 ppm advanced inland by 2.5 km. In the case of 0.5 m rise, the concentration line 35,000 intruded further inland the aquifer by 1.5 km and the concentration line 1,000 ppm advanced inland by a distance of 9.0 km. Sherif et al. [11] used the 2D-FED model to simulate the saltwater intrusion in the Nile Delta aquifer on the vertical direction to assess the effect of sea level rise due to global warming. A 20 cm rise in the seawater level of the Mediterranean Sea caused the equi-concentration line 35,000 ppm intruded inland to a distance of 2.0 km. A 50 cm rise in the seawater level of the

Mediterranean Sea causes about 4.5 km inland intrusion of equi-concentration line 35,000 ppm. Nofal et al. [14] used a numerical solute transport model (MODFLOW and SEAWAT) to simulate the impact of saltwater and freshwater interface behavior in the Nile Delta aquifer with different scenarios of sea level rise. The effect of three values of sea level rise 0.2, 0.5, and 1.0 m on groundwater heads and salinity distribution was evaluated. It was concluded that the sea level rise affects the aquifer to a certain limit. Also, that groundwater heads after 30 years will increase ranging from 0.0 to 0.5 m according to the increase in sea level rise. On the other hand, the groundwater salinity will be marked southward at a distance of 7.0 km from the sea. Abdelaty et al. [12] used a three-dimensional model (SEAWAT) to study the seawater intrusion in the Nile Delta aquifer considering the increase of sea level rise. Three different scenarios of sea level rise were considered 0.25, 0.5, and 1.0 m to the model. The results showed that an increase of sea level by 0.25, 0.5, and 1.0 m would lead the equi-concentration 35,000 ppm at the section in the middle of the Delta to intrude inland into the Nile Delta aquifer to distances of about 66.75, 67.0, and 67.75 km, respectively. On the other hand, the increase of sea level by 0.25, 0.50, and 1.0 m will cause the equi-concentration 1,000 ppm to move to a distance of 97.75, 96.25, and 97.0 km, respectively, at the same cross section. Abd-Elhamid et al. [13] used a coupled transient (2D–FEST) finite element model to simulate the saltwater intrusion in the Nile Delta aquifer under the impacts of climate change. The results showed that a rise of 100 cm in the water level of the Mediterranean Sea would lead to an additional saltwater intrusion into the Nile Delta aquifer with about 10 km.

3 Groundwater Recharge in the Nile Delta Aquifer

The main sources of groundwater recharge in the Nile Delta aquifer are recharge from the Nile River, seepage from irrigation canals and drains, excessive irrigation water, and groundwater recharge from rainfall.

3.1 Groundwater Recharge from Rainfall

Building an integrated groundwater model of the Nile Delta aquifer needs input data related to long-term groundwater recharge for accurate representation and simulation. Integrated groundwater modeling of the Nile Delta is essential for assessing the effect of climate change (the rise in sea level and hydrological conditions). Also, it helps in assessing the related impacts of increasing development in terms of managing the groundwater resources of the Nile Delta aquifer to quantify the variations of groundwater level, its water budget, and groundwater salinity conditions [24]. Infiltration recharges the Nile Delta aquifer from a network of irrigation canals, excess irrigation water, and precipitation that percolates through the upper clay layer contributing to the groundwater recharge [8]. Groundwater recharge from rainfall in the Nile Delta occurs only during the winter months. It is an important mechanism in the coastal zones, especially in the Mediterranean coastal aquifer where coastal dunes overlay the aquifer [22, 26]. Sherif and Sighn [8], Sefelnasr and Sherif [27], and Abdelaty et al. [12] built a groundwater model of the Nile Delta for studying saltwater intrusion and the impact of climatic changes on water resources in the Nile Delta aquifer. Mabrouk et al. [24] stated that the rainfall recharge in the Nile Delta is neglected in most groundwater modeling studies.

3.2 Groundwater Recharge from Excessive Irrigation Water

The main source of recharge in the Nile Delta aquifer is the percolation from agriculture and infiltration from irrigation and drainage networks canals. Recharge varies according to hydrogeological conditions. The pattern of infiltration/drainage may be reversed according to the season [28]. Shamrukh et al. [29] reported that the agricultural recharge to the Quaternary aquifer ranged between 0.8 and 1.1 mm/day for old lands and between 1.9 and 2.1 mm/day for the reclaimed area. DRI [30] reported the average percolation to the Quaternary aquifer in the Nile Delta at about 0.8 mm/day. Warner et al. [31] reported that the percolation rate in the central and southern parts of the Nile Delta ranges between 0.25 and 0.8 mm/day. The percolation rate depends on the type of soil and irrigation and drainage practices. This value increases in the desert areas to the west for furrow irrigation from 1.0 to 1.5 mm/day and decreases to values range from 0.1 to 0.5 mm/day for drip and sprinkler irrigation. The percolation rates in the Eastern parts of the Nile Delta ranged between 0.2 and 5 mm/day in the large reclamation projects [24, 32]. El Ramly [33] mentioned that the groundwater aquifer system in the Nile Delta is recharged from deep percolation of subsurface drainage water and seepage from the main irrigation canals crossing the Nile Delta region. The average rate of recharge to groundwater from the excess irrigation water varies from 0.25 to 1.1 mm/day. Morsy [22] reported that the rate of drainage surplus equals to 0.25 mm/day at the northern part of the Nile Delta, 1 mm/day in the middle part, and 1.1 mm/day in the southern part of the Delta. These values decrease in the old lands to equal 0.29 and 0.5 mm/day in the eastern and western parts of the Nile Delta, respectively. In the newly reclaimed lands in the western deserts, drainage surplus rate ranges from 0.87 to 3.2 mm/day.

3.3 Groundwater Recharge from Irrigation Canals Network

Another significant influence on the recharge of the main aquifer comes from the water levels in the irrigation canals. These water levels are also a significant factor in groundwater modeling, because they influence the surface water-groundwater interaction. According to Mabrouk et al. [24], most modeling studies were represented with a constant average water level value along the canals (e.g., [34]). On the other hand, water levels of the canals vary from 1 month to another and throughout different sectors of the canals, which needs to be taken into account for more accurate representation of the interactions between the aquifer and the surface water in the delta. The delta receives 35×10^9 m^3/year of surface water from the Nile for irrigation, industry, and water supply. This recharges the aquifer through the infiltration of excess irrigation water through seepage from an extensive net of canals and drains [2, 4]. Previous studies considered in their account the groundwater recharge from 10 canals for simulation of the saltwater intrusion in the Nile Delta aquifer despite the actual irrigation canal network consisting of more than 200 canals.

4 Water Budget of the Nile Delta Aquifer

Kashef [4] reported that the total annual abstraction from the Nile Delta aquifer equals to 2.42 km^3/year and the balance between the inputs and outputs equals to 3.98 km^3/year; this value represented the future water surplus for the future development. Kashef [4] mentioned that the balance which is about 3.98 km^3/year is approximately assumed correct and indicated the annual gain to the Nile Delta aquifer. Kashef [4] concluded that if this value is repeated annually, a detrimental situation will develop, leading to waterlogging. On the other hand, if the total amount of abstraction increased more than the present level, the waterlogging would be reduced, leading to better land production, while the danger of an increase in saltwater would develop. Dahab [20] concluded that the difference between the inputs and outputs of the Nile Delta aquifer system equals to +2,813.984 million m^3/year; this value is lost through downward infiltration from main drains and excess irrigation water.

5 Data Limiting About Irrigation Canal Networks in the Nile Delta Region

The Nile Delta canals receive about 35.5 billion m^3 of surface water annually, which is used for agriculture, industrial activities, and water supply. The intensive irrigation canal networks recharge the aquifer in two ways: infiltration of excess irrigation water and seepage from canals and drains [2, 4]. The main source of recharge to the Nile Delta aquifer is percolation from agriculture and infiltration from irrigation and drainage network canals [28]. Mabrouk et al. [24] stated that the water levels of the irrigation canals have a significant influence on the groundwater recharge of the Nile Delta aquifer system. The major portion of annual overall groundwater recharge of Nile Delta aquifer is provided by the direct seepage from the Nile River, the huge network of irrigation canals, and the excess irrigation water [4]. In that study, water budget concept was applied, despite the lack of irrigation canal bed and bank levels, to estimate recharge from irrigation canals and irrigation water, which was about 6.4 km^3/year. The annual seepage of surface water to groundwater system in the Nile Delta aquifer was estimated to equal to 2.5 km^3/year [20]. The effect of freshwater recharge on seawater intrusion in the Nile Delta aquifer is also observed in the upper layer around the Nile River and its branches [11]. The seawater intruded toward the Nile Delta aquifer with the decrease of water levels of canals [12]. However, those previous studies considered groundwater recharge based on up to ten canals in groundwater modeling despite the high density of irrigation network in Nile Delta region that consists of more than 200 canals [35]. There was obviously a shortage in the data of bed and water levels of the Nile Delta canals, although this data of bed, bank level, and bed slope is necessary for more accurate representation and simulation of groundwater modeling to understand the interaction between surface water and groundwater in the Nile Delta aquifer system. Determination of water level in canals requires information of bank levels which are not available for irrigation canal network of the Nile Delta in the previous studies. Nowadays, many different tools are available to estimate upper canal width and bank levels of rivers. Ssegane and Tollner [36] used Google Earth to derive elevations in the Nzoia basin (Kenya) and to calculate the differences between Google Earth and GPS. It could be concluded that there was a shortage of bank level, water level, and upper canal width data for irrigation canal network in the Nile Delta, while it was needed for building an integrated groundwater modeling of the Nile Delta aquifer and accurate simulation of the interaction between surface water and groundwater.

6 Estimation of Groundwater Recharge from Rainfall in the Nile Delta Aquifer

Armanuos and Negm [37] described the variations between the values of WetSpass input parameter which represents the Nile Delta region and the default values in the model. An extensive sensitivity analysis is required for accurate determination of the WetSpass input parameters based on the crop pattern in the Nile Delta region. Armanuos et al. [38] used WetSpass hydrological model to estimate groundwater recharge from rainfall for the years 1970, 1980, 1990, 1991, 2000, and 2010. ENVI software was used to generate the land use/land cover maps for the Nile Delta. The input data for WetSpass model was collected from the Nile Delta region which includes topography, soil map, land use, groundwater level, rainfall, wind speed, temperature, and evapotranspiration. The simulated values of groundwater recharge by WetSpass model were compared to the estimated values by using the water balance model in calibration process for 25 points for the years 1991 and 2000. The following subsections describe the analysis of WetSpass input data.

6.1 Data Analysis

6.1.1 Topography

Figure 2 shows the topography map of the Nile Delta according to EGSA (Egyptian General Survey and Mining) [39]. In the central part of the Nile Delta, the land level ranges between 18 m + MSL at the south near Cairo city and decreases toward the Mediterranean Sea in the north to be less than 1.0 m. In the eastern part, the level of the Delta land ranges between less than 1.0 m + MSL in the northern border near Port Said and increases toward the south to reach 35.0 m + MSL near Ismailia Canal. It increases in the southeast of the Nile Delta to reach 112.0 m + MSL. In the western part of the Nile Delta, the land level ranges between 112 m + MSL at the south and decreases toward the north near the Mediterranean border to be less than 1.0 m, except Wadi El Natrun area where the land level decreases to 15 m −MSL.

6.1.2 Land Use

Rapid imbalance changes were observed among the five land cover classes (urban, desert, water, fish farms, and agricultural land). Figure 3 shows the classification of land use in the Nile Delta for 1972 and 2009. The percentage of agriculture land increases gradually from 51% in 1972 to 68% in 2009. The temporal data for the five images consists of four Landsat scenes, which were mosaicked to cover the entire Nile Delta. A support vector machine (SVM), which is a supervised classification algorithm, was implemented to produce classification maps in a thematic form. A total number of 1,319 points, obtained by the ENVI software, was used as training points, which were randomly generated under the ArcGIS environment for each image. Another 750 points were selected as validation points for accuracy assessment. Table 1 shows the percentage of training, validation points, and accuracy assessment for each class. The overall accuracy of land classification was 88.83%. Reclaimed land areas were observed in the southwestern part of the Nile Delta. For desert land, it decreases gradually from 33% in 1972 to 12% in 2009. Urban land increases from 6.5% in 1972 to 14% in 2009, as shown in Fig. 4.

6.1.3 Soil

According to the distribution of the soil map of the Nile Delta region [41, 42], the most common type of soil in the ND is Nile silt and clay distributed in the central part. Sand dunes are observed in the northern border near the Mediterranean Sea. Coarse sand

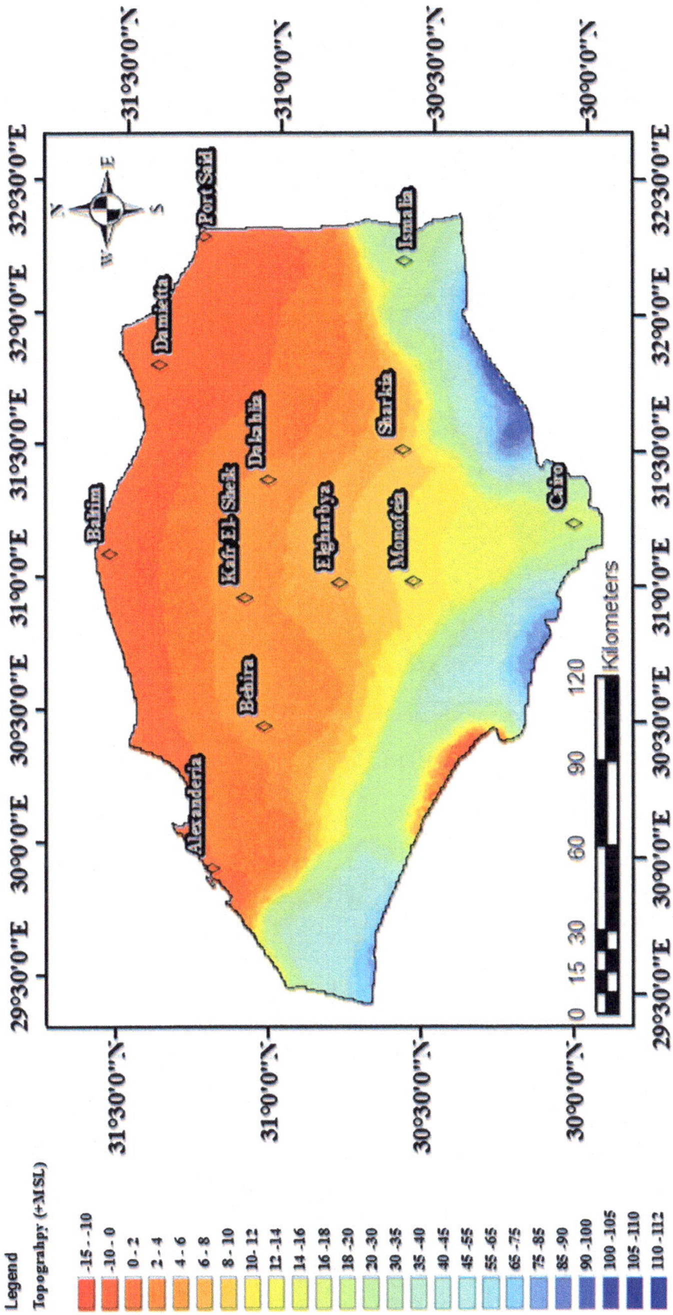

Fig. 2 Topographic map of the Nile Delta region and location of data collection stations

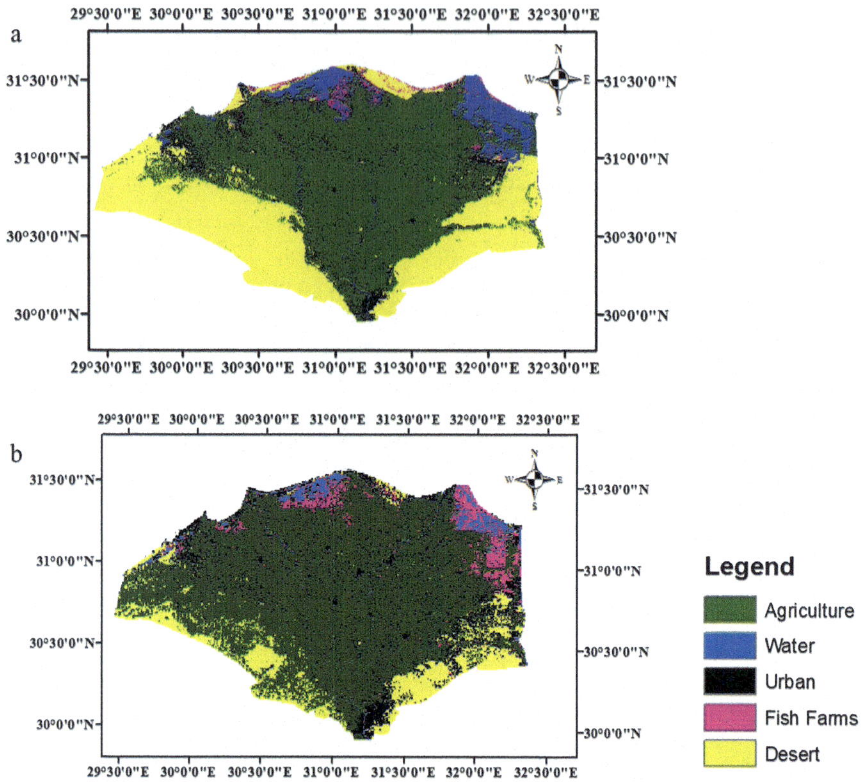

Fig. 3 Land use/land cover changes in the Nile Delta region, Armanuos et al. [38, 40]. (**a**) 1972, (**b**) 2009

Table 1 Total number of the training, validation points, and accuracy assessment per class for the five images

Class name	Total number of the training points	Training points percentage per class	Total number of the validation points	Validation points percentage per class	Assessment accuracy percentage
Agriculture	972	63.69	400	53.33	100
Water	128	9.70	100	13.33	100
Fish farms	65	4.93	50	6.67	75
Urban	67	5.08	100	13.33	71.16
Desert	87	16.60	100	13.33	98

and gravel, mixed with clay, are observed in the southeast and southwest parts. Desert soil crust is found in northern-west parts and in the border of Manzala and Borolus

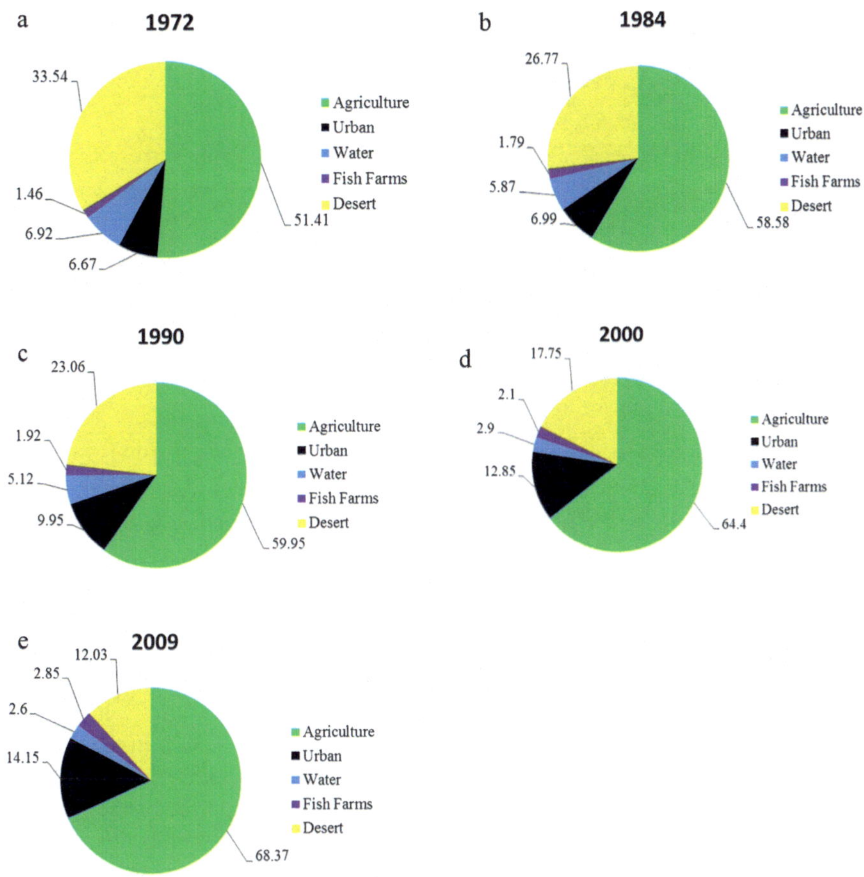

Fig. 4 Land use/land cover class percentage in the Nile Delta region at 1972, 1984, 1990, 2000, and 2009. (**a**) 1972, (**b**) 1984, (**c**) 1990, (**d**) 2000, (**e**) 2009

lakes in the north. Sabkha deposit soil type is observed in the northern part of ND near the north border, as shown in Fig. 5.

6.1.4 Groundwater Level

In 1974, in the central part and the eastern part, the groundwater levels range between 18.0 m amsl at the southern border and 0 m at the northern border, while in the northwestern part, it increases to reach 18 amsl. On the other hand, it decreases to 18.0 m bmsl in the southwestern part of the Nile Delta. In 1980, the groundwater level ranges between 16.0 m amsl near Cairo city at the south and decreases toward the north to 0.0 m amsl. The groundwater level increases toward the west to reach 14 amsl, while it is still below the mean sea level by 15.0 m in the

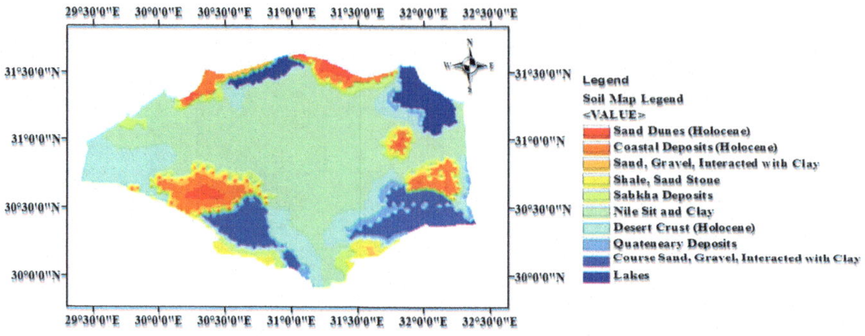

Fig. 5 Soil map in the Nile Delta

southwest. The groundwater level increased slightly in the year 1990 to reach 18.0 m amsl in the south and also increased in the northern-west part to reach about 25.0 m amsl. In 1992, the groundwater level decreased slightly to about 16 amsl near Cairo and 20+ MSL in the west and remained at 15 bmsl in the southwest. In the year 2000, it decreases again to reach 15 amsl near the south border, while it remained at still the same level in the west and in the southwest. The groundwater level remained at 15 amsl in the south border and also at 15 bmsl in the southwest, while it increased to 30.0 m amsl in the northwestern part of the Nile Delta, as shown in Fig. 6. The depth to groundwater level in the Nile Delta ranges from 1 to 5 m in the central part and increased to 50 m in the south. In the eastern part, it ranges from 1 to 3 m in the north and increased in the south, reaching 100 m, while in the western part, it ranges from 1 to 5 m, increased in the central part to 15 m, and reached more than 50 m in the south. The depth to groundwater level is higher than the root depth in the Nile Delta region; the available near maps were used to present the selected studied years.

6.1.5 Rainfall

The amount of rainfall in Alexandria fluctuated through the period at around 198.3 as an average value. Minimum rainfall values were 7 mm and 55 mm in 1999 where maximum values equaled to 165 and 410 mm observed in 1991 for Cairo and Alexandria, respectively. Rainfall in the Nile Delta increased from the south near Cairo toward the northern areas near the Mediterranean border at Alexandria. Rainfall in 1970 increased from 16.7 mm at Cairo to 192.9 mm at Alexandria. Since 1970–1980, rainfall decreased slightly below the average values. In 1980, rainfall increased in Alexandria to 195.5 mm and decreased to 10.5 mm in Cairo. Since 1980–1990, rainfall slightly exceeds the average values. Minimum rainfall was 15 mm in Cairo, and it increased toward the north to 189 mm. Since 1990–2000, rainfall fluctuated around the average values where the maximum rainfall is recorded in 1991. In 2000, it reached 17 and 200 mm in Cairo and

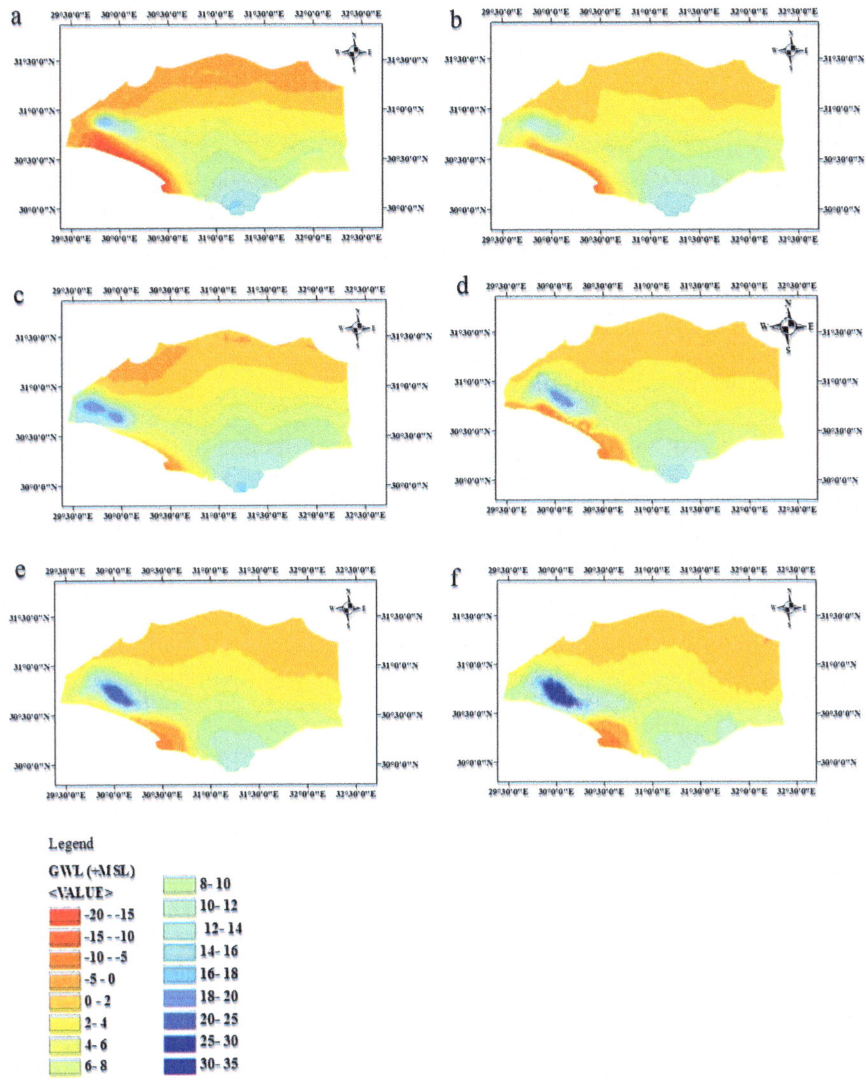

Fig. 6 Groundwater level in the Nile Delta aquifer. (**a**) 1974, (**b**) 1980, (**c**) 1990, (**d**) 1992, (**e**) 2002, (**f**) 2008

Alexandria, respectively. From 2000 to 2010, the rainfall was near the average value reaching 20 and 225 mm in Alexandria and Cairo, respectively. Based on an analysis of rainfall, 6 years were selected to present the rainfall distribution in the Nile Delta, as shown in Fig. 7.

6.1.6 Temperature

It is observed from Fig. 8a–f that temperature ranged from 13 to 17.2°C in the Nile Delta through the study period. The distribution of temperature recorded minimum values in the north in Alexandria; Egypt increased toward the east in Port Said and reached maximum values in Cairo in the south part of the Nile Delta. In 1970 and 1980, it ranged from 13.5 to 16.33°C and in 1990 ranged from 13.0 to 14.67°C. It recorded an increase in the year 2000 to reach maximum values in Cairo about 17.2°C while still in a range from 14.2 to 16°C in 1991 and 2010, as shown in Fig. 8.

6.1.7 Potential Evapotranspiration

It can be observed from the spatial and temporal distribution of potential evapotranspiration that PET ranged from 97.8 and 121.5 mm in the winter season, as shown in Fig. 9a–e. The potential evapotranspiration values were observed in Alexandria where minimum temperature was recorded, increasing to the east in Port Said and reaching maximum values toward the south in Cairo where maximum temperature was observed. Figure 9 presents the PET values in Alexandria, Port Said, and Cairo from 1970 to 2010. It ranged from 113.3 to 117.3 mm in 1970. It increased slightly in 1980 where maximum value was observed in Cairo about 121.5 mm. From 1990 to 2010, it ranged from 97.8 to 113.5 mm, while the minimum value of PET was observed in Alexandria.

6.1.8 Wind Speed

It can be observed from Fig. 10a–f that the wind speed ranged from 3 to 5.5 m/s in the Nile Delta through the study period. The distribution of wind speed recorded minimum values in Cairo, increased toward the north in Alexandria, and reached maximum values in Port Said in the northeastern part of the Nile Delta. In 1970, it ranged from 3.72 to 5.5 m/s and decreased slightly in 1980 and 1990 ranging from 3.63 to 5.25 m/s. It recorded an increase in the year 2000 to reach maximum values in Port Said about 5.5 m/s while still in a range from 3 to 5 m/s in 1991 and 2010, as shown in Fig. 10.

Values of groundwater recharge from rainfall ranged from 0.0 to 304 mm in the winter season in the year 1991, which equates to about 0.0–74% from the values of precipitation. Low values of groundwater recharge were observed in the eastern, northeastern, and southern areas in the Nile Delta at Gharbia, Ismailia, Kafr El Sheikh, and Port Said. The values of groundwater recharge increased toward the western areas of Beheira and Baltim cities. High values of recharge were observed in the northwestern part of Alexandria city, which ranged from 0.0 to 99 mm (49.5%) for 1970 and 1980. The groundwater recharge values decreased in 1990

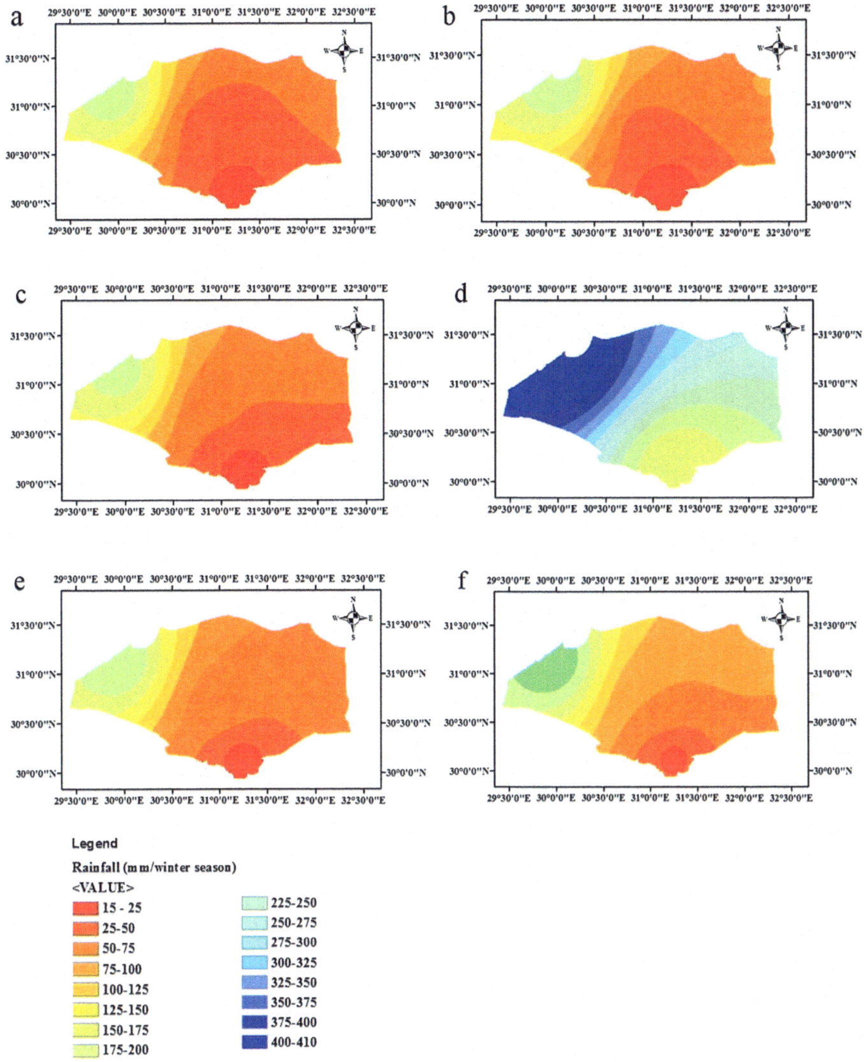

Fig. 7 Annual rainfall (mm) in the Nile Delta. (**a**) 1970, (**b**) 1980, (**c**) 1990, (**d**) 1991, (**e**) 2000, (**f**) 2010

to reach a maximum of 89 mm (44.5%). Groundwater recharge increased in 2000 and 2010 to reach a maximum of 134 mm (59.6%), as shown in Fig. 11.

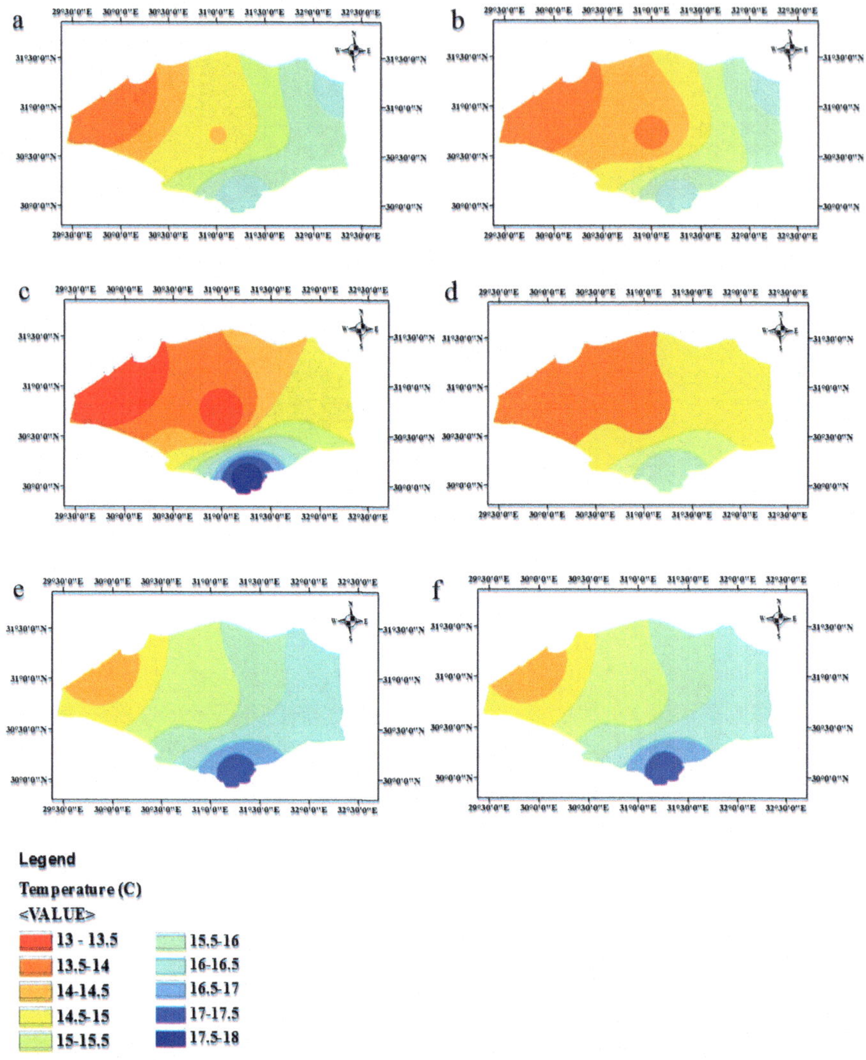

Fig. 8 Average winter temperature (°C) in the Nile Delta. (**a**) 1970, (**b**) 1980, (**c**) 1990, (**d**) 1991, (**e**) 2000, (**f**) 2010

7 Estimation of Top Width and Bank Levels of Irrigation Canal Network in the Nile Delta Region

Armanuos et al. [40] used Google Earth Pro software based on the available Landsat imagery of Nile Delta to estimate the ban level and top width of irrigation canal network in the Nile Delta region. The method is first applied to Damietta

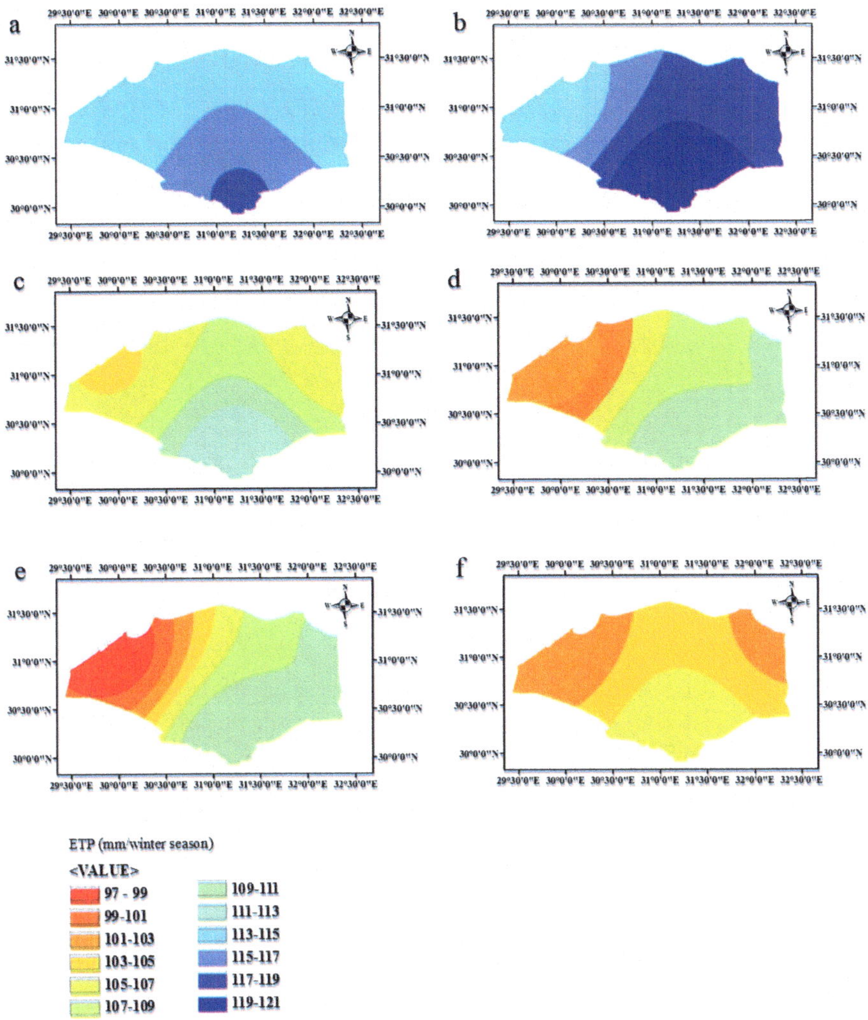

Fig. 9 Potential evapotranspiration in the Nile Delta mm/winter season. (**a**) 1970, (**b**) 1980, (**c**) 1990, (**d**) 1991, (**e**) 2000, (**f**) 2010

branch where the estimated values for bank level and top width for Damietta branch are compared with the measured values by the Hydraulics Research Institute (HRI) of Egypt [43]. The results confirmed that the difference between the measured and estimated values of bank level for Damietta branch ranged from 0.01 to 0.80 m [40]. The R^2 values of bank level and top width for Damietta branch showed a strong correlation between the estimated and measured values. The method was applied to all irrigation canal network in the Nile Delta region. The application results of the proposed method are very useful for accurate groundwater simulation

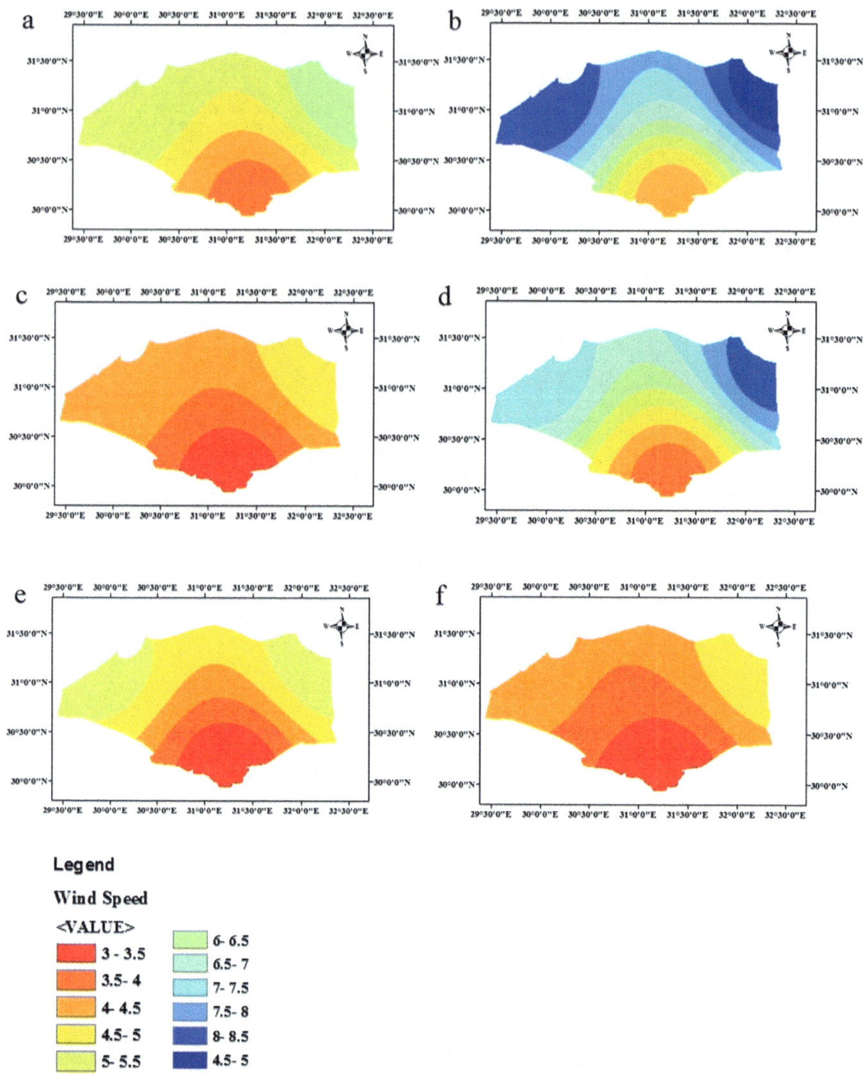

Fig. 10 Average winter wind speed (m/s) in the Nile Delta. (**a**) 1970, (**b**) 1980, (**c**) 1990, (**d**) 1991, (**e**) 2000, (**f**) 2010

of the Nile Delta aquifer and interaction between surface water and groundwater in the Nile Delta aquifer. Tables 2, 3, and 4 summarize the bank level, length, and top width of irrigation canals in central, eastern, and western parts of the Nile Delta region, respectively.

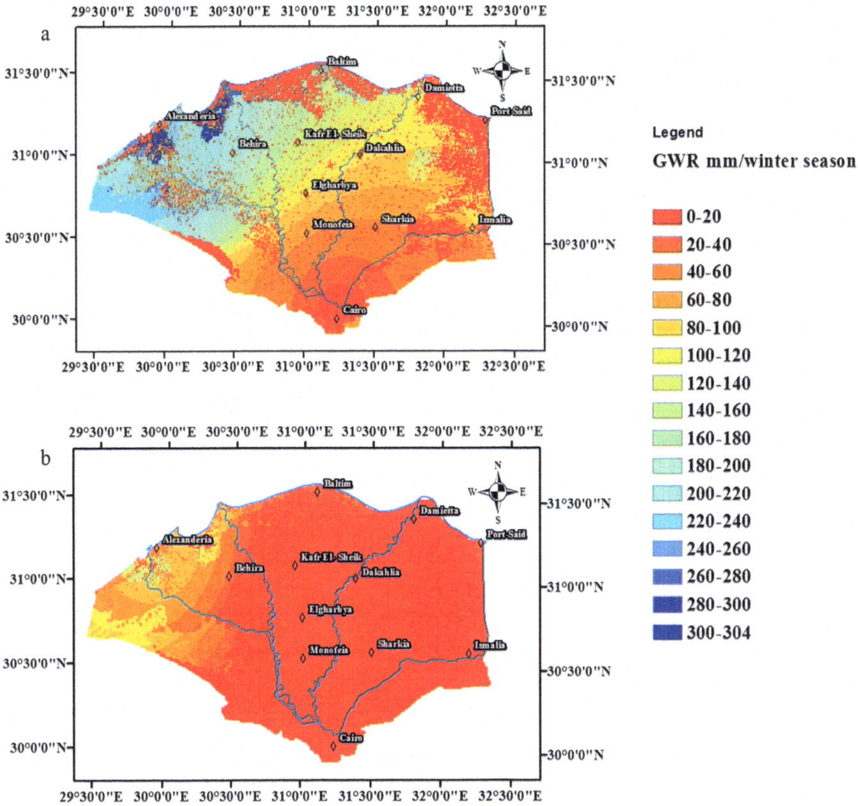

Fig. 11 Groundwater recharge from rainfall in the Nile Delta region for the years 1991 and 2010, Armanuos et al. [38]. (**a**) 1991, (**b**) 2010

8 Building Integrated Groundwater Model of the Nile Delta Aquifer

Nile Delta aquifer is considered to be among the largest underground freshwater reservoirs all over the world [9]. The Nile Delta region is about 25,000 km^2, located between latitudes 30°05 and 31°30/N and longitudes 29°50 and 32°15/E in the northern area of Egypt [44]. The Nile Delta is the bounded by the Mediterranean Sea in the north, by the Nile River in the south, by Suez and Ismailia canals in the east, and by El Nobaria canal in the west [44]. In the Nile Delta region, the Nile River is divided into nine large canals. Its main branches are Rosetta and Damietta branches. Rosetta branch is located in the west with a length of about 239 km, whereas Damietta branch is located in the east and about 245 km in length [4]. The groundwater system in the Nile Delta aquifer is a complex system. The Nile Delta Quaternary aquifer is a semi-confined aquifer; a clay layer covers its top [3, 45]. The aquifer thickness ranges from 200 m in the south and increases to the

Table 2 Bank level, length, and top width of irrigation canals (central Nile Delta)

Name of canals	Bank level + MSL	Length (km)	Top width (m)
Central Nile Delta			
Abou Ghalib canal	2.4, 1.5	3.8	9.9
Abou Halal canal	3.8, 2.2	8.6	9.5
Abou Ismail canal	1.7, 0.0	6.0	8.8
Aisha canal	10.5, 6.6	14.0	7.0
Amr Beek canal	10, 7.3	24.0	11.8
Aremion canal	6.6, 5.2	8.3	13.8
Atf canal	14.2, 12.0	32.2	34.1
Bagoria canal	16.6, 2.0	173	36
Bahe Semela	11.0, 8.0	14.0	7.0
Bahr El Samar El Ayman	4.0, 3.2	15.0	9.8
Bahr Assei canal	10.4, 6.8	21	17
Bahr Basondia	5.7, 4.0	36.0	26
Bahr Bemam	12, 7.5	20.0	12.9
Bahr El Bashma	6.6, 3.8	14.0	21.4
Bahr El Batahla canal	4.0, 0.0	25.0	11.0
Bahr El Bonwat	4.0, 1.2	25.0	16.6
Bahr El Malah	7.4, 4.0	33.0	11.5
Bahr Hafeir Shab El Dein	8.3, 3.0	29.8	36.0
Bahr Loktei	7.0, 5.4	15.0	12.1
Bahr Nashret	8.5, 6.0	20.0	20.4
Bahr Nemra	6.8, 4.9	20.0	20.4
Bahr Seief	10.6, 7.5	20.0	12.8
Bahr Shebin	7.2, 7.0	16.8	49
Bahr Shershaba	10.0, 8.0	9.0	9.0
Bahr Tiro	7.1, 2.0	65.0	43
Balkena canal	7.4, 4.5	12.0	7.7
Baseis canal	7.8, 1.0	19.0	15.4
Bashbesh canal	5.6, 4.5	5.3	8.03
Beshara canal	3.6, 2.4	5.3	14.8
Canal Mareis el Gamal	6.0, 2.5	12.5	12.5
Damat canal	8.0, 7.0	7.50	10.3
Deiama canal	3.8, 2.6	6.0	9.5
Dia el kom canal	13.8, 11.5	6.5	6.8
Draz canal	3.0, 2.0	3.50	5.2
Ebraheem El Gendi canal	12.3, 11.2	11.0	5.1
Ebshan canal	3.6, 2.5	13.0	5.5
Ebto canal	5.0, 1.5	16.0	8.3
El Aalaf canal	4.4, 4.0	9.0	7.7
El Amreia canal	7.4, 5.3	10.0	9.8
El Bahr El Sehediei	5.5, 3.0	22.8	24.6
El Baradehi canal	8.4, 6.6	6.0	3.3
El Bashawat canal	5.0, 3.0	10.0	11.0
El Betanonia canal	3.5, 1.7	26.0	9.0

(continued)

Table 2 (continued)

Name of canals	Bank level + MSL	Length (km)	Top width (m)
El Daramallei canal	1.5, 1.0	10.0	13.3
El Darfeil canal	2.5, 1.0	10.0	12.8
El Doon canal	2.1, 1.2	18.5	16.4
El Eslah canal	5.0, 2.0	15.0	21.1
El Farameli canal	3.0, 1.3	6.0	11.3
El Fazrei canal	13.0, 10.0	12.0	7.53
El Gafria canal	11, 7.5	20.0	12.8
El Galban canal	5.5, 1.0	12.0	10.0
El Gamoos canal	14.5, 12.2	10.0	7.0
El Ganeim canal	10, 9.0	5.8	4.3
El Halwanei canal	5.0, 1.0	19.50	12.4
El Henawei canal	4.8, 3.0	12.0	10.0
El Hessa canal	4.5, 3.0	7.3	6.30
El Hosha canal	4.6, 2.8	8.2	8.4
El Kadei canal	10.6, 7.8	6.0	5.0
El Kaleig canal	5.0, 1.0	17.5	15.5
El Kaleig El Abassei	8.0, 5.2	7.8	19.3
El Karadwa canal	3.5, 2.5	9.0	9.7
El Khaial canal	5.0, 0.0	11	13.3
El Khawaled canal	4.0, 1.5	9.0	3.0
El Sawakei canal	8.2, 3.5	10.00	7.80
El Serbawia canal	15.5, 10.2	31.6	16.0
El Sharkia canal	3.5, 3.0	5.0	7.50
El Sharkwa canal	3.1, 2.3	12.50	15.8
El Shawamei canal	4.2, 2.3	15.0	7.50
El Shiek Ahmed canal	2.5, 1.0	6.50	9.10
El Taiefa canal	6.3, 2.3	12.0	18.14
El Tebia canal	7.0, 6.0	11.0	11.7
El Teien canal	3.0, 2.0	7.00	11.9
El Wastanei canal	3.4, 2.0	11.00	10.50
El Zawia canal	9.1, 2.5	27.0	43
Fawwa canal	3.0, −1.0	7.00	8.10
Ganabiat Shobra Ell Namla	10.5, 8.0	12.00	9.00
Kabreit canal	3.0, 2.0	11.00	15.3
Kafr Seliman canal	3.4, 1.5	7.30	8.20
Kafr Tanbadei canal	14.8, 10.5	16.00	8.00
Ketah El Zalat canal	3.4, 1.5	6.50	9.90
Ketai canal	11, 8.0	14.00	8.30
Kom El Farahein canal	4.8, 2.8	5.00	10.36
Kotor canal	7.5, 6.0	8.00	6.90
Mansour canal	5.4, 2.5	8.00	4.50
Meit El Serag canal	6.0, 5.0	12.00	13.80
Mit Yazeid canal	9.5, 1.5	71.50	46.6
Moftah canal	3.0, 2.0	10.00	8.80

(continued)

Table 2 (continued)

Name of canals	Bank level + MSL	Length (km)	Top width (m)
Moharam canal	2.0, 1.3	3.00	6.70
Mounof canal	16.0, 11.0	36.0	29.52
Nasha canal	8.0, 5.0	11.8	13.80
Nesheel canal	6.0, 5.0	13.00	8.40
New senhour canal	6.0, 3.7	7.00	7.30
Om Ameik canal	11.0, 9.5	13.50	18.16
Om Kaleifa canal	12.5, 10.0	15.00	7.50
Om Sherief canal	15.8, 13.7	10.00	6.30
Qwesena canal	12.0, 10.0	13.00	10.70
Raes el Kalieg canal	4.0, 2.5	9.00	11.80
Sahel El Garaber canal	10.2, 8.0	13.50	5.10
Salmon canal	11, 5.5	16.00	12.50
Samdar el baharei canal	15.0, 9.5	21.00	6.90
Samed canal	4.6, 3.3	8.00	9.0
Samool canal	9.4, 5.2	14.00	6.90
Sandioon canal	2.3, 1.0	2.60	10.80
Sandseic canal	7.0, 4.0	10.00	5.80
Seleem canal	12.2, 12	5.00	3.80
Senara canal	5.5, 5.0	14.00	9.10
Shakeib canal	4.3, 3.0	6.00	4.30
Shalma canal	7.0, 2.5	9.00	17.0
Shemei canal	5.2, 4.0	8.00	8.00
Sheshna canal	9.0, 7.0	13.00	6.60
Shobrareic canal	8.9, 7.8	4.00	4.00
Tanta canal	16.0, 5.0	96.50	21.0
Yuosseif afandei canal	5.0, 1.0	15.00	18.79
Zeil El Kaseed canal	5.5, 1.0	22.00	10.20
El Kodrat canal	12.5, 12.0	13.0	6.9
El Konaisa canal	5.0, 3.5	9.30	11.8
El Kordei canal	5.9, 5.0	7.00	10.1
El Manaifa canal	5.8, 2.8	15.0	19.3
El Marbat canal	2.4, 0.0	15.0	9.0
El Mestalia canal	3.6, 2.0	8.00	9.5
El Monshaa canal	2.0, 1.5	10.00	11.5
El Mostafa canal	6.8, 4.8	15.00	11.8
El Nafqa canal	3.2, 2.2	7.9	19.7
El Nagar canal	18.0, 13.1	32.0	12.5
El Nasher canal	9.5, 7.4	3.8	7.80
El Nehnaeia canal	5.0, 1.0	10.0	14.4
El Neshrein canal	5.0, 4.2	10.0	9.5
El Nile canal	5.4, 2.0	25.0	21.6
El Ramla canal	6.5, 3.6	12.0	8.1
El Salab canal	3.2, 1.0	13.0	18.0
El Salmia canal	4.5, 1.5	3.0	9.7
El Samda canal	1.5, −1.5	5.0	9.0

Table 3 Bank level, length, and top width of irrigation canals (eastern Nile Delta)

Name of canals	Bank level + MSL	Length (km)	Top width (m)
Bahr Mois canal	16.0, 1.6	133.0	25.0
Bahr Sageir	8.5, 1.3	65.0	35.0
Bahr Tanah canal	9.6, 3.5	60.0	14.50
Boaheia canal	12.8, 6.0	50.0	21.50
El Raiah El Twafiq	20.5, 2.0	160.0	49.0
Ismalia canal	19.0, 7.0	130.0	67.0
Salehia canal	13.4, 7.5	55.0	18.0
Sayedia canal	12, 6.5	60.0	18.0
Sharkawia canal	19.0, 10.0	50.0	43.0
Wadei canal	10.4, 1.5	80.0	65.10

Table 4 Bank level, length, and top width of irrigation canals (western Nile Delta)

Name of canals	Bank level + MSL	Length (km)	Top width (m)
Abou Deiab El Aahla canal	9.6, 4.2	32.0	18.50
Bloktor canal	3.4, −1.5	10.0	11.00
Canal project naser	3.4, −2.0	11.50	19.00
East Kahnat canal	8.6, 5.0	44.0	38.0
El Farwania canal	5.8, −1.5	25.0	9.00
El Kanobia canal	3.5, 2.0	15.0	16.50
El Nobaria canal	11, 3.0	110.0	79.0
El Qanawia canal	3.0, −2.0	17.00	7.50
El Rahiah El Naserie	19.0, 3.0	83.50	52.3
El Raiah El Beheriei	19.0, 12.5	80.00	87.0
Fareash canal	7.5, 1.0	41.50	16.0
Hagar canal	6.5, −1.5	74.30	28.0
Hamed meneis canal	4.5, −2.0	14.00	9.30
Mahmoudia canal	6.5, 1.3	55.00	41.0
Rosetta branch	16. 6, 1.5	235.0	180
Sahel Balmoon canal	9.0, 3.0	74.00	20.0
Sahel Morcous canal	10.0, 5.0	83.50	35.0
West Roshedila canal	7.4, 4.0	40.00	22.10

north to reach 1,000 m [21]. The upper clay layer acts as an aquitard in the south and ranges from 5 to 25 m in thickness, while in the north, its thickness increases to reach more than 50 m acting as an aquiclude [46]. In 2000 and 2010, groundwater recharge from rainfall in the Nile Delta aquifer ranged between 0.0 and 134 mm in the winter at Cairo and Alexandria, respectively [38].The average rate of percolation to the Nile Delta Quaternary aquifer is 0.8 mm/day DRI [30]. The percolation rate depends on the soil type, irrigation, and drainage systems [31]. It ranges from 0.25 to 0.8 mm/day in the central and southern part of the delta. The maximum rates are from 1.0 to 1.5 mm/day found in the western desert areas where furrow irrigation is used. The minimum rates are from 0.1 to 0.5 mm/day observed in drip and sprinkler system regions [32]. The names and directions of the irrigation canal networks in the eastern and western parts of the Nile Delta were identified based on Roest [47], and those in the central part of the Nile Delta were identified based on maps from the National Water Research Center (NWRC), MWRI [35], and NWRP [48].

The Nile Delta aquifer has been extensively studied by many geological, hydrochemical, and hydrological researchers for different purposes. These researches identified the characteristics of groundwater aquifer in the Nile Delta [24]. Table 5 shows the vertical and horizontal hydraulic conductivity values of the top clay layer estimated by different studies [57]. Table 6 summarizes the estimations of hydraulic parameters of Quaternary aquifer by various studies. Different values of the hydraulic conductivity and storage coefficient of the Nile Delta aquifer were reported by Farid [3] and Armanuos et al. [57]. The hydraulic conductivity of the Nile Delta aquifer increases toward the south and west directions. An effective porosity that equals to 0.3 is considered to represent the aquifer medium. The storage coefficient and transmissivity values were estimated by RIGW [26] to be 2.5×10^{-3} and 5,000 m/day, respectively. The longitudinal dispersivity and the transverse dispersivity values for the Nile Delta were estimated to be 100 m and 10 m, respectively, while the value of diffusion coefficient (D^*) equals to 10^{-4} m^2/day [5]. The depth from the ground surface to the groundwater head level in the Nile Delta aquifer increases from the north toward the south. It varies between 1 and 2 m in the northern area, increases to range between 3 and 4 m at the middle part, and reaches the maximum value of 5 m at southern parts [10, 22]. RIGW in Egypt

Table 5 The hydraulic properties of the top clay layer in the Nile Delta

Study	K_h (m/day)	K_v (m/day)
Farid [3]	0.2160	0.0025
Farid [49]	–	0.05
Anon [2]	–	0.0025
RIGW/IWACO [50]	0.2505	0.0484
Wolf [51]	0.1037	0.0011
Sherif et al. [5]	–	0.05
Warner et al. [31]	0.2160	0.0073
RIGW [21]	0.05–0.5	0.0025
Dahab [20]	–	0.01–10.0
Arlt [52]	0.3800	0.0046
Sherif [19]	–	0.0025
Sakr and Mabrouk [53]	10.0	0.15–16
El-Arabi [54]	0.1–0.25	0.01–0.025
Fadlelmawla and Dawoud [55]	–	0.05–0.1
Sherif et al. [11]	–	0.0005–0.005
Abdelaty et al. [12]	0.1–0.25	0.01–0.025
Nofal et al. [14]	0.05–0.5	0.0025
Morsy and El-Fakharany [56]	0.1–0.25	0.01–0.025

Table 6 The hydraulic parameters of the Quaternary aquifer

Study	K (m/day)	T (m^2/day)	S (storage coefficient)	Effective porosity (%)
RIGW [21]	35–100	15,000–75,000	10^{-4}–10^{-3}	–
UNDP [58]	55–103	20,000–103,000	10^{-4}–10^{-3}	–
Anon [2]	100	–	–	–
Farid [3]	112	72,000	2.53×10^{-3}	32–35
Zaghloul [59]	119	–	10^{-4}–10^{-3}	20–28
Shahien [60]	50	2,500–25,900	10^{-5}–10^{-4}	23–25
Laeven [61]	150	10,350–59,800	–	–
Shahin [62]	35	–	–	–
RIGW/IWACO [50]	35–75	25,000	–	–
Mabrook et al. [63]	72–108	–	0.72×10^{-4} -2.3×10^{-4}	21–30
Sollouma and Gomaa [64]	23–65	–	–	–
Shata and El Fayoumy [65]	86	71,800	1.1×10^{-3}	–
Bahr [66]	75		1.1×10^{-3}	18
Dahab [20]	50–240	3,000–15,000	9.0×10^{-4} -1×10^{-1}	11.5–19
RIGW/IWACO [6]				
Southern ND	50–100	5,000–25, 000	–	25–30
Northern ND	Less than 50	–	–	More than 30
Sherif [17]	70–100	–	–	12–19
Sherif and Al-Rashed [9]	100	–	10^{-4}–10^{-3}	30
Sherif [19]	100	–	10^{-4}–10^{-3}	30
El-Arabi [54]	5–100	–	5×10^{-4} -1×10^{-3}	20–60
El Tahlawi [67]	35–75	500–25,000	–	–
Sherif et al. [11]	36–240	2000–15,000	–	25–40
Morsy and El-Fakharany [56]	20–55	–	2.35×10^{-3}	25
Abdelaty et al. [12]	5–100	–	5×10^{-4} -1×10^{-3}	20–60
Mazi et al. [16]	100–120	–	–	–
Elshinnawy et al. [68]	23.0–180.0	–	–	–

carried the inventory of the abstraction wells from the Nile Delta region for the years 1992, 1995, 1997, 2002, and 2008. The total abstraction from the groundwater system in the Nile Delta aquifer at 2008 equals to 3.48×10^9 m^3/year [12].

A three-dimensional model was built for the Nile Delta aquifer by using MODFLOW software. The governing equations of the groundwater flow are derived from a mathematical combination between the water balance equation and Darcy's law [69]. The MODFLOW model describes groundwater flow in a nonhomogeneous isotropic medium according to the following equation [70, 71]:

$$\frac{\mathrm{d}}{\mathrm{d}x}\left(K_{xx}\frac{\mathrm{d}h}{\mathrm{d}x}\right)+\frac{\mathrm{d}}{\mathrm{d}y}\left(K_{yy}\frac{\mathrm{d}h}{\mathrm{d}y}\right)+\frac{\mathrm{d}}{\mathrm{d}z}\left(K_{zz}\frac{\mathrm{d}h}{\mathrm{d}z}\right)-W=S_s\frac{\mathrm{d}h}{\mathrm{d}t} \tag{1}$$

where K_{xx}, K_{yy}, and K_{zz} are values of hydraulic conductivity in x, y, and z directions (LT^{-1}), h is the piezometric head (L), W is a volumetric flux per unit volume of aquifer representing sources and/or sinks of water (T^{-1}), S_s is the specific storage of the porous material (L^{-1}), and t is the time (T).

SEAWAT program [72] is used to simulate saltwater intrusion for the barrier wall and freshwater injection experiments. SEAWAT program combines both MODFLOW and MT3DMS. The program is used to solve the coupled groundwater flow and solute transport equations. SEAWAT performs the variable density groundwater flow calculations. SEAWAT code was widely tested by solving various benchmark problems including Simpson and Clement [73, 74] problems. In the variable-density flow (VDF) process of SEAWAT program, the well-established MODFLOW is used as familiar tool to solve the equations of variable density groundwater flow system by the meaning of finite-difference method (FDM) [75]. In the current study, the equations of solute transport which related to the Integrated MT3DMS Transport (IMT) process were solved by using MT3DMS, and the following variable-density groundwater flow equation was solved by VDF process in SEAWAT program [75]:

$$\begin{aligned}&\nabla\left[\rho\times\frac{\mu_o}{\mu}\times k_o\left(\nabla\times h_o+\frac{\rho-\rho_f}{\rho_f}\times\nabla Z\right)\right]\\&=\rho\times S_{S,O}\frac{\partial h_o}{\partial t}+\theta\times\left(\frac{\partial\rho}{\partial C}\right)\left(\frac{\partial C}{\partial t}\right)-\rho_s\times q_s'\end{aligned} \tag{2}$$

The IMT process in SEAWAT program solved the following solute transport equation [75]:

$$\left(1+\frac{\rho_b\times k_d^k}{\theta}\right)\frac{\partial(\theta\times C)}{\partial t}=\nabla\left(\theta D\times\nabla C^K\right)-\nabla\left(q\times C^K\right)-\left(q_s'\times C_S^K\right) \tag{3}$$

where ρ_f is the density of fluid [ML^{-3}]; ρ_s is the saline groundwater density [ML^{-3}]; μ_o is the fresh groundwater dynamic viscosity [$ML^{-1}T^{-1}$]; μ is the saline groundwater dynamic viscosity [LT^{-1}]; k is the hydraulic conductivity [L]; h is the hydraulic head [L]; S is the specific storage [L^{-1}]; T is the time [T]; θ is the porosity; C is the concentration of salt, a source or a sink of fluid with density [ML^{-3}]; q_s' is a source or a sink of fluid with density [ML^{-3}]; ρ_b is the bulk density [ML^{-3}]; k_d^k is the distribution coefficient of the species [L^3M^{-1}]; D is the hydrodynamic dispersion coefficient [L^2T^{-1}]; q is the specific discharge [LT^{-1}]; and C_S^K is the source or sink concentration of species k [ML^{-3}].

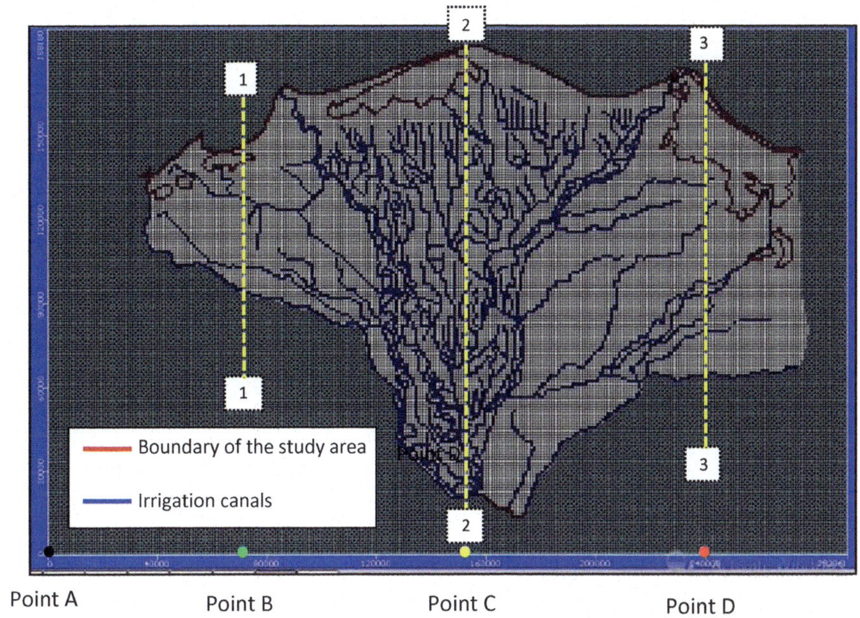

Fig. 12 Model geometry and discretization

The 3D model of the Nile Delta aquifer was built by grid system consisting of 292 columns and 190 rows with cell dimension of 1.0 × 1.0 km with variation in depth from 1,000 m at the shoreline of the Mediterranean Sea in the north to 200 m in the south. Figure 12 shows the model boundary condition and the canal distribution of the Nile Delta aquifer model [57]. The built model is divided into 11 layers; the upper clay layer is represented by layer number one; the following layers from two to eleven represent the quaternary layer. The thickness of the upper clay layer is 25 m in the south and increases towards the north to reach about 50 m. The thickness of the quaternary aquifer is 200 m near Cairo in the southern part of the Nile Delta and increases toward the Mediterranean Sea to reach about 1,000 m. The groundwater flow in the Nile Delta aquifer is from the south to the north toward the Mediterranean Sea with an average value of hydraulic gradient about 11.0 cm/km, and the gradient is mild in the northern area about 7.0 cm/km.

Figure 13a–c shows the three vertical cross sections 1–1, 2–2, and 3–3 in the Nile Delta model, respectively. Based on the literature, the south boundary condition of the Nile Delta model defined a constant head that equals to 16.96 amsl [12, 22]. The upper boundary condition in the north defined a zero value along the shoreline of the Mediterranean Sea. The east boundary is left as free where the Suez Canal. In the southeast, the model is bounded by Ismailia canal where the water level starts by 16.17 amsl in the south and ends by 7.01 amsl in the east. In the southwest, the model is bounded by El Rayah, El Behery, and El Nubaria canals where the water level in the field starts by 16.00 m amsl in the south and ends by 0.50 m amsl in the

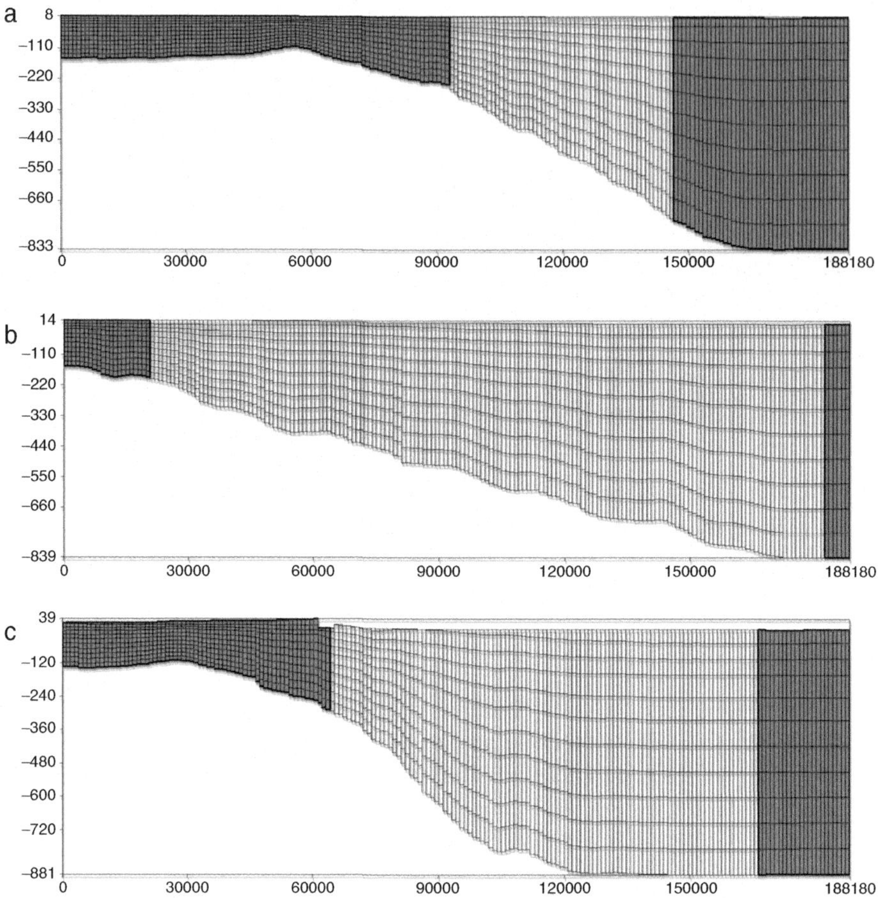

Fig. 13 Vertical cross sections in the Nile Delta aquifer model: 1–1, 2–2 and 3–3. (**a**) cross section 1–1, (**b**) cross section 2–2, (**c**) cross section 3–3

north. The four main lakes of the Nile Delta are Maryut, Idku, Burullus, and Manzala and were assigned to the model as a constant head, as they are directly connected to the Mediterranean Sea. The river package in the MODFLOW was used to assign the hydraulic properties of Damietta and Rosetta branches and also El Rayah and the main canals of the model. The hydraulic properties of the river package in MODFLOW include river stage level, bottom level, riverbed vertical hydraulic conductivity, and vertical hydraulic conductance. The drain package in the MODFLOW was used to simulate the main drains of the Nile Delta. Morsy [22] reported the water level of El Rayah, main canals, and drains that were used as an input to the model. The bank level and top width for irrigation canals network which were estimated by Armanuos et al. [40] were used in building the model.

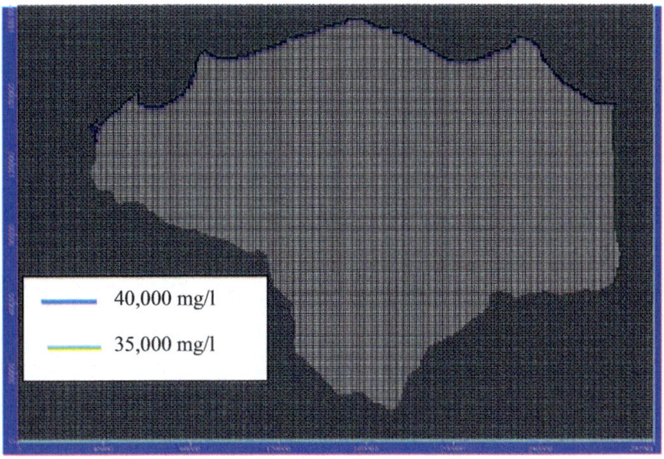

Fig. 14 Concentration boundaries in the Nile Delta model

The groundwater recharge from excess irrigation water depends on the soil type, irrigation, and drainage practices assigned to the model based on reported values by Morsy [22]. The concentration boundary conditions are defined to the model domain as 40,000 mg/L in the north domain (the shoreline of the Mediterranean Sea) and 35,000 mg/L to the east boundary domain (along with the coastal line of the Suez Canal), as shown in Fig. 14. The initial groundwater concentration is set to be zero mg/L. The annual extraction rate per governorate in the Nile Delta region in the year 2008 was assigned to the model, where the total volume rate of extraction from the total studied area of the Nile Delta was 2.78×10^8 m^3/year in the year 2008 [22]. The freeboard between the bank level and water level in irrigation canals according to the Egyptian code of water resources and irrigation works equals to 0.5, 0.75, and 1.0 m for distributed, branch, and main canals consequently [76].

In the calibration process, the model was run many times to minimize the difference between the measured hydraulic head by RIGW in 2008 and the simulated head by the model. A 60 distributed observation wells have been selected in the study area for the calibration process until the maximum difference between the measured and simulated hydraulic head level became 0.60 m. For model validation, the observed hydraulic heads were compared to the simulated ones for other 60 records. The RMSE between the observed and simulated groundwater head was constant for different values of horizontal hydraulic conductivity of the first clay layer. The difference in value between the vertical and horizontal hydraulic conductivity equals to 10%. The calibrated values of the vertical hydraulic conductivity of the first layer vary between 0.01 and 0.025 m/day. The values of the calibrated horizontal hydraulic conductivity of the second layer vary from 50 to 240 m/day. Figure 15a, b presents the observed and simulated groundwater levels. The comparison between the observed and simulated groundwater levels for 60 records from different observation wells by RIGW in 2008 [22] was shown in

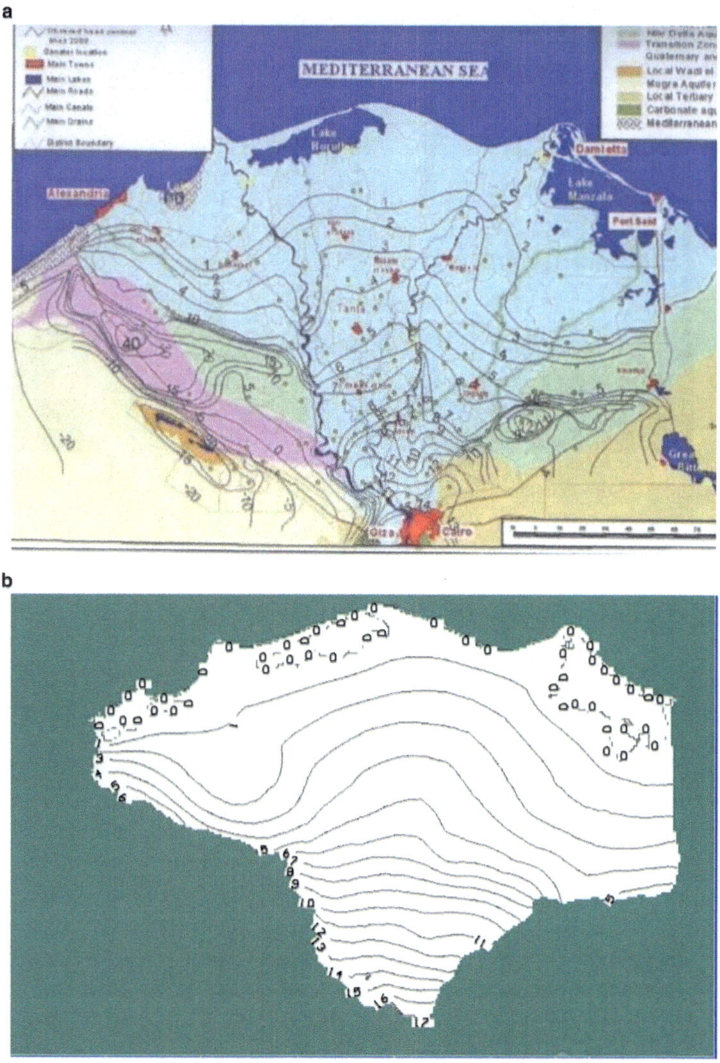

Fig. 15 Groundwater head in the Nile Delta aquifer 2008, (**a**) Observed groundwater level, Morsy [22] and (**b**) Simulated groundwater level

Fig. 16a. The correlation factor R^2 and RMSE between the observed and simulated groundwater levels were 0.9844 and 0.60 m, respectively [57]. Figure 16b shows the comparison between the observed and simulated groundwater levels for other 60 well records were chosen for model validation. The correlation coefficient R^2 between the simulated and observed head equals to 0.981, and the RMSE equals to 0.60 m [57].

SEAWAT program based on variable density modeling approach was used to recognize the dispersion zone between the freshwater and saltwater in the Nile

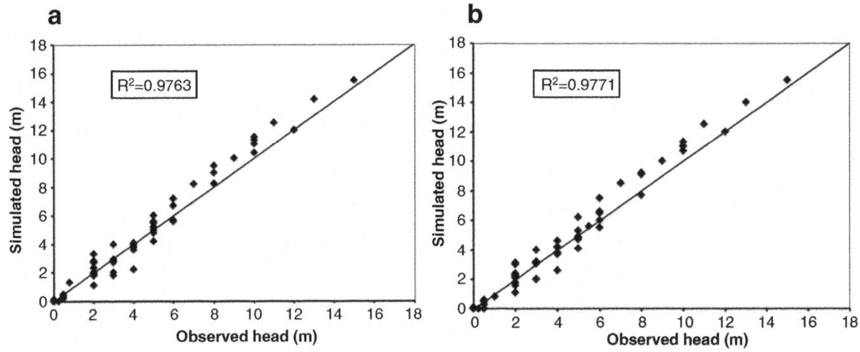

Fig. 16 Relationships between observed and simulated groundwater levels for (**a**) Calibration and (**b**) Validation process

Delta aquifer. Total dissolved solids distribution in the Nile Delta aquifer was identified in horizontal views and vertical cross sections. Three vertical cross sections were identified with depth range varying from the south portion (landside) to the northern part (shoreline) [77]. As explained previously and shown in Fig. 13a–c, cross sections (1–1), (3–3), and (6–6) are located in the western part, middle part, and eastern part of the Nile Delta, respectively. Consecutively, these cross sections have depths of 240, 200, and 200 m at the south portion (landside) and of 900 m in the northern part (shoreline). As illustrated in Fig. 13a–c, the lengths of cross sections (1–1), (2–2), and (3–3) are 70 km, 150 km, and 120 km, respectively. The validation process of the current model considered the year 2008 as a base case.

9 Simulation of Saltwater Intrusion in the Nile Delta Aquifer

For the base case, at cross section 1, the concentration line 35,000 mg/L intruded inland at a distance of about 40.00 km from the shoreline, and the concentration line 1,000 mg/L intruded inland at a distance of 66.00 km from the shoreline where the width of the transition zone equals to 26.00 km [77] as shown in Fig. 17a. At cross section 2, the concentration line 35,000 mg/L intruded inland at a distance of about 56.00 km from the shoreline, and the concentration line 1,000 mg/L intruded inland at a distance of 105.00 km from the shoreline where the width of the transition zone equals to 49.00 km [77] as shown in Fig. 17b. At cross section 3, the concentration line 35,000 mg/L intruded inland at a distance of about 75.00 km from the shoreline, and the concentration line 1,000 mg/L intruded inland at a distance of 100.00 km from the shoreline. In cross section 3, the width of the transition zone between the concentration line 35,000 mg/L (front of the saltwater) and the concentration line 1,000 mg/L (front of the freshwater) equals to 25.00 km [77]

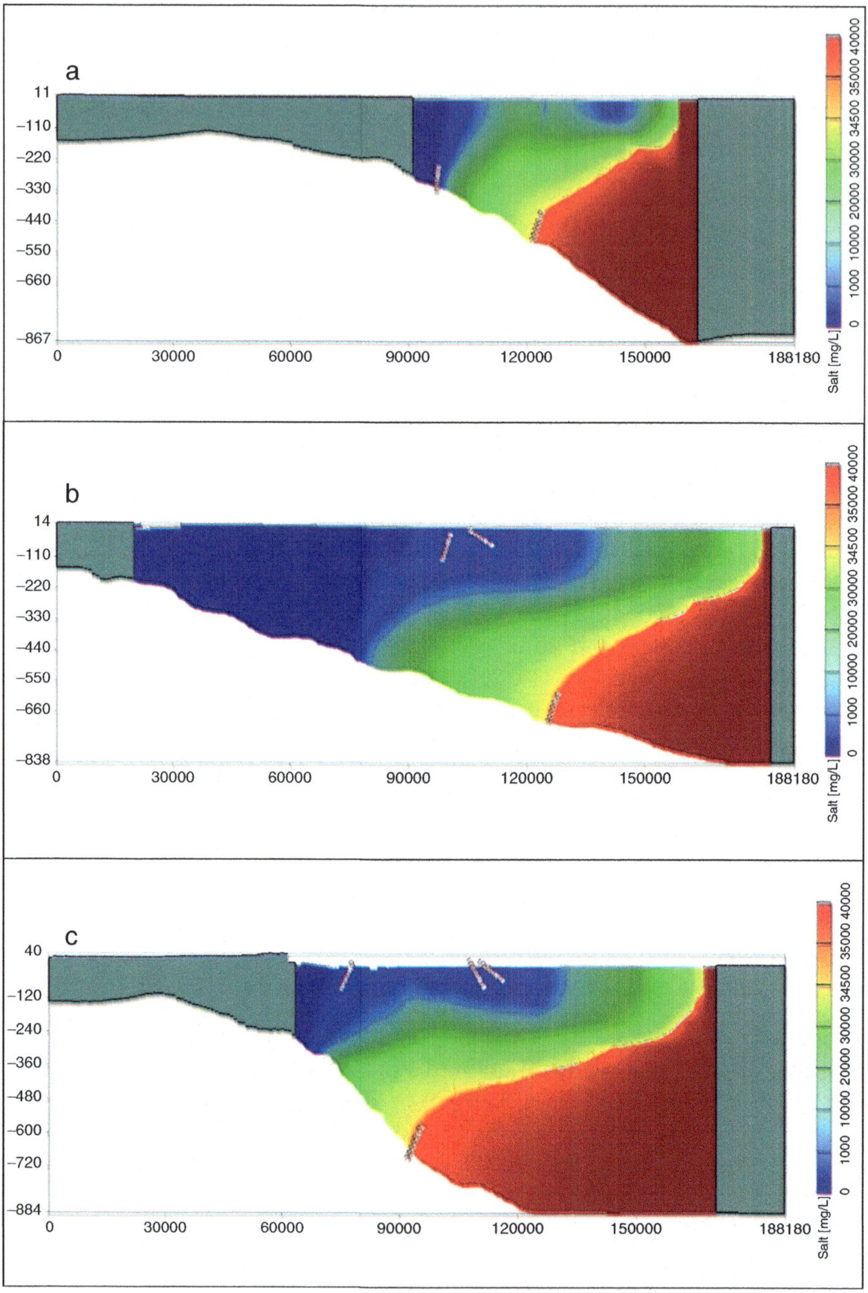

Fig. 17 Vertical distribution of salt concentration in the Nile Delta aquifer at cross sections 1, 2, and 3. (**a**) cross section 1 [West], (**b**) cross section 2 [Middle], (**c**) cross section 3 [East]

as shown in Fig. 17c. IPCC predicts that in the year 2100, the global warming will cause a rise in sea water level between 0.11 and 0.88 m [78, 79]. In scenario 1, the SEAWAT model was used to simulate the effect of sea level rise by 100 cm on the saltwater intrusion in the Nile Delta aquifer in the year 2100. The total annual groundwater abstraction from the Nile Delta aquifer has increased dramatically during the last 30 years. The Research Institute for Groundwater (RIGW) in Egypt reported its increase from 1.6×10^9 m^3/year in 1980 to 3.5×10^9 m^3/year in 2003 and reached 4.6×10^9 m^3/year in 2010 [24, 34]. It is expected that the annual abstraction rate will exceed 0.20×10^9 m^3 per year [24]. In scenario 2, the SEAWAT model was used to predict the saltwater intrusion in the Nile Delta aquifer under additional groundwater pumping where the groundwater table in the upper boundary declined by 100 cm in the year 2100. In scenario 3, the SEAWAT model was run to study the saltwater intrusion in the Nile Delta aquifer under the combination of the first and second scenarios.

9.1 Impact of Sea Level Rise on Saltwater Intrusion in the Nile Delta Aquifer

The current model was run for different climate change scenarios including sea level rise, the decline of groundwater head in the lower boundary due to additional groundwater pumping and the combining between sea level rise and groundwater head decline. In scenario 1, the model was used to study the effect of sea level rise on saltwater intrusion in the Nile Delta aquifer by 100 cm after 100 years. Increasing the sea level by 100 cm caused the concentration line 35,000 mg/L to intrude inland to distances of 44.65 km, 60.25 km, and 82.20 km in cross sections 1–1, 2–2 and 3–3 consequentially. Also, the concentration line 1,000 mg/L intruded inland into the aquifer to distances of 68.84 km, 108.2 km, and 106.5 km from the shoreline in cross sections 1–1, 2–2, and 3–3, respectively, as shown in Fig. 18. The concentration line 35,000 mg/L advanced inland into the aquifer by a distance of 8.2 km with sea level rise by 100 cm compared with the base case on 2008. The concentration line 1,000 mg/L advanced inland into the Nile Delta aquifer by a distance of 6.5 km compared with the base case with sea level rise by 100 cm [77].

9.2 Impact of Decreasing Head Due to Additional Pumping on Saltwater Intrusion in the Nile Delta Aquifer

In scenario 2, the model was used to predict the saltwater intrusion in the Nile Delta aquifer under additional groundwater pumping where the groundwater head in the south boundary declined by 100 cm after 100 years. Declining the groundwater head by 100 cm leads the concentration line 35,000 mg/L to intrude inland to

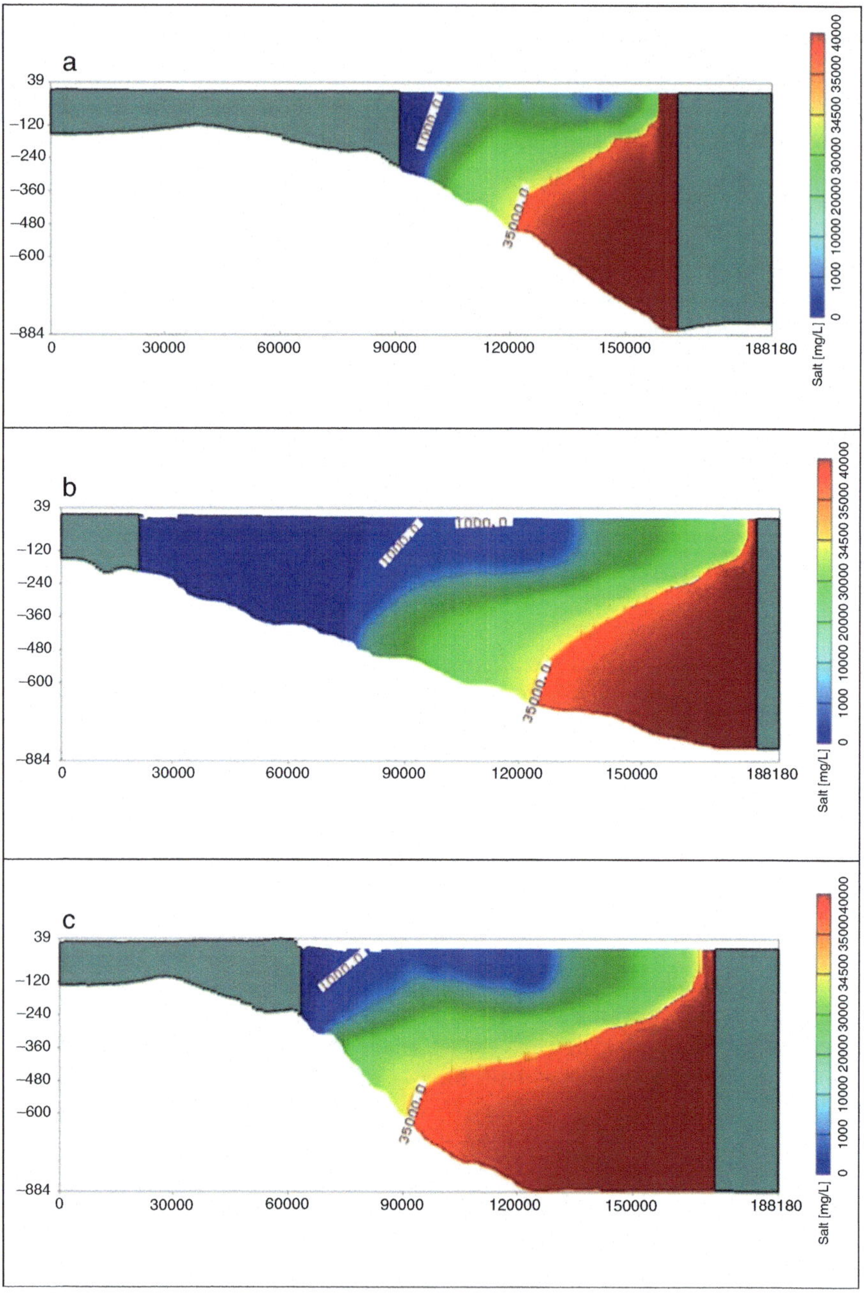

Fig. 18 Vertical distribution of salt concentration in the Nile Delta aquifer at SLR by 100 cm. (**a**) Sec 1–1: West, (**b**) Sec 2–2: Central, (**c**) Sec 3–3: East

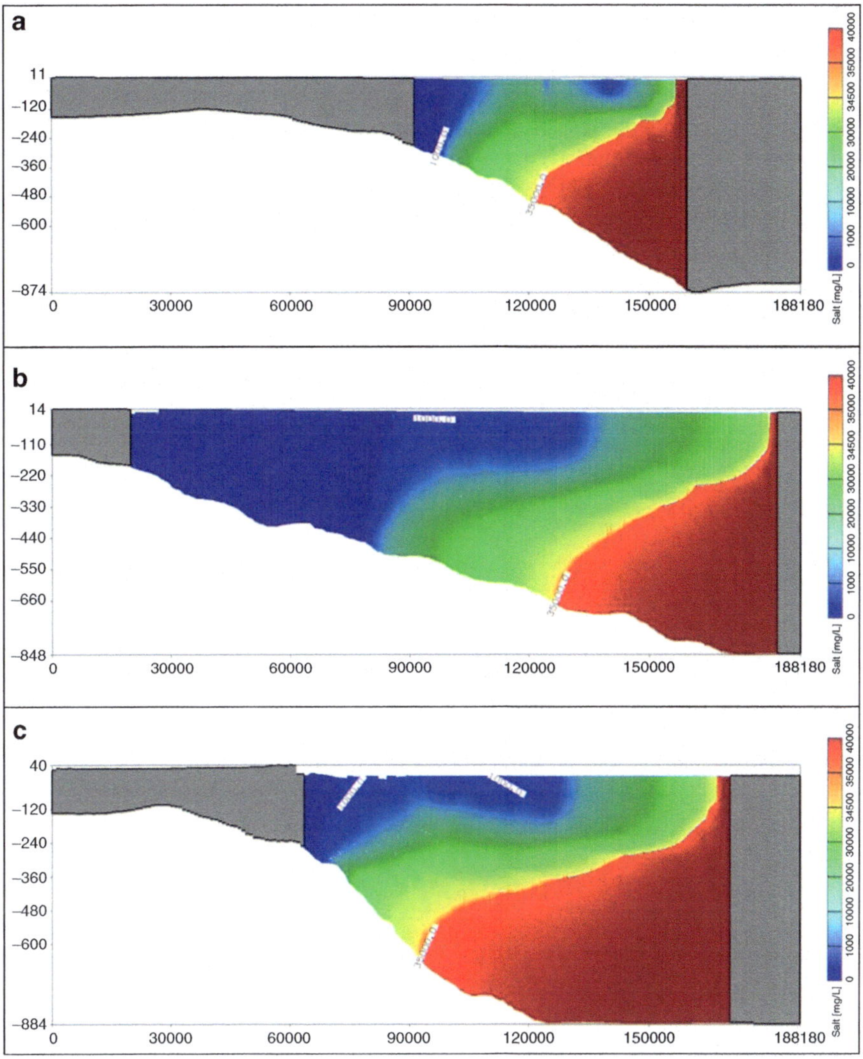

Fig. 19 Vertical distribution of salt concentration in the Nile Delta aquifer at decreasing head by 100 cm. (**a**) Sec 1–1: West, (**b**) Sec 2–2: Central, (**c**) Sec 3–3: East

distances of 44.27 km, 58.74 km, and 80.75 km in cross sections 1–1, 2–2, and 3–3 consequentially. Also, the concentration line 1,000 mg/L intruded inland the aquifer to distances of 68.02 km, 106.4 km, and 105.5 km in cross sections 1–1, 2–2, and 3–3, respectively, from the shoreline, as shown in Fig. 19. The concentration line 35,000 mg/L advanced inland into the aquifer by a distance of 6.75 km with decreasing the boundary head due to additional groundwater pumping by 100 cm, while the concentration line 1,000 mg/L advanced inland into the aquifer by a distance of 5.5 km compared with the base case [77].

9.3 Impact of Combining Between Sea Level Rise and Decreasing Groundwater Head

In scenario 3, the model was run to study the saltwater intrusion in the Nile Delta aquifer under the combination of the first and second scenarios, sea level rise by 100 cm and groundwater head decline by 100 cm after 100 cm. Increasing sea level by 100 cm and declining the groundwater head by 100 cm caused the concentration line 35,000 mg/L to intrude inland to distances of 44.64 km, 60.50 km, and 84.20 km in cross sections 1–1, 2–2, and 3–3 consequentially. Also, the concentration line 1,000 mg/L intruded inland the aquifer to distances of 69.00 km, 109.8 km, and 107.0 km in cross sections 1–1, 2–2, and 3–3, respectively, from the shoreline, as shown in Fig. 20. The concentration line 35,000 mg/L advanced inland into the aquifer by a distance of 10.20 km with combining sea level rise by 100 cm with decreasing boundary head due to additional groundwater pumping by 100 cm, while the concentration line 1,000 mg/L intruded inland into the aquifer by a distance of 7.0 km compared with the base case [77].

Including the actual irrigation canal networks in the current model of the Nile Delta aquifer using SEAWAT code causes the saltwater to intrude inland into the aquifer to distances quite less than the results got by Sherif and Al-Rashed [9], Abdelaty et al. [12], and Abd-Elhamid et al. [13]. The comparison is considered at the same cross sections due to different climate change scenarios including sea level rise and decreasing the groundwater head at the bottom boundary. As all previous studies had ignored the actual irrigation canal networks in the simulation of saltwater intrusion of the Nile Delta aquifer using different codes 2D-FED, 2D-FEST, and SEAWAT, the corresponding groundwater recharge could be less than the actual one, which is included in the current model [77].

10 Management of Pumping Discharges to Control Saltwater Intrusion in the Nile Delta Aquifer

Table 7 describes the studied scenarios to select the best pumping scenarios to control saltwater intrusion in the Nile Delta aquifer. The first scenario represents the reduction of pumping from all areas of the Nile Delta region. The second, third, and fourth scenarios represent the reduction of pumping discharges from western, central, and eastern parts only in the Nile Delta, respectively. The fifth, sixth, and seventh introduce the reduction of pumping discharges from western, middle, and eastern parts of the Nile Delta only and increase in the other parts [77]; see Table 9.

Seven pumping scenarios were introduced to select the best scenario to control the saltwater intrusion in the Nile Delta aquifer. Table 8 represents the results of saltwater intrusion lengths for scenario 1. Scenario 1 represents decreasing the pumping discharges in all areas of the Nile Delta by 10, 20, 30, 40, and 50%. Reduction of the pumping discharges from 10 to 50% leads to retreat the saltwater

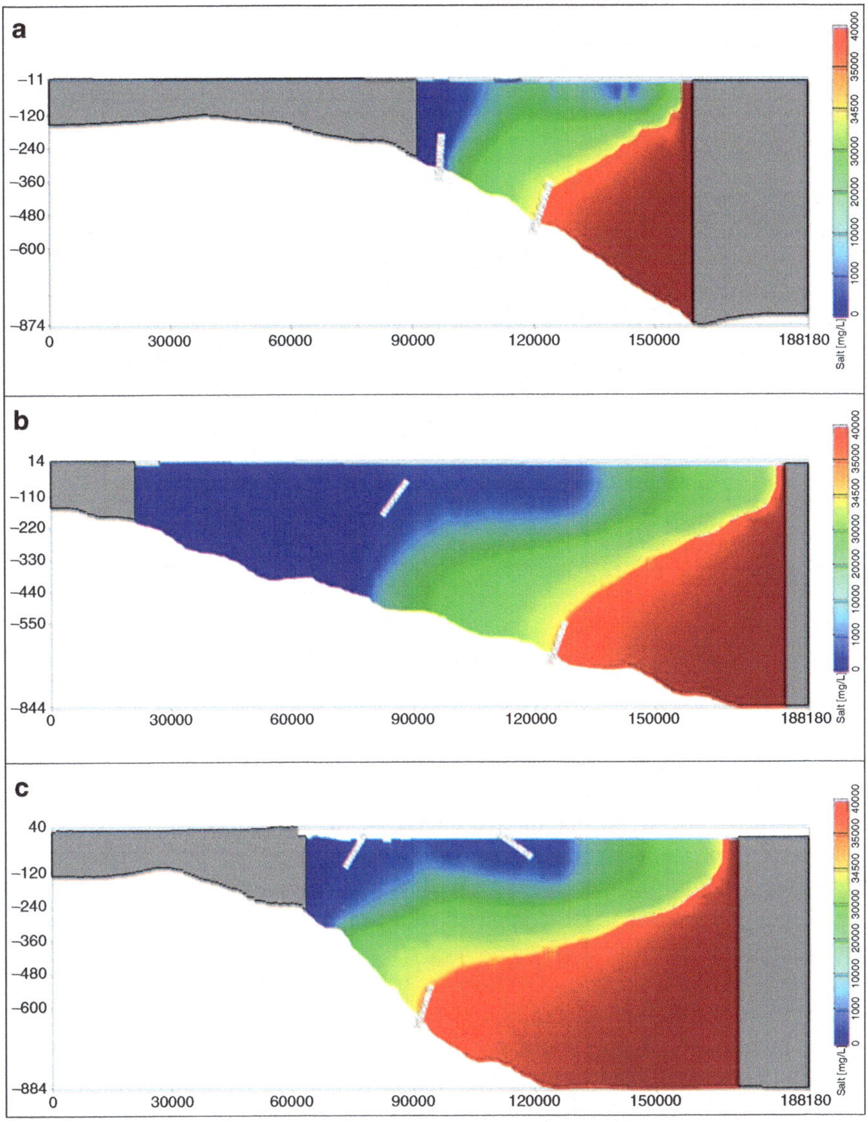

Fig. 20 Vertical distribution of salt concentration in the Nile Delta aquifer at SLR by 100 cm and decreasing head by 100 cm. (**a**) Sec 1–1: West, (**b**) Sec 2–2: Central, (**c**) Sec 3–3: East

lines from 32.2 to 31.33 km measured from the shoreline, while the freshwater line retreated from 56.15 to 55.29 km in the western part of the Nile Delta. In the middle part of the Nile Delta, the saltwater and freshwater lines are expected to retreat to 53.01 and 93.01 km if the pumping discharges reduced by 50%, respectively. In the eastern part, the saltwater and freshwater lines retreated from 72.57 and 99.0 km to

Table 7 Scenarios of decreasing pumping discharges to control saltwater intrusion in the Nile Delta aquifer

Scenario	Definition
S-1	Decreasing the pumping discharges in all areas of the Nile Delta by 10, 20, 30, 40, and 50%
S-2	Decreasing the pumping discharges in the western area of the Nile Delta by 10, 20, 30, 40, and 50%
S-3	Decreasing the pumping discharges in the middle area of the Nile Delta by 10, 20, 30, 40, and 50%
S-4	Decreasing the pumping discharges in the eastern area of the Nile Delta by 10, 20, 30, 40, and 50%
S-5	Decreasing the pumping discharges in the western area of the Nile Delta by 10, 20, 30, 40, and 50% and increasing the pumping discharges in middle and eastern area by 5, 10, 15, 20, and 25%
S-6	Decreasing the pumping discharges in the middle area of the Nile Delta by 10, 20, 30, 40, and 50% and increasing the pumping discharges in the western and eastern area by 5, 10, 15, 20, and 25%
S-7	Decreasing the pumping discharges in the eastern area of the Nile Delta by 10, 20, 30, 40, and 50% and increasing the pumping discharges in the western and middle area by 5, 10, 15, 20, and 25%

54.05 and 94.12 km with reduction of the pumping discharges to 10 and 50%, respectively.

Table 9 shows the results of saltwater intrusion lengths in western, middle, and eastern parts of the Nile Delta in case of scenario 2. Scenario 2 represents decreasing the pumping discharges in the western area of the Nile Delta by 10, 20, 30, 40, and 50%. Reduction of the pumping discharges in the western part of the Nile Delta only has a significant impact observed only in the western area. In the western area, the saltwater and freshwater lines retreated to 34.2 km and 57.00 km with decreasing the pumping rates by 50% compared with 44.51 and 68.00 km consequentially without controlling method.

Table 10 shows the results of saltwater intrusion lengths in western, middle, and eastern parts of the Nile Delta in case of scenario 3. Scenario 3 represents decreasing the pumping discharges in the middle area of the Nile Delta by 10, 20, 30, 40, and 50%. Decreasing the pumping discharges in the middle part of the Nile Delta leads to retreat the saltwater intrusion in the Nile Delta. However, its impact is less than the impact of the first scenario on retreat of the saltwater and freshwater lines toward the sea.

The intruded lengths of saltwater and freshwater for scenario 4 are shown in Table 11. Scenario 4 represents decreasing the pumping discharges in the eastern area of the Nile Delta by 10, 20, 30, 40, and 50%. Reduction of the pumping discharges in the eastern area of the Nile Delta only leads to retreat the saltwater and freshwater lines in all the areas of the Nile Delta. The impact of this scenario is significant only in the eastern parts, while its impacts are very limited in other areas compared with first scenarios.

Table 8 Results of saltwater intrusion lengths for scenario 1

	Group 1. Reduction of pumping discharges in the whole area of the Nile Delta										Situation in 2050 without controlling method (SLR = 0.50 m)	
	10%		20%		30%		40%		50%			
Sec.	Equi-line 35 (km)	Equi-line 1 (km)	Equi-line 35 (km)	Equi-line 1 (km)	Equi-line 35 (km)	Equi-line 1 (km)	Equi-line 35 (km)	Equi-line 1 (km)	Equi-line 35 (km)	Equi-line 1 (km)	Equi-line 35 (km)	Equi-line 1 (km)
West	32.20	56.15	32.00	56.00	31.75	55.75	31.50	55.50	31.33	55.29	44.51	68.00
Middle	56.00	103.25	55.50	100.85	54.81	98.16	54.05	95.38	53.01	93.01	59.05	107.4
East	72.57	99.00	68.63	97.46	63.62	96.31	58.39	95.05	54.05	94.12	81.10	105.6

Table 9 Results of saltwater intrusion lengths for scenario 2

	Group 2. Part A. Reduction of pumping discharges in the western area of the Nile Delta										Situation in 2050 without controlling method	
	10%		20%		30%		40%		50%			
Sec.	Equi-line 35 (km)	Equi-line 1 (km)	Equi-line 35 (km)	Equi-line 1 (km)	Equi-line 35 (km)	Equi-line 1 (km)	Equi-line 35 (km)	Equi-line 1 (km)	Equi-line 35 (km)	Equi-line 1 (km)	Equi-line 35 (km)	Equi-line 1 (km)
West	34.45	57.98	34.40	57.75	34.35	57.50	34.35	57.25	34.20	57.00	44.51	68.00
Middle	56.6	105.8	56.77	105.9	56.90	106	57.16	106.13	57.29	106.25	59.05	107.4
East	74.00	100.00	74.33	100.35	74.66	100.69	74.98	101.04	75.31	101.38	81.10	105.6

Table 10 Results of saltwater intrusion lengths for scenario 3

Sec.	Group 2. Part B. Reduction of pumping discharges in the central area of the Nile Delta										Situation in 2050 without controlling method	
	10%		20%		30%		40%		50%			
	Equi-line 35 (km)	Equi-line 1 (km)	Equi-line 35 (km)	Equi-line 1 (km)	Equi-line 35 (km)	Equi-line 1 (km)	Equi-line 35 (km)	Equi-line 1 (km)	Equi-line 35 (km)	Equi-line 1 (km)	Equi-line 35 (km)	Equi-line 1 (km)
West	34.31	57.91	34.28	57.85	34.25	57.80	34.20	57.75	34.16	57.75	44.51	68.00
Middle	56.00	104.26	55.8	103.19	55.70	101.97	55.5	100.65	55.40	99.24	59.05	107.4
East	75.15	101.17	75.00	100.25	74.70	100.30	74.50	99.90	74.31	99.50	81.10	105.6

Table 11 Results of saltwater intrusion lengths for scenario 4

Sec.	Group 2. Part C. Reduction of pumping discharges in the Eastern area of the Nile Delta										Situation in 2050 without controlling method	
	10%		20%		30%		40%		50%			
	Equi-line 35 (km)	Equi-line 1 (km)	Equi-line 35 (km)	Equi-line 1 (km)	Equi-line 35 (km)	Equi-line 1 (km)	Equi-line 35 (km)	Equi-line 1 (km)	Equi-line 35 (km)	Equi-line 1 (km)	Equi-line 35 (km)	Equi-line 1 (km)
West	34.25	57.85	34.22	57.80	34.20	57.73	34.18	57.68	34.15	57.60	44.51	68.00
Middle	55.82	104.18	55.11	102.305	54.39	101.06	53.52	100.20	52.69	99.25	59.05	107.4
East	73.10	99.42	69.16	97.78	64.61	96.61	59.55	95.61	55.45	94.61	81.10	105.6

Table 12 shows the results of saltwater intrusion lengths in western, middle, and eastern parts of the Nile Delta in case of scenario 5. Scenario 5 represents decreasing the pumping discharges in the western area of the Nile Delta by 10, 20, 30, 40, and 50% and increasing the pumping discharges in middle and eastern areas by 5, 10, 15, 20, and 25%. Reduction of pumping discharges in the western part and increasing the pumping discharges on middle and eastern parts lead to retreat of the saltwater and freshwater lines in the western parts. On the other hand, the freshwater and the saltwater intruded more inland into the Nile Delta aquifer in the middle and eastern areas.

The intruded lengths of saltwater and freshwater for scenario 6 are shown in Table 13. Scenario 6 represents decreasing the pumping discharges in the middle area of the Nile Delta by 10, 20, 30, 40, and 50% and increasing the pumping discharges in the western and eastern areas by 5, 10, 15, 20, and 25%. Reduction of pumping discharges rates in the middle part of the Nile Delta only and increasing the pumping on the other parts lead to retreat of the saltwater and freshwater lines only in the middle parts, while scenario 6 indicated that the saltwater intruded more inland into the Nile Delta aquifer.

Table 14 shows the results of saltwater intrusion lengths in western, middle, and eastern part of the Nile Delta in case of scenario 7. Scenario 7 represents decreasing the pumping discharges in the eastern area of the Nile Delta by 10, 20, 30, 40, and 50% and increasing the pumping discharges in the western and middle areas by 5, 10, 15, 20, and 25%. Reduction of pumping scenarios in the eastern part only of the Nile Delta and increasing the pumping discharges on the other area lead to retreat of the saltwater only in the eastern area. On the other hand, the saltwater intruded more on the other parts of the Nile Delta.

11 Conclusion

This chapter presented an assessment of land use/land cover changes in the Nile Delta region using ENVI software. Moreover, it discussed the estimation of groundwater recharge from rainfall in the Nile Delta aquifer using WetSpass model. Also, building an integrated 3D groundwater model for the Nile Delta aquifer using MODFLOW and simulation of the saltwater intrusion using SEAWAT code are combined with the Nile Delta aquifer model. Different climate change scenarios including sea level rise, decreasing groundwater head due to additional groundwater pumping, and combination of the first and the second scenarios are simulated. On the other hand, seven scenarios were tested by SEAWAT code for management of pumping discharges to control saltwater intrusion in the Nile Delta region. The results of the WetSpass hydrological model for 1991 indicated that the minimum and maximum recharge values were, respectively, 0.0 and 304 mm and were 0.0 and 134 mm for 2000. The WetSpass model results indicate a good agreement with the manually calculated values by using the water balance equation. The results from groundwater recharge confirmed that the different amounts of rainfall cause a large

Table 12 Results of saltwater intrusion lengths for scenario 5

	Group 3. Part A. Redistribution of pumping discharges (western area of the Nile Delta)										Situation in 2050 *without controlling* method	
	10%		20%		30%		40%		50%			
Sec.	Equi-line 35 (km)	Equi-line 1 (km)	Equi-line 35 (km)	Equi-line 1 (km)	Equi-line 35 (km)	Equi-line 1 (km)	Equi-line 35 (km)	Equi-line 1 (km)	Equi-line 35 (km)	Equi-line 1 (km)	Equi-line 35 (km)	Equi-line 1 (km)
West	34.42	57.85	34.35	57.75	34.25	57.65	34.12	57.55	34.00	57.45	44.51	68.00
Middle	56.65	107.20	56.89	108.79	57.03	110.14	57.01	111.65	56.89	113.24	59.05	107.4
East	76.29	102.58	76.98	103.53	77.54	104.57	78.00	105.12	78.56	105.20	81.10	105.6

Table 13 Results of saltwater intrusion lengths for scenario 6

Sec.	Group 3. Part B. Redistribution of pumping discharges (central area of the Nile Delta)										Situation in 2050 without controlling method	
	10%		20%		30%		40%		50%			
	Equi-line 35 (km)	Equi-line 1 (km)	Equi-line 35 (km)	Equi-line 1 (km)	Equi-line 35 (km)	Equi-line 1 (km)	Equi-line 35 (km)	Equi-line 1 (km)	Equi-line 35 (km)	Equi-line 1 (km)	Equi-line 35 (km)	Equi-line 1 (km)
West	34.15	57.78	34.22	57.69	34.28	57.95	34.34	58.00	34.40	58.11	44.51	68.00
Middle	56.32	104.77	56.22	103.86	56.08	102.84	55.89	101.43	55.73	100.28	59.05	107.4
East	76.11	102.06	76.70	102.45	77.13	102.85	77.44	103.15	77.74	103.31	81.10	105.6

Table 14 Results of saltwater intrusion lengths for scenario 7

Sec.	Group 3. Part C. Redistribution of pumping discharges (eastern area of the Nile Delta)										Situation in 2050 without controlling method	
	10%		20%		30%		40%		50%			
	Equi-line 35 (km)	Equi-line 1 (km)	Equi-line 35 (km)	Equi-line 1 (km)	Equi-line 35 (km)	Equi-line 1 (km)	Equi-line 35 (km)	Equi-line 1 (km)	Equi-line 35 (km)	Equi-line 1 (km)	Equi-line 35 (km)	Equi-line 1 (km)
West	33.97	57.82	34.05	57.90	34.13	57.98	34.20	58.10	34.30	58.15	44.51	68.00
Middle	55.73	104.74	57.23	106.24	58.73	107.74	60.23	109.24	61.73	110.74	59.05	107.4
East	73.18	99.54	69.44	98.11	68.85	98.07	60.24	96.06	57.37	95.82	81.10	105.6

variation in recharge, while the modeling trend results were similar to the trend of observed rainfall values. The different soil classes in the Nile Delta were shown to cause a large variation in groundwater recharge. High values of recharge were observed in the northern part related to the combination of the sand soil type and the high amount of rainfall. The difference between the estimated and measured bank level for Damietta branch varied from 0.01 to 0.80 m for bank level. The RMSE and R^2 values of top width and bank level for Damietta branch showed a fairly good correlation between the estimated and measured values. A good and consistent agreement was found regarding the TDS distribution and the intruded length of saltwater between the current model using SEAWAT code and previous studies. The results of saltwater intrusion problem in the Nile Delta aquifer showed that the concentration line 35,000 mg/L intruded inland into the aquifer to a distance ranging from 40.0 to 74.0 km measured at the aquifer bottom boundary from the shoreline, while the concentration line 1,000 mg/L intruded inland into the aquifer to a distance ranging from 60.0 to 100.0 km measured at the aquifer bottom boundary from the shoreline. The results of the current model showed that the increase of sea level rise has a significant impact on the saltwater intrusion in the Nile Delta aquifer and the location of the transition zone. A rise of sea level by 1.0 m leads to that the concentration line 35,000 mg/L advanced inland by a distance of 8.2 km measured at the bottom boundary compared with the base case (2008). The impact of increasing the sea level on saltwater intrusion in the Nile Delta is more significant than decreasing the groundwater head at the south boundary due to additional groundwater pumping. Decreasing of the groundwater head due to additional groundwater pumping by 1.0 m leads to that the concentration line 35,000 mg/L to intrude further inland by a distance of 6.75 km measured at the aquifer bottom boundary compared with the base case. Scenario 3 presented the worst case where the saltwater intruded inland further than the first and second scenarios. A rise in sea level by 1.0 m and decreasing the groundwater head by 1.0 m lead the concentration line 35,000 mg/L to advance further inland into the aquifer by a distance of 10.20 km, respectively, measured at the aquifer bottom boundary compared with the base case. Including the actual irrigation canal networks in the current model causes the saltwater to intrude inland into the Nile Delta aquifer to distances less than the results got by previous studies at the same cross sections due to different climate change scenarios. Reduction of the pumping scenarios from the all areas of the Nile Delta represents the best scenario as the saltwater and freshwater lines retreated toward the sea compared with the results of the other scenarios.

12 Recommendations

Decreasing the pumping discharges of wells from all the areas in the Nile Delta is highly recommended to enhance the head gradient and the flow from the Nile Delta aquifer toward the seaside and retreat the saltwater intrusion. Including the actual drainage network with the actual canal network is recommended for future studies in groundwater modeling for the Nile Delta for accurate representation and simulation of the interaction between the surface water and the groundwater system. Update the numerical simulation of saltwater intrusion by including the recent data of groundwater level, salinity measurements, water levels of canals, and pumping rate records. The results of the current model could be used in the identification of the most vulnerable areas to the saltwater intrusion in the Nile Delta aquifer and management of groundwater resources in the Nile Delta region. The integrated built model is a useful predictive tool for more understanding of the hydrological process of the Nile Delta aquifer and the saltwater intrusion process. The future strategies for the adaption and mitigation for the saltwater intrusion in the Nile Delta aquifer should be based on an integrated built groundwater model for the Nile Delta aquifer. Future studies should also link between the results of the numerical model for simulating the saltwater intrusion in the Nile Delta aquifer and the socioeconomic effect of the recent development in the Nile Delta especially the coastal area near to the Mediterranean Sea border. Authorities and stakeholders should use the built integrated model with another analysis tools as a decision support system to determine the best location and the maximum abstraction rates from groundwater wells in the Nile Delta region.

References

1. Abd-Elhamid HF (2010) Saltwater intrusion investigation and control methods. PhD thesis, University of Exeter
2. Anon (1980) Investigation project of sustained yield of aquifers. Part 1. Groundwater Research Institute, Cairo, p 342
3. Farid MS (1980) Nile Delta groundwater study. MSc thesis, Faculty of Engineering, Cairo University, Giza
4. Kashef (1983) Salt-water intrusion in the Nile Delta. Groundwater 21(2):160–164. https://doi.org/10.1111/j.1745-6584.1983.tb00713.x
5. Sherif MM, Singh VP, Amer AM (1988) A two-dimensional finite element model for dispersion (2D-FED) in coastal aquifer. J Hydrol 103:11–36
6. RIGW/IWACO (1999) Environmental management of groundwater resources (EMGR). Final technical report TN/70.0067/WQM/97/20, Research Institute for Groundwater, Kanater El-Khairia
7. Soliman M, EL-Mongy A, Talaat M, Hassan A (1991) Groundwater quality model with applications to various aquifers. Environ Geol Water Sci 17(3):201–208
8. Sherif M, Sighn VP (1991) Effect of climate change on sea water intrusion in coastal aquifers. Hydrol Process 13:1277–1287
9. Sherif MM, Al-Rashed MF (2001) Vertical and horizontal simulation of seawater intrusion in the Nile Delta Aquifer. In: First international conference on saltwater intrusion and coastal aquifersó monitoring, modeling, and management, Essaouira, 23–25 Apr
10. RIGW (2002) Nile Delta groundwater modeling report. Research Institute for Groundwater, Kanater El-Khairia
11. Sherif MM, Sefelnasr A, Javadi A (2012) Incorporating the concept of equivalent freshwater head in successive horizontal simulations of seawater intrusion in the Nile Delta aquifer, Egypt. J Hydrol 464–465:186–198. https://doi.org/10.1016/j.jhydrol.2012.07.007

12. Abdelaty M, Hany FA, Maha RF, Gamal MA (2014) Investigation of some potential parameters and its impacts on saltwater intrusion in Nile Delta aquifer. J Eng Sci 42(4):931–955
13. Abd-Elhamid H, Javadi A, Abdelaty I, Sherif M (2016) Simulation of seawater intrusion in the Nile Delta aquifer under the conditions of climate change. Hydrol Res 47(5):1–14. https://doi.org/10.2166/nh.2016.157
14. Nofal ER, Amer MA, El-Didy SM, Fekry AM (2015) Sea water intrusion in Nile Delta in perspective of new configuration of the aquifer heterogeneity using the recent stratigraphy data. J Am Sci 11:281–292
15. Wassef R, Schüttrumpf H (2016) Impact of sea-level rise on groundwater salinity at the development area western delta, Egypt. Groundwater Sustain Dev 2–3:85–103. https://doi.org/10.1016/j.gsd.2016.06.001
16. Mazi K, Koussis AD, Destouni G (2014) Seawater intrusion risks and controls for safe use of coastal groundwater under multiple change pressures. Department of Physical Geography and Quaternary Geology, Stockholm University
17. Sherif MM (1999) The Nile Delta aquifer in Egypt. In: Bear J, Cheng A, Sorek S, Ouazar D, Herrera A (eds) Seawater intrusion in coastal aquifers, concepts methods and practices. Theory and application of transport in porous media, vol 14. Kluwer Academic, Dordrecht, pp 559–590
18. Sherif MM (2001) Simulation of seawater intrusions in the Nile Delta aquifer. In: First international conference on saltwater intrusion and coastal aquifers monitoring, modeling and management, Essaouira
19. Sherif MM (2003) Assessment, modeling and management of seawater intrusion in the Nile Delta aquifer. IGME, Madrid
20. Dahab KA (1993) Hydrological evaluation of the Nile Delta after the high dam construction. PhD thesis, Faculty of Science, Monofia University, Egypt, p 30
21. RIGW (1992) Hydrogeological map of Nile Delta. Scale 1: 500,000. 1st edn. ND
22. Morsy WS (2009) Environmental management to groundwater resources for Nile Delta region. PhD thesis, Faculty of Engineering, Cairo University, Cairo
23. Serag El Din HM (1989) Geological, hydrochemical and hydrological studies on the Nile Delta quaternary aquifer. PhD thesis, Faculty of Science, Mansoura University, Mansoura
24. Mabrouk MB, Jonoski A, Solomatine D, Uhlenbrook S (2013) A review of seawater intrusion in the Nile Delta groundwater system – the basis for assessing impacts due to climate changes and water resources development. Hydrol Earth Syst Sci Discuss 10:10873–10911
25. El-Arabi N (2012) Environmental management of groundwater in Egypt via artificial recharge extending the practice to soil aquifer treatment (SAT). Int J Environ Sustain 1(3):66–82
26. RIGW (1980) Projected of the safe yield study for groundwater aquifer in the Nile Delta and Upper Egypt, part 1 (in Arabic). Ministry of Irrigation, Academy of Scientific Research and Technology, and Organization of atomic Energy, Cario
27. Sefelnasr A, Sherif M (2014) Impacts of seawater rise on seawater intrusion in the Nile Delta aquifer, Egypt. Groundwater 52(2):264–276
28. Wahaab R, Badawy MI (2004) Water quality assessment of the River Nile system: an overview. Biomed Environ Sci 17:87–100
29. Shamrukh M, Corapcioglu Y, Hassona F (2001) Modeling the effect of chemical fertilizers on ground water quality in the Nile Valley aquifer, Egypt. Ground Water 39:59
30. DRI Drainage Research Institute (1989) In: Amer MH, de Ridder NA (eds) Land drainage in Egypt. Drainage Research Institute, Cairo
31. Warner JW, Gates TG, Attia FA, Mankarious WF (1991) Vertical leakage in Egypt's Nile Valley: estimation and implications. J Irrig Drain Eng 117:515–533
32. RIGW/IWACO (1990) Hydrological inventory and groundwater development plan western Nile Delta region. TN77, 01300-9-02, Research Institute for Groundwater, Kanater El-Khairia
33. El Ramly M (1997) Hydrogeological and water quality characteristics of the saturated zone beneath the various land uses in the Nile Delta region, Egypt, Freshwater Contamination. In: Proceedings of Rabat symposium S4, IAHS Publishing No. 243, Apr–May 1997

34. RIGW (2003) Monitoring of groundwater microbiological activities in the Nile Delta aquifer. A study completed for the National Water Quality and Availability Management project (NAWQAM), Kanater El-Khairia
35. MWRI (1954) Irrigation and drainage network Nile Delta map, Scale 1:200,000. Prepared by Ministry of Water Resources and Irrigation
36. Ssegane H, Tollner EW (2007) Tools for remotely assessing riparian buffers protecting streams rom sediment pollution. In: Proceedings of the 2007 Georgia water resources conference, University of Georgia, Georgia, 27–29 Mar 2007
37. Armanuos AM, Negm A (2016) Assessment of the variations of local parameters of Wetspass model: case study Nile Delta aquifer. Proc Eng 154:276–283. https://doi.org/10.1016/j.proeng.2016.07.475
38. Armanuos AM, Negm A, Yoshimura C, Valeriano OCS (2016) Application of WetSpass model to estimate groundwater recharge variability in the Nile Delta aquifer. Arab J Geosci 9:553. https://doi.org/10.1007/s12517-016-2580-x
39. EGSA (Egyptian General Survey and Mining) (1997) Topographical maps cover Nile Delta, scale 1:1500000 and 1:10000
40. Armanuos AM, Negm A, Yoshimura C, Valeriano OCS (2016) Estimation of bed and bank levels of an irrigation canal network towards accurate groundwater modeling of the Nile Delta aquifer. Int Water Technol J 6(1):74–84
41. FAO-UNESCO (1988) Soil map of the world. Revised Geographies Informatica. Mexico, Legend, Reprinted with corrections. World Soil Resources report 60. FAO, Rome, pp 818–833
42. Elewa H (2010) Potentialities of water resources pollution of the Nile River Delta, Egypt. Open Hydrol J 4:1–13
43. HRI (Hydraulic Research Institute), NWRC (2000) Final Report, Studying the capacity of Damietta branch for different discharges, pp 17–22
44. MWRI (Ministry of Water Resources and Irrigation) (2013) Proposed climate change adaptation strategy for the ministry of water resources and irrigation in Egypt, pp 4–5
45. Wilson J, Townley LR, Da Costa AS (1979) Mathematical development and verification of a finite element aquifer flow model AQUIFEM-1. Technology adaptation program, report no. 79–2, MIT, Cambridge
46. Said R (1962) The geology of Egypt. Elsevier, Amsterdam
47. Roest J (1999) Regional water distribution in the Nile Delta of Egypt. In: ILRI workshop: water and food security in (semi-) arid areas. Proceedings of the Wageningen water workshop, Wageningen, pp 61–82
48. NWRP (2005) National water resources plan of Egypt – 2017. Arab Republic of Egypt, Ministry of Water Resources and Irrigation, Planning Sector, Cairo, pp 6–7
49. Farid MS (1985) Management of groundwater system in the Nile Delta. PhD thesis, Faculty of Engineering, Cairo University, Cairo
50. RIGW/IWACO (1992) Hydrogeological map of the Nile Delta scale 1:500,000. Cairo
51. Wolf P (1987) The problem of drainage and its solution in the Nile valley and Nile Delta. Nat Res Dev 25:62–73
52. Arlt HD (1995) A hydrogeological study of the Nile Delta aquifer with emphasis on saltwater intrusion in the Northern Area, Mitteilung/Institut für Wasserbau und Wasserwirtschaft, Technische Universität Berlin, Nr. 130, OCLC No. 636899992, pp 291–302
53. Sakr MA, Mabrouk BM (2006) Invited paper in the environmental assessment and controls of groundwater quality El-Gharbiya Governorate-Egypt. In: Workshop on: a better environment in delta area, Tanta University
54. El-Arabi M (2007) Environmental impact of new settlements in groundwater in a region in the Nile Delta. MSc thesis, Faculy of Engineering, Zagazig University
55. Fadlelmawla AA, Dawoud MA (2006) An approach for delineating drinking water well head protection areas at the Nile Delta, Egypt. J Environ Manag 79:140–149

56. Morsy WS, El-Fakharany Z (2012) Predicting the impact of surface wastewater on groundwater quality in Quesna industrial area. J Am Sci 8:772–781
57. Armanuos AM, Ibrahim MG, Mahmod WE, Abdelazim N, Yoshimura C, Takemura J, Zidan B (2017) Evaluation the potential impact of grand Ethiopian renaissance dam and pumping scenarios on groundwater level in the Nile Delta aquifer. Water Sci Technol Water Supply 17(15):ws2017037. https://doi.org/10.2166/ws.2017.037
58. UNDP (1981) Competitive use of water by major field crops in Egypt, Water Master Plan, Cairo, Egypt. Soil and Water Research Institute, Ministry of Irrigation/UNDP/IBRD, p 28
59. Zaghloul MG (1958) Flow distribution through Groundwater aquifer of the Nile Delta. MSc thesis, Faculty of Engineering, Alexandria University, Alexandria
60. Shahien M (1987) Hydrology of the Nile basin development in water science, vol 21. Elsevier Science, B.V, Amsterdam, p 575
61. Leaven MT (1991) Hydrogeological study of the Nile Delta and adjacent desert areas, Egypt, with emphasis on hydrochemistry and isotope hydrology. Thesis, RIGW/IWACO as technical note TN 77.01300-91-01, Free University, Amsterdam
62. Shahin M (1991) Assessment of groundwater resources in Egypt, IHE report series 23. International Institute for Hydraulics and Environmental Engineering, Delft
63. Mabrook B, Swailem F, El Sheikh R, El Dairy F (1983) Shallow aquifer parameters and its influence on groundwater flow, Nile Delta Egypt. In: Australian Water Resources Council conference, series 8, pp 187–197
64. Sollouma M, Gomaa MA (1997) Groundwater quality in the Miocene aquifers east and west of the Nile Delta and the north west desert, Egypt. Sci J Fac Sci 35
65. Shata A, El Fayoumy LF (1997) Remarks on the hydrology of the Nile Delta, UAR. In: Proceeding of the bucharest symposium, IASHIAIHS-UNESCO
66. Bahr R (1995) Groundwater management and saltwater intrusion in Nile Delta aquifer. Berlin TU, Berlin
67. El Tahlawi MR, Farrag AA, Ahmed SS (2008) Groundwater of Egypt: an environmental overview. Environ Geol 55:639–652
68. Elshinnawy H, Zeidan BA, Ghoraba SM (2015) Impact of hydraulic conductivity uniformity on seawater intrusion in the Nile Delta aquifer, Egypt. https://www.researchgate.net
69. Anderson MP, Woessner WW (1992) Applied groundwater modeling: simulation of flow and advection transport. Academic Press, San Diego, p 318
70. Bear J (1979) Hydraulics of groundwater. McGraw-Hill, New York
71. Bear J, Verruijt A (1987) Modeling groundwater flow and pollution. D. Reidel Publishing, Dordrecht, p 414
72. Guo W, Langevin CD (2002) User's guide to SEAWAT: a computer program for simulation of three-dimensional variable-density ground-water flow. U.S. Geological Survey, Reston
73. Simpson MJ, Clement TP (2003) Theoretical analysis of the worthiness of Henry and elder problems as benchmarks of density-dependent groundwater flow models. Adv Water Resour 26:17–31. https://doi.org/10.1016/S0309-1708(02)00085-4
74. Simpson MJ, Clement TP (2004) Improving the worthiness of the Henry problem as a benchmark for density-dependent groundwater flow models. Water Resour Res 40:W01504. https://doi.org/10.1029/2003WR002199
75. Langevin CD, Daniel T, Thorne J, Alyssa M, Dausman MC, Weixing G (2008) SEAWAT version 4: a computer program for simulation of multi-species solute and heat transport. Techniques and methods book 6. US Department of the Interior, US Geological Survey
76. NWRC (2003) Egyptian code of water resources and irrigation works. In: Horizontal expansion, planning and designing of irrigation and drainage canals, pp 1–7
77. Armanuos AM, Ibrahim MG, Mahmod WE, Yoshimura C, Takemura J, Fujii M (2017) Experimental and numerical study of saltwater intrusion in the Nile Delta aquifer, Egypt. PhD thesis, E-JUST

78. Intergovernmental Panel of Climate Change (IPCC) (2001) Climate change 2001: impacts, adaptations, and vulnerability. In: Contribution of Working Group II to the third assessment report of the Intergovernmental Panel on Climate Change. Cambridge University Press, Cambridge
79. Intergovernmental Panel of Climate Change (IPCC) (2007). Climate change 2007: impacts, adaptations, and vulnerability. In: Contribution of Working Group II to the fourth assessment report of the Intergovernmental Panel on Climate Change. Cambridge University Press, Cambridge

Groundwater Modelling and Assessment Under Uncertain Hydrological Conditions for Egyptian Sahara

Wael Elham Mahmod

Abstract Lack of hydrogeological data is the main reason for the difficulty of groundwater management, especially in arid zones. Egypt's Sahara Desert is located in Western Egypt and is lacking hydrogeological data. Recent development of the Egyptian Sahara is mainly due to the Nubian Sandstone Aquifer System (NSAS) as a unique source of water there. NSAS covers a great part of Egypt, Sudan, Chad and Libya and is considered as a main source of freshwater. During the last two decades, excess pumping of groundwater at the Egyptian Sahara brought about a significant drawdown of the groundwater table. This chapter will discuss a new technique that was developed to overcome the uncertainty from data gaps to facilitate the implementation of numerical models to improve strategies for optimal groundwater management. The core of this developed method is to understand the temporal and spatial variation of groundwater table. In the Egyptian Sahara, the hydrogeological data needed for groundwater simulation are lacking, thereby introducing a problem for numerical models calibration and validation. A newly developed model named the modified grey model (MGM) was proposed to analyse groundwater flow. At its core it is a finite element method (FEM) with a new developed modified genetic algorithm (MGA) to obtain the goodness of fit with observations. The MGM is an attempt to determine a selection process of the best input models' trends with the appropriate values of input parameters for achieving acceptable fitting to the measured data. Kharga Oasis was selected as a case study for application of the developed MGM in groundwater flow analysis. The MGM simulation results clearly show that the future groundwater table will face a severe

W.E. Mahmod (✉)
Environmental Engineering Department, School of Energy Resources, Environment and Chemical and Petrochemical Engineering, Egypt-Japan University of Science and Technology, E-JUST, Alexandria, Egypt

Civil Engineering Department, Faculty of Engineering, Assiut University, Assiut 71515, Egypt
e-mail: wael.elham@ejust.edu.eg; wdpp2006@aun.edu.eg

A. M. Negm (ed.), *Groundwater in the Nile Delta*,
Hdb Env Chem (2019) 73: 545–584, DOI 10.1007/698_2017_84,

drawdown in the northeastern part of the study area compared with that in the southwestern part. On the other hand, by 2060, the hydraulic head difference between these two parts will reach 140 m. Considering the uncertainty and lack of available data, the MGM produced more realistic results compared with those obtained from only FEM. Three development scenarios of groundwater withdrawal were proposed. These scenarios include either expanding the current extraction rate or redistributing the groundwater withdrawal over the recent working production wells (RWPWs). The results concluded that, for the northern part of the oasis, the groundwater table could be temporally recovered to an economical piezometric level; however, the table in the southern part is severely decreased. Conclusively, the MGM could be applied to other cases with similar data limitations.

Keywords Egyptian Sahara, Kharga Oasis, Modified grey model, Nubian Sandstone Aquifer System, Uncertainty analysis

Contents

1 Water Resources in Egypt

Egypt located in the northeast of Africa has a total area of 1 million km^2. Around 96% of Egypt is a desert plateau (desert area) dissected by the Nile Valley and its Delta, which occupy about 4% of the total country area (green area). For the green area, the River Nile is considered the main source of freshwater since ancient times until present for all types of human activities. For the desert area, human activities remain confined to a few valleys, oases and depressions, where groundwater is available through springs and seepage zones [1] (see Fig. 1). The River Nile water budget is around 55.5 billion m^3/year under the 1959 Nile Waters Agreement between Egypt and Sudan. Egypt is a water-scarce country, with around 600 m^3 of available freshwater/year/capita. The rate of water usage in Egypt is approximately 200 L/ person/day. On the other hand, water needs for typical Middle East and North African diet is 294 L/person/day [2]. The annual average rainfall is 51 mm [3].

Groundwater exists in the fringes of the Nile Valley, Nile Delta, Western Desert and Sinai Peninsula. Groundwater aquifers which are located in Nile Valley, Nile Delta and Sinai Peninsula replenished from the river Nile by seepage from canals and deep percolation from irrigation application [5]. Groundwater in the Western Desert is the only available resource for interdisciplinary development in the desert

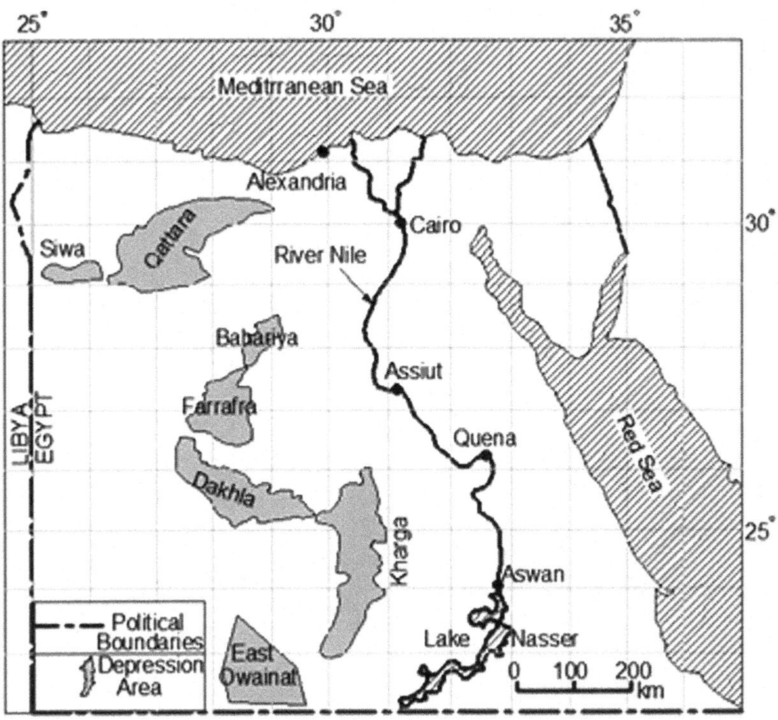

Fig. 1 Egypt map showing the major oasis in the New Valley Governorate [4]

area, and it is a non-renewable water resource [4, 6]. Developing strategies of the desert area depend on pumping costs of fossil water and potential economic return over a fixed time period. For coastal zones, groundwater reservoirs are recharged from local rainfall [7]. The largest groundwater deposit is located underneath the eastern part of the African Sahara and is shared between Egypt, Sudan, Libya and Chad as shown in Fig. 2 [8, 9].

Fig. 2 Schematic distribution of Nubian Sandstone under Egypt, Libya, Sudan and Chad [9]

2 Hydrogeology of Aquifer Systems in Egypt

The hydrogeological framework of Egypt comprises seven aquifer systems [1, 2, 5, 7, 10–12], as shown in Fig. 3:

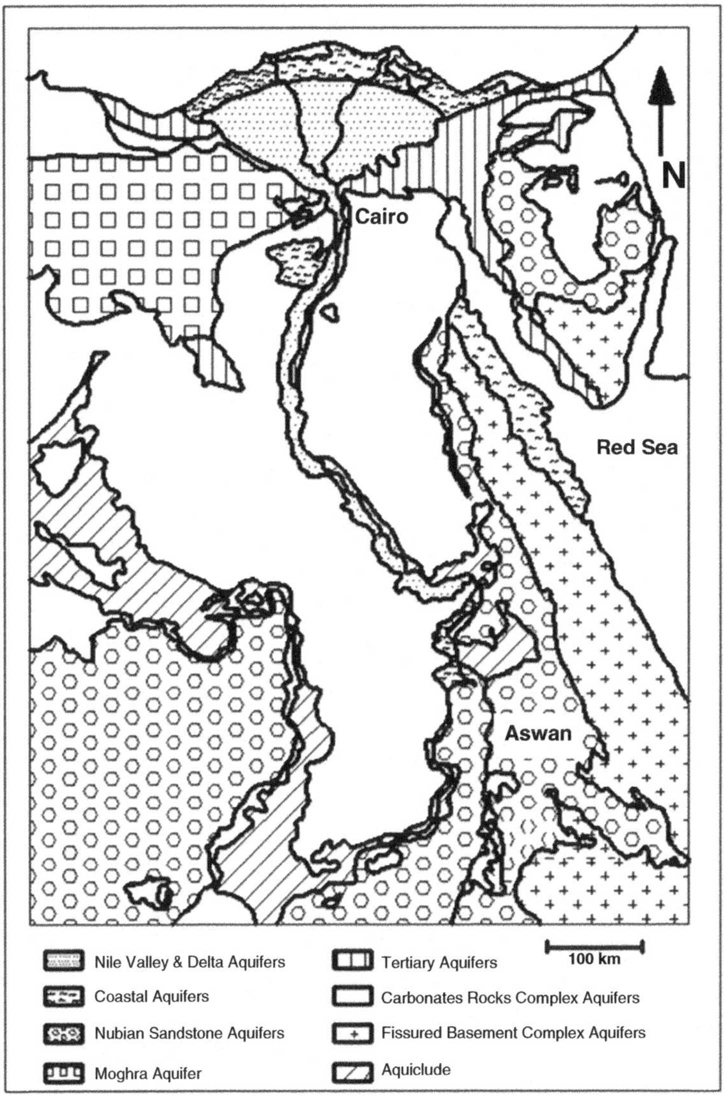

Fig. 3 Hydrogeological map of Egypt [1]

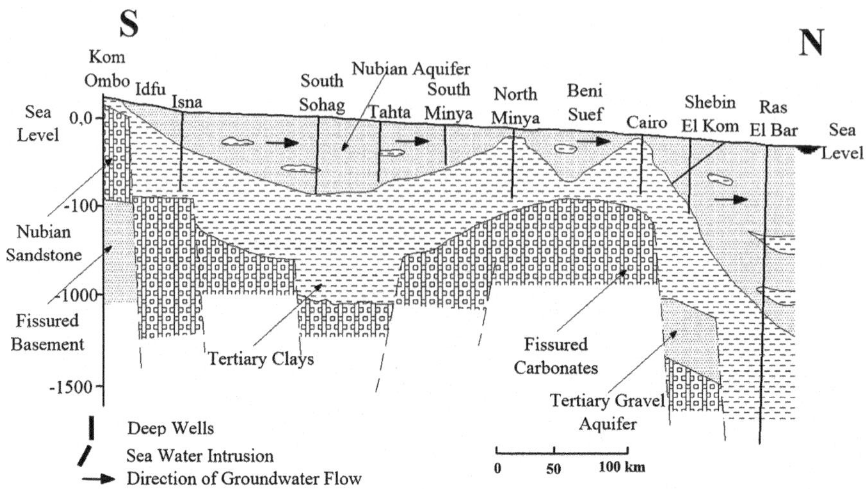

Fig. 4 Hydrogeological profile through the Nile Valley and Delta basins [6, 13]

2.1 *The Nile Valley Aquifer Systems (NVAS)*

The aquifer assigned to the Quaternary and Late Tertiary is located under the Nile flood plain and desert fringes and consists of a thick layer of graded sand and gravel covered by a clay layer in its major part. NVAS thickness varies from 300 m in Middle Egypt (Sohag Governorate) and decreases to only a few metres to the north towards Cairo and in the south direction up to Aswan as shown in Fig. 4. Below this aquifer, there are clayey deposits that were formed during the Pliocene, which are impermeable and prevent the connection between this aquifer and the Nubian Sandstone aquifer. However, there could be a connection between the Quaternary deposits and the surrounding limestone. The water is mainly used for domestic purposes and irrigation; its salinity is less than 1,500 ppm. The aquifer system is renewable, and the main recharge source is the infiltration from the excess water application for agriculture and seepage from the irrigation water distribution system [1]. The saturated thickness of NVAS aquifer is in the range of 10–200 m. On the other hand, its hydraulic conductivity ranges from 50 to 70 m/day.

2.2 *The Nile Delta Aquifer Systems (NDAS)*

Four regional aquifers are present in the NDAS in addition to six subregional and local aquifers [1] as shown in Fig. 5. Tahlawi et al. [1] mentioned that the NDAS assigned to the Pleistocene age consists mainly of graded sand and gravel. However, to the north, it consists of fine sand and clay. NDAS is considered as semi-confined aquifer. It is overlain by Holocene silt, clay and fine sand in the floodplain

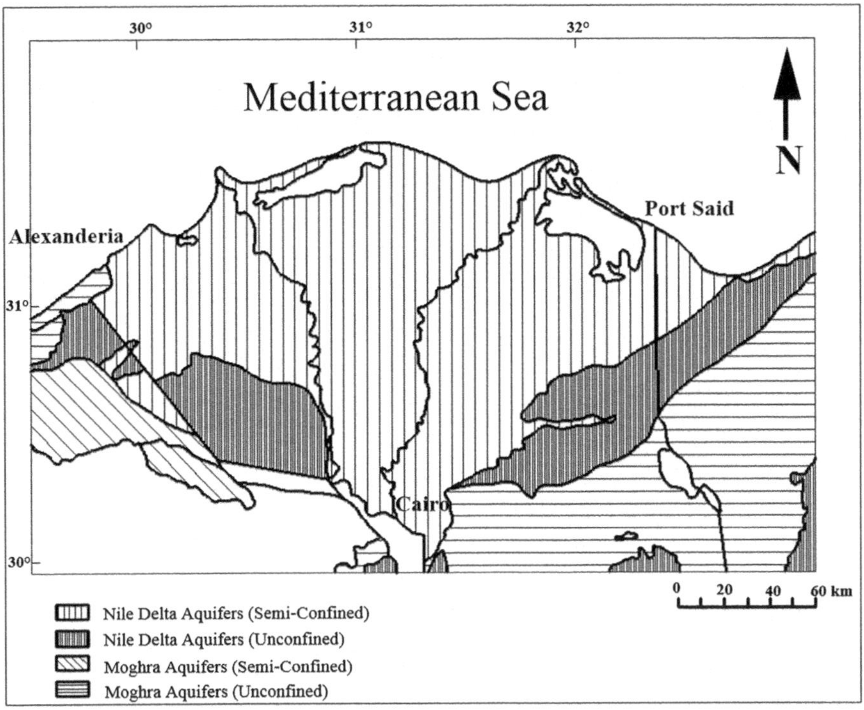

Fig. 5 Groundwater aquifers in the Nile Delta [1]

of the Nile. A calcareous loamy layer acts as a semi-confining zone outside the floodplain in the north-western part of the Delta with a thickness ranging between 0 and 20 m. However, to the west, there is a hydraulic connection between the Quaternary aquifer with the Moghra aquifer. This connection could be declared by the westward continuity of the piezometric gradient (Fig. 6). In the desert fringes, the semi-confined layer is missing, and phreatic conditions prevail [1]. The thickness of the layers holding the groundwater is different from one place to another; however, its average thickness is 190 m, increasing gradually to the north until it reaches 350 m at Tanta City. At the western part of the Delta, the thickness of the layer holding water varies between 120 and 220 m, decreasing gradually towards the east. The total thickness of the aquifer increases from Cairo northward to about 1,000 m along the Mediterranean coast [7, 12]. The depth to the groundwater level in the Nile Delta aquifer increases from the north towards the south. It varies between 1 and 2 m at the north, increases to vary from 3 to 4 m at the middle and reaches maximum value 5 m at the south [14]. The Nile Valley and Nile Delta aquifer are expected to be separated as the aquifer thickness is reduced to a few metres at Cairo. The quality and quantity of the southern part of the aquifer are much better than the northern part as TDS does not exceed 1,000 ppm. The aquifer is continuously recharged by infiltration of irrigation water [1, 7]. The permeability

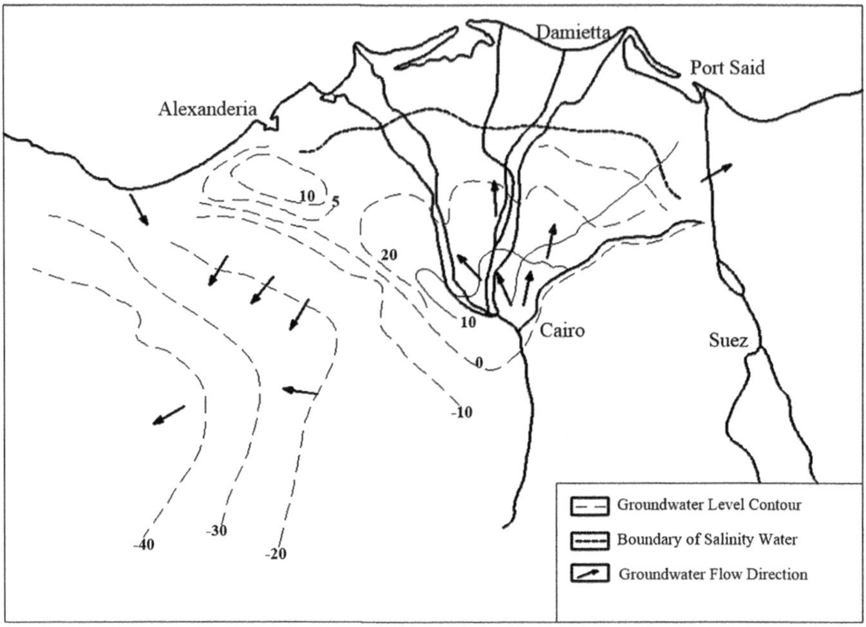

Fig. 6 Regional groundwater levels and trends in the Nile Delta and its fringes from hydrogeological map of Egypt [1]

ranges between 35 and 75 m/day, and it decreases near the coastline due to increase in clay content. The horizontal hydraulic conductivity in the Quaternary Nile Delta aquifer ranges between 0.05 and 0.5 m/day. The storage coefficient range is in the range of 10^{-4} to 10^{-3}, while 0.3 represents the porosity of the aquifer medium. The transmissivity ranges between less than 500 m^2/day at the edges in the desert fringe to more than 25,000 m^2/day in the apex of the Nile Delta. The characteristics of the northern portion of that system are considerably different. The aquifer becomes less productive, containing mainly brackish or saline water due to seawater intrusion.

2.3 The Coastal Aquifer Systems (CAS)

The aquifer belongs to the Quaternary and Late Tertiary types. The systems are found along the Mediterranean and Red Sea coasts. The CAS is in the form of scattered pockets. Groundwater in CAS aquifer is in the form of thin lenses over the saline water. The aquifer is considered as renewable aquifer, and it is recharged from rainfall [9].

The CAS aquifer could be divided into two zones, the Mediterranean and Red Sea littoral zones. For the Mediterranean littoral zone, it occupied an area about 10,000 km^2 . It is characterized by relatively high rainfall up to 200 mm/year. On the other hand, many local aquifer systems are located in the Mediterranean littoral zone. These aquifers are represented by the oolitic limestone aquifer dominating the area to the west of the Nile Delta and by the complex fluviatile sandy gravels and the shallow marine calcareous limestone in North Sinai, near El Arish. On the west of the Nile Delta, the oolitic limestone aquifer has a thickness of 40 m. The aquifer is recharged from winter rainfall and exists under phreatic conditions. It forms a thin layer (± 1.0 m thick) floating on the saline water body resulting from seawater intrusion [1].

At El Arish in northern Sinai, the aquifer is characterized by a top sandy aquifer, a middle fluviatile gravelly aquifer and a lower calcareous sandstone aquifer. For the first, it is known as Thaimail, and the groundwater is recharged by local rainfall and occurs at shallow depth from the surface (± 2.0). For the middle fluviatile gravelly aquifer, its thickness is about 30 m; the groundwater there is under semi-confined conditions and found at a depth of about 15 m from the surface and is capped by a clayey layer. The last aquifer is the lower calcareous sandstone aquifer which is known also as Kurkar. It is of shallow marine origin and has a wide geographical distribution in northeastern Sinai. It has a thickness of 40 m and extends inland for a distance of about 15 km from the coast. This aquifer directly underlies the fluviatile gravelly aquifer and is connected hydrologically. It is recharged in the foot slope area of Sinai highland, by local rainfall, by runoff water and presumably also by the upward leakage from the high pressure water in the Nubian Sandstone Aquifer System underlying the area. The salinity of the water varies between 3,000 and 5,000 ppm.

For the Red Sea coast aquifers, it extends also into Sinai, comprising essentially the quaternary fluviatile aquifer and the tertiary aquifers. The groundwater in the quaternary fluviatile aquifer is under phreatic conditions with a hydraulic head level same as sea level. For the tertiary aquifers, they are recharged by runoff water, by infiltration from the quaternary aquifers and locally by upward leakage from deep-seated aquifers. The salinity is about 2,000–2,500 ppm.

2.4 The Nubian Sandstone Aquifer System (NSAS)

The NSAS is a regional system and considered as a non-renewable aquifer system. It extends into Libya, Sudan and Chad and western Saudi Arabia with a total area of about two million kilometres. The groundwater volume of NSAS is about 150,000 km^3 [15]. NSAS in Egypt could be divided into five subsystems with respect of their locations: at the Western Desert, at Nile Valley, at the Eastern Desert, at Gulf of Swiz and at Sinai. For the NSAS at the Western Desert, it is exploited particularly in the New Valley area, where intensive deep drilling was carried out during the past five decades. There are more than 500 wells that have been drilled to depths from 500 to 1,000 m. Groundwater flow in this aquifer (under New Valley) is towards the northeast with a gradient of 0.5 m/km. The hydraulic parameters are determined in various parts of the New Valley: Kharga, Dakhla, El Bahariya (El Bauity), El Farafra and southward east of Oweinat area. At Kharga and Dakhla Oasis, the salinity decreases from 600 in the upper layers to 200 ppm in the lower layers. On the other hand, NSAS at the Nile Valley extends from the north to the south up to Aswan. West of Cairo, groundwater has been found at a depth of 1 km with high salinity. However, the hydrogeological information about this aquifer is insufficient. The springs of Helwan are mainly connected to this aquifer. At Eastern Desert, NSAS appears in many zones such as El Laqeita area and east of Qena. At El Laqeita area, the groundwater is flowing freely, and the piezometric level is up to 112 m above mean sea level (MSL). The salinity varies in the Eastern Desert between 1,000 and 10,000 ppm. For the NSAS that lies under Gulf of Swiz, its water originates from the watershed in Sinai with salinity about 100,000 ppm. In Sinai Peninsula, NSAS water is basically a fossil water (Late Pleistocene). It is recharged from the uptake areas in Southern Sinai, where the rainfall rate is about 120 mm/year. The flow pattern is in the northward direction. Sinai is bounded by two major rift valley basins, and the local groundwater flow direction is towards these basins: westward in Western Sinai and eastward in the East. Consequently, springs are present near the Gulfs of Suez and Aqaba. In Central Sinai, the piezometric level is about 200 m above sea level, and at the springs of Oyun Moussa (south of Suez), it is close to sea level. The salinity of the water in Central and Southern Sinai is of the order of 1,500 ppm, but it increases to more than 5,000 ppm in North and West Sinai.

2.5 The Moghra Aquifer System (MAS)

The aquifer assigned to the Lower Miocene occupies mainly the western edge of the Delta up to the Qattara Depression. The MAS outcrops on the surface in Wadi El Natrun and Wadi El Farigh [1, 6] illustrate that MAS aquifer has an average thickness of 300 m and is considered as a non-renewable aquifer system. The hydraulic gradients have a value less than 0.2 m/km. The groundwater in the MAS aquifer is a mixture of fossil and renewable water. Discharge of the aquifer is through evaporation in the Qattara and Wadi El Natrun depressions and through the lateral seepage into carbonate rocks in the western part of the Qattara Depression. The base of the aquifer slopes from ground level near Cairo to 1,000 m below mean sea level west of Alexandria. The saturated thickness is between 70 and 700 m. Permeability ranges between 25 m/day east of Wadi El Farigh to less than 1 m/day in the Qattara area and near the coast. Transmissivity ranges between 500 and 5,000 m/day. The salinity of the water changes from less than 1,000 ppm in the east, i.e. close to the main recharging area, to more than 5,000 ppm in the west, i.e. close to the main discharging area.

2.6 *The Karstified Carbonate Aquifer System (KCAS)*

KCAS is assigned to the Eocene and to the Upper Cretaceous. It predominates essentially in the north and middle parts of the Western Desert. Although the fissured and karstified carbonate aquifer complex occupies at least 50% of the total area of the country, it is the least explored and exploited in Egypt. The recharge to this aquifer is essentially from the upward leakage from the underlying Nubian Sandstone aquifer, and El Tahlawi et al. [1] mentioned that the carbonate complex is generally divided into three horizons: lower horizon assigned to Upper Cretaceous, middle horizon assigned to Lower and Middle Eocene and upper horizon assigned to Middle Miocene. These three horizons are separated by Esna Shale ($\pm$100 m) and Daba Shale (200 m). KCAS generally overlies the Nubian Sandstone complex. The hydrogeology of the fissured limestone is not well understood. In Siwa Oasis, the fissured limestone complex has a thickness of about 650 m (Upper Cretaceous to Middle Miocene) and is lying unconformably on the Nubian Sandstone complex. Over 200 natural springs have a total flow of about 200,000 m^3/day pumping out water from the top portion of the fissured limestone (Middle Miocene). Groundwater salinity there is from 1,500 to 7,000 ppm. On the other hand, test drilling in Siwa area points to the occurrence of water with a salinity of about 200 ppm in the underlying limestone, which belongs to the Eocene and the Cretaceous.

2.7 *The Fissured Hard Rock Aquifer System (FHRAS)*

The Fissured Hard Rock Aquifer System is mainly located in the Eastern Desert and Sinai and assigned to the Pre-Cambrian type. This aquifer is influenced by many factors including, namely, tectonic, morphotectonic and lithologic. Figure 7 shows the distribution of the outcropping and subsurface basement rocks.

3 Nubian Sandstone Aquifer System in Western Desert

The aquifer assigned to the Paleozoic–Mesozoic occupies a large area in the Western Desert and parts of the Eastern Desert and Sinai. Groundwater can be found at very shallow depths, where the water-bearing horizon is exposed, or at very large depths (up to 1,500), where the aquifer is semi-confined. The NSAS is a regional system. It extends into Libya, Sudan and Chad, and it is a non-renewable aquifer system Mahmod et al. [4], Mahmod and Watanabe [6], Ebraheem et al. [17] illustrate that the Nubian Sandstone Aquifer System (NSAS) is considered to be the largest fossil groundwater aquifer system in the world and is located under the eastern end of the Sahara Desert. NSAS groundwater is shared among Egypt, Libya, Sudan and Chad (see Fig. 2). In Egypt, the NSAS extends under the New Valley

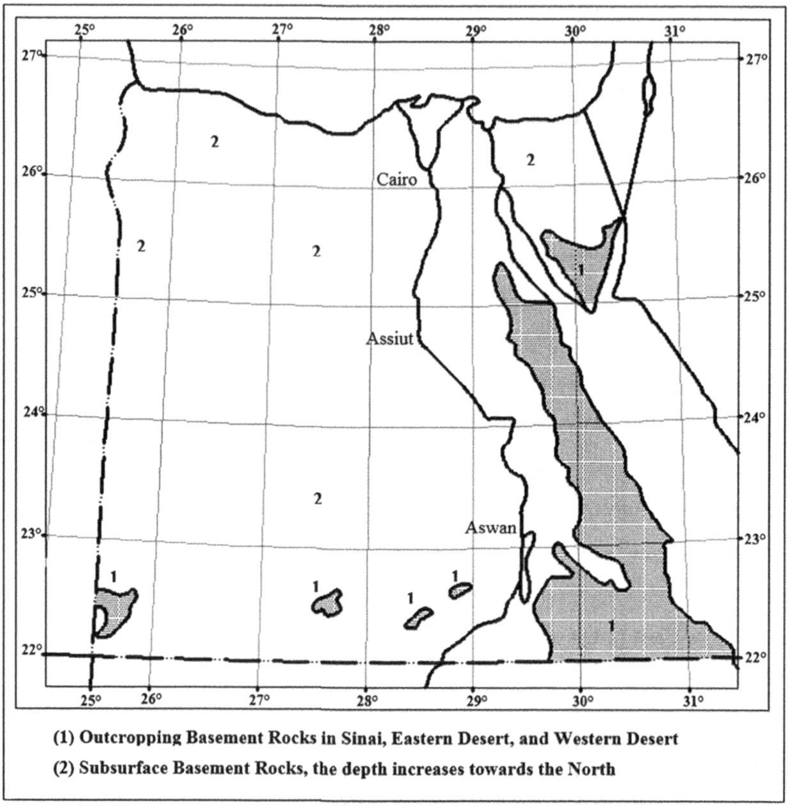

Fig. 7 Distribution of basement rocks in Egypt [1, 16]

Governorate area, where it acts as a local major water source. The total area of the aquifer is approximately 630,000 km^2 and at some areas reaches to a depth of 3.5 km below the ground surface [18]. The recharge of the NSAS aquifer is thought to be generally negligible [19]. Due to overuse of NSAS groundwater for industrial and agricultural applications during the last two decades, a significant drawdown in the groundwater table has been observed. The origin of groundwater in the NSAS has been extensively discussed by several authors since approximately 1920; two possibilities for the origin have been proposed. The older postulated concept is that the groundwater is supplied from precipitation in the southern mountainous parts of the NSAS. This concept was suggested primarily by Ball [20] and supported by Sanford [21]. The newer postulate by Sonntag [22] suggested that the bulk of the groundwater mass within the aquifer was originally supplied in situ during the humid pluvial periods (up to 30,000 years B.P.). There are three possible ways of groundwater recharge: regional groundwater influx from areas with modern groundwater recharge, local infiltration through precipitation during wet periods in the past and finally post-High Dam construction seepage of Nile water [24]. In addition, [20, 21, 23] noted that the general flow direction in the NSAS is from the southwest to the northeast Thorweihe and Heinl [25] found that the amount of the current recharge to the NSAS from the south is not significant based on the groundwater velocity, aquifer resistance time and climatic conditions.

The geology of the Western Desert of Egypt, including the Kharga Oasis, has been well documented by many geologists [21, 26–28]. The NSAS is a confined aquifer that is overlaying an impermeable crystalline rock that comprises the lower confining layer. The upper confining layer is an impermeable variegated shale and clay overlying the Nubian Sandstone aquifer. The NSAS is composed of a series of mostly unfossiliferous formations. In the Kharga Oasis as an example, NSAS formations are comprised of sands, sandstones, clays and shales (see Fig. 8). The NSAS formation changes gradually from primarily continental sandy facies in the southern region (Sudan) to intercalations of sandstones and clays of alternating continental and shallow marine facies in the central region (e.g. the Kharga Oasis). The formation then changes to marine facies in the northern region (Qattara Depression), where it consists of thick beds of clays intercalated with beds of limestone, dolomites and sandstones. It is generally thought that the geological age of the NSAS ranges from the Cambrian to the Upper Cretaceous period [29].

Lamoreaux et al. [23] and Thorweihe and Schandelmeier [30] reported that the thickness of the Nubian Sandstone aquifer beneath the Kharga Oasis ranged from 300 to 1,000 m. Shata [31] noted that in the Kharga Oasis, 32 water-bearing horizons had been recognized within the Nubian Sandstone succession. Shata [31] grouped these water-bearing horizons into three aquifers that include (1) a top layer with an average thickness of 200 m, (2) a middle layer with an average thickness of 200 m and (3) a lower layer with a thickness of approximately 250 m. Lamoreaux et al. [23] mentioned that the NSAS consists of thick sequences of

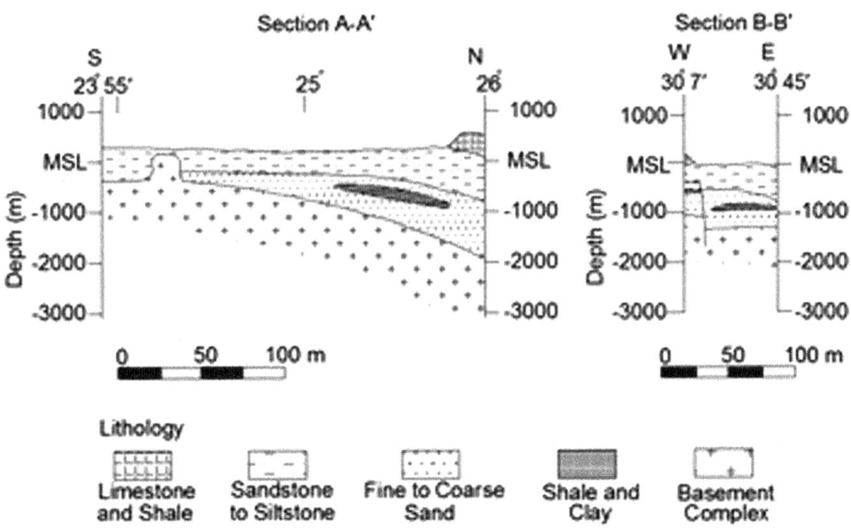

Fig. 8 Lithostratigraphic sections across the study area [4, 6, 9]

coarse, clastic sediments of sandstone and sandy clay interbedded with shale and clay beds. These clayey impermeable beds restrict the vertical movement of water among water-bearing layers. These beds of lower permeability, however, are lenticular and discontinuous; thus, regionally, the Nubian sandstone forms a single aquifer complex.

4 Groundwater Modelling and Uncertainty

4.1 Model Types

A model is a simulation of areal and temporal properties of a system, or one of its parts, in either a physical or mathematical way. The difficulties of building physical model are obvious in how feasible is the water table (hydraulic head) in a multilayer aquifer exposed to various stresses such as precipitation, surface runoff and leakage from deep underlying strata and then change some of its geometric and hydrogeological properties as needed. For these difficulties, the application of real physical models has been limited to educational and demonstration purposes. Mathematical models are those equations used to analyse groundwater flow. Three mathematical models could be identified; depending upon the nature of equations involved, these models can be empirical, probabilistic and deterministic models. Empirical models are derived from experimental procedures until relations are fitted to some mathematical function. Although empirical models are limited in scope, they can be an important part of more complex problems. Probabilistic (stochastic) models are based on statistics such as probability. These models either have various simple probability distributions of a hydrogeological information or complicated stochastic, time-dependent models. The main problem of the stochastic models, for a wider use in hydrogeology, is that they need large data sets for parameter identification. Moreover, it is difficult to predict the future behaviour of groundwater flow under any change of water table stress. Deterministic models are those models that are predetermined by physical laws governing groundwater flow such as Theis equation and Darcy's law. These types of models are used to solve complicated groundwater problems as a multiphase flow through a multilayered, heterogeneous, anisotropic aquifer system. There are two categories of deterministic models: analytical and numerical models. Analytical models are used to solve one equation of groundwater flow at a certain time for one point or line of points. Analytical models are easy to be applied to one point with homogeneous aquifer property. Calculation time is increased if the model is applied for many points and if the aquifer is quite heterogeneous. If the situation gets really complicated, such as when there are several boundaries, more pumping wells and several hydraulically connected aquifers, the feasible application of analytical models terminates. Numeric models describe groundwater flow for many points as specified by the user considering heterogeneity of the aquifer. This is achieved by dividing the area of interest into small subdivisions named cells or elements. Then, a basic groundwater flow equation is solved for each cell considering its water balance. The outputs are a distribution of hydraulic heads at points representing individual cells. These points can be placed at the centre of the cell, at intersections between adjacent cells or elsewhere. These models are coded by solving the basic differential flow equation for each cell approximately to finally get an algebraic equation that can be easily solved through an iterative process (numeric) solution. The two most widely applied numeric models are the finite difference model (FDM) and finite element model (FEM). The difference between FDM and FEM is mainly at approximation of the differential flow equation until getting the algebraic equation to be solved then by iterative process.

4.2 Source of Uncertainty

Groundwater management in many arid areas in the world has been difficult due to the limited availability of hydrogeological data. The lack of basic data often limits the implementation of the numerical models for groundwater flow analysis, as well as introduces a problem for model calibration and validation. A deterministic approach is useful to overcome the uncertainty because it depends on variables or parameters with values determined by taking into account previously collected historical information. Although this type of approach has some limitations, it can be very useful in the decision-making process. It is relatively simple and allows for a detailed exploration of other aspects that are sometimes neglected in more sophisticated models.

There are many sources of uncertainty in groundwater analysis. The most common sources are:

- Natural variability of hydrologic properties of the groundwater aquifer
- Limited and incorrect information on hydrogeological variables and parameters that characterize the groundwater flow
- Lack of information on future groundwater exploitation rate because of temporal changes of anthropogenic activities, land use, land cover, and climate change

Sensitivity analysis is a straightforward technique that allows the identification of the parameters and variables that have more influence on system performance and allows evaluation of the scale of this influence. Uncertainty in constraints can also be modelled using several different techniques. When uncertainty is small, it can be done in terms of expected values. Chance-constrained models are very useful in reliability analysis (reliability expresses the probability of a system operating without failure). Chance-constrained models describe constraints in mathematical models in the form of probability levels of attainment. The chance constraints can be converted into deterministic equivalents and can therefore be easily incorporated into deterministic programming models. From the previous two sections, it could be concluded that it is difficult to develop a proper groundwater management system given such uncertainties and data limitations when using ordinary numerical models.

5 Overcoming Uncertainty

To implement an optimal groundwater management, temporal and spatial variation of groundwater table should be investigated first as an indispensable factor in groundwater flow analysis. Temporal and spatial variation of groundwater table is an important factor for implementing the optimal management of groundwater resources. A model is suggested to make clear this spatial distribution of groundwater flow by analysing spatial relationships among OWs considering the limited availability of data. This model is linear regression model. Genetic algorithm (GA) was adapted as a technique to estimate the optimized values of the linear parameters considering robustness and high conversion of this method as one of the soft computing techniques (SCTs). This mentioned technique quantifies the spatial structure of the data and produces a prediction of pore pressure change at arbitrary points. A newly developed model named the grey model (GM) was adopted to analyse groundwater flow. This model combines a finite element method (FEM) with GA model to try to obtain the best-fit piezometric level trends compared to observations. On the other hand, the GM has no clear theories that exist for selecting the number of input model trends and the most suitable values of input parameters. Therefore, a modification of the GM is newly proposed and called the modified grey model (MGM) in an attempt to determine a process for selecting the best input models' trends with the appropriate values of input parameters to achieve acceptable fitting to observations.

5.1 Grey Model (GM)

GM is a combination of a 2D-FEM with genetic algorithms (GA) as the NM and SCT, respectively. FEM was used to analyse groundwater flow under the assumption of hydrogeological and boundary conditions to produce approximate solutions for temporal piezometric level changes. The GA model was used to fit the measured temporal piezometric level changes.

Although the GM has been proposed by some hydrogeologists, it has not yet been widely applied to groundwater problems. The main concept of the GM is combining a numerical model (NM) with a soft computing technique (SCT) Mohammed et al. [32] proposed a GM as a combination of a 3D-FEM with an artificial neural network (ANN) as the NM and SCT, respectively Mohammed et al. [32] used a GM to predict the piezometric level change in the fractured rock mass at Mizunami Underground Research Laboratory (MIU), Japan. In the current study, an FEM and a linear regression model based on genetic algorithms (GAs) were used. FEM was used to analyse groundwater flow under the assumption of hydrogeological and boundary conditions to produce approximate solutions for temporal piezometric level changes. The GA model was used to fit the measured temporal piezometric level changes.

5.1.1 Model Structure

Due to the uncertainty in the hydraulic parameters for water-bearing layers and lack of information about the pumping rates from each layer, it was difficult to properly construct a 3D-FEM model. In addition, the OWs do not provide piezometric level data for specific aquifer layers but rather for the entire deep aquifer (layers B, C and D). Because of these considerations, a 2D-FEM was applied in this study. The governing equation for saturated groundwater flow can be obtained by combining a special form of Darcy's law (derived from the water phase momentum balance) and the continuity equation written for the water phase [33, 34], as given in the equation

$$\frac{\partial}{\partial x}\left(T_{xx}\left(\frac{\partial h}{\partial x}\right)\right)+\frac{\partial}{\partial y}\left(T_{yy}\left(\frac{\partial h}{\partial y}\right)\right)=S\left(\frac{\partial h}{\partial t}\right)-W$$

where T_{xx} and T_{yy} are the transmissivities along the x and y directions, respectively, h is the hydraulic head, S is the storativity, W is the groundwater volume flux per unit area (positive for outflow and negative for inflow), x and y are the Cartesian coordinates and t is the time.

The FEM calculations were carried out assuming n different 2D hydrogeological models. For each model, the grid geometry and number of elements were identical. However, different sets of hydrogeological conditions, such as horizontal transmissivity in the x and y directions (T_{xx} and T_{yy}) and specific storage and boundary conditions were assumed for each model. Then, n piezometric level trends ($H_1(t)$, $H_2(t)$,. . .,$H_n(t)$) were calculated by FEM. These simulated piezometric level trends were not representative of the measured piezometric level trends. The linear model was used to reconstruct the measured piezometric level trend by combining FEM output piezometric level trends ($H_1(t)$, $H_2(t)$,. . ., $H_n(t)$) as given in the equation

$$H_s(t)=\sum_{i=1}^{i=n}\alpha_i\cdot H_i(t),\quad \sum\alpha=1$$

where $H_s(t)$ is the reconstructed piezometric level trend at a certain location, $H_i(t)$ are the 2D-FEM simulated piezometric level trends, and α_i are weight parameters for the linear regression model. The best-fitting parameters for the linear model are estimated by a GA model. The weight parameters can also be used for future prediction of temporal piezometric level change at any time (t). The structure of the GM is schematically illustrated in Fig. 9.

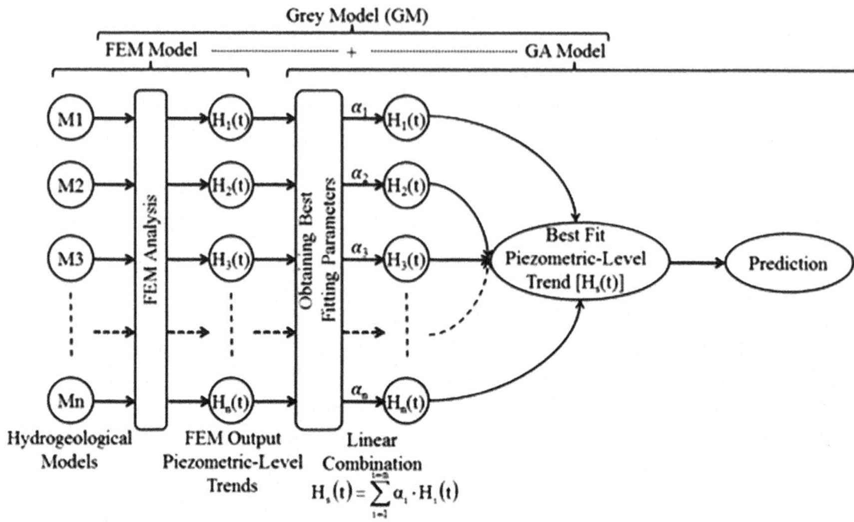

Fig. 9 Structure of the developed grey model [4]

The basic process of the GA used in this study is as follows (see Fig. 10):

1. Generate N different chromosomes by giving random values in the range of 0.0–1.0 as genes. The gene values are adjusted to fit the criterion that the sum of the genes in each chromosome should equal 1.0. The number of chromosomes N is called the population size, and all chromosomes are put in the initial (1st) pool representing the first generation.
2. The previous equation is applied to obtain the reconstructed piezometric level trend ($H_s(t)$). Then, the root mean square error (RMSE) and fitness value (F_i) between the measured ($H_o(t)$) and the reconstructed piezometric level trends ($H_s(t)$) for every chromosome are calculated. Chromosomes in the first pool are rearranged based on the probability value of each chromosome (rating process):

$$F_i = \sqrt{\frac{N}{\sum_{i=1}^{i=N} (H_S(t) - H_o(t))^2}} = \frac{1}{\text{RMSE}}$$

$$P_i = \frac{F_i}{\sum_{i=1}^{i=N} F_i}$$

3. Based on the rating process, the best-fit chromosomes are selected and sent to fill the 2nd pool. The selection of chromosomes was performed by using a deterministic sampling method [35, 36]. In this method, the average fit of the

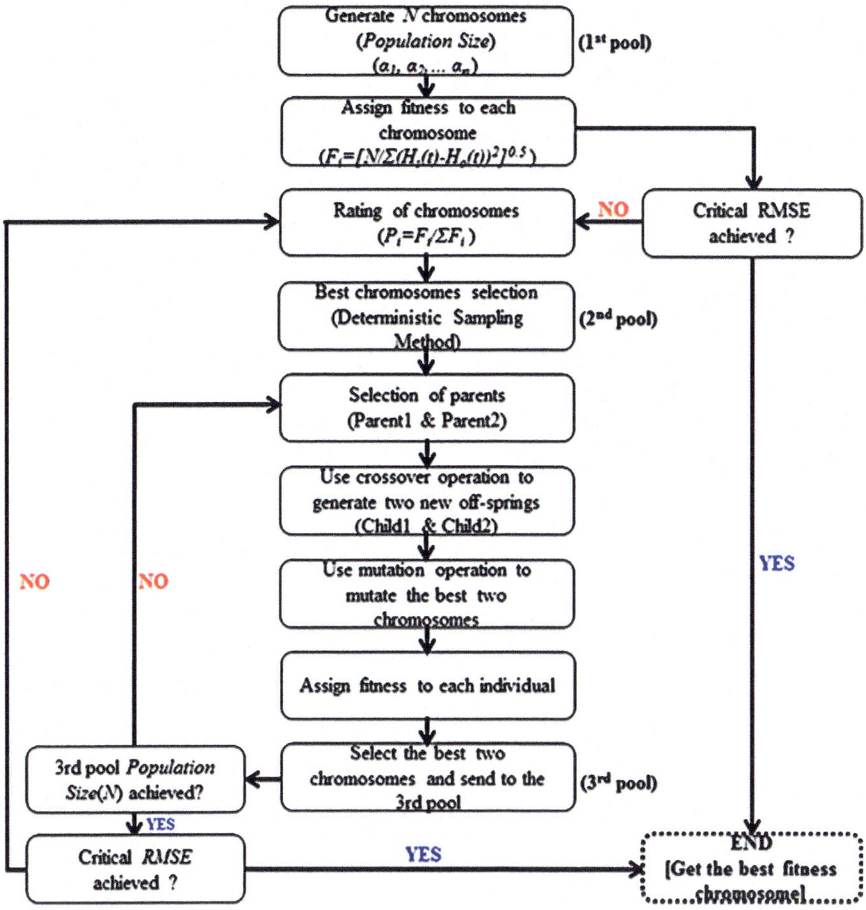

Fig. 10 Flow chart describing the GA optimization process [4]

population is calculated. Then, the fitness for each chromosome is divided by the average fitness, but only the integer part of this operation is used to select individuals to send to 2nd pool. The entire fraction is used to sort the individuals. The remaining free places in the new population are filled with the chromosomes that have the largest fraction values in the sorted list.

4. Two chromosomes are selected randomly from the second pool and are called parent chromosomes. Two offsprings are created from the parents by using a crossover operation. A crossover is a process to mix the genes of parent chromosomes. In this study, a two-point crossover technique was used [37, 38]. By using the chromosomes of parents and children, four temporal piezometric level trends are calculated, and then two of these trends with a large value of F_i are selected. The ratio between the number of chromosomes thus created and the population size is called the crossover probability.

5. The selected two chromosomes constructed and calculated by the crossover process were mutated by changing some gene values using random numbers between 0.0 and 1.0. After mutation, the sum of the gene values should equal 1.0. Among the four chromosomes, the two chromosomes that have the best-fitting values are sent to the 3rd pool. The ratio between the mutated gene and all genes of one chromosome is called the Mutation Probability. Crossover and mutation are repeated until the population size of the 3rd pool is filled; this process is referred to as the second generation.
6. The process from step 2 to step 5 is repeated for a fixed number of generations or until the error between measured and calculated piezometric level trends falls below an assumed critical RMSE value.

All steps from 1 to 6 are applied to a portion of the time series data to train the model (training/learning phase). The other part of the data is used for validation of the resulting weights (prediction/test phase). Due to the limitation of data and to guarantee good results in the prediction phase and future assessments, the majority of the data series is used in training, while the rest is left for the prediction phase.

5.2 *Modified Grey Model (MGM)*

In the GM, the used GA model has no clear concept for selecting the number of input piezometric level trends and values of input parameters. The input piezometric level trends are the calculated piezometric level trends from the assumed hydrogeological models. On the other hand, the input parameters are the parameters which used to fit the measured observations such as population size, crossover probability, mutation probability and number of generations. For simplicity, in the current study, the number of input piezometric level trends will be named as input models' trends.

In the current study, this concept is processed, and new model named as modified genetic algorithm (MGA) is proposed trying to select the best input models' trends for fitting to the measured piezometric level trends. Moreover, the effect of number of input models' trends on the values of input parameters was studied by meaning of sensitivity analysis for both models GA and MGA. By combining this new proposed model MGA with the FEM model, a new model is produced and named as modified grey model (MGM).

In the current study, the MGM was newly developed for the analysis of groundwater flow. Broadly speaking, the main concept of the MGM is quite similar to the GM because it combines a numerical model (NM) with a soft computing technique (SCT). The GM has not yet been widely applied to groundwater problems. In the GM, the number of input models' trends and the values of the input parameters are selected hypothetically before obtaining the best-fitting values [4, 32]. The main problem in the GM is that there is no clear concept in the GA model for the selection of either the number of input models' trends or the values of input parameters, such as population size, crossover probability, mutation probability and number of generations (fitting calculation time). For this purpose, a modification of the GA model was developed to select the input models' trends that achieve the best fit to the observations. Then, the effects of the best selected models' trends on the values of GA input parameters were studied. The GA model after modification will be called the modified genetic algorithm (MGA) model. The combination of the best models' trends will be called the best models combination (BMCk, where k is the number of selected models' trends). The MGM is the combination of FEM and MGA. As an application for MGM, this model is used to evaluate three different future exploitation/development plans under consideration for the aquifer system to explore the hydrological feasibility of these plans. The MGM simulation is applied for the period from the year 1960 to the year 2100.

5.2.1 Model Structure

The FEM calculations were carried out assuming n different 2D hydrogeological models. Then, n piezometric level trends ($H_1(t)$, $H_2(t)$,. . .,$H_n(t)$) were calculated by FEM. These simulated piezometric level trends were not representative of the measured piezometric level trends. The linear model (GA model) was used to reconstruct the measured piezometric level trend by combining the calculated piezometric level trends ($H_1(t)$, $H_2(t)$,. . .,$H_n(t)$), as given in the equation below. For MGA, only the best models' trends were selected to be combined together by the best model combinations (BMC$_k$ ($R_1(t)$, $R_2(t)$,. . .$R_k(t)$, where R_k is the selected models' trends from H_n models' trends) to match the target measured trend:

$$R_s(t) = \sum_{j=1}^{k} \alpha_j \cdot R_j(t), \qquad \sum \alpha_j = 1, \qquad k \leq n$$

where $R_s(t)$ are the reconstructed piezometric level trends at a certain location estimated by MGM; $R_j(t)$ is the 2D-FEM simulated piezometric level trends; and α_j are weight parameters. The produced weight parameters can also be used for the future prediction of temporal piezometric level changes at any time (t). The structure of the MGM is schematically illustrated in Fig. 11.

Many researchers reported that the GA model provides good performance for estimating the optimum combination of weight parameters [4]. This combination of weight parameters is called a chromosome (α_1, α_2,.. α_n), and each weight parameter is called a gene. However, the optimum number of inputs (hydrogeological models) and the values of the input parameters remain a problem. For this reason, the GA model has been modified by adding a selection process for finding the optimum input models' trends. This newly developed model is called the modified genetic algorithm (MGA) model. The selection process is used for selecting the best inputs that achieve an appropriate goodness of fit with the measured piezometric level trends. The number of selected inputs is "k," and the combination of these inputs is called the best models combination (BMC$_k$). The concept of the proposed model (MGA) could be described in two stages, as follows (see Fig. 12) [4, 39]:

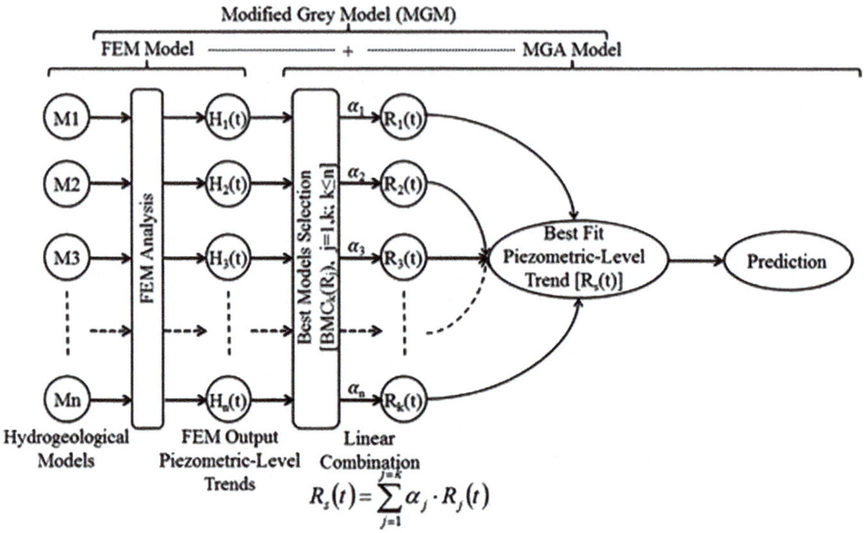

Fig. 11 Structure of the developed grey model [6]

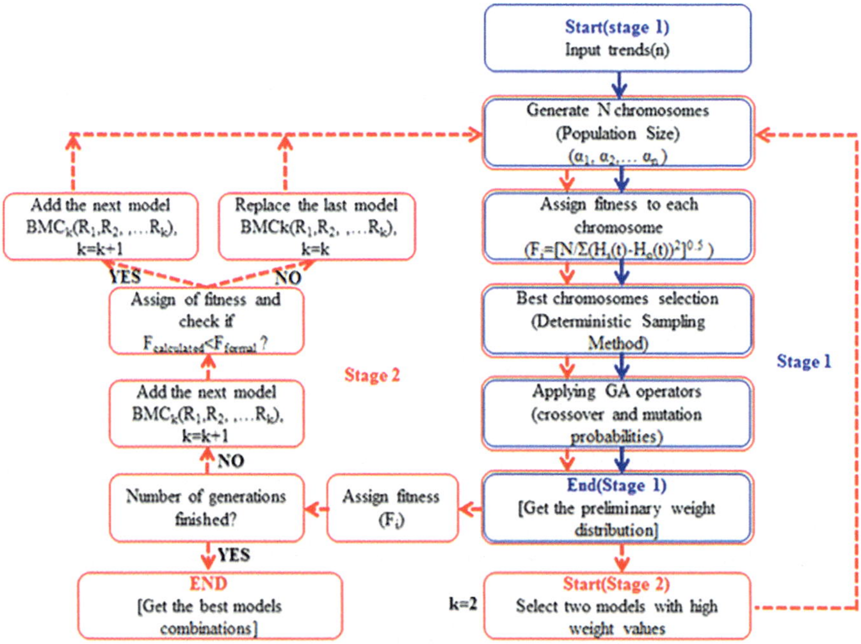

Fig. 12 Flow chart describing the MGA optimization process [6]

Stage 1 (Preliminary weight distribution): The process in this stage is the same as in the ordinary GA model [4]; however, the target of this stage is not determining the optimum weight parameters. This stage is used to obtain a preliminary weight distribution for the inputs with any value for number of generations. For this reason, using a small number of generations is enough to obtain this type of distribution. This stage contains the following main steps:

1. After simulating groundwater flow using the hydrogeological model explained previously, n temporal trends of hydraulic heads ($H_1(t)$, $H_2(t)$,. . .,$H_n(t)$) are obtained and used as inputs to the GA model. Then, N different chromosomes are generated by giving random values in the range of 0.0–1.0 as genes. The gene values are adjusted to fit the criterion that the sum of the genes in each chromosome should equal 1.0. The number of chromosomes N is called the population size.
2. As the second step of the basic process of the GA model, the reconstructed piezometric level trend ($H_s(t)$) is obtained. Then, the fitness values (F_i) between the measured ($H_o(t)$) and the reconstructed piezometric level trends ($H_s(t)$) for every chromosome are calculated.
3. The best-fit chromosomes are selected using a deterministic sampling method. In this method, the average fit of the population is calculated. Then, the fitness for each chromosome is divided by the average fitness, but only the integer part of this operation is used to select individuals. Then, the entire fraction is used to sort the individuals. The remaining free places in the new population are filled with the chromosomes that have the largest fraction values in the sorted list.
4. These "n" models' trends are processed by GA processes (crossover and mutation processes) to obtain the final weight parameters (α_1, α_2,.. α_n). At the initial stage, the weight parameters are given as random numbers. Then, the weight parameters are pursued through the repetition of crossover and mutation processes until completing the selected number of generations. Crossover and mutation processes are represented by two operators: crossover probability and mutation probability.

Stage 2 (Obtaining the $BMC_k(R_j(t))$; $j = 1,k$; $k \leq n$): Regarding stage 1, the weights of each input for fitting the observed trend were recognized. In this stage, the inputs are renamed according to their weights from $R_1(t)$ for the highest weight input to $R_n(t)$ for the lowest weight input. Hereafter, the following steps take place:

1. The first two inputs ($BMC_k(R_j(t))$; $j = 1,k$; $k = 2$) are combined using the GA model (stage 1 procedures). Then, $R_s(t)$ is calculated, and the fitness with the observed trend $R_o(t)$ is calculated.
2. The third input $R_3(t)$ is added to the formal combination to become $BMC_3(R_1(t)$, $R_2(t)$, $R_3(t))$, and then the fit is recalculated.
3. If the calculated fit is less than the formal fit ($F_{\text{calculated}} < F_{\text{formal}}$), then the input will be added to the combination to become $BMC_4(R_1(t), R_2(t), R_3(t), R_4(t))$. On the contrary, if the calculated fit is larger than the formal fit, the last input in the combination $BMC_3(R_1(t), R_2(t), R_3(t))$ will be replaced by its next input to

become $BMC_3(R_1(t), R_2(t), R_4(t))$. This process is continued until the selected number of generations has been completed.
4. The best models combination $(BMC_k((R_1(t),R_2(t),...R_k(t))$ is obtained.

The two stages explained above are applied on a part of the time series data as training for the model (training phase). The other part of the data is used for the validation of the resulting weights (test phase). Due to the limited data and to guarantee good results at the prediction phase and in future assessment, a large percentage of the data is used in the training phase, and the other remains for the prediction phase.

6 Application of MGM on the NSAS Under Kharga Oasis

As applications of the new proposed model MGM, future assessment of piezometric level change was analysed by developing three exploratory scenarios of groundwater withdrawal that involved either expanding the present extraction rate or redistributing the groundwater withdrawal over the recent working production wells (RWPWs).

The Kharga Oasis is one of the depressions located in the New Valley in the southern part of the Western Desert of Egypt (Fig. 1). This oasis is bounded by the Eocene limestone plateau on the east and north, where steep cliffs form a sharp boundary. This limestone plateau stretches along Middle and Upper Egypt at an elevation up to 550 m above mean sea level (amsl) (approximately 400 m above the depression floor of the oasis). The studied area lies between the latitudes of 23° 55′ to 26° N and the longitudes 30° 7′ to 30° 45′ E, as shown in Fig. 13. The calculated area shown in Fig. 13 is used in the numerical analysis, as explained below.

The Kharga Oasis is situated in the tropical arid climate zone. The maximum daytime temperature reaches up to 45–50 °C in the summer; however, the temperature drops below zero °C at night in the winter. The Kharga Oasis is one of the driest areas of the Eastern Sahara, which thus makes it one of driest areas on Earth [39]. The average potential evaporation rate reaches approximately 18.4 mm/day, and the mean annual value of the relative humidity is approximately 39%. The annual precipitation is extremely low and is almost insignificant (1 mm/year) [40].

Thorweihe and Heinl [25] found that the amount of the current recharge to the NSAS from the south is not significant based on the groundwater velocity, aquifer resistance time and climatic conditions Heinl and Thorweihe [19] estimated that the flow velocity in the deep aquifer is approximately 1 m/year, which means the groundwater would require over 100,000 years crossing the distance from either tropical infiltration areas or the River Nile to reach the Kharga Oasis part of the NSAS. Because of the known precipitation and potential evaporation conditions in the study area, recharge of the Kharga Oasis part of the NSAS was neglected in the current study.

Lamoreaux et al. [23] and Ebraheem et al. [18] reported that the thickness of the Nubian Sandstone aquifer beneath the Kharga Oasis ranged from 300 to 1,000 m Shata [31] noted that in the Kharga Oasis, 32 water-bearing horizons had been recognized within the Nubian Sandstone succession Shata [31] grouped these

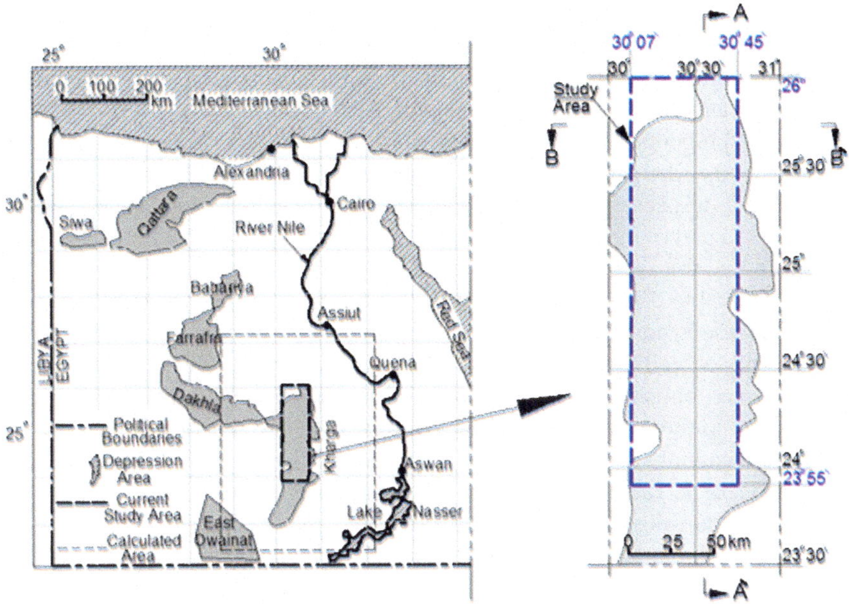

Fig. 13 Location map of the study area showing observed sites [6]

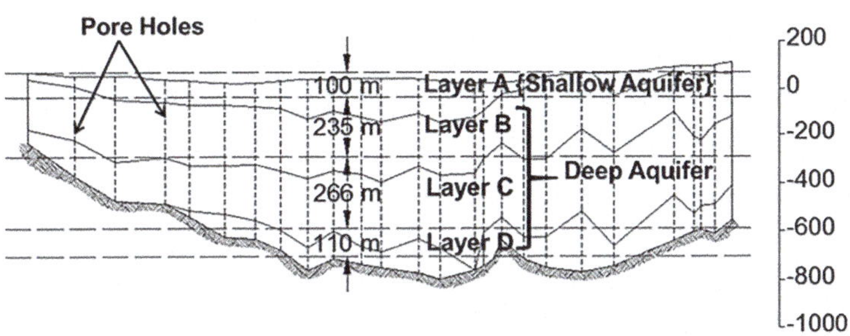

Fig. 14 Classification of water-bearing layers compiled from Lamoreaux et al. (1985) [9]

water-bearing horizons into three aquifers that include (1) a top layer with an average thickness of 200 m, (2) a middle layer with an average thickness of 200 m and (3) a lower layer with a thickness of approximately 250 m Lamoreaux et al. [23] mentioned that the NSAS consists of thick sequences of coarse, clastic sediments of sandstone and sandy clay interbedded with shale and clay beds. These clayey impermeable beds restrict the vertical movement of water among water-bearing layers. These beds of lower permeability, however, are lenticular and discontinuous; thus, regionally, the Nubian sandstone forms a single aquifer complex.

At the Kharga Oasis, [23] classified these water-bearing layers of the average thickness of approximately 700 m into four layers as shown in Fig. 14. In the current study and following [23], the classification and the average thickness of each layer are as follows: (1) layer A with an average thickness of 100 m (top layer), (2) layer

B with an average thickness of 235 m, (3) layer C with an average thickness of 266 m and (4) layer D with an average thickness of 110 m (bottom layer). On the basis of this classification, Layer A refers to the shallow aquifer, while the others (layers B, C and D) refer to the deep aquifer.

Shata [31] reported that the transmissivity and storage coefficient values for the NSAS under the Kharga Oasis were in the range of 100–2,000 m^2/day and 1.5×10^{-4} to 1.5×10^{-2}, respectively Hesse et al. [41] reported that the hydraulic conductivity of the Kharga part of the NSAS has an average value of 2.16 m/day Ebraheem et al. [18] noted that the average hydraulic conductivity for the regional NSAS is in the range of 0.0864 to 0.864 m/day, with an average storage coefficient of 1.0×10^{-4}. However, these hydraulic properties have not been confirmed for the entire aquifer.

Two types of wells are considered in this study, recent working production wells (RWPWs) and observation wells (OWs). In the current study, the whole area is divided into northern and southern parts. Based on distribution of settlements and RWPW allocations, the studied area was subdivided into eight subdivisions, as shown in Table 1. Table 1 summarizes the name and the available RWPW data for every subdivision. Figure 15 illustrates the locations and distributions of the RWPWs and OWs over the study area. The RWPWs and OWs are distributed over 14% and 45% of the study area, respectively. From Fig. 15, it could be concluded that RWPWs are highly concentrated over two regions: the middle area of the northern part (Area 3–Area 5) and Area 7 in the southern part. For the current study, these two regions were named the M-region and the SN region, respectively. Recent screen depths of the RWPWs range from 76 to 813 m, while the recent depths of the OWs range from 200 to 950 m below the ground surface. Recent groundwater extraction has only been from the deep aquifer (layers B, C and D).

Q is the discharge from each well

6.1 Hydrogeological Model

Based on the hydrogeological properties of the Kharga Oasis part of the NSAS reported by previous researchers, *n* hydrogeological models were constructed. For each model, transmissivities (T_{xx} and T_{yy}) and storativity values were assumed to be

Table 1 Summary of RWPW data throughout subdivisions of the study area

Area no.	Area name	No. of RWPWs	Well depth (m)	Q m^3/h
1	Monera	10	101–720	14–117
2	Sherka	11	125–703	14–114
3	Kharga	41	88–706	28–177
4	Jenah	14	89–804	31–329
5	Bulaq	14	163.5–813	46–216
6	Garmashin	12	217–673	20–205
7	Paris	44	76–600	14–229
8	Darb El-Arbeen	41	91.3–485.4	44–56

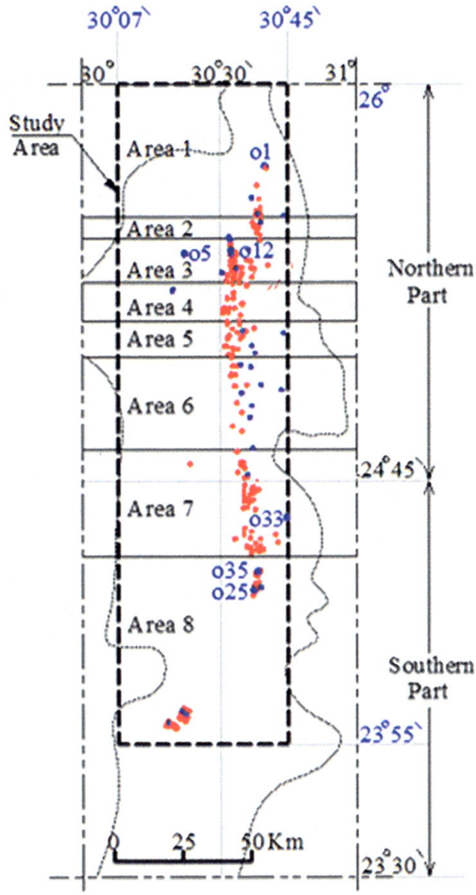

Fig. 15 Distribution of RWPWs and OWs over the Kharga Oasis; RWPWs shown in *red* and OWs in *blue* [6]. *Dotted line* indicates the schematic boundary of the Kharga Oasis. *Dashed line* indicates the current study area

in average ranges of 50–500 m^2/day and 3.28×10^{-3} to 3.28×10^{-2}, respectively. Although the total discharge extracted from the Kharga Oasis part of the NSAS is known for the period from 1960 to date (Fig. 16), the discharge for each production well within the same period is unknown. For this reason, a scenario of groundwater withdrawal of each RWPW was assumed considering that discharge since 2005 has been almost steady. Groundwater withdrawal for each RWPW was assumed to follow the same groundwater withdrawal trend as the total discharge curve given in Fig. 16. The assumed scenario of groundwater withdrawal started from the year 1960 and increased gradually until the year 2005. Then, it continued steady until 2009. Beyond 2009, three development scenarios of groundwater withdrawal were proposed that involved either expanding the present extraction rate or redistributing the groundwater withdrawal over the recent working production wells (RWPWs) as will be explained later.

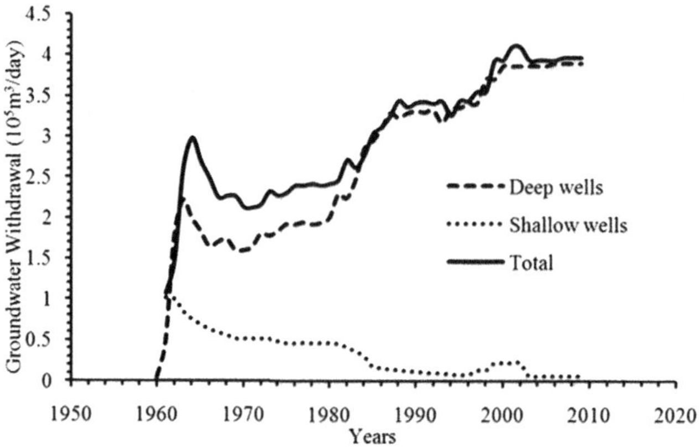

Fig. 16 History of groundwater withdrawal from deep and shallow aquifers [4]

6.2 Initial and Boundary Conditions

The NSAS in the Kharga Oasis area is only a small part of the entire NSAS in Egypt. Groundwater flow in the NSAS is governed by the conditions at the boundaries of the regional system. These conditions are not well defined for the Kharga Oasis area. Because of that, the boundary condition was assumed as a fixed boundary head with the flow direction from the southwest to the northeast. With that assumption, the boundary conditions were expressed as hydraulic head values given at the four corners (Z1–Z4) of the calculated area indicated in Fig. 13. Hydraulic heads along each boundary line were assumed to change linearly as shown in Fig. 17. Values of Z1, Z2, Z3 and Z4 were roughly estimated by considering the goodness of fit between the calculated and measured piezometric level trends. In this study, 16 hydrogeological models were assumed with different sets of hydrogeological and boundary conditions as summarized in Table 2.

To reduce the direct effects of boundary conditions on the results, the calculated area was extended beyond the study area boundary, as shown in Fig. 13. This area was subdivided into 24,069 rectangular elements and 24,396 nodes with the RWPWs and OWs located on mesh nodes. The elements around the RWPWs and OWs were set as the smallest elements with the sizes gradually increasing towards the boundary of the calculated area, as illustrated in Fig. 17. The minimum and maximum lengths of the element sides were 377 m and 7,000 m, respectively.

6.3 Proposed Future Scenarios for Groundwater Withdrawal

Although the total discharge extracted from the Kharga Oasis part of the NSAS is known from 1960 to 2009 (Fig. 16), the discharge for each production well in the same period is unknown. For this reason, a scenario of groundwater withdrawal of

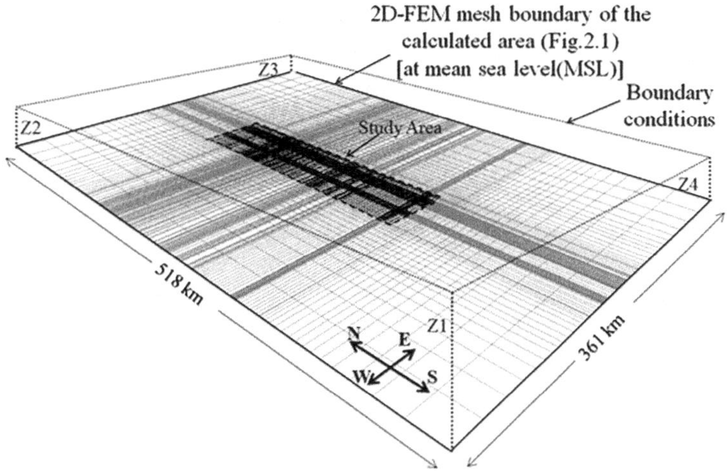

Fig. 17 Adopted 2D-FEM mesh geometry for the calculated area, and the study areas. Z1–Z4 are the assumed hydraulic heads at the corners of the calculated area (after six)

each RWPW was assumed to follow the same groundwater withdrawal trend as the total discharge curve given in Fig. 16, considering that the discharge was almost steady within the period 2005–2009. From 2009 to the present, groundwater withdrawal was assumed to continue being steady. Hereafter, for each RWPW, three scenarios of groundwater withdrawal are assumed, considering that no new production wells will be constructed. The main criterion in the three proposed scenarios is to keep the sum of groundwater withdrawal from the current study area constant. These scenarios are explained as follows:

Scenario 1: The present groundwater withdrawal is increasing and kept constant until the final simulation year of 2100.

Scenario 2: In this scenario, the groundwater withdrawal is redistributed among the RWPWs in the M-region and the southern subdivision Area 8. Area 8 is the highest piezometric level area of the study area [4, 23]. In this scenario, the groundwater withdrawal in the M-region is equal to that in the 1st scenario reduced by 20%. Then, the sum of the groundwater withdrawal reduced from the RWPWs of the M-region is distributed over the RWPWs in Area 8, while keeping the ratio between the RWPW discharges constant.

$$\sum Q_{\text{M-region}} = 0.80 \times \sum Q_{\text{M-region(Scenario1)}}$$

$$\sum Q_{\text{Area7}} = \sum Q_{\text{Area8(Scenario1)}} + 0.20 \times \sum Q_{\text{M-region(Scenario1)}}$$

Scenario 3: In this scenario, groundwater withdrawal is redistributed among the RWPWs in the northern part and the southern subdivision Area 8. The groundwater withdrawal of the northern part of this scenario is equal to that in the 1st scenario reduced by 5%. Then, the sum of the groundwater withdrawal reduced from the

Table 2 Proposed hydrogeological models

	Corner hydraulic heads						
Model name	Z1 (m)	Z2 (m)	Z3 (m)	Z4 (m)	T_{xx} (m/day)	T_{yy} (m/day)	$S \times 10^{-3}$
M1	87	220	150	105	50	50	32.8
M2	37	170	100	55	50	50	3.28
M3	27	160	90	45	50	500	3.28
M4	37	170	100	55	5	50	3.28
M5	37	170	100	55	500	500	32.8
M6	85	85	85	85	50	500	3.28
M7	37	170	100	100	50	500	3.28
M8	−282	484	123	−111	50	500	3.28
M9	−282	484	123	−90	50	500	3.28
M10	−282	490	123	−111	50	500	3.28
M11	−300	484	123	−111	50	500	3.28
M12	−297	495	108	−126	50	500	3.28
M13	−282	484	123	−111	138	500	3.28
M14	−282	484	123	−111	50	50	3.28
M15	−232	534	123	−61	97	500	3.28
M16	−282	484	123	−111	138	500	32.8

RWPWs in the northern part is distributed among the RWPWs in Area 8 while keeping the ratio between the RWPW discharges constant:

$$\sum Q_{\text{Northern-part}} = 0.95 \times \sum Q_{\text{Northern-part(Scenario1)}}$$

$$\sum Q_{\text{Area7}} = \sum Q_{\text{Area8(Scenario1)}} + 0.05 \times \sum Q_{\text{M-region(Scenario1)}}$$

for which Q is the groundwater withdrawal of each RWPW.

6.4 Sensitivity Analysis

The input parameters population size, crossover probability, mutation probability and number of generations should be properly determined for the proposed MGA model. Moreover, the comparison between the goodness-of-fit values of the MGA and GA models should be analysed under the different values for input parameters. For this purpose, a sensitivity analysis was performed to properly select the appropriate values of these parameters. In this analysis, the measured piezometric level trend at OW o1 was used as a target, and the 16 calculated piezometric level trends calculated by FEM were used as input data for the MGA and GA models. For the GA model, Fig. 18a, b display the relationships among RMSE, number of generations and population size and among RMSE, crossover probability and mutation probability, respectively, and the same for MGA in Fig. 19a, b. From Fig. 18a, it is found

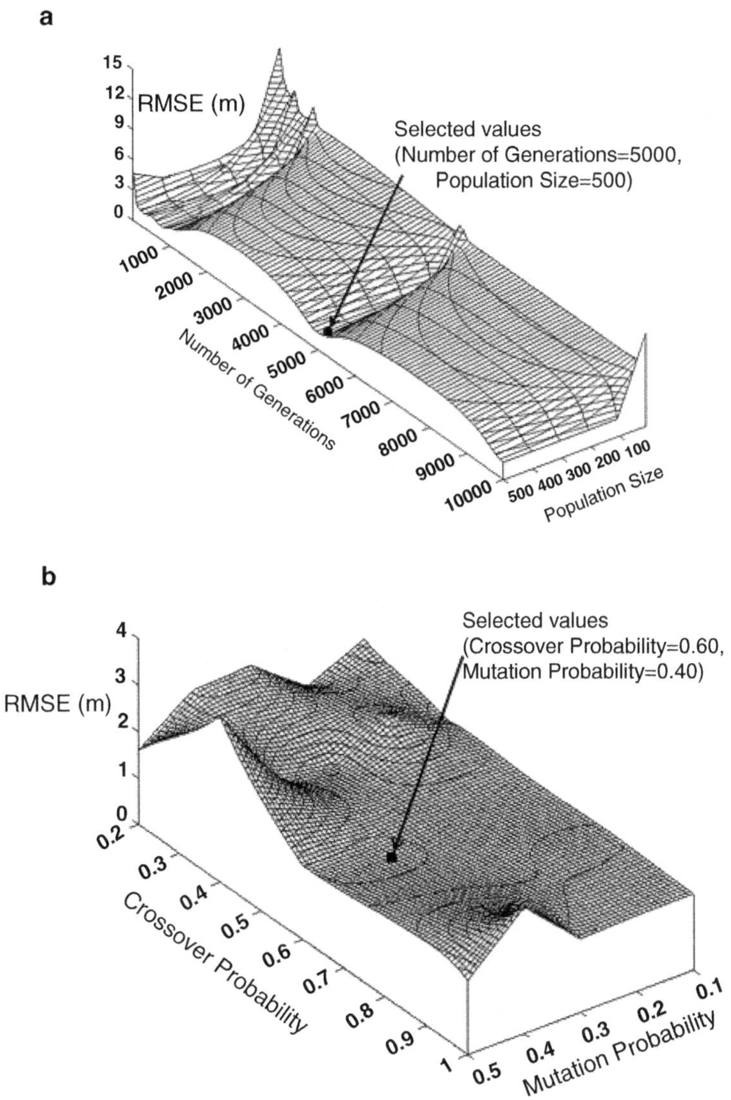

Fig. 18 Sensitivity range analysis for the GM: (**a**) number of generations and population size; (**b**) crossover and mutation probabilities [6]

that the RMSE value became small with the number of generations = 5,000 and population size ≥100. The relationships among RMSE, crossover probability and mutation probability are more complicated, as displayed in Fig. 18b. It could be concluded that the RMSE became small in the range of 0.70 > crossover probability >0.60 and mutation probability >0.35. On the contrary, Fig. 19a, b show high stability of results in a wide range of parameter values. Considering these results,

a

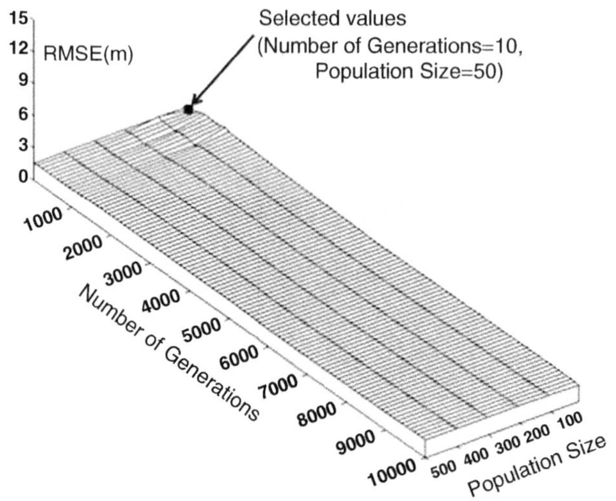

b

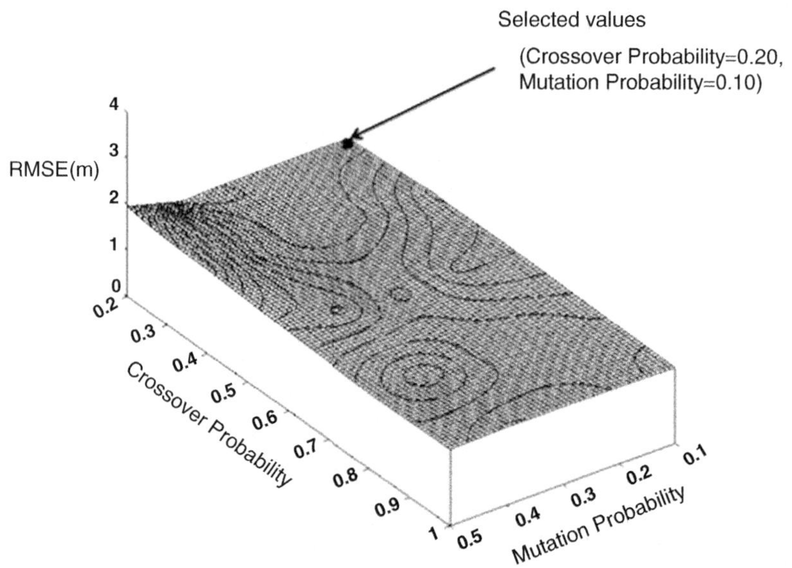

Fig. 19 Sensitivity range analysis for the MGM: (**a**) number of generations and population size; (**b**) crossover and mutation probabilities [9]

the selected values of the input parameters for the GA and MGA models are summarized and displayed in Table 3. Consequently, MGA produces a goodness of fit between measured and calculated piezometric levels within a wide range of input parameter values.

Table 3 Setting of GA and MGA models parameters [6]

	For minimum RMSE		Selected parameters	
Parameters	GA	MGA	GA	MGA
Number of generations	5,000	10–10,000	5,000	10
Population size	≥100	10–500	500	50
Crossover probability	0.60–0.70	0.2–1.0	0.60	0.20
Mutation probability	≥0.35	0.1–0.4	0.40	0.10

6.5 Future Assessment of Temporal Piezometric Level Change

The good agreement between the measured and calculated piezometric level trends in the short period simulation (1979–2005) was considered to be a verification of the MGM presented in this study. Thus, the MGM can be used to predict piezometric level trends and drawdown of the groundwater table over long periods, as well as to evaluate the hydrological feasibility of the proposed groundwater withdrawal plans (scenarios 1, 2 and 3). With regards to the proposed scenarios for groundwater withdrawal described previously, the simulated declines in hydraulic head at the target OWs were calculated for the period 1960 to 2100.

For the proposed scenarios of groundwater withdrawal (scenario 1, scenario 2 and scenario 3), Figs. 20, 21 and 22 show the simulated piezometric levels of the target OWs, respectively. For scenario 1, Fig. 20 shows the piezometric level trends simulated by the MGM and that previously simulated by the GM [4] for the target OWs. From this figure, both models simulating the piezometric level trends are highly correlated, which confirms the validity of the newly developed MGM for future assessment of hydraulic heads. However, the MGM input parameters values are smaller than those of GM. The results indicate that the MGM produced satisfactory validation results compared to the ordinary GM considering the use of a small number of inputs and the low values of the input parameters. Moreover, the time required for calculations, represented by the number of generations parameter, is reduced from 5,000 for the GM to 10 for the MGM, which is a considerably large reduction of 99.98%.

For scenario 2, Fig. 21 shows that the hydraulic heads of OWs o1, o5 and o33 are recovered by 32.0, 37.5 and 25.0 m, respectively, compared to those in scenario 1. On the contrary, the hydraulic head of OW o35 is severely decreased by 77.8 m. Comparing scenario 3 with scenario 1, Fig. 22 shows that the hydraulic heads of OWs o1 and o5 are recovered by 104.0 and 64.5 m, respectively, to reach their economical piezometric levels by the years 2039 and 2018. On the other hand, the hydraulic heads of OWs o33 and o35 are decreased by 13 and 165 m, respectively. For both scenarios 2 and 3, the high decrease in hydraulic head of OW o35 as a representative of Area 8 is due to the high groundwater withdrawal that added to the RWPWs in this area. Therefore, it is recommended to drill new production wells in the southern part of the study area to reduce groundwater withdrawal over the

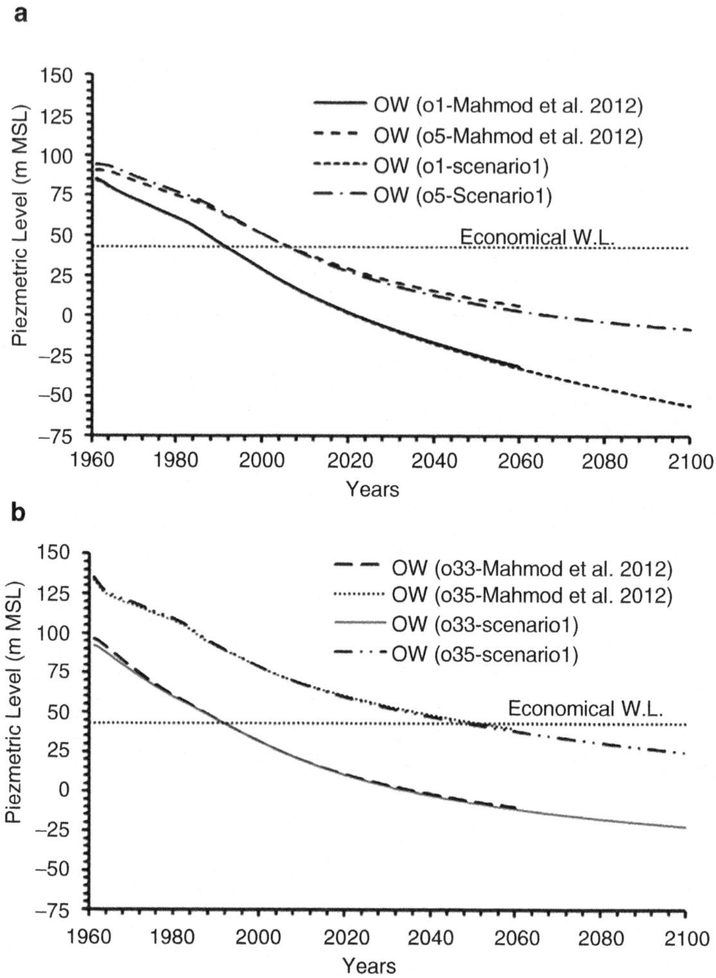

Fig. 20 Comparison of simulated piezometric levels for MGM and GM [39] until the year 2100: (**a**) OWs o1 and o5; (**b**) OWs o33 and o35 [6]

RWPWs. For the three scenarios, Table 4 shows the differences in hydraulic heads between the economical piezometric head levels and the target OWs' piezometric levels beyond 2100. A positive sign indicates that the piezometric level of the OW is above the economical one and vice versa.

The MGM shows a better fitting performance than the ordinary GM, for which the best input models' trends are recognized by means of the best models combination (BMC_k). On the other hand, the MGM produced a wide range of input parameter values, which guarantees goodness of fit with any hypothetical value of these parameters. Moreover, a high reduction in required calculation time was observed for reaching the best fit to the observations.

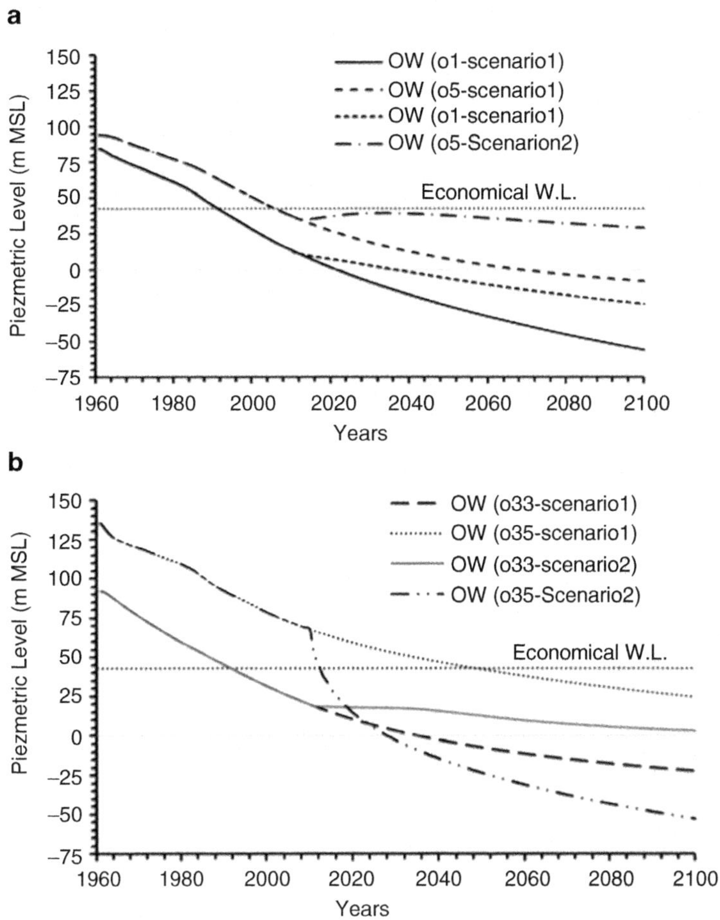

Fig. 21 Comparison of MGM-simulated piezometric levels for scenario 2 and scenario 1 until the year 2100: (**a**) OWs o1 and o5; (**b**) OWs o33 and o35 [6]

Table 4 Hydraulic head differences in the economical piezometric levels for the target OWs

	Difference from the economical piezometric level (43 m) by 2100		
OW	Scenario 1	Scenario 2	Scenario 3
o1	−98.88	−66.89	4.98
o5	−51.16	−13.58	13.27
o33	−65.22	−39.76	−78.26
o35	−18.33	−96.08	−183.62

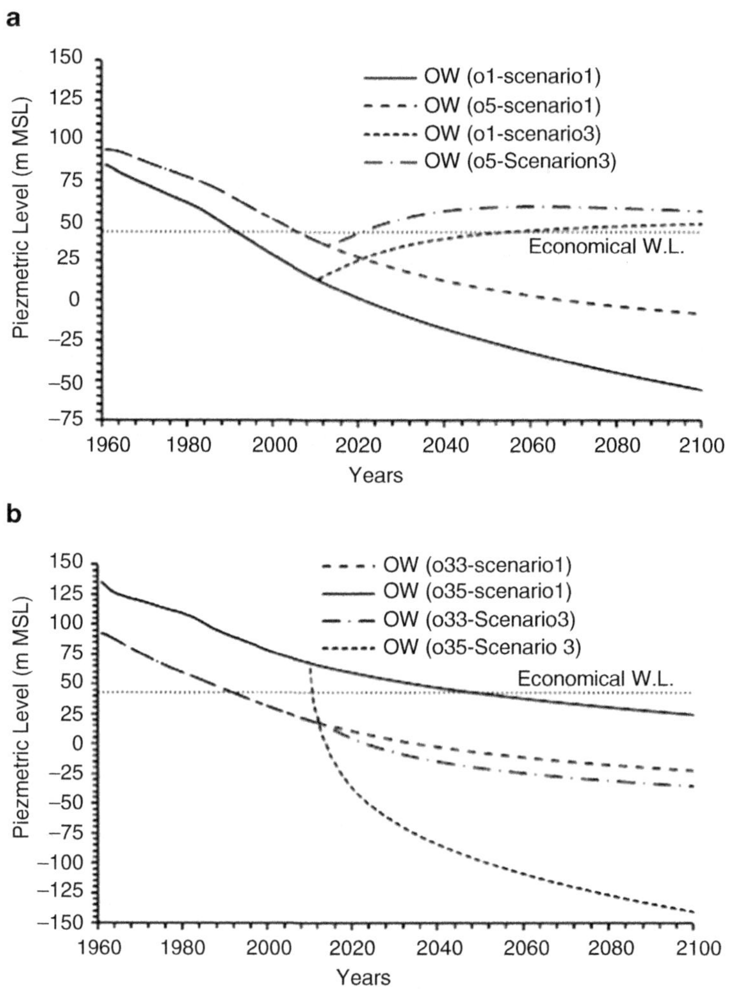

Fig. 22 Comparison of MGM-simulated piezometric levels for scenario 3 and scenario 1 until the year 2100: (**a**) OWs o1 and o5; (**b**) OWs o33 and o35 [6]

6.6 Evaluation of the Proposed Groundwater Withdrawal Scenarios

As a trial to evaluate the proposed scenarios of groundwater withdrawal, a contour map of hydraulic head distributions was constructed to understand the future hydrogeological conditions of groundwater flow in the study area. Figure 23a–c show the flow direction and hydraulic head distributions for the three proposed scenarios, scenario 1, scenario 2 and scenario 3 for year 2100, respectively. From these figures, it could be concluded that the hydraulic heads of the northern part

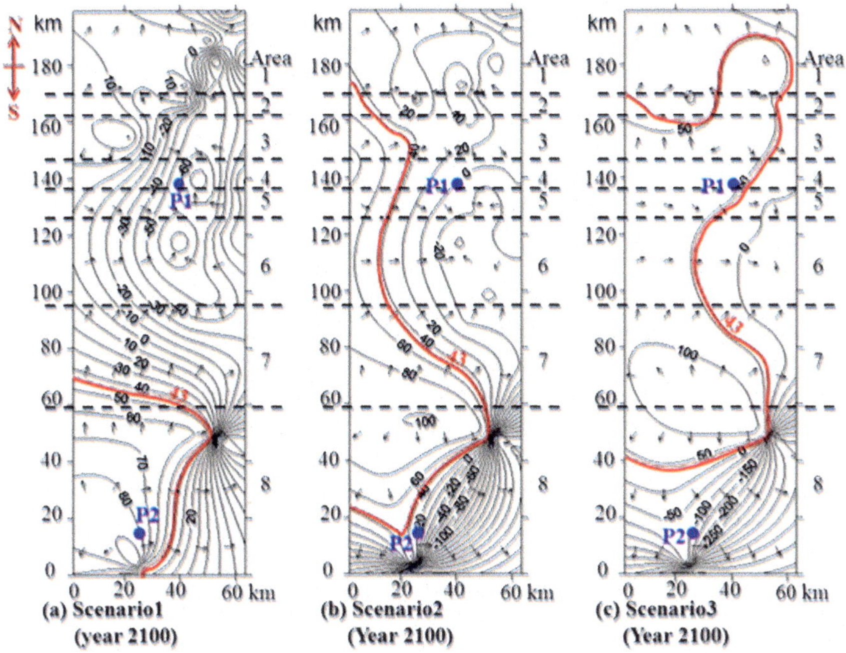

Fig. 23 Comparison between the simulated piezometric level contours along with groundwater flow vectors in the study area for the proposed scenarios [9]

could be recovered as well as the hydraulic head of point P1 for the three scenarios; scenario 1, scenario 2 and scenario 3 have the values of −60, 0, and 50 m, respectively. On the other hand, for the southern part, the hydraulic head of point P2 for the same three scenarios is 80, 20, and −100 m respectively. The drawdown which occurred at the southern part (Area 8) was due to the small number of RWPWs that were used in the redistribution process of groundwater withdrawal. By comparing the hydraulic heads at points P1 and P2, which are representative of the northeastern and southwestern parts of the study area, respectively, it could be concluded that the differences in hydraulic heads between these two points were 140, 20, and −150 m for the three scenarios: scenario 1, scenario 2 and scenario 3, respectively. The negative sign here is a result of a decrease in the piezometric level of point P2 comparing with that of point P1 because of small number of RWPWs in the southern part (Area 8). The percentage of the whole currently affected study area for the three scenarios scenario 1, scenario 2 and scenario 3 at year 2100 are 80%, 70% and 50%, respectively. In general, a recovery of the hydraulic head distribution could be achieved by redistribution of the groundwater withdrawal of the RWPWs. Consequently, the northern part of the study area could be highly recovered for the three proposed scenarios, as well as a high recovery percentage of the whole area was achieved by applying scenario 3 (50% recovered) with a recommendation of increase in the number of RWPWs at the southern part of the study area (Area 8).

7 Conclusions

In this study, groundwater flow analysis was performed using the developed modified grey model (MGM) which combines the finite element method (FEM) and the new developed modified genetic algorithm (MGA). The analysis of observed data shows enormous pore pressure decrease by 50 m in the northeastern part of the study area within the period 1979–2005. The MGM produced stable results for piezometric level simulations while using a wide range of input parameters values compared to the ordinary grey model (GM). Moreover, the time required for calculations to achieve proper convergence with the measured piezometric level trends was reduced by 99.8% compared to the GM. MGM used a smaller number of input models' trends than the GM for achieving goodness of fit, with a percentage of reduction in the range of 68.75–81.25%. The RMSE was in the ranges of 0.63–2.936 m and 0.439–2.665 for the MGM and the GM, respectively. The results from the various scenarios of groundwater withdrawal demonstrate that if the present extraction rate is expanded (scenario 1), the groundwater piezometric level continuously declines and drops below the economical piezometric level until the year 2100 at the end of the simulation period. To avoid groundwater depletion in the Kharga Oasis, the extraction rate in the M-region is lowered by 20% and added to Area 8 (scenario 2). The planned extraction rate is not feasible for the future and will have a negative impact on the piezometric level of Area 8. However, a recovery in the piezometric levels for all other subdivisions was observed. For the 3rd scenario of groundwater withdrawal, the recovery head for the northern area increased more than in scenario 2, as well as reached above the economical piezometric level. However, a large drawdown of the piezometric level of the southern part was observed.

8 Recommendations

To protect the groundwater table from being severely drawn down, it is recommended that new production wells be constructed in the southern part of the current study area far enough from the RWPWs. For reliable piezometric measurements, a network of monitoring wells should be drilled that cover the whole area. Regular water-level measurements (at least monthly) would provide the basic data needed for time series compilations in the future, which could give feedback on the constructed model. For the future, it is suggested that the MGM be used for simulating the total piezometric level surface of the whole area and hence confirm the MGM performance. The proposed model (MGM) is significant for the decision-makers to have knowledge about groundwater resources in their regions. For that, MGM could be useful to study the effect of increased pressure on the finite water resource on the sustainability of agricultural production and livelihoods as well as the social cohesion within new settler communities. Moreover, a new developed monitoring system could be proposed to reduce the monitoring cost based on giving up the idea of constructing new observation wells.

References

1. El Tahlawi MR, Farrag AA, Ahmed SS (2008) Groundwater of Egypt: "an environmental overview". Environ Geol 55:639–652
2. Elbeih SF (2015) An overview of integrated remote sensing and GIS for groundwater mapping in Egypt. Ain Shams Eng J 6:1–15
3. UNESCO (2013) Joint programme for climate change risk management in Egypt – proposed climate change adaptation strategy for the Ministry of Water Resources & Irrigation Egypt. Ministry of Water Resources and Irrigation
4. Mahmod WE, Watanabe K, Zahr-Eldeen AA (2013) Analysis of groundwater flow in arid areas with limited hydrogeological data using the Grey model: a case study of the Nubian Sandstone, Kharga Oasis, Egypt. Hydrgeol J 21:1021–1034
5. El Arabi N (2012) Environmental management of groundwater in Egypt via artificial recharge extending the practice to soil aquifer treatment (SAT). Int J Environ Sustain 1(3):66–82
6. Mahmod WE, Watanabe K (2014) Modified Grey model and its application to groundwater flow analysis with limited hydrogeological data: a case study of the Nubian Sandstone, Kharga Oasis, Egypt. Environ Monit Assess J 186:1063–1081
7. Idris H, Nour S (1990) Present groundwater status in Egypt and the environmental impacts. Environ Geol Water Sci 16(3):171–177
8. Alnaggar D (2003) Water resources management and policies for Egypt. Workshop on policies and strategies options for water management in Islamic countries, Tehran; December
9. Mahmod WE (2016) Modeling of complex groundwater flow systems using MGM. LAP Lambert Academic Publishing, 128 p
10. RIGW/IWACO (1988) Hydrogeological mapping of Egypt, scale 1:2,000,000. 1st edn. TN 70.120-88-03, http://earthwise.bgs.ac.uk/index.php/Hydrogeology_of_Egypt
11. Allam AR, Ele-Jan S, Dawoud MA (2002) Desalination of brackish groundwater in Egypt. Desalination 152:19–26
12. Dawoud MA (2004) Design of national groundwater quality monitoring network in Egypt. Environ Monit Assess 96:99–1180
13. Hefny K, Shata A (2004) Underground water in Egypt. Ministry of water supplies and irrigation, Cairo, Egypt, p 295 (in Arabic)
14. Armanuos AM, Ibrahim MG, Mahmod WE, Negm, A, Yoshimura C, Jiro Takemura J, Zidan BA (2017) Evaluation the potential impact of Grand Ethiopian Renaissance Dam and pumping scenarios on groundwater level in the Nile Delta aquifer, Water Science & Technology: Water Supply. doi:10.2166/ws.2017.037
15. Ambroggi RP (1966) Water under the Sahara. Sci Am 214:21–29
16. Hefny K, Shata A (1995) Strategies for planning and management of groundwater in the Nile Valley and Delta in Egypt. Strategic research program-working paper series no. 31-1. Environmental and Natural Resources Policy and Training Project (EPAT)
17. Ebraheem AM, Garamoon HK, Wycisk P, Seif El Nasr AM (2003) Numerical modeling of groundwater resource management options in the East Oweinat area, SW Egypt. Environ Geol 44:433–447

18. Ebraheem AM, Riad S, Wycisk P, Seif El Nasr AM (2002) Simulation of impact of present and future groundwater extraction from the non-replenished Nubian Sandstone Aquifer in SW Egypt. Environ Geol 43:188–196
19. Heinl M, Thorweihe U (1993) Groundwater resources and management in SW Egypt. In: Meissner B, Wycisk P (eds) Geopotential and ecology. Catena Suppl 26:99–121
20. Ball J (1927) Problems of the Libyan desert. Geogr J 70:21–38, 105–128, 209–224
21. Sanford KS (1935) Source of water in the northern-western Sudan. Geogr J 85:412–431
22. Sonntag C (1986) A time-dependent groundwater model for the Eastern Sahara. Berl Geowiss Abh A72:124–134
23. Lamoreaux PE, Memon BA, Idris H (1985) Groundwater development, Kharga Oasis, western desert of Egypt: along-term environment concern. Environ Geol Water Sci 7:129–149
24. Masoud MH, Schneider M, El Osta MM (2013) Recharge flux to the Nubian Sandstone aquifer and its impact on the present development in southwest Egypt. Journal of African Earth Science 85:115–124
25. Thorweihe U, Heinl M (2002) Groundwater resources of the Nubian aquifer system, NE-Africa. Synthesis, Observatoire du Sahara et du Sahel, Paris
26. Knetsch G, Yallouze M (1955) Remarks on the origin of the Egyptian Oasis depressions. Bull Soc Géogr Égypte 28:21–33
27. Said R (1962) The geology of Egypt. Elsevier, Amsterdam
28. Said R (1990) Geomorphology. In: Said R (ed) The geology of Egypt. Tailor & Francis, Rotterdam
29. Thorweihe U, Schandelmeier H (1993) Geoscientific research in Northeast Africa. Proceedings of international conference on geoscience. A.A. Balkema, Rotterdam, 776 p
30. Thorweihe U (1990) The Nubian aquifer system. In: Said R (ed) The geology of Egypt, Balkema, Lisse, The Netherlands, pp 601–614
31. Shata AA (1982) Hydrogeology of the great Nubian Sandstone basin, Egypt. Q J Eng Geol 15:127–133
32. Mohammed M, Watanabe K, Takeuchi S (2010) Grey model for prediction of pore pressure change. Environ Earth Sci 60:1523–1534
33. Anderson MP, Woessener WW (1992) Applied groundwater modeling-simulation of flow and advective transport. Academic Press Inc, San Diego
34. Pinder GF, Gray WG (1977) Finite element simulation in surface and subsurface hydrology. Academic Press Inc, New York
35. Sivaraj R (2011) A review of selection methods in genetic algorithm. Int J Eng Technol 3:3792–3797
36. Goldberg DE, Deb K (1991) A comparative analysis of selection schemes used in genetic algorithms. In: Rawlins GJE (ed) Foundations of genetic algorithms. Morgan Kaufmann Publishers Inc, San Francisco, CA
37. Mitchell M (1999) An introduction to genetic algorithms, 5th printing. Massachusetts Institute of Technology, London
38. Coley DA (1999) An introduction to genetic algorithms for scientists and engineers. World Scientific Publishing Co Pte Ltd, Singapore
39. Mahmod WE, Watanabe K (2012) A preliminary modification and validation of the Grey Model for groundwater flow analysis of arid regions, VII international conference on environmental hydrology, Egypt
40. Salman AB, Howari FM, El-Sankary MM, Wali AM, Saleh MM (2010) Environmental impact and natural hazards on Kharga Oasis monumental sites, Western Desert of Egypt. J African Earth Sci 58:341–353
41. Hesse K, Hissne A, Kheir O, Schnäcker E, Schneider M, Thorweihe U (1987) Hydrogeological investigations in the Nubian sandstone aquifer system, Eastern Sahara. Berl Geowiss Abh A75:397–4641

Groundwater Potential in the New Valley South West of the Nile Delta in Egypt

Abeer M.M. Soliman and Mostafa M. Solimsn

Abstract Egypt is continuously facing a decrease in water share per capita due to a decline in available water resources.

The objective of the present study is to evaluate and manage the groundwater resources of the Nubian Sandstone Aquifer System (NSAS) in the New Valley area. It is located in the middle part of Egypt's Western Desert. It lies between latitudes of 24°–28° N and longitudes of 27°–31.5° E. It covers an area of 440 km long by 460 km wide where the total area is about (202,400 km^2). A detailed review of the Nubian Sandstone Aquifer in the western desert is also introduced.

A finite difference model using "Visual MODFLOW" was applied on the Nubian Sandstone Aquifer of Dakhla Basin. It was adapted to simulate groundwater flow in such aquifer. The simulation was calibrated with available groundwater head data from CEDARE Report (2002). An optimum solution is established for the safe groundwater mining in the study area.

The scenario applications could allow for an increase in reclamation at Dakhla Oasis by 15%, with the condition of safe drawdown values less than 60 m for the period of 100 years.

The study provides the benefits of applying the modeling techniques. Numerous valuable inputs for the national development plan in Egypt are presented. The study found that it is important to seek an alternative water resource to compensate for the groundwater depletion.

A.M.M. Soliman (✉)
Research Institute for Groundwater (RIGW), National Water Research Center, MWRI, El Qanater El Khayreya, Egypt
e-mail: mostrhmn@yahoo.com

M.M. Solimsn
Irrigation and Hydraulics Department, Faculty of Engineering, Ain Shams University, Cairo, Egypt
e-mail: msoliman1@hotmail.com

A. M. Negm (ed.), *Groundwater in the Nile Delta*,
Hdb Env Chem (2019) 73: 585–622, DOI 10.1007/698_2017_62,

Keywords CEDARE, GIS, Groundwater, Modeling, MODFLOW, Nubian, RIGW, Sandstone Aquifer

Contents

1 Introduction

Groundwater is one of the most important resources of water in Egypt, ranking second after the River Nile. Although in terms of quantity the contribution of groundwater to the total water supply in Egypt is very moderate. Groundwater is the only available source of water for people living in the desert area. In terms of quality, groundwater, unless contaminated, is generally of better quality than surface water. The quality of groundwater depends on two main factors, namely the origin of the water and the type of rocks bearing it. Groundwater is obtained in many areas from fissures, and joints of the igneous and metamorphic rocks. It may be found in layers of weathered basement rocks. It can also be extracted from alluvial deposits and aquifers in coastal areas.

Demand for water has increased during the last decade particularly due to the declining availability of freshwater. This has expanded the development of searching and extracting groundwater resources. An evaluation of the renewable groundwater resources is essential for any Integrated Water Resource Management (IWRM) plans that are sustainable. In this study, the evaluation of the renewable groundwater resources in Egypt from water quality and quantity perspective is discussed.

2 Groundwater System in Egypt

The aquifers systems in Egypt are distributed over the country and the hydrogeological framework of Egypt. It can be characterized into eight hydrogeological units as shown in Fig. 1 [3, 4]. These units are:

1. Nile Valley and Delta aquifers,
2. Coastal aquifers,
3. Moghra aquifer,
4. Aquiclude rocks,
5. Carbonate rocks complex aquifers,
6. Nubian Sandstone Aquifer,
7. Fissured basement complex aquifers, and
8. Tertiary aquifers

They differ in general characteristics, including extension, hydraulic conductivity, transmissivity, storage, and recharge.

The above-mentioned aquifers are classified into two groups based on the water-bearing rocks as given in Table 1:

1. Granular water-bearing rocks and
2. Fissured and karstified water-bearing rocks

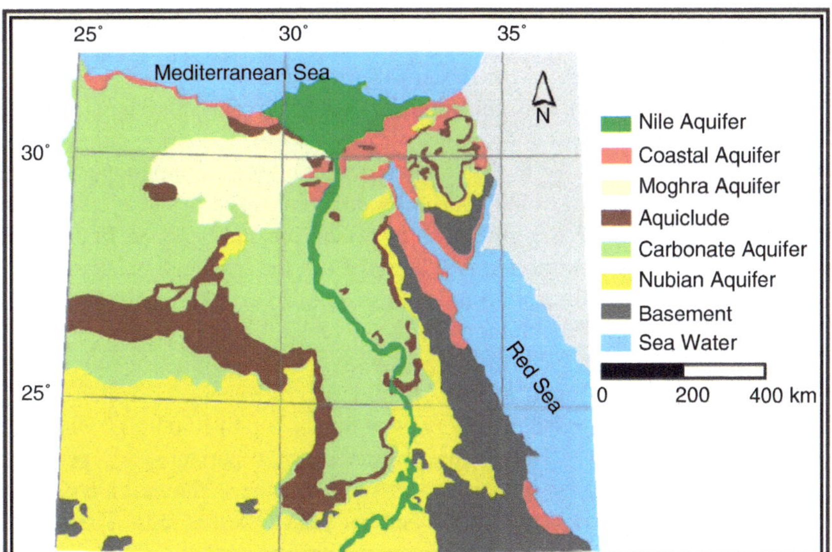

Fig. 1 The aquifer systems of Egypt. Modified after RIGW [1, 2]

Table 1 The main hydrogeological units in Egypt [3]

Lithology rock type	Unit	Recharge Surface	Recharge Sub-surface	Distribution	Productivity
Granular material	Nile Valley and Delta	Continues	Continues	Extensive	High
	Coastal aquifer	Occasional	Limited	Local	High
	Nubian Sandstone	None	Limited	Extensive	High

The Nubian Sandstone Aquifer is selected for further study in this respect since it is considered the main groundwater resource in the New Valley in the Western Desert of Egypt.

3 Regional Nubian Sandstone Aquifer System

The Nubian Sandstone Aquifer system is a regional system. It extends into Libya, Sudan, and Chad. It is a nonrenewable aquifer system [5]. The Nubian aquifer system (Fig. 2), one of the largest groundwater systems of the Sahara, is formed by two major basins: the Kufra Basin in Libya, north-eastern Chad and northwestern Sudan, and the Dakhla Basin of Egypt. In addition, the aquifer area includes the southernmost strip of the northwestern Basin of Egypt and the Sudan Platform 7, Brinkman et al. [8].

The total area is about two million square kilometers. The Nubian Sandstone Aquifer in Egypt is assigned to the Paleozoic–Mesozoic. It occupies a large area in the Western Desert and parts of the Eastern Desert and Sinai.

"In spite of its vast area, it is considered as a broadly closed system, as it has natural boundaries to the east and southeast formed by the mountain ranges of the Nubian Shield and is bounded to the south and west by the mountainous outcrops of Kordofan Block, Ennedi, and Tibesti. To the southwestern part of the basin, the aquifer is bounded by the groundwater divide located between Ennedi and Tibesti Mountains. The natural northern boundary of the Nubian Sandstone Aquifer System is set to the so-called Saline-Freshwater Interface, whose location is considered spatially stable, although slight movement is believable," Sefelnasr [9, 10].

"Groundwater can be found at very shallow depths, where the water-bearing formation (Horizon) is exposed or at very large depths (up to 1,500 m)," where the aquifer is semi-confined. "The deepest water-bearing horizons are generally encountered in the North," as in the Siwa Oasis, while the shallowest are encountered in the southern portion of Kharga and the East Oweinat area. The aquifer transmissivity is generally medium to low, varying from 1,000 to 4,000 m^2/day, Sefelnasr [10].

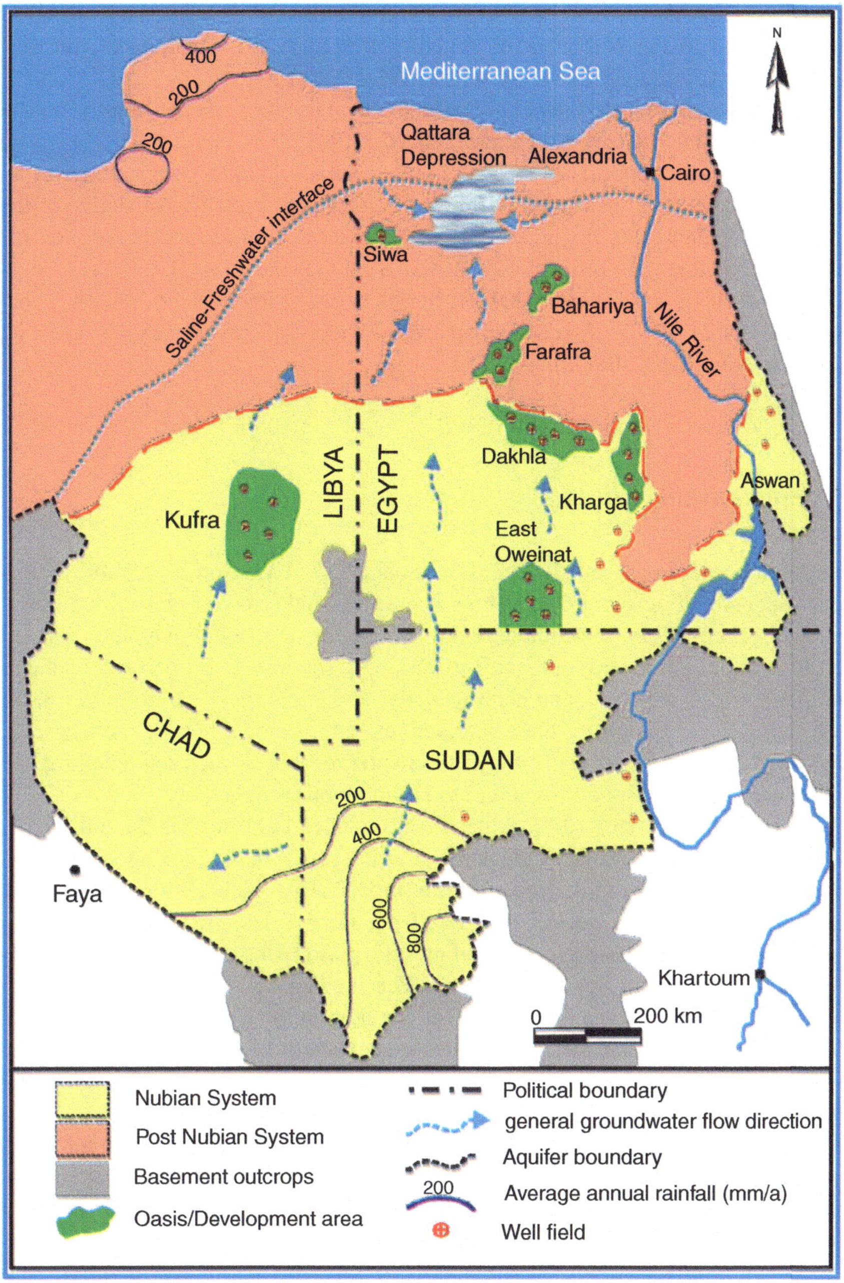

Fig. 2 Regional hydrogeological overview of the Nubian Sandstone Aquifer System. Compiled from: CEDARE [6] and Bakhbakhi [7]

CEDARE/IFAD [6] in a program for the development of a regional strategy for the utilization of the Nubian Sandstone Aquifer System has estimated the fresh groundwater volume within the system at 372,950 km^3.

"Groundwater quality is generally good in the major part, except near the coastal regions and Sinai," RIGW [11] and Sefelnasr [10].

The following section is discussing in more detail the most recent study for the groundwater potential in the New Valley area of Egypt [12]. The objective of this study is to evaluate and manage the use of the valuable groundwater resources of the Nubian Sandstone Aquifer System (NSAS) in the study area. It is located in the middle part of Egypt's Western Desert. It lies between latitudes of 24°–28° N and longitudes of 27°–31.5° E. It covers an area of about 440 km long by 460 km wide (about an area of 202,400 km^2).

4 Conceptual Model

The hydrogeological groundwater flow modeling is based on a valuable tool to better understanding groundwater flow in aquifers and to better managing groundwater resources. Numerical models are one of the few tools available that can consider a complex array of aquifer variables (hydraulic properties, recharge, pumping, rivers, structure, and heterogeneity) and allow these variables to interact with themselves. Exploring these interactions with a model can be used to make predictions important for managing groundwater resources, such as predicting how water levels might respond to increased pumping or drought.

The model study area of "Dakhla Basin includes Farafra, Dakhla and Kharga Oases." The Dakhla basin includes the upper Post Nubian Aquifer, separated from the Nubian by an aquitard. However, the model study includes the main deeper aquifer since the upper post-Nubian aquifer is almost declining and that it is not feasible to include it in the study. For this reason, the upper aquifer will be considered inactive in this model study. The Nubian aquifer in this respect still contains a huge groundwater storage that can help in the development of the New Valley region, provided that good water management plan is required since this Nubian system is not well recharged and should be treated as groundwater mining aquifer.

Based on the hydrogeological map of Egypt (Fig. 3a) and the schematic cross sections shown in (Fig. 3b, c), they were schematized as the series of layers. The upper layers are considered as zone 1and the Nubian Sandstone Aquifer is considered as zone 2, as follows:

- The oldest and most extended aquifer comprises the Paleozoic and Mesozoic continental deposits (zone 2). It is older than the Upper Cinemania (zone 1). A large part of the lower aquifer is exposed south of the 26° N parallel and is in unconfined condition.

- More recent groundwater reservoir includes the Tertiary Carbonate rocks in western Desert of Egypt. However, this aquifer is heavily depleted, and most of the wells are becoming dry which forced the natives in the area abandon those wells and replace them with deeper wells drilled in the deep Nubian aquifer.
- The two zones are separated by an impermeable shale layer (aquitard or confining bed) above which the Upper Cretaceous deposit is located.
- According to the aforementioned discussion, it is only feasible to develop the lower deep Nubian aquifer. In this situation, the upper aquifer in zone 1 will be discarded because most of the springs and wells taping the upper limestone aquifer (PNSA) are becoming dry. This is beside the very low conductive layer that separates the two zones. A conceptual numerical model domain and its extensions are given below.

a

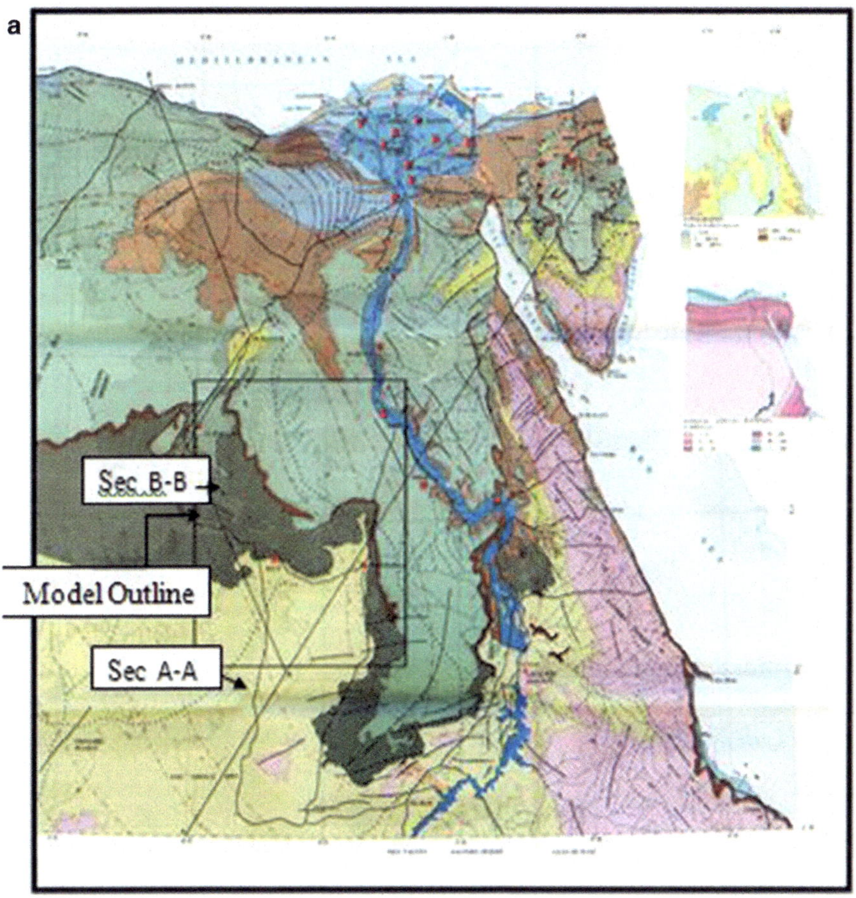

Fig. 3 (**a**) Hydrogeological map of Egypt by RIGW [13] showing the study area. (**b**) Cross section A-A through the study area. (**c**) Cross section B-B through the study area

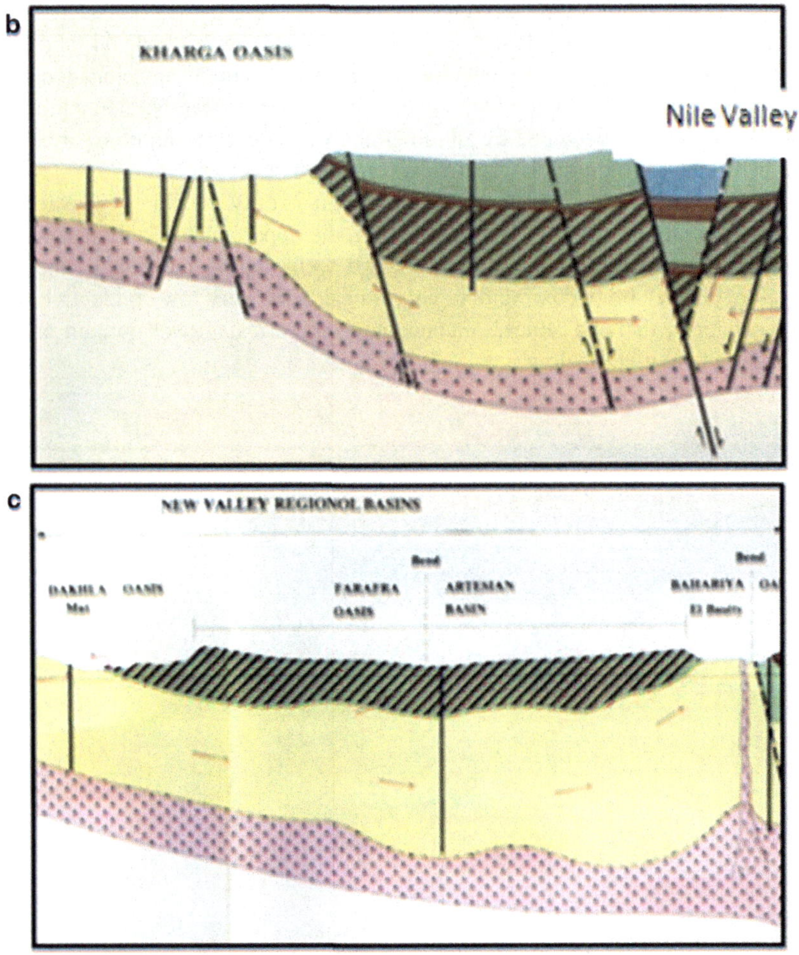

Fig. 3 (continued)

5 Numerical Simulation of the Study Area

5.1 Hydrologic Setting of the Study Area Model

The study area behavior of the aquifer system is acquired mainly from the hydrogeological map and other reports and papers available about the Nubian Aquifer system. It must be noted that CEDARE Report 2002 will be considered as the main source of data in this study. The modeling was carried out as follows:

- In order to have more confidence in the hydrologic parameters GIS with Visual Basics model is used in order to fill the gaps in those parameters needed for the Nubian main model. Each Oasis will have a separate local model [14].
- A second model is set for the entire model domain using the MODFLOW model software. The objective of the second model is to set the required aquifer boundary conditions.
- Upon setting the model properties and its boundary conditions from the last models, a third final model domain was furnished including all wells replacement until the year 2010 and their discharges. It must be noted that water supply wells are added to the final model in order to have an actual groundwater management plan.
- At this stage, simulating the development of the different scenarios is followed for the prediction of the related aquifer response using the latest well groups. The model will run for 100 years (2010–2110) in order to set the future water management plan for the three Oases.

5.2 *Mathematical Formulation (Governing Equations)*

The Governing equation that suits the Nubian confined aquifer is the two-dimensional groundwater partial differential equation as given below:

$$\partial\left(T_{\mathrm{xx}}\frac{\partial h}{\partial x}\right)+\partial\left(T_{\mathrm{yy}}\frac{\partial h}{\partial y}\right)+Q=S\frac{\partial h}{\partial t} \tag{1}$$

where the transmissivity and the horizontal conductivity (HC) are defined as follows:

$T_{\mathrm{xx}} = K_{\mathrm{xx}} \times$ Saturated thickness of layer in x direction (L^2/T)
$T_{\mathrm{yy}} = K_{\mathrm{yy}} \times$ Saturated thickness of layer in y direction... (L^2/T)

And the storage coefficient S is given as:

$S = S_s \times$ Saturated thickness of layer
K_{xx}, K_{yy} (L/T) are values of the hydraulic conductivity along the horizontal principal axes in the layer. In case of the Nubian aquifer, it was found from many research works and CEDARE Report that almost the horizontal HC $K_{\mathrm{xx}} = K_{\mathrm{yy}} = K$
S_s is the specific storage coefficient of the aquifer (L^{-1})
h is the piezometric head at any point in the model domain (L)
Q is the pumping rate in the model (L^3/T)
S_y is the specific yield or effective porosity of the unconfined aquifer.

On the other hand, for the unconfined aquifer, the three-dimensional groundwater flow equation becomes:

$$\frac{\partial}{\partial x}\left(K_{xx}\frac{\partial h}{\partial x}\right)+\frac{\partial}{\partial y}\left(K_{yy}\frac{\partial h}{\partial y}\right)+\frac{\partial}{\partial z}\left(K_{z}\frac{\partial h}{\partial z}\right)+Q=S_{y}\frac{\partial h}{\partial t} \tag{2}$$

where z is the third dimension and K_z is the vertical hydraulic conductivity.

5.3 The Model Package of the MODFLOW

Visual MODFLOW model is the software package developed by Waterloo Hydrogeological Inc. It is used to solve three- or two-dimensional groundwater flow and contaminant transport simulations with the finite difference approach.

This package combines MODFLOW and MT3D with the most intuitive and powerful graphical interface available.

The logical menu structure and easy to use graphical tools allow the following:

- Dimensioning the model domain and selecting units.
- Conveniently assigning model properties and boundary conditions.
- Running the model simulations.
- Calibrating the model using manual or automated techniques.
- Optimizing the pumping well rates and locations.
- Visualizing the results using 2D or 3D graphics.

The MODFLOW software used the finite difference approach for its simulation principles. Complete explanations about using the finite difference techniques are given in [9].

5.4 Procedures for MODFLOW Implementation

5.4.1 The Model Domain

As explained, the extent of the model domain for the Dakhla Basin covers an area between longitudes (27° E and 31.5° E) and between latitudes (24° N and 28° N). Since the MODFLOW supports only both the foot unit and the metric unit systems. Therefore, the metric unit system is used in our model domain. The model dimensions, therefore cover 440 × 460 km. The shorter lengths are the eastern and western boundaries of the model parallel to longitudes 27° E and 31.5° E, respectively. On the other hand, the longer lengths are located at the southern and northern boundaries of the model parallel to the latitudes 24° N and 28° N, respectively. The model grid has a uniform cell size all over the model domain. The cell size of 4 × 4 km was found suitable for the problem at hand. The model grid, in this case,

Table 2 Relation between (geographical) and model coordinates

Coordinates	Geographical coordinates	Model dimension in m
X_{min}	At Long 27° E	0
Y_{min}	At Lat 24° N	0
X_{max}	At Long 31.5° E	460,000
Y_{max}	At Lat 28° N	440,000

The model origin (0,0) is located, therefore, at the intersection of Long. 27° E and Lat 24° N

will have 115 columns and 110 rows. The model coordinate systems related to the geographical coordinates are given in Table 2. It must be remembered that the model domain was drawn on Google earth. An error of 4% was found between the coordinates obtained from both the hydrological map and the Google Earth data. This error is considered agreeable to all geological survey Departments.

According to the conceptual model, the MODFLOW model will have two zones. The model first zone will simulate the depleting limestone layer and the non-conductive shale layer. The model second zone will simulate the main and productive Nubian aquifer. It is a deep aquifer where its lower surface boundary is located at the basement complex surface. The model thickness will simulate the actual vertical thickness which is between the maximum ground surface elevation (little less than 700 m (amsl)) and the minimum elevation of −3,300 m (bmsl) at the basement complex surface boundary.

By placing the three surfaces, the upper ground surface, the intermediate boundary surface, and the basement complex surface using the MODFLOW processing the model domain can be formulated. Figures 4, 5, and 6 show the plan of the ground surface contours, the intermediate surface contours, and the basement complex surface boundary contours. Figure 7 shows an intermediate cross section through Dakhla Oasis in the model domain located at Lat 26° N at a distance $Y = 220{,}000$ m from the model origin.

5.4.2 Model Properties

The model properties include the hydraulic conductivity (HC), storage or specific storage coefficient (S_s), and the initial head. Specific storage (1/L) will be used. In spite of the several pumping tests that were done in the three Oases as given in Table 3 (obtained from CEDARE report), yet some data for S_s are missing. This can be overcome by using local models and in the same time adjusting the available data for the model properties. A regional calibration model which includes the entire Oasis together using the MODFLOW model would adjust the boundary conditions.

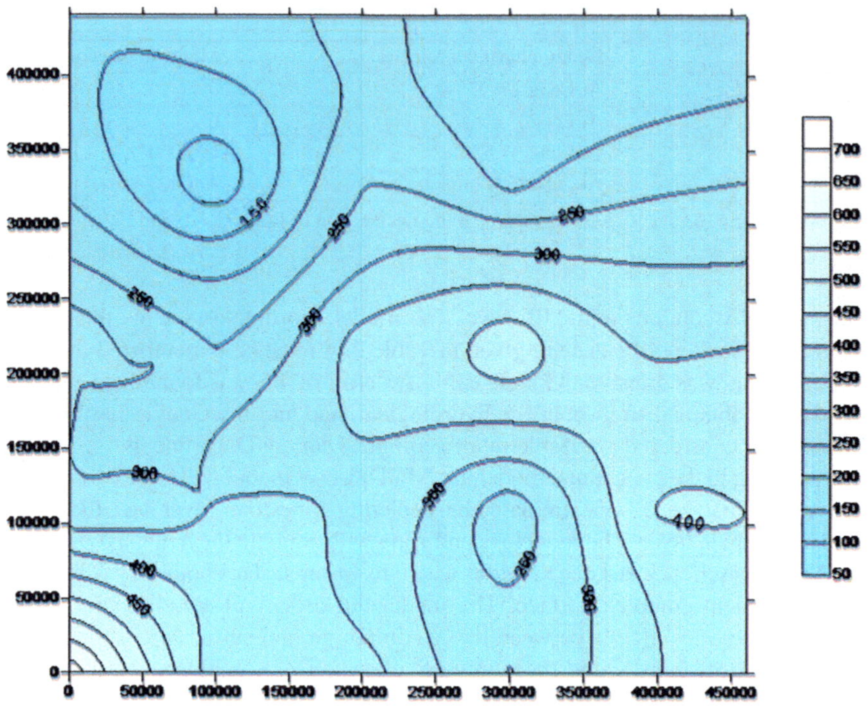

Fig. 4 Topographical map of the model area

5.5 *Local Models*

In this research work, the local models were used to further adjust both HC and S_s and find any missing data as S_s. The method used for adjustment of both properties is by using the GIS with the visual basic model which was achieved by Gad and Soliman [14]. The method was used for adjustment or calibration process of both the HC and S_s.

The adjustment started by using data shown in Table 3 of HC and S (obtained by CEDARE from pumping tests [6]) and comparing it with calculated data from the drawdown by Eq. (3). This equation was introduced to the data code of the visual basic (VB) which is in the GIS extension tool. Upon getting the drawdown values for the entire wells at selected network stations obtained by the VB, the GIS can draw the drawdown contours through the spatial analyst in the GIS tools as given below.

Table 4 which was obtained by CEDARE includes the number of pumping tests in each Oasis, the average aquifer thicknesses, the HC, and the storativity for each Oasis. Table 5 shows the total optimum discharges for each Oasis for specific years that was optimized by CEDARE model (Fig. 8), which was also developed by CEDARE and was helpful in getting the relation between the piezometric head with

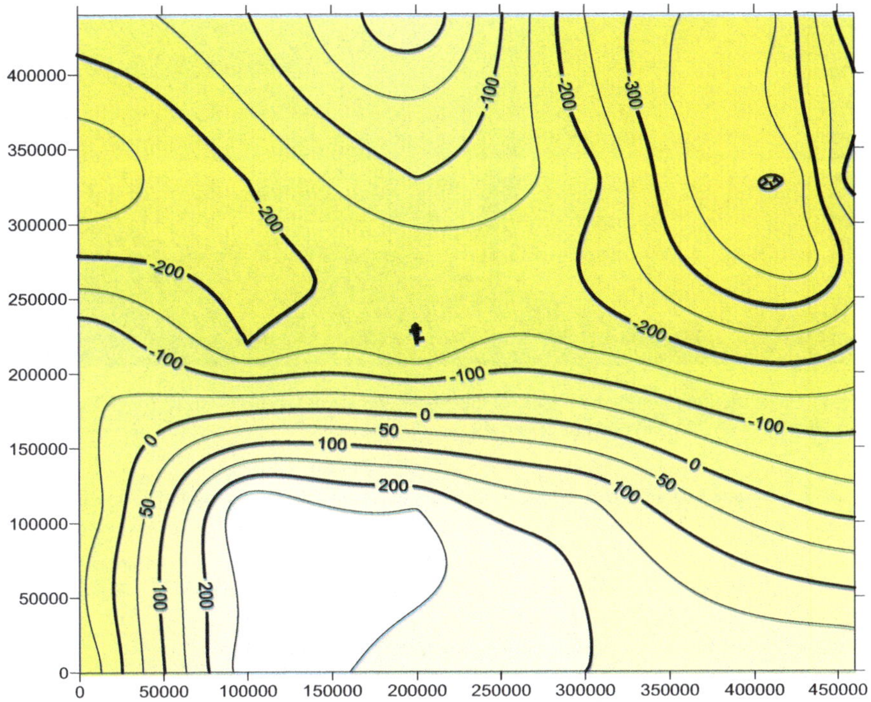

Fig. 5 Contour lines of zone 1 bottom

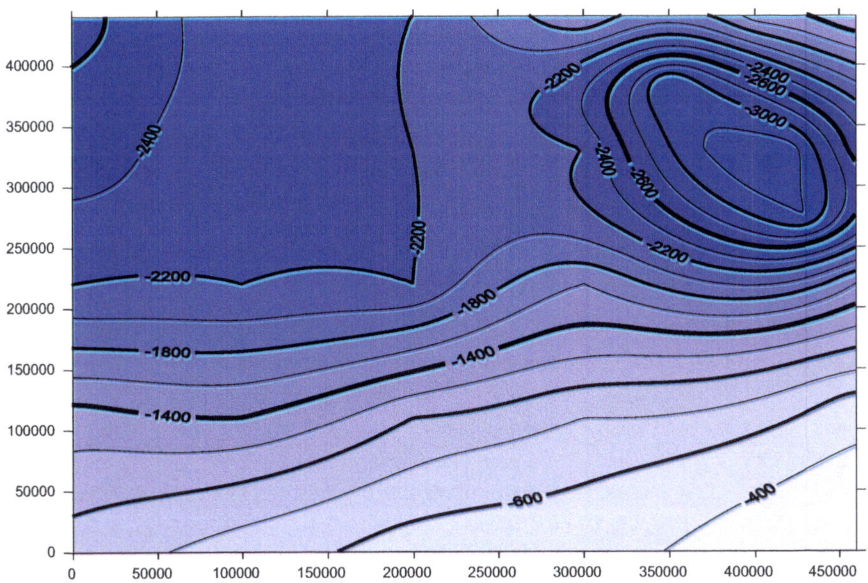

Fig. 6 Basement complex bottom layer 2

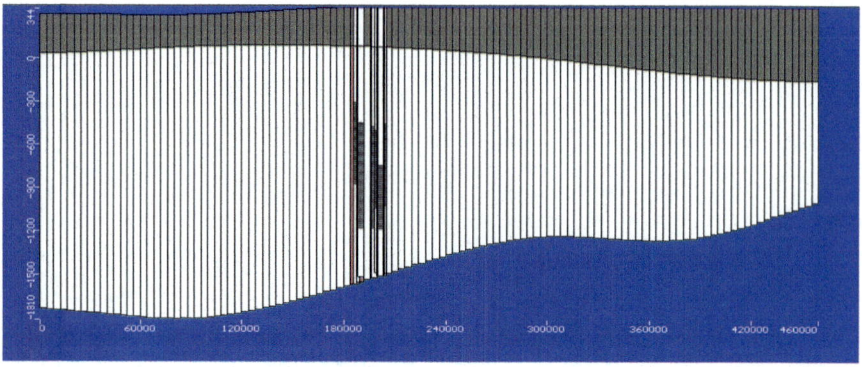

Fig. 7 Cross section through the middle of the model domain at Dakhla Oasis (through Lat 26° N) with distance $y = 220{,}000$ m from the origin

Table 3 Average hydraulic parameters of the Nubian Aquifer System (refer to CEDARE Report [15])

Region	No of Pumping tests	Saturated thickness (m)	Transmissivity (m^2/s)	Hydraulic cond. (HC) (m/s)	Storage coefficient(s) (*S*)
Farafra	8	2,600	1.20E-02	5.4×10^{-5}	?
Abu minqar	3	2,500	5.469E-02	2.18×10^{-4}	?
Dakhla	21	1,850	1.1E-02	6.1E-05	6.35×10^{-4}
Kharga	59	1,280	5.9E-03	2.90E-05	2.84×10^{-4}

Note: ? = missing data in CEDARE Report. All missing data was achieved by VB-GIS model as stated in the literature

Table 4 New Valley extractions (1998–2010) (CEDARE Report [15])

	1998		2000		2005		2010	
Area	m^3/s	Mm^3/year	m^3/s	Mm^3/year	m^3/s	Mm^3/year	m^3/s	Mm^3/year
Farafra	5.05	159	2.85	90	3.5	110	4.5	142
Dakhla	8.75	276	8.75	276	10.3	325	11.85	374
Kharga	3.96	125	3.96	125	3.96	125	5.00	158

Table 5 Drawdowns (DD) calculated and observed for Farafra parameters

Oasis name	Q (m^3/s)	Observation well locations	Pumping time	Model DD (m)	Observed DD (m)	Note
Farafra	2.85	NF9 (Obs1)	5 years (1995–2000)	2.8	2.5	Av error 6.5%
	3.5	$X = 60{,}000$ m	5 years (2005–2010)	3.8	3.4	$K = 5$ m/day
	5.05	$Y = 308{,}000$ m	10 years (1990–2000)	6.4	6.2	$S_s = 2.00E\text{-}05$

Note: The data for Q and observed drawdown (DD) at Obs1 were obtained from Fig. 9 and Table 4 which were adopted from CEDARE Report

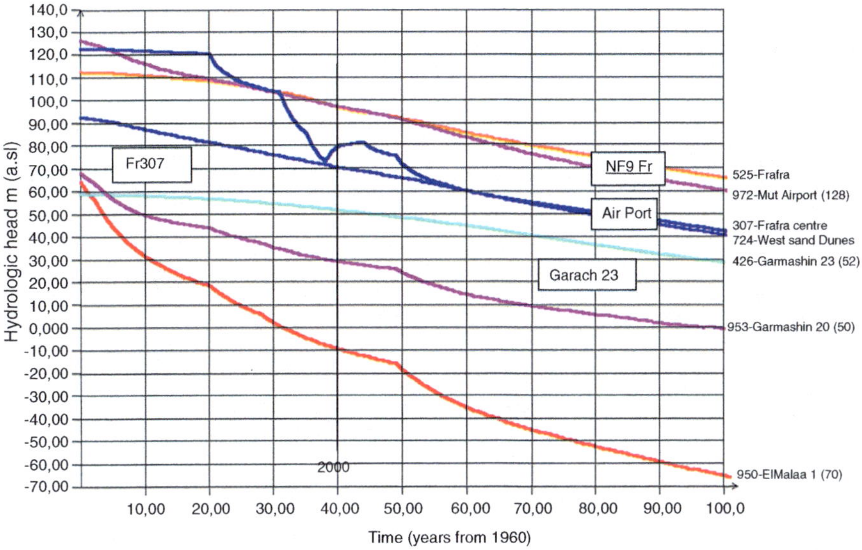

Fig. 8 Piezometric head at various locations of wells at the New Valley [2]

respect to time in years for the observed wells in the three Oases. This figure was helpful to extract the observed drawdown at the shown wells for a selected short period with its related discharges. The reason for selecting the short periods with the local models is to try to get the zero drawdown in each Oasis as shown in the following sections. The GIS method follows the following steps:

1. Locate first the well group for each Oasis in a point shapefile using the selected coordinates in model domain as shown in Figs. 29, 30, and 31 in Appendix 1
2. Once the well groups were located, and the well discharges are assigned to each well then the Visual Basic Code (VB Code) is designed to make a grid of points to stand for the observation points using the same coordinates of the well groups. There will be two coordinate systems $(\mathrm{xw}, \mathrm{yw})_i$ for the well (i) and other coordinates for the observation points $(\mathrm{xp}, \mathrm{yp})_k$ (i for the well number and k for the grid point number).
3. In order to calculate the drawdown at any grid or observation point, Modified Theis Equation is used in VB Code as,

$$d_k = (2.3Q_i/4T\pi)\left[\log\left(2.25Tt/R_i{}^2S\right)\right] \quad (3)$$

where

d_k = drawdown at an observation point k (m)

Q_i = well discharge for well i (m^3/day)

T = average transmissivity of the aquifer = K – Saturated aquifer thickness of the selected Oasis as given in Table 5 (m^2/day)

K = the hydraulic conductivity (HC) (m/day) of the Oasis

t = selected pumping time in days

R_i = the distance between the well (i), and the observation point (k) $=[(x(k) - x(i))^2 + (y(k) - y(i))^2]^{0.5}$

4. For each well the drawdown was calculated and summed at each observation point k. This was obtained by VB code and located at the respective point. It must be noted that any observation point located at any well is discarded by the Code. The GIS will then draw the drawdown contours as shown for the three Oases (Figs. 9, 11, and 13). Notice that the drawdown zero contours were located inside the Oases boundaries. This condition was found in this respect when the pumping time will not exceed 10 years for any Oasis in order to prevent any groundwater interference between Oases. Figures 29, 30, and 31 in Appendix 1 give both Geographic coordinates and the model coordinates system for each well at Farafra, Dakhla, and Kharga Oases, respectively.
5. The adjustment process started by selecting the drawdown values for one or two wells near the center of each Oasis. Then compare the results with the observed and calculated values as mentioned before. In this research, it was found that Storage Coefficient S was much more sensitive on the drawdown than the H.C. K, so S_s was selected in the calculations for its variation to try to narrow the gaps between both observed and calculated drawdown values (Figs. 9, 10, 11, 12, 13, and 14).

Tables 5, 6, and 7 include the three observation well locations used for correlations and the results after the adjustment processing. Figures 10, 12, and 14 show the correlations between the observed and the calculated drawdown.

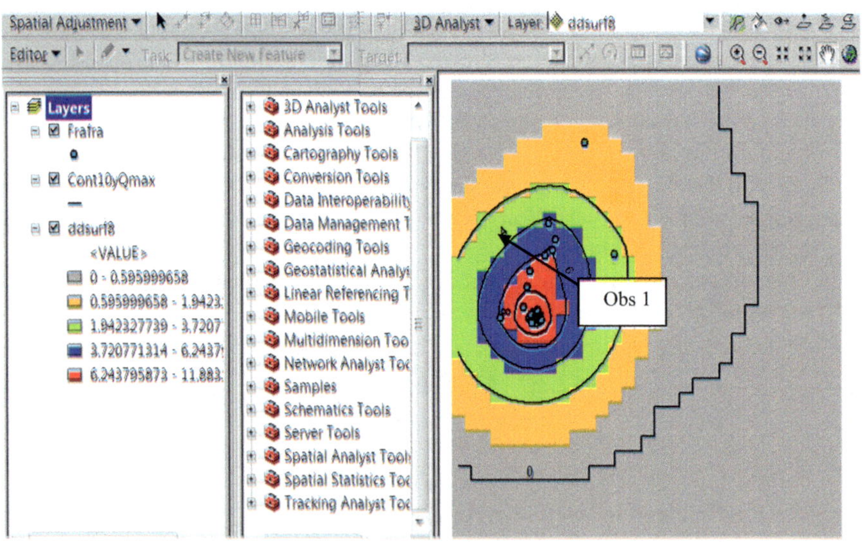

Fig. 9 Drawdown contours for Farafra. Total $Q = 5.05$ m^3/s run for a maximum period of 10 year as given in Table 5

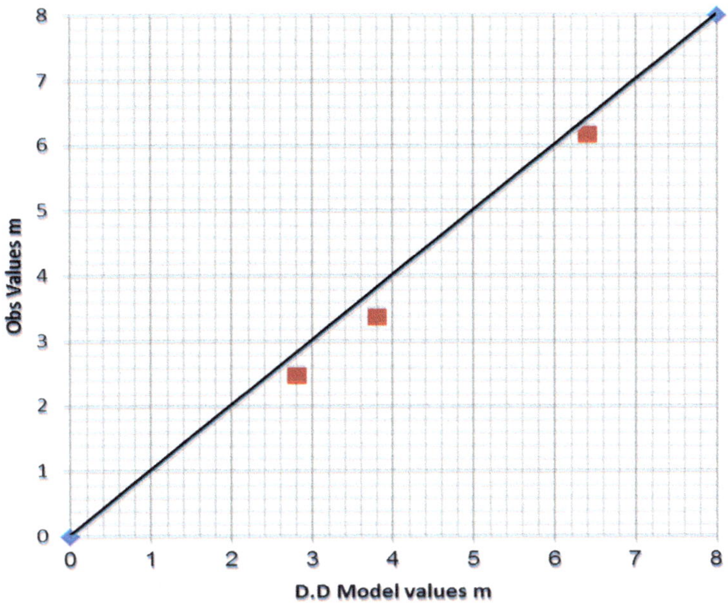

Fig. 10 Farafra aquifer correlation values (refer to Table 5)

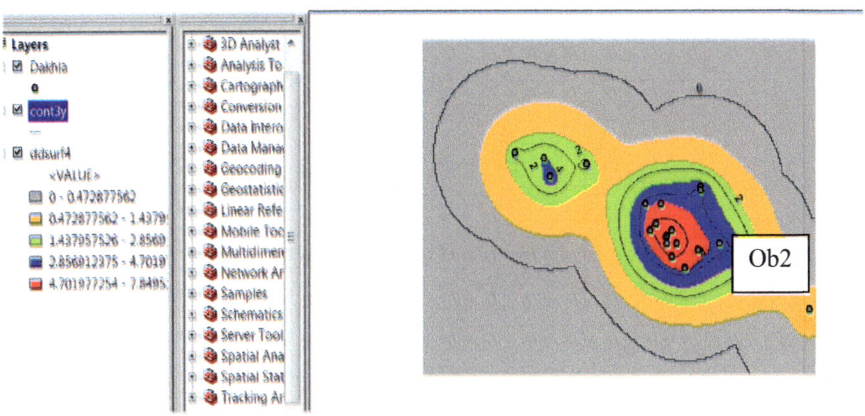

Fig. 11 Drawdown contours of Dakhla Oasis using GIS and VBC (observation well locations Obs2 is at Mut airport)

Regarding the local models of the three Oases, one can state that the difference between the HC values obtained by the GIS/VB and CEDARE values is close. However, the storativity values were adjusted, and the missing storage values were obtained. For this reason, Soliman [12] found that the local model results gave more confidence in the obtained model properties. However, a regional model for the three Oases is set to calibrate the model boundary conditions.

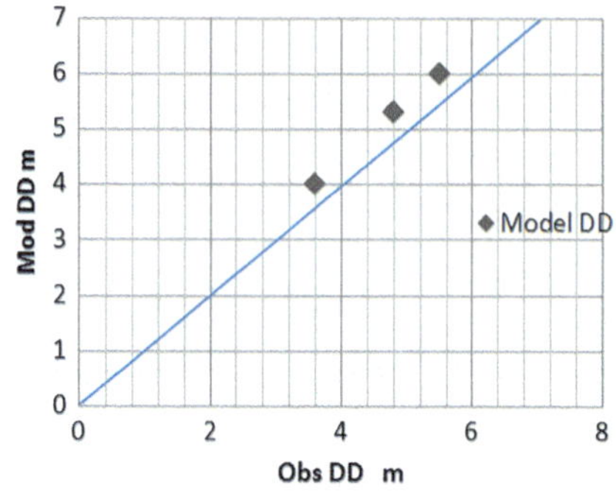

Fig. 12 Relations between observed and model data for DD in Dakhla Oasis. See for observed well locations Obs2

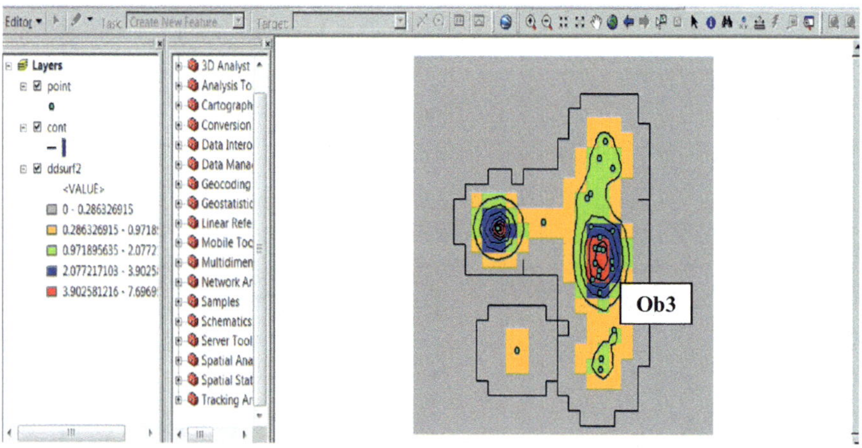

Fig. 13 Drawdown for 5Y contours for Kharga (*Obs3 stands for well* observation 3)

5.6 *Regional Model*

5.6.1 Model Preparations

The hydraulic conductivity and specific storage adjustments using the GIS with the Visual Basic extension program were found successfully for each Oasis. The regional model using the MODFLOW domain was built particularly for the calibration stage in order to adjust the model boundary conditions. The model domain would contain the well network and its discharges for the period of years (2000–2020). The well discharges are as given in Table 4 which was obtained from CEDARE Report. The well network location is found in Figs. 29, 30, and 31 in Appendix 1.

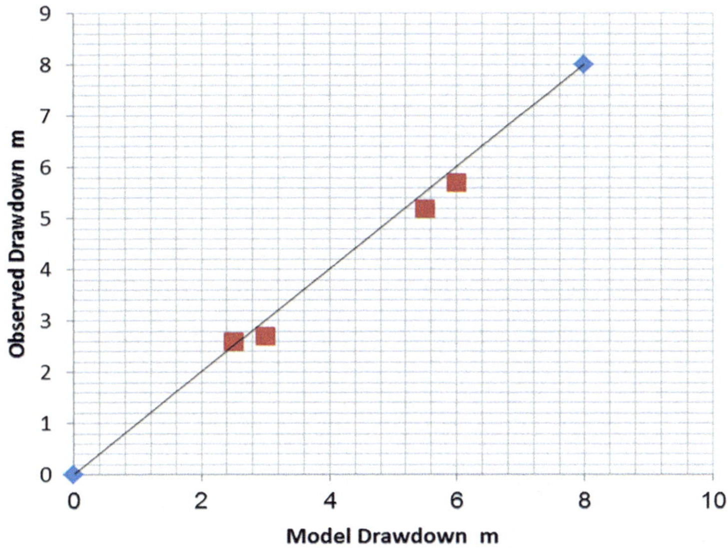

Fig. 14 Kharga Oasis Parameter adjustment

Table 6 Drawdowns (DD) calculated and observed for Dakhla Parameters

Oasis	Observation well with its model coordinates	Q (m^3/s)	Pumping time	Model DD (m)	Observed DD (m)	Note
Dakhla	Airport Mut (Obs2)	10.3	5 years (2005–2010)	6.0	5.5	Av error 5%
	$X = 200{,}170$ m		3 years (2005–2008)	4	3.6	$K = 8$ m/day
	$Y = 156{,}354$ m	8.75	5 years (2000–2005)	5.3	4.8	S_s 8.00E-06

Note: The data of Q and observed drawdown (DD) were obtained from Fig. 8 which were adopted from CEDARE Report

Table 7 Drawdowns (DD) calculated and observed for Kharga Parameters

Oasis name	Q (m^3/s)	Observation well with its model coordinates	Pumping time	Model DD (m)	Observed DD (m)	Note
Kharga	3.96	Garmashin 23 (Obs 3) $X = 360{,}900$ m $Y = 104{,}600$ m	5 years (2000–2005)	3	2.7	Av error 7.5% $K = 3.25$ m/day $S_s = 2.40$E-05
			10 years (2000–2010)	6	5.7	
		Garmashin 20 (Obs4) $X = 362{,}000$ m $Y = 93{,}170$ m	5 years (2000–2005)	2.5	2.6	
			10 years (2000–2010)	5.5	5.2	

Note: The data of Q and observed drawdown (DD) were obtained from Fig. 8 which were adopted from CEDARE Report

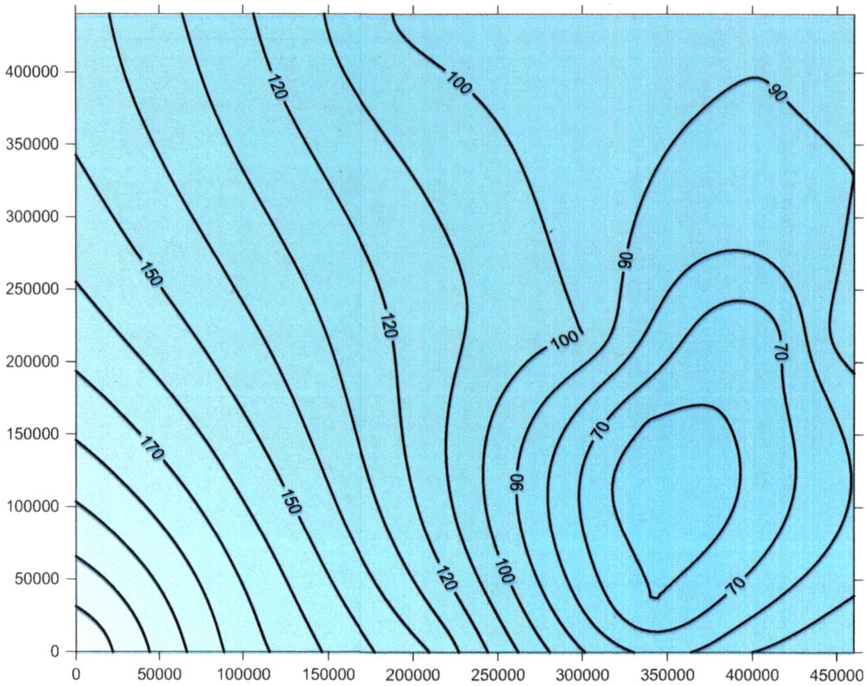

Fig. 15 Initial head contours for Dakhla Basin developed from the Hydrogeological Map by RIGW

Table 8 The final adjusted values for model properties

Location	S_s (1/L)	K (cm/s)
Farafra	2.00E-05	6.00E-05
Dakhla	8.00E-06	8.00E-05
Kharga	2.40E-05	4.00E-05

The initial head needed for the model as given in Fig. 15 was adopted from the hydrogeological map obtained by RIGW 1999/2000. Three observation wells (vis Obs1, Obs2, and Obs3) were selected for calibration process, one observation well in each Oasis.

Those wells are chosen to fit the drawdown data in Fig. 8 and as located in the GIS model. The benefit of using this stage is also to get the piezometric head at year 2010 which will be used in the final model as its initial head. Table 8 gave the final adjustment for model properties resulted from the local models. Those values were distributed in the model domain as given in Figs. 16 and 17.

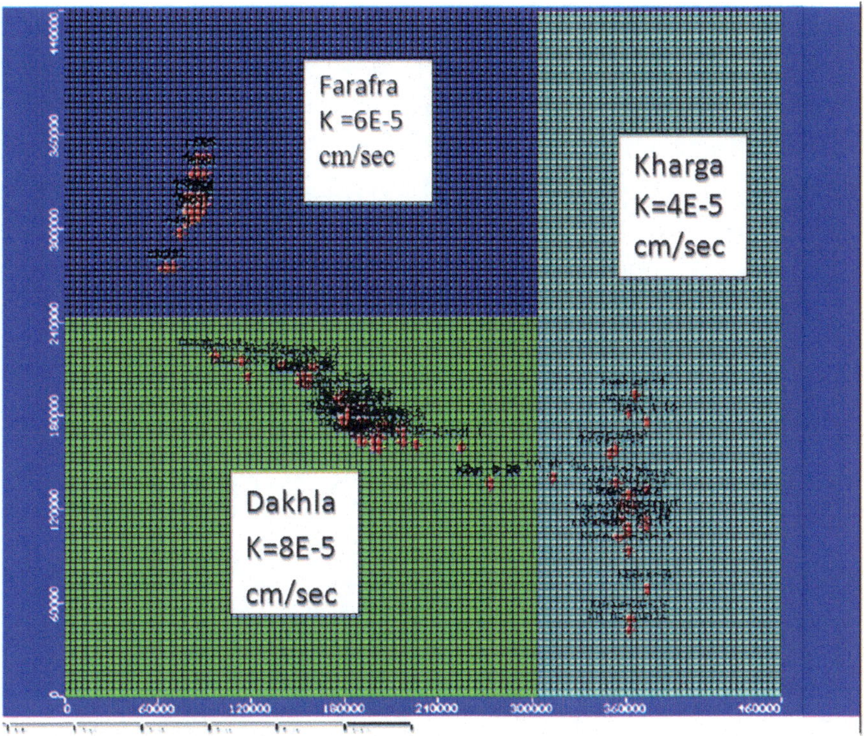

Fig. 16 HC > distribution in the model domain

5.6.2 Boundary Conditions

As a first glance to the model area, the boundary of Assiut Governorate area at the upper eastern corner of the model domain can be taken as a constant head boundary due to the presence of some seepage flow as reported by the groundwater Institute in this location. This boundary is also formed from series of faults between the Nile Basin alluvial material and the Nubian aquifer as seen in Fig. 3b. In spite of both aquifers are different from each other, yet there is a possibility that both aquifers are hydraulically connected as reported also by the groundwater Institute in this location.

Since the groundwater level fluctuates around 50 m (amsl) in Assiut alluvial aquifer, therefore a constant head boundary of 50 m can be taken at the Nile Basin boundary at north east of the model domain.

Upon looking at the initial head map, two other sources of boundaries may be postulated. These are the southwestern corner (BC2) and the southeastern corner (BC3) of the model domain (Fig. 21 where more details are given). The rest of the model boundaries are approximated as flow lines.

The aforementioned boundary conditions were slightly increased or decreased cell by cell several times until successful calibration or an optimum condition gave

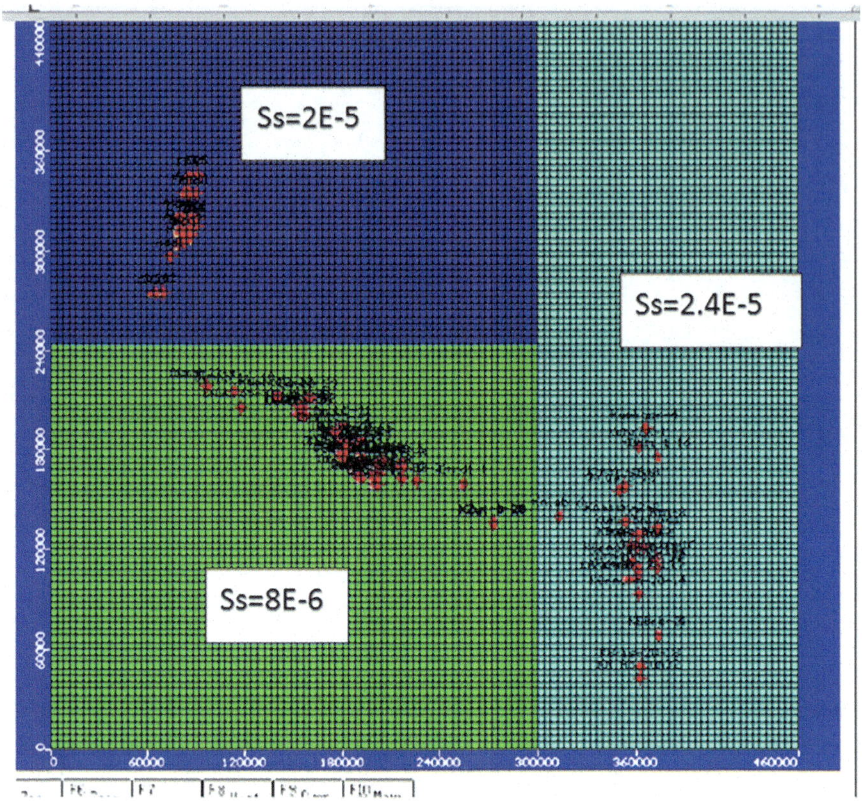

Fig. 17 Specific storage distribution in the model domain

95% confidence as given in Figs. 18 and 19. Those figures were obtained from the model output for 20-year period (year 2000 to year 2020) as given in Fig. 20.

Figure 21 shows the observation wells, the final boundary condition after calibration, and the initial head at the year 2000 as given below:

1. For observation well coordinates (Far obs1), (D obs2), and (Kh obs3) see Tables 5, 6, and 7.
2. Boundary condition BC1 is set at the Nile Basin boundary south of Asyut Governorate as mentioned before.
3. Boundary conditions BC2 and BC3 which took L section are set as follows:
   (a) The number of cells of BC2 = 11 cells in the vertical direction and 8 cells in the horizontal direction. The average potential head was found from the calibration trials to be 180 m (amsl)
   (b) Also from the calibration trials, the L section BC3 boundary was found to have four vertical cells and three horizontal cells. The average head value was also found to be =75 m (amsl) resulted from the calibration process.

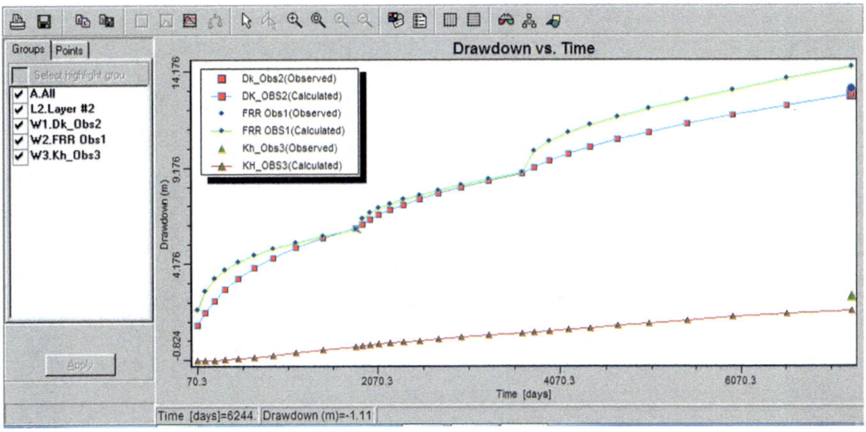

Fig. 18 DD vs time at the tutee Oasis

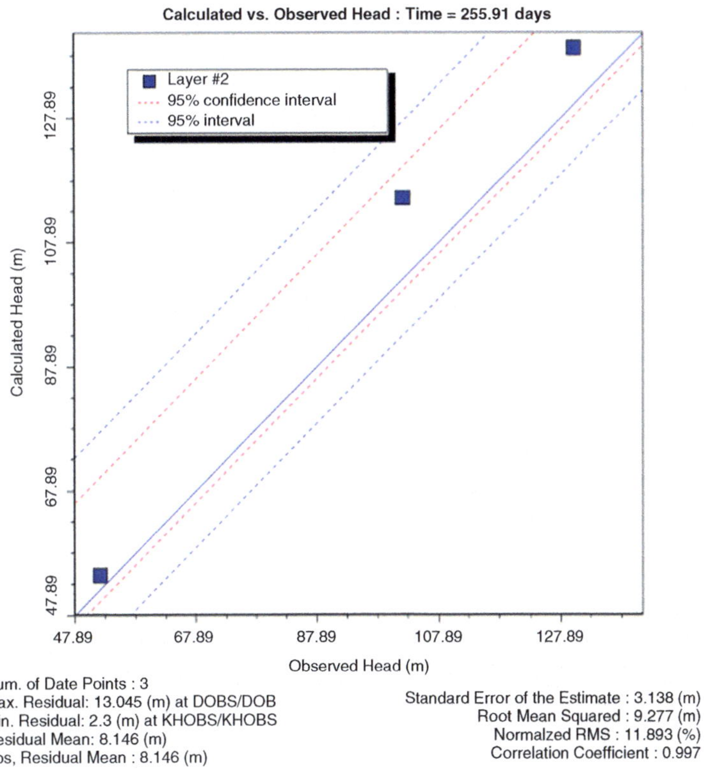

Fig. 19 Calculated vs observed head at the observed wells in the three Oases. The observed data is from CEDARE drawdown data

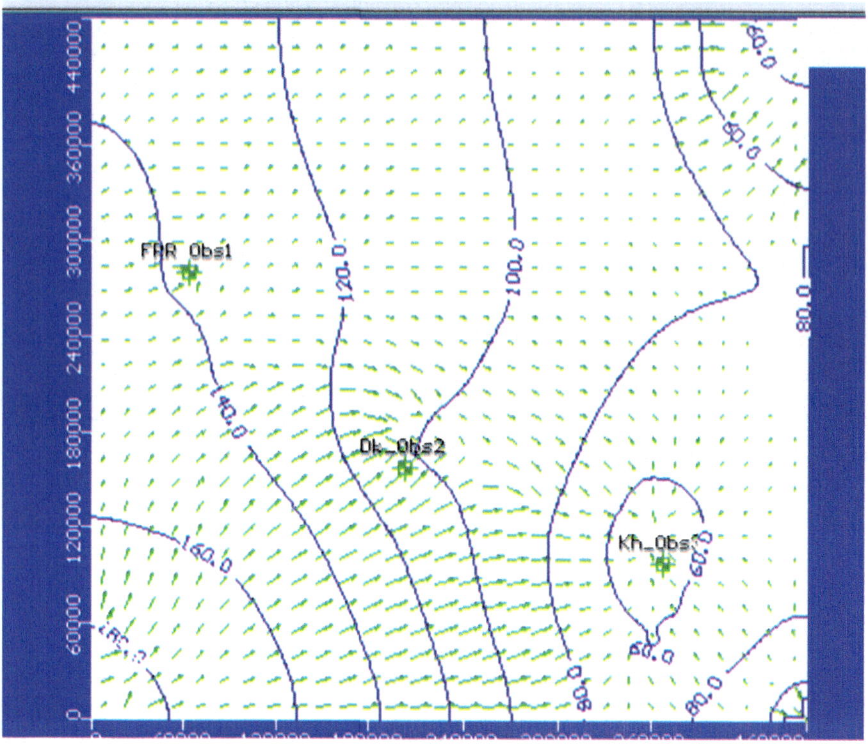

Fig. 20 Piezometric head map at 2020 resulted from the calibration process

Figure 22 shows the potential head at the year 2010 which was selected to be the initial head for the final model.

5.7 Final Model and Analysis

The final model domain is arranged to include all the wells which were replaced and constructed recently (September 2010) in the three Oases. Most of the new wells were drilled during 2008–2010. The boundary condition was already installed as mentioned in the regional model study.

Tables 9, 10, and 11 (see Appendix 2) include all the wells and its locations that have been obtained from the Groundwater Sector in the Ministry of Water Resources and Irrigation. It must be noted that all the water supply wells were also included in the tables. Two scenarios are adopted to run the final model as given below.

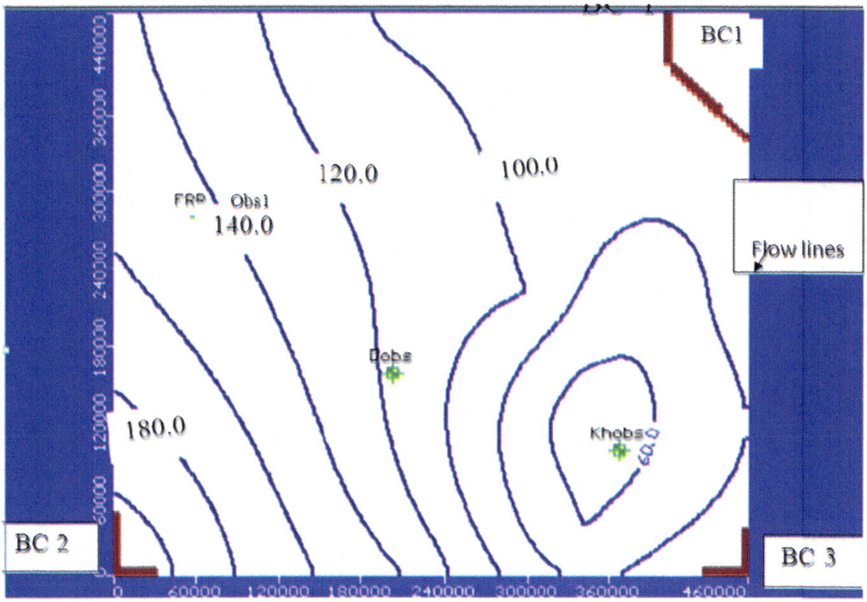

Fig. 21 Initial head in the year 2000, boundary conditions and the three observation wells used in the model domain

Scenario 1

In order to study the groundwater management in Dakhla Basin, an optimization study should be done. The optimization procedure of the Groundwater Sector of the Ministry of Irrigation is used. The process first of all deals with the Nubian aquifer in Dakhla Basin as a nonrenewable groundwater reservoir.

In this kind of aquifer, groundwater development should follow the groundwater mining system rules. This means that any groundwater exploitation should postulate constrictions for development. The Groundwater Sector of the Ministry adopted the main constriction in the groundwater management plan is that the maximum drawdown of any well field in such kind of aquifer should not be greater than 60 m resulted from running the pumping wells for 100-year period plus the historical drawdown value that exists before the 100 year period. This postulation took mainly into consideration the economic and the timescale factors. The timescale factor is postulated to have the 100-year period as enough to investigate for an additional water resources system to cover the lack of the water requirement of the area.

The final model was run first for the 100-year period (from the year 2010 to 2110) in order to find the maximum drawdown for each Oasis. The second step was to find the historical drawdown before the year 2010. Figure 8 developed by CEDARE was used to get the historical drawdown for the period between the

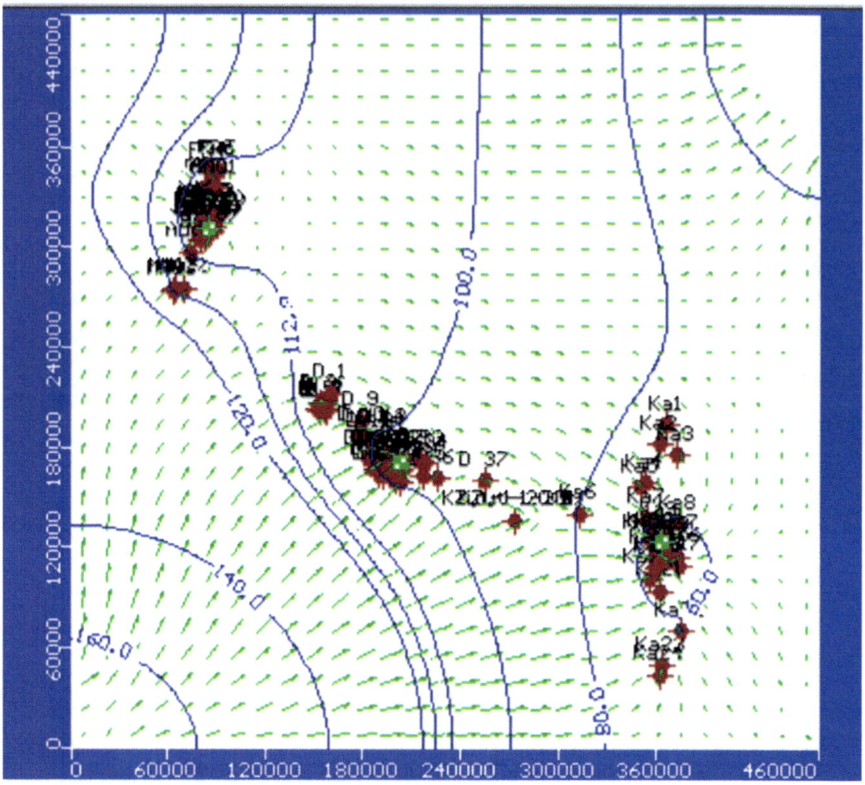

Fig. 22 Piezometric head map at year 2010 used as initial head of final model and showing also the three observation well locations

year 1960 (the time at which the New Valley project started) and the year 2010. The three wells that were selected to find the historical drawdown for the three Oases are

- Well number 307 in Farafra was used to get the historical drawdown for Farafra Oasis.
- Well Mut Airport was used to get the historical drawdown for Dakhla Oasis.
- Garmashine well 23 was used to get the historical drawdown for Kharga Oasis.

Figure 23 shows the final model results of the head contours in the model area at the year 2110. The total drawdown for the same period is given in Fig. 24. Notice in the figure the zero drawdown contours separates the Khrga Oasis well field from both Farafra and Dakhla Oases well fields.

Notice also that the drawdown in Kharga Oasis is denoted by negative sign since it is below zero drawdown line. On the other hand, Fig. 25 shows the drawdown as recorded by the observation wells. It is clear from the figure that Farafra Oasis gave the highest drawdown. Figure 26 shows the maximum and minimum drawdown

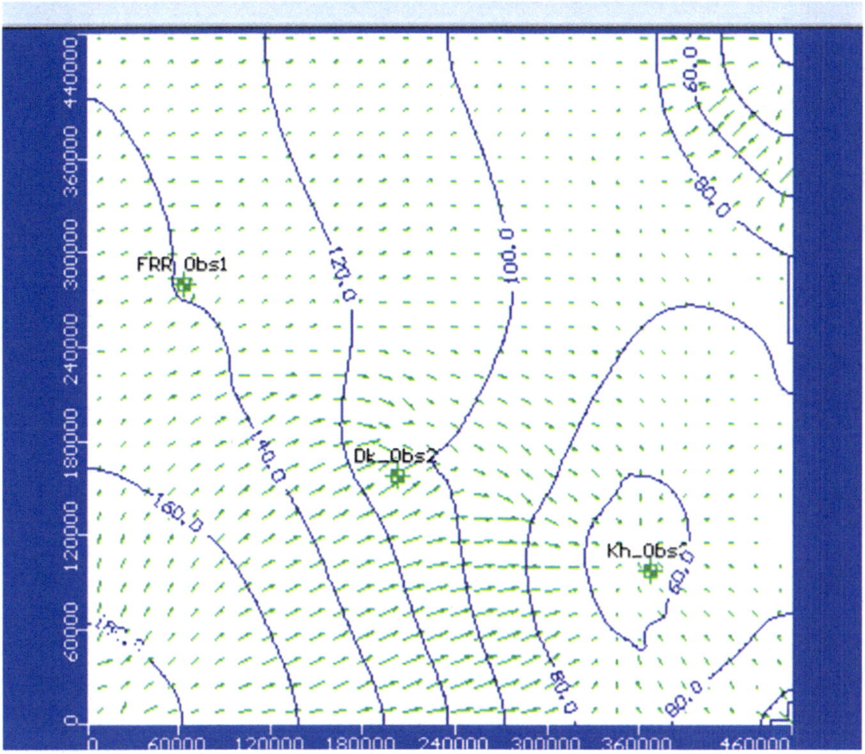

Fig. 23 Model outcome of head contours of 100-year period

values obtained by the MODFLOW. It also gave an absolute drawdown of 55 m for the model area of Dakhla Basin.

The value of drawdown −20 m denotes the maximum drawdown in Kharga Oasis whereas the drawdown value of 35 m denotes the maximum drawdown near Farafra area. The negative sign in the drawdown was only denoted algebraically by the model. This is also due to the presence of the 0.0 drawdown contour where any value above 0.0 is positive, and any value below 0.0 is negative. The 0.0 contour can be resembled by an island encountered by water, postulated as 0.0 level, and any drawdown inside the island is considered as having a negative value. It can be noticed that the flow inside the 0.0 contour approaches the steady state condition.

This can be visualized by subtracting the piezometric head in the year 2010 minus the piezometric head at the year 2110. The reason for that is due to the reduction of the well discharges in Dakhla Oasis in the year 2010 by 50% as given in Table 10 in Appendix 2. The model study considered this reduction until the year 2110.

According to optimization procedure for Kharga Oasis the maximum drawdown was found to be 20 m in 100-year period. The historical drawdown value started

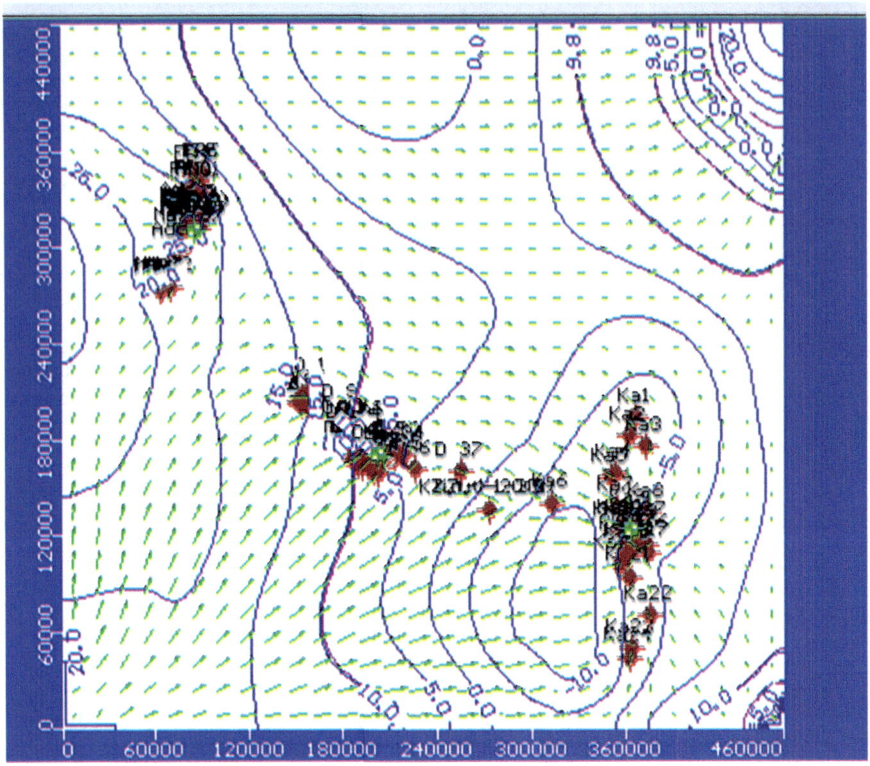

Fig. 24 Drawdown contours for 100-year period

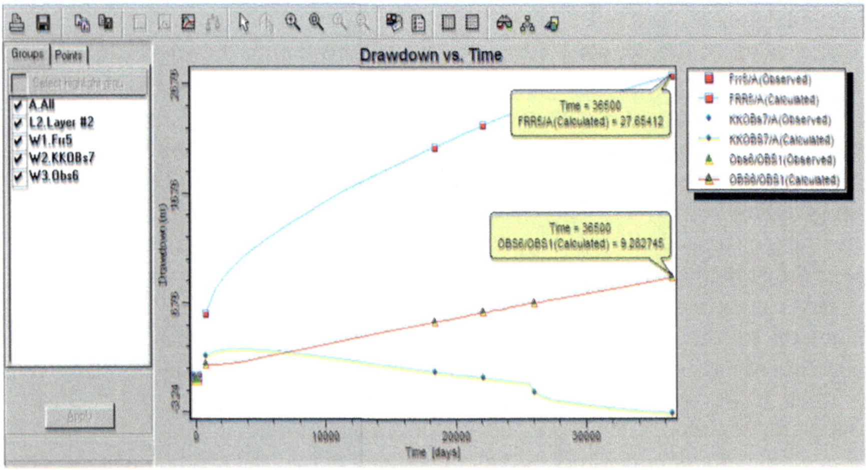

Fig. 25 Drawdown vs time at the three observation wells in the three Oases

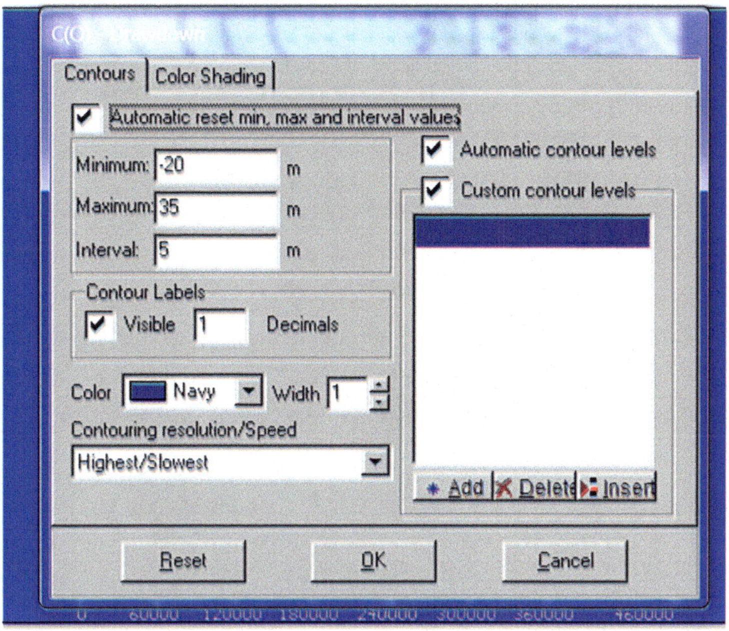

Fig. 26 Maximum and minimum drawdowns at the year 2110

from the year 1960 which was considered as the beginning of New Valley development may be adopted from Fig. 8.

The Garmashine well 23 in Fig. 8 (see its location in Fig. 31 in Appendix 1) was used to give the historical drawdown. This value was found to be 42 m.

Therefore, the total drawdown in this respect until the year 2110 would be 62 m which is considered as critical drawdown value by the Ministry of Irrigation optimization process.

The historical drawdown in Farafra Oasis using Well 307-Farafra gave 26 m (see Fig. 8). On the other hand, the 100-year maximum drawdown in Farafra was found to be 35 m (Fig. 26). Therefore, the total drawdown became 61 m which do not favor any discharge increase. This is according to the Groundwater Sector postulation as mentioned before.

Concerning Dakhla Oasis, the 100-year maximum drawdown was found to be 24 m (Fig. 24). On the other hand, the historical drawdown of Mut Air port well gave 34 m. Accordingly, the total drawdown gave 58 m. Therefore, Dakhla Oasis allows further expansion which is discussed in the Scenario 2.

Scenario 2

In this scenario, the final model was run with higher discharge for Dakhla Oasis than that in Scenario 1. This was proved possible since the total maximum drawdown of Dakhla Oasis was found to be 58 m which is less than the critical drawdown (60 m as mentioned before). However, one should be very cautious to the well discharge increase. After a series of trials 15% increase of the well

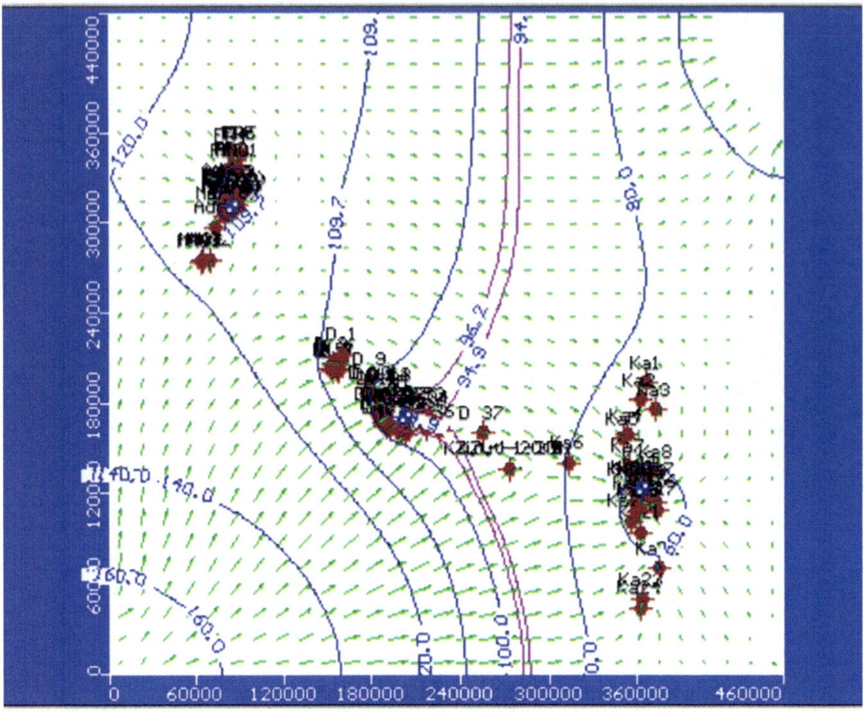

Fig. 27 Piezometric head map of the model output at the year 2110 after increasing the well discharge by 15%

discharge (Q) was found reasonable in Dakhla Oasis. Figure 27 shows the expected piezometric head at the year 2110 with the increased well discharge. The new well discharges for Dakhla Oasis are also given in Table 10 in Appendix 2.

In order to calculate the maximum drawdown in Dakhla Oasis in this scenario, the piezometric map of Fig. 27 was zoomed near the center of the Oasis as given in Fig. 28 which gave the lowest contour value as 94.4 m. The initial contour that was passing through this area as given in Fig. 22 is 120 m. Therefore the drawdown in Dakhla center would be 25.6 m. Accordingly, by adding 25.6 m to the historical drawdown found in Scenario 1 of 34 m resulted total maximum drawdown of 59.6 m which considered safe and little less than critical drawdown. This fact should be taken into consideration when discharge increase for Dakhla basin development is planned.

6 Conclusions

The study area in this research is located in the middle part of Egypt's Western Desert. It lies between latitudes 24°–28°N and longitudes of 27°–31.5° E. It includes the three Oases, Farafra, Dakhla, and Kharga, which are part of the New

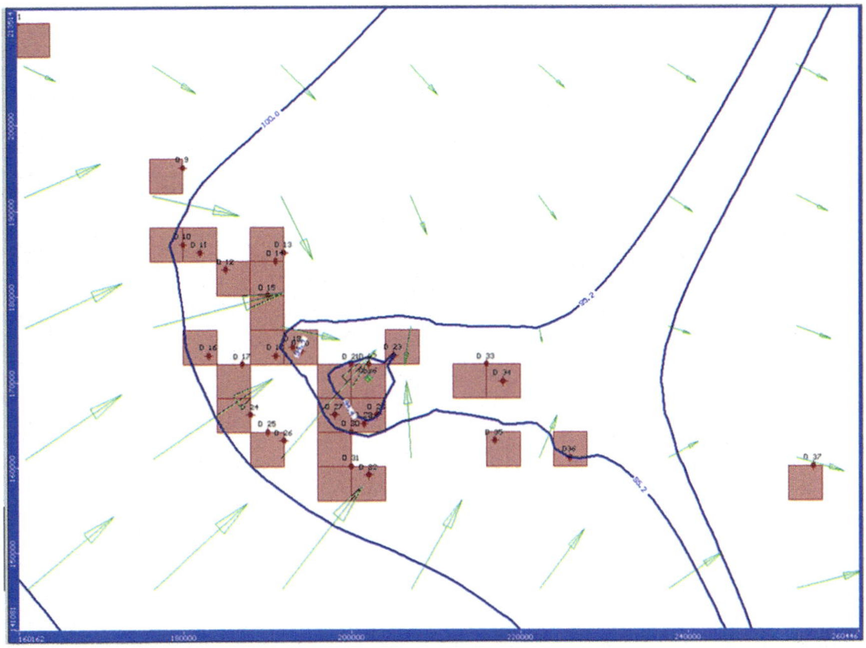

Fig. 28 Zooming of piezometric head map of Fig. 24 at Dakhla Oasis Center

Valley area. It covers an area of about 440 km long by 460 km wide. It is equivalent to an area of 202,400 km^2 which is named as the Dakhla Basin. The western limestone plateau bound it to the east and north of Kharga and Dakhla region.

The main objective of this research is to evaluate the groundwater resources and its potentiality for development in the study area. The aquifer of this groundwater resource is mainly the deep Nubian Sandstone Aquifer which is located in the three Oases. Due to the scarcity of the groundwater recharge in the Nubian aquifer, it is hydrogeologically treated as groundwater mining aquifer [12].

Three groundwater models were used in this respect. A local model using GIS accompanied by the visual basic was first prepared to adjust the model properties. A second model using MODFLOW was furnished to calibrate the boundary conditions of the main model. The third model was using the main model as a final model to find the main objectives which are the groundwater potentiality and its management in Dakhla Basin The model was run for a 100-year period to suit the groundwater mining conditions.

As an outcome of the model study for the optimum management strategy in Dakhla Basin, two scenarios were postulated. The objective is to maintain the resultant drawdown in the development areas within an acceptable range for economic purposes besides obtaining the most appropriate area for development in Dakhla Basin.

The salient fact from this study showed that Dakhla Oasis is the only place in the New Valley area where land reclamation development can safely take place.

Appendix 1

FID	Shape	name	Long	Lat	X	Y	Q
0	Point	AbuMin 1	27.6575	26.512778	65750	276405.58	20777
1	Point	AbuMin 3	27.666667	26.496389	66666.7	274602.79	20777
2	Point	AbuMin 5	27.6275	26.541389	62750	279552.79	20777
3	Point	AbuMin 6	27.596389	26.510556	59638.9	276161.16	20777
4	Point	AbuMin 7	27.6075	26.493333	60750	274266.63	20777
5	Point	AbuMin 9	27.693333	26.497222	69333.3	274694.42	20777
6	Point	AbuMin 11	27.638611	26.495278	63861.1	274480.58	20777
7	Point	AbuMin 12	27.625833	26.489167	62583.3	273808.37	20777
8	Point	AbuMin 13	27.619444	26.480556	61944.4	272861.16	20777
9	Point	AbuMin 14	27.3775	26.511667	37750	276283.37	20777
10	Point	Frf- Dalla (F1)	28.25	26.906	125000	319660	20777
11	Point	Frf-Qasr frf F4	28.02	27.6	102000	396000	20777
12	Point	Frf-Frf 3-11 F5	27.8	27	80000	330000	20777
13	Point	Frf-KfahF6	27.7776	26.9336	77760	322696	20777
14	Point	Frf-Mrzq SbhF7	27.7452	27.0955	74520	340505	20777
15	Point	Frf-Brqa (frf)F8	27.72	26.9	72000	319000	20777
16	Point	Frf-Qrwn NF9	27.6	26.8	60000	308000	20777
17	Point	Frf-Qrwn S F10	27.58	26.7	58000	297000	20777
18	Point	Frf-Abu MinqrF11	27.5509	26.5772	55090	283492	20777
19	Point	Frf-Enma wls(minqr)F12	27.66	26.56	66000	281600	20777
20	Point	Frf-Bustan minqrF13	27.611	26.528	61100	278080	20777

Fig. 29 Attribute table for Farafra well coordinates (X, Y in m and Q in m^3/day)

FID	Shape	Name	F2	X	Y	Q
0	Point		D1	114180	215600	34363
1	Point	Dkla-Mouhob 1-17	D2	117800	205810	34363
2	Point	Dkl-west mouhob 1-40	D3	140120	212630	34363
3	Point	Dkla-Bdkhul 1-14	D4	175780	191180	34363
4	Point	Dkla-El Qasr 1-21	D5	185500	191180	34363
5	Point	Dkla-El Rashda 1-14	D6	182260	180510	34363
6	Point	Dakla-El Gdid 1-22	D7	179020	176880	34363
7	Point	Dkla-El Qlamun 1-17	D8	188740	169840	34363
8	Point	Dkla-wali 1-6	D9	189000	173800	34363
9	Point	Dakla-El Hindew 1-14	D10	191990	176880	34363
10	Point	Dakhla-Mut 1-36	D11	209600	200000	34363
11	Point	Dkla-Ismant1-10	D12	195230	169840	34363
12	Point	Dakhla well	D13	191990	162690	34363
13	Point	Dkla-Masara 1-20	D14	208190	166287	34363
14	Point	Dkla-Balat 1-27	DS1	221160	169862	34363
15	Point	Dkla-Tneda- N	D16	230890	166287	34363
16	Point	Dkla-Zyat	D17	276270	134189	34363
17	Point	Mut airport	D18	200170	156354	34363
18	Point	Mawhoub W-15	D19	117830	205788	34363
19	Point	Qour El Malek	D20	96610	218438	34363
20	Point	South Mut(new)	D21	209580	198000	34363
21	Point	South Teneida	D22	230000	165000	34363

Fig. 30 Attribute table for Dakhla well coordinates (X, Y in m and Q in m^3/day)

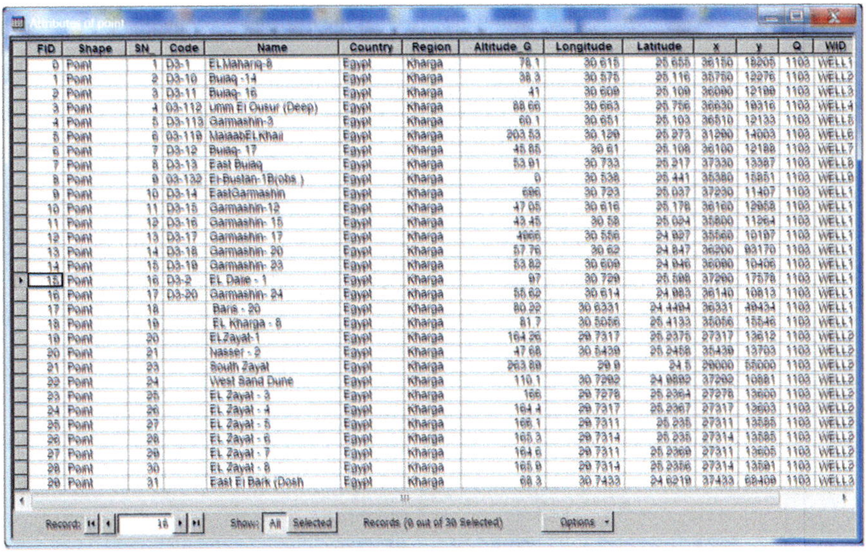

Attributes of point

FID	Shape	SN_	Code	Name	Country	Region	Altitude_G	Longitude	Latitude	x	y	Q	WID
0	Point	1	D3-1	ELMaharig-8	Egypt	Kharga	78.1	30.615	25.655	36150	18205	1103	WELL1
1	Point	2	D3-10	Buiaq -14	Egypt	Kharga	38.3	30.575	25.116	35750	12276	1103	WELL2
2	Point	3	D3-11	Buiaq- 16	Egypt	Kharga	41	30.609	25.109	36090	12199	1103	WELL3
3	Point	4	03-112	umm El Ousur (Deep)	Egypt	Kharga	88.66	30.663	25.756	36630	19316	1103	WELL4
4	Point	5	D3-113	Garmashin-3	Egypt	Kharga	60.1	30.651	25.103	36510	12133	1103	WELL5
5	Point	6	03-119	MalaabELKhail	Egypt	Kharga	203.53	30.129	25.273	31290	14003	1103	WELL6
6	Point	7	D3-12	Buiaq- 17	Egypt	Kharga	45.85	30.61	25.108	36100	12188	1103	WELL7
7	Point	8	D3-13	East Buiaq	Egypt	Kharga	53.91	30.733	25.217	37330	13387	1103	WELL8
8	Point	9	03-132	El-Bustan-1B(obs.)	Egypt	Kharga	0	30.538	25.441	35380	15851	1103	WELL9
9	Point	10	D3-14	EastGarmashin	Egypt	Kharga	696	30.723	25.037	37230	11407	1103	WELL1
10	Point	11	D3-15	Garmashin-12	Egypt	Kharga	47.05	30.616	25.178	36160	12958	1103	WELL1
11	Point	12	D3-16	Garmashin- 15	Egypt	Kharga	43.45	30.58	25.024	35800	11264	1103	WELL1
12	Point	13	D3-17	Garmashin- 17	Egypt	Kharga	4966	30.556	24.927	35560	10197	1103	WELL1
13	Point	14	D3-18	Garmashin- 20	Egypt	Kharga	57.76	30.62	24.847	36200	93170	1103	WELL1
14	Point	15	D3-19	Garmashin- 23	Egypt	Kharga	53.82	30.609	24.946	36090	10406	1103	WELL1
15	Point	16	D3-2	EL Daie - 1	Egypt	Kharga	97	30.729	25.598	37290	17578	1103	WELL1
16	Point	17	D3-20	Garmashin- 24	Egypt	Kharga	55.62	30.614	24.983	36140	10813	1103	WELL1
17	Point	18		Baris - 20	Egypt	Kharga	80.22	30.6331	24.4494	36331	49434	1103	WELL1
18	Point	19		EL Kharga - 8	Egypt	Kharga	81.7	30.5056	25.4133	35056	15546	1103	WELL1
19	Point	20		ELZayat-1	Egypt	Kharga	164.26	29.7317	25.2375	27317	13612	1103	WELL2
20	Point	21		Nasser - 2	Egypt	Kharga	47.68	30.5439	25.2458	35439	13703	1103	WELL2
21	Point	23		South Zayat	Egypt	Kharga	263.89	29.9	24.5	29000	55000	1103	WELL2
22	Point	24		West Sand Dune	Egypt	Kharga	110.1	30.7292	24.9892	37292	10881	1103	WELL2
23	Point	25		EL Zayat - 3	Egypt	Kharga	166	29.7278	25.2364	27278	13600	1103	WELL2
24	Point	26		EL Zayat - 4	Egypt	Kharga	164.4	29.7317	25.2367	27317	13603	1103	WELL2
25	Point	27		EL Zayat - 5	Egypt	Kharga	166.1	29.7311	25.235	27311	13585	1103	WELL2
26	Point	28		EL Zayat - 6	Egypt	Kharga	165.3	29.7314	25.235	27314	13585	1103	WELL2
27	Point	29		EL Zayat - 7	Egypt	Kharga	164.6	29.7311	25.2369	27311	13605	1103	WELL2
28	Point	30		EL Zayat - 8	Egypt	Kharga	165.9	29.7314	25.2356	27314	13591	1103	WELL2
29	Point	31		East El Bark (Dosh	Egypt	Kharga	68.3	30.7433	24.6219	37433	68409	1103	WELL3

Record: 16 Show: All Selected Records (0 out of 30 Selected) Options

Fig. 31 Attribute table showing Kharga well locations and its coordinates

Appendix 2

Table 9 Farafra well group for final Model

No		*X* (m)	*Y* (m)	*Q* (m^3/day)
1	MZ1	88,000	312,000	13,300
2	MZ2	80,000	304,000	18,700
3	MZ3	81,000	311,000	17,050
4	Mz4	80,000	315,000	15,650
5	Mz5	86,000	311,000	18,000
6	Mz6	87,000	310,000	19,800
7	MZ7	85,000	308,000	18,000
8	MZ8	90,000	315,000	12,500
9	MZ9	90,000	319,000	8,900
10	MZ10	88,000	317,000	15,250
11	Mz11	83,500	306,000	18,000
12	MZ12	83,000	307,000	15,600
13	MZ13	84,000	310,000	17,000
14	MZ14	85,000	313,000	9,200
15	FR1	85,500	335,000	16,430
16	FR2	83,000	335,000	18,095
17	FR3	85,000	336,000	16,720
18	FR4	86,000	345,000	22,572
19	FR5	91,000	346,000	12,932
20	FR6	92,000	346,000	18,216
21	MNQ1	90,000	335,000	15,050
22	MINQ2	70,000	275,000	15,050
23	MNQ3	64,000	274,000	15,824
24	MNQ4	65,000	275,000	11,650
25	Add1	74,000	297,000	16,450
26	Add2	86,000	310,000	14,950
27	Add3	87,000	320,000	13,000
28	Add4	80,000	321,000	12,300
29	Add5	80,000	312,000	13,500
30	Add6	82,000	320,000	13,000
			Total (m^3/day)	462,689
			Total (m^3/s)	5.355197

Table 10 Dakhla well group for final model

No		Longitude	Latitude	*X* (m)	*Y* (m)	*Q* (m^3/day)	1.15 × *Q*
1	Teneda1	29.250	25.455	225,000	160,000	12,407	14,268.05
2	Teneda2	29.260	25.464	226,000	161,000	11,874	13,655.1
3	Balat1	29.180	25.545	218,000	170,000	10,786	12,403.9
4	Balat2	29.170	25.482	217,000	163,000	8,581	9,868.15
5	Balat3	29.160	25.564	216,000	172,000	12,944	14,885.6
6	Asmant1	29.050	25.573	205,000	173,000	14,636	16,831.4
7	Masara1	29.020	25.564	202,000	172,000	14,947	17,189.05
8	Masara2	29.015	25.500	201,500	165,000	15,303	17,598.45
9	Shekwali	29.030	25.509	203,000	166,000	10,855	12,483.25
10	Mute1	28.900	25.491	190,000	164,000	11,678	13,429.7
11	Mute2	29.000	25.455	200,000	160,000	13,797	15,866.55
12	Mut3	29.020	25.445	202,000	159,000	12,906	14,841.9
13	Mut4	28.920	25.482	192,000	163,000	14,466	16,635.9
14	Mut5	29.000	25.491	200,000	164,000	14,465	16,634.75
15	Hend1	28.980	25.509	198,000	166,000	15,289	17,582.35
16	Owena1	29.000	25.564	200,000	172,000	16,176	18,602.4
17	Rashda1	28.910	25.573	191,000	173,000	16,382	18,839.3
18	Rashda2	28.930	25.582	193,000	174,000	17,196	19,775.4
19	Bedkhlo1	28.940	25.577	194,000	173,500	16,692	19,195.8
20	Kalamon1	28.880	25.509	188,000	166,000	15,570	17,905.5
21	ElGedida1	28.870	25.564	187,000	172,000	14,025	16,128.75
22	ElGedida2	28.830	25.573	183,000	173,000	15,526	17,854.9
23	Mosh1	28.900	25.636	190,000	180,000	16,544	19,025.6
24	Kasr1	28.910	25.673	191,000	184,000	17,960	20,654
25	Kasr2	28.920	25.682	192,000	185,000	14,856	17,084.4
26	Kasr3	28.850	25.664	185,000	183,000	16,844	19,370.6
27	Mawhub1	28.820	25.682	182,000	185,000	15,807	18,178.05
28	Mawhub2	28.800	25.691	180,000	186,000	15,507	17,833.05
29	Mawhub3	28.800	25.773	180,000	195,000	16,693	19,196.95
30	Mawhub4	28.550	25.864	155,000	205,000	14,396	16,555.4
31	Mawhub5	28.480	25.891	148,000	208,000	17,800	20,470
32	Mawhub6	28.520	25.836	152,000	202,000	17,277	19,868.55
33	Mawhub7	28.570	25.827	157,000	201,000	15,012	17,263.8
34	Mawhub8	28.540	25.827	154,000	201,000	16,750	19,262.5
35	Mawhub9	28.560	25.836	156,000	202,000	17,979	20,675.85
36	Mawhub10	28.570	25.864	157,000	205,000	14,431	16,595.65
37	Mawhub11	28.540	25.873	154,000	206,000	10,640	12,236
				Total (m^3/day)		544,997	626,746.55
				Total (m^3/s)		6.3078	7.2540

Table 11 Kharga wells for final model

No	Name	Longitude	Latitude	*X* (m)	*Y* (m)	*Q* (m^3/day)
1	ELMahariq-8	30.615	25.655	361	182	13,900
2	Bulaq-14	30.575	25.116	357	122	13,900
3	Bulaq-16	30.609	25.109	360	121	13,900
4	Umm El Ousur	30.663	25.756	366	193	13,900
5	Garmashln-3	30.651	25.103	365	121	13,900
6	MalaabELKhail	30.129	25.273	312	140	13,900
7	Bulaq-17	30.61	25.108	361	121	13,900
8	East Bulaq	30.733	25.217	373	133	13,900
9	El-Bustan-1 B	30.538	25.441	353	158	13,900
10	EastGarmashln	30.723	25.037	372	114	13,900
11	Garmashln-12	30.616	25.178	361	129	13,900
12	Garmashln-15	30.58	25.024	358	112	13,900
13	Garmashln-17	30.556	24.927	355	101	13,900
14	Garmashln-20	30.62	24.847	362	931	13,900
15	Garmashln-23	30.609	24.946	360	104	13,900
16	EL Dalie-1	30.729	25.598	372	175	13,900
17	Garmashln-24	30.614	24.983	361	108	13,900
18	Baris-20	30.6331	24.4494	363	494	13,900
19	EL Kharga-8	30.5056	25.4133	350	155	13,900
20	ElZayat-1	29.7317	25.2375	273	136	13,900
21	Nasser-2	30.5439	25.2458	354	137	13,900
22	South Zayat	29.9	24.5	290	550	13,900
23	West Sand Dune	30.7292	24.9892	372	108	13,900
24	EL Zayat-3	29.7278	25.2364	272	136	13,900
25	EL Zayat-4	29.7317	25.2367	273	136	13,900
26	EL Zayat-5	29.7311	25.235	273	135	13,900
27	EL Zayat-6	29.7314	25.235	273	135	13,900
28	EL Zayat-7	29.7311	25.2369	273	136	13,900
29	EL Zayat-8	29.7314	25.2356	273	135	13,900
30	East El Bark	30.7433	24.6219	374	684	13,900
31	North Abu Bayan	30.6261	24.3864	362	425	13,900
				Total (m^3/day)		430,900
				Total (m^3/s)		4.9873

References

1. RIGW (1998) Hydrogeological map of Egypt, 1:2,000,000 with explanatory note. Academy of Scientific Research and Technology, Cairo
2. RIGW (1998) Hydrogeological map of Luxor, 1:5,00,000 with explanatory note. Academy of Scientific Research and Technology, Cairo
3. RIGW, Research Institute for Groundwater (1988) Major regional aquifer in North East Africa. Int. report, Giza

4. RIGW (1993) The Hydrogeological Map of Egypt. Scale 1:2,000,000. Research Institute for Ground Water, Cairo
5. RIGW (2001) Southern Egypt development project. Research Institute for Ground Water, unpublished internal report
6. CEDARE (2002) Regional Strategy for the Utilization of the Nubian Sandstone Aquifer System. Draft final report, Centre for Environment and Development for the Arab Region and Europe, Heliopolis Bahry, Cairo. https://www.iaea.org/sites/default/files/sap180913.pdf
7. Bakhbakhi M (2006) Nubian Sandstone Aquifer System. IHP-VI, series on groundwater, 10:75–81. UNESCO, Paris
8. Brinkman PJ, Heinl M, Hollander R, Reich G (1989) Retrospective simulation of groundwater, flow and transport in the Nubian Aquifer System. Berliner Geowiss, Berlin
9. LaMoreaux PE, LaMoreaux JW, Soliman MM, Memon BA, Assaad FA (2008) Environmental hydrogeology, 2nd edn. CRC Press, New York. ISBN 9781420054859 - CAT# 54856
10. Sefelnasr AM (2007) Development of groundwater flow model for water resources management in the development areas of the western desert, Egypt. MSc in hydrogeology, submitted to the Faculty of Natural Sciences III of the Martin Luther University Halle-Wittenberg. https://sundoc.bibliothek.uni-halle.de/diss-online/07/07H178/t5.pdf
11. RIGW (1999) Updating of hydrogeological map of Frafra Oasis, 1:5,00,000 with explanatory note. Academy of Scientific Research and Technology, Cairo
12. Soliman AMM (2016) Groundwater resources management in terms of quantity and quality in western desert of Egypt in the Nubian sandstone aquifer system. PhD thesis, Faculty of Engineering, Cairo University, Cairo
13. RIGW (1999) A plan for the development and management of deep groundwater in the Oases. Internal strategy report, Research Institute for Ground Water, Cairo
14. Gad M, Soliman MM (2015) A GIS spatial interpolation method for determining the drawdown field in the vicinity of multiple wells. Int Water Technol J 5(4):284–293
15. CEDARE (2001) Regional Strategy for the Utilization of the Nubian Sandstone Aquifer System. Volume II. Centre for the Environment and Development for the Arab Region and Europe, Hydrogeology, Cairo

Part IX
Groundwater Usage and Groundwater Quality Assessment

Hydrogeochemistry and Quality Assessment of Groundwater Under Some Central Nile Delta Villages, Egypt

Zenhom E. Salem, Gamal Elsaiedy, and Abdelaziz ElNahrawy

Abstract The target of this work is to assess the impact of the human activities on the hydrochemistry and quality of the groundwater under Nile Delta villages. Sixteen groundwater samples were collected during 2016. Hydrochemical analyses including major and trace elements were done. Spatial distribution of the element concentrations according to WHO guidelines, WQI (drinking water quality index), and statistical analysis was used for assessment. The sampled groundwater showed variable quality from one village to another and was classified into three clusters. Cluster A is characterized by higher concentrations of total dissolved solids (TDS), electrical conductivity (EC), potassium, magnesium, sodium, calcium, sulfate, chloride, bicarbonate, Mn, Zn, P, NH_4, Ba, and unfit WQI values. Low water quality of this sample is related to the effect of El-Gharbia main drain and seawater intrusion. Cluster B included samples 4, 8, 11, and 15 and had the moderately mean ion concentrations and higher concentrations of Fe, Sr, and Si compared to the other two clusters. The latter cluster major ion concentration arrangement is sodium > calcium and chloride > bicarbonate. WQI of this cluster varies from poor, very poor to unfit.

Samples 1, 2, 3, 5, 6, 7, 9, 10, 12, 13, and 14 belong to cluster C which had the lowest ion concentrations and dominated by Ca > Na > Mg and HCO_3 > Cl ion concentration arrangement. WQI of most of these samples is good. Undesirable concentrations of arsenic and ammonia are indications of direct infiltration from septic tanks and/or seepage from the drains. Finally, water treatment should be done before usage of groundwater from under the Nile Delta villages because of human-induced contamination.

Z.E. Salem (✉) and A. ElNahrawy
Geology Department, Faculty of Science, Tanta University, Tanta, Egypt
e-mail: zenhomsalem@yahoo.com

G. Elsaiedy
Ministry of Environmental Affairs, Cairo, Egypt

A. M. Negm (ed.), *Groundwater in the Nile Delta*,
Hdb Env Chem (2019) 73: 625–646, DOI 10.1007/698_2017_111,

Keywords Nile Delta, Rural groundwater contamination, Water quality index (WQI)

Contents

1 Introduction

The guidelines for drinking water are attributed to two main criteria: (1) the absence of unacceptable taste, odor, and color and (2) the absence of substances that harm the physiological functions. In this way, water needs to get together with certain physical, chemical, and microbiological principles, that is, it must be free from illnesses delivering microorganisms and concoction substances, hazardous to well-being before it can be named consumable.

Most of the villages of the developing countries currently have a water supply, yet few of them have wastewater collection and treatment system [1–3]. There is a need to construct proper sewer frameworks in the small communities; otherwise, individuals will depend on on-site sanitation. In such septic tank, sewage water infiltrates to the underlying aquifer forming the biggest volumetric source of pathogenic organisms to the groundwater. Wastewater and sludge are also dumped in the closest water body or directly on agrarian fields.

Bishnoi and Arora [1] found that village's wastewater effluent was the major source of fluoride in the potable groundwater in Haryana, India. Village's environs in central India were the source of nitrate contamination in basaltic groundwater aquifer [2]. Gao et al. [4] monitored the groundwater quality in drinking water wells in a village belonging to Shanghai City (China). They found that the groundwater in the residential areas was highly contaminated by NH4-N without significant seasonal patterns and most water samples are unsuitable for human consumption. The work of Sajidu et al. [5] revealed high levels of fluorides (>1.5 mg/l) in some villages of Southern Malawi, and a positive correlation was observed between the pH of the water and fluoride concentrations. Geriesh and EL-Rayes [6] studied the municipal contamination of the shallow groundwater under south Ismailia City villages (Egypt). They concluded that the major source of the groundwater contamination was the seepage from the poorly constructed septic tanks; therefore, the human activity was the main source of higher nitrate and heavy metals concentrations and high bacterial count in the shallow groundwater. Emara et al. [7] deduced that the industrial activities, wastewater drains, and fertilizers are the main sources of groundwater pollution in the water-bearing formation of the Quaternary aquifer under some rural areas in Giza Governorate of Greater Cairo, Egypt. In the central part of the Nile Delta, most of the villages are unsewered; therefore, the consumed water for domestic purposes is stored in the septic tanks and artificially recharges the groundwater causing water logging and groundwater pollution. Salem [3] integrated the water level, sedimentological setting, stable isotopes, subsurface temperature, and hydrochemistry data to investigate and evaluate the problem.

The Nile Delta is facing great challenges, with extremely dense housing, high pressure on the cultivated land, high water requests, and rapid population growth. Groundwater is an essential source of drinking water in the countryside regions in Egypt. Each needs about 2 l of clean drinking water every day. Hence, it is necessary to evaluate the quality of water proposed to be utilized for drinking. Indeed, the chemical pollutants seriously affect the human health. Natural processes and anthropogenic activities affect the water quality. The quality of water is portrayed by water parameters (physical, chemical, and microbiological), and human well-being is at hazard if values surpass satisfactory limits [8].

Water quality index (WQI) is the best strategy for measuring water quality. Various water quality parameters are incorporated into a mathematical equation to rate water quality, estimating the suitability of water for drinking [9]. WQI was initially created by [10] to gauge water quality by utilizing ten most routinely used water parameters. Subsequently, this method was modified by various specialists. The utilized water quality parameters fluctuate by number and types. The weights in every parameter depend on its separate guidelines, and the assigned weight shows the parameter's importance and effects on the index. WQI enables comparison between various samples. The index is simplifying a complex dataset into easily estimated, usable data and understandable even by lay people. Groundwater was sampled from 16 villages (Fig. 1a, b) in the central part of the Nile Delta to evaluate the effect of human activities in the some Nile Delta villages on groundwater hydrogeochemistry and quality. Many researchers have investigated the water quality in the Nile Delta among them [3, 11–26].

2 Geology of the Study Area

The geology of the Nile Delta has been extensively discussed by some researchers over several decades [27–34]. Mit Ghamr Formation composed of sand and gravels with thin clay intercalations. It is assumed to be Pliocene to Quaternary age. The whole sequence of Mit Ghamr is capped by Bilqas formation of the Holocene age. The depositional environment of this formation is probably shallow marine to

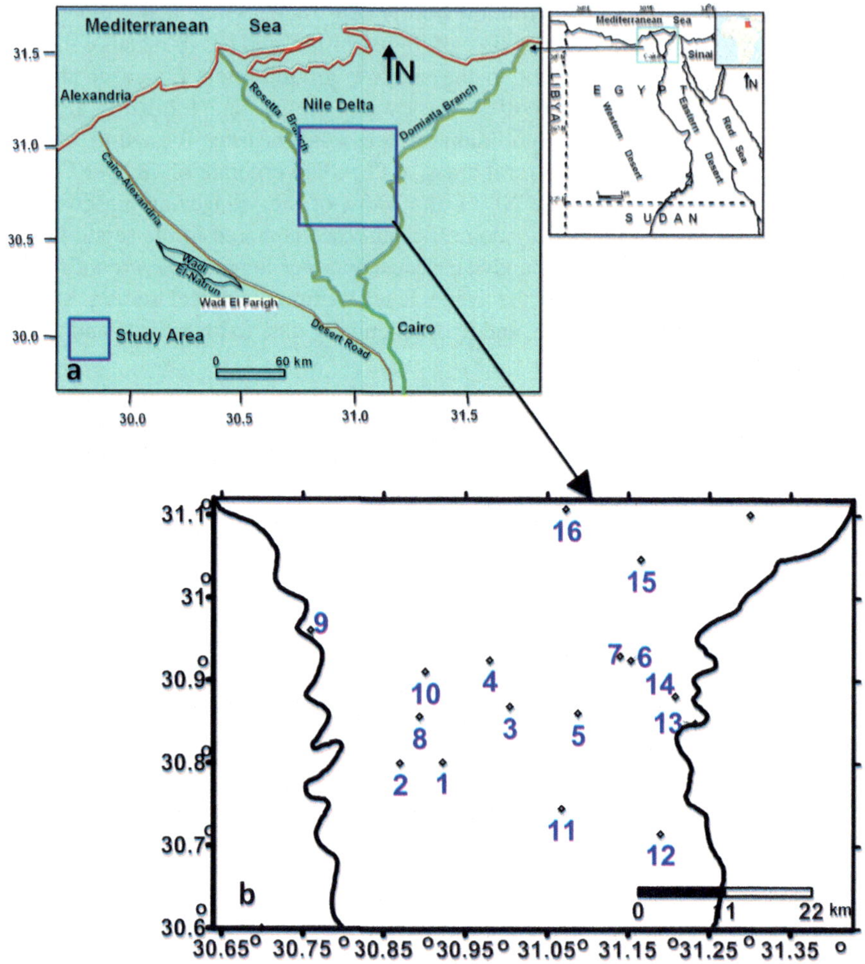

Fig. 1 Location map of the study area (**a**) and the locations of the collected groundwater samples (**b**)

fluvial. On both sides of the present delta, these deposits form a series of gravel terraces at various heights [35]. Bilqas formation is the top cover of the Delta area and consists mostly of clay. Fine-grained sands, silts, plant remains, and peat deposits are frequent. According to Salem [3] "These sediments appear to have been deposited under continental, lagoonal, fluvatile and beach environments. They represent the advent of a third Holocene sea transgression phase advancing almost from the N and NE direction."

In the study area, according to the shallow cross section given in Fig. 2, Bilqas formation consists mainly of fine detrital materials ranging from the two classical end members, clay, and silt including some sand tracks, particularly in the north central part. The clay thickness increases from south to north and from west to east.

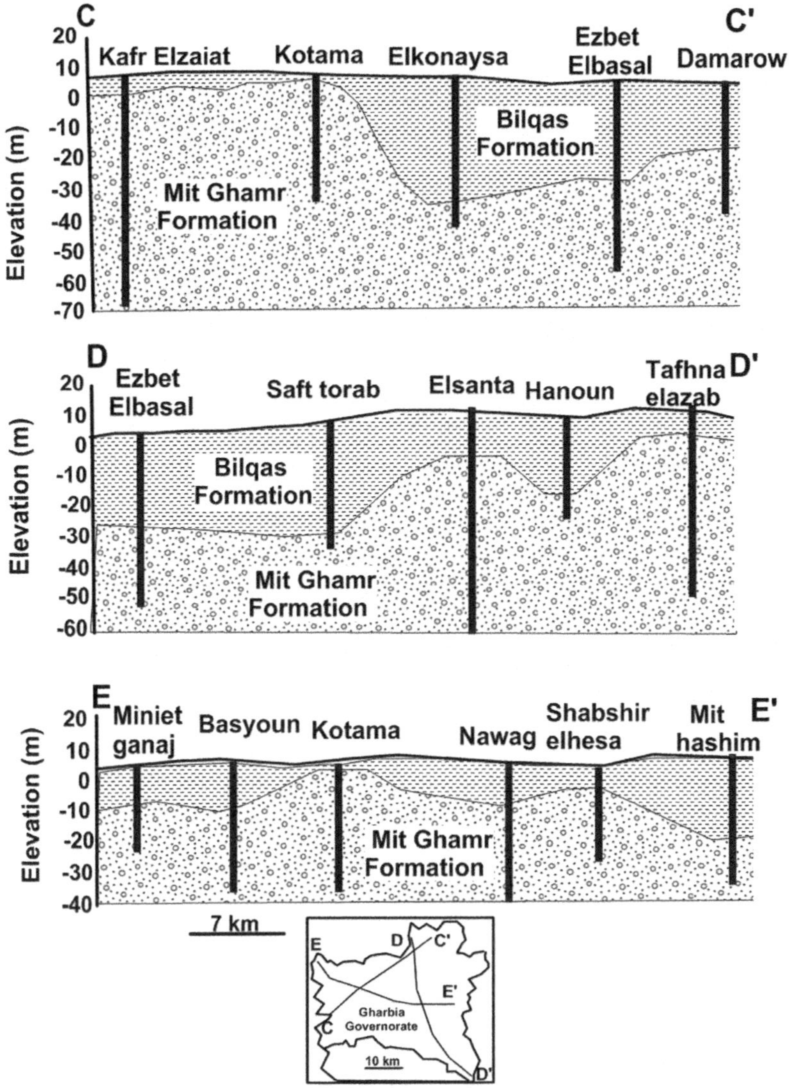

Fig. 2 Geohydrologic sections through the groundwater aquifer in the study area [36]

The lowest clay thickness appears at Kotamma location to the west of the study area.

3 Materials and Methods

Groundwater samples were collected from 16 villages in the study area during 2016 (Fig. 1b). Samples were obtained from wells range in depth from 9 to 46 m.

3.1 Hydrochemical Analysis

Samples were gathered in prewashed clean polyethylene bottles. Temperature, pH, DO (dissolved oxygen), ORP, TDS (total dissolved solids), and electrical conductivity (EC) of the water samples were measured in site. Samples were kept at 4 °C and chemically analyzed after a short time of sampling to reduce the physicochemical changes [37]. Alkalinity and chloride concentrations were measured by titration. Sulfate, nitrate, and ammonia were measured by a spectrophotometer. Concentrations of cations were measured utilizing the ICP-Perkin Elmer/Optima 7000DV (inductively coupled plasma) at the Central Lab of Tanta University, Egypt. Complete hydrochemical analysis of the collected water samples is listed in Table 1.

Values of desirable and maximum allowable limits for drinking of the measured parameters are listed in Table 2 according to WHO [8]. Each parameter is assigned a weight according to its relative importance for the quality of water for drinking purposes. Maximum weight of 5 is assigned to total dissolved solids (TDS), EC, (NO_3), (Pb), and As; weight of 4 is assigned to (SO_4), TH, (Mn), and (Cr); weight of 3 is assigned to pH, (Cl), (Na), and (Cd); and weight of 2 is assigned to (K), (Mg), (Ca), (CO_3), (HCO_3), (Fe), (Cu), and (Zn). The water quality parameters are selected based on the guidelines as shown in Table 2 [38]. WQI is calculated as follows:

$$\mathrm{WQI} = \Sigma\, Qi \times Wi \tag{1}$$

Qi is the ith quality rating and is given by Eq. (2), and Wi is the ith relative weight of the parameter I and is given by Eq. (3).

$$Qi = (Ci/Si) \times 100 \tag{2}$$

where Ci is the ith concentration of water quality parameter and Si is the ith drinking water quality standard according to the guidelines of WHO [8] in mg/l.

$$Wi = wi/wini = 0 \tag{3}$$

where wi is the weight of ith parameter and n is the number of chemical parameters.

Table 1 Hydrochemical characteristics of the analyzed water samples

S. No		1	2	3	4	5	6	7	8	9	10	11	12	13	14	15	16
Village name		Shubra El-Namla	Abiar	Nawag	Damatt	Shabsheer Elhessa	Shenta Ayaash	Bolqina	Pirma	Saa Elhagr	Kotama	Shubra Qaas	Nahttai	Mit Badr Halawa	Mit Habib	Damro	Dokhmees
Param.	Units																
Depth	m	35	27	44	15	46	45	38	28	28	30	9	25	33	32	21	16
Temp.	°C	25.3	25.4	25.3	25.1	25	25.5	24	23.7	23.7	24.1	20.2	un	un	un	21.6	22.4
TDS	mg/l	430	440	690	1,550	520	770	330	2,580	680	350	1,500	350	330	500	2,500	7,420
EC	us/cm	880	880	1,400	3,320	1,060	1,055	580	5,180	1,370	740	3,075	717.5	676.5	1,025	5,020	14,870
ORP	mv	−69	−69	150	−0.6	135	28	57	−9.6	−25.5	−50.1	−29	−26.5	−75	−75	−48.9	−17.3
PH		7.64	7.42	7.77	7.29	7.64	7.74	8.45	7.01	7.3	7.72	7.34	7.3	8.1	7.96	7.83	7.26
DO	mg/l	0.5	0.5	0.7	0.7	0.7	0.6	0.7	un	un	un	un	un	un	un	2.3	un
TH	mg/l	241.6	279.6	350.8	724.18	311.2	425.8	168.4	1,376	405.9	161.28	328.9	222	178.72	259.2	557.84	3898.6
K	mg/l	4.725	3.81	5.71	20.037	12.57	3.126	5.96	95.18	3.52	2.696	1.79	2.288	2.6	3.13	33.34	118.5
Na	mg/l	36	42	56	200.79	41	71	34	216.66	64.27	39.8	435.5	45.2	33.9	74.2	698	1,087
Mg	mg/l	26	31	38	49.8	32	38	24	160	49	15.8	29	20	14.2	12	32.4	446
Ca	mg/l	54	61	78	208	72	108	28	288	82	38.6	84	56	48.2	84	170	828
Cl	mg/l	105.8	98	157.6	516	82	113.4	82.5	697.1	163	58.8	613.5	57	49.9	166.7	486	2,566
NO_3	mg/l	1.93	1.69	0.6	1.87	2.23	2.14	2.01	6.63	4.83	1.97	1.7	1.83	1.68	4.64	28	3.8
SO_4	mg/l	23.24	56.66	72.47	72.5	52.27	63.67	25.29	188.2	39	15.6	72.3	65.2	0.3	4	244.53	220.68
HCO_3	mg/l	205	215	236	81.646	266	386	146	817.8	321.4	206	386	201	242.5	210	860	2,563
CO_3	mg/l	0	0	12	13	12	12	11	17.2	3.6	8	10	14	12	2.1	62	40
Fe	mg/l	0.286	0.025	0	0.43	0	0	0	1.144	1.824	0.0056	0	0	1.231	0.527	0	0
Mn	mg/l	1.087	0.722	0.645	0.774	0.51	0.324	0.2	0.863	1.616	0.644	1.85	0.387	0.25	0.217	0.299	1.496
Zn	mg/l	0.007	0.088	0.0072	0.007	0.007	0.0071	0	0.008	0.067	0.005	0.01	1.025	1.45	0.174	0.048	2.9
N	mg/l	1.93	1.69	2.07	1.87	2.23	2.14	1.52	6.63	4.83	0.122	5.155	4.09	4.14	3.46	4.53	3.8
P	mg/l	0.598	0.401	0.398	0.38	0.373	0.376	0.365	0.13	0.64	0.11	0.48	0.44	0.53	0.64	0.53	0.57
AS	mg/l	0.15	0.184	0.182	0.182	0.177	0.227	0.01	0.24	0.008	0.186	0.01	0.008	0.006	0.012	0.011	0.11
NH_4	mg/l	0.57	0.42	0.6	0.47	0.61	0.52	0.6	12.45	1.49	1.02	2.82	0.79	2.03	5.83	6.97	9.78
Ba	mg/l	0.177	0.145	0.027	0.025	0.02	0.02	0.01	0.024	0.025	0.067	0.074	0.068	0.066	0.091	0.314	0.114
Sr	mg/l	0.748	0.579	1.187	1.34	0.71	0.631	0.42	0.794	0.919	0.569	1.246	0.505	0.452	0.36	1.039	0.76
Si	mg/l	42.55	28.22	31.56	35.75	35.59	29.96	14	50.66	26.58	42.16	14.38	12.69	10.99	14.04	17.35	13.31

Note: un means unmeasured

Table 2 Desirable limits of parameters and its assigned relative weight

Parameters	WHO desirable limits	Weight (*wi*)	Relative weight
TDS	500 mg/l	5	0.08621
PH	6.5–8.5	3	0.05172
EC	1,500 us/cm	5	0.08621
TH	300 mg/l	4	0.06897
Ca	200 mg/l	2	0.03448
Na	200 mg/l	3	0.05172
Mg	125 mg/l	2	0.03448
K	12 mg/l	2	0.03448
Cl	250 mg/l	3	0.05172
CO_3	350 mg/l	2	0.03448
SO_4	250 mg/l	2	0.03448
NH_4	0.5 mg/l	4	0.06897
HCO_3	350 mg/l	2	0.03448
NO_3	50 mg/l	5	0.08621
Fe	0.3 mg/l	3	0.05172
Mn	0.4 mg/l	4	0.06897
Zn	5 mg/l	2	0.03448
As	0.01 mg/l	5	0.08621
		$\sum W = 58$	$\sum W = 1$

3.2 Statistical Analysis

Multivariate statistical techniques, such as CA and PCA, are often used as "unbiased methods" to summarize associations between samples and variables. Such associations based on similar magnitudes and variations in the chemical and physical composition may reveal the effects of climate or human activity on water quality. Agglomerative hierarchical clustering approach options include single linkage, complete linkage, weighted and unweighted pair group average linkage, and centroid linkage. Ward's method was used for this analysis because it is commonly applied in the literature germane to this study [39, 40]. One of the reasons it is widely applied is that it tends to yield spherical clusters of the same size (i.e., classification variables over all classes essentially have the same variance).

PCA is an effective method for pattern estimation that used to clarify the difference of an extensive arrangement of intercorrelated variables. It demonstrates the relationship between variables by decreasing the dimensionality of the dataset [41]. The covariance matrix of the original variables is used to extract the eigenvalues and eigenvectors. PCA changes informational collection into an uncorrelated one [42]. The first PC is the projections of the given points onto the extreme fluctuations, the second PC has the change subject to being orthogonal to PC1; the third PC has the greatest difference subject to being orthogonal to the first and second PCs [43, 44]. The multivariate statistical analysis relates variables into PCs depending on their common relationship coefficients and these affiliations might be used to estimate the mineralization, lithology and ecological procedures [45].

4 Results

The physical and chemical properties of the groundwater samples from the study area were given in Table 1.

4.1 Description and Spatial Distribution

The pH values were acceptable in the range of 7.01–8.45. Electrical conductivity reached its minimum value (580 μs/cm) in sample 7 and its maximum (14,870 μs/cm) in sample 16. According to WHO [8], groundwater samples 1, 2, 7, 10, 12, and 13 are under the desirable limit; samples 3, 5, 6, 9, and 14 are of permissible values; and samples 4, 8, 11, 15, and 16 are unfit for drinking. As shown in Table 1 and Fig. 3a, TDS content is variable where samples 2, 7, 10, 12, and 13 have TDS $<$ 500 mg/l and fall within the desirable range for drinking category; samples 3, 5, 6, 9, and 14 have TDS range from 500 to 1,000 mg/l exceeding the desirable limit and permissible for drinking; and samples 4, 8, 11, 15, and 16 have TDS $>$ 1,000 mg/l and fall within the unfit category for drinking. The water salinity reaches its maximum value (TDS equal 7,420 mg/l) at Dokhmees (sample 16) in the north-eastern part of the area.

The values of alkalinity in all samples ranged from 157 mg/l (sample 7) to 2,603 mg/l (sample 16). Except samples 6, 8, 11, 15, and 16, other samples are of the desirable limit of HCO_3 (Fig. 3b). Total hardness concentrations range from 161.28 mg/l (sample 10) to 3898.6 mg/l (sample 16). According to Sawyer and McCarty [46] common hardness scale, the groundwater samples 1, 2, 7, 10, 12, 13, and 14 are under the hard category, but all other samples are very hard water (Fig. 3c). The lowest concentration of the chloride ion is 49.9 mg/l (sample 13), and the highest level is 2,566 mg/l was detected in sample 16. Chloride levels in most samples are below WHO permissible limits (Fig. 3d) except that of samples 4, 8, 11, 15, and 16. Nitrate concentrations were found to be in a range of 0.6 mg/l (samples 3) to 6.63 mg/l (sample 8). Therefore, all samples are under the desirable limits (Fig. 3e). Concentrations of sulfate are below the WHO permissible limits (Table 1 and Fig. 3f) and range from 0.3 mg/l (samples 13) to 244.53 mg/l at sample 16.

Concentrations of calcium and magnesium in the collected water samples, respectively, range from 28 and 12 mg/l (samples 7 and 14, respectively) to 828 and 446 mg/l (samples 16). Calcium concentrations (Fig. 4a) are compatible with the desirable range in most samples except samples 4, 8, and 16 which are of unfit values. Magnesium levels are of desirable range in most samples and unfit in

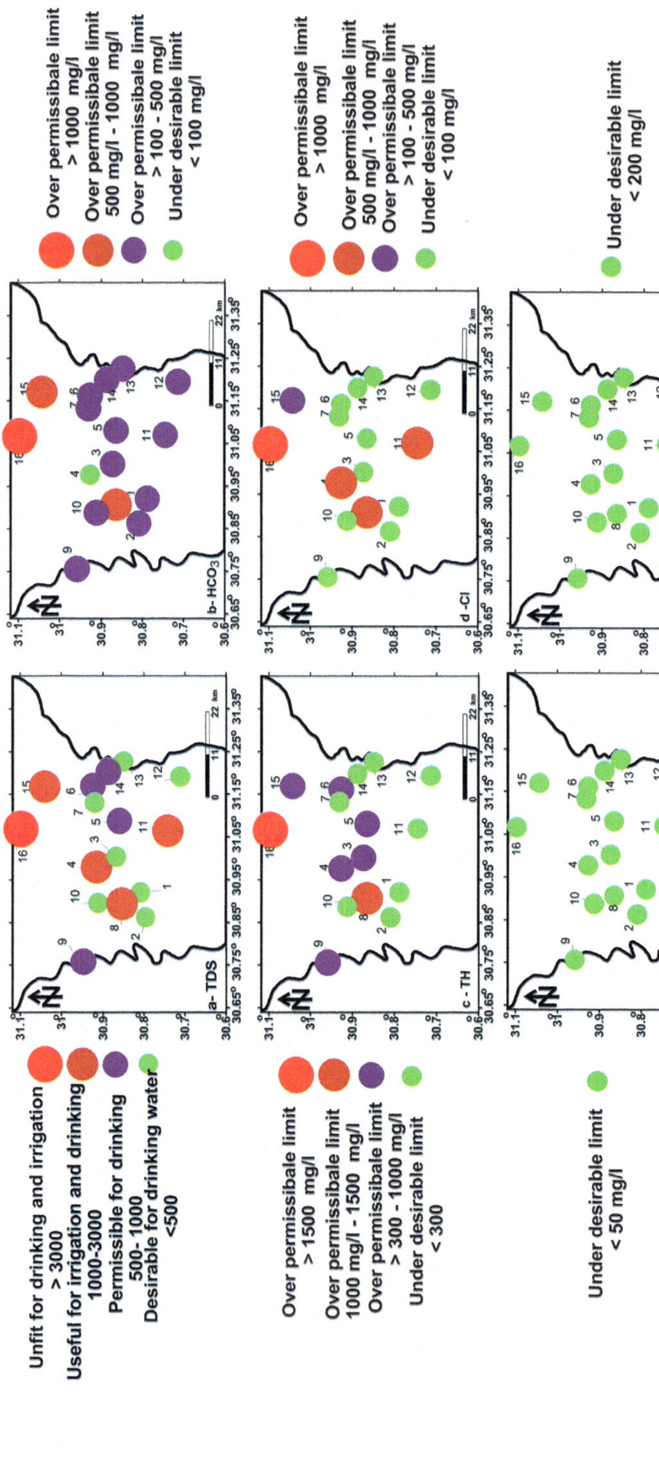

Fig. 3 Spatial distribution maps of TDS (**a**), HCO_3 (**b**), TH (**c**), Cl (**d**), NO_3 (**e**), and SO_4 (**f**) concentrations according to WHO [8] drinking water guidelines

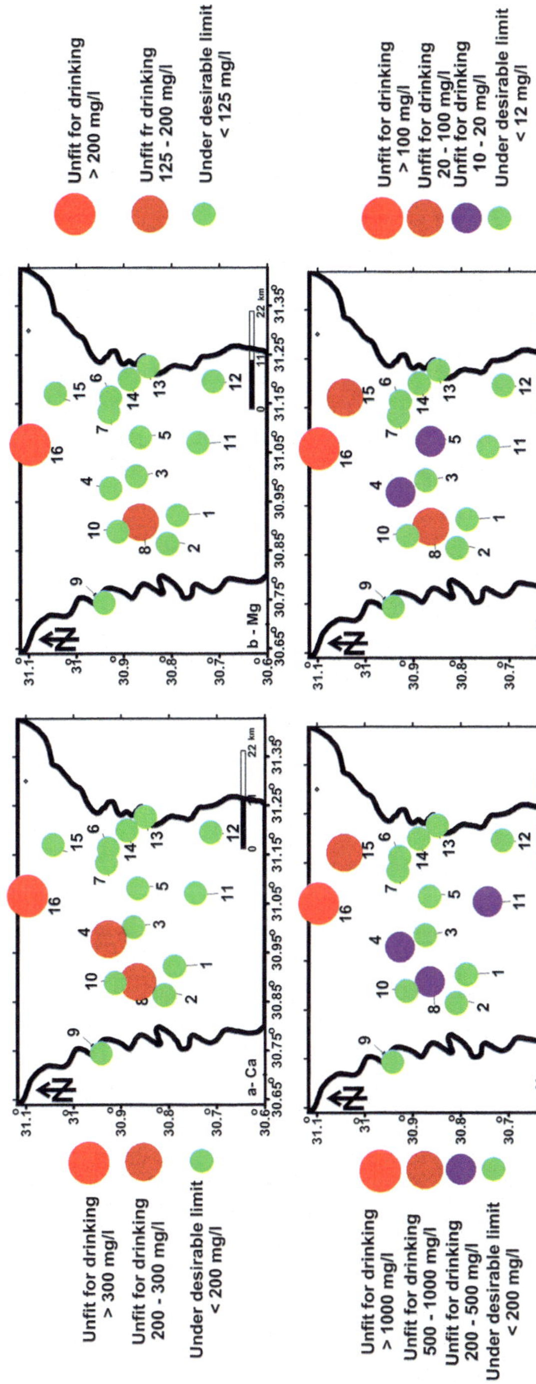

Fig. 4 Concentrations spatial distribution maps of Ca (**a**), Mg (**b**), Na (**c**), and K (**d**) according to WHO [8] drinking water guidelines

samples 8 and 16 (Fig. 4b). Water with unfit sodium levels has a bad taste. Ranges of sodium and potassium levels in the collected samples are, respectively, 33.9 mg/l (samples 13) to 1,087 mg/l (samples 16) and from 1.79 mg/l (samples 11) to 118.5 mg/l (samples 11). Sodium concentrations are compatible with the desirable range in samples 1, 2, 3, 5, 6, 7, 9, 10, 12, and 13 and of unfit values in samples 4, 8, 11, 15, and 16 (Fig. 4c). Potassium concentrations are of desirable range in samples 1, 2, 3, 6, 7, 9, 10, 11, 12, 13, and 14 and of unfit value in samples 4, 5, 8, 15, and 16 (Fig. 4d).

Trace element concentrations are listed in Table 1. Iron concentrations range from 0 to 1.824 mg/l. Samples 4, 8, 9, 13, and14 are above the acceptance WHO [8] limit which is 0.3 mg/l (Fig. 5a). Concentrations of zinc in all the samples are under WHO desirable limit (5 mg/l) and range from 0.005 to 1.45 mg/l (Fig. 5b). The content of manganese is lower than the desirable limit which is 0.4 mg/l in samples 6, 7, 12, 13, 14, and 15 (Fig. 5c). The levels of silicon in the samples were in the range of 10.99 mg/l in sample 13 to 50.66 mg/l in sample 8. Strontium highest concentration (1.34 mg/l) was recorded in sample 4, and sample 14 has the lowest

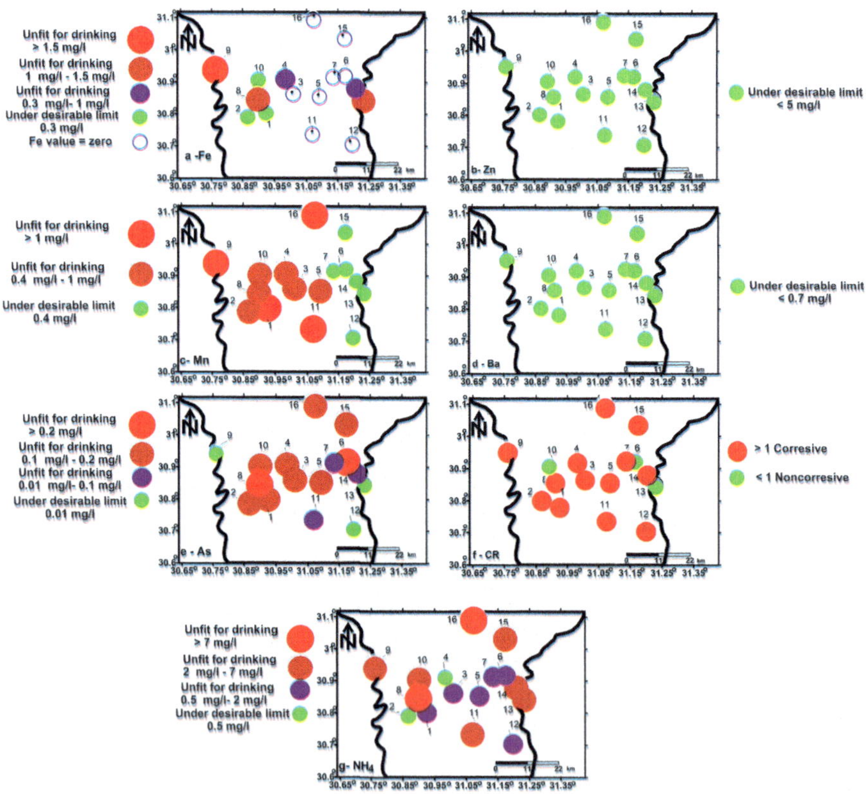

Fig. 5 Spatial distribution maps of Fe (**a**), Zn (**b**), Mn (**c**), Ba (**d**), As (**e**), CR (**f**), and NH_4 (**g**) concentrations according to WHO [8] drinking water guidelines

concentration (0.36 mg/l). Barium concentrations were ranging from 0.01 mg/l in sample 7 to 0.314 mg/l in sample 15 (Fig. 5d). Concentrations of barium in all the samples are under WHO desirable limit (0.7 mg/l). The levels of arsenic in the samples were in the range of 0.006 in sample 13 to 0.24 mg/l in sample 8 (Fig. 5e) and exceeded the WHO guideline (0.01 mg/l) in all samples except samples 9, 12, and 13. Phosphorus concentrations were ranging from 0.11 mg/l in sample 10 to 0.64 mg/l in sample 14. Calculated corrosive ratio (CR) values reveal that most of the samples fall within corrosive category (13 wells), and just 3 wells are of noncorrosive properties (Fig. 5f). Concentrations of ammonia in all samples are ranging from 0.42 mg/l in sample 2 to 12.45 mg/l in sample 8. Samples 2 and 4 are under the acceptance WHO limit which is 0.5 mg/l, and all other are above the WHO limit (Fig. 5g).

4.2 Statistical Analysis

Analysis of correlation between the measured physicochemical parameters (Table 3) reveals that strong relationships were recognized between TDS-K, TDS-Na, TDS-Mg, TDS-Ca, TDS-Cl, TDS-SO_4, TDS-HCO_3, K-Na, K-Mg, K-Ca, K-Cl, K-SO_4, K-HCO_3, K-NH_4, Na-Mg, Na-Ca, Na-Cl, Na-SO_4, Na-HCO_3, Mg-Ca, Mg-Cl, Mg-SO_4, Mg-HCO_3, Mg-Zn, Ca-Cl, Ca-HCO_3, Ca-Zn, Ca-SO_4, Cl-SO_4, Cl-HCO_3, NO_3-CO_3, NO_3-Ba, SO_4-HCO_3, SO_4-NH_4, HCO_3-Zn, and As-Si. The intermediate relationship is shown between TDS-Zn, TDS-NH_4, K-Zn, Na-Zn, Na-NH_4, Mg-NH_4, Ca-NH_4, Cl-Zn, Cl-NH_4, NO_3-SO_4, SO_4-CO_3, HCO_3-NH_4, CO_3-Ba, Fe-N, and Mn-Sr. These very strong to intermediate relationships suggest that the increase in TDS was accompanied by an increase in the concentrations of the all major ions as well as the increase in NH_4, NO_3, Zn, and As. This indicates that water under the villages receives water from the domestic wastes due to bad or absence of sewage networks.

Using HCA dendrogram, collected samples were classified according to the abundance of the elements into three groups A, B, and C (Fig. 6). The elements' average concentrations are shown in Table 4. Cluster A includes sample 16 which is dominated by higher concentrations of TDS, EC, potassium, magnesium, sodium, calcium, sulfate, chloride, bicarbonate, Mn, Zn, P, NH_4, and Ba compared to the other two clusters. However, its pH values (7.26) were the lowest compared to cluster B (7.40) and cluster C (7.73). This sample was obtained from Dokhmees village which is located in the northeastern part of the study area and near to El-Gharbia main drain. Therefore, such concentrations could be related to the effect of the drain water and/or seawater intrusion.

Cluster B includes samples 4, 8, 11, and 15. As for chemical elements, this water type had the moderately mean concentrations among the three clusters (Table 4), and it had higher ion concentrations of Fe, Sr, and Si than the other two clusters and dominated major ion concentrations arranged as follows: sodium > calcium and chloride > bicarbonate. They also had relatively high concentrations of potassium,

Table 3 Correlation matrix for the chemical constitutions present in the collected groundwater samples

	TDS	K	Na	Mg	Ca	Cl	NO_3	SO_4	HCO_3	CO_3	Fe	Mn	Zn	N	P	AS	NH_4	Ba	Sr	Si
TDS	1																			
K	**0.91**	1.00																		
Na	**0.93**	**0.77**	1.00																	
Mg	**0.95**	**0.90**	**0.78**	1.00																
Ca	**0.98**	**0.91**	**0.85**	**0.98**	1.00															
Cl	**0.99**	**0.88**	**0.90**	**0.96**	**0.98**	1.00														
NO_3	0.26	0.23	0.48	0.02	0.13	0.12	1.00													
SO_4	**0.93**	**0.88**	**0.91**	**0.83**	**0.87**	**0.86**	*0.52*	1.00												
HCO_3	**0.99**	**0.89**	**0.91**	**0.96**	**0.98**	**0.98**	0.25	**0.91**	1.00											
CO_3	0.14	0.13	0.38	−0.11	0.01	0.01	**0.90**	0.44	0.12	1.00										
Fe	−0.10	0.05	−0.22	−0.03	−0.06	−0.11	−0.02	−0.17	−0.09	−0.15	1.00									
Mn	0.42	0.36	0.40	0.43	0.39	0.49	−0.18	0.28	0.38	−0.32	0.20	1.00								
Zn	*0.71*	*0.56*	*0.60*	**0.76**	**0.75**	*0.74*	−0.08	*0.54*	**0.75**	−0.13	0.00	0.17	1.00							
N	0.34	0.48	0.37	0.27	0.27	0.30	0.36	0.41	0.30	0.30	*0.52*	0.32	0.21	1.00						
P	0.12	−0.13	0.24	0.07	0.11	0.15	0.16	0.03	0.15	−0.06	0.20	0.21	0.31	0.23	1.00					
AS	−0.01	0.17	−0.22	0.12	0.07	−0.02	−0.32	0.01	−0.01	−0.21	−0.27	−0.14	−0.29	−0.49	*−0.70*	1.00				
NH_4	*0.72*	**0.86**	*0.65*	*0.66*	*0.68*	*0.66*	0.46	**0.75**	*0.69*	0.30	0.21	0.17	0.37	*0.67*	−0.05	−0.07	1.00			
Ba	0.22	0.09	0.46	0.03	0.10	0.14	**0.76**	0.39	0.23	*0.57*	−0.28	−0.08	0.07	0.06	0.36	−0.34	0.24	1.00		
Sr	0.23	0.16	0.33	0.07	0.15	0.22	0.20	0.29	0.15	0.30	−0.04	*0.51*	−0.24	0.16	0.00	0.04	0.01	0.02	1.00	
Si	−0.20	0.05	−0.38	−0.11	−0.16	−0.23	−0.17	−0.14	−0.22	−0.14	0.07	0.02	−0.55	−0.31	−0.65	**0.81**	−0.04	−0.26	0.15	1.00

Values listed in bold font are strong positive relationship. Values listed in italic font are intermediate positive relationship

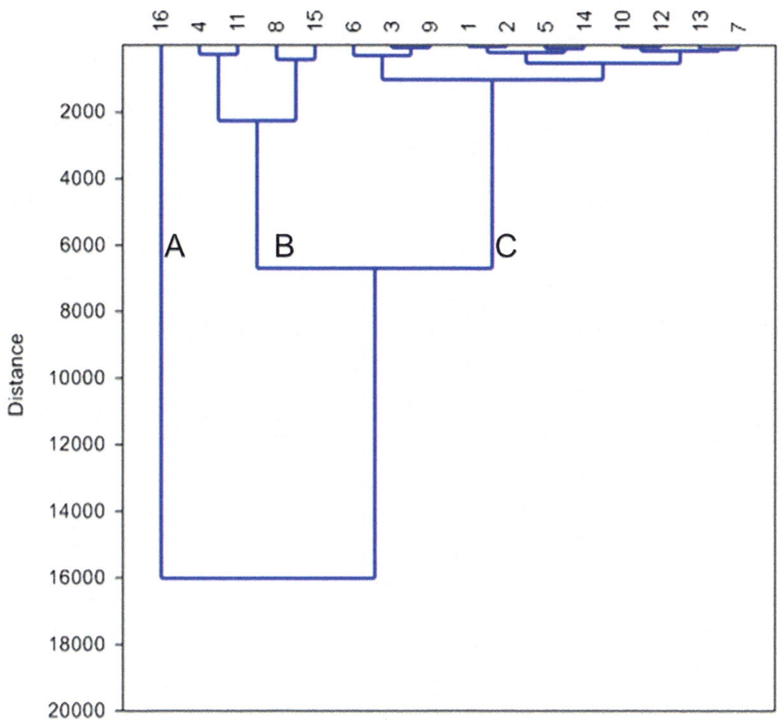

Fig. 6 Dendrogram based on Ward's method shows three groups of the collected water samples according to the abundance of the hydrochemical composition

magnesium, and sulfate. Finally, the water samples of cluster C (samples 1, 2, 3, 5, 6, 7, 9, 10, 12, 13, and 14) were dominated by Ca > Na > Mg and HCO_3 > Cl ion concentration arrangement (Table 4). The average concentrations of the physico-chemical parameters of this cluster are the lowest among the three groups.

PCA scatter diagram (Fig. 7) showed that K, Mg, Ca, Cl, HCO_3, SO_4, Mn, Sr, EC, and TDS were the most effective in samples 8 and 16. Na, NH_4, Zn, Fe, NO_3, CO_3, Ba, N, and P were the most effective in samples 11 and 15. pH was effective on samples 9, 12, 13, and 14. Si and As were the most effective elements in samples 1, 2, 3, 4, 5, 6, 7, and 10.

4.3 Drinking Water Quality

Drinking water quality index (WQI) of the collected samples (Fig. 8) shows that good water is represented in some localities in the eastern side of the area. Those samples include 5, 6, 7, 10, 12, 13, and 14 (villages Shabsheer Elhessa, Shenta Ayaash, Bolqina, Kotama, Nahttai, Mit Badr Halawa, and Mit Habib, respectively). On the other hand, samples 1, 2, 3, and 4 (villages Shubra El-Namla, Abiar, Nawag,

Table 4 Hydrochemical compositions of clusters A, B, and C

	Unit	Cluster A (sample 16)	Cluster B (samples 4,11, 8, and 15)			Cluster C (samples 1, 2, 3, 5, 6, 7, 9, 10, 12, 13, and 14)		
			Min	Max	Average	Min	Max	Average
TDS	mg/l	7,420	1,500	2,580	2032.5	330	770	490
EC	μS/cm	14,870	3,075	5,180	4148.75	580	1,400	944
PH		7.26	7.01	7.83	7.3675	7.300	8.450	7.731
K	meq/l	3.03	0.48	2.44	1.07	0.059	0.322	0.117
Na	meq/l	47.27	8.73	30.36	16.86	1.474	3.227	2.085
Mg	meq/l	36.71	2.39	13.17	5.58	0.988	4.033	2.245
Ca	meq/l	41.40	4.20	14.40	9.38	1.400	5.400	3.226
Cl	meq/l	72.41	14.55	32.44	20.99	1.407	4.701	2.909
NO_3	mg/l	3.80	1.70	28.00	9.55	0.600	4.830	2.323
SO_4	meq/l	7.29	0.58	3.91	1.88	0.006	1.507	0.790
HCO_3	meq/l	42.03	6.33	13.41	8.77	2.394	6.330	3.928
CO_3	meq/l	0.66	0.16	1.02	0.42	0.000	0.230	0.129
Fe	mg/l	0.01	0.01	1.14	0.40	0.002	1.824	0.365
Mn	mg/l	1.496	0.30	1.85	0.95	0.217	1.616	0.611
Zn	mg/l	2.9	0.01	0.05	0.02	0.005	1.450	0.259
N	mg/l	3.8	1.87	6.63	4.55	0.122	4.830	2.610
P	mg/l	0.57	0.13	0.53	0.38	0.110	0.640	0.444
AS	mg/l	0.11	0.01	0.24	0.11	0.006	0.227	0.120
NH_4	mg/l	9.78	0.47	12.45	5.68	0.420	5.830	1.316
Ba	mg/l	0.114	0.02	0.31	0.11	0.020	0.177	0.069
Sr	mg/l	0.76	0.79	1.34	1.10	0.360	1.187	0.653
Si	mg/l	13.31	14.38	50.66	29.54	10.990	42.550	27.545

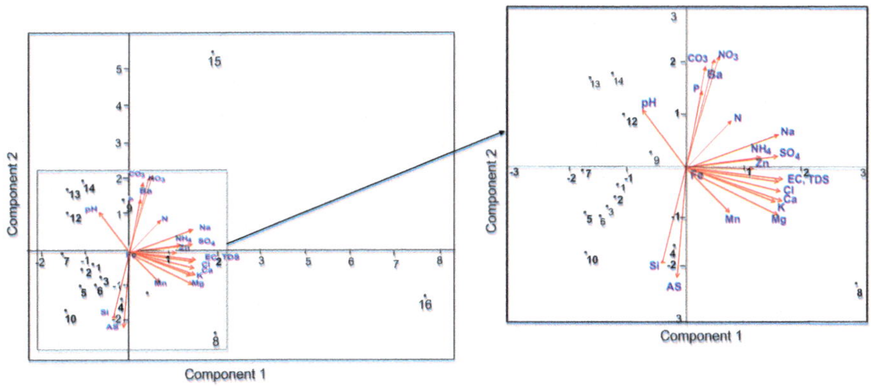

Fig. 7 PCA scatter diagram showing the relationships between samples and the water physicochemical characteristics

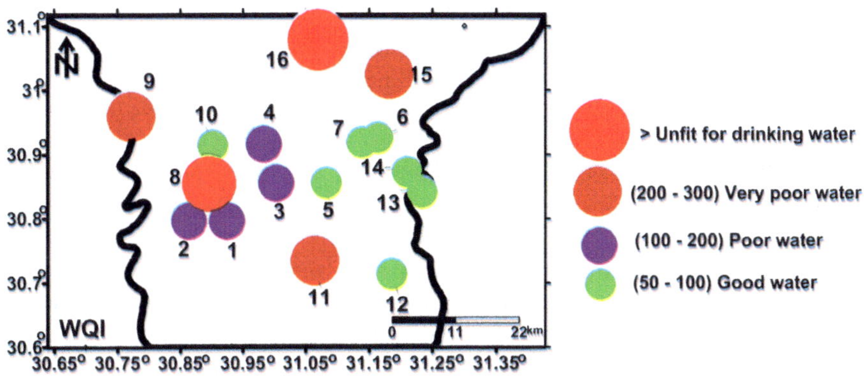

Fig. 8 Spatial distribution map of drinking water quality index (WQI)

and Damatt, respectively) are of poor groundwater quality and located in the central part of the study area. Water quality decreases toward northern villages where very poor water quality occurs in Saa Elhagr (sample 9), Damro (sample 15), as well as Shobra Qaas village (sample 11) in the southwest. In villages of Pirma and Dokhmees, groundwater samples (8 and 16, respectively) are of unfit character for drinking.

5 Discussion

Egyptians living in the villages directly use groundwater for various purposes. Nile Delta villages have no or badly constructed sewage networks. Therefore, groundwater under these villages might receive pollutants. Physicochemical parameters of

16 groundwater samples (one sample for one village) were measured to evaluate this problem. The quality of the sampled groundwater varied greatly from one village to another. Using cluster analysis, the collected samples were classified into three groups. Cluster A includes sample 16 which is dominated by higher concentrations of TDS, EC, potassium, magnesium, sodium, calcium, sulfate, chloride, bicarbonate, Mn, Zn, P, NH_4, Ba, and unfit WQI values. Bad water quality of this group might be related to the effect of El-Gharbia main drain and seawater intrusion.

Cluster B includes samples 4, 8, 11, and 15. This group had the moderately mean ion concentrations among the three clusters, and it had higher ion concentrations of Fe, Sr, and Si than the other two clusters and dominated major ion concentrations arranged as follows: sodium > calcium and chloride > bicarbonate. They also had relatively high concentrations of potassium, magnesium, and sulfate. WQI of this group varies from poor characters in Damatt (4), very poor in Shobra Qaas village (11) and Damro (15), to unfit in Pirma (8). Groundwater of this group might receive a seepage from the domestic wastes.

Samples of cluster C (1–3, 5–7, 9, 10, and 12–14) were dominated by Ca > Na > Mg and HCO_3 > Cl ion concentration arrangement (Table 4). The average concentrations of the physicochemical parameters of this cluster are the lowest among the three groups. WQI of these samples varies from good as in samples 5, 6, 7, 10, 12, 13, and 14, poor (1–3, and) to very poor (sample 9). A wide range of WQI in this group is assumed to be related to the slight seepage from domestic wastes which increase the values of minor elements like NH_4 but did not affect the major ion concentrations.

The presence of undesirable concentrations of arsenic and ammonia in most of the samples is related to the direct seepage from the domestic wastes such that of the drains and septic tanks. Higher concentrations of arsenic and ammonia sometimes accompanied with higher TDS and major element concentrations due to the higher leakage from septic tanks under the village like that of Pirma (sample 8) and both the effect of seepage from the main drains and seawater intrusion as in sample 16 (village of Dokhmees).

6 Conclusions

Groundwater was sampled from 16 villages during 2016. Spatial distribution of the elements' concentrations according to WHO guidelines, WQI (drinking water quality index), and statistical analysis was used to evaluate the groundwater pollution under Central Nile Delta villages. The studied groundwater was affected by the seepage from the septic tanks and/or drains in most of the villages. The groundwater affected by drains showed high TDS, EC, potassium, magnesium, sodium, calcium, sulfate, chloride, bicarbonate, Mn, Zn, P, NH_4, Ba, and unfit WQI values. Samples directly affected by seepage from the septic tanks had moderately mean ion concentrations and higher concentrations of Fe, Sr, and Si. It also had WQI values range from poor, very poor to unfit. Other samples were affected by neither drains nor septic tanks and had the lowest TDS concentrations and good WQI. Arsenic and ammonia undesirable concentrations in most of the collected groundwater samples are indications of direct infiltration from septic tanks and/or seepage from the drains. These two chemicals are of severe hazardous effect on the human health.

7 Recommendations

Sewage networks should be constructed in the Egyptian villages to prevent the groundwater contamination due to the use of septic tanks. The state and civil society organizations should inform the residents of the villages of the need to analyze the groundwater of their wells and consult the concerned authorities to determine the validity of this water for different uses. Also, they should be aware of the health hazards resulting from the use of contaminated water. The state should also extend the villages with clean water networks so that the need for underground water is reduced. Detailed hydrochemical and the microbial survey should be done to evaluate the groundwater contamination under the Egyptian village.

References

1. Bishnoi M, Arora S (2007) Potable groundwater quality in some villages of Haryana, India: focus on fluoride. J Environ Biol 28(2):291–294
2. Reddy DV, Nagabhushanam P, Peters E (2011) Village environs as source of nitrate contamination in groundwater: a case study in basaltic geo-environment in central India. Environ Monit Assess 174:481–492. https://doi.org/10.1007/s10661-010-1472-x
3. Salem ZE (2009) Natural and human impacts on the groundwater under an Egyptian village, central Nile Delta – a case study of Mehallet Menouf. Thirteenth International Water Technology Conference (IWTC 13), 12–15 Mar 2009. Hurghada, Egypt, 3:1397–1414
4. Gao Y, Yu G, Luo C, Zhou P (2012) Groundwater nitrogen pollution and assessment of its health risks: a case study of a typical village in rural-urban continuum, China. PLoS One 7(4): e33982. https://doi.org/10.1371/journal.pone.0033982
5. Sajidu SMI, Masamba WRL, Thole B, Mwatseteza JF (2008) Groundwater fluoride levels in villages of Southern Malawi and removal studies using bauxite. Int J Phys Sci 3(1): 001–0112008
6. Geriesh MH, EL-Rayes AE (2001) Municipal contamination of shallow groundwater beneath south Ismailia villages. 5th International conference on geochemistry. Alex. Univ., Egypt, pp 241–253. Accessed 12–13 Sept
7. Emara MM, El Sabagh I, Kotb A, Turkey AS, Husseen D (2007) Evaluation of drinking groundwater for the rural areas adjacent to the nearby desert of Giza Governorate of greater Cairo, Egypt. In: Linkov I, Kiker GA, Wenning RJ (eds) Environmental security in harbors and coastal areas. NATO security through science series (series C: environmental security). Springer, Dordrecht
8. WHO (2011) Guideline for drinking water quality, 4th edn. World Health Organization, Geneva

9. Ochuko U, Thaddeus O, Oghenero OA, John EE (2014) A comparative assessment of water quality index (WQI) and suitability of river Ase for domestic water supply in urban and rural communities in Southern Nigeria. Int J Human Soc Sci 4(1):234–245
10. Horton RK (1965) An index number system for rating water quality. J Water Pollut Control Fed 37:300–305
11. Elewa HH (2010) Potentialities of water resources pollution of the Nile River Delta, Egypt. Open Hydrol J 4:1–13
12. Gemail K, El-Shishtawy AM, El-Alfy M, Ghoneim MF, El-Bary MHA (2011) Assessment of aquifer vulnerability to industrial waste water using resistivity measurements. A case study, along El-Gharbyia main drain, Nile Delta, Egypt. J Appl Geophys 75:140–150
13. Morsy W, El-Fakharany Z (2012) Predicting the impact of surface wastewater on groundwater quality in Quesna Industrial Area. J Am Sci 8:772–781
14. Salema MG, El-Awady MH, Amine E (2012) Enhanced removal of dissolved iron and manganese from nonconventional water resources in Delta District, Egypt. Energy Procedia 18:983–993
15. Ghoraba SM, Zyedan BA, Rashwan IMH (2013) Solute transport modeling of the groundwater for quaternary aquifer quality management in Middle Delta, Egypt. Alex Eng J 52:197–207
16. Khalil MA, Salem ZE, Gheda SF, El-Sheekh MM (2013) Quality assessment of drinking water in Tanta City, Egypt. J Environ Sci Eng B 2:257–275
17. El Bedawy R (2014) Water resources management: alarming crisis for Egypt. J Manag Sustain 4:108–124
18. Bennett PC, El Shishtawy AM, Sharp JM, Atwia MG (2014) Source and migration of dissolved manganese in the Central Nile Delta aquifer, Egypt. J African Earth Sci 96:8–20
19. Elkafoury A, Dawoud W, Negm A, Bady M, Aly MH (2014) Integrated framework for evaluating the impact of urban transportation gaseous emissions on groundwater quality. Int Water Technol J 4:114–124
20. Fattah MK, Ragab EG (2014) Assessment of groundwater vulnerability to pollution in the southern part of Nile Delta, Egypt. Standard Sci Res Essays 2:725–738
21. El-Kowrany SI, El-Zamarany EA, El-Nouby KA, El-Mehy DA, Ali EA, Othman AA, Salah W, El-Ebiary AA (2016) Water pollution in the middle Nile Delta, Egypt: an environmental study. J Adv Res 7:781–794
22. Negm AM, Armanuos AM (2017) GIS-based spatial distribution of groundwater quality in the western Nile Delta, Egypt. In: Negm AM (ed) The Nile Delta, Hdb Env Chem, Springer International Publishing Switzerland. Doi: https://doi.org/10.1007/698_2016_66
23. Negm AM, Eltarabily MGA (2017) Modeling of fertilizer transport through soil, case study: Nile Delta. In: Negm AM (ed) The Nile Delta, Hdb Env Chem, Springer International Publishing Switzerland. Doi: https://doi.org/10.1007/698_2016_66
24. Salem ZE, Al Temamy AM, Salah MK, Kassab M (2016) Origin and characteristics of brackish groundwater in Abu Madi coastal area, Northern Nile Delta, Egypt. Estuar Coast Shelf Sci 178:21–35
25. Sharaky AM, El Hasanein AS, Atta SA, Khallaf KM (2017) Nile and groundwater interaction in the Western Nile Delta, Egypt. In: Negm AM (ed) The Nile Delta, Hdb Env Chem, Springer International Publishing Switzerland. Doi: https://doi.org/10.1007/698_2016_66
26. Salem ZE, Osman OM (2017) Use of major ions to evaluate the hydrogeochemistry of groundwater influenced by reclamation and seawater intrusion, West Nile Delta, Egypt. Environ Sci Pollut Res 24:3675–3704
27. Attia MI (1954) Deposits in the Nile Valley and the Delta "Geological survey Egypt, Cairo," vol 12, pp 147–165
28. Hurst HE (1952) Long-term storage capacity of reservoirs. Trans Am Soc Civ Eng 116: 770–808
29. Kashef AI (1981a) The Nile-one River and nine countries. J Hydrol 53:53–71
30. Kashef AI (1981b) Technical and ecological impacts of the High Aswan Dam. J Hydrol 53: 73–84

31. Said R (1962) The geology of Egypt, Elsevier, Amsterdam, The Netherlands
32. Said R (1993) The Nile River: geology, hydrology, and utilization. New York
33. Sestini G (1989) Nile Delta; a review of depositional environments and geological history. Whateley, MKG, Pickering KT (eds) Geol Soc Spec Pub No. 40, pp 99–127
34. Shata AA, El-Fayoumy IF (1969) Remarks on the hydrogeology of the Nile Delta. Proceedings of the Bucharest symposium in Deltas
35. Sandford KS, Arkell WJ (1939) Paleolithic man and the Nile Fazuiom divide Chicago. Univ. Oriental Inst. Pub., I., pp 1–77
36. Habib MM (1996) Hydrogeological studies on El-Gharbia Governorate, Central Nile Delta, Egypt. MSc Thesis, Tanta University
37. IMFO (1996) Resource and ecological assessment of San Pedro Bay, Philippines, Technical Report for Fishery Sector Program, U.P. Visayas Foundation, Inc. and Institute of Marine Fisheries and Oceanography
38. Abbasi TA, Abbasi SA (2012) Water quality indices, Elsevier BV, pp 19–28
39. Cloutier V, Lefebvre R, Therrien R, Savard MM (2008) Multivariate statistical analysis of geochemical data as indicative of the hydrogeochemical evolution of groundwater in a sedimentary rock aquifer system. J Hydrol 353:294–313
40. Guler C, Thyne GD, McCray JE, Turner AK (2002) Evaluation of graphical and multivariate statistical methods for classification of water chemistry data. Hydrgeol J 10:455–474
41. Ravikumar P, Somashekar RK (2017) Principal component analysis and hydrochemical facies characterization to evaluate groundwater quality in Varahi river basin, Karnataka state, India. Appl Water Sci 7:745–755
42. Ishaku JM, Kaigama U, Onyeka NR (2011) Assessment of groundwater quality using factor analysis in Mararaba-mubi area, Northeastern Nigeria. J Earth Sci Geotech Eng 1:9–33
43. Hotelling H (1933) Analysis of a complex of statistical variables into principal components. J Educ Psychol 24:417–441
44. Gnanadesikan R (1977) Methods for statistical data analysis of multivariate observations. Wiley, New York
45. Nton ME, Adejumo SA, Elueze AA (2007) Hydrogeochemical assessment of surface water and groundwater quality in Agbowo-Orogun area of Ibadan, Southwestern Nigeria. Global J Geol Sci 5:13–23
46. Sawyer CN, McCarty DL (1967) Chemistry of sanitary engineers, 2nd edn. McGraw-Hill, New York, 518 p

Assessment of the Groundwater Quality for Drinking and Irrigation Purposes in the Central Nile Delta Region, Egypt

Zenhom E. Salem, Gamal Elsaiedy, and Abdelaziz ElNahrawy

Abstract One hundred sixty-nine groundwater samples were collected, chemically analyzed, and classified into shallow, intermediate, and deep zones to evaluate the vertical and lateral change in groundwater quality in the central part of the middle Nile Delta. To estimate the groundwater suitability for drinking, parameter's concentrations were evaluated according to WHO drinking water guidelines to delineate the samples of desirable and undesirable range in every zone. According to the computed WQI, most part of the shallow groundwater is unsuitable for drinking [unfit (8 wells, 14.55%), very poor (3 wells, 5.45%), and poor drinking quality (26 wells, 47.3%)]. Intermediate groundwater zone is mostly suitable [excellent (4 wells, 8.9%) and good (24 wells, 53.3%)]. The deep groundwater quality is classified into unfit (3 wells, 4%), very poor (5 wells, 7%), poor water (27 wells, 40%), good quality (30 wells, 45%), and excellent (2 wells, 3%).

Groundwater suitability was also evaluated using TDS, Na%, SAR, RSC, Cl, KI, PI, MH, CAI, and CR. Irrigation water quality index (IQW) was also used as an integrated method. The studied groundwater is mostly of medium suitability where a number of samples which fall within this class are 36 (65.5%), 29 (64%), and 34 (51%) for the shallow, intermediate, and deep groundwater. Water samples have good irrigation quality which increases downward where 15 (27.2%), 13 (29%), and 32 (48%) samples are recorded in this class, respectively. Samples belonging to the poor quality class are mostly located in the northern part, and its sample numbers are 4 (7.3%), 3 (7%), and 1 (1%), respectively.

Z.E. Salem (✉) and A. ElNahrawy
Geology Department, Faculty of Science, Tanta University, Tanta, Egypt
e-mail: zenhomsalem@yahoo.com

G. Elsaiedy
Ministry of Environmental Affairs, Cairo, Egypt

A. M. Negm (ed.), *Groundwater in the Nile Delta*,
Hdb Env Chem (2019) 73: 647–684, DOI 10.1007/698_2017_137,

Keywords Drinking water quality index, Groundwater pollution, Irrigation water quality index, Nile Delta, Water quality assessment

Contents

1 Introduction

Resources of freshwater distribution are uneven all around the world, and the freshwater accessibility is gradually becoming rare and attributable to population growth and varies human activities. The lack of fresh surface water leads to misuse of groundwater to meet the demand required by different areas. Groundwater quality is similarly essential as its amount is attributable to the appropriateness of water for different targets. Spatial differences in groundwater quality in certain areas are a component of physical, chemical, and biological factors that are significantly affected by aquifer geology and human activities [1, 2]. Groundwater became the main water supply for residential, industrial, and irrigation divisions of numerous nations. Spatial change in the quality of groundwater because of geologic setup and anthropogenic elements warrants the assessment of the groundwater quality for any use including that for human utilization. Evaluation of the water quality for drinking purposes includes the determination of the chemical composition of groundwater and the remedial measures for the restoration of the water quality in case of its deterioration demand and the estimation of probable sources of the groundwater pollution [3]. About 33% of the world's population use groundwater for drinking [4, 5].

Low quality of water unfavorably influences human health and plant development. In developing nations like Egypt, about 60% of all infections are specifically occurring due to the poor quality of the drinking water [6]. The spatial distribution of water quality and significant components should be compared with the geology and area using land use/land cover spread maps in GIS environment [7, 8]. Therefore the chemical processes of water and the methods of their acquisition could be clearly understood [9]. Egyptian population is basically situated in the narrow valley and delta of the Nile. This high population density is accompanied with a continuous growing size of the population, wastewater, development of industrialization, living standards, the expanded utilization of chemicals in agriculture, the lack of real control on the transfer of unsafe waste materials, and in addition the absence of ecological open mindfulness. This study aims to evaluate the quality characteristics of the groundwater in the central Nile Delta part wherein the people of the terrain are mostly dependent on the groundwater for their needs and to assess the groundwater suitability for drinking and irrigation purposes. Many researchers have investigated the water quality in the Nile Delta, among them Salem [10], Elewa [11], Gemail et al. [12], Morsy and El-Fakharany [13], Salema et al. [14], Ghoraba et al. [15], Khalil et al. [16], El Bedawy [17], Bennett et al. [18], Elkafoury et al. [19], Fattah and Ragab [20], El-Kowrany et al. [21], Negm and Armanuos [22], Negm and Eltarabily [23], Salem et al. [24], Sharaky et al. [25], and Salem and Osman [26].

No encountered research work on the central Nile Delta used the integrated method of water quality indices to evaluate the groundwater for irrigation and drinking purposes as will be shown in this study.

2 Study Area

The Nile Delta, one of the biggest depocenters in the Mediterranean, is currently mainly a man-altered beachfront plain where the Nile sediments have stopped accumulating in the Mediterranean and locally are retreating. The Nile Delta area is around 22,000 km^2 and records for 66% of Egypt's agrarian surface. Its smooth coastline is 225 km long and lies 160 km north of Cairo (Fig. 1). Land elevation diminishes gradually northward, from 18 m masl (meter above ocean level) close to Cairo to less than 1.0 m close to the coast [27]. The Nile water is flowing from the Aswan High Dam to the delta. Thirty-three percent of this water volume is lost by evapotranspiration and penetration to groundwater aquifers, and the other two third flows slowly through the intense system of irrigation channels and drains. Agricultural activities are overwhelming in the Nile Delta because of the soil salinity [27, 28] and a watering system framework setup. The seawater intrusion might strongly influence the quality of the groundwater in the area. The rise in the sea level as well as the changes of Nile River flow leads to salinity increase [28, 29] and water logging in the northern part of the Nile Delta [30]. Furthermore, the present and future anthropogenic activities, particularly intensive unplanned groundwater pumping, are bringing about the disintegration of the accessible groundwater resources.

The study area occupied the central part of Nile Delta (Fig. 1) and is located between the Damietta branch (in the east) and Rosetta branch (in the west). It lies between latitudes 30° 06$'$ and 31° 10$'$ north. The study area extends from the northern part of Monufia Governorate into the southern part of Kafr El Sheikh Governorate and including the whole area of Gharbia Governorate.

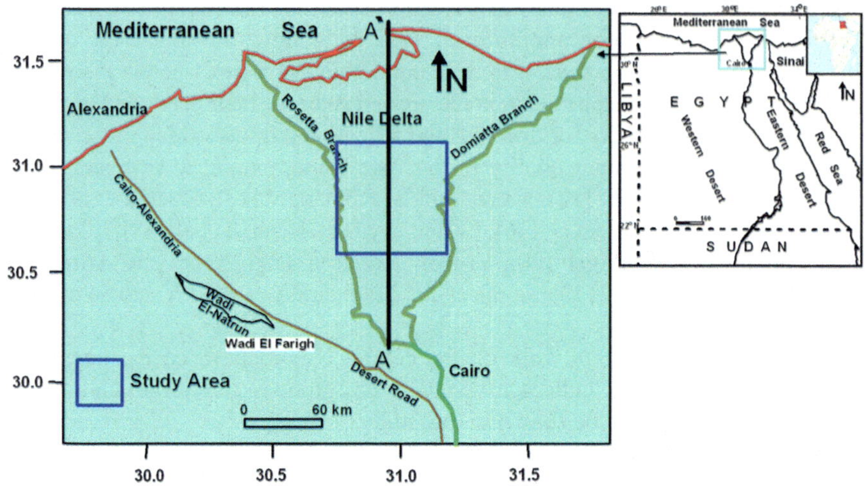

Fig. 1 Location map of the study area in the central part of the middle Nile Delta. A–A′ is the location of the cross section shown in Fig. 2

3 Hydrogeological Setting

Several studies dealing with the geology, hydrogeology, and hydrogeochemistry of the Nile Delta quaternary aquifer were done. The significant part of the annual recharge of the Nile Delta aquifer is gotten from the immediate seepage from the watering system trenches and from the infiltration through soils due to irrigation activities. The yearly general groundwater recharge to the aquifer is evaluated at 6.70 km^3/year [24, 31–34].

The Nile Delta quaternary aquifer is of semi-confined type [29, 35]. It occupies the entire Nile Delta. This aquifer changes in thickness from 200 m in the south where Cairo is located to 1,000 m in the northern direction [36] (Fig. 2). The depth to water in this aquifer ranges between 1 and 2 m in the north, 3–4 m in the middle (study region), and 5 m in the south [29]. Distinctive evaluated groundwater depth has been accounted for by RIGW [37] and Morsy [38].

Bilqas formation is a thin clay layer that covers the top of the quaternary aquifer of the Nile Delta [24, 29, 39, 40]. This clay layer is responsible for the semi-confining characters of the aquifer. Bilqas formation's thickness differs from 5 to 20 m in the southern and the central part of the delta and achieves 50 m in the northern part [24, 29, 41]. The lithological and thickness characteristics of the Bilqas formation greatly affect the level of interaction between the surface water and the groundwater [42]. Mit Ghamr formation is the main aquifer in the Nile Delta which is formed by quaternary sediments [24, 29] (Fig. 2). The variable hydraulic parameters and water salinity of the aquifer are related to the various deltaic depositional conditions [39]. These sediments show various aggradations and degradation cycles that were typically produced by changes in sea level [41]. These quaternary sediments made

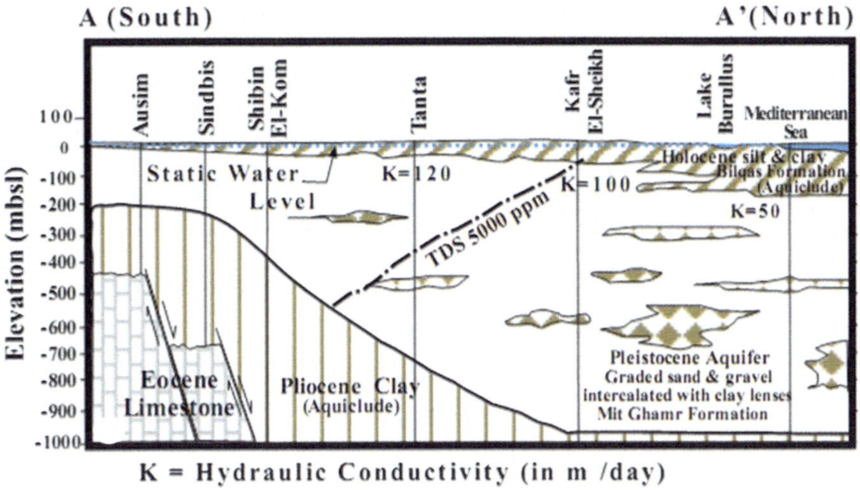

Fig. 2 Hydrogeological cross section from south to north in the Nile Delta showing the groundwater regime [11]

the Nile Delta aquifer as a large water reservoir that is recharged by the Nile water flowing from south to north through an intensive network watering system [43]. The quaternary Nile Delta aquifer is not hydraulically connected with the underlying tertiary formations where the latter rocks act as an aquiclude [42].

4 Methodology

The performed working methodology of this work is shown in Fig. 3. One hundred sixty-six groundwater samples were gathered during field inventory in 2014–2015. Samples were collected from the drinking and irrigation water wells with depths ranging from around 15 to 120 m. Samples were gathered in newly washed plastic bottles. The performed field measurements and laboratory major ions analysis were discussed in Salem et al. [24] and Salem and El-horiny [44], where electrical conductance (EC, mS/cm), pH (hydrogen ion activities), temperature (°C), and TDS (total dissolved solids, mg/l) were measured in situ. Hach's portable EC/TDS meter and portable Consort pH meter (model P 314) were utilized [45]. Chemical analysis of the water samples was carried out by the central laboratory of the ministry of environmental affairs in Tanta City and included the determination of the major ions (i.e., K, Na, Ca, Mg, Cl, HCO_3, CO_3, and SO_4) and trace elements (Fe, Mn, Cu, and Zn). The chemical analysis was finished within few days of water sample gathering. The laboratory analyses of the major ions were done utilizing the standard analytical techniques depicted by Hach [45]. Precipitation with barium chloride method was used for sulfate measuring by spectrophotometer. Chloride, bicarbonate, calcium, and magnesium were measured utilizing a digital titrator. Titration method

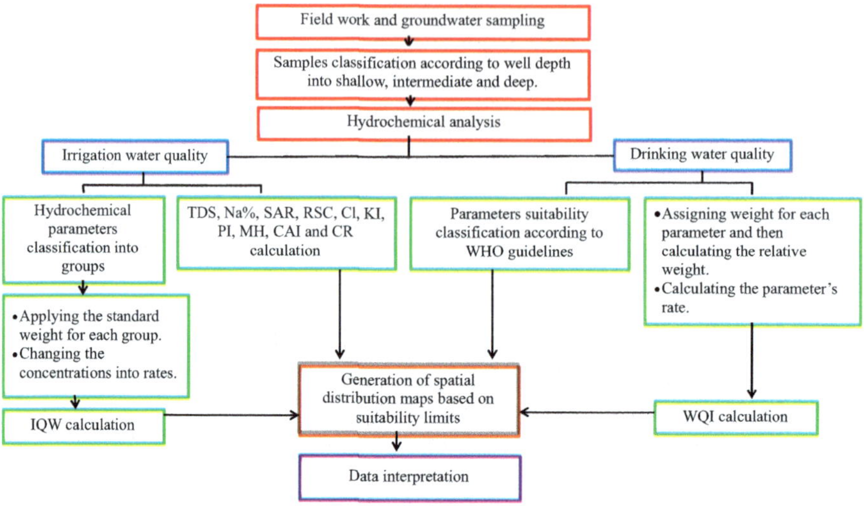

Fig. 3 Flow chart shows the working methodology

using a standard solution of mercuric nitrate and diphenyl carbazone reagent powder was used for chloride ion measurements. Measurements of bicarbonates were done by titration using standard sulfuric acid solution. Concentrations of calcium and magnesium were measured through TH test (total hardness) and calcium hardness test. Atomic absorption was utilized for measuring Fe, Mn, Cu, and Zn. To have a definite depiction of the groundwater quality of the study zone, the gathered samples are classified into shallow groundwater (Fig. 4a), intermediated groundwater (Fig. 4b), and profound groundwater tests (Fig. 4c). The statistical assessment of the water chemical analysis for drinking and irrigation objects is listed in Tables 1 and 2.

4.1 Estimation of Drinking Water Quality Index (WQI)

Chemically based drinking water guidelines according to WHO [46, 47] including the desirable limits of the measured parameters are listed in Table 3. Depending on the relative importance of each parameter for drinking water quality, parameter's weight is assigned. TDS, EC, NO_3, and Pb each has a maximum weight of 5. SO_4, TH, and Mn assigned a weight of 4. 3 is used as the weight of pH, Cl, and Na. K, Mg, Ca, HCO_3, Cu, Fe, and Zn weight is 2 [48]. The calculating method of WQI is as follows:

$$\mathrm{WQI} = \sum Qi \times Wi \tag{1}$$

$$Qi = (Ci/Si) \times 100 \tag{2}$$

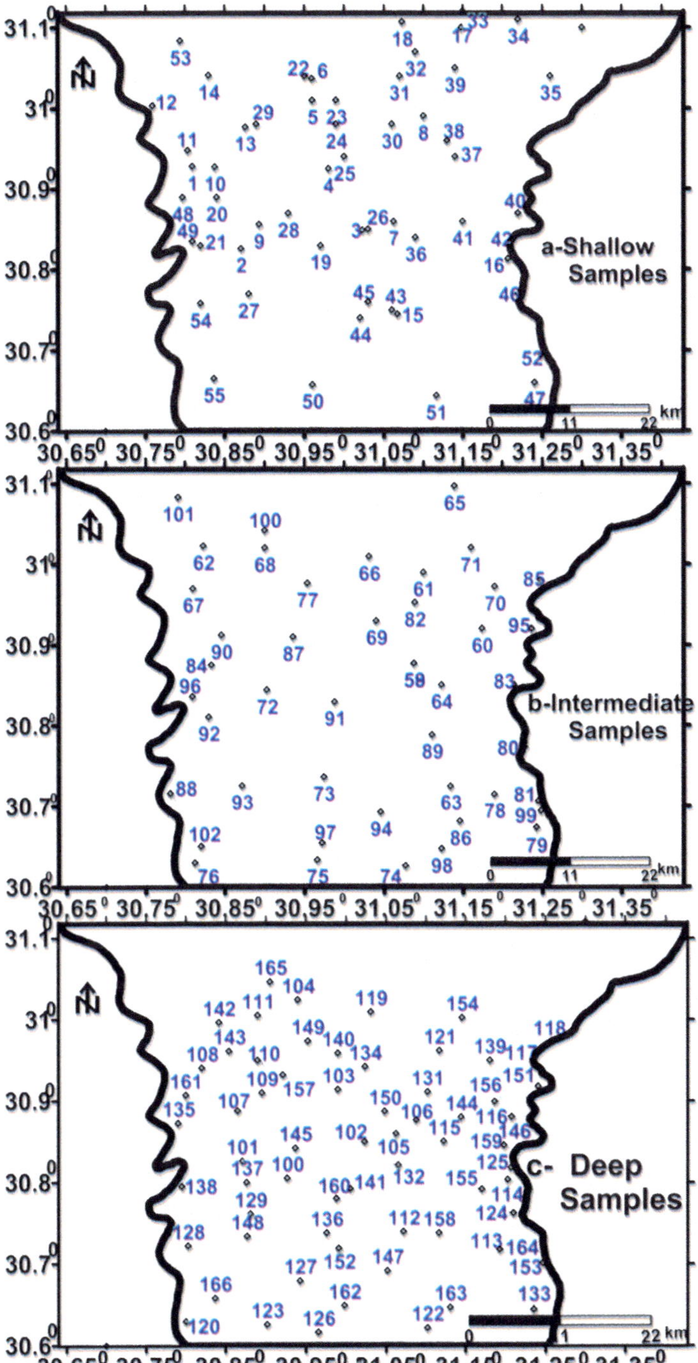

Fig. 4 Location map of the collected groundwater samples in (**a**) shallow, (**b**) intermediate, and (**c**) deep zones

Table 1 Statistical evaluation and water suitability for drinking according to WHO [46, 47]

Classification pattern	Statistical analysis				Classification according to WHO guidelines						
					Desirable limit (mg/l)	Shallow		Intermediate		Deep	
	Range	Shallow	Intermediate	Deep		No	%	No	%	No	%
pH	Min	6.8	7.34	7.3	6.5–8.5	0 up	0	0 up	0	0 up	0
	Max	8.31	8.40	8.45		54 down	100	44 down	100	67 down	100
	Mean	7.74	7.75	7.65							
Total dissolved solids (TDS)	Min	210	173.00	180	500	36 up	65.5	26 up	42	39 up	58.2
	Max	8,820	6,430.00	1,940		19 down	34.5	19 down	58	28 down	41.8
	Mean	1,165.7	804.43	635.11							
Total hardness (TH)	Min	136.55	2.40	91.562	300	18 up	32.7	0 up	0	26 up	38.8
	Max	1,946.38	141.46	1,026		37 down	67.3	45 down	100	41 down	61.2
	Mean	342.1	23.12	306.24							
Sodium (Na)	Min	24.2	15.20	17.20	200	11 up	20	5 up	11.11	5 up	7.5
	Max	2,160	1,249.5	427.7		45 down	80	40 down	88.99	62 down	92.5
	Mean	244.4	135.40	98.41							
Potassium (K)	Min	0.060	1.1	0.15	12	0 up	0	0 up	0	0 up	0
	Max	11.40	3.20	2		55 down	100	45 down	100	67 down	100
	Mean	1.70	0.78	0.66							
Calcium (Ca)	Min	27	8.4	30.70	200	3 up	5.5	2 up	4.4	1 up	1.49
	Max	620	750	230		52 down	94.5	43 down	95.6	66 down	98.51
	Mean	78	85.40	75							

Magnesium (Mg)	Min	10.50	4	9.72	125	1 up	1.8	1 up	2.2	0 up	0
	Max	206.56	141.4	110		54 down	98.2	44 down	97.8	67 down	100
	Mean	33	39.90	28.92							
Bicarbonates (HCO_3)	Min	137.20	80	87.80	350	16 up	29.1	14 up	31.1	17 up	25.4
	Max	2,562.81	542.70	904.3		39 down	70.9	31 down	68.9	50 down	74.6
	Mean	378	288.40	303.7							
Sulfate (SO_4)	Min	0.20	0.48	0.31	250	0 up	0	0 up	0	2 up	3
	Max	246.20	189	252.90		55 down	100	45 down	100	65 down	97
	Mean	43.30	60.10	43.30							
Chloride (Cl)	Min	50	16	17	250	12 up	21.8	6 up	13.3	8 up	11.9
	Max	3,166	3,610	748.30		43 down	78.2	39 down	86.7	59 down	88.1
	Mean	362.80	256.40	160.30							
NO_3	Min	0.022	0.01	6E-05	50	0 up	0	0 up	0	0 up	0
	Max	42	42.00	30.8		55	100	45 down	100	67down	100
	Mean	3.05	3.90	2.4							
Iron (Fe)	Min	0	0.00	0	0.3	34 up	61.8	16 up	64.4	25 up	37.3
	Max	3.42	2.81	1.76		21 down	38.2	29 down	35.6	42 down	62.7
	Mean	0.96	0.69	0.382							
Manganese (Mn)	Min	0.042	0.07	0.036	0.4	25 up	45.5	21 up	53.3	34 up	50.7
	Max	2.76	1.95	2.89		30 down	54.5	24 down	46.7	33 down	49.3
	Mean	0.36	0.65	0.524							

(continued)

Table 1 (continued)

Classification pattern	Statistical analysis				Classification according to WHO guidelines						
					Desirable limit	Shallow		Intermediate		Deep	
	Range	Shallow	Intermediate	Deep	(mg/l)	No	%	No	%	No	%
Cu	Min	0.002	0.00	0.003	2	0 up	0	0 up	0	0 up	0
	Max	1.15	1.28	1.95		55 down	100	45 down	100	67 down	100
	Mean	0.91	0.25	0.62							
Zn	Min	0.006	0.01	0	5	14 up	25.5	11 up	24.4	14 up	20.9
	Max	2.63	1.54	1.37		41 down	74.5	34 down	75.6	53 down	79.1
	Mean	0.314	0.36	0.283							

Table 2 Groundwater suitability for irrigation

Classification pattern	Categories	Ranges	Shallow		Intermediate		Deep	
			No	%	No	%	No	%
Total dissolved solids (TDS)	Desirable	<500	23	41.80	18.00	40.00	28.00	41.79
	Permissible	500–1,000	21	38.20	21.00	46.70	31.00	46.27
	Useful for irrigation	1,000–3,000	6	10.90	4.00	8.90	8.00	11.94
	Unfit for irrigation	>3,000	5	9.10	2.00	4.44		0.00
Sodium per-cent sodium (Na%)	Excellent	0–20	0	0.00	3.00	6.67	2.00	2.90
	Good	20–40	32	58.20	28.00	62.22	44.00	65.70
	Permissible	40–60	12	21.80	12.00	26.70	15.00	22.40
	Doubtful	60–80	11	20.00	2.00	4.40	3.00	4.50
	Unsuitable	>80	0	0.00		0.00	3.00	4.50
Sodium absorption ratio (SAR)	Excellent	0–10	45	81.80	43.00	95.60	65.00	97.00
	Good	18	7	12.70	2.00	4.40	2.00	3.00
	Doubtful	18–26	3	5.50	0.00	0.00	0.00	
	Unsuitable	>26	0	0.00	0.00	0.00	0.00	
Residual sodium car-bonate (RSC)	Good	<1.25	48	87.30	42.00	93.30	63.00	94.00
	Doubtful	1.25–2.5	2	3.60	2.00	4.50	2.00	3.00
	Unsuitable	> 2.5	5	9.10	1.00	2.20	2.00	3.00
Permeability index (PI)	Class I	>75	13	23.60	7.00	15.60	9.00	13.40
	Class II	25–75	42	76.40	38.00	84.40	58.00	86.60
Chloro-alka-line Indices (CAI)	Base exchange	Negative	25	45.50	25.00	55.60	35.00	52.20
	Cation–anion	Positive	30	54.50	20.00	44.40	32.00	47.80
Chloride (Cl^-)	Extremely fresh	<0.14		0.00			0.00	0.00
	Very fresh	0.14–0.85	0	0.00	2.00	4.40	4.00	6.00
	Fresh	0.85–4.23	34	61.80	30.00	66.70	38.00	56.70
	Fresh–brackish	4.23–8.46	9	16.40	10.00	22.30	17.00	25.40
	Brackish	8.46–28.21	6	10.90	1.00	2.20	8.00	11.90
	Brackish–salt	28.21–282.06	6	10.90	2.00	4.40	0.00	0.00
	Salt	282.06–564.13	0	0.00	0.00	0.00	0.00	0.00
	Hypersaline	>564.13	0	0.00	0.00	0.00	0.00	0.00
Magnesium hazard (MH)	Unsuitable	> 50%	9	16.40	16.00	35.60	7.00	10.45
	Suitable	< 50%	46	83.60	29.00	64.40	60.00	89.55
Kelley's index (KI)	Unsuitable	> = 1	14	25.50	10.00	22.20	11.00	16.40
	Suitable	<1	41	74.50	35.00	77.80	56.00	83.60
Corrosive ratio (CR)	Noncorrosive	<1	17	30.90	11.00	24.40	22.00	32.80
	Corrosive	>1	38	69.10	34.00	75.60	45.00	67.20

Table 3 Desirable limits and weightings of the parameters used in WQI calculations

Parameters	WHO desirable limits (mg/l)	Weight (*wi*)	Relative weight (mg/l)
TDS	500 (mg/l)	5	0.1000
PH	6.5–8.5	3	0.0600
EC	1,500 us/cm	5	0.1000
TH	300 (mg/l)	4	0.0800
Ca	200 (mg/l)	2	0.0400
Na	200 (mg/l)	3	0.0600
Mg	125 (mg/l)	2	0.0400
K	12 (mg/l)	2	0.0400
Cl	250 (mg/l)	3	0.0600
SO_4	250 (mg/l)	4	0.0800
Cu	2.0 (mg/l)	2	0.0400
HCO_3	350 (mg/l)	2	0.0400
NO_3	50 (mg/l)	5	0.1000
Fe	0.3 (mg/l)	2	0.0400
Mn	0.4 (mg/l)	4	0.0800
Zn	5 (mg/l)	2	0.0400
		$\sum W = 50$	$\sum W = 1$

$$Wi = wi / \sum_{i=0}^{n} wi \tag{3}$$

As stated by Armanuos et al. [49], "Qi is the ith quality rating and is given by Eq. (2), Wi is the ith relative weight of the parameter I and is calculated by Eq. (3), Ci is the ith concentration of parameter and Si is the ith drinking water guideline according to WHO guidelines in mg/l, wi is the weight of ith parameter, and n is the number of chemical parameters."

4.2 Estimation of Irrigation Water Quality (IWQ) Index

Five parameter groups are utilized to compute the IWQ. The used parameters were classified into groups according to the rules introduced by Ayers and Westcost [50] and Simsek and Gunduz [51] (Tables 4 and 5). The IWQ index calculating method in the current study was done according to what was stated in Simsek and Gunduz [51] and Spandana et al. [52] as follows:

$$\text{IWQ} = \sum\nolimits_{i=1}^{5} Gi \tag{4}$$

where i is an incremental index and G is the contribution of each one of the five hazard groups that are essential to assess the quality of water resource for irrigation. Salinity hazard as EC value represents the first category and is determined as:

Table 4 IWQ's parameter classification [50, 51]

Hazard	Weight	Parameter	Range	Rating	Suitability
Salinity hazard	5	Electrical conductivity (μs/cm)	EC < 700	3	High
			700 ≤ EC ≤ 3,000	2	Medium
			EC > 3,000	1	Low
Infiltration and permeability hazard	4	See table IV for details			
Specific ion toxicity	3	Sodium adsorption ratio	SAR <3.0	3	High
			3.0 ≤ SAR ≤ 9.0	2	Medium
			SAR >9.0	1	Low
		Boron (mg/l)	B < 0.7	3	High
			0.7 ≤ B ≤ 3.0	2	Medium
			B > 3.0	1	Low
		Chloride (mg/l)	Cl < 140	3	High
			140 ≤ Cl ≤350	2	Medium
			Cl > 350	1	Low
Trace element toxicity	2	See Simsek and Gunduz [51]			
Miscellaneous effects to sensitive cops	1	Nitrate nitrogen (mg/l)	NO3-N < 50	3	High
			5.0 ≤ NO_3-N ≤ 30.0	2	Medium
			NO_3-N > 30.0	1	Low
		Bicarbonate (mg/l)	HCO_3 < 90	3	High
			90 ≤ HCO_3 ≤ 500	2	Medium
			HCO3 > 500	1	Low
		pH	7.0 ≤ pH ≤ 8.0	3	High
			6.5 ≤ pH <7.0 and 8.0 < pH ≤ 8.5	2	Medium
			pH < 6.5 or pH > 8.5	1	Low

$$G_1 = w_1 r_1 \tag{5}$$

where w is the group weighting, and r is the parameter rating value. EC–SAR combination is the second category and represents the infiltration and permeability hazard that is computed as:

$$G_2 = w_2 r_2 \tag{6}$$

where w and r are the group weight and the parameter rating value, respectively. The third group is the specific ion toxicity of SAR, chloride, and boron ions in the water and is calculated as a weighted average of the three ions:

Table 5 Classification for infiltration and permeability hazard for irrigation water [51]

						Rating	Suitability
SAR	<3	3–6	6–12	12–20	>20		
EC	>700	>1,200	>1,900	2,900	5,000	3	High
	700–200	1,200–300	1,900–500	2,900–1,300	5,000–2,900	2	Medium
	<200	<300	<500	1,300	2,900	1	Low

$$G_3 = \frac{W_3}{3} \sum_{j=1}^{3} r_j \tag{7}$$

where j is an incremental index, w is the weighting of the third group, and r is the rating value of each parameter. The fourth group is the trace element toxicity that is calculated as a weighted average of all the ions available for analysis:

$$G_4 = \frac{W_4}{N} \sum_{k=1}^{N} r_k \tag{8}$$

where k is an incremental index, N is the total number of trace element available for the analysis, w is the weight value of this group, and r is the rating value of each parameter. The fifth and the final category is the miscellaneous effects to sensitive crops that are represented by nitrate–nitrogen and bicarbonate ions and the pH of the water and is formulated as a weighted average:

$$G_5 = \frac{W_5}{3} \sum_{m=1}^{3} r_m \tag{9}$$

where m is an incremental index, w is the weight value of this group, and r is the rating value of each parameter.

5 Results and Discussion

5.1 *Evaluation of Groundwater Quality for Drinking*

Drinking water characteristics are soft, low TDS, and has no dissolved toxic components. WHO [47] drinking water guidelines are taken as the basis for the groundwater quality assessment. The statistical analysis of the measured chemical parameters which are used to assess the studied groundwater drinking quality is listed in Table 1.

5.1.1 Shallow Groundwater

The shallow groundwater pH values range from 6.8 to 8.31 with an average value of 7.74. One hundred percent of the samples of the shallow zone are within the desirable limits. TDS vary in the range from 210 to 8,820 mg/l, and the average value is 1,165.7 mg/l. Based on WHO [47] for water suitability according to TDS values, 65.5% of the wells (36 well) are unsuitable water for drinking, and 34.5% of wells are suitable [53]. Spatially the samples have unsuitable TDS values, occupying most of the study area except some areas to the east and the west (Fig. 5a). TH values range from 136.55 to 1,946.4 mg/l with 342.1 mg/l average value. 67.3% (37 wells) of the groundwater samples have soft characters; 32.7% (18 wells) are of hard to very hard category which is beyond the suggested safe limit [54]. The latter unsuitable groundwater samples occupy the northern and southeastern parts of the area (Fig. 5b). NO_3 concentration varies from 0.022 to 42 mg/l with an average value of 3.05 mg/l. One hundred percent of the shallow groundwater samples have NO_3 concentrations under the permissible limit (Table 1) [55]. Higher NO_3 values are recognized in the southern and northwestern parts of the area.

Na concentrations range from 24.2 to 2,160 mg/l with an average of 244.4 mg/l. Based on WHO [46] guidelines, 11 samples (20%) are above the limit and spatially located in the northern parts (Fig. 6a). One hundred percent of the shallow groundwater samples have K concentration below the desirable limit and ranges from 0.060 to 11.40 mg/l with an average value of 1.70 mg/l (Table 1). Ca concentrations range from 27 to 620 mg/l with an average of 78 mg/l. About 5.5% (three wells) of the shallow samples have Ca concentration which exceeded the desirable limit and is located in the northern part (Fig. 6b). Mg concentrations range from 10.50 to 206.56 mg/l, and its average value is 33 mg/l. Most of the samples (98.2%, 54 wells) have Mg concentrations within the desirable range (Fig.6c). The concentration of HCO_3 ranges from 137.20 to 2,562.81 mg/l with an average value of 378 mg/l. Thirty-nine wells (about 70.9%) of the total wells have HCO_3 concentrations lower than the desirable limit and occupy most of the study area. 29.1% of wells are above HCO_3 limit and mostly located in the northern parts of the area (Fig. 6d). It was found that concentrations of SO_4 ions range from 0.20 to 246.20 mg/l with an average of 43.30 mg/l. One hundred percent of the samples are within the desirable limit (Table 1). Chloride concentrations vary from 50 to 3,166 mg/l with an average of 362.80 mg/l. Forty-three wells (78.2%) are within the desirable range and located in the southern and central parts. The samples of the northern part of the area (12 wells, 21.8%) are above the permissible limit (Fig. 6e). The concentration of Fe ranges from 0 to 3.42 mg/l with an average of 0.96 mg/l. Twenty-one wells (about 38.2%) are within the desirable range and represented at different localities in the study area. Thirty-four wells (61.8%) are above the permissible limit and cover the most part of the study area (Fig. 7a). Mn concentrations vary from 0.042 to 2.76 mg/l with an average of 0.677 mg/l. Twenty-five wells (45.5%) of the shallow groundwater samples have unsuitable Mn concentrations and occupy the most part of the study area. Thirty wells (54.5%) of the samples are within the desirable range and occupy the area located to the northeastern and southwestern parts (Fig. 7b). A higher concentration of Mn is toxic and is usually bad in terms of taste, odor, and discoloration of food. It was found that the amount of Cu ions ranges from 0.002 to 1.15 mg/l with an average of 0.91 mg/l. One hundred percent of wells are within the desirable limit with concentration increase toward NE parts as a general trend (Table 1). The concentration of Zn varies from 0.006 to 2.63 mg/l with an average of

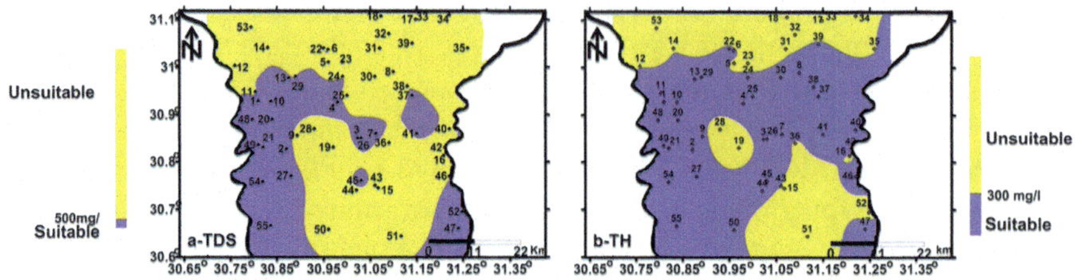

Fig. 5 Spatial distribution maps of the shallow groundwater suitability for drinking based on (**a**) TDS and (**b**) TH concentrations

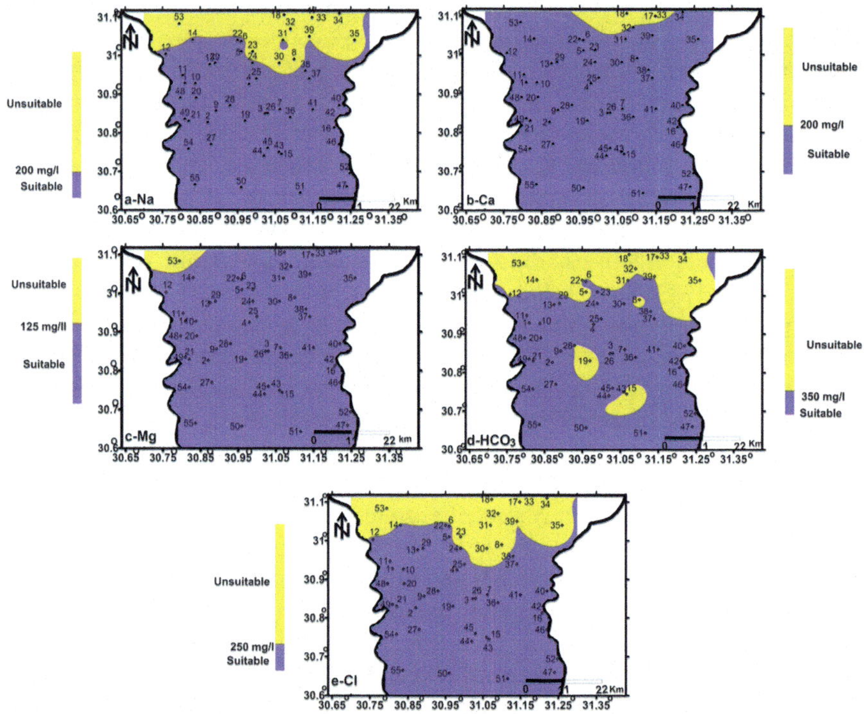

Fig. 6 Spatial distribution maps of the shallow groundwater suitability for drinking based on (**a**) Na, (**b**) Ca, (**c**) Mg, (**d**) HCO_3, and (**e**) Cl concentrations

0.314 mg/l. Most of the wells (41 wells, 74.5%) are within desirable limit. Fourteen wells (25.5%) exceeded the permissible limit and are located to the northern and eastern directions (Fig. 7c).

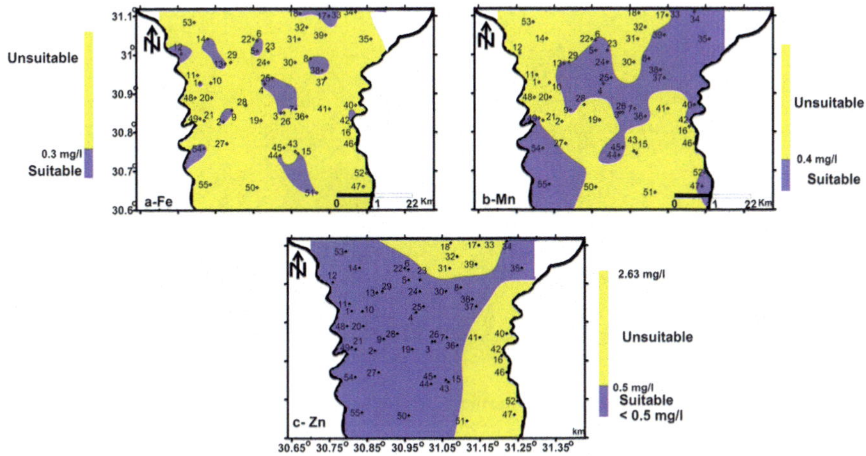

Fig. 7 Spatial distribution maps of the shallow groundwater suitability for drinking based on (**a**) Fe, (**b**) Mn, and (**c**) Zn concentrations

5.1.2 Intermediate Zone

The pH values of the intermediate groundwater range from 7.34 to 8.40 with an average value of 7.75. One hundred percent of the wells in the study area are within the desirable pH limits. TDS of the intermediate groundwater vary in the range from 173 to 6,430 mg/l with an average value of 804.43 mg/l. According to WHO [46], 58% (26 wells) of the wells are unsuitable water for drinking, and 42% (19 wells) are of suitable water for drinking. The wells above the desirable limit are located at the northern and the southern parts of the study area (Fig. 8). TH as $CaCO_3$ in the intermediate groundwater ranges from 2.4 to 141.5 mg/l with an average value of 23.43 mg/l (Table 1). All the groundwater wells fall in the soft category (100%), and the concentration increases toward the northern direction as a general trend. NO_3 concentration ranges from 0.0124 to 42 mg/l with an average value of 3.9 mg/l. All the samples of this zone have NO_3 concentrations under the permissible limit with an increasing trend toward the northwestern direction (Table 1).

Na concentration ranges from 15.20 to 1,249.5 mg/l with an average of 135.40 mg/l and increases toward northern parts as a general trend (Fig. 9a). Five wells (11.11%) are above the desirable limit and are located in northern parts. K concentration ranges from 1.1 to 3.20 mg/l with an average value of 0.78 mg/l, all samples are under the desirable limit (Table 1), and concentrations increase toward the northern direction. Ca concentrations range from 8.4 to 750 mg/l with an average of 85.40 mg/l. About 95.6% (43 wells) of the samples are within the permissible limit and show an increase toward the northern direction (Fig. 9b). The concentrations of Mg range from 4 to 141.4 mg/l with an average value of 39.9 mg/l. Most of the Mg concentrations (44 wells, 97.8%) are within the desirable limit and increase toward the northern parts as a general trend (Fig. 9c). The concentration of HCO_3 in the intermediate groundwater ranges from 80 to 542.7 mg/l with an average value of 288.40 mg/l.

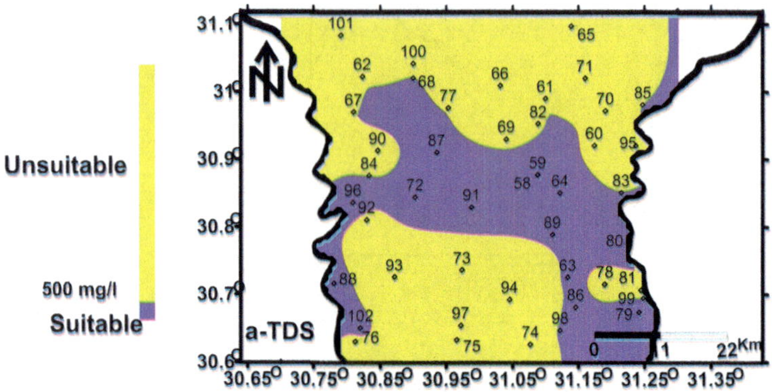

Fig. 8 Spatial distribution maps of the intermediate groundwater suitability for drinking based on TDS concentrations

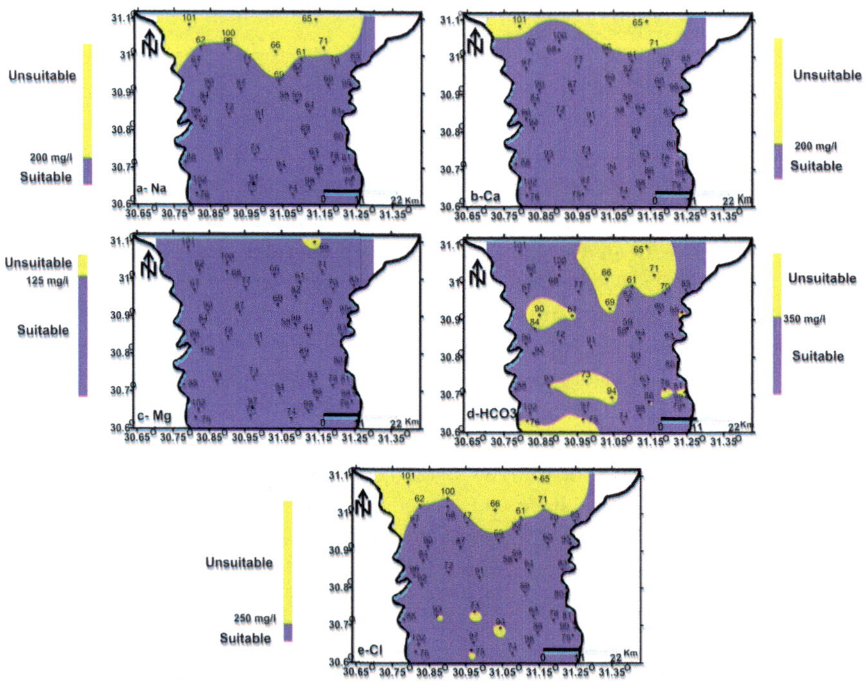

Fig. 9 Spatial distribution maps of the intermediate groundwater suitability for drinking based on (**a**) Na, (**b**) Ca, (**c**) Mg, (**d**) HCO_3, and (**e**) Cl concentrations

Thirty-one samples (about 68.9%) of the total wells are under the desirable limit and occupy most of the study area, while the wells above the permissible limit are located in the northern parts and some localities at central and southern parts (Fig. 9d). It was found that 100% of wells are within the desirable limit of SO_4 concentrations and range from 0.48 to 189 mg/l with an average of 60.10 mg/l (Table 1). The concentrations of Cl ion vary from 16 to 3,610 mg/l with an average of 256.40 mg/l. Thirty-nine wells (86.7%) are within the desirable limit and increase toward the northern parts as a general trend (Fig. 9e).

The concentration of Fe ranges from 0 to 2.81 mg/l with an average of 0.693 mg/l. Sixteen samples (about 35.6%) are within the desirable range, and concentrations increase toward the northwest and northeast. The wells above the desirable limit (29 wells, 64.4%) are located in the northern, western, and southwestern parts of the study area (Fig. 10a). The concentration of Mn varies from 0.068 to 1.95 mg/l with an average of 0.65 mg/l. Most of the wells (24 wells, 53.3%) are within the desirable range, and the unsuitable samples (21 wells, 46.7%) are located at the southern and northern parts (Fig. 10b). It was found that concentrations of Cu ion range from 0 to 1.28 mg/l with an average of 0.246 mg/l. One hundred percent of wells are within the desirable limit (Table 1) with concentrations increasing toward the northern parts as a general trend. The concentration of Zn varies from 0.006 to 1.54 mg/l with an average of 0.361 mg/l. Most of the wells (34 wells, 75.6%) are within the permissible limit, and 11 wells (24.4%) located mostly in the central and southeastern parts exceed the permissible limit (Fig. 10c).

5.1.3 Deep Zone

pH values of the deep groundwater range from 7.30 to 8.45 with an average value of 7.65. One hundred percent of the deep groundwater samples are within the desirable limits with higher values in some locations in the northern and southern direction. The TDS values vary in the range from 180 to 1,940 mg/l with an average value of 635.11 mg/l. Thirty-nine wells (58.2%) are unsuitable, and 28 wells (41.8%) are of suitable water for drinking. The suitable water samples are located in the eastern and western directions (Fig. 11a). TH in this zone ranges from 91.96 to 1,026 mg/l with an average value of 306.24 mg/l. Forty-one wells (61.2%) of the groundwater wells fall in the soft category; 26 wells (38.8%) are located in the central part of the area and fall in hard to very hard category (Fig. 11b). NO_3 concentration ranges from 0.006 to 30.4 mg/l with an average value of 2.7 mg/l. All the samples of this zone have NO_3 concentrations under the permissible limit with an increasing trend toward the south and southeast (Table 1).

Na concentrations show a range from 17.20 to 427.7 mg/l with an average of 98.41 mg/l. Most of the deep groundwater has Na concentrations within the desirable range except the five wells (7.5%) which are mostly located in the northeastern part (Fig. 12a).All the deep samples have K concentration under the desirable limit and range from 0.15 to 2 mg/l with an average value of 0.66 mg/l (Table 1). The concentration of Ca ranges from 30.70 to 230 mg/l with an average of 75 mg/l. Sixty-six wells (98.51%) of the groundwater samples are within the desirable range except one sample located in the southern part (Fig. 12b). The concentrations of Mg range from 9.72 to 110 mg/l with an average value of 28.92 mg/l. All of the Mg

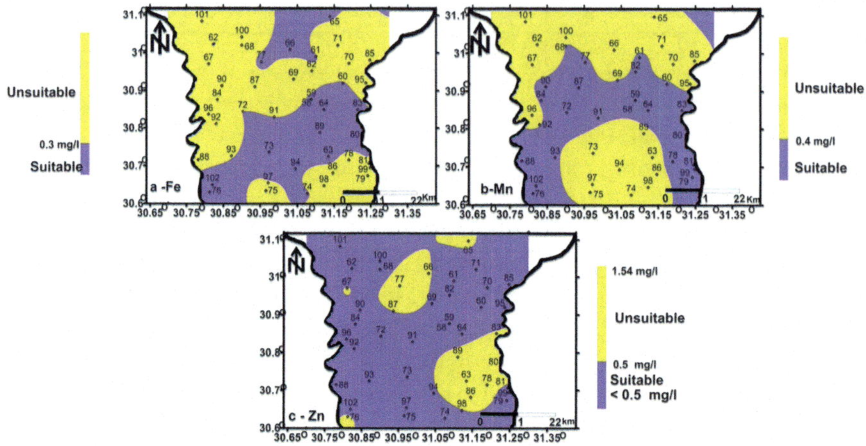

Fig. 10 Spatial distribution maps of the intermediate groundwater suitability for drinking based on (**a**) Fe, (**b**) Mn, and (**c**) Zn concentrations

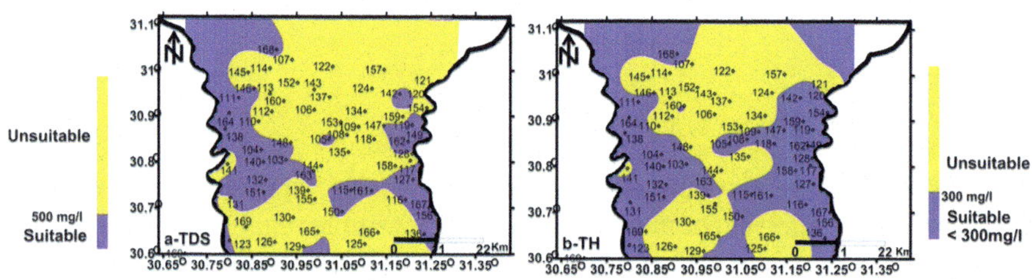

Fig. 11 Spatial distribution maps of the deep groundwater suitability for drinking based on (**a**) TDS and (**b**) TH concentrations

concentrations (100%) are within the desirable limit of 125 mg/l (Table 1). The concentration of HCO_3 in this zone ranges from 87.92 to 904.3 mg/l with an average value of 303.7 mg/l. Seventeen wells (25.4%) are above the desirable limit. These samples are recognized in different localities within the study area (Fig. 12c). It was found that the amount of SO_4 ion ranges from 0.31 to 252.90 mg/l with an average of 17 mg/l and about 65 wells (97%) are within the desirable limit of 250 mg/l (Fig. 12d). The concentrations of Cl varies from 17 to 748.30 mg/l with an average 160.30 mg/l. Fifty-nine wells (88.1%) are within the desirable range, and eight wells (11.9%) are above the limit and located mostly at the northern parts (Fig. 12e).

The concentrations of Fe range from 0 to 1.76 mg/l with an average of 0.38 mg/l. Forty-two samples (62.7%) are under the desirable limit (0.3 mg/l), and 25 wells (37.3%) are unsuitable (Fig. 13a). The unsuitable Fe-related groundwater samples are represented mostly in the northwestern direction and some localities in the southeastern and southwestern directions. The concentrations of Mn have ranged from 0.036 to 2.89 mg/l with an average of 0.524 mg/l. 49.3% of the samples are

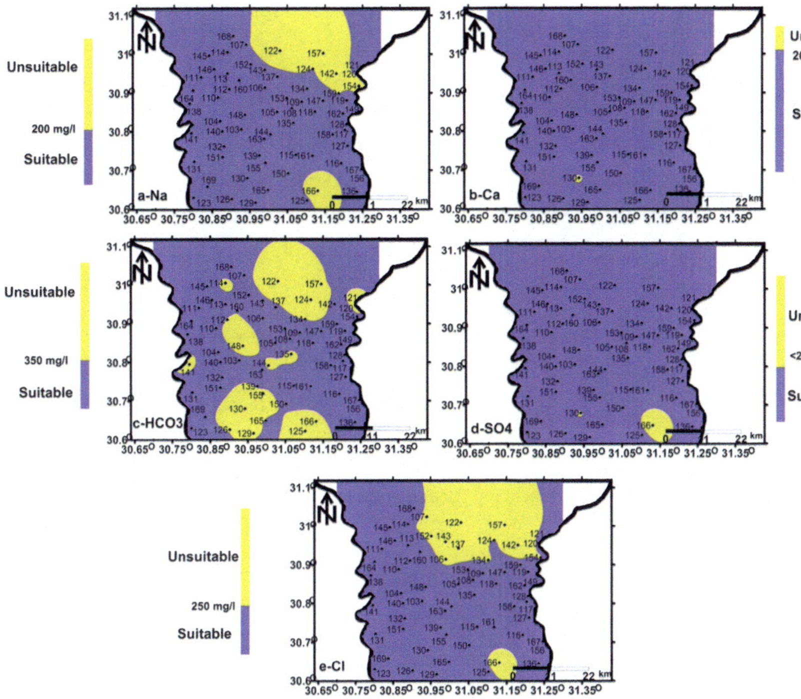

Fig. 12 Spatial distribution maps of the deep groundwater suitability for drinking based on (**a**) Na, (**b**) Ca, (**c**) HCO_3, (**d**) SO_4, and (**e**) Cl concentrations

within the desirable range and 50.7% are above the limit. The unsuitable groundwater extended along the southeastern–northwestern direction (Fig. 13b). All wells are within the desirable limit of Cu concentration with a range from 0.003 to 1.95 mg/l and an average value equal to 0.62 mg/l (Table 1). The concentration of Zn varies from 0 to 1.37 mg/l with an average of 0.283 mg/l. Most of the wells (53 well, 79.1%) are within the desirable range (0.5 mg/l), and 14 wells (20.9%) exceed the limit (Fig. 13c). The unsuitable wells are located in the southwestern corner of the area.

5.1.4 Drinking Water Quality Index (WQI)

Drinking water quality index of the shallow, intermediate, and deep groundwater zones (Table 6, Fig. 14a–c, respectively) and spatial distribution have different patterns. Most of the area in the shallow zone (Fig. 14a) is characterized by poor drinking quality (26 wells, 47.3%) with patches of good quality (18 wells, 32.7%). The northern and southern parts of the area show very poor (3 wells, 5.45%) to unfit (8 wells, 14.55%) groundwater quality. The groundwater of the intermediate zone has excellent (4 wells, 8.9%) to good (24 wells, 53.3%) water quality in the central part of the area. On the other hand, the samples from the southern part have poor

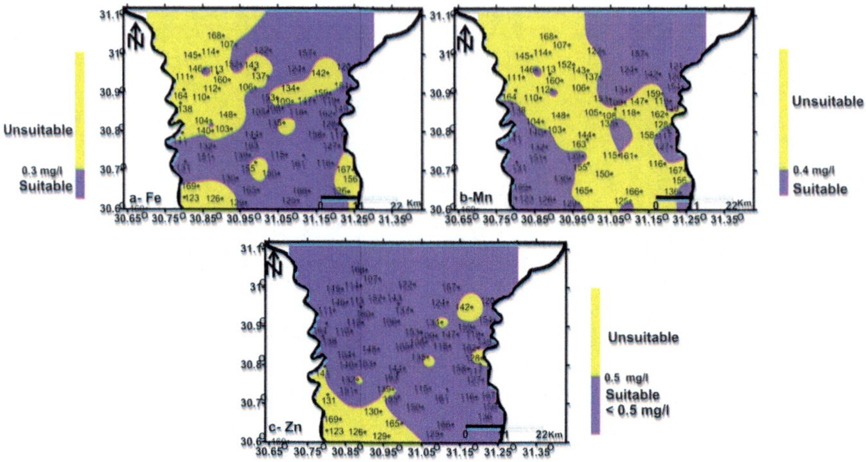

Fig. 13 Spatial distribution maps of the deep groundwater suitability for drinking based on (**a**) Fe, (**b**) Mn, and (**c**) Zn concentrations

Table 6 Samples classification according to WQI

Type of water	Range	Shallow zone		Intermediate zone		Deep zone	
		No. of wells	% of wells	No. of wells	% of wells	No. of wells	% of wells
Excellent water	<50	0	0	4	8.9	2	3
Good water	50–100	18	32.7	24	53.3	30	45
Poor water	100–200	26	47.3	13	29	27	40
Very poor water	200–300	3	5.45	2	4.4	5	7
Unfit for drinking water	>300	8	14.55	2	4.4	3	4

quality (13 wells, 29%), and in the northern part, it has very poor (2 wells, 4.4%) to unfit (2 wells, 4.4%) characters (Fig. 14b). In the deep zone (Fig. 14c), the poor water quality class (27 wells representing 40%) covers most of the study area. Its water quality changed into very poor (5 wells, 7%) to unfit (3 wells, 4%) in the southern direction and excellent (2 wells, 3%) to good quality (30 wells, 45%) to the eastern and western directions.

The correlation matrix (Table 7) shows the relationships between WQI and different water quality parameters. As the correlation coefficient increases, the elements' bad effects on the drinking water quality increase. In the shallow groundwater, there is a strong positive relationship between water quality deterioration (higher WQI) and TDS, EC, TH, Ca, Na, Mg, K, Cl, HCO_3, and Zn. It means all these parameters increase with increasing TDS which are the mean factor for quality deterioration. Mn is moderately affecting the groundwater quality, but SO_4, NO_3, Fe, and Cu show weak to a very weak relationship with WQI. This means the shallow

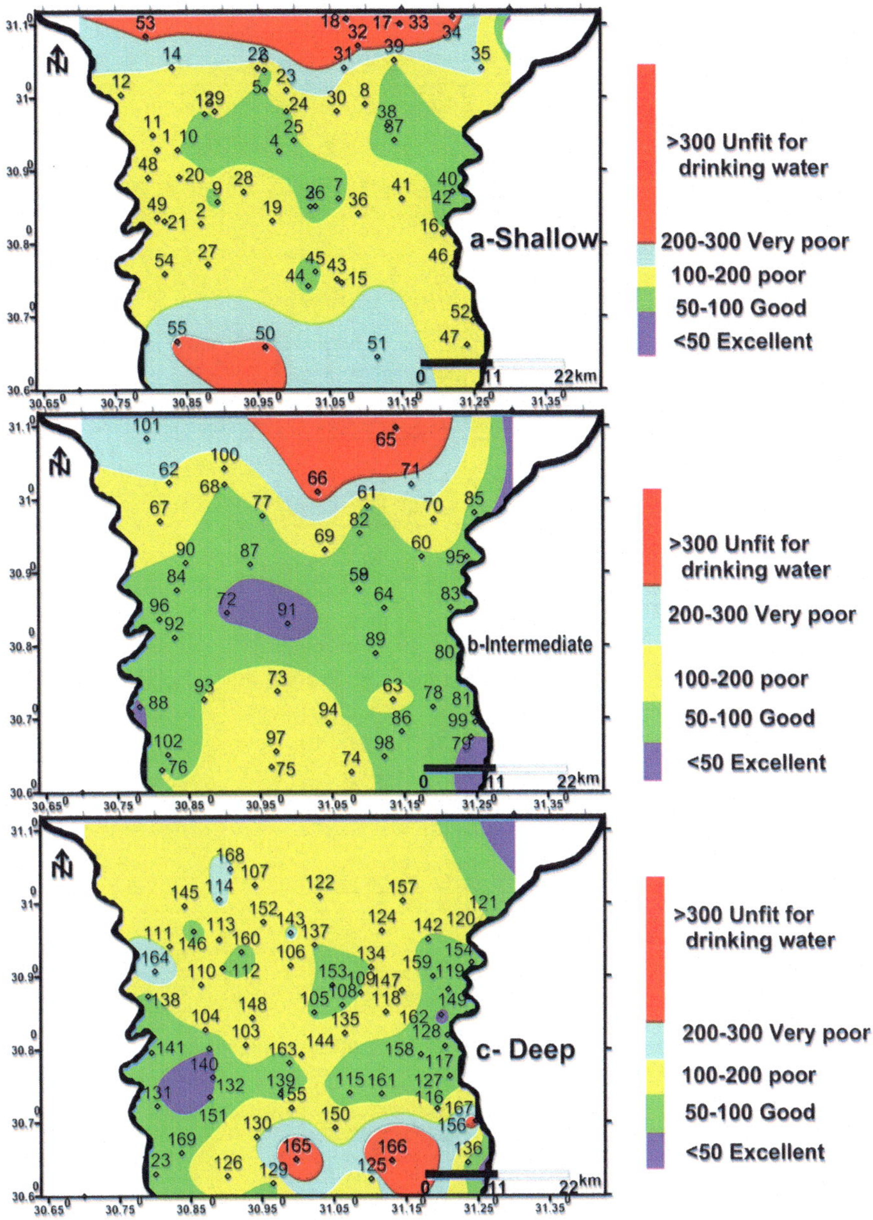

Fig. 14 Spatial distribution maps of the groundwater suitability according to the calculated drinking water quality index (WQI) in the (**a**) shallow, (**b**) intermediate, and (**c**) deep zones

groundwater is not affected by seawater intrusion because SO_4 does not increase linearly with TDS and NO_3, Fe, and Cu concentrations are not related to salinity increase, but it could be related to groundwater–sediment interaction.

Table 7 Correlation matrix between WQI and IQW and the governing parameters

WQI (drinking)	TDS	pH	EC	TH	Ca	Na	Mg	K	Cl	HCO_3	SO_4	NO_3	Fe	Mn	Cu	Zn
Shallow	0.9	−0.18	0.933	0.91	0.88	0.88	0.74	0.87	0.87	0.88	0.38	0.15	−0.01	0.5	−0.04	0.76
Intermediate	0.78	−0.15	0.804	0.67	0.67	0.86	0.85	0.63	0.8	0.19	0.67	0.12	0.38	0.63	−0.17	0.01
Deep	0.62	0.024	0.665	0.63	0.42	0.53	0.6	0.33	0.51	0.33	0.58	0.55	0.179	0.85	−0.04	−0.2
IQW (irrigation)	EC	SAR	Cl	NO_3	HCO_3	PH	Fe	Mn	Zn	Cu						
Shallow	0.70	0.91	0.72	−0.11	0.65	−0.01	−0.03	0.03	0.38	0.41						
Intermediate	0.69	0.83	0.60	0.02	−0.14	−0.04	0.18	0.20	0.36	0.41						
Deep	0.27	0.83	0.21	0.30	0.05	0.18	0.45	0.35	0.56	0.59						

Similar to the shallow water, the intermediate zone WQI shows a strong to an intermediate positive relationship with seawater intrusion-related parameters which are TDS, EC, Na, Mg, Cl, as well as SO_4. In contrast, the freshwater-related water quality parameters (Ca, K, HCO_3, and NO_3) show a weak relationship with WQI which could be related to a higher freshwater fraction. As the deep samples generally have lower salinity, therefore WQI has intermediate positive relationships (>0.5) with seawater-related parameters and weak relationship with the freshwater-related parameters. Iron and Mn show an increasing effect on water quality deterioration in the intermediate zone for iron and deep zone for manganese. In contrast, Cu is highly affecting the water quality in the shallow zones and nearly with no effect on the intermediate and deep zones.

5.2 *Evaluation of Groundwater Quality for Irrigation*

In the Nile Delta, groundwater is the second water resource for irrigation. Therefore, TDS, Na%, SAR, RSC, Cl, KI, PI, TH, MH, CAI, and CR (Table 2) and irrigation water quality index (IQW) were computed to survey the suitability of this groundwater for irrigation purposes. Sodium is a vital element since it indicates the soluble alkali/sodium effect to soils. Since sodium decreases soil penetrability which badly affects cultivation process, SAR (sodium adsorption ratio) was calculated using the following equation:

$$\mathrm{SAR} = \frac{\mathrm{Na}}{\sqrt{\frac{(\mathrm{Ca+Mg})}{2}}} \tag{10}$$

Sodium percent is a factor likewise calculated to assess the appropriateness of water for irrigation purposes [56, 57]. This factor is computed by the accompanying equation:

$$\mathrm{Na\%} = \frac{(\mathrm{Na} + \mathrm{K})}{(\mathrm{Ca} + \mathrm{Mg} + \mathrm{Na} + \mathrm{K})} \times 100 \tag{11}$$

The relative wealth of sodium regarding alkaline earths and boron and the amount of HCO_3 and CO_3 in abundance of soluble earths additionally impact the appropriateness of water for irrigation (RSC) [58]. RSC (residual sodium carbonate) is computed as follows:

$$\mathrm{RSC} = (\mathrm{CO_3} + \mathrm{HCO_3}) - (\mathrm{Ca} + \mathrm{Mg}) \tag{12}$$

The soil penetrability is influenced by water use for a long time. Na, Mg, Ca, and HCO_3 component in the soil impact it. Permeability index (PI) was developed by Doneen [59] to evaluate the appropriateness of water for irrigation where

$$PI = \frac{(Na + \sqrt{HCO_3})}{(Ca + Mg + Na)} \times 100 \quad (13)$$

Measured Na versus Mg and Ca was used by Kelly [60] and Paliwal [61] as Kelley index. This factor is computed as shown in this equation:

$$KI = \frac{Na}{(Ca + Mg)} \quad (14)$$

Magnesium ratio (MH) was suggested by Szabolcs and Darab [62] to assess the irrigation water by the given formula:

$$MH = \frac{Mg}{(Ca + Mg)} \times 100 \quad (15)$$

CAI (chloro-alkaline index) is an important factor and characterized as the ion-exchange characters between the groundwater and aquifer sediments [63]. CAI is computed according to the following equation:

$$CAI = \frac{(Cl - (Na + K))}{Cl} \quad (16)$$

CR (corrosivity ratio) is essential to assess the ability to transfer water in metallic pipes. Water with CR < 1 is safe, while >1 shows corrosive characters and thus not to be transferred in metal pipes. The CR is computed utilizing this equation:

$$CR = \frac{\frac{Cl}{35.5} + 2\left(\frac{SO_4}{96}\right)}{2\left(\frac{HCO_3 + CO_3}{100}\right)} \quad (17)$$

5.2.1 Shallow Zone

The classification of the shallow groundwater suitability for irrigation using the abovementioned parameters is listed in Table 2, and the spatial distribution is shown in Figs. 15 and 16. According to Jain et al. [64], most parts of the study area (50 wells, 90.9%) have desirable TDS range less than 3,000 mg/l except the northern parts which have unfit TDS values (five samples) (Fig. 15a). Na% in the shallow groundwater zone ranged between 19.4 and 80.1%, with an average of 44.7% (Table 1). It is observed that most of the shallow groundwater samples located in the southeastern and western parts fall into the category of good (58.2%, 32 sample). Groundwater samples located in the northern part of the study area have permissible (21.8%, 12 samples) and doubtful (20%, 11 samples) ranges (Fig. 15b). The calculated value of SAR ranges from 0.83 to 21.4 and has been classified mostly as excellent (45 samples, 81.8%) and good (7 samples, 12.7%) for irrigation.

Fair category for groundwater (3 samples, 5.5%) is represented in the northeastern part of the study area (Fig. 15c). RSC reveal that most of the samples fall within good category (48 sample, 87.3%) suitable for irrigation. Two samples (3.6%) and five samples (9.1%) restricted to the northeastern part fall within the category of doubtful and unsuitable, respectively (Fig. 15d). Permeability index reveals that most of the shallow samples fall within class II category (42 sample, 76.4%) except the northeastern part where 13 wells (23.6%) are classified as class I (Fig. 15e). Based on CAI, 30 wells (54.5%) of groundwater samples have cation–anion exchange reaction characters and occupy various locations in the study area. Samples belong to the base-exchange reaction category are represented by 25 wells (45.5%) and have no definite spatial distribution pattern (Fig. 15f). According to chloride concentrations, 34 wells (61.8%) fall within fresh category suitable for irrigation, 9 wells (16.4%) are considered fresh–brackish, 6 wells (10.9%) are brackish, and 6 wells (10.9%) are located in the northern part and fall within brackish–salt range (Fig. 16a). Magnesium hazard (MH) shows that 46 wells (83.6%) occupy most of the study area and are under the suitable category. Nine wells (16.4%) are unsuitable and located as spots in the southeastern, northeastern, and northwestern parts (Fig. 16b). According to KI, most of the shallow samples fall within suitable category (41 samples, 74.5%) which occupy most of the study area. Fourteen wells (25.5%) are classified as unsuitable and located in the northern parts (Fig. 16c). The calculated CR values reveal that most of the shallow samples fall within corrosive category (38 wells, 69.1%). Noncorrosive category (17 wells, 30.9%) is represented mostly in the western part of the area (Fig. 16d).

5.2.2 Intermediate Zone

The classification of the intermediate groundwater suitability for irrigation is shown in Table 2, and the spatial distribution is presented in Figs. 17 and 18. Forty-three wells (95.6%) and two samples (4.4%) have suitable ($<$3,000 mg/l) and unfit TDS values for irrigation, respectively. The unfit groundwater is represented in the northern part (Fig. 17a). Calculated Na% reveals that most of the samples fall into the category of excellent (28 wells, 62.2%) and good water (3 wells, 6.7%). Permissible (12 wells, 26.7%) and doubtful (2 wells, 4.4%) categories are represented mostly in the northern parts (Fig. 17b). Intermediate groundwater is classified as suitable for irrigation ranging from excellent to good according to SAR-calculated values (Fig. 17c). Most of the groundwater samples have RSC values falling within good category (42 wells, 93.3%) and occupying most of the study area. Doubtful (2 wells, 4.5%) and unsuitable (1 well, 2.2%) categories are represented as scattered few locations (Fig. 17d). Class II is the predominant PI category (38 wells, 84.4%) and represented in all parts of the study area while 7 wells (15.6%) are related to class I and represented as three scattered small locations (Fig. 17e). Base-exchange reactions are the common CAI category (25 wells, 55.4%) and occupying an area extending from the southwestern to the northeastern and northern directions (Fig. 17f). Cation–anion exchange reactions (20 wells, 44.6%) mostly characterize

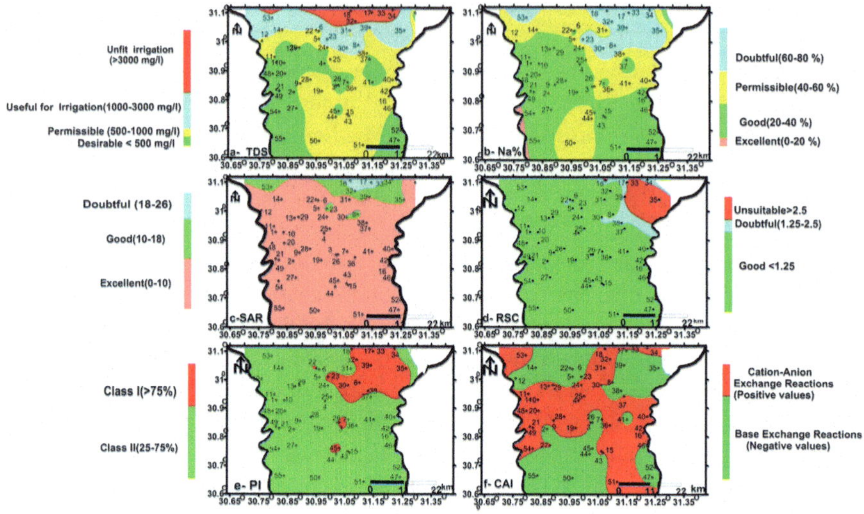

Fig. 15 Spatial distribution maps of the shallow groundwater suitability for irrigation based on (**a**) TDS, (**b**) Na%, (**c**) SAR, (**d**) RSC, (**e**) PI, and (**f**) CAI values

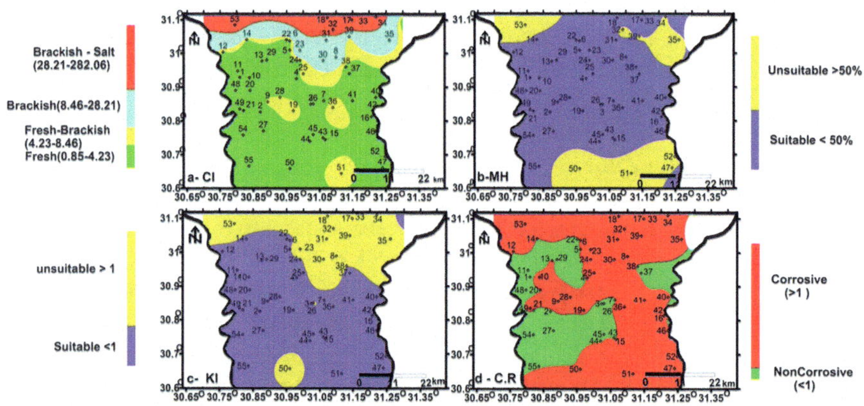

Fig. 16 Spatial distribution maps of the shallow groundwater suitability for irrigation based on (**a**) Cl, (**b**) MH, (**c**) KI, and (**d**) CR values

the water of the southeastern and northwestern parts of the area. Chloride concentrations reveal that most of the groundwater samples fall within fresh category (30 wells, 66.7%) and occupy the central parts, two samples (4.4%) are very fresh, and ten wells (22.3%) fell within the fresh–brackish category and are located in the southern and the northern parts. Brackish (1 well, 2.2%) and brackish–salt (2 wells, 4.4%) groundwater are located in the northern parts as shown in Fig. 18a. According to the calculated magnesium hazard (MH), 29 wells (64.4%) are under suitable

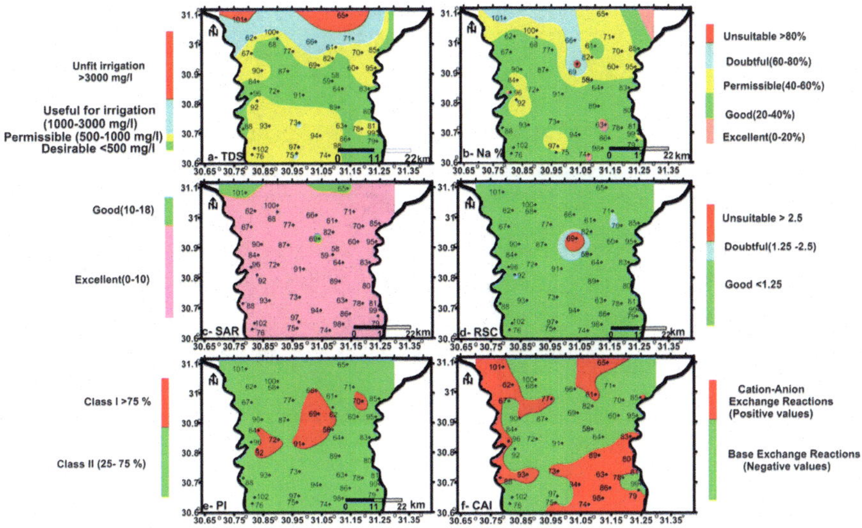

Fig. 17 Spatial distribution maps of the intermediate groundwater suitability for irrigation based on (**a**) TDS, (**b**) Na%, (**c**) SAR, (**d**) RSC, (**e**) PI, and (**f**) CAI values

category and occupy most of the study area, and 16 wells (35.6%) fell within the unsuitable category and are represented mostly in the southwestern part of the area (Fig. 18b). Most of the samples of the intermediate zone fall within the suitable KI category (35 wells, 77.8%), while ten wells (22.2%) are unsuitable and located in the northern parts (Fig. 18c). As shown in Fig. 18d, the calculated CR values reveal that most of the intermediate samples fall within corrosive category (34 wells, 75.6%). Eleven samples (24.4%) are of noncorrosive properties and located in the central part of the area and other few locations.

5.2.3 Deep Zone

Parameters used for examining of the deep groundwater suitability for irrigation were listed in Table 2 and spatially distributed in Figs. 19 and 20. Based on TDS values, all samples are suitable for irrigation (TDS < 3,000 mg/l) (Fig. 19a). In respect to Na%, 44 wells (65.7%) fall within the category of good and occupy most of the area, and 15 wells (22.4%) are of permissible values and located in the southeastern and the northeastern directions. Few samples fall in excellent category (2 wells, 2.9%), three wells (4.5%) are in doubtful category, and three wells (4.5%) are unsuitable (Fig. 19b). The calculated values of SAR reveal that most of the samples fall within excellent category (65 wells, 97%), and the rest of the samples (2 wells, 3%) fall into the good category (Fig. 19c). Calculated RSC reveal that 63 wells (94%) fall into good category and 4 wells (6%) located in the northeastern part fall within categories of doubtful and unsuitable (Fig. 19d). Class II is the predominant permeability index

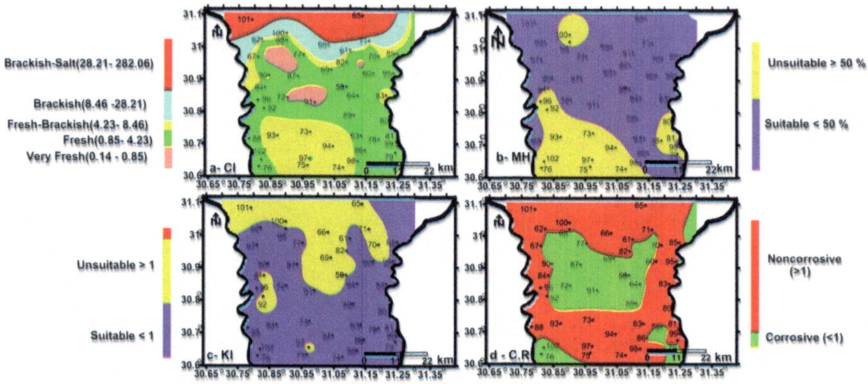

Fig. 18 Spatial distribution maps of the intermediate groundwater suitability for irrigation based on (**a**) Cl, (**b**) MH, (**c**) KI, and (**d**) CR values

category (58 wells, 86.6%), while 9 wells (13.4%) located in the eastern part fall within class I (Fig. 19e). The calculated CAI has no definite trend where 32 wells (47.8%) are under cation–anion exchange reaction category, and 35 wells (52.2%) fall within base-exchange reaction category (Fig. 19f). Chloride concentrations show that 4 wells (6%) and 46 wells (56.7%), respectively, fall within very fresh and fresh category and occupy most of the study area. Seventeen wells (25.4%) have fresh–brackish characters, and eight wells (11.9%) are brackish and located mostly in the northeastern direction (Fig. 20a). Calculated magnesium hazard (MH) values reveal that 60 wells (89.55%) are under suitable category and occupy most of the study area, and 7 wells (10.45%) are unsuitable and mostly located in the northwestern direction (Fig. 20b). KI reveal that most of the deep samples fall within suitable category (56 wells, 83.6%), and 11 wells (16.4%) are unsuitable and characterize small areas in the south and northeastern parts (Fig. 20c). The calculated CR has no definite trend, and most of the samples fall within corrosive category (45 wells, 67.2%), while 22 samples (32.8%) are characterized by noncorrosive properties (Fig. 20d).

The binary relationships between salinity hazards and SAR (Fig. 21) and total ion concentrations and Na% (Fig. 22) reveal that most of the shallow, intermediate, and deep groundwater samples fall within ranges from good to permissible. Some samples are of the unsuitable category. It is obvious that the quality of groundwater for irrigation in the southern and middle parts is good to permissible, but the groundwater in the northern parts is of doubtful to unsuitable characters.

5.2.4 Irrigation Water Quality (IQW) Index

Irrigation water quality index of the shallow, intermediate, and deep groundwater zones (Table 8, Fig. 23a–c, respectively) has different patterns. IQW of the shallow zone (Fig. 23a) shows that 4 wells (7.3%) fell within poor category and are located in

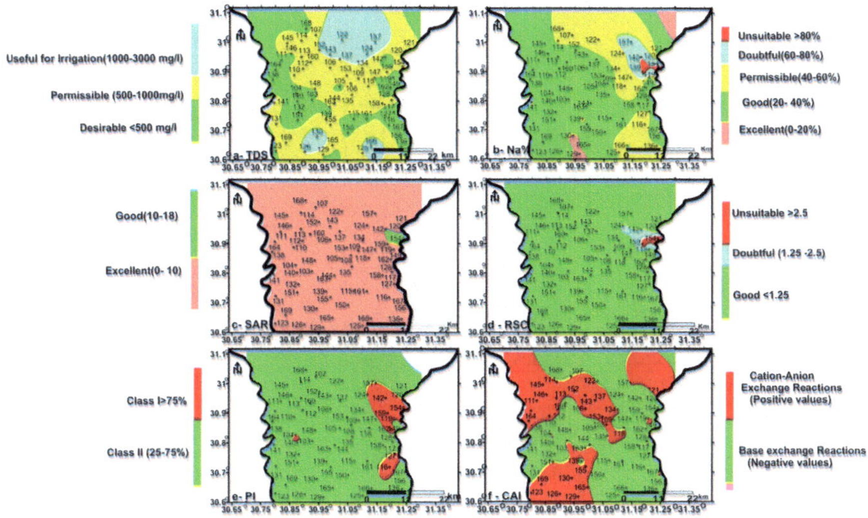

Fig. 19 Spatial distribution maps of the deep groundwater suitability for irrigation based on (**a**) TDS, (**b**) Na%, (**c**) SAR, (**d**) RSC, (**e**) PI, and (**f**) CAI values

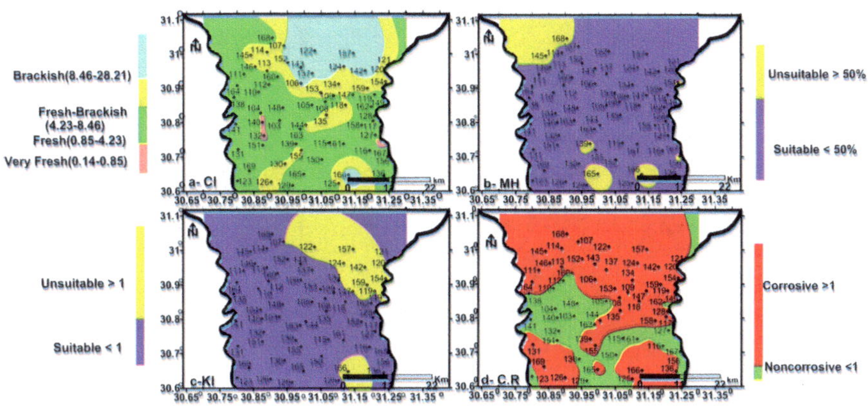

Fig. 20 Spatial distribution maps of the deep groundwater suitability for irrigation based on (**a**) Cl, (**b**) MH, (**c**) KI, and (**d**) CR values

the northeastern and northwestern parts, and 15 wells (27.2%) are of good water quality. The medium water quality (36 wells, 65.5%) is the predominant category and occupies most of the study area. The intermediate groundwater (Fig. 23b) is predominantly of good (13 wells, 29%) to medium (29 wells, 64%) water quality and changes to low quality to the northern part (3 wells, 7%). The deep zone (Fig. 23c), in general, looks similar to the intermediate zone with medium (34 wells, 51%) to good (32 wells, 48%) water quality changed to low quality (1 wells, 1%) in the northern part.

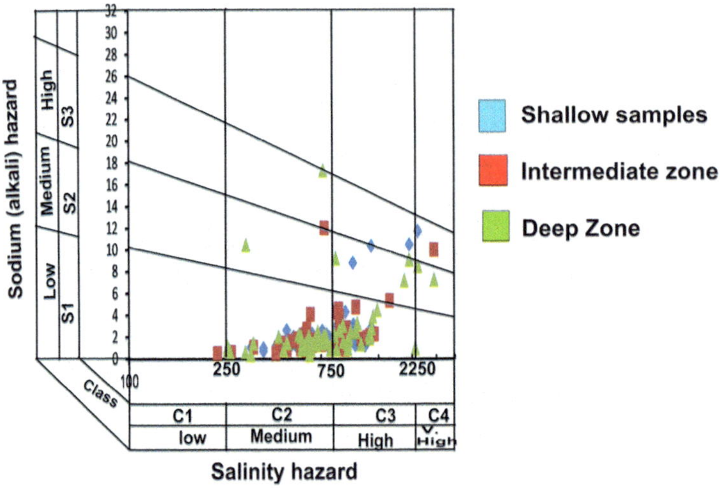

Fig. 21 Classification of the collected groundwater samples according to salinity hazards and SAR values

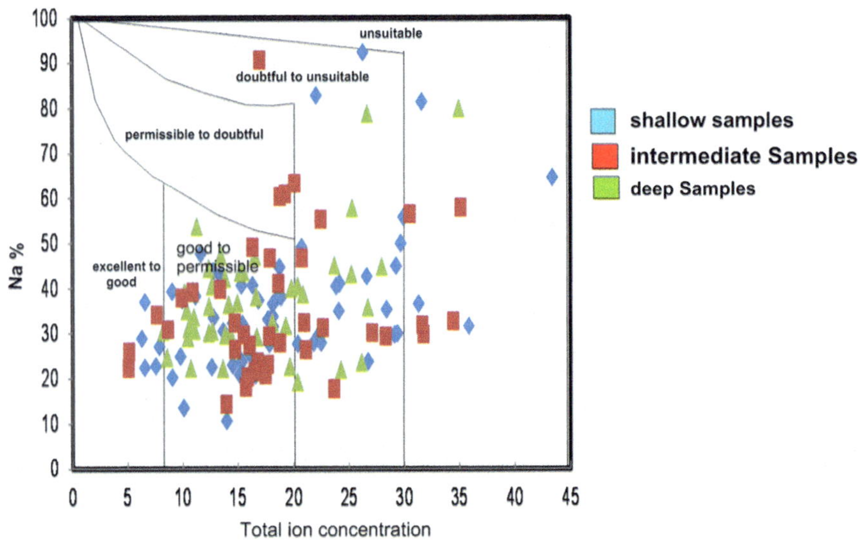

Fig. 22 Wilcox diagrams for groundwater classification for irrigation use

The correlation matrix (Table 7) shows the relationships between IQW and different water quality parameters. In the shallow groundwater, EC, SAR, Cl, and HCO_3 are the most effective parameters in IQW where the correlation coefficients range from 0.65 to 0.91. Weak correlation values between IWQ and NO_3, pH, Fe, Mn, Zn, and Cu are noticed. As the soil hydrogeochemical processes decrease in the

Table 8 Classification of groundwater according to its suitability for irrigation

Type of water	Range	Shallow zone		Intermediate zone		Deep zone	
		No. of wells	% of wells	No. of wells	% of wells	No. of wells	% of wells
Low water quality	<22	4	7.3	3	7	1	1
Medium water quality	22–37	36	65.5	29	64	34	51
High water quality	>37	15	27.2	13	29	32	48

intermediate and deep groundwater, the effect of the major parameters like EC, SAR, and Cl on the IQW decreases downward. In contrast to the shallow groundwater, correlation coefficient values of EC, Cl, and HCO_3 with IQW decrease downward reaching weak class in the deep groundwater. On the other hand, the effect of the minor elements on the irrigation water quality index increases downward reaching 0.3, 0.45, 0.35, 0.56, and 0.59 for NO_3, Fe, Mn, Zn, and Cu, respectively. This could be related to the predominance of the water–rock interaction in the intermediate and deep groundwater zones.

6 Conclusions

Based on WHO [46, 47] guidelines for drinking water, Na, K, Mg, Ca, SO_4, Cl, Cu, and NO_3 concentrations are lower than the desirable limits in around 80% or more of the collected samples. Mostly, the samples have undesirable concentrations of these ions which are located in the northern part of the study area. Around 68–75% of the samples have HCO_3 desirable concentration. TDS, iron, and manganese are the most effective individual factors for quality evaluation of this groundwater. 65%, 42%, and 58% of the samples have TDS higher than 500 mg/l for the three zones, respectively. Regarding iron concentrations, 61%, 64%, and 37% of the wells have undesirable concentrations. Around half of the collected water samples have undesirable manganese concentrations. Ninety-eight percent of the intermediate groundwater samples have desirable TH concentrations, while around 67% and 61% are recorded for the shallow and deep zones, respectively.

According to WQI, most part of the area in the shallow zone is characterized by poor quality (26 wells, 47.3%), and 18 wells (32.7%) fall in good categories. The northern and southern parts of the area show very poor (3 wells, 5.45%) to unfit (8 wells, 14.55%) groundwater. The intermediate zone has excellent (4 wells, 8.9%) to good (24 wells, 53.3%) water quality in the central part of the area. While in the southern part, it has poor quality (13 wells, 29%), and in the northern part, it has very poor (2 wells, 4.4%) to unfit (2 wells, 4.4%) characters. In the deep zone, the poor (27 wells, 40%) water quality class covers most of the study area. The latter class changed into very poor (5 wells, 7%) to unfit (3 wells, 4%) in the southern

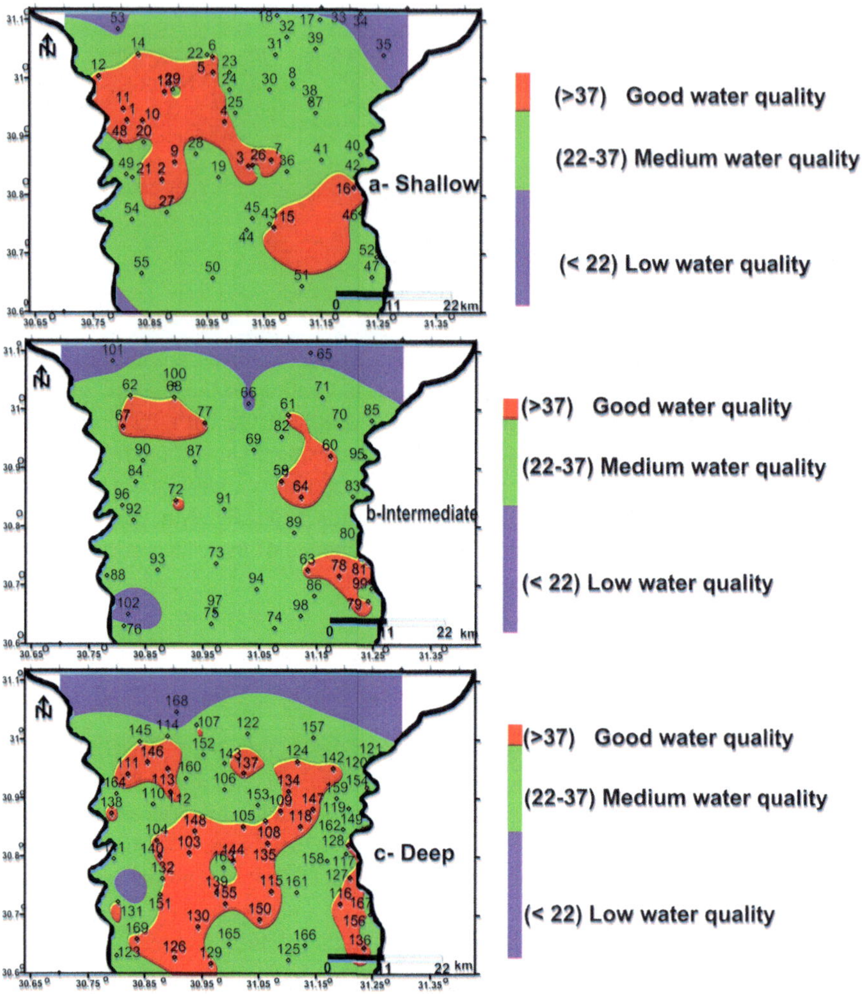

Fig. 23 Spatial distribution maps of the groundwater suitability according to the calculated irrigation water quality (IWQ) index in the (**a**) shallow, (**b**) intermediate, and (**c**) deep zones

direction and excellent (2 wells, 3%) to good quality (30 wells, 45%) to the eastern and western directions.

Ten individual parameters were utilized to assess the quality of studied groundwater and its reasonableness for irrigation. These parameters are TDS, SAR, Na%, Cl, RSC, PI, KI, CAI, C.R, and MH. Groundwater wells were ordered concurring TDS as around 90, 99, and 100% of wells for shallow, intermediate, and deep zone separately have TDS under 3,000 mg/l. Subsequently groundwater is reasonable for irrigation. Concerning Na%, the vast majority of the groundwater samples fall into the classification of good to permissible. The computed estimation of SAR varies from 0.83 to 21.37 for the shallow zone, 0.57 to 12.1 for the intermediate zone, and

0.7 to 17.38 for the deep zone and has been classified as about 82% for shallow, 92% for intermediate, and 100% for deep zones of groundwater wells are suitable for irrigation. Most samples have good RSC category, class II (76–86%, PI), fresh and fresh–brackish characters (Cl), and suitable KI and MH. Samples that have base-exchange and cation-exchange characters are mostly present with equal percentages in the three zones. A considerable number of samples are of corrosive properties, 69%, 75%, and 67%, respectively, for the shallow, deep, and intermediate zones.

Few samples located in the northern part of the area have IWQ values less than 22 and are considered to be of poor irrigation quality. Groundwater of such regions could decrease soil quality and led to yield misfortune. As a general guideline, water pumping from such regions must be stopped. It was estimated that the quality of the groundwater in the studied aquifer is fairly accepted and in the most parts of the study area is reasonable for irrigation. Such spatial distribution of the factors controlling the studied groundwater quality for drinking and irrigation in the three zones is a reasonable tool to be used in future domestic and rural administration plans and in deciding the most appropriate site for drilling wells and for evaluating the general groundwater suitability.

7 Recommendations

The Nile Delta groundwater aquifer is influenced by many environmental factors like human activity on the surface, sea water intrusion, and the type of aquifer sediments. Therefore, the decision-maker must take into consideration the lateral and vertical changes in groundwater quality when digging wells for different uses in order to achieve the highest utilization of wells. Microbial contamination should also be studied in spatial form to have a complete image about the Nile Delta aquifer. The researchers should not study this aquifer as a single homogenous body, but they should consider the vertical and horizontal change of groundwater chemistry. To complete this topic, the authors will conduct a hydrogeochemical study in the three dimensions of the aquifer.

Acknowledgments The authors are grateful to Tanta University for the financial support offered by the project number "TU-01-12-03" during the course of this research work. The authors thank the editor Prof. Dr. Abdelazim Negm for his constructive remarks.

References

1. Singh CK, Shashtri S, Mukherjee S, Kumari R, Avatar R, Singh A, Singh RP (2011) Application of GWQI to assess effect of land use change on groundwater quality in lower shiwaliks of Punjab: remote sensing and GIS based approach. Water Resour Manag 25:1881–1898. https://doi.org/10.1007/s11269-011-9779-0

2. Subramaniam T, Elango L, Damodarasamy SR (2005) Groundwater quality and its suitability for drinking and agricultural use Chithar River basin, Tamil Nadu, India. Environ Geol 47:1099–1110
3. Shabbir R, Ahmad SS (2015) Use of geographic information system and water quality index to assess groundwater quality in Rawalpindi and Islamabad. Arab J Sci Eng 40:2033. https://doi.org/10.1007/s13369-015-1697-7
4. Toumi N, Hussein BHM, Rafrafi S, El Kassas N (2015) Groundwater quality and hydrochemical properties of Al-Ula region, Saudi Arabia. Environ Monit Assess 187:84–100. https://doi.org/10.1007/s10661-014-4241-4
5. UNEP (United Nations Environment Program) (1999) Global environment outlook 2000. Earthscan, London
6. Vincy V, Brilliant R, Pradeepkumar AP (2015) Hydrochemical characterization and quality assessment of groundwater for drinking and irrigation purposes: a case study of Meenachil River basin. Western Ghats, Kerala, India. Environ Monit Assess 187:4217. https://doi.org/10.1007/s10661-014-4217-4
7. Bruce BW, Becker MF, Pope LM, Gurdak JJ (2003) Ground-water quality beneath irrigated agriculture in the central high plains aquifer, 1999–2000. U.S. Geological Survey Water-Resources Investigations Report 03–4219, U.S. Geological Survey, Reston, VA, p 39
8. Zhang W, Kinniburgh D, Gabos S (2013) Assessment of groundwater quality in Alberta, Canada Using GIS Mapping. Third international conference on medical, biological and pharmaceutical sciences (ICMBPS'2013), Bali, pp199–203
9. Arumaikkani GS, Chelliah S, Gopalan M (2017) Mapping the spatial distributions of water quality and their interpolation with land use/land cover using GIS and remote sensing in Noyyal River basin, Tamil Nadu, India. J Geosci Environ Protection 5:211–220
10. Salem ZE (2009) Natural and human impacts on the groundwater under an Egyptian village, central Nile Delta e a case study of Mehallet Menouf. In: 13th International Water Technology Conference (IWTC), 12–15 March 2009, Hurghada, 3, pp 1397–1414
11. Elewa HH (2010) Potentialities of water resources pollution of the Nile River Delta, Egypt. Open Hydrol J 4:1–13
12. Gemail K, El-Shishtawy AM, El-Alfy M, Ghoneim MF, El-Bary MHA (2011) Assessment of aquifer vulnerability to industrial waste water using resistivity measurements. A case study, along El-Gharbyia main drain, Nile Delta, Egypt. J Appl Geophys 75:140–150
13. Morsy W, El-Fakharany Z (2012) Predicting the impact of surface wastewater on groundwater quality in quesna industrial area. J Am Sci 8:772–781
14. Salema MG, El-Awady MH, Amine E (2012) Enhanced removal of dissolved iron and manganese from nonconventional water resources in Delta District, Egypt. Energy Procedia 18:983–993
15. Ghoraba SM, Zyedan BA, Rashwan IMH (2013) Solute transport modeling of the groundwater for quaternary aquifer quality management in Middle Delta, Egypt. Alex Eng J 52:197–207
16. Khalil MA, Salem ZE, Gheda SF, El-Sheekh MM (2013) Quality assessment of drinking water in Tanta City, Egypt. J Environ Sci Eng B 2:257–275
17. El Bedawy R (2014) Water resources management: alarming crisis for Egypt. J Manag Sustain 4:108–124
18. Bennett PC, El Shishtawy AM, JM S, Atwia MG (2014) Source and migration of dissolved manganese in the central Nile Delta aquifer, Egypt. J Afr Earth Sci 96:8–20
19. Elkafoury A, Dawoud W, Negm A, Bady M, Aly MH (2014) Integrated framework for evaluating the impact of urban transportation gaseous emissions on groundwater quality. Int Water Technol J 4:114–124
20. Fattah MK, Ragab EG (2014) Assessment of groundwater vulnerability to pollution in the southern part of Nile Delta, Egypt. Standard Sci Res Essays 2:725–738
21. El-Kowrany SI, El-Zamarany EA, El-Nouby KA, El-Mehy DA, Ali EA, Othman AA, Salah W, El-Ebiary AA (2016) Water pollution in the middle Nile Delta, Egypt: an environmental study. J Adv Res 7:781–794

22. Negm AM, Armanuos AM (2016) GIS-based spatial distribution of groundwater quality in the western Nile Delta, Egypt. In: Negm AM (ed) The Nile Delta. The handbook of environmental chemistry. Springer, Cham. https://doi.org/10.1007/698_2016_66
23. Negm AM, Eltarabily MGA (2016) Modeling of fertilizer transport through soil, case study: Nile Delta. In: Negm AM (ed) The Nile Delta. The handbook of environmental chemistry. Springer, Cham. https://doi.org/10.1007/698_2016_66
24. Salem ZE, Al Temamy AM, Salah MK, Kassab M (2016) Origin and characteristics of brackish groundwater in Abu Madi coastal area, Northern Nile Delta, Egypt. Estuar Coast Shelf Sci 178:21–35
25. Sharaky AM, El Hasanein AS, Atta SA, Khallaf KM (2016) Nile and groundwater interaction in the western Nile Delta, Egypt. In: Negm AM (ed) The Nile Delta. The handbook of environmental chemistry. Springer, Cham. https://doi.org/10.1007/698_2016_66
26. Salem ZE, Osman OM (2017) Use of major ions to evaluate the hydrogeochemistry of groundwater influenced by reclamation and seawater intrusion, West Nile Delta, Egypt. Environ Sci Pollut Res 24:3675–3704
27. Negm AM, Saavedra O, El-Adawy A (2016) Nile Delta biography: challenges and opportunities. In: Negm A (ed) The Nile Delta. The handbook of environmental chemistry, vol 55. Springer, Cham
28. Dawoud M (2004) Design of national groundwater quality monitoring network in Egypt. J Environ Monit Assess 96:99–118
29. Mabrouk MB, Jonoski A, Solomatine D, Uhlenbrook S (2013) A review of seawater intrusion in the Nile Delta groundwater system – the basis for assessing impacts due to climate changes and water resources development. Hydrol Earth Syst Sci Discuss 10:10873–10911
30. Barrocu G, Dahab K (2010) Changing climate and saltwater intrusion in the Nile Delta, Egypt. In: Taniguchi M, Holman IP (eds) Groundwater response to changing climate. CRC Press, Boca Raton, pp 11–25. https://doi.org/10.1201/b10530-3
31. Kashef AI (1983) Salt water intrusion in the Nile Delta. Ground Water 21:160–167
32. Mikhailova M (2001) Hydrological regime of the Nile Delta and dynamics of its coastline. Water Resour 28:477–490
33. Sefelnasr A, Sherif MM (2014) Impacts of seawater rise on seawater intrusion in the Nile Delta aquifer, Egypt. Groundwater 52:264–276
34. Sherif MM, Sefelnasr A, Javadi A (2012) Incorporating the concept of equivalent freshwater head in successive horizontal simulations of seawater intrusion in the Nile Delta aquifer. Egypt. J Hydrol 464:465
35. Ball J (1939) Contribution to the Geography of Egypt, Survey, Cairo, pp 23–84
36. RIGW (1992) Hydrogeological map of Nile Delta, scale 1:500,000, 1st edn. Nile Delta
37. RIGW (2002) Nile Delta groundwater modeling report. Research Institute for Groundwater, Kanater El- Khairia, Egypt
38. Morsy S (2009) Environmental management to groundwater resources for Nile Delta region. PhD thesis, Faculty of Engineering, Cairo University, Egypt
39. Farid M (1980) Nile Delta groundwater study. MSc thesis, Cairo University, Egypt
40. Wilson J, Townley R, SaDa Costa A (1979) Mathematical development and verification of a finite element aquifer flow model AQUIFEM-1. Technology adaptation program, Report No. 79–2, Cambridge, MA
41. Diab S, Dahab K, El Fakharany M (1997) Impacts of the paleohydrological conditions on the groundwater quality in the northern part of Nile Delta, the geological society of Egypt. J Geol 4112B:779–795
42. Saleh F (1980) Some hydrological and hydrochemical studies on the Nile Delta. MSc. thesis, Faculty of Science, Ain Shams University, Egypt
43. Abdel Maged H (1994) Water logging phenomena in the north of the delta region. MSc thesis, Faculty of Science, Cairo University, Egypt
44. Salem ZE, El-horiny MM (2014) Hydrogeochemical evaluation of calcareous eolianite aquifer with saline soil in a semiarid area. Environ Sci Pollut Res 21:8294–8314

45. HACH (1990) Chemical procedures explained. Hach Technical Center for Applied Analytical Chemistry, Colorado
46. WHO (World Health Organization) (1996) Guidelines for drinking-water quality, 2nd edn
47. WHO (World Health Organization) (2011) Guidelines for drinking-water quality, 4th edn
48. Abbasi T, Abbasi A (2012) Water quality indices. Elsevier, Amsterdam, pp 19–28
49. Armanuos AM, Negm A, Valeriano OCS (2015) Groundwater quality investigation using water quality index and ArcGIS: case study: western Nile Delta aquifer, Egypt. Proceedings of the 18th water technology conference (IWTC18), Sharm ElSheikh, 12–14 March
50. Ayers S, Westcost W (1985) Water quality for agriculture, FAO irrigation and drainage paper. Food Agriculture Organization, Rome, p 29
51. Simsek C, Gunduz O (2007) IWQ index: a GIS-integrated technique to assess irrigation water quality. Environ Monit Assess 128:277–300
52. Spandana MP, Suresh KR, Prathima B (2013) Developing an irrigation water quality index for Vrishabavathi command area. Int J Eng Res Technol 2:821–830
53. Srinivas Y, Oliver DH, Raj AS, Chandrasekar N (2013) Evaluation of groundwater quality in and around Nagercoil town, Tamilnadu, India: an integrated geochemical and GIS approach. Appl Water Sci 3:631. https://doi.org/10.1007/s13201-013-0109-y
54. Ramesh K, Seetha K (2013) Hydrochemical analysis of surface water and groundwater in tannery belt in and around Ranipet, Vellore district, Tamil Nadu, India. Int J Res Chem Environ 3:36–47
55. Brindha K, Neena Vaman KV, Srinivasan K, Sathis Babu M, Elango L (2014) Identification of surface water-groundwater interaction by hydrogeochemical indicators and assessing its suitability for drinking and irrigational purposes in Chennai, southern India. Appl Water Sci 4:159. https://doi.org/10.1007/s13201-013-0138-6
56. Tiwari N, Manzoor A (1988) River pollution in Kathmandu valley (Nepal), suitability of river water for irrigation. Indian J Environ Protection 8:269–274
57. Wilcox V (1948) The quality of water for irrigation use. US Department of Agriculture, Technical Bulletin, Washington
58. Richards A (1954) Diagnosis and improvement of saline and alkali soils. USDA Hand Book
59. Doneen D (1954) Salination of soil by salts in the irrigation water. Am Geophys Union Trans 35:943–950
60. Kelly P (1940) Permissible composition and concentration of irrigation waters. Proc Am Soc Civ Eng 66:607–613
61. Paliwal V (1967) Effect of gypsum application on the quality of irrigation waters. Madras Agric J 59:646–647
62. Szabolcs I, Darab C (1964) The influence of irrigation water of high sodium carbonate content on soil. In: Szabolics I (ed) Proceedings of the 8th international congress soil science, sodics soils. Res Inst Soil Sci, Agric Chem, Hungarian Acad Sci, ISSS Trans II, pp 802–812
63. Schoeller H (1967) Geochemistry of ground water. An international guide for research and practice, vol 15. UNESCO, Paris, pp 1–18
64. Jain K, Kumar P, Sharma K (2003) Ground water qualities of ghataprabha command area Karnataka. Indian J Environ Ecoplann 7:251–262

Part X
Groundwater Management for Sustainability

Groundwater Management for Sustainable Development East of the Nile Delta Aquifer

Mohamed Galal A. Eltarabily and Abdelazim M. Negm

Abstract Due to the progressive increase in land reclamation projects in Egypt, especially east on Nile Delta, efficient water resources management plans and accurate land cover maps are highly vital. In this chapter, the hydrogeological characteristics of the Quaternary groundwater aquifer east of the Nile Delta are presented. This description includes the geomorphology, lithostratigraphy, geological Structure, surface hydrogeology, and the recharge and discharge of the aquifer. The change in land use maps between 1990 and 2004 shows a significant increase in land reclamation which is critical for informing efficient and sustainable policies for groundwater management. MODFLOW software was used to model groundwater flow in three dimensions based on integration with ArcGIS dataset. The stratigraphy of the aquifer was mapped using the solids approach. The groundwater head distribution was calculated in 1990 and between 1990 and 2004 after model calibration for steady and transient states, respectively. The land use for the year 2004 was used to run the predictive transient model to simulate the groundwater flow and budget analysis for the upcoming 8 years (until 2025). Budget analysis results showed that Ismailia Canal and Damietta Branch are the primary recharge components of the aquifer, while the production wells are the main discharge element. Different strategies for well operation can be implemented to control the withdrawal based on the storage capacity and aquifer yield as well as the rising water table in the low elevated lands.

M.G.A. Eltarabily (✉)
Civil Engineering Department, Faculty of Engineering, Port Said University, Port Said 42523, Egypt
e-mail: eng_mohamad1987@hotmail.com; eng_m_trabily@eng.psu.edu.eg

A.M. Negm
Water and Water Structures Engineering Department, Faculty of Engineering, Zagazig University, Zagazig 44519, Egypt
e-mail: amnegm85@yahoo.com; amnegm@zu.edu.eg

A. M. Negm (ed.), *Groundwater in the Nile Delta*,
Hdb Env Chem (2019) 73: 687–708, DOI 10.1007/698_2017_102,

Keywords Groundwater modelling, Land use/land cover, Nile Delta, Quaternary aquifer, Sustainable development

Contents

Abbreviations and Symbols

ω	Volumetric flux
ENE	East-northeast
FAO	Food and Agriculture Organization of the United Nations
K_{xx}	Hydraulic conductivity along x axis
K_{yy}	Hydraulic conductivity along y axis
K_{zz}	Hydraulic conductivity along z axis
LULC	Land use/land cover maps
MSL	Mean sea level
S_s	Specific storage
T	Transmissivity

1 Introduction

Groundwater is the second most abundant traditional water resource, after the Nile River. Recently groundwater development and management have become crucial. The continuous population growth in Egypt leads to increasing demand for groundwater since the surface water from the River Nile is not sufficient to meet demands. Due to this increase, successive governments have encouraged agricultural development expansion projects. When initiating such major agricultural projects, the great challenge is to maintain sustainable water resources. Thus, the demand for accurate land cover change information is increasing. In the East Nile Delta, the land reclamation projects have drawn attention in the last few decades. These projects require particular effort to eliminate or at least reduce the negative hydrogeological impacts of large-scale reclamations such as water logging, drawdown/rising water table, and soil salinity. Based on the land use/land cover maps

(LULC) for different years during the time frame of the agricultural expansion projects, the accurate quantifying of the changes must be detected, and spatial patterns should be delineated. Such information is critical for making efficient and sustainable policy decisions for resource management. After this, the three-dimensional groundwater flow model will be developed to outline the future framework for groundwater management in the East Nile Delta where a continuous rise in head associated with agricultural development is reported.

2 Groundwater Aquifer East Nile Delta

The East Nile Delta region is enclosed between latitudes 30° 00′ and 31° 30′ N and longitudes 31° 00′ and 32° 30′ E, approximately. It covers an area of about 14,000 km^2. It is bounded on the north by El Manzala Lake, to the east by Suez Canal, to the south by Ismailia Canal, and the west by Damietta Branch (Fig. 1). The region slopes towards the north to east-northeast (ENE). A dense network of irrigation canals and drains constitute the traditionally cultivated land. Within the study area, a subregion is distinctly characterized by greater agricultural development and shows different conditions within the Quaternary aquifer.

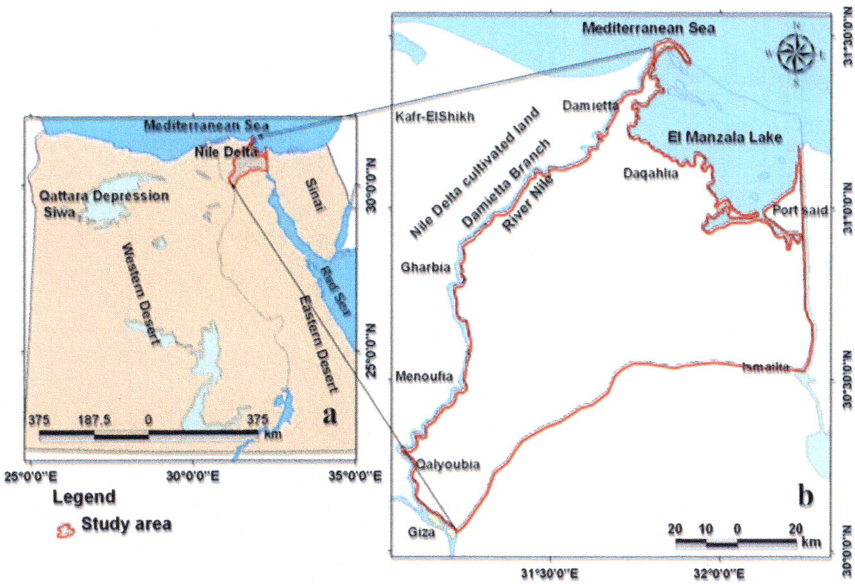

Fig. 1 Study area location map of East Nile Delta

2.1 Physical Setting

2.1.1 Geomorphology

The topography of the study area decreases in elevation northwards with relatively low topography with gentle slope to the north of Ismailia Canal, whereas the southern provinces have high features with steeper slopes related to intensive geological structure. Physical and chemical weathering substantially modified the morphology of the land surfaces, especially the physical weathering due to the prevailing arid conditions. Some geomorphologic studies [1–3] have been conducted, particularly those related to groundwater development where discharge drainages, collection basins, and watersheds have been investigated. In light of these studies, eight physiographic provinces can be recognized in the study area (Fig. 2).

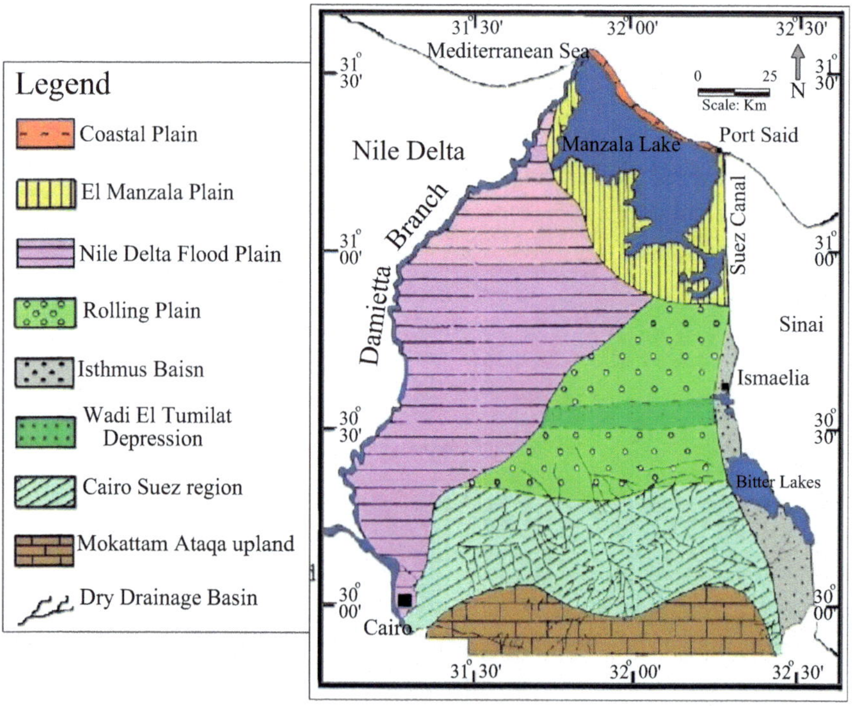

Fig. 2 Physiographic projection, modified after Shatta et al. [3]

2.1.2 Lithostratigraphy

The study area is covered with Tertiary and Quaternary sediments with sporadic mid-Tertiary basalt flows. The deposits form a vast succession in the subsurface (~5,000 m) with several formations having excellent hydraulic properties that act as groundwater aquifers. The lithostratigraphic succession in the study area can be compiled as shown in Fig. 3. In discussing lithostratigraphic section, particular emphasis shall be given to the water-bearing units of the Quaternary sediments. Quaternary deposits have a great extension over the land surface and unconformably overlie the Tertiary units. The Pleistocene deposits act as the main aquifer in the study area and directly rest on the Pliocene clay. The Pleistocene deposits can be subdivided into two units: the Early and Late Pleistocene sediments. The Holocene deposits consist of coastal deposits, lagoonal clay and lake deposits (sabkha), and alluvial deposits. Alluvial deposits are present in the form of silty and sandy clay layers. The silty layer occupies the present flood plain of the Nile River and acts as a semi-confined stratum in the Nile Delta [5].

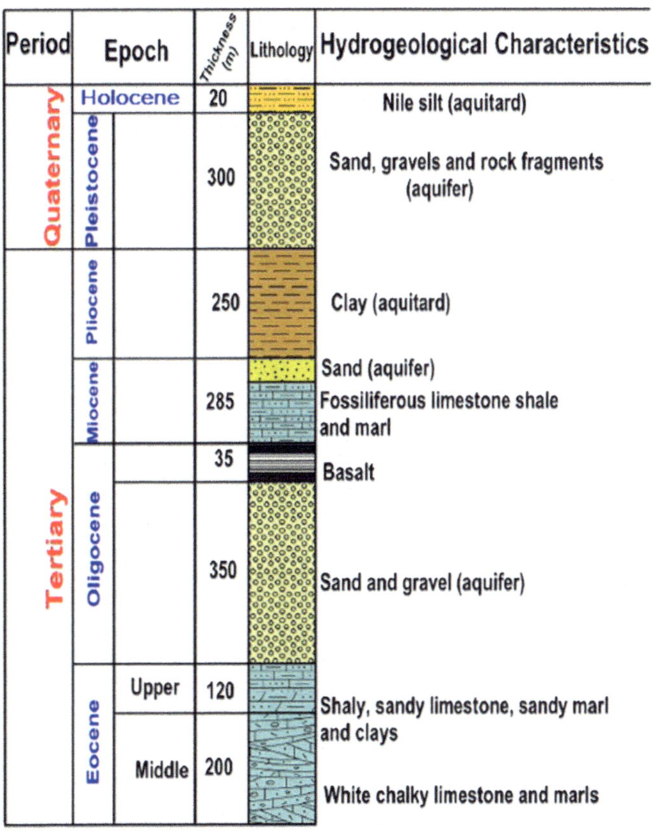

Fig. 3 Stratigraphic section of the area east of Nile Delta, modified after Schlumberger [4]

2.1.3 Geological Structure

Faults and folds represent the main features of the geological map of the study area. Faulting is noted in the topographic features of the southern portion of the study area, while folding may show local imprints. According to this tectonic-structural map (Fig. 4), the study area can be divided into two tectonic zones, the northern and the southern, separated by F1–F1′ and F2–F2′ faults. Unconformity surfaces may

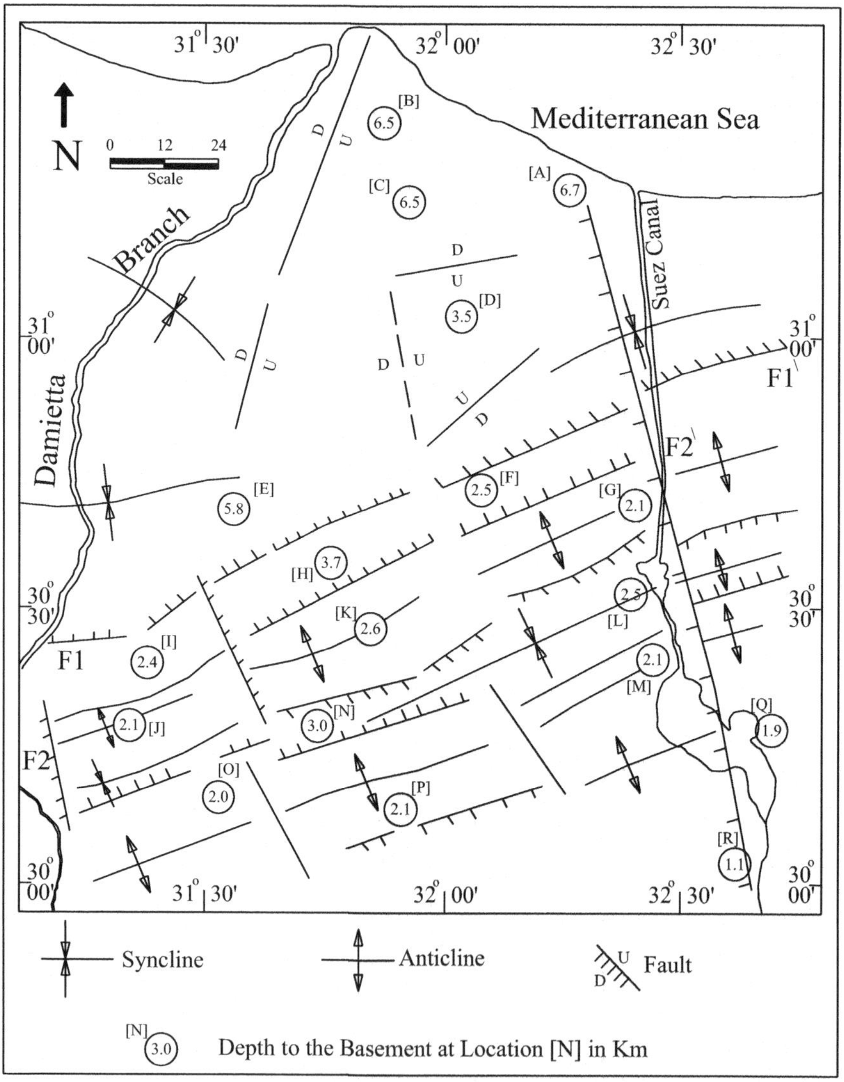

Fig. 4 Tectonic-structural map of the area east of Nile Delta, modified after El-Dairy [6]

influence the occurrence and movement of groundwater, especially when combined with dramatic changes in sedimentary facies of the adjacent Pleistocene sediments.

2.2 *Surface Hydrology*

Groundwater is strongly influenced by the surface water systems except for the perfectly confined parts of the Quaternary aquifer. The rate of recharge, flow direction, and groundwater heads are spatially and quantitively controlled by the distribution of the surface water system. Surface hydrology includes surface water system, seasonal precipitation, infiltration rates, and evaporation and transpiration.

2.2.1 Surface Water System

Surface water system includes all water bodies such as canals, drains, lakes, and tributaries existing in the study area. Suez Canal and El Manzala Lake are saltwater, and Damietta Branch and Ismailia Canal represent the main freshwater bodies. The dimensions and capacity of the main water bodies in the study area are shown in Table 1.

2.2.2 Seasonal Precipitation and Infiltration Rates

The study area falls within the arid climate zone, and the rainfall intensity is very low. Average annual precipitation is minimal in Cairo ~25 mm/year, but the maximum values are recorded in the coastal areas ~133 mm/year [8]. The rainfall increases from south to north, and its effect on groundwater recharge is limited except for the Ismailia region. Infiltration tests on the different surface sediments at various localities in the study area were performed by FAO [9], and the results are shown in Table 2.

Table 1 Main irrigation canals, drains, and lakes in the study area [7]

Water body	Total length	Bottom width (m)	Mean depth (m)	Range of surface water levels (m)
Damietta Branch	245 km	200:500	6	16.0:−1.0
Ismailia Canal	136 km	10:30	5	16.0:0.0
Bahr El Baqar Drain	85 km	100:200	5	–
Suez Canal	173 km	–	16	–
El Manzala Lake (area)	1,500 km^2	–	1.2	1.0:0.0

Table 2 Infiltration rates of the dominant sedimentary facies at different locations

Location	Infiltration rates (cm/h)			Sedimentary facies
	From	To	Average	
El Manzala	4	9.6	6.1	Silty clay, loam to clay
El Salheya	2.7	19.4	10.18	Coarse to very coarse sand
El Ismailia	1	23.8	9.6	Sandy clay
El Tumilat	3.6	20.8	12.65	Sand and clay

Table 3 Average values of the surface evaporation and evapotranspiration (mm/year) at different locations in the study area [7]

Location	Free surface evaporation	Evapotranspiration
El Mansoura	149	104
El Zagazig	145	102
Damietta	153	107
Port Said	180	126
Suez	200	140
Cairo	189	132
El Ismailia	175	123

2.2.3 Evaporation and Evapotranspiration

Evaporation and evapotranspiration have substantial impacts on groundwater, especially in regions that fall in the arid zone. Evaporation rates decrease towards the north and east with the low value recorded in El Mansoura (44 mm/year) and a high rate in Cairo (142 mm/year). Average values of the surface evaporation and evapotranspiration at different locations in the study area are presented in Table 3. The evapotranspiration increases southwards with a maximum value at Suez and high values at Port Said. This can be attributed to the wind speed and relatively high temperature in these locations. Due to the high evaporation rates and limited precipitation in the study area, the local rainfall has negligible replenishment to groundwater.

2.3 Hydrogeological Characteristics of the Quaternary Aquifer

The Quaternary aquifer is the most important water-bearing deposit (~900 m thick) of the Nile Delta. Two cross sections were constructed to reveal the subsurface Quaternary aquifer (Fig. 5), where it was classified based on the lithological facies variation. In the Nile Delta floodplain, the aquifer is confined; however, in the coastal plain and the northern part of the Rolling Plains, it is semi-confined [11]. The unconfined condition is common in the south of the Rolling Plains and

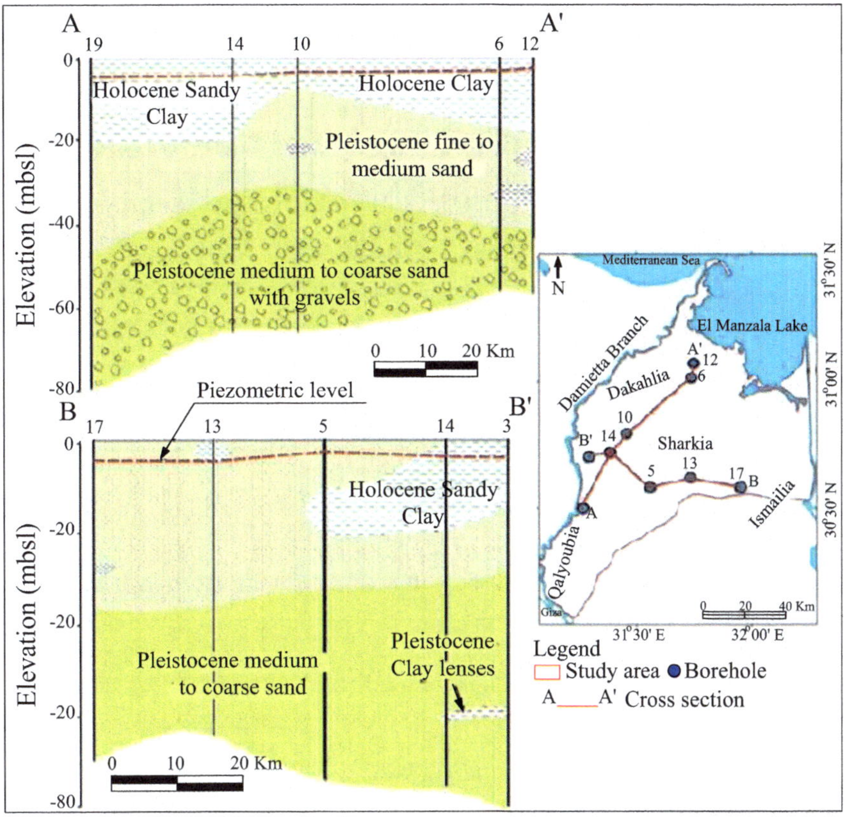

Fig. 5 Hydrogeological cross sections of the Quaternary aquifer modified after Attwa et al. [10]

extends parallel to the course of Ismailia Canal, but variable conditions are dominant in the aquifer at the Isthmus basin. The aquifer is underlain by Pliocene plastic clay, which acts as an aquiclude, particularly in the floodplain region and entirely isolates the base of the aquifer from the effects of the older formation. Towards the eastern and southern regions, the Quaternary aquifer directly overlies the fissured carbonate aquifers of Miocene and Oligocene age (Fig. 6). The saturated thickness of the Quaternary aquifer increases gradually in the north-west direction to reach ~900 m (Fig. 7).

2.3.1 Hydraulic Parameters

Hydraulic conductivity (K) and transmissivity (T) showed variation within the same sedimentary unit at various locations and depths. This difference is attributed to the significant change in the nature of the water-bearing formations laterally and vertically due to the clay layer interbedding [13]. Other factors include the

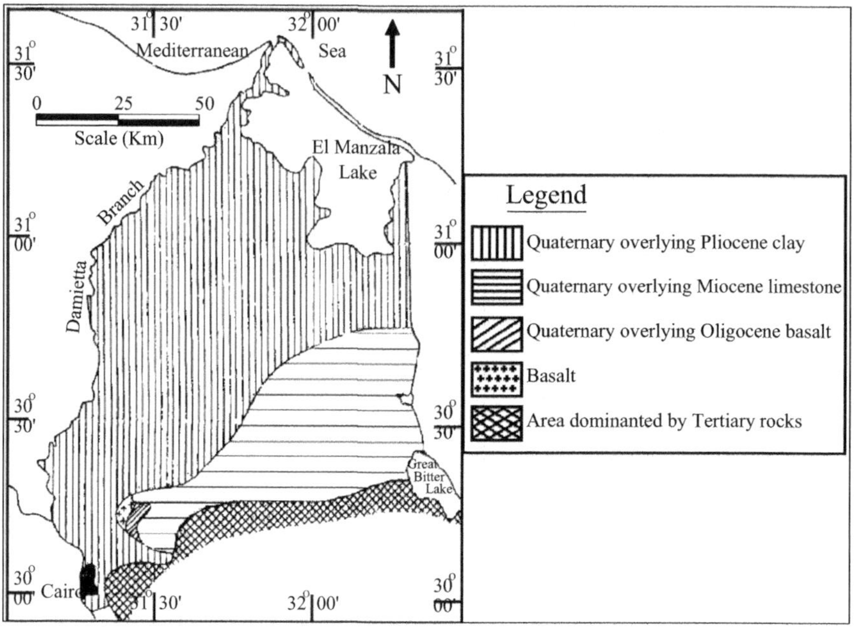

Fig. 6 Distribution and relationship between Quaternary aquifer and Tertiary rocks, modified after El-Fayoumy [12]

unaccounted leakage from deeper aquifers in contact with the Quaternary aquifer and partial penetration of pumping wells to the sedimentary layer. Therefore, it is extremely difficult to assign average values for these parameters in any direction. The hydraulic conductivity ranges from 10 to 76 m/day (Fig. 8). From the interpolated values, it could be concluded that the *K* values increase from east to west of the study area.

The southern part of the Quaternary aquifer is characterized by 200 m thickness, 20% porosity, 75 m/day horizontal conductivity (K_{hal}), 25 m/day vertical conductivity (K_{val}), and 10,000–20,000 m^2/day transmissivity [15]. The Quaternary aquifer can be described as highly productive. The change in *K* values is associated with the change in clay content, while the change in *T* values is attributed to the change in the aquifer thickness. Table 4 shows the average value of the estimated *K* and *T* values.

The average specific yield for different sedimentary of the Quaternary aquifer was estimated [13]. Nile silt, sandy clay, and clayey sand sediments show a gradual decrease in the specific yield values (12–0%) towards both El Manzala Lake and Damietta Branch. These changes were attributed to the increased clay bed thickness. Fine to medium sand deposits had specific yield values between 10 and 35% with lower values encountered where clay cap or fully saturated layers exist. Higher values are reported at Wadi El Tumilat where the clay cap is missing. Coarse sand

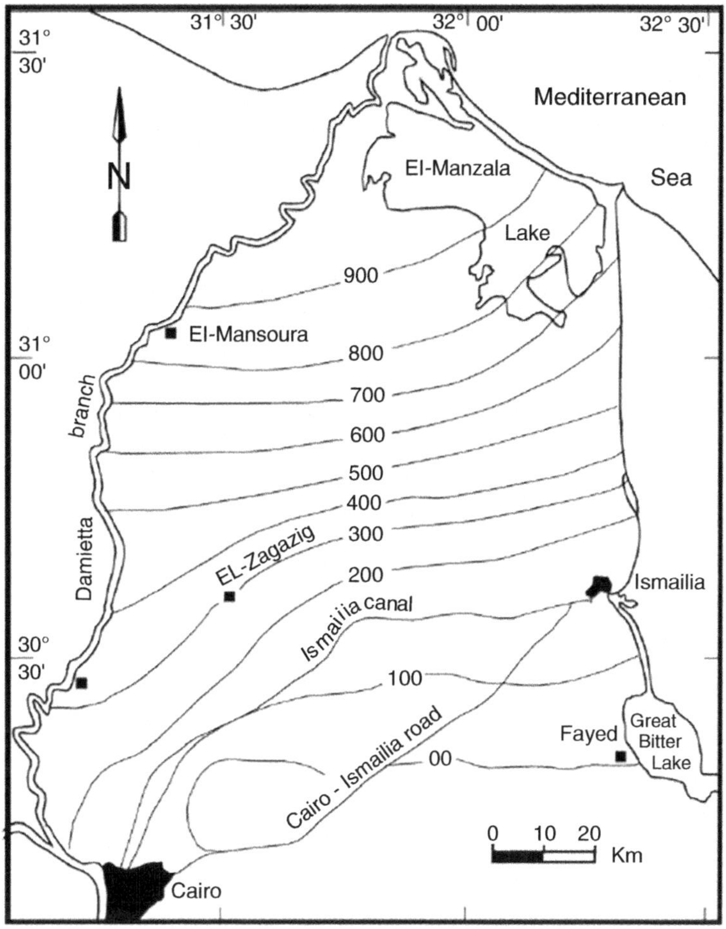

Fig. 7 Thickness (meters) of the Quaternary aquifer east of Nile Delta [7]

and gravel showed slight lateral and vertical changes in facies which enable relatively constant specific yield values (~35%) over the entire area. The storativity for these three formations is 2.3×10^{-3}, 1.9×10^{-3}, and 6.5×10^{-3}, respectively [16].

2.3.2 Groundwater Movement

Flow patterns can be investigated by the depth to water and water level maps (Fig. 9a, b). The majority of piezometers had a slight hydraulic head (<7 m), but the depth to water increased progressively to reach more than 35 m on the newly reclaimed land. Water levels ranged between 14.25 m above MSL near Cairo and

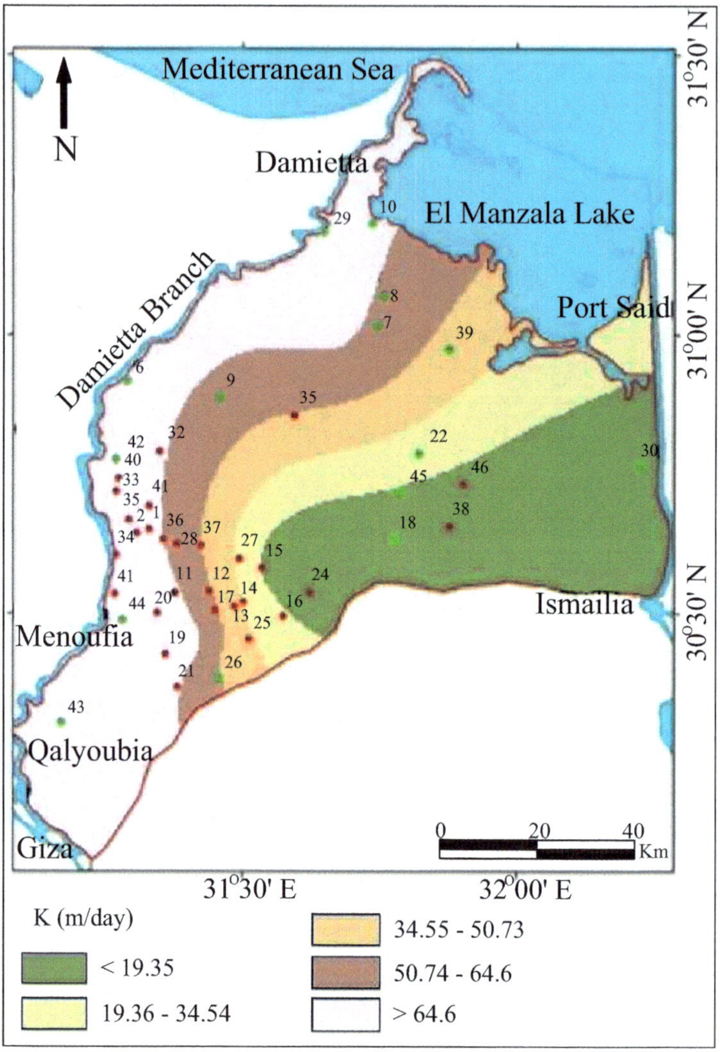

Fig. 8 Spatial distribution of hydraulic conductivity modified after Elewa et al. [14]

Table 4 Summary of the hydraulic parameters in some selected areas within the study region

Region	Aquifer thickness (m)	Transmissivity (m^2/day)	Hydraulic conductivity (m/day)
El Zagazig	100–500	10,000–20,000	40–100
Ismailia	200–400	10,000–20,000	25–40
Belbies	150–200	5,000	20–25

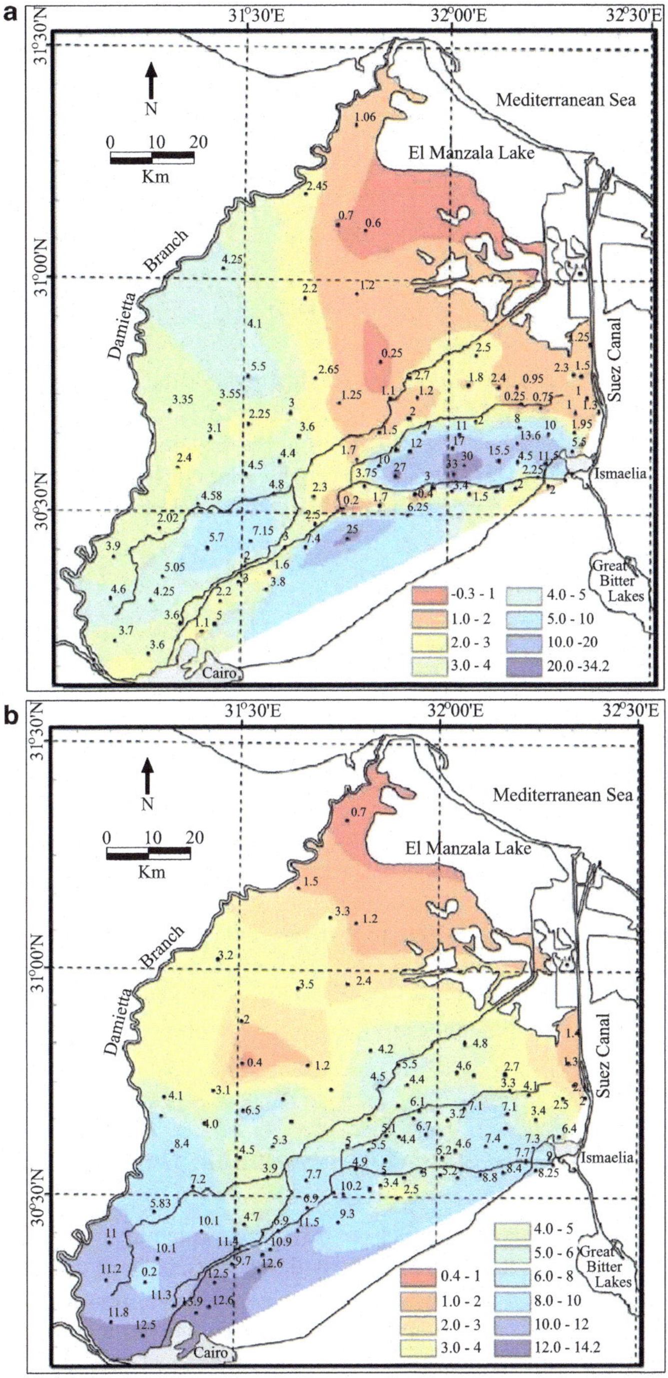

Fig. 9 (**a**) Depth to water table map; (**b**) groundwater levels distribution in the year 2005 after Ismael [17]

Table 5 Extraction of groundwater from the Quaternary aquifer in the study area after RIGW/IWACO [18]

Area	Governorate	Drinking water (million m^3/year)	Irrigation water (million m^3/year)	Total extraction (million m^3/year)
Flood plain	Cairo	0.7	126.8	127.5
	Qalubia	56.4	219	275.4
	El Sharqiya	70	197	267
	Ismailia	–	49.8	49.8
Desert fringe	El Sharqiya	310	7.5	317.5
	Ismailia	–	145	145
Total		**436.4**	**618.3**	**1,054.7**

<1 m near El Manzala Lake. Groundwater flows towards the northeast to ENE with marked local flow directions at the newly reclaimed lands. This perturbation in water level and groundwater flow directions at the reclaimed land is attributed to local changes in the water level as a result of the variations in agricultural irrigation practices, particularly the irrigation methods that include flooding, drip, and sprinkler.

2.3.3 Recharge and Discharge of the Aquifer

The Quaternary aquifer is recharged by the seepage of freshwater from the Damietta Branch and Ismailia Canal, the percolation of return irrigation water especially in the newly reclaimed lands with sandy soil, the infiltration of excess storage water in the overlying aquitard, and the inter-aquifer flow in particular between the Nile Delta aquifer and Miocene aquifer. Groundwater is discharged naturally through evaporation/evapotranspiration and the natural outflow to discharge sites (e.g. El Manzala Lake and northern coastal aquifers). It discharged artificially through agricultural tile drains and by withdrawal from wells. Table 5 shows the estimated extraction rates of groundwater from the Quaternary aquifer. Table 6 shows the groundwater balance in the Quaternary aquifer [19], in which the runoff and infiltration from agricultural lands represent a significant recharge component.

3 Groundwater Modelling for Sustainable Development

Most studies of groundwater have focused on aquifer-stream interactions with little emphasis on the effect of the intermediate confining units and land use. The changes in confinement conditions are crucial to the groundwater flow in the aquifer, especially when a significant change in land use is involved and a considerable volume of water is added or abstracted from the system [20]. The complex nature of the lithology and hydrogeology requires the use of numerical tools capable of

Table 6 Groundwater balance of eastern Nile Delta after RIGW/IWACO [19]

Component	Recharge (m^3/year)	Discharge (m^3/year)
Subsurface drainage	1,900,000	–
Seepage from the Nile and main canals	400,000	–
Inflow across boundaries	100,000	–
Discharge by tile drains		700,000
Inflow to Damietta		600,000
Discharge into drains		400,000
Groundwater withdrawals		400,000
Evaporation in barren areas		200,000
Outflow across boundaries		100,000
Balance	**2,400,000**	**2,400,000**

simulating groundwater flow in such heterogeneous media [21]. Boundary conditions, hydraulic properties, and initial conditions are classic prerequisite parameters for the aquifer domain, and the accuracy of the final output is highly dependent on these parameters. To achieve an accurate simulation, the spatial distribution of the aquifer hydraulic properties conditioned with physical measurement [e.g. hydraulic conductivity (K), porosity (φ), storativity (S), and dispersivity] should be determined [21].

Despite all these difficulties and challenges, the numerical model remains an important technique for understanding the hydrogeological phenomenon and identifying certain aquifer parameters. It is a computer-based method for solving flow and transport equations given the hydraulic parameters of the system in an array pattern. Numerical approaches deal efficiently with the heterogeneity of the aquifer hydraulic parameters as well as transport coefficients. Also, numerical models can simulate several aquifer systems or layers in two or three dimensions and are extremely flexible in dealing with boundary conditions.

MODFLOW is a well-known public domain, 3D, cell centred, finite difference, and saturated flow model developed by the United States Geological Survey [22]. It can perform both steady-state and transient analyses and has a wide variety of boundary conditions and input options. To model groundwater of the Quaternary aquifer associated with the agricultural development in the area north Ismailia Canal in the east of Nile Delta region, a three-dimensional GIS-based model that simulates the dominant hydrogeologic conditions of the aquifer and major land cover was developed. This model is expected to provide a management tool and implementation guide to evaluating the possible water policies that mitigate the expected deteriorations until the year 2025. Also, the model is capable of showing the estimated water budget of the system for each defined time step in the simulation, which helps to better understand the dominant hydrological processes that control the groundwater flow.

3.1 Objectives

In the Nile Delta, groundwater provided the principal water source for the development projects which undoubtedly impinged on the quality and quantity of groundwater reserves. As one of the most dynamic development regions, the East Nile Delta was selected as an ideal case study for planning groundwater resource management based on 3D modelling. In this research, a groundwater modelling approach was utilized to test several groundwater development strategies that include episodes of intense pumping stress to evaluate the potential of the system during drought and episodes of medium to low pumping for sustainable water use.

3.2 Methodology

In this study, a three-dimensional finite difference model was developed to simulate groundwater flow in the Quaternary aquifer of the area north Ismailia Canal, East Nile Delta region. MODFLOW was selected to numerically solve the governing flow equation based on water balance with a fully implicit finite different approximation [23]. The governing equation is written as follows:

$$\frac{\partial}{\partial x}\left(K_{xx}\frac{\partial h}{\partial x}\right)+\frac{\partial}{\partial y}\left(K_{yy}\frac{\partial h}{\partial y}\right)+\frac{\partial}{\partial z}\left(K_{zz}\frac{\partial h}{\partial z}\right)-\omega=S_s\frac{\partial h}{\partial t}$$

where K_{xx}, K_{yy}, and K_{zz} are hydraulic conductivities values (m^3/day) along x, y, and z axis, h is the hydraulic head, ω is a source and sink term (volumetric flux m^3/day), S_s is specific storage (dimensionless), and t is time (day). Mapping stratigraphy and assigning hydraulic parameters are always a great challenge and important factors that determine the model accuracy. The available 93 borehole data was utilized to map the stratigraphy, and the hydraulic conductivity and other hydraulic parameters were assigned based on the results of lab measurements and pumping tests. All calculations are based on water mass balance equations where in–budget = out–budget ± storage. All model boundaries were assigned as specified head boundary conditions in which an unlimited volume of water enters or leaves the system based only on hydraulic properties and constraints and aquifer stress. The selected boundaries which include Damietta Branch, Ismailia Canal, Suez Canal, and the Mediterranean Sea were assigned as the constant head is justified. The specified head in Damietta Branch and Ismailia Canal boundaries can be updated in the case of future dramatic changes in the rate of water release from Aswan Dam to maintain the acceptable accuracy of the results.

Other water bodies such as drains, irrigation canals, and El Manzala Lake were mapped in different ways (Fig. 10). Drains can only gain water from the system if the groundwater level was higher than the drain level and the infiltration is controlled by the drain bed conductance. Irrigation canals may gain or lose water based on the water level in the canal relative to groundwater level. El Manzala Lake

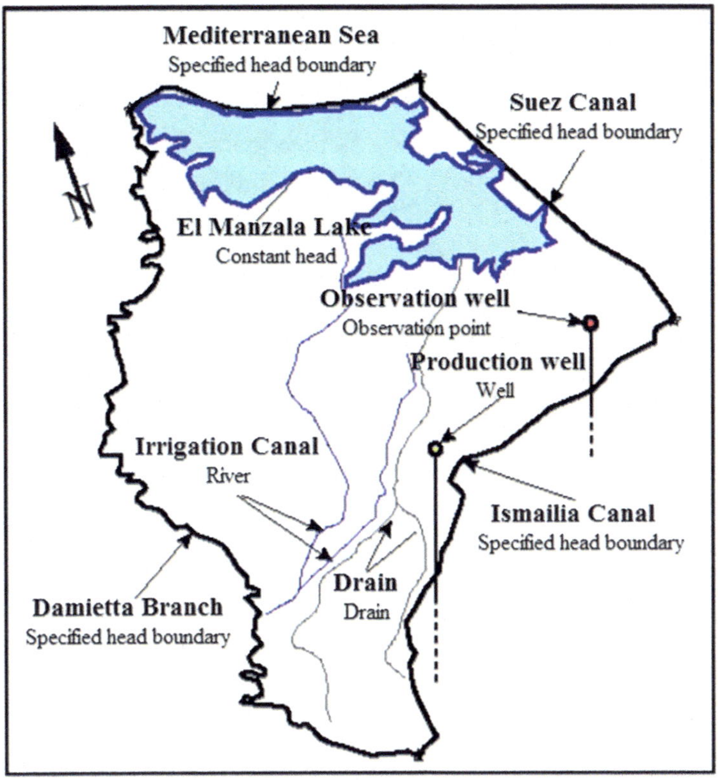

Fig. 10 Simplified sketch of the main model components

represents the significant discharge for the surface water system, but this setting may vary in case of groundwater due to the change in hydraulic head and the lake bottom bed conductance. Therefore, El Manzala Lake was mapped as a general head to swiftly manipulate the variations in hydraulic head and lake bed conductance. Infiltration from agricultural fields can be simulated as recharge zone in which one can relatively estimate the volume of water that reaches the saturated area of the groundwater system.

3.2.1 3D Stratigraphic Model

Solids developed from 3D borehole data can be used arbitrarily to map such complex stratigraphy of the aquifer. The relationship between the hydrogeological units of 93 boreholes selected in the study area (Fig. 11) was first checked to confirm a relative spatial continuation of the layers in the model domain. This is an essential requirement before assigning horizon IDs for each hydrogeological unit that will be subsequently used for mapping the stratigraphy into solids.

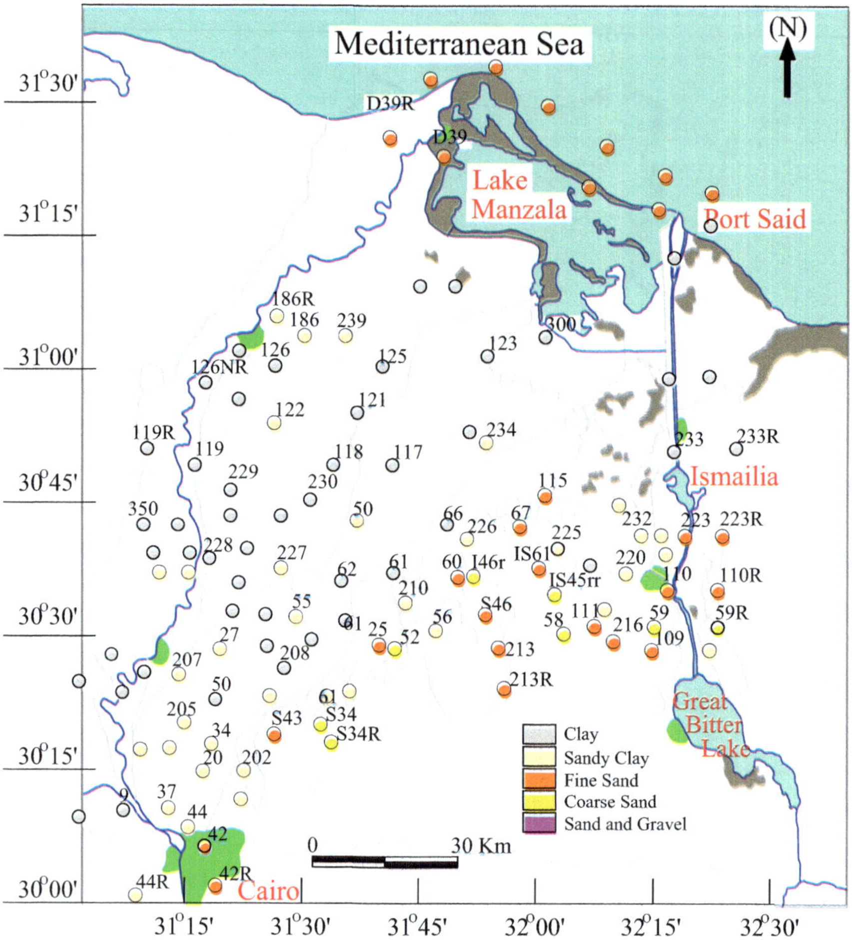

Fig. 11 Location of borehole dataset east of Nile Delta; colours represent the surface lithology

3.2.2 Building the GIS-Based Conceptual Model

In a conceptual model approach, the model parameters are designed as sources and sinks, boundary conditions, and observation points in the map module using the GIS tools. Once this is completed, the model can be exported to the MODFLOW, in such a way that these hydraulic data are automatically arranged in arrays. Since the upper 100 m of the Quaternary aquifer is modelled, the effect of the bottom layer is negligible, and the assumption made is that the bottom layer is deeply extended. Assuming a relatively constant water level throughout the study period, Damietta Branch, Ismailia Canal, Suez Canal, and the Mediterranean Sea were mapped as specified head boundary conditions. El Manzala Lake was mapped as general head,

Table 7 Calibrated parameters for the steady-state model simulation

Hydraulic parameters	Material	Value (m/day)
K_{xx} (m/day)	Clay	0.0015
	Sandy clay	0.85
	Fine sand	1.1
	Coarse sand	85
	Sand and gravel	375
Recharge rate	Traditional agriculture	0.00000582:0.00005
	Reclamation land	0.00005:0.0029456
Conductance	Irrigation canals	0.001:0.03
	Drains	0.01:0.12

while irrigation canals and drains were mapped and assigned attributes using, respectively, river packages and drain package. Both the production and observation wells were assigned to the conceptual model using, respectively, well and observation packages.

3.2.3 Model Execution and Calibration in Steady State

For model execution, the PCG2 package in MODFLOW was used to set the criteria for the iteration process used in solving the flow equation. The first run of the model was successfully executed using the initial input parameters to define the head distribution in the unstressed steady state. Model calibration is an essential part of adjusting the input parameters until the calculated and observed head in observation points reaches acceptable matching. The sets of these parameters were systematically adjusted starting with *K* values, recharge rates, and finally conductance values. Table 7 shows the final input parameters that were used in the calibrated unstressed steady-state model. Figure 12a shows the distribution of the calculated head obtained from the calibrated unstressed steady state.

3.2.4 Model Calibration in Transient State

Transient simulations were used to analyse time-dependent starting with the calibrated steady-state condition. Managing transient simulations involves large amounts of transient datasets such as pumping well data, recharge data, and observation data. In MODFLOW, the computed water budget consists of two parts: the inflow and outflow budgets. The inflow budget includes five components, namely, constant head, general head, recharge, infiltration from irrigation canals, and in-storage (stored water in the aquifer). The outflow budget consists of six components, namely, constant head, general head, production wells, drains, infiltration to irrigation canals, and out-storage. The groundwater head spatial distribution from steady-state model in 1990 and transient state simulation between

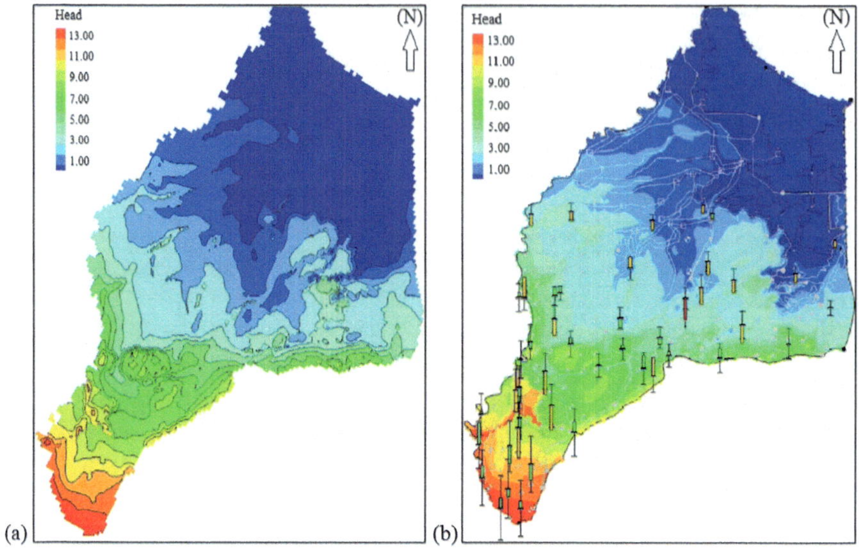

Fig. 12 (**a**) Groundwater head distribution by the calibrated steady-state model in 1990; (**b**) head distribution resulting from the interpretive transient model 1990–2004 (*bars* represent the accuracy reported in observation wells, *green*: ≤0.5 m, 0.5 m < *yellow* ≤1.0, and *yellow* >1.0 m)

1990 and 2004 are shown in Fig. 12a, b. The statistical analysis of groundwater flow rates shows an increase in the recharge rate of the groundwater aquifer during this period.

3.2.5 Prediction and Future Development

Using groundwater spatial distribution information as input data for the predictive transient simulation and assuming no changes in hydrogeological conditions and land use in the modelled area over the next 8 years; the head distribution by the year of 2025 was predicted in the study area. The spatial distribution of the predicted changes in the head is valuable information that helps planning efficient and sustainable groundwater development strategies such as rate and distribution of withdrawal stress from production wells.

4 Conclusion

This chapter summarizes the hydrogeological characteristics of the Quaternary groundwater aquifer east of Nile Delta. The integration of GIS and GMS models played a significant role in the construction of a 19 layer groundwater flow model and effectively enabled smooth management to the required database. The budget analysis of groundwater flow simulation in the three-dimensional model using the MODFLOW code indicated that pumping discharge from production wells efficiently controls the rising water level in the confined part of the Quaternary aquifer and the infiltration from agricultural fields is the key factor in the unconfined part. The primary recharge of the aquifer is the infiltration from the Ismailia Canal and Damietta Branch which significantly increases during the high pumping stress from wells. The up-to-date land use maps should be provided and interpolated to the GIS model for accurate model simulation. The results of the present model exercise constitute a foundation for sustainable water resource management in the East Nile Delta, and the calibrated model can be used to evaluate the efficiency of multidisciplinary policies shortly.

5 Recommendations

For better groundwater management for sustainable development, it is recommended that the strategies for the production well operations are controlled. Setting up a GIS database for groundwater resources in the study area is required. The GIS database includes the active and abandoned groundwater wells, their pumping rates and screen depths, irrigation canals, drains, seepage surfaces, soil type, and agricultural fields. This information is important for monitoring programmes and efficiently helps to obtain the optimum results from groundwater models.

Implementing monthly monitoring of groundwater levels, especially in the northern confined aquifer and near the saltwater interface, verifies the validity of the resulting management plan of this study and updating the model as required for improving the accuracy of the predictions. Implementation of a licensing system for new groundwater well construction strictly adheres to the guidelines of drilling depth, screening interval, and the distance between wells based on the permissible withdrawal rates and the capacity of drawdown recovery.

References

1. El-Shazly EM, Abd El-Hady MA, El-Shazly MM, El-Kassas IA, El-Ghawaby MA, Salman AB, Morsi MA (1975a) Geology and groundwater potential studies of El Ismailia master plan study area. Remote Sensing Research Project, Academy of Scientific Research and Technology, Cairo, 24 pp
2. Shatta AA, El-Fayoumy IF (1970) Remarks on the hydrogeology of the Nile Delta, UAR. In: Proceedings of the international symposium on the hydrogeology of deltas, Bucharest, UNESCO, vol II, pp 385–396
3. Shatta AA, Abdel Salam AA, Harga AA et al (1979) Soil map of the Eastern Delta Region, El Tumilat-Suez. 4th report, internal report, Academy of Scientific Research Technology, Cairo, pp 89–105

4. Schlumberger (1995) Well evaluation conference of Egypt. Schlumberger Technical Editing Service, Chester, 87 pp
5. Geriesh MH, Balke KD, El-Rayes AE, Mansour BM (2015) Implications of climate change on the groundwater flow regime and geochemistry of the Nile Delta, Egypt. J Coast Conserv 19 (4):589–608. https://doi.org/10.1007/s11852-015-0409-5
6. El-Dairy MD (1980) Hydrogeological studies on the eastern part of the Nile Delta using isotope techniques. MSc thesis, University of Zagazig, Zagazig
7. El Haddad IM (2002) Hydrogeological studies and their environmental impact on future management and sustainable development of the new communities and their surroundings, east of the Nile Delta. PhD thesis, University of Mansoura, Mansoura
8. Egyptian Meteorological Authority (2006) Climatic atlas of Egypt. Arab Republic of Egypt, Ministry of Transportation and Communications, Cairo
9. FAO, Food and Agriculture Organization of the United Nations (1966) High dam soil survey project, vol 3. Ministry of Agriculture, Cairo, pp 1–348
10. Attwa M, Basokur A, Akca I (2014) Hydraulic conductivity estimation using direct current (DC) sounding data: a case study in East Nile Delta, Egypt. Hydrogeol J 22(5):1163–1178. https://doi.org/10.1007/s10040-014-1107-3
11. Arnous MO, Green DR (2015) Monitoring and assessing waterlogged and salt-affected areas in the Eastern Nile Delta region, Egypt, using remotely sensed multi-temporal data and GIS. J Coast Conserv 19(3):369–391. https://doi.org/10.1007/s11852-015-0397-5
12. El-Fayoumy IF (1968) Geology of ground water supplies in the region east of the Nile Delta and its extension in the north Sinai. PhD thesis, Cairo University, Cairo
13. Sallouma M (1983) Hydrogeological and hydrogeochemical studies east of the Nile Delta. PhD thesis, University of Ain Shams, Cairo
14. Elewa HH, Shohaib RE, Qaddah AA, Nousir AM (2013) Determining groundwater protection zones for the Quaternary aquifer of northeastern Nile Delta using GIS-based vulnerability mapping. Environ Earth Sci 68(2):313–331. https://doi.org/10.1007/s12665-012-1740-x
15. RIGW/IWACO (1994) Environmental management of groundwater resources project. Inception report
16. El-Shazly MM, Sanad S, Rofail NH (1975b) Hydrogeological and hydrological investigation of the site of proposed funnel at Kantara (Suez Canal). Technical report, Engineering and Geological Consulting Office, Prof. Dr. M. Abdel Hady, Cairo, pp 1–44
17. Ismael AMA (2007) Application of remote sensing, GIS, and groundwater flow modelling in evaluating groundwater resources: two case studies – East Nile Delta, Egypt and Gold Valley, California. PhD thesis, University of Texas at El Paso, El Paso, ProQuest, p 402. ISBN: 0549355057-9780549355052
18. RIGW/IWACO (1991) Groundwater potential in the Nile Valley and Delta. Internal report, Cairo
19. RIGW/IWACO (1990) Assessment of groundwater pollution from agricultural activities. TN.77. 01300-90-05, Cairo
20. Urbano L, Waldron B, Larsen D, Shook H (2006) Groundwater-surface water interactions at the transition of an aquifer from unconfined to confined. J Hydrol 321:200–212. https://doi.org/10.1016/j.jhydrol.2005.08.001
21. Pasquier P, Marcotte D (2006) Steady- and transient state inversion in hydrogeology by successive flux estimation. Adv Water Resour 29(12):1934–1952. https://doi.org/10.1016/j.advwatres.2006.02.001
22. McDonald MG, Harbaugh AW (1988) A modular three-dimensional finite-difference ground-water flow model. US Geological Survey techniques of water resources investigations, book 6, chapter A1, US Geological Survey, Reston, 586 pp
23. Wang HF, Anderson MP (1982) Introduction to groundwater modelling: finite difference and finite element methods. Academic Press, California, p 256

Groundwater Management for Sustainable Development Plans for the Western Nile Delta

Mohamed Galal A. Eltarabily and Abdelazim M. Negm

Abstract Egypt is considered an arid country and the primary water resource is the River Nile. The limited availability of renewable freshwater for agriculture and urban development is a major constraint. The role of groundwater is steadily increasing and will cover 20% of the total water supply in the coming decades especially in the reclaimed areas of the Western Nile Delta. Serious environmental problems are emerging in the groundwater aquifer in Western Nile Delta such as waterlogging, soil salinity, and the risk of saline water intrusion to the north aquifer. An efficient integrated and sustainable management plan for groundwater resources is needed to avoid the deterioration of the groundwater aquifer in Western Nile Delta. In this chapter, a brief description of the groundwater aquifer in Western Nile Delta and a review of previous studies on groundwater hydrology were presented. GIS and MODFLOW models were integrated to simulate the groundwater flow in the studied Quaternary aquifer. The developed model was calibrated for steady-state and transient conditions for groundwater heads till 2002. The groundwater potentiality was evaluated, and different management scenarios were analyzed for groundwater prediction. The results of the current situation of groundwater showed that groundwater aquifer in Western Nile Delta is susceptible to significant water table reduction especially in the unconfined parts for the case of overstress discharge. The net aquifer recharge was increased for the case of reducing the surface water inflow while increasing the annual abstraction and improving the irrigation system. The annual aquifer potentiality was increased by the construction of a new canal to feed the aquifer towards the northwest direction. Therefore, efficient integrated and sustainable management of

M. G. A. Eltarabily (✉)
Civil Engineering Department, Faculty of Engineering, Port Said University, Port Said, Egypt
e-mail: eng_mohamad1987@hotmail.com; eng_m_trabily@eng.psu.edu.eg

A. M. Negm
Water and Water Structures Engineering Department, Faculty of Engineering, Zagazig University, Zagazig, Egypt
e-mail: amnegm85@yahoo.com; amnegm@zu.edu.eg

A. M. Negm (ed.), *Groundwater in the Nile Delta*,
Hdb Env Chem (2019) 73: 709–728, DOI 10.1007/698_2018_247,

groundwater resources relies on a comprehensive database that represents the characteristics of the aquifer and modeling software to achieve the impacts of decision alternatives.

Keywords Groundwater management, MODFLOW, Sustainable development, Western Nile Delta

Contents

1 Introduction

With the increase in the demand for freshwater, the attention of nonconventional water resources is drawn. Nile water alone is no longer adequate for the increasing water requirements for the different development activities in Egypt. Groundwater is playing an essential role in water supply because of its importance for the domestic, agricultural, and industrial use. However, many cautions and worries are increasingly being voiced about the dangers that surround the groundwater resources. These concerns are related to the groundwater depletion as a result of overabstraction [1] and also, quality deterioration brought by many modes of contamination. Intensive expansion in land reclamation, increased population, and construction of new industrial projects cause changes in water demand and the groundwater regime. Ministry of Water Resources and Irrigation [2] introduced the West Delta water conservation and irrigation rehabilitation project aiming to the implementation of a pipe network to transfer surface water to irrigate about 107,000 ha (264,403 feddan) located in the South of the West Delta. This will be reflected in rising of groundwater levels accompanied by the formation of local groundwater mound sand logging problems.

The Nile Delta aquifer is considered an essential source of groundwater abstraction as it represents 87% of the total groundwater abstraction in Egypt [3]. The primary source of the Nile aquifer recharge is the seepage from the surface water irrigation/drainage systems and infiltration from the cultivated land. The conjunctive use of surface water and groundwater plays an active role during the period of peak irrigation demand. The amount of water abstraction from the Nile aquifer (Delta and Valley) was estimated in 2010 at 6.2 BCM, which is within the safe yield (8.4 BCM) that was evaluated by MWRI [4].

In the Western Nile Delta region, groundwater is the primary source for domestic, industrial, and agriculture use. With the expansion of developing activities in this area, it is essential to develop a groundwater management strategy to avoid any environmental impacts on the aquifer system due to the extensive future abstraction of groundwater. For better management of groundwater resources, it is crucial to have enough data about the physical and hydrogeological settings for the study area. Physical parameters include land use, meteorological data, topography, soil classification, and drainage as well as irrigation systems. Hydrogeological setting comprises the aquifer system, boundary conditions, hydraulic parameters for all aquifer layers, and monitored groundwater levels.

Data used in groundwater modeling consists of the aquifer system stress factor, the aquifer system geometry, and the hydrogeological parameters [5]. Stress factors for groundwater flow include effective recharge, pumping volumes, water surface flow exchanges, etc. Appropriate aquifer system geometry can be determined using geological information (maps and cross sections), topographic maps, as well as contour maps of the upper and lower limits for the aquifer strata and aquitards. Links can be organized between MODFLOW, groundwater model, and the GIS. The GIS software is used to preprocess and postprocess the spatial data. Recently, the use of GIS has grown rapidly in groundwater assessment and management researches such as Arshad and Zulfiqar [6], Kharad et al. [7], Sarma and Saraf [8], and Singh and Prakash [9].

2 Study Area West Nile Delta

Western Nile Delta region occupies the area between Cairo and Alexandria. The southern boundary is near Cairo, the eastern boundary is the Rosetta Branch, and the north is bounded by the coast of the Mediterranean Sea. It extends westward to the desert area from the west of Wadi El-Natrun up to the eastern edge of the Qattara Depression. The modeled area covers about 15,170.6 km^2 (Fig. 1).

3 Physical Setting

3.1 Climate Condition

Table 1 shows some data from meteorological stations situated in the Western Nile Delta region. These data include air temperature, relative humidity, evaporation, and

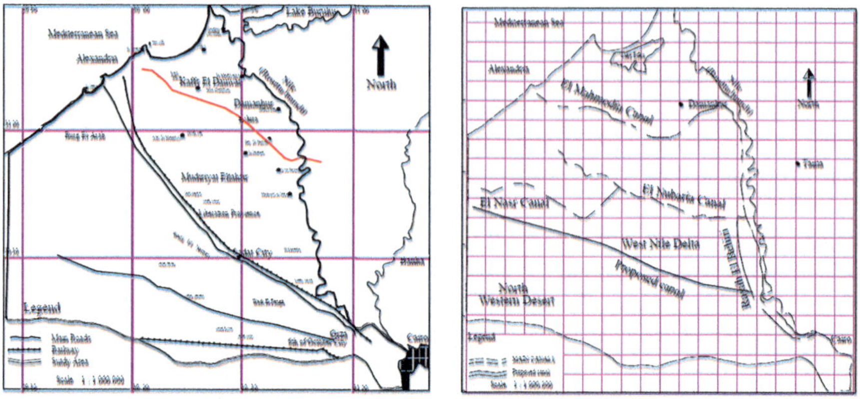

(a) (b)

Fig. 1 (**a**) Location map of Western Nile Delta area; (**b**) surface water system of Western Nile Delta [10]

rainfall. The annual mean of the evaporation intensity decreases from South to North and from summer to winter. Rainfall increases in the northward direction. Given the available rainfall records, a 100 mm amount of rainfall may take place intermittently and at intervals hardly exceed a couple of hours.

3.2 Geology and Geomorphology

West Nile Delta consists of a sedimentary succession which ranges in age from the Late Cretaceous to Quaternary. The eastern area of the studied region is covered with Holocene clay while the western part is covered with Quaternary sediments [11]. West Nile Delta area can be classified into four forms: young alluvial plains, old alluvial plains, fanglomerates, and dunes [12]. The young alluvial plains dominate the cultivated lands bordering Rosetta branch. Irrigation canals and drains dissect these plains, sloping from South to North. The ground elevation ranges between 18 m (+MSL) in the South, to about 4 m (+MSL) in the North by an average slope value of 1 m/10 km [13]. The old alluvial plains expand over the western fringe of the Nile Delta and are occupied by most of the newly reclaimed areas. Fanglomerates are considered to represent the Wadi downwash brought by the drainage lines depositing their loads before reaching the Nile. They are covered by relatively thick conglomerates and loose sand deposits. They are mainly composed of fine-grained sand.

Table 1 Summary of meteorological data for Western Nile Delta region

Station	Temperature (°C)			Relative humidity	Evaporation	Rainfall	
	Annual average	Max.	Min.	Annual average (%)	Annual average (mm/day)	Annual (mm/year)	Max. (mm/day)
El-Tahrir	20	34.5	5.80	64.0	6.6	34.4	8.50
Cairo	21.2	34.0	7.50	56.0	8.7	17.62	4.60
Alexandria	20.1	26.4	13.5	68.7	5.0	190.9	53.5
Damanhur	19.6	25.8	12.3	67.7	4.4	90.2	24.2
Wadi el-Natrun	20.9	27.4	13.2	55.3	9.6	38.5	11.2

3.3 Land Use

The traditionally cultivated lands are predominating in the eastern part of the Western Nile Delta region. Reclaimed areas are present in the western part of the region that depends on surface water or groundwater for agriculture. Drainage water from open drains is partly reused for irrigation to cover the shortages of the irrigation water [14].

3.4 Surface Water and Drainage System

The main surface water channels are Rosetta branch, Rayah El Beheiry, Rayah El Nasseri, and El Nubariya canals (Fig. 1b). These water channels are mainly cutting through sands. Therefore, a direct connection between the surface water and groundwater exists [3]. Table 2 lists the average monthly surface water levels and inflow through the surface water system. The Rosetta branch (239 km in length) acts as a drain for the groundwater [15]. Rayah El Beheiry starts from Delta barrage and extends for a distance of 65 km. Rayah El Nasseri begins from Delta barrage and extends for a total length about 68 km running northward in the beginning and then northwestward until it joins with El Nubariya canal. Rayah El Beheiry, Rayah El Nasseri, and El Nubariya canals in their first reaches act as influent streams and are considered the main source for groundwater recharge in the study area [11]. El Nubariya canal, in the end reaches, acts as a discharging effluent stream for groundwater. The open drains are intensively distributed in the old lands of the eastern part of the study area. These drains discharge large quantities of drainage water, but their effects are local. In most of the reclaimed areas in the western desert fringes, there is no tile drainage system. More open drains must be planned and constructed to solve many problems of water logging in these areas.

Table 2 Monthly average surface water inflow to the Western Nile Delta aquifer (in million m^3/day) (after [14])

Month	Rosetta branch		Rayah El Beheiry		Rayah El Nasseri		El Nubariya canal	
	Water level (m)	Discharge	Water level (m)	Discharge	Water level (m)	Discharge	Water level (m)	Discharge
August	12.76	20.56	16.12	26.76	16.03	9.50	9.55	13.85
September	13.06	13.14	18.53	24.57	15.54	7.19	9.42	14.46
October	13.05	11.82	15.36	20.76	14.78	4.17	9.06	19.95
November	12.98	8.16	15.11	18.85	14.78	5.54	8.90	21.04
December	12.94	5.84	14.43	17.83	14.65	3.87	8.41	19.44
January	12.68	9.06	14.44	11.88	14.66	4.09	8.12	21.70
February	12.99	8.94	14.79	16.20	14.90	4.99	8.69	20.18
March	13.03	11.07	15.13	18.69	15.24	5.64	8.82	16.54
April	13.09	12.52	15.31	20.18	15.24	5.59	9.00	16.41
May	13.11	16.09	15.72	23.18	15.16	5.42	9.31	12.30
June	13.20	22.52	16.06	25.53	15.73	7.40	9.51	9.66
July	13.18	21.07	16.12	26.50	15.95	9.02	9.54	11.76
Average	13.01	13.40	15.37	20.91	15.22	6.03	9.03	16.44

4 Hydrogeological Conditions

4.1 Groundwater Quaternary Aquifer

Quaternary aquifer, the Pliocene aquifer, and the Miocene aquifer are groundwater aquifers reusable for exploitation in the Western Nile Delta region (Fig. 2). The Quaternary aquifer is the main water-bearing formation in the study area. The most important one is the Quaternary aquifer that consists of Pleistocene sand and gravel, and clayey facies in the North. This aquifer is overlaid by a clay cap which is considered as a semi-confining layer with a thickness varying from 10 m in the South to 30 m in the North. The thickness of the Quaternary aquifer ranges between 50 m along the desert fringes in the West and 800 m in the North. Tables 3 and 4 show the hydraulic conductivity of the Quaternary aquifer and clay cap layer. The groundwater salinity is about 800 ppm in the South, and it reaches about 359,000 ppm in the North due to the effect of seawater intrusion [25]. The piezometric lines show radial flow towards the depression of Wadi El-Natrun and the groundwater is discharged through seepage zones into small lakes, ponds, and sabkhas, where the water evaporates in amounts up to 70×10^6 m^3/year.

4.2 Groundwater Levels

The depth from the ground surface to the groundwater is less than 3 m in the eastern part of the Western Nile Delta region. It decreases northward and to the northeast. It ranges from 3 to 8 m in northeastern part and increases in the southwest direction ranging from 20 m to more than 50 m near western desert fringes (high topography area) (Fig. 3). The piezometric head levels are decreasing within the Western Nile Delta region from more than 15 m (+MSL) in Cairo to 1 m (+MSL) near the coast. The piezometric contour lines in the hydrogeological maps of the Western Nile Delta show a local increase in heads due to the effects of old reclamation projects based on surface water using the traditional flood irrigation method where the infiltration rate is very high. Also, local depression appeared due to extensive groundwater abstraction where many recent private reclamation projects have been developed.

4.3 Recharge and Discharge of Groundwater

The groundwater in Quaternary aquifer in Western Nile Delta exists mainly under free water table conditions (unconfined). Semi-confined conditions are locally present where clay cap is found (northeast of the study area). The aquifer in the floodplain (southern and central portions) is continuously recharged by the leakage of excess surface irrigation water; thus, the aquifer acts as a storage reservoir. It is also recharged by the

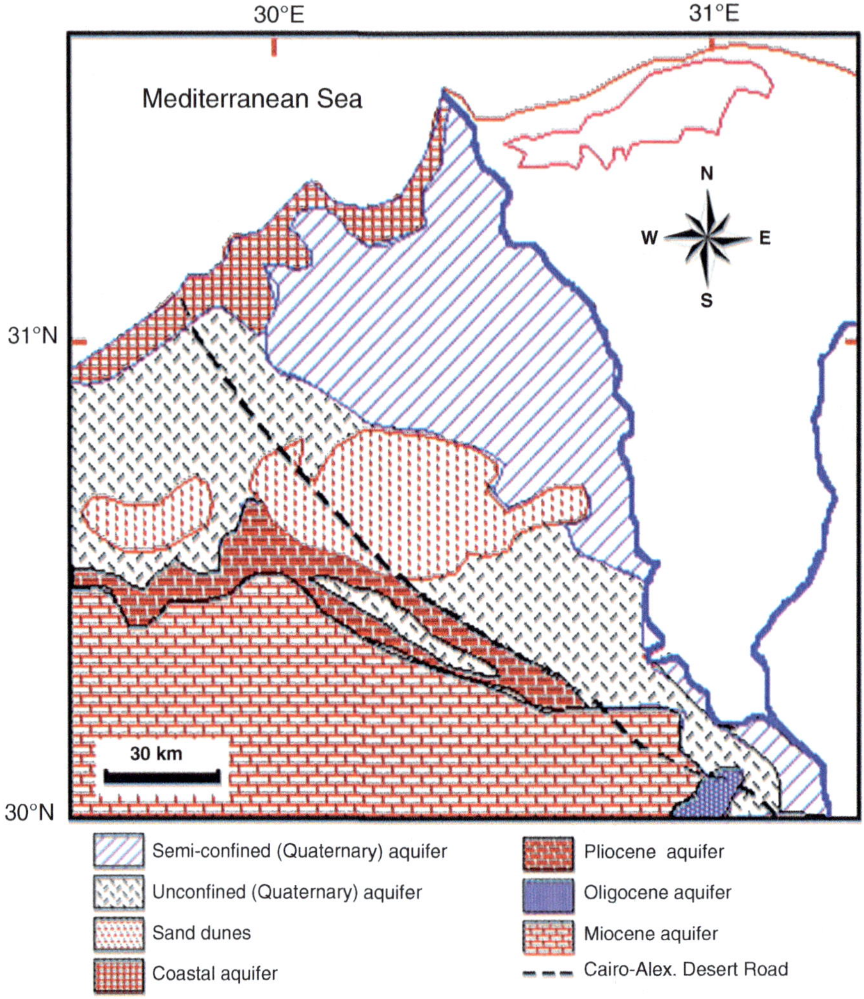

Fig. 2 The groundwater aquifers in Western Nile Delta (after [16])

seepage from canals, and infiltration of rainfall that takes place only during the winter months. In the southern and central parts of the floodplain, the downward leakage towards the aquifer ranges between 0.25 and 0.8 mm/day, depending on irrigation, and drainage practices, and the soil type. In the desert areas, relatively high leakage rates are observed for basin, furrow, and sprinkler irrigation (1.0–1.5 mm/day) with much lower rates for drip and central pivot irrigation (0.1–0.5 mm/day). Discharge of groundwater takes place naturally through the outflow into the Rosetta branch and drainage system, evapotranspiration, and inter-aquifer flow of groundwater, and artificially through direct abstraction. Groundwater return flow to Rosetta branch and drains and extraction by production wells are the main discharge components.

Table 3 Hydraulic parameters of the Quaternary aquifer in Western Nile Delta

Study	*K* (m/day)	*T* (m^2/day)	*S*	S_y	Porosity (%)	Effective porosity (%)
Zaghloul [17]	119.00		10^{-4}–10^{-3}	0.15	30	
Shahin [18]	50.00	2,500–25,900	10^{-5}–10^{-4}	0.20	25	23.25
Laeven [19]	150.00	10,350–59,800			25–30	
RIGW [20]	75.00	15,000–75,000	10^{-4}–10^{-3}		25–40	
Bahr [21]	75.00		1.1×10^{-3}		25	18

K hydraulic conductivity (m/day), *T* transmissivity (m^2/day), *S* storage coefficient, *Sy* specific yield

Table 4 Hydraulic conductivity of the semi-confining clay cap layer

Study	K_h (m/day)	K_v (m/day)
Farid [22]	0.2160	0.0025
RIGW/IWACO [23]	0.2505	0.0484
Warner et al. [24]	0.2160	0.0073

5 Groundwater Flow Modeling

In the present study, TRIWACO a steady-state, three-dimensional, finite-difference groundwater flow model has been employed to simulate flow and get the budget of groundwater in the study area. TRIWACO is a numerical program package for three-dimensional simulation of groundwater flow under the steady-state and transient conditions by finite element technique. MODFLOW has been used through the comprehensive packages of the program. It has been used in the present study as tools for every phase of the groundwater simulation including area characterization, model development, postprocessing, calibration, and visualization [27].

5.1 Conceptual Model Development

To develop a conceptual model, the locations of rivers, canals, drains, wells, layer parameters such as hydraulic conductivity, and model boundaries for the simulation have been defined at the conceptual model level. Once these data are completed, the conceptual model is converted to the grid model, and the program performs all of the cell-by-cell assignments automatically. The model consists of the multilayer aquifer system. The first layer represents the semi-confining layer (upper clay layer) that is modeled as an aquifer in which horizontal and vertical flow is simulated. All surface water features such as Rayah El Beheiry, Rayah El Nasseri, and El Nubariya canals and main open drains are included as "rivers" in the model. Groundwater abstraction occurs from Quaternary and Miocene aquifers, while storage changes (confined or

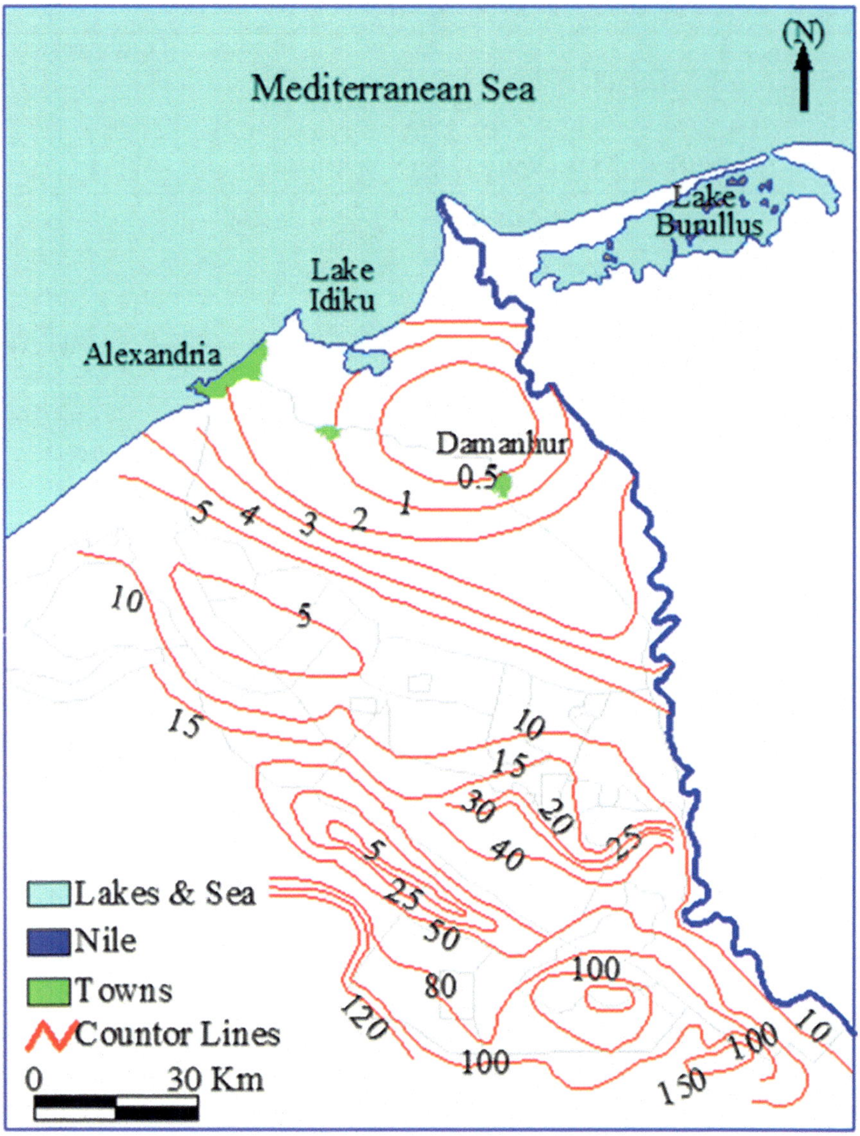

Fig. 3 Average depth to groundwater in the Quaternary aquifer in Western Nile Delta in 2008 [26]

unconfined, depending on the groundwater level relative to the top of the layer) are simulated when the model is running in the transient state (Fig. 4).

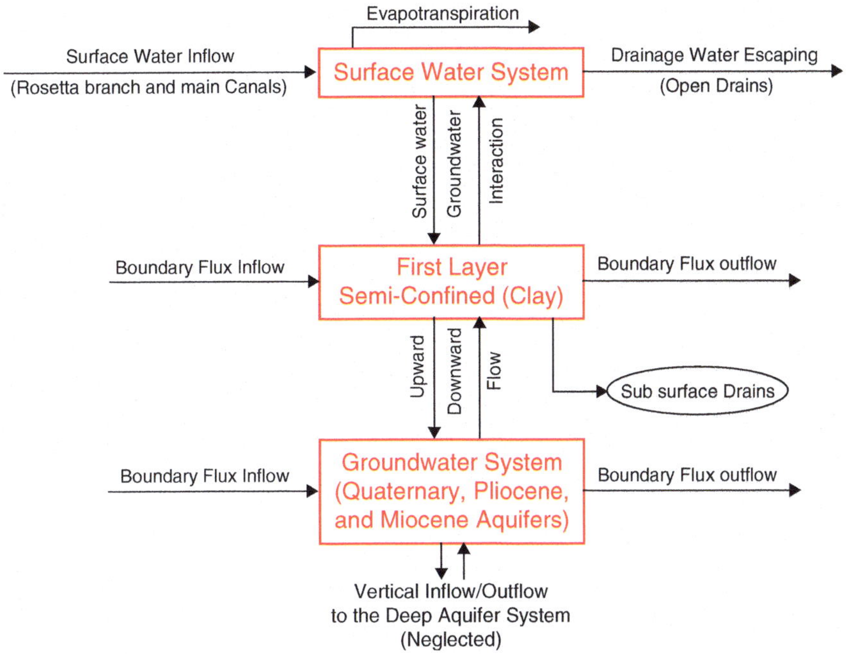

Fig. 4 Multilayer conceptual model for the study area

5.2 *Numerical Simulation*

After initializing the numerical model data, interpolation is used to define the top and bottom elevations of the layer and the starting (observed or initial) head levels of groundwater. The conceptual model is then converted from a high-level feature object-based to a grid-based MODFLOW numerical model. GIS was used to manage the spatially distributed input parameters and outputs of the model [28]. Primarily, the hydrogeological properties of the aquifer (geological layers, hydraulic conductivity, and porosity), and also maps of land use, the location of pumping wells, and surface water bodies were imported to the database of the ArcGIS. Ground elevation and topography map was generated by the interpolation of the points' data from boreholes over the area. Two aquifer layers covering an area of about 15,170.6 km^2 have been simulated in the computational 3D grid that consists of 32,800 nodes and 64,870 equilateral elements incorporating the triangular elements size that ranges from 500 m in the main area of interest to 1,000 m in the remainder of the area and the desert fringes outside the area.

5.3 *Boundary Conditions*

For the southwestern border where faults exist, and no flow from the western desert enters the aquifer system, no flow boundary was specified (Neumann conditions). In this case, the derivatives of the head (flux) across the boundary are set to zero. Water levels in Rosetta branch were almost constant. Therefore, the piezometric heads are constant (Fig. 5) and do not change with simulation time. This eastern border is specified across the model as a fixed head boundary (Dirichlet conditions). Rayah El Beheiry to the east and El Nubariya canal to the north are defined as stream flow-routing (River) package that simulates the interaction between surficial streams and the groundwater. Water can move from the stream to the aquifer or from the aquifer to the stream depending on the relative differences in the stream stage and the water table elevations. The recharge zones depend on the land use and hydrogeological conditions for each zone. The influx to the system is assumed primarily through recharge due to infiltration from canals, excess irrigation water, and limited rainfalls.

5.4 *Model Calibration*

The model was calibrated against the available average annual groundwater heads of about 60 observed wells during the period from 1990 to 2002. The calibration is by steady-state and transient nonlinear conditions using trial and error method which requires several trials by adjusting the hydraulic conductivity, specific storage, and recharge. Model is repeatedly run within an acceptable level of accuracy (± 0.5 m) until the computed solution is matched with the observed values. Figure 6 presents a map of the calibrated piezometric surface for the Western Nile Delta area.

5.5 *Groundwater Assessment*

To assess the groundwater resources within the study area, the calibrated model has been used to calculate the water balance and the groundwater aquifer potentiality. The surface water inflow to the study area is about 9,375 million m^3, and the drainage water is approximately 4,545 million m^3, out of this volume about 875 million m^3 is reused for irrigation. The calculated water balance indicated that the annual net recharge to the aquifer system is estimated at 1,950 million m^3 and the annual total aquifer potentiality at about 468 million m^3. Figure 7 shows the schematic diagram for the calibrated water balance for the Western Nile Delta.

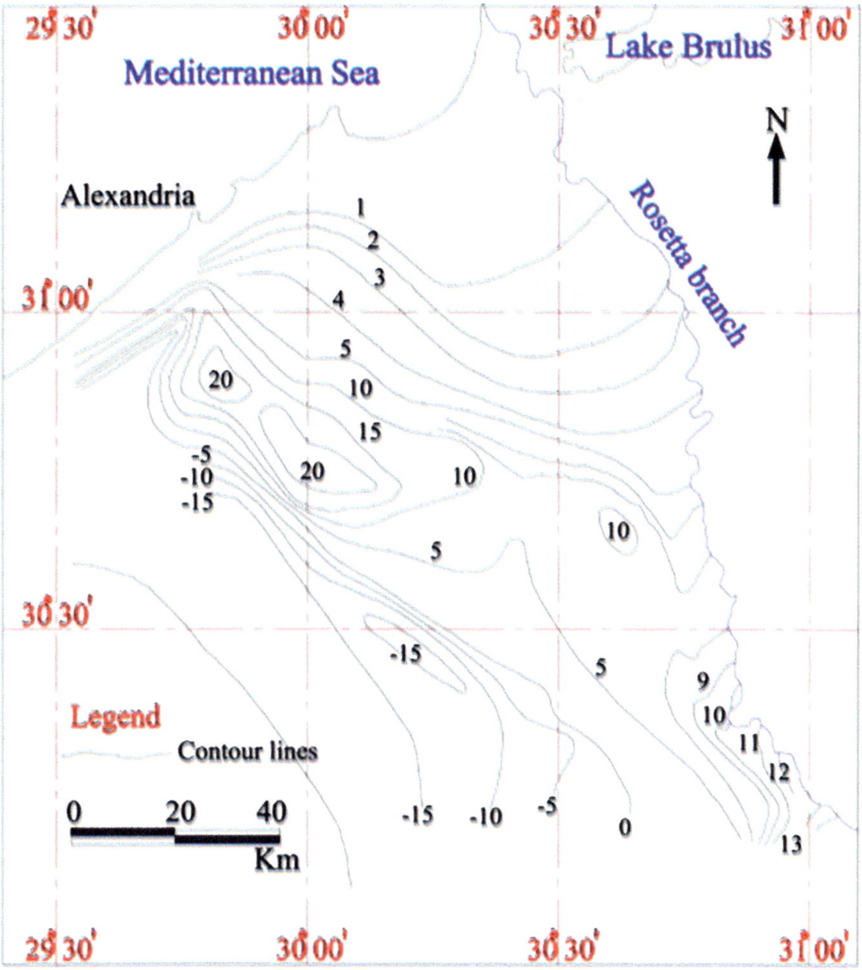

Fig. 5 Constant heads for Western Nile Delta [26]

6 Groundwater Management Plans

To achieve the sustainability plan of water resources developments of the Ministry of Water Resources and Irrigation [4] at Western Nile Delta, it is clear that the management plans are required to provide the water requirements for the newly proposed reclaimed areas. Therefore, water budgets analysis till the reference year should be adequately calculated for accurate prediction of the groundwater heads and tested with different alternative scenarios. Previous researchers studied the groundwater management for sustainable developments of Western Nile Delta (e.g., [10, 29, 30]).

El Molla et al. [10] studied different management scenarios for the Western Nile Delta aquifer to satisfy the need for 465,000 feddan of new lands reclamation. Water

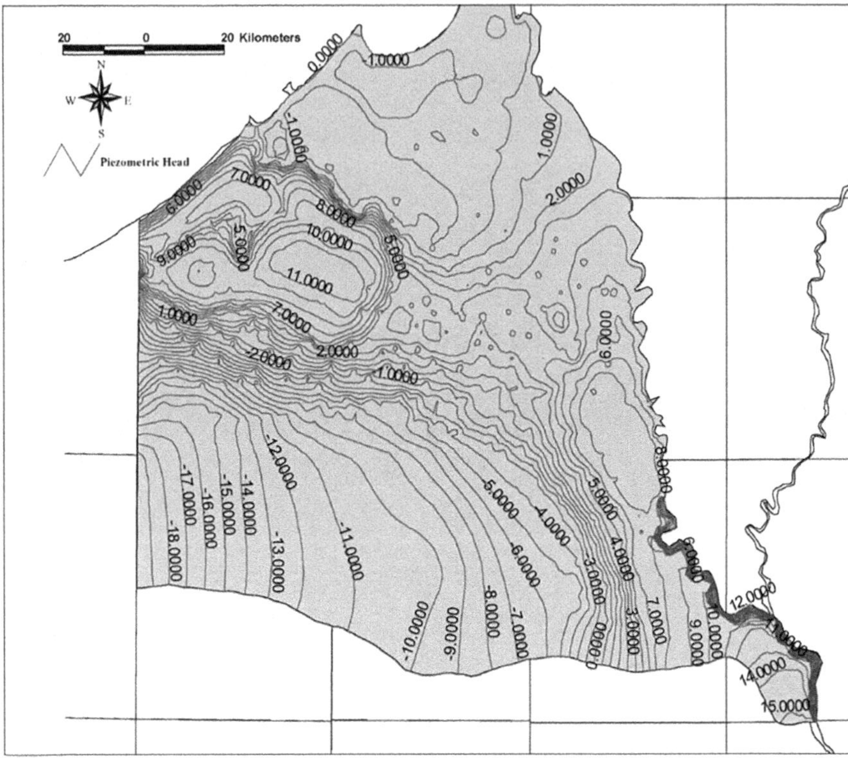

Fig. 6 Calibrated groundwater piezometric heads for Western Nile Delta [29]

demand of 7,000 m^3/year/feddan as an average for Western Nile Delta is assumed [3]. Therefore, the excess inflow required for irrigation water supply is nearly 3.25 bm^3. Recharge seepage down to the aquifer after full reclamation is 1.2 mm/day per feddan. Safe abstraction from the groundwater aquifer in the year 2017 was calculated after reclaiming all areas for cultivation in Western Nile Delta. Three different scenarios were studied with alternative conjunctive uses for available water resources in Western Nile Delta (surface water, groundwater, and drainage water reuse) to prevent aquifer depletion.

The first scenario is the construction of a new canal with 2.1 BCM of surface water taken from Rayah El Beheiry as proposed by MWRI [3]. The increase of safe groundwater abstraction to avoid aquifer exploitation, as calculated by the model, is about 1.0 BCM/year. Groundwater piezometric heads for the first scenario are shown in Fig. 8a. It is clear that observed excess recharge has affected the piezometric head in the southern part of the study area, that is raised from -15 to 3 m. These changes caused moving back of the contour lines in the northern area especially in the part near to the Rosetta branch. Contour lines in the southern section near Rosetta branch were decreased by about 1 m. It is observed that the amount of abstraction by wells was approximately equal to the net recharge of the aquifer. Therefore, finally no changes

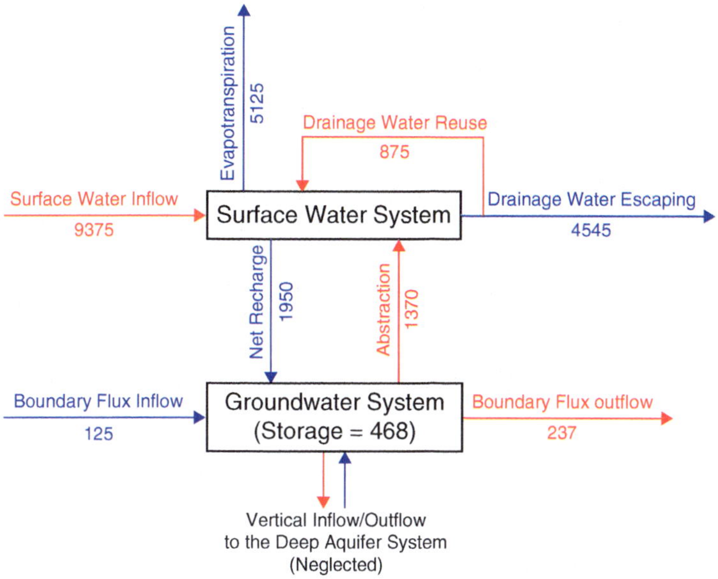

Fig. 7 Schematic diagram for the water balance of Western Nile Delta [29]

in piezometric levels except for local conditions are attributed to the vicinity of the modeled abstraction wells. Water budget analysis shows an increase in recharge with 1.0 BCM according to the increase in the cultivated area. Constant head outflow increased by 0.05 BCM, and the influx changed only with 0.001 BCM.

The second scenario is the drainage water reuse of the drainage water that flows into the sea. The drainage water reuse accounts 4.50 BCM, 5.50 BCM, 13.50 BCM, and 15.50 BCM in year 2005, 2007, 2015, and 2017, respectively [2, 10, 31, 32]. This scenario was processed by using the same input data as in the first scenario, but without the proposed canal therefore extra 2.25 BCM was gained from the use of drainage water. Quantitatively, this situation is similar to the first one except for the additional amount of freshwater needed for mixing with drainage water and also salt concentrations are increased.

The third scenario is the reduction of Rosetta branch discharge to account for Toshka project. Water requirements for Toshka project is 5.5 BCM per year. In 2017, the Ministry of Water Resources and Irrigation was organizing a plan to decrease 4 BCM from delta region by changing the crop patterns and 1.5 BCM from current discharges of Upper Egypt [3]. This reduction of 4 BCM will be divided between western, middle, and eastern delta with a ratio according to the current recharge. The current water distribution for Western Nile Delta is 10.98 BCM (32.3% of the delta). This scenario has an effect on the total water resources availability Western Nile Delta. Rosetta branch flow will decrease to 9.69 BCM, which is not enough for the existing cultivation area.

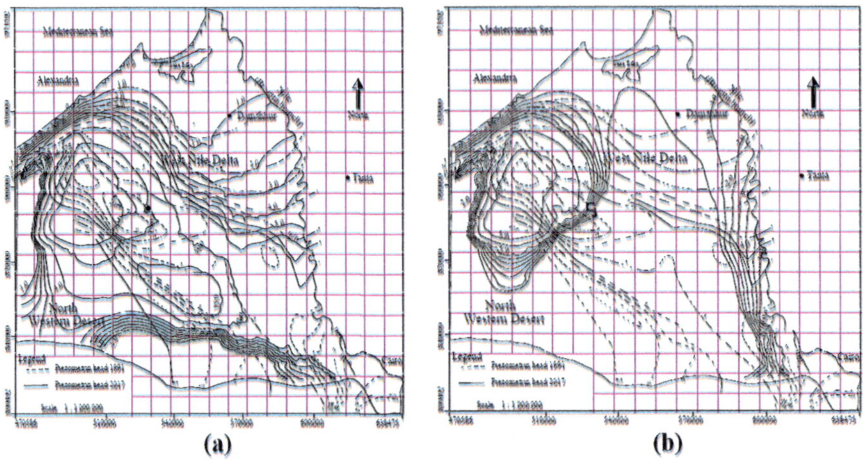

Fig. 8 Groundwater heads, (**a**) for the first scenario; (**b**) for the third scenario [10]

Groundwater levels are declined, and the highest groundwater level is about 6 m. This lowering in the piezometric head would increase the saltwater intrusion from the Mediterranean Sea. Water budget for the groundwater aquifer indicated an apparent shortage in the recharge values in comparison to the abstraction. The water budget analysis for the study area in 2017 is listed as follows [10]:

Available water quantities:

- Rosetta branch, 9.69 BCM
- Abstraction from groundwater aquifer, 2.08 BCM

Required water quantities:

- Old cultivated areas, 10.9 BCM
- Newly cultivated lands, 3.25 BCM

Shortage of about 2.38 BCM/year will occur if this scenario is applied. It is therefore recommended to either decrease cultivated area (do not reclaim all the planned regions) or change the crop pattern to decrease irrigation water application or increase the canals' network discharge by the 2.38 BCM.

Dawoud et al. [29] developed a GIS-based model for the Western Nile Delta aquifer to evaluate groundwater potentiality. Two scenarios were simulated and tested using the calibrated groundwater model in 2002. The first scenario considers a reduction of the surface water inflow to Western Delta region. The problem of shortage in irrigation water could, in this case, be partly solved by increasing the annual abstraction from the groundwater by about 450 million m^3 and also by improving the tradition flooding irrigation system. The surface water is hydraulically connected with the Quaternary aquifer in Western Nile Delta. So, any change in the surface water inflow would have a direct impact on the groundwater heads. Also, irrigation system improvement could reduce the rate of water application and then minimize the amount of recharge to the aquifer system.

The new water levels in the surface water bodies after the reduction of surface water inflow have been calculated along with the new rate of recharge after the irrigation system improvement. After running the model, it was found that although recharge input to the model has been reduced by 10%, the net recharge calculated from the model is increased by about 5.7%. This is due to the effect of river/aquifer interaction. The increase in net recharge is coming from the seepage from the surface water bodies. But, it was also found that the aquifer potentiality has been reduced by about 91%.

The second scenario has evaluated the effect of constructing a new canal that diverts the water from Rayah El Beheiry to the western part of the study area. It could help to feed the newly reclaimed lands and to minimize the overpumping of groundwater in the stressed areas. Although the path of this new canal is not yet defined along with the water levels, a preliminary test has been evaluated and shows that the annual aquifer potential could be increased by about 23%.

Morsy [30] studied the impact of groundwater development plan from the year 1992 to 2008 for the Western Nile Delta aquifer on the water table and highlighted the waterlogging problems. A drawdown of groundwater levels in the southern part of the western fringes of the Nile Delta (between 5 and 15 m) due to extensive exploitation of groundwater and lack of recharge was recorded.

7 Conclusion and Recommendations

This chapter summarizes the physical and hydrogeological settings of the Quaternary aquifer in Western Nile Delta including land use, meteorological data, topography, soil classification, drainage as well as irrigation systems, boundary conditions, and hydraulic parameters. The groundwater aquifer is mainly recharged by Rayah El Beheiry, Rayah El Nasseri, and El Nubariya canals in their first reaches. Discharge from the aquifer takes place through the outflow into the Rosetta branch and drainage system, evapotranspiration, inter-aquifer flow of groundwater, and direct abstraction by production wells. Groundwater aquifer in Western Nile Delta is in direct connection with Rosetta branch and main canals. A GIS-based model has been developed with the conjunction of MODFLOW groundwater model to simulate the water resources in the Western Nile Delta. The model has provided a useful tool to evaluate the groundwater resources status and to test the different proposed management alternatives in the region of West Nile Delta. The model was calibrated against the average annual groundwater heads of about 60 observation wells during the period from 1990 to 2002.

Results showed that, in case of planned reclamation completion for 460,000 feddan, the regional water balance indicated that the extra abstraction must not increase more than 1.0 BCM/year to avoid aquifer exploitation. Water budget in Western Nile Delta shows the need to increase the surface water quantity by 2.1

BCM through the newly proposed canal to fulfill the required irrigation demands for the new reclamation area. Results also clearly indicate that reducing the surface water inflow and increasing the dependency on groundwater abstraction and irrigation system improvement would decrease the annual aquifer potentiality by about 91%. These results illustrate the need for a more detailed analysis concerning the effect of irrigation improvement and thus could be considered as preliminary regional evaluation for testing the alternative water management scenarios in Western Nile Delta area.

It is strongly recommended to study the effect of saltwater intrusion on the groundwater quality for a better understanding of the water management scenarios especially if drainage water reuse is implemented. To avoid groundwater deterioration at Western Nile Delta, it is recommended to apply well-controlled license system of wells abstraction to update the wells inventory database. Some baseline data is available, but it is full of inconsistency due to several reasons among them the unregistered or unofficial abstraction from wells. Updating this information would be a first step and additional research will be needed for the optimum operation and maintenance of the whole groundwater aquifer system in Western Nile Delta.

References

1. Soliman SM, Fattah MK, Ahmed MG (2014) Sustainable development and management of water resources in West Nile Delta, Egypt. Int J Sci Eng Res 5(11):1296–1310
2. MWRI (Ministry of Water Resources and Irrigation) (2007) Environmental and social impacts assessment and framework management plan, the West Delta conservation and irrigation rehabilitation project. Ministry of Water Resources and Irrigation, Giza
3. MWRI (Ministry of Water Resources and Irrigation) (2005) National water resources plan for Egypt 2017
4. MWRI (Ministry of Water Resources and Irrigation) (2012) Strategy of water resources of Egypt till 2050
5. Gogu RC, Carabin G, Hallet V, Peters V, Dassargues A (2001) GIS-based hydrogeological databases and groundwater modeling. Hydrogeol J 9:555–569
6. Arshad A, Zulfiqar A (2012) Integration of groundwater flow modeling and GIS. In: Nayak P (ed) Water resources management and modeling. InTech, London, pp 239–262
7. Kharad SM, Srinivas Rao K, Rao GS (2002) GIS-based groundwater assessment model. Available at the GIS Development hope page: http://www.gisdevelopment.net/application/nrm/water/ground
8. Sarma B, Saraf AK (2002) Study of land use – groundwater relationship using an integrated remote sensing and GIS approach. http://www.gisdevelopment.net/application/nrm/water/ground
9. Singh AKr, Prakash SR (2003) An integrated approach of remote sensing, geophysics, and GIS to evaluation of groundwater potentiality of Ojhala Subwatershed, Mirzapur District, India. http://www.gisdevelopment.net/application/nrm/water/ground
10. El Molla AM, Dawoud MA, Hassan MS, Ewea HA, Mohamed RF (2005) Integrated management of water resources in Western Nile Delta, management scenarios. In: Proceedings of the fourth international symposium on environmental hydrology, American Society of Civil Engineering, Cairo

11. El Abd AE (2005) The geological impact on the water bearing formations on the area Southwest Nile Delta, Egypt. PhD thesis, Faculty of Science, Menoufia University, Cairo
12. Diab MS, Mohammed MA, Rizk ZS (1995) The role of geology, hydrogeology, and human activities in the contamination of shallow water resources northwest of the Rosetta Nile Branch, Egypt. J Fac Sci United Arab Emirates Univ 8(2):260–291
13. Saleh MF (1980) Some hydrogeological and hydrochemical studies on the Nile Delta. MSc thesis, Faculty of Science, Ain Shams University, Cairo
14. Sobeih M, El-Arabi EN, El-Deen YH, Awad BS (2017) Management of water resources to control groundwater levels in the southern area of the Western Nile Delta, Egypt. Water Sci 31(2):137–150
15. Nour El Din S (2009) Environmental impacts on the surface and groundwater system in the area surrounding the southern part of the Rosetta branch – Nile Delta – Egypt. MSc thesis, Faculty of Science, Menoufia University, Menofia Governorate
16. RIGW/IWACO (1991) Monitoring and control groundwater pollution in the Nile Delta and adjacent desert areas. El Kanater El Khairia, TN 77.01300-91-12
17. Zaghloul MG (1985) Flow distribution through groundwater aquifer of the Nile Delta. MSc thesis, Faculty of Engineering, Alexandria University, Alexandria
18. Shahin M (1985) Hydrology of the Nile Basin. Development in water science 21. Elsevier Science, Amsterdam, p 575
19. Laeven MT (1991) Hydrogeological study of the Nile Delta and adjacent desert areas in Egypt with emphasis on hydrochemistry and isotope hydrology. MSc thesis, Free University, Amsterdam
20. RIGW (1992) Hydrogeological map for the Nile Delta area. Scale 1: 500000. Research Institute for Groundwater, El Kanter El Khairia
21. Bahr B (1995) Nile Delta aquifer with emphasis on saltwater intrusion in the northern area. MSc thesis, Technical University of Berlin, Institute for Applied Geoscience, Berlin
22. Farid MS (1985) Management of groundwater system in the Nile Delta. PhD thesis, Faculty of Engineering, Cairo University
23. RIGW/IWACO (1990) Development and management of groundwater resources in the Nile Valley and Delta: assessment of groundwater pollution from agricultural activities. Research Institute for Groundwater, El Kanater El Khairia
24. Warner JW, Gates TG, Attia FA, Mankarious WF (1991) Vertical leakage in Egypt's Nile Valley: estimation and implications. J Irrig Drain Eng 117(4):515–533
25. RIGW/IWACO (1998) Environmental management of groundwater resources (EMGR): identification, priority setting and selection of area for monitoring groundwater quality. Technical report TN/70.00067/WQM/97/20, Research Institute for Groundwater (RIGW), Cairo
26. Eltarabily MG, Negm AM, Yoshimura C, Saavedra OC (2017) Modeling the impact of nitrate fertilizers on groundwater quality in the southern part of the Nile Delta, Egypt. Water Sci Technol Water Supply 17(2):561–570
27. TRIWACO (1992) Groundwater modelling system. Manual for version 6.0. IWACO, Rotterdam
28. ESRI (1992) ARC/INFO data model, concepts, and key terms. ESRI, Redlands
29. Dawoud MA, Darwish MM, El-Kady MM (2005) GIS-based groundwater management model for western Nile delta. Water Resour Manag 19:1–20
30. Morsy WS (2009) Environmental management to groundwater resources for Nile Delta region. PhD thesis, Faculty of Engineering, Cairo University, Cairo
31. Eltarabily MG (2017) Experimental and numerical study of fertilizers impact on Nile Delta groundwater aquifer. PhD thesis, Egypt Japan University of Science and Technology, Alexandria
32. Omar Mohie El Din M, Moussa Ahmed MA (2016) Water management in Egypt for facing the future challenges. J Adv Res 7:403–412

Part XI
Conclusions

Update, Conclusions, and Recommendations for Groundwater in the Nile Delta

Abdelazim M. Negm, El-Sayed E. Omran, and Sommer Abdel-Fattah

Abstract The current Egyptian situation is framed by land and water scarcity, which are under severe stress. The Nile Delta is well known as one of the most densely populated deltas in the world. On the one hand, soil and water resources are at the center of sustainable development and are critical for socioeconomic development. On the other hand, groundwater is considered the second main source of water supply in Egypt after the Nile River, although it represents less than 3% of the total water supply. The Nile Delta aquifer is among the largest underground freshwater reservoirs in the world, and it has been extensively utilized and conjunctively used with the Nile water to cope with the increased demands due to implementing economic development plan in Egypt. The major challenge facing the Nile Delta aquifer is it receives its water (recharging) from the Nile River which is threatened nowadays by the construction and most probably the improper operation of the GERD particularly over the long term. This chapter encapsulates the key groundwater sustainability (in terms of conclusions and recommendations) of the existing main agri-food system and presents insights derived from the cases in the volume. Also, some (update) findings from a few recently published research related to the sustainability covered themes. This chapter presents the main current challenges facing the groundwater aquifer with the set of recommendation to protect the Nile

A. M. Negm (✉)
Water and Water Structures Engineering Department, Faculty of Engineering, Zagazig University, Zagazig, Egypt
e-mail: Amnegm@zu.edu.eg; Amnegm85@yahoo.com

E. E. Omran
Soil and Water Department, Faculty of Agriculture, Suez Canal University, Ismailia, Egypt
e-mail: ee.omran@gmail.com

S. Abdel-Fattah
McMaster University, Hamilton, ON, Canada
e-mail: abdelfa@mcmaster.ca; sommerab@gmail.com

A. M. Negm (ed.), *Groundwater in the Nile Delta*,
Hdb Env Chem (2019) 73: 731–752, DOI 10.1007/698_2018_335,

Published online: 13 September 2018

Delta aquifer to its sustainability to supply water to the Nile Delta populations and farmers.

Keywords Assessment, Delineation, Egypt, Geophysical methods, Groundwater, Hydrogeochemical, Management, Modeling, Nile Delta, Quality, Salinization, Seawater intrusion, Sedimentology, Sustainability

Contents

1 Introduction

Water resources in Egypt are the Nile, deep groundwater, rainfall, and desalinated water complemented by shallow groundwater in the Nile Delta and both reused agricultural drainage and treated wastewater as non-conventional water resources [1]. About 86% of that water comes from the Nile, and around 11% from underground sources, 2% is recycled [2]. With a population density of 1,500 inhabitants per square kilometer when including the capital [3], the Nile Delta is one of the most densely populated deltas in the world. Groundwater is a precious and most widely distributed resource of the earth, which is required for agriculture, industry, and domestic purposes [4]. Groundwater plays a vital role in water supply everywhere throughout Egypt. Therefore, this chapter will present a general idea of the sustainable agriculture and its importance for Egypt and the researchers. In designing sustainable agricultural production systems, it is necessary to give due consideration to the characteristics of groundwater resources used, which render the resultant production system unsustainable. Therefore, the intention of the volume is to addresses the following main theme based on the latest research findings:

- Overview of the groundwater in the Nile Delta
- Groundwater use
- Sedimentology and hydrogeophysical characteristics
- Groundwater investigations and aquifer characterization
- Groundwater contamination and degradation
- Saltwater intrusion
- Delineation of groundwater flow and seawater intrusion
- Groundwater modeling
- Groundwater usage and groundwater quality assessment
- Groundwater management for sustainability

The next section presents a brief of the important findings of some of the recent (updated) published studies on the groundwater in the Nile Delta, then the main conclusions and recommendations of the volume chapters extracted mainly from the presented chapters.

2 Update

The following is the major update for the book project based on the main book theme.

Overview Groundwater is an important natural resource. Many agricultural, domestic, and industrial water users rely on groundwater as the main (and probably the sole) source because of its low cost and high quality. However, in recent years it has become clear that human activities and climate change have a negative impact on both quantity and quality of groundwater resources. The depletion of groundwater may occur due to excessive pumping and contamination of the groundwater by waste disposal or other activities. The Mediterranean region is characterized by a strong development of coastal areas with a high concentration of water-demanding human activities, resulting in weakly controlled withdrawals of groundwater, which accentuate the saltwater intrusion phenomenon [5]. Zeidan et al. [6] studied the groundwater quality management in the Middle Nile Delta through a combination of laboratory and numerical modeling work. Environmental isotope techniques were used to investigate the recharge sources and the nitrogen compound pollution sources. MODFLOW and MT3DMS were employed numerically by applying the method of finite differences for solving the three-dimensional problem [4, 7]. The obtained results include the prediction of water levels and solute concentration values distribution in the area at different times.

However, with a population density of 1,500 inhabitants per square kilometer when including the capital [3], the Nile Delta is one of the most densely populated deltas in the world. On the one hand, soil and water resources are at the center of sustainable and socioeconomic development. On the other hand, water and soil resources are under threat because of the way we treat it. High Dam has freed the Egyptian Nile Valley and Delta from floods but also dramatically confines sediment transport and water availability. Since the 1960s, the Egyptian Nile is completely controlled by the High Dam and a series of barrages along its course to the Mediterranean Sea. Dams most prominently the Renaissance Dam in Ethiopia will put additional pressure on the territory [8]. As our soil and water resources come under increasing pressure, hard decisions should be made so that resources are not degraded or tipping points reached. Water is an irreplaceable resource, and it is only renewable if well managed [9]. As the demand on freshwater increases, consideration regarding non-conventional water resources consequently increases in Egypt. The demand for adequate and safe supplies of water is becoming crucial especially in the overpopulated Egypt.

Groundwater Use The irrigation demand of the agricultural sector already amounts to “more than 70% of global water-supply withdrawals and about 85% of global water-resource consumption – and it is estimated that groundwater sources provide 43% of all water used for irrigation” [10]. Food production requires the largest quantity of water, with groundwater resources providing more than 40% of all water used globally for irrigated agriculture [10, 11]. Groundwater is stored in aquifers, which are water-bearing rock formations that hold water in the inter-particle pore space and cracks within rock material. Shallow groundwater in the Nile aquifer cannot be considered a separate source of water. The aquifer is recharged only by seepage losses from the Nile, the irrigation canals and drains and percolation losses from irrigated land. Hence, its yield must not be added to Egypt's total water resources. Therefore, it is considered as a reservoir in the Nile River system with a huge capacity, but only 7.5 BCM/year is rechargeable live storage.

Furthermore, Groundwater is an essential source of freshwater. Groundwater abstraction could significantly contribute to agriculture directly as it used in irrigation mainly or in areas of supplemental irrigation. Also, groundwater could use in irrigation via indirect way through control drainage system, so water table will decrease and thus, contribute to crop water requirement. In arid and semiarid regions, coastal aquifers represent an essential source of freshwater. In these regions, groundwater resources are overexploited to meet the development and urbanization of coastal regions [12]. In Egypt, to face the challenge of high population density and a constant freshwater resource from the Nile River, Egypt adopted many policies to establish new agricultural communities outside the overpopulated Nile Delta and Nile valley. These newly reclaimed areas depend almost exclusively on groundwater as a water resource. The growth of irrigation activities, urbanization, and industrialization leads to rapid growth of groundwater abstraction.

Sedimentology and Hydrogeophysical Characteristics These two basic information are discussed in the following. First, basic sedimentological information should shape some portion of any evaluation of possibly polluted destinations and part of examinations concerning the scattering and catching of contaminants in fluvial frameworks [13]. This information is additionally required for sound ecological administration to guarantee that planning policies are good with common environmental limitations. When the aquifer is unconfined and shallow, there is a high possibility of contaminants whatever source and kind, particularly when they are near a highly populated area with high anthropogenic effect [14]. Omorogieva and Imasuen [15] estimated that aquifer deposits come from the close surface, igneous outcrops, and volcanic and sedimentary rocks.

Second are the hydrogeophysical characteristics of the central Nile Delta aquifer. Well logs are used widely in the exploration of mineral and hydrocarbon resources because they provide detailed and reliable information about the geometrical and petrophysical characteristics of the geological structures. It is aimed to estimate the spatial variability of the formation lithology, porosity, permeability, groundwater salinity, and hydraulic conductivity. In hydrogeophysics, the main target of well log analysts is to estimate the layer thickness, water saturation, groundwater salinity, effective porosity, clay content, and aquifer hydraulic conductivity accurately as possible [16]. As the Nile Delta aquifer is so important freshwater source for the highly populated area in Egypt, determining the spatial change in its petrophysical, hydrogeological, and hydrogeochemical characteristics is of great importance.

Groundwater Investigations and Aquifer Characterization Two areas are used to investigate the aquifer. The first is the resistivity characterization of the aquifer in coastal semiarid areas. Geoelectrical methods are increasingly popular in detecting and characterizing aquifers as the resistivity of sediment and rocks depends mainly on their water and material contents [17]. One of such electrical methods is direct current resistivity (DCR) technique. In DCR method, the potential difference is measured via injecting DC into the ground with electrodes at the ground surface. The DCR method has received great attention because of its potential applications in the field of hydrogeology and saltwater intrusion tomography [18]. In the arid and semiarid region, the DCR method is popularly used for groundwater exploration and aquifer mapping [19]. Usually, DCR method is preferred, but the resistivity data interpretation suffers from a major flaw: the inability to discriminate between surface and bulk conductivity. This causes unrealistic interpretation in hydrogeophysics. Such main problem can be resolved by using spectral induced polarization (SIP) method.

The second is hydrogeophysical investigations at El-Nubariya-Wadi El-Natrun area. An adequate supply of freshwater is one of the prerequisites for every type of developmental programs. Water conservation is very important because usable water is a limited resource. As water supplies dwindle, water shortages in many areas of the world constitute a major problem for both agriculture and nations. Groundwater is the main source for domestic, industrial, and agriculture uses in most of the newly reclaimed areas in the western Nile Delta region [20]. Geophysics, especially geoelectric techniques, have been successfully used to detect the freshwater/saltwater interface in coastal aquifers. Resistivity surveys are often used to search for groundwater in both porous and fissured media (e.g., Tarabees and El-Qady) [21]. The methods provide detailed information about the geometry, source, and amount of contamination.

Groundwater Contamination and Degradation Two groundwater contaminations and degradation issues are updated. The first issue related to the salinization and origin of the coastal shallow groundwater aquifer, northwestern Nile Delta, Egypt. There are several processes causing groundwater salinization. They include seawater intrusion due to intense aquifer exploitation, interactions with deep saline palaeowaters, water-rock interaction [22]. Other sources that cause salinization include the dissolution of evaporites [23], the evaporation of freshwater, and pollution by untreated wastewater. The Nile Delta aquifer is the principal groundwater source for Egypt. The annual rate of groundwater withdrawal from the Nile Valley and Delta aquifers increased from 5.5 billion m^3 in 2000, 6.13 billion m^3 to more than 7 billion m^3 in 2016 [24]. Over-pumping is the most severe issue of the groundwater mainly in the Nile Delta aquifers that are followed by salinization through saltwater intrusion. The coastal aquifer, northwestern Nile Delta, is characterized by the presence of brackish water that endangers, often irreversibly, the future of water resources in the area. Over-pumping of groundwater in the coastal aquifer has severely degraded water quality due to seawater intrusions [25, 26].

The second issue is the soil aquifer treatment (SAT) system design equation for organic micropollutant removal. SAT system is considered attractive unconventional water resources for Egypt, which is suffering water sacristy. The use of SAT can provide treatment for the wastewater and recharge in groundwater aquifers. While guidelines are available for the use of SAT system in Egypt for removal of nitrogen and organic matter, no guidelines are available for the SAT removal potential of organic micropollutants. a prediction model for the organic micropollutants (OMPs) removal in SAT system was provided which is based on previous works on soil aquifer treatment system and analysis models [27–30].

Saltwater Intrusion Two basic factors to investigate and control of saltwater intrusion are identified. First is the investigation of saltwater intrusion in coastal aquifers. The problems of saltwater intrusion into groundwater have become a considerable concern in many countries with coastal areas [26]. New methods to control saltwater intrusion in coastal aquifers are presented and discussed in details; the advantages and disadvantages of each method were highlighted. Finally, investigation and control of saltwater intrusion in Egypt, especially in the Nile Delta aquifer, are discussed. The possibility of applying new methods to control saltwater intrusion in Egypt is presented.

Second is the control of saltwater intrusion in coastal aquifers. A number of methods had been adopted to control seawater intrusion to protect groundwater reserves in coastal aquifers [25]. A number of numerical models had been developed and used to help to understand the relevant process that causes saltwater intrusion in coastal aquifers and identifies suitable methods of control. Extensive research has been carried out to investigate saltwater intrusion in coastal aquifers. However, only limited amount of research has been directed to study the control of saltwater intrusion. The coastal aquifers' management requires careful planning of withdrawal strategies for control of saltwater intrusion. Therefore, efficient control of seawater intrusion is very important to protect groundwater resources from depletion. From the literature, a number of control methods can be applied in Egypt. The selection of the method is very important and should consider environmental, social, and economic issues. The most suitable methods that can be applied to control saltwater intrusion in the Nile Delta are optimization of abstraction rates, pump and treat, and recharge and abstraction. These three techniques have been applied in different locations of the world (e.g., Gaza, Florida USA, India). These three methods are capable of preventing the intrusion of saltwater into the Nile Delta aquifer, but the cost of applying these techniques may be high and may have some environmental and social impacts.

Delineation of Groundwater Flow and Seawater Intrusion Three techniques are used to delineate of groundwater and seawater intrusion. The first technique is using 1D subsurface temperature profiles to characterize the groundwater flow system in the northwestern part of the Nile Delta, Egypt. Changes in climatic conditions, particularly surface warming, are influencing the subsurface temperature and recorded as an inversion in the subsurface temperature profiles. All the investigations deal with the relationship between the subsurface temperature and the groundwater flow system expect the temperature profile in the well is illustrative of the temperature in the aquifer. In open or screened boreholes, the difference in water levels of regional scale flow system that interface with a well makes vertical flow inside the borehole. Recently, some studies dealing with the relation between the subsurface temperature and the groundwater flow systems in Egypt were performed. Salem and Osman [31] and Salem and El Bayumy [32] modeled the vertical 2D groundwater flow and heat transport to estimate the vertical groundwater flow velocities in the study area and east Wadi El-Natrun region, respectively.

The second technique is using of geoelectrical resistivity to delineate the seawater intrusion in the northwestern part of the Nile Delta, Egypt. Seawater intrusion becomes a severe problem in arid and semiarid regions where the groundwater constitutes the main freshwater resource. Mixing of only 3% seawater with freshwater in a coastal aquifer would render the freshwater resource unsuitable for human consumption [33]. Diverse methodologies have been received to evaluate seawater intrusion. Salem and Osman [34] have utilized geochemical strategies in view of modeling technique, stable isotopes, and hydrochemical data to evaluate the seawater intrusion.

The third technique is integrated subsurface thermal regime and hydrogeochemical data to delineate the groundwater flow system and seawater intrusion in the Middle Nile Delta, Egypt. The Nile Delta aquifer system is considered a leaky aquifer in the southern and middle parts and a free aquifer in the western and eastern borders, where the thickness of the top Holocene deposits reaches its minimum value. The integrated tracer technique between subsurface temperature and water chemistry was a good technique for tracing the groundwater flow system and seawater in the Nile Delta Quaternary aquifer. The most dangerous information given from this research is that the groundwater in the Nile Delta is not affected only by seawater intrusion but also affected by hypersaline brine water inland propagation. Seawater affects the upper 200–250 m, but the hypersaline water was indicated in Motobes well which is of 420 m depth. The upward seawater intrusion flux rates in Motobes wells were 2.8 m/year which is much higher than the groundwater recharge flux at Tala well which was 0.8 m/day.

Groundwater Modeling The objective of this part is to model and manage the groundwater resources using different techniques. First is the integrated groundwater modeling for simulation saltwater intrusion in the Nile Delta aquifer, Egypt. Abd-Elhamid et al. [12] conducted a coupled transient (2D FEST) finite element model in order to simulate the fluid flow and the solute transport in both saturated and unsaturated zone for studying the saltwater intrusion in the Nile Delta aquifer under the impacts of climate change. The results of the model indicated that the Isoline 35 intruded inland into the Nile Delta aquifer by a distance of about 64 km from the shoreline, while the Isoline 1 moved inland into the aquifer at a distance of 112 km at a cross section in the central part of the Nile Delta. Nofal et al. [35] used the three-dimensional finite difference model SEAWAT to simulate the saltwater intrusion in the Nile Delta aquifer with considering the available heterogeneity data and the variation of groundwater density values in the recently drilled boreholes. The model results showed that the appreciable agreement related to the flow fluxes and piezometric head. The model delineated the saltwater intrusion in the Nile Delta aquifer. Wassef and Schüttrumpf [36] built a three-dimensional finite element models in the western area of the Nile Delta by using FEFLOW software to study the saltwater intrusion under different climate change scenarios. The results showed that by the year 2100, it is expected that the interface of the seawater will reach a maximum of about 43 km according to RCP 2.6 scenarios, while it is expected to reach 57 km according to RCP 8.5 scenario. The groundwater overexploitation will cause an increase in the salinity concentration to about 5,000 mg/l.

Second is the groundwater potential in the New Valley South West of the Nile Delta in Egypt. Three groundwater models were used in this respect. A local model using GIS accompanied by the visual basic was first prepared to adjust the model properties. A second model using MODFLOW was furnished to calibrate the boundary conditions of the main model. The third model was using the main model as a final model to find the main objectives, which are the groundwater potentiality and its management in Dakhla Basin. The scenario applications could allow for an increase in reclamation at Dakhla Oasis by 15%, with the condition of safe drawdown values less than 60 m for 100 years. The study provides the benefits of applying the modeling techniques. Numerous valuable input for the national development plan in Egypt is presented. The study found that it is important to seek an alternative water resource to compensate for the groundwater depletion.

Groundwater Usage and Groundwater Quality Assessment Two strategies are used for groundwater quality assessment. First is the hydrogeochemistry and quality assessment of groundwater under some central Nile Delta Villages, Egypt. Water quality index (WQI) is the best strategy for measuring water quality. Various water quality parameters are incorporated into a mathematical equation to rate water quality, estimating the suitability of water for drinking [37]. WQI enables comparison between various samples. The index is simplifying a complex dataset into easily estimated, usable data and understandable even by lay people. Many researchers have investigated the water quality in the Nile Delta, among them Negm and Armanuos [38], Negm and Eltarabily [39], Sharaky et al. [20], and Salem and Osman [34].

Second is the assessment of the groundwater quality for drinking and irrigation purposes in the central Nile Delta Region, Egypt. The Nile Delta groundwater aquifer is influenced by many environmental factors like human activity on the surface, seawater intrusion, and the type of aquifer sediment. Therefore, the decision-maker must be taken into consideration the lateral and vertical changes in groundwater quality when digging wells for different uses to achieve the highest utilization of wells. Low quality of water unfavorably influences human health and plant development. In developing nations like Egypt, about 60% of all infections are specifically occurring due to the poor quality of the drinking water. The spatial distribution of water quality significant components should be compared with the geology and area land use/land cover spread maps in a GIS environment. Therefore the chemical processes of water and the methods of their acquisition could be understood [40].

Groundwater Management for Sustainability To manage groundwater, two plans were developed. The first plan is related to groundwater management for sustainable development East of the Nile Delta aquifer. Integration of GIS and groundwater modeling system (GMS) played a significant role in the construction of a 19-layer groundwater flow model and effectively enabled smooth management to the required database. From the budget analysis of groundwater flow simulation in the three-dimensional model using the MODFLOW code indicated that pumping discharge from production wells efficiently controls the rising water level in the confined part of the Quaternary aquifer and the infiltration from agricultural fields is the key factor in the unconfined part. The up-to-date land use maps should be provided and interpolated to the GIS model for accurate model simulation [41]. The results of the model exercise constitute a foundation for sustainable water resource management in the East Nile Delta, and the calibrated model can be used to evaluate the efficiency of multidisciplinary policies shortly.

The second is the groundwater management for sustainable development plans for the Western Nile Delta. In the Western Nile Delta region, groundwater is the primary source for domestic, industrial and agriculture use. With the expansion of developing activities in this area, it is essential to develop a groundwater management strategy to avoid any environmental impacts on the aquifer system due to the future extensive abstraction of groundwater. For better management of groundwater resources, it is crucial to have enough data about the physical and hydrogeological settings for the study area. Physical parameters include land use, meteorological data, topography, soil classification, and drainage as well as irrigation systems. Hydrogeological setting comprises the aquifer system, boundary conditions, hydraulic parameters for all aquifer layers, and monitored groundwater levels. Data used in groundwater modeling consist of the aquifer system stress factor, the aquifer system geometry, and the hydrogeological parameters. Stress factors for groundwater flow include effective recharge, pumping volumes, water surface flow exchanges, etc. Appropriate aquifer system geometry can be determined using geological information (maps and cross sections), topographic maps, as well as contour maps of the upper and lower limits for the aquifer strata and aquitards. Links can be organized between MODFLOW, groundwater model, and the GIS [41]. The GIS software is used to preprocess and post-process the spatial data. Recently, the use of GIS has grown rapidly in groundwater assessment and management researches.

3 Conclusions

Throughout the current volume, we were able to reach several conclusions, which have been drawn from this volume's chapters. Besides methodological insights, this chapter originates key lessons from the cases in the volume, in particular, the promising characteristics of both the historical and current local food system. These conclusions are important to increase sustainable food supply in Egypt as the agricultural food production is mainly dependent on water. The following conclusions could be stated based on the materials presented in all chapters of this volume:

1. The main sources of the Nile Delta aquifer come from Nile River branches (Rosetta and Damietta). The direct seepage from irrigation and drainage systems and the irrigated and cultivated land are the main source of the Nile Delta aquifer recharge. Recharge from rainfall is minor, but it could be significant over other areas outside the Nile Delta where flash floods take place.
2. Use of groundwater in agriculture activity is effective in the newly reclaimed land in the deserts where the Nile water is not available or accessible. Most of the groundwater in Egypt are nonrenewable water except the shallow groundwater in the Nile valley and delta land and its fringes in addition to some famous depression and oasis-like Wadi El-Natrun in West Delta and Siwa Oasis south the northwest coast of the Mediterranean.
3. It is expected that groundwater will cover about 20% of the total water supply in the upcoming decades especially in the newly reclaimed areas along the desert fringes of the Nile Delta and Valley. Among the main six aquifers in Egypt, the Nile aquifer represents 87% of the total groundwater pumping.
4. Egypt is obliged to reuse of both agricultural drainage water and treated sanitary wastewater, in addition to recycling of treated industrial wastewater, to offer about 17 BCM which is helping to minimize the water shortage to be 11 BCM instead of 28 BCM.
5. Reuse of marginal poor quality water in Egypt was established and will be continuous as a fixed water policy in the future to face the water scarcity as an obligation.
6. The Nile Delta aquifer suffers from the apparent decline of its piezometric surface due to the increase in the pumping and an increase in water salinity as well.
7. Agriculture activities in Egypt are making up more than 80–90% of groundwater withdrawals due to lack of sufficient precipitation and a shortage of surface water to grow crops leading the farmers to use water from the underground to irrigate. Consequently, the amount of nonrenewable groundwater used for irrigation was doubled in Nile Valley and Delta.
8. The annual groundwater abstraction in the Nile aquifer system and fringes is about 4.6 billion m^3. Another 0.5 billion m^3 is abstracted from the desert aquifers and the coastal areas. Groundwater abstraction is expected to increase to 11.4 billion m^3. Groundwater could be used in irrigation directly as usual practice or in an indirect way through controlling drainage system, so water table will decrease and thus, contribute to crop water requirement, except where shallow groundwater exists which negatively affect agriculture production as secondary soil salinization and waterlogging. These two challenges could be overcome through using surface and subsurface drainage systems.

9. It is expected that groundwater would be in a deficit in the future compared to the situation in the past. Modeling technique could tell us the approximate deficit. More details about the groundwater and its usage in Egypt and particularly in the Nile Delta are presented in Part II of this volume.
10. The majority of sediments belong to the river sands and river processes with multi-directional depositional currents. The main clay minerals in the intercalated clay lenses of the aquifer are smectite and kaolinite. The essential carbonate minerals include calcite and dolomite, whereas noncarbonate minerals are represented by quartz, feldspar, hematite, and gypsum.
11. The use of the electrical resistivity and gamma-ray logs for 34 wells revealed that the Nile Delta aquifer in its central part consists of 2 main formations. The upper is the Belqas Formation, which is of a clay nature, and the lower formation is of a sandy nature and called Mit Ghamr Formation. The petrophysical, chemical, and hydrogeological characteristics of the Nile Delta aquifer in its central part were identified and presented in Salem et al. [42] in this volume.
12. The results of applying the direct current resistivity (DCR) method with the available boreholes, as an integrative approach over the East Nile Delta, proved the efficiency of the DCR method for hydrogeological evaluation in coastal semiarid areas. Also, the validity of the DCR method to predict the TDS and K using the empirical relationships for this area is approved. For further details, the reader is advised to consult in Attwa and Ali [43].
13. DCR technique is found to be very appropriate in the determining subsurface layers at shallow depths with a high resolution. The time domain electromagnetic (TEM) technique is more effective than DCR technique in detection more details of subsurface layers which could not be detected by vertical electrical soundings (VES) method, especially in the deep depths. The application of these techniques to the area extending from Wadi El-Natrun city to El-Nubariya city along the Alexandria-Cairo desert road indicated that the underground formation is divided into Pleistocene and Pliocene aquifers. The hydrogeological details of these two aquifers and the main findings of using TEM and VES are reported in Ibraheem and El-Qady [44] in this volume.
14. The origin of groundwater in coastal aquifer northwestern Nile Delta is meteoric in origin (Nile River and local rainfall), which is mixed with marine water due to seawater intrusion. The groundwater of the shallow coastal aquifer, northern Nile Delta, is brackish and is unsuitable for drinking purposes, but it may be used safely for irrigation of some suitable crops. Moreover, the coastal aquifer is recognized to be at high risk of increasing salinization. The qualitative and quantitative analysis of the hydrogeochemical can be found in details in Sharaky et al. [45] in this volume.
15. Primary sources of excess salinization of the groundwater are seawater intrusions where the review of the studies related to seawater intrusion revealed that the seawater intrusion to the Nile Delta aquifer represents a severe risk to groundwater resources particularly in the coastal part of the aquifer. The seawater intrusion in the Nile Delta aquifer has extended to a distance of more than 100 km from the Mediterranean coast. The second source of salinization of

groundwater is the aquifer rocks that contain carbonates and gypsum. On the one hand, the concentration of the examined significant ions is higher than maximum contaminant level (MCL), on the highest allowable concentrations of a contaminant in drinking water, which is set by the U.S. Environmental Protection Agency. The nutrient content such as nitrates is also higher than the standard values, which are mainly produced from rural sources. On the other hand, several methods can be used to control the seawater instruction. These methods include (1) abstraction of saline water and recharge using surface ponds; (2) abstraction, desalination, and recharge (ADR); and (3) treatment, recharge, abstraction, and desalination (TRAD).

16. Moreover, to limit the pollution of the aquifer by the organic micropollutants (OMPs), one of the most important methods to remove OMPs at the water and wastewater treatment plants is discussed which utilizes soil aquifer treatment (SAT). The mechanism of organic micropollutants removal using extreme learning machine (ELM) method was evaluated. Analysis of numerical model results indicated the acceptable accuracy of ELM models in prediction of the pollutant removal efficiency during the SAT operation. Interested reader who would like to know how this conclusion is derived can consult [46] in this volume.
17. Making use of the connection between the subsurface temperature distribution and the groundwater stream framework, the investigated area in northwestern part of the Nile Delta is considered a case for the abnormal thermal system. It is observed that the subsurface temperatures in the desert within the study area, the recharge territory, have higher values while in the discharge region of the old agricultural land have low values. Also, the geothermal gradient is found to be superior to the thermal system for the groundwater stream framework in such region. It is estimated that the recharge flow flux varies among the recharge profiles. Groundwater discharge flux is also variable. Seawater intrusion probably affects the thermal regime in the discharge area to the northern part. This estimation is based on the occurrence of a colder temperature zone in the northern parts of the area and a downward reduction in the discharge rate of the temperature profiles.
18. Using of geoelectrical resistivity technique in the northwestern part of the Nile Delta has indicated that the area has four geoelectrical layers as follows: first, the surface layer composed of sand, clay, and silt; second, the aquifer layer composed of sand (fine sand with intercalations of clay and medium to coarse sand); third, the clay layer with a thickness from 2 to 69 m; and fourth, the brackish to saltwater intrusion zone with depth range from -35 m bmsl in the northwest to -100 m bmsl in the south.
19. The resistivity of the aquifer layer is increased toward the south and east directions where both the freshwater quality and thickness are increased. However, the resistivity of the aquifer layer is decreased toward the northwest where the effect of seawater intrusion is decreased. The reader is advised to read in [47] for more details.

20. Using the subsurface temperature and water chemistry as integrated tracer technique indicated that the Nile Delta Quaternary aquifer has two flow systems, one is regional and the other is local. Both flow systems are affected by seawater intrusion. The regional one is recharged south Tala and discharges northward in the area from kafrelarab (south Tanta city) until El Karada (south Kafr El Sheik city). The estimates of the flux indicate that fresh groundwater starts to move vertically upward southern Tanta city. The subsurface thermal regime and the hydrogeochemical data in the study area gave an image about the spatial extent of seawater intrusion in the Nile Delta. The most critical conclusions are that the groundwater in the Nile Delta is not affected only by seawater intrusion but also affected by hypersaline brine water inland propagation.
21. The recharge of the Nile Delta aquifer from both rainfall and seepage from the canals network was estimated. In addition, an integrated 3D groundwater model for the Nile Delta aquifer using MODFLOW and simulation of the saltwater intrusion using SEAWAT code combined with the Nile Delta aquifer model are introduced. Different climate change scenarios are included in the modeling. The modeling results revealed that a rise in sea level by 1.0 m and decreasing the groundwater head by 1.0 m increased the saltwater intrusion and led the concentration line 35,000 mg/l to advance further inland into the aquifer by a distance of 10.20 km measured at the aquifer bottom boundary compared with the base case. The interested reader is advised to consult [48] in this volume.
22. The use of the modified gray model and the modified genetic algorithms on Sahara desert indicated that if the present extraction rate is expanded, the groundwater piezometric level continuously declines and drops below the economical piezometric level until the year 2100 at the end of the simulation period leading to possible depletion of groundwater in the Kharga Oasis. The full set of conclusions and the scenarios results are available in Mahmod [49].
23. Modeling results via GIS and MODFLOW for the middle part of Egypt's Western Desert including the three oases, Farafra, Dakhla, and Kharga, as a part of the New Valley area revealed that Dakhla Oasis is the only place in the New Valley area where land reclamation development can safely take place. More details on how these conclusions were obtained; the reader is advised to read Soliman and Solimsn [50].
24. Assessment of groundwater quality using water quality index (WQI) based on samples collected from 16 villages in central Nile Delta during 2016 indicated that the water quality ranged between good and unfit with arsenic and ammonia undesirable concentrations in most of the collected groundwater samples. It was estimated that the quality of groundwater in the studied aquifer is fairly accepted and in the most parts of the study area is reasonable for irrigation except for the most northern part of the Nile Delta where pumping should stopped there. It is highly recommended to read Salem [51, 52] for the detailed analysis and additional conclusions.

25. Groundwater quaternary aquifer western Nile Delta is mainly recharged by Rayah El Beheiry, Rayah El Nasseri, and El-Nubariya canal in its first reaches. Discharge from the aquifer takes place through the outflow into the Rosetta branch and drainage system, evapotranspiration, inter-aquifer flow of groundwater, and direct abstraction by production wells. For the Eastern part of the aquifer, the primary recharge of the aquifer is the infiltration from the Ismailia Canal and Damietta branch which significantly increases during the high pumping stress from wells.
26. Modeling using GIS and MODFLOW for the Western Nile Delta aquifer showed that, in case of planned reclamation completion for 460,000 feddans, the regional water balance indicated that the extra abstraction must not increase more than 1.0 BCM/year to avoid aquifer exploitation. Water budget in western Nile Delta shows the need to increase the surface water quantity by 2.1 BCM through the newly proposed canal to fulfill the required irrigation demands for the new reclamation area. Results also clearly indicate that reducing the surface water inflow increasing the dependency on groundwater abstraction and irrigation system improvement would decrease the annual aquifer potentiality by about 91%. These results illustrate the need for a more detailed analysis concerning the effect of irrigation improvement and thus could be considered as preliminary regional evaluation for testing the alternative water management scenarios in the Western Nile Delta area.
27. Similarly, modeling the Eastern Nile Delta aquifer and the interpretation of the budget analysis of groundwater flow simulation indicated that pumping discharge from production wells efficiently controls the rising water level in the confined part of the Quaternary aquifer and the infiltration from agricultural fields is the key factor in the unconfined part. The up-to-date land use maps should be provided and interpolated to the GIS model for accurate model simulation. The results of the present model exercise constitute a foundation for sustainable water resource management in the East Nile Delta, and the calibrated model can be used to evaluate the efficiency of multidisciplinary policies shortly. More details can be found in Eltarabily and Negm [53] in this volume.

4 Recommendations

A key aspect of groundwater sustainability is the ability to adapt to future challenges. We argue that sustainable systems need built-in flexibility to achieve this goal. Throughout this volume, we noted some areas that could be explored to further improvement. Based on the authors' findings and conclusions, this section offers a set of recommendations providing suggestions for future researchers in exceeding the scope of this book.

1. It is recommended, that the concerned authority in Egypt, to identify all the locations where groundwater wells are operating both official and nonofficial to make their management easier. Also, evaluation of the efficiency of these wells

should be undertaken seriously. Therefore, the concerning authority can take the needed measure to regulate the use of the groundwater wells and find the suitable measure to control the excessive use of the Nile Delta groundwater. This will help to protect the Nile Delta aquifer against severe saltwater intrusion.

2. The future study should be focused on cases where underground reservoirs or aquifers are overused.
3. It is highly recommended to monitor the surface and groundwater regularly to take the needed measures to improve groundwater quality, avoid saline groundwater supply, and prevent groundwater pollution in the Nile Delta region.
4. Future exploration of the groundwater could benefit from lithological, grain size, and mineralogical analysis of the clastic sediments, which are presented in this volume.
5. The join use of the conventional and non-conventional direct current resistivity (DCR) inversion techniques is recommended for hydrogeological evaluation in East Nile Delta. Further research including pumping tests is required to verify the prediction of hydraulic conductivity using the geoelectrical empirical relationships in Nile Delta. Other techniques such as TEM and DC resistivity methods could be applied too.
6. The indicators presented in the volume of the water quality could be used as a guide in case of drilling new boreholes for drinking or agricultural purposes with an update whenever possible.
7. It is worthy to have a detailed investigation to delineate the extension of the brackish water zone.
8. We also recommend installing a tile drainage system to overcome the problem of trapping surface water near El-Nubariya city and are having similar situations.
9. To maintain groundwater resources in the Nile Delta for sustainable use, an integrated management plan should be prepared. The plan should include the following characteristic: preparing digital maps for groundwater distribution and depth and defining the safe extraction; controlling or forbidding drilling of wells especially in North Nile Delta close to the Mediterranean Sea; increasing ground recharge and/or reducing ground abstraction; treatment of sewage and drainage water before using in irrigation; integrated management of applying fertilizers in agriculture areas to prevent groundwater pollution; and applying controlled drainage in agricultural areas in Nile Delta to reduce the transport of agricultural pollutants as pesticides, herbicides, nutrients, and some heavy metals in addition to enhancing water use efficiency.
10. Sustainability of groundwater is a vital issue in Egypt, and it is only possible if water managers, monitoring and characterizing of groundwater resources, local communities, management, and hydrogeologists work together to devise measures and policies by backcasting. Moreover, adapt future measures in achieving the long-term sustainable targets.
11. Governmental regulation of the pumping process from the Quaternary Nile Delta aquifer is urgently needed as the groundwater is not just facing the pollution from surface human activities and seawater intrusion but also suffers

from the deeper hypersaline brine groundwater inland propagation. Integration between subsurface thermal and hydrogeochemical data is a good tool in recognition of the groundwater flow system and seawater intrusion.

12. Subsurface temperature is a good method for tracing the groundwater flow system, where it can be used for quantitative determination of the recharge and discharge rates as well as it can estimate the spatial circulation of the groundwater in an aquifer. Therefore, borehole temperature is recommended to be used widely in Egypt to help for solving some complicated groundwater.
13. Geoelectrical resistivity was successful for imaging the seawater-freshwater relationship and classifying the aquifer lithology in the target area. Therefore, the results of this work are recommended to be used by the decision-makers for groundwater management planning in the study area.
14. For reliable piezometric measurements, a network of monitoring wells should be drilled that cover the whole area. Regular water-level measurements (at least monthly) would provide the basic data needed for time series compilations in the future, which could give feedback on the constructed model. Modeling tools that are significant for the decision-makers to have knowledge about groundwater resources could be useful to study the effect of increasing pressure on the finite water resource on the sustainability of agricultural production and livelihoods as well as the social cohesion within new settler communities.
15. Including the actual drainage network with the actual canals network is recommended for future studies in groundwater modeling for the Nile Delta for accurate representation and simulation of the interaction between the surface water and the groundwater system. Update the numerical simulation of saltwater intrusion by including the recent data of groundwater level, salinity measurements, water levels of canals, and pumping rates records. Integrated modeling results of the current model could be used in the identification of the most vulnerable areas to the saltwater intrusion in the Nile Delta aquifer and management of groundwater resources in the Nile Delta region. The integrated built model is a useful predictive tool for more understanding of the hydrological process of the Nile Delta aquifer and the saltwater intrusion process. Future studies should also link between the results of the numerical model for simulating the saltwater intrusion in the Nile Delta aquifer and the socioeconomic effect of the recent development in the Nile Delta especially the coastal area near to the Mediterranean Sea border. The built integrated model could be used with another analysis tools as a decision support system to determine the best location and the maximum abstraction rates from groundwater wells in the Nile Delta region for the benefits of the stakeholders and the concerned authorities.
16. For better groundwater management for sustainable development, it is recommended that the strategies for the production wells operations are controlled. Setting up a GIS database for groundwater resources in the study area is required. The GIS database includes the active and abandoned groundwater wells, their pumping rates and screen depths, irrigation canals, drains, seepage surfaces, soil type, and agricultural fields. This information is important for

monitoring programs and efficiently helps to obtain the optimum results from groundwater models.

17. Implementing monthly monitoring of groundwater levels, especially in the northern confined aquifer and near the saltwater interface, to verify the validity of the resulting management plan of this study and updating the model as required for improving the accuracy of the predictions. Implementation of a licensing system for new groundwater wells construction to strictly adhere to the guidelines of drilling depth, screening interval, and the distance between wells based on the permissible withdrawal rates and the capacity of drawdown recovery.
18. Because Egypt suffers from freshwater supply to meet the increasing demand, the people in the rural area used untreated groundwater for drinking purposes. Some rural areas do not have sewage networks, and they are using septic tanks. Therefore, people in rural areas should be aware of the possible contamination of the groundwater due to leakage from the septic tanks. It is essential to let the people know that regular analysis of water samples from their wells is vital to protect their health from being drinking unsuitable water. It is highly recommended to supply the villages with clean freshwater networks and detailed hydrochemical, and the microbial survey should be done to evaluate the groundwater contamination under the Egyptian village. Also, constructing sewage networks for the rural areas which are using septic tanks are a fundamental step to reduce the groundwater contamination. A 3D hydrogeochemical study of the aquifer is highly needed for the Nile Delta aquifer to investigate both the vertical and horizontal groundwater chemistry to have a complete spatial microbial contamination image of the groundwater Nile Delta aquifers to enable the decision-makers to achieve the optimal possible utilization of groundwater wells.
19. The integration of GIS and groundwater modeling system could play a significant role in the construction of all the needed layer groundwater flow model and effectively enabled smooth management to the required database. The budget analysis of groundwater flow using 3D modeling indicated that pumping discharge from production wells efficiently controls the rising water level in the confined part of the Quaternary aquifer and the infiltration from agricultural field domain is the key factor in the unconfined part. The primary recharge of the aquifer is the infiltration from the Ismailia Canal and Damietta branch, which significantly increases during the high pumping stress from wells. The results of such modeling system could constitute a foundation for sustainable groundwater resource management and to evaluate the efficiency of multidisciplinary policies shortly.
20. It is recommended to apply well-controlled license system of wells abstraction to update the wells inventory database to avoid degradation of the groundwater updating the available information, which could be a good baseline, would be a first step to improve the quality of groundwater research for the optimum operation and maintenance of whole groundwater aquifer system Nile Delta.

21. It is highly recommended that water managers, planners, and decision- and policy-makers continue to look widely for ways to improve water management and augment water supplies such as rainfall harvesting, water use efficiency, water recycling, and desalinated water for domestic and industrial purposes.
22. The demand is hardening because of new governmental development projects in conjunction with the private expansion in both agriculture and urbanization. There is a definite need for detailed monitoring of the groundwater parameters to ensure sustainable development of water resources.
23. It is highly recommended to put water labels, along with the lines of food labels, for creating public awareness to show how much water is used domestically and internationally in the production and whether these water amounts are from sustainable or non-sustainable sources. Water labels and caps on extraction, linked to agricultural trade, are among measures proposed tackling the trend.

References

1. El-Din MMN (2013) Climate change risk management in Egypt proposed climate change adaptation strategy for the Ministry of Water Resources and Irrigation in Egypt. UNESCO Office, Cairo
2. CAPMAS (2012) Central Agency for Public Mobilization and Statistics. http://www.capmas.gov.eg/
3. Tamburelli P, Thill O (2013) The Nile metropolitan area. Berlage-Institute, TU Delft, Delft
4. Pathak R, Awasthi MK, Sharma SK, Hardaha MK, Nema RK (2018) Ground water flow modelling using MODFLOW – a review. Int J Curr Microbiol Appl Sci 7(2):83–88
5. De Filippis G, Giudici M, Negri S, Margiotta S, Cattaneo L, Vassena C (2014) Numerical modeling of groundwater flow in the coastal aquifer system of Taranto (southern Italy). In: Geophysical research abstracts, vol 16. EGU2014-393-1
6. Zeidan BA, Aly AI, Rashwan IM, Ahmed MA, Ghoraba SM (2015) Scenarios for groundwater remediation using N15 in Nile Delta. In: 18th international water technology conference, IWTC2015, Sharm El-Shiekh, 12–14 Mar
7. Khayyun TH (2018) Simulation of groundwater flow and migration of the radioactive Cobalt-60 from LAMA nuclear facility-Iraq. Water 10:176. https://doi.org/10.3390/w10020176
8. Kantoush SA (2013) The downstream impacts of Ethiopia's cascade dams in the upper blue Nile on Egypt. In: Proceedings of regional sustainable building conference SB13. Cairo Fairmont Towers Hotel, Cairo
9. Chao Z, Song X, Feng X (2018) Concept and connotation of water resources carrying capacity in water ecological civilization construction. In: IOP conference series: earth and environmental science, vol 111, p 012003
10. IAH (2015) Food security and groundwater. International Association of Hydrogeologists strategic overview series. www.iah.org
11. UNESCO (2015) Water for a sustainable world. World water development report 2015. ISBN 978-92-3-100071-3. ePub ISBN 978-92-3-100099-7
12. Abd-Elhamid H, Javadi A, Abdelaty I, Sherif M (2016) Simulation of seawater intrusion in the Nile Delta aquifer under the conditions of climate change. Hydrol Res 47(5):1–14. https://doi.org/10.2166/nh.2016.157
13. Imasuen OI, Omorogieva OM, Nwokoloh NJ (2016) Grain size and heavy mineral analyses of two boreholes in recent to miocene aquifer in benin formation. Niger J Technol 35:979–986

14. Isikhuemen MI, Omorogieva OM (2015) Hydrogeochemical and biophysical characterization of groundwater in eastern Nigeria: a case study of onisha and environ. Niger J Technol 34:875–882
15. Omorogieva OM, Imasuen OI (2016) Factors contributing to the concentration of heavy metals in stream sediment along Ikpoba River tributary in Oluku (upstream) to Ikpoba River dam (downstream) and their implication. Niger J Appl Sci 34:187–193
16. Szabó NP, Anett K, Anett H (2015) Hydrogeophysical characterization of groundwater formations based on well logs: case study on cenozoic clastic aquifers in East Hungary. Geosci Eng 4(6):45–71
17. Ronczka M, Hellman K, Günther T, Wisén R, Dahlin T (2017) Electric resistivity and seismic refraction tomography: a challenging joint underwater survey at Äspö Hard Rock Laboratory. Solid Earth Sci 8:671–682
18. Goebela M, Adam P, Rosemary K (2017) Resistivity imaging reveals complex pattern of saltwater intrusion along Monterey coast. Hydrol Sci 551:746–755
19. Muhammad M, Khalid P (2017) Hydrogeophysical investigations for assessing the groundwater potential in part of the Peshawar basin Pakistan. Arab J Sci Eng Sci 42:327–337
20. Sharaky AM, El Hasanein AS, Atta SA, Khallaf KM (2017) Nile and groundwater interaction in the western Nile Delta, Egypt. In: Negm AM (ed) The Nile Delta. Handbook of environmental chemistry, vol 55. Springer, Cham, pp 33–62. https://doi.org/10.1007/698_2016_127
21. Tarabees E, El-Qady G (2016) Seawater intrusion modeling in Rashid area of Nile Delta (Egypt) via the inversion of DC resistivity data. Am J Clim Chang 5:147–156. https://doi.org/10.4236/ajcc.2016.52014
22. Armandine Les Landes A, Aquilina L, Davy P, Vergnaud-Ayraud V, Le Carlier C (2015) Timescales of regional circulation of saline fluids in continental crystalline rock aquifers (Armorican Massif, western France). Hydrol Earth Syst Sci 19:1413–1426
23. Merchán D, Auqué LF, Acero P, Gimeno MJ, Causapé J (2015) Geochemical processes controlling water salinization in an irrigated basin in Spain: identification of natural and anthropogenic influence. Sci Total Environ 502:330–343
24. MWRI (Ministry of Water Resources and Irrigation, Egypt) (2016) Water scarcity in Egypt: the urgent need for regional cooperation among the Nile Basin countries, 5 pp. http://www.mfa.gov.eg/SiteCollectionDocuments/Egypt%20Water%20Resources%20Paper_2014.pdf
25. Maity PK, Das S, Das R (2018) Remedial measures for saline water ingression in coastal aquifers of South West Bengal in India. MOJ Eco Environ Sci 3(1):00061. https://doi.org/10.15406/mojes.2018.03.00061
26. Alfarrah NID, Walraevens K (2018) Groundwater overexploitation and seawater intrusion in coastal areas of arid and semi-arid regions. Water 10:143. https://doi.org/10.3390/w10020143
27. Atieh M, Taylor G, Sattar AM, Gharabaghi B (2017) Prediction of flow duration curves for ungauged basins. J Hydrol 545:383–394
28. Sattar AM, Gharabaghi B, Sabouri F, Thompson AM (2017) Urban stormwater thermal gene expression models for protection of sensitive receiving streams. Hydrol Process 31 (13):2330–2348. https://doi.org/10.1002/hyp.11170
29. Gharabaghi B, Sattar AM (2017) Empirical models for longitudinal dispersion coefficient in natural streams. J Hydrol. https://doi.org/10.1016/j.jhydrol.2017.01.022
30. El-Hakeem M, Sattar AM (2017) Explicit solution for the specific flow depths in partially filled pipes. J Pipeline Syst Eng Pract 8(4):06017004. https://doi.org/10.1061/(ASCE)PS.1949-1204.0000283
31. Salem ZE, Osman MO (2016) Shallow subsurface temperature in the environs of El-Nubaria canal, northwestern Nile Delta of Egypt: implications for monitoring groundwater flow system. Environ Earth Sci 75:1241. https://doi.org/10.1007/s12665-016-6046-y
32. Salem ZE, El Bayumy DA (2016) Use of the subsurface thermal regime as a groundwater-flow tracer in the semi-arid western Nile Delta, Egypt. Hydrogeol J 24(4):1001–1014. https://doi.org/10.1007/s10040-016-1377-z

33. Salem ZE, Al Temamy AM, Salah MK, Kassab M (2016) Origin and characteristics of brackish groundwater in Abu Madi coastal area, northern Nile Delta, Egypt. Estuar Coast Shelf Sci 178:21–35
34. Salem ZE, Osman OM (2017) Use of major ions to evaluate the hydrogeochemistry of groundwater influenced by reclamation and seawater intrusion, West Nile Delta, Egypt. Environ Sci Pollut Res 24:3675–3704. https://doi.org/10.1007/s11356-016-8056-4
35. Nofal ER, Amer MA, El-Didy SM, Fekry AM (2015) Sea water intrusion in Nile Delta in perspective of new configuration of the aquifer heterogeneity using the recent stratigraphy data. J Am Sci 11:281–292
36. Wassef R, Schüttrumpf H (2016) Impact of sea-level rise on groundwater salinity at the development area western delta, Egypt. Groundwater Sustain Dev 2–3:85–103. https://doi.org/10.1016/j.gsd.2016.06.001
37. Anomohanran O (2014) Assessment of groundwater potential in Ozoro, Delta State, Nigeria using the electrical resistivity method. Appl Phys Res 6(5):116. https://doi.org/10.5539/apr.v6n5p116
38. Negm AM, Armanuos AM (2017) GIS-based spatial distribution of groundwater quality in the western Nile Delta, Egypt. In: Negm AM (ed) The Nile Delta. Handbook of environmental chemistry. Springer, Cham. https://doi.org/10.1007/698_2016_66
39. Negm AM, Eltarabily MGA (2017) Modeling of fertilizer transport through soil, case study: Nile Delta. In: Negm AM (ed) The Nile Delta. Handbook of environmental chemistry. Springer, Cham. https://doi.org/10.1007/698_2016_66
40. Arumaikkani GS, Chelliah S, Gopalan M (2017) Mapping the spatial distributions of water quality and their interpolation with land use/land cover using GIS and remote sensing in Noyyal River basin, Tamil Nadu, India. J Geosci Environ Protection 5:211–220
41. Ng G-HC, Wickert AD, Somers LD, Saberi L, Cronkite-Ratcliff C, Niswonger RG, McKenzie JM (2018) GSFLOW-GRASS v1.0.0: GIS-enabled hydrologic modeling of coupled groundwater–surface-water systems. Geosci Model Dev Discuss. https://doi.org/10.5194/gmd-2017-321
42. Salem ZE, Negm AM, Nahrawy A (2017) Hydrogeophysical characteristics of the central Nile Delta aquifer. Handb Environ Chem. https://doi.org/10.1007/698_2017_75
43. Attwa M, Ali H (2018) Resistivity characterization of aquifer in coastal semi arid-areas: an approach for hydrogeological evaluation. Handb Environ Chem. https://doi.org/10.1007/698_2017_210
44. Ibraheem IM, El-Qady G (2017) Hydrogeophysical investigations at El-Nubariya-Wadi El-Natrun area, west Nile Delta, Egypt. Handb Environ Chem. https://doi.org/10.1007/698_2017_154
45. Sharaky AM, El Hassanein AS, Atta SA, Khallaf KMA (2017) Salinization and origin of the coastal shallow groundwater aquifer, northwestern Nile Delta, Egypt. Handb Environ Chem. https://doi.org/10.1007/698_2017_183
46. Abdel Sattar AM, Bonakdari H, Negm A, Gharabaghi B, Elhakeem M (2017) Soil aquifer treatment system design equation for organic micropollutant removal. Handb Environ Chem. https://doi.org/10.1007/698_2017_136
47. Salem ZE, Osman OM (2017) Use of geoelectrical resistivity to delineate the seawater intrusion in the northwestern part of the Nile Delta, Egypt. Handb Environ Chem. https://doi.org/10.1007/698_2017_175
48. Armanuos AM, Negm A (2018) Integrated groundwater modeling for simulation saltwater intrusion in the Nile Delta aquifer, Egypt. Handb Environ Chem. https://doi.org/10.1007/698_2017_184
49. Mahmod W (2017) Groundwater modeling and assessment under uncertain hydrological conditions for Egyptian Sahara. Handb Environ Chem. https://doi.org/10.1007/698_2017_84
50. Soliman AMM, Solimsn MM (2017) Groundwater potential in the new valley south west the Nile Delta in Egypt. Handb Environ Chem. https://doi.org/10.1007/698_2017_62

51. Salem ZE, Elsaiedy G, ElNahrawy A (2017) Hydrogeochemistry and quality assessment of groundwater under some central Nile Delta villages, Egypt. Handb Environ Chem. https://doi.org/10.1007/698_2017_111
52. Salem ZE, Elsaiedy G, ElNahrawy A (2017) Assessment of the groundwater quality for drinking and irrigation purposes in the central Nile Delta region, Egypt. Handb Environ Chem. https://doi.org/10.1007/698_2017_137
53. Eltarabily MGA, Negm AM (2017) Groundwater management for sustainable development east of the Nile Delta aquifer. Handb Environ Chem. https://doi.org/10.1007/698_2017_102

Index

A. M. Negm (ed.), *Groundwater in the Nile Delta*,
Hdb Env Chem (2019) 73: 753–760, DOI 10.1007/978-3-319-94283-4,

X

Z

Printed by Printforce, the Netherlands